Combinatorics of Symmetric Designs

The aim of this book is to provide a unified exposition of the theory of symmetric designs with emphasis on recent developments. The authors cover the combinatorial aspects of the theory giving particular attention to the construction of symmetric designs and related objects. The last five chapters of the book are devoted to balanced generalized weighing matrices, decomposable symmetric designs, subdesigns of symmetric designs, non-embeddable quasi-residual designs, and Ryser designs. Most results in these chapters have never previously appeared in book form. The book concludes with a comprehensive bibliography of over 400 entries.

Researchers in all areas of combinatorial designs, including coding theory and finite geometries, will find much of interest here. Detailed proofs and a large number of exercises make this book suitable as a text for an advanced course in combinatorial designs.

YURY J. IONIN is a professor of mathematics at Central Michigan University, USA.

MOHAN S. SHRIKHANDE is a professor of mathematics at Central Michigan University, USA.

New Mathematical Monographs

For information about Cambridge University Press mathematics publications visit http://www.cambridge.org/mathematics

Combinatorics of Symmetric Designs

YURY J. IONIN and MOHAN S. SHRIKHANDE
Central Michigan University

CAMBRIDGE
UNIVERSITY PRESS

CAMBRIDGE UNIVERSITY PRESS
Cambridge, New York, Melbourne, Madrid, Cape Town, Singapore, São Paulo

Cambridge University Press
The Edinburgh Building, Cambridge CB2 2RU, UK

Published in the United States of America by Cambridge University Press, New York

www.cambridge.org
Information on this title: www.cambridge.org/9780521818339

First published 2006

Printed in the United Kingdom at the University Press, Cambridge

A catalog record for this publication is available from the British Library

ISBN-13 978-0-521-81833-9 hardback
ISBN-10 0-521-81833-8 hardback

To Irina, Tania, and Timur

To Neelima, Aditi, and Sean

Contents

Preface

Design theory is a well-established branch of combinatorial mathematics. The origins of the subject can be traced back to statistics in the pioneering works of R. A. Fisher, F. Yates, and R. C. Bose. From the very beginning, one of the central objects of design theory has been symmetric designs. The prototype of a symmetric design is a finite projective plane, and the theory of symmetric designs borrows its methods and ideas from finite geometries, group theory, number theory, and linear algebra.

It is notoriously difficult to construct an infinite family of symmetric designs or even a single symmetric design. However, in recent years new ideas in constructing symmetric designs have been discovered and new infinite families have been found. The central role in these constructions is played by balanced generalized weighing matrices. These matrices generalize the notion of a symmetric design but until recently they were often regarded as a rather obscure combinatorial object. Now they seem to be a useful tool in unifying different construction methods that have been developed since the 1950s.

This book is primarily a research monograph which aims to give a unifying exposition of the theory of symmetric designs with emphasis on these new developments. The book covers the combinatorial aspects of the theory with particular attention to constructing symmetric designs and related objects. Recent results that have never previously appeared in book format are developed mainly in the last five chapters. These chapters are devoted to balanced generalized weighing matrices, decomposable symmetric designs, subdesigns of symmetric designs, non-embeddable quasi-residual designs, and Ryser designs. The preceding chapters on finite geometries, Hadamard matrices, resolvable designs, t-designs, strongly regular graphs, and difference sets emphasize relations between these objects and symmetric designs.

We believe that this book can also be used as a text for a course in combinatorial designs. We begin with a brief introduction to combinatorial set theory, including such beautiful results as Fisher's Inequality, the Ray-Chaudhuri–Wilson Inequality, and the Ryser–Woodall Theorem. The proofs of these theorems are elementary, but we hope they may be of interest even to the expert. Both Fisher's Inequality and the Ryser–Woodall Theorem allow us to introduce the notion of a symmetric design even before the formal definition is given in Chapter 2. Chapters 2–4 and 6–9 contain basic material on combinatorial designs, finite geometries, Hadamard matrices, strongly regular graphs, difference sets, and codes. We have included many examples and exercises and presented the proofs of many theorems in a manner suitable for graduate and advanced undergraduate students. Every chapter of the book is concluded by notes containing comments, references, and historical material. We suggest that the following chapters and sections could form a course in combinatorial designs: Chapter 1, Chapter 2 (without Sections 2.7 and 2.8), Chapter 3 (without Section 3.7), and also Sections 4.1, 4.2, 4.3, 6.1, 6.2, 6.3, 6.5, 7.1, 7.2, 9.1, and 9.2. A standard course of linear algebra and the basic notions of combinatorics and abstract algebra should form a sufficient background for this book.

The numbering of theorems, definitions, remarks, and examples is consecutive within each section and includes the chapter and section numbers, so, for instance, Theorem 3.7.10 can be found in Section 3.7. However, equations are numbered consecutively within each chapter. The last two sections of every chapter are **Exercises** and **Notes**. The **Appendix** contains the list of parameters of all known symmetric designs, which are combined into 23 series and 12 sporadic designs. We conclude the book with an extensive **References** section of over 400 entries, all of which are cited in the book.

We would like to acknowledge people and institutions who through their help, financial support, and hospitality made this work possible. Our particular thanks are due to Alphonse Baartmans, Dieter Jungnickel, Hadi Kharaghani, Vassili Mavron, Gary McGuire, Damaraju Raghavarao, Dijen Ray-Chaudhuri, S. S. Shrikhande, and Vladimir Tonchev for their comments and encouragement during various stages of preparation of this book.

We thank O. Abu Ghnaim, T. Al-Raqqad, J. R. Angelos, T. Ionin, D. Levi, A. Sarker, and K. W. Smith for help and comments and also the students of three classes at Central Michigan University who had to use imperfect drafts of the book as their textbooks.

Our own research that is included in this book, and the writing of the book were done at Central Michigan University, with extensive use of its facilities. The university has also supported us with sabbaticals and numerous travel grants. We are especially thankful to Central Michigan University for two Research Professorship grants awarded to each of us. We would also like to acknowledge the hospitality and financial support of the following institutions: Mathematisches Forschungsinstitut, Oberwolfach, Germany; Michigan Technological University, Houghton, Michigan, USA; Ohio State University, Columbus, Ohio, USA; University of Lethbridge, Lethbridge, Alberta, Canada; Temple University, Philadelphia, Pennsylvania, USA; University of Wales, Aberystwyth, Wales, UK.

We thank Roger Astley and the staff of Cambridge University Press for their superb assistance during preparation and production of this book.

Finally, we would like to thank our wives for their unwavering support, patience, and understanding.

1
Combinatorics of finite sets

A number of advances in combinatorics originated in the following problem: given a finite set and a property of families of subsets of this set, estimate the size of a family with this property and then explore families of maximum or minimum size.

In this chapter we will discuss three problems of this kind:

(i) given a nonempty finite set V, estimate the size of a family $\mathcal{F}$ of subsets of V such that $|A \cap B|$ is the same for all distinct $A, B \in \mathcal{F}$;
(ii) given a nonempty finite set V and positive integers k and s, estimate the size of a family $\mathcal{F}$ of k-subsets of V such that $|A \cap B|$ takes at most s values for distinct $A, B \in \mathcal{F}$;
(iii) given a nonempty finite set V, estimate the size of a family $\mathcal{F}$ of subsets of V such that the cardinality of the symmetric difference of A and B is the same for all distinct $A, B \in \mathcal{F}$.

This discussion will lead us to *symmetric designs*, the central object of study in this book.

1.1. Fisher's Inequality

When we consider families of subsets of a finite set V of cardinality v, it is convenient to think of V as the set $\{1, 2, \ldots, v\}$ and associate with every subset X of V a (0, 1)-string $(x_1, x_2, \ldots, x_v)$ of length v where $x_i = 1$ if $i \in X$ and $x_i = 0$ if $i \notin X$.

We now introduce a simple but useful idea. In order to estimate the size of a family $\mathcal{F}$ of subsets of V, we will select a suitable finite-dimensional vector space P over the rationals and associate an element of P with each element of

$\mathcal{F}$. If the set of vectors associated with the elements of $\mathcal{F}$ is linearly independent, then the cardinality of $\mathcal{F}$ does not exceed the dimension of P.

As the first application of this idea, we take P to be the vector space of linear polynomials $a_0 + a_1x_1 + a_2x_2 + \cdots + a_vx_v$ in v variables with rational coefficients. Clearly, $\dim P = v + 1$. We will now give a proof of the following result:

Theorem 1.1.1 (Nonuniform Fisher's Inequality). *Let V be a nonempty finite set and $\mathcal{F}$ a family of subsets of V such that the cardinality of the intersection of any two distinct members of $\mathcal{F}$ is the same positive integer. Then $|\mathcal{F}| \le |V|$.*

Proof. Let $\mathcal{F}$ be a family of subsets of the set $V = \{1, 2, \ldots, v\}$. Assume there exists a positive integer λ such that $|A \cap B| = \lambda$ for any distinct A and B in $\mathcal{F}$.

Suppose first that there exists $A \in \mathcal{F}$ such that $|A| \le \lambda$. Then $|A| = \lambda$ and the intersection of any two distinct members of $\mathcal{F}$ is the set A. By subtracting A from each member of $\mathcal{F}$, we obtain a family of pairwise disjoint subsets of the set $V \setminus A$. Since the cardinality of such a family does not exceed $|V \setminus A| + 1$, we obtain that $|\mathcal{F}| \le v - \lambda + 1 \le v = |V|$.

From now on, we assume that $|A| > \lambda$ for any $A \in \mathcal{F}$. With each $A \in \mathcal{F}$, we associate the linear polynomial $f_A = \sum_{i \in A} x_i - \lambda$. Then $f_A(X) = |A \cap X| - \lambda$ for any $X \subseteq V$ (regarded as a (0, 1)-string). In particular, for any $A, B \in \mathcal{F}$,

$$f_A(B) = \begin{cases} 0 & \text{if } B \ne A, \\ |B| - \lambda & \text{if } B = A. \end{cases} \tag{1.1}$$

We claim that the subset $\{f_A \colon A \in \mathcal{F}\}$ of the vector space P is linearly independent. Indeed, if $\sum_{A \in \mathcal{F}} \alpha_A f_A = 0$ for some (rational) coefficients α_A, then, applying both sides of this equation to an arbitrary $B \in \mathcal{F}$ and using (1.1), we obtain that $\alpha_B(|B| - \lambda) = 0$, so $\alpha_B = 0$.

Suppose that the constant polynomial 1 is spanned by the polynomials f_A, $A \in \mathcal{F}$, i.e.,

$$1 = \sum_{A \in \mathcal{F}} \alpha_A f_A. \tag{1.2}$$

for some coefficients α_A. Then, applying both sides of (1.2) to $B \in \mathcal{F}$ and using (1.1), we obtain that $\alpha_B(|B| - \lambda) = 1$, so

$$1 = \sum_{A \in \mathcal{F}} \frac{1}{|A| - \lambda} f_A.$$

Applying both sides of this equation to the empty set, we obtain

$$1 = \sum_{A \in \mathcal{F}} \frac{-\lambda}{|A| - \lambda},$$

a contradiction, since the right-hand side of the last equation is negative.

Thus, the set $\{f_A : A \in \mathcal{F}\} \cup \{1\}$ of linear polynomials is linearly independent. Since $\dim P = v + 1$, we obtain that $|\mathcal{F}| + 1 \le v + 1$, i.e., $|\mathcal{F}| \le v = |V|$. □

The bound given by Fisher's Inequality is sharp. If $\mathcal{F}$ is the family of all $(v-1)$-subsets of the v-set V, then $|A \cap B| = v - 2$ for all distinct $A, B \in \mathcal{F}$ and $|\mathcal{F}| = v$.

1.2. The First Ray-Chaudhuri–Wilson Inequality

If A and B are distinct elements of a family $\mathcal{F}$ of subsets of a set V, the number $|A \cap B|$ is called an *intersection number* of $\mathcal{F}$. In the previous section, we considered families of subsets with one intersection number. In this section, we will consider families with s intersection numbers. To estimate the size of such a family, we will use the vector space P_s of multilinear polynomials of total degree s or less in v variables.

Definition 1.2.1. Let Q_s be the vector space of all polynomials in variables $x_1, x_2, \ldots, x_v$ of total degree $\le s$ with rational coefficients. For each $I \subseteq \{1, 2, \ldots, v\}$, let $x_I = \prod_{i \in I} x_i$ (with the convention that $x_\varnothing = 1$). A polynomial $f \in Q_s$ is called *multilinear* if it can be represented as a linear combination of the polynomials x_I with $|I| \le s$. For every polynomial f in variables $x_1, x_2, \ldots, x_v$, let f^* be the multilinear polynomial obtained by replacing each occurrence of x_i^k by x_i (for $k \ge 2$ and $i = 1, 2, \ldots, v$).

Multilinear polynomials form a subspace P_s of Q_s, and the polynomials x_I with $|I| \le s$ form a basis of P_s. Therefore, $\dim P_s = \sum_{i=0}^{s} \binom{v}{i}$.

With every subset X of $\{1, 2, \ldots, v\}$, we again associate a (0, 1)-string $(x_1, x_2, \ldots, x_v)$ of length v where $x_i = 1$ if $i \in X$ and $x_i = 0$ if $i \notin X$. Then, for any polynomial f in v variables, we have $f(X) = f^*(X)$.

Theorem 1.2.2 (The First Ray-Chaudhuri–Wilson Inequality). *Let $\mathcal{F}$ be a family of subsets of a set V of cardinality v. Let M be a set of non-negative integers, $|M| = s$. Suppose that $|A| = k$ is the same for all $A \in \mathcal{F}$, $|A \cap B| \in M$ for any distinct $A, B \in \mathcal{F}$, and $k > m$ for all $m \in M$. Then $|\mathcal{F}| \le \binom{v}{s}$.*

Proof. Let $V = \{1, 2, \ldots, v\}$ and let $\mathcal{F}$ be a family of k-subsets of V satisfying the conditions of the theorem. With each $A \in \mathcal{F}$, we associate the polynomial

$$g_A = \prod_{m \in M} \Big(\sum_{i \in A} x_i - m\Big),$$

and the multilinear polynomial g_A^*. Then

$$g_A^*(X) = \prod_{m \in M} (|A \cap X| - m)$$

for any $X \subseteq V$, and $g_A^*(B) = 0$ for any distinct $A, B \in \mathcal{F}$. Note that $g_A^*(A) > 0$ for any $A \in \mathcal{F}$. We also put $h(x_1, x_2, \ldots, x_v) = \sum_{i=1}^{v} x_i - k$. Then $h(X) = |X| - k$ for any subset X of V, so $h(A) = 0$ for any $A \in \mathcal{F}$.

We claim that the set

$$\{g_A^* \colon A \in \mathcal{F}\} \cup \{(x_I h)^* \colon I \subseteq V, |I| \leq s-1\}$$

of multilinear polynomials is linearly independent. Since all these polynomials are in P_s, this would imply that

$$|\mathcal{F}| + \sum_{i=0}^{s-1} \binom{v}{i} \leq \dim P_s,$$

so $|\mathcal{F}| \leq \binom{v}{s}$.

Assume that

$$\sum_{A \in \mathcal{F}} \alpha_A g_A^* + \sum_{\substack{I \subseteq V \\ |I| \leq s-1}} \beta_I (x_I h)^* = 0,$$

for some rational coefficients α_A, β_I. Applying both sides of this equation to $B \in \mathcal{F}$, we obtain that $\alpha_B g_B^*(B) = 0$, so $\alpha_B = 0$. Therefore,

$$\sum_{\substack{I \subseteq V \\ |I| \leq s-1}} \beta_I (x_I h)^* = 0. \tag{1.3}$$

We will show by induction on $|I|$ that $\beta_I = 0$.

Note that for $J \subseteq V$, we have

$$x_I(J) = \begin{cases} 1 & \text{if } I \subseteq J, \\ 0 & \text{otherwise.} \end{cases} \tag{1.4}$$

Applying both sides of (1.3) to the empty set and using (1.4), we obtain $\beta_\varnothing = 0$. Let $1 \leq u \leq s-1$ and let $\beta_I = 0$ whenever $|I| \leq u-1$. Then we have

$$\sum_{\substack{I \subseteq V \\ u \leq |I| \leq s-1}} \beta_I (x_I h)^* = 0.$$

Applying both sides of this equality to a subset J of V of cardinality u and using (1.4), we obtain that $\beta_J = 0$. This completes the induction and the proof of the theorem. □

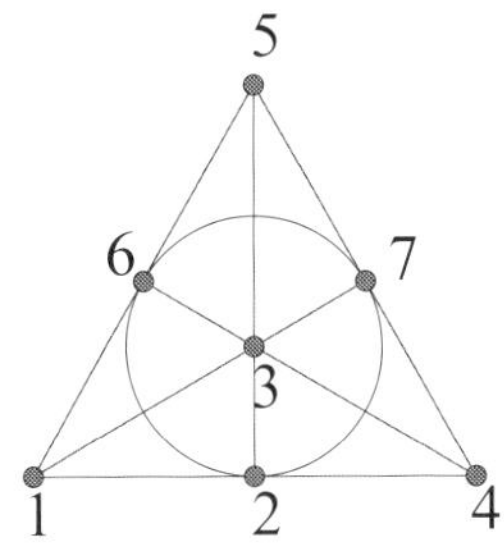

Figure 1.1 Fano Plane.

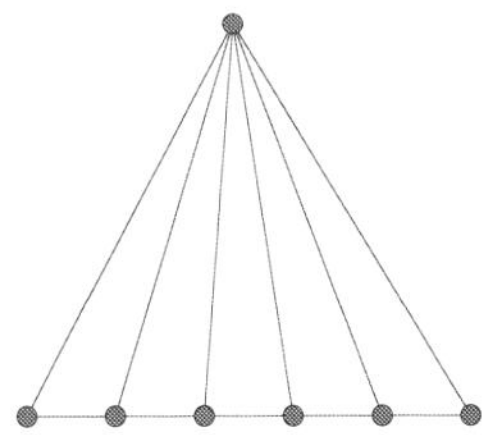

Figure 1.2 Pencil.

If $\mathcal{F}$ is the family of all s-subsets of the v-set V, then $|A \cap B| \in \{0, 1, \ldots, s-1\}$ for any distinct $A, B \in \mathcal{F}$ and $|\mathcal{F}| = \binom{v}{s}$, so the Ray-Chaudhuri–Wilson bound is sharp.

1.3. Symmetric designs and Ryser designs

By Fisher's Inequality (Theorem 1.1.1), the cardinality of a family of subsets of a v-set with one (nonzero) intersection number does not exceed v. In this section, we will consider families attaining this bound. The set of all $(v-1)$-subsets of a v-set is an example of such a family. We will give several less trivial examples.

Example 1.3.1. Let $V = \{1, 2, 3, 4, 5, 6, 7\}$ and let $\mathcal{F} = \{\{1, 2, 4\}, \{2, 3, 5\}, \{3, 4, 6\}, \{4, 5, 7\}, \{5, 6, 1\}, \{6, 7, 2\}, \{7, 1, 3\}\}$. Then $|\mathcal{F}| = |V|$ and $|A \cap B| = 1$ for any distinct $A, B \in \mathcal{F}$. This configuration is known as the *Fano Plane*. In Fig. 1.1, triples of points on lines or on the circle represent elements of the family $\mathcal{F}$. All these triples are regarded as *lines* in the Fano Plane.

Example 1.3.2. Let V be a finite set. Fix $x \in V$ and define $\mathcal{F}$ to be the family consisting of the set $V \setminus \{x\}$ and all 2-subsets of V containing x. Then $|\mathcal{F}| = |V|$ and $|A \cap B| = 1$ for any distinct $A, B \in \mathcal{F}$. Such a configuration is called a *pencil* (Fig. 1.2).

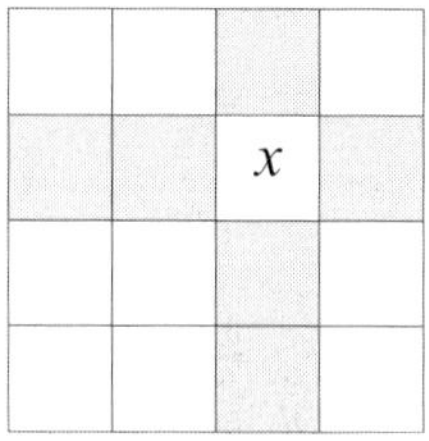

Figure 1.3 Block B_x.

Example 1.3.3. Arrange the elements of a set V of cardinality 16 in a 4×4 array. For each $x \in V$, define a subset B_x of size 6 by taking the elements of V, other than x, which occur in the same row or column as x. It is easy to see that $|B_x \cap B_y| = 2$ for any distinct $x, y \in V$.

Let $V = \{1, 2, \ldots, v\}$ be a set of cardinality v. Let λ be a positive integer and let $\mathcal{F}$ be a family of subsets of V such that $|A \cap B| = \lambda$ for any distinct $A, B \in \mathcal{F}$. For each $A \in \mathcal{F}$, denote by f_A the linear polynomial

$$f_A = \sum_{i \in A} x_i - \lambda. \tag{1.5}$$

In the proof of Theorem 1.1.1, we have shown that the set $\{f_A \colon A \in \mathcal{F}\} \cup \{1\}$ is linearly independent in the vector space P of linear polynomials in variables $x_1, x_2, \ldots, x_v$ (over the rationals).

Suppose now that the family $\mathcal{F}$ is of maximum size, i.e., $|\mathcal{F}| = v$. Then this set of polynomials is a basis of P. By expanding monomials x_i in this basis we will attempt to extract information which can be used to obtain a crude classification of the extremal case. For the next theorem we introduce the notion of the *replication number* that will be used throughout the book.

Definition 1.3.4. Let $\mathcal{F}$ be a family of subsets of a finite set V. For any $x \in V$, the number of elements of $\mathcal{F}$ which contain x is called the *replication number of x in $\mathcal{F}$*.

Theorem 1.3.5 (The Ryser–Woodall Theorem). *Let v and λ be positive integers and let $\mathcal{F}$ be a family of v subsets of a v-set V such that $|A \cap B| = \lambda$ for any distinct $A, B \in \mathcal{F}$. Then either all elements of V have the same replication number or they have exactly two distinct replication numbers r and r^* and $r + r^* = v + 1$. In the latter case, if $|A| > \lambda$ for all $A \in \mathcal{F}$, then $2 \le r \le v - 1$ and $2 \le r^* \le v - 1$.*

Proof. Let $V = \{1, 2, \ldots, v\}$. If there is $A \in \mathcal{F}$ such that $|A| \le \lambda$, then $|A| = \lambda$ and $B \cap C = A$ for any distinct $B, C \in \mathcal{F}$. Therefore, each element of A has replication number $r = v$ and each element of $V \setminus A$ has replication number $r^* = 1$. Thus we have $r + r^* = v + 1$. From now on, we assume that $|A| > \lambda$

for each $A \in \mathcal{F}$. Then the set $\{f_A : A \in \mathcal{F}\} \cup \{1\}$ where the polynomials f_A are defined by (1.5), is a basis of the vector space P of linear polynomials in variables $x_1, x_2, \ldots, x_v$ over the rationals. We will expand the monomials x_i in this basis:

$$x_i = \sum_{A \in \mathcal{F}} \alpha_A^{(i)} f_A + \beta_i.$$

Applying both sides of this equation to $B \in \mathcal{F}$ and using (1.1), we obtain that $\alpha_B^{(i)} = (1-\beta_i)/(|B|-\lambda)$ if $i \in B$ and $\alpha_B^{(i)} = -\beta_i/(|B|-\lambda)$ if $i \notin B$. Therefore,

$$x_i = (1-\beta_i)\sum_{A \ni i} \frac{f_A}{|A|-\lambda} - \beta_i \sum_{A \not\ni i} \frac{f_A}{|A|-\lambda} + \beta_i. \tag{1.6}$$

Applying both side of (1.6) to the empty set and to the singleton $\{i\}$, we obtain:

$$0 = (1-\beta_i)(-\lambda)\sum_{A \ni i} \frac{1}{|A|-\lambda} - \beta_i(-\lambda) \sum_{A \not\ni i} \frac{1}{|A|-\lambda} + \beta_i, \tag{1.7}$$

$$1 = (1-\beta_i)(1-\lambda)\sum_{A \ni i} \frac{1}{|A|-\lambda} - \beta_i(-\lambda) \sum_{A \not\ni i} \frac{1}{|A|-\lambda} + \beta_i. \tag{1.8}$$

Subtract (1.7) from (1.8) to obtain that $\beta_i \neq 1$ and

$$\sum_{A \ni i} \frac{1}{|A|-\lambda} = \frac{1}{1-\beta_i}. \tag{1.9}$$

Equations (1.7) and (1.9) imply that $\beta_i \neq 0$ and

$$\sum_{A \not\ni i} \frac{1}{|A|-\lambda} = \frac{1}{\beta_i} - \frac{1}{\lambda}. \tag{1.10}$$

Adding (1.9) to (1.10) yields

$$\frac{1}{\lambda} + \sum_{A \in \mathcal{F}} \frac{1}{|A|-\lambda} = \frac{1}{\beta_i(1-\beta_i)}. \tag{1.11}$$

We can reduce (1.11) to a quadratic equation in β_i, whose coefficients do not depend on i. Therefore, β_i can have at most two distinct values, β and $\beta^* = 1-\beta$. If $\beta_i = \beta$, then applying both sides of (1.6) to the set V yields

$$1 = (1-\beta)r_i - \beta(v - r_i) + \beta,$$

where r_i is the replication number of i. This equation implies that $r_i = \beta(v-1)+1$. Similarly, if $\beta_i = \beta^*$, we obtain that $r_i = \beta^*(v-1)+1$. Thus, if all β_i are the same, then all points $i \in V$ have the same replication number. If β and

β^* are the two distinct values of β_i, then the elements of V have two distinct replication numbers r and r^*. Since $\beta + \beta^* = 1$, we have $r + r^* = v + 1$.

Since $r + r^* = v + 1$, we have $r \geq 1$. If $r = 1$, then $r = \beta(v-1) + 1$ implies $\beta = 0$ which is not the case. Therefore, if the family $\mathcal{F}$ has two replication numbers and $|A| > \lambda$ for all $A \in \mathcal{F}$, then the replication number of each element of V is greater than 1 and less than v. □

Let us now discuss the two possibilities that arise from the Ryser–Woodall Theorem.

Suppose first that $\mathcal{F}$ is a family of v subsets of a v-set V such that $|A \cap B| = \lambda$ for any distinct $A, B \in \mathcal{F}$ and all elements of V have the same replication number r. Fix $A \in \mathcal{F}$ and count in two ways pairs (x, B) with $B \in \mathcal{F}$, $B \neq A$, and $x \in A \cap B$. We obtain that $|A|(r-1) = \lambda(v-1)$. Therefore, if $\lambda > 0$, then all $A \in \mathcal{F}$ have the same cardinality. In this case, we will say that $(V, \mathcal{F})$ is a *symmetric* (v, k, λ)*-design*, where $k = |A|$ for all $A \in \mathcal{F}$. Counting in two ways pairs (x, A) with $A \in \mathcal{F}$ and $x \in A$ yields $k = r$. Examples 1.3.1 and 1.3.3 describe a symmetric $(7, 3, 1)$-design and a symmetric $(16, 6, 2)$-design, respectively. The precise definition and many other examples of symmetric designs will be given in the next chapter.

The second possibility arising from the Ryser–Woodall Theorem leads to the notion of a *Ryser design*.

Definition 1.3.6. Let v and λ be positive integers. A *Ryser design of index* λ *on* v *points* is a pair $(V, \mathcal{F})$ where V is a set of cardinality v and $\mathcal{F}$ is a family of v subsets of V (blocks) such that

(i) $|A \cap B| = \lambda$ for any distinct $A, B \in \mathcal{F}$;
(ii) $|A| > \lambda$ for all $A \in \mathcal{F}$;
(iii) there are blocks A and B such that $|A| \neq |B|$.

Example 1.3.2 describes a Ryser design of index 1 on v points. As will be shown in Section 14.1, pencils are the only possible Ryser designs of index 1 on v points.

1.4. Equidistant families of sets

We will now consider a distance function on the set of subsets of a finite set. It will measure how different two subsets are. The following definition introduces the famous *Hamming distance*.

Definition 1.4.1. Let V be a finite set. For any $X, Y \subseteq V$, define the Hamming distance $d(X, Y)$ to be the cardinality of the symmetric difference $X \triangle Y$ of X and Y.

The Hamming distance has the following properties that can be easily verified:

(i) $d(X, Y) \geq 0$; $d(X, Y) = 0$ if and only if $X = Y$;
(ii) $d(X, Y) = d(Y, X)$;
(iii) $d(X, Y) + d(Y, Z) \geq d(X, Z)$.

Definition 1.4.2. A family $\mathcal{F}$ of subsets of the set V is called *equidistant* if there exists a positive integer d such that $|A \triangle B| = d$ for any distinct A and B in $\mathcal{F}$.

In this section we will first find the maximum cardinality of an equidistant family of subsets of a v-set.

Theorem 1.4.3. *If $\mathcal{F}$ is an equidistant family of subsets of a finite set V of cardinality v, then $|\mathcal{F}| \leq v + 1$.*

Proof. Let $\mathcal{F}$ be an equidistant family of subsets of the set $V = \{1, 2, \ldots, v\}$, $|\mathcal{F}| \geq 2$, and let $d = |A \triangle B|$ for any distinct A and B in $\mathcal{F}$. With each $A \in \mathcal{F}$ we associate the following linear polynomial f_A in variables $x_1, x_2, \ldots, x_v$:

$$f_A = \sum_{i \notin A} x_i - \sum_{i \in A} x_i + |A| - d. \tag{1.12}$$

Then, for any subset X of V (regarded as a $(0, 1)$-string),

$$f_A(X) = |A \triangle X| - d. \tag{1.13}$$

This implies that for any $A, B \in \mathcal{F}$,

$$f_A(B) = \begin{cases} 0 & \text{if} \quad B \neq A, \\ -d & \text{if} \quad B = A. \end{cases} \tag{1.14}$$

We claim that the set $\{f_A : A \in \mathcal{F}\}$ of linear polynomials is linearly independent (over the rationals). Indeed, if $\sum_{A \in \mathcal{F}} \alpha_A f_A = 0$ for some rational coefficients α_A, then, applying both sides of this equality to $B \in \mathcal{F}$ and using (1.14), we obtain that $\alpha_B(-d) = 0$, so $\alpha_B = 0$. Since the dimension of the vector space of linear polynomials in the variables $x_1, x_2, \ldots, x_v$ equals $v + 1$, it follows that $|\mathcal{F}| \leq v + 1$. □

Hadamard matrices provide examples of maximum cardinality equidistant families.

Definition 1.4.4. A *Hadamard matrix* is a square matrix with all entries equal to ± 1 and with any two distinct rows orthogonal.

For example,

$$\begin{bmatrix} 1 & 1 & 1 & 1 \\ 1 & 1 & -1 & -1 \\ 1 & -1 & 1 & -1 \\ 1 & -1 & -1 & 1 \end{bmatrix}$$

is a Hadamard matrix of order 4.

Hadamard matrices arise in different areas of combinatorics. The order of a Hadamard matrix is 1 or 2 or a multiple of 4. One of the most famous open conjectures in combinatorics is that there exists a Hadamard matrix of every order that is divisible by 4. We will discuss Hadamard matrices at length in Chapter 4.

Example 1.4.5. Let $V = \{1, 2, \ldots, v\}$, and let $H = [h_{ij}]$ be a Hadamard matrix of order $v + 1$ with all entries in the last column equal to 1. For $i = 1, 2, \ldots, v + 1$, let $A_i = \{j \in V : h_{ij} = 1\}$. Then the family $\mathcal{F} = \{A_i : 1 \leq i \leq v + 1\}$ is equidistant. It is called a *Hadamard family*.

We will now show that this is the only possible example of a maximum size equidistant family.

Theorem 1.4.6. *Let $\mathcal{F}$ be an equidistant family of subsets of a v-set V. If $|\mathcal{F}| = v + 1$, then $\mathcal{F}$ is a Hadamard family.*

Proof. Let $|\mathcal{F}| = v + 1$, $|A \triangle B| = d$ for any distinct $A, B \in \mathcal{F}$, and let polynomials f_A be defined by (1.12). It was shown in the proof of Theorem 1.4.3 that the set $\{f_A : A \in \mathcal{F}\}$ of linear polynomials is linearly independent. Since $|\mathcal{F}| = v + 1$, this set is a basis of the vector space P of linear polynomials in $x_1, x_2, \ldots, x_v$. Expand the constant polynomial 1 in this basis:

$$1 = \sum_{A \in \mathcal{F}} \alpha_A f_A$$

for some rational coefficients α_A. Applying both sides of this equality to $B \in \mathcal{F}$, we derive that $\alpha_B(-d) = 1$, so $\alpha_B = -1/d$ for any $B \in \mathcal{F}$. Therefore, we have

$$\sum_{A \in \mathcal{F}} f_A = -d. \tag{1.15}$$

Applying both sides of (1.15) to the empty set and the set V, we obtain:

$$\sum_{A \in \mathcal{F}} (|A| - d) = -d$$

and

$$\sum_{A \in \mathcal{F}} (v - |A| - d) = -d.$$

Adding these equalities yields $(v+1)(v-2d) = -2d$, which implies $d = \frac{v+1}{2}$. Let $\mathcal{F} = \{A_1, A_2, \ldots, A_{v+1}\}$. Define the following square matrix $H = [h_{ij}]$ of order $v+1$:

$$h_{ij} = \begin{cases} 1 & \text{if} \quad j = v+1, \\ 1 & \text{if} \quad 1 \le j \le v \quad \text{and} \quad j \in A_i, \\ -1 & \text{if} \quad 1 \le j \le v \quad \text{and} \quad j \notin A_i. \end{cases} \tag{1.16}$$

Since $|A \triangle B| = d = \frac{v+1}{2}$ for any distinct A and B in $\mathcal{F}$, the inner product of any two distinct rows of H is equal to 0, i.e., H is a Hadamard matrix and therefore $\mathcal{F}$ is a Hadamard family. □

We will return to equidistant families of sets (regarded as binary equidistant codes) in Section 4.7.

Exercises

(1) Let $\mathcal{F}$ be a set of pairwise disjoint subsets of a v-set V.
   (a) Prove that $|\mathcal{F}| \le v+1$.
   (b) Prove that if $|\mathcal{F}| = v+1$, then $\mathcal{F}$ consists of the empty set and all singletons.

(2) For any positive integer n, $\pi(n)$ denotes the number of primes that do not exceed n. Let X be a subset of the set $\{1, 2, \ldots, n\}$ such that the product of all elements of any nonempty subset Y of X is not a square (in particular, no element of X is a square). Prove that $|X| \le \pi(n)$.

(3) Let $\mathcal{F}$ be a set of subsets of a v-set V such that for any distinct $A, B \in \mathcal{F}$, $A \cup B \ne V$. Prove that $|\mathcal{F}| \le 2^{v-1}$. Give an example of a set $\mathcal{F}$ of cardinality 2^{v-1} having this property.

(4) Let $\mathcal{F}$ be a set of subsets of a v-set V such that $A \cap B \ne \varnothing$ for all $A, B \in \mathcal{F}$. Prove that if $|\mathcal{F}| < 2^{v-1}$, then there exists $X \subseteq V$ such that $X \notin \mathcal{F}$ and $X \cap A \ne \varnothing$ for all $A \in \mathcal{F}$.

(5) Let V be a v-set with $v \ge 3$. Prove that there is a set $\mathcal{F}$ of subsets of V such that $A \cap B \ne \varnothing$ for all $A, B \in \mathcal{F}$, $|\mathcal{F}| = 2^{v-1}$, and $\bigcap_{A \in \mathcal{F}} A = \varnothing$.

(6) Let $V = \{1, 2, 3, 4, 5, 6, 7\}$ and

$$\mathcal{B} = \{\{1,2,4\}, \{2,3,5\}, \{3,4,6\}, \{4,5,7\}, \{5,6,1\}, \{6,7,2\}, \{7,1,3\}\}.$$

   Let $\mathcal{F}$ be the set of all subsets of V which contain at least one member of $\mathcal{B}$.
   (a) Find $|\mathcal{F}|$.
   (b) Prove that $A \cap B \ne \varnothing$ for all $A, B \in \mathcal{F}$.

(7) Let $\mathcal{F}$ be a set of subsets of a finite set V such that $|A \cap B|$ is the same for all distinct $A, B \in \mathcal{F}$. Fix $C \in \mathcal{F}$ and define $\mathcal{G} = \{C\} \cup \{A \triangle C \colon A \in \mathcal{F}, A \ne C\}$. Prove that $|A \cap B|$ is the same for all distinct $A, B \in \mathcal{F}$.

(8) Let $(V, \mathcal{F})$ be a symmetric (v, k, λ)-design. Let X be a subset of V such that $|X \cap A|$ is the same for all $A \in \mathcal{F}$. Prove that $X = \varnothing$ or $X = V$.
   Hint: Expand the polynomial $\sum_{i \in X} x_i$ in the basis introduced in the proof of the Ryser–Woodall Theorem.

(9) Let $(V, \mathcal{F})$ be a Ryser design and let X be a subset of V such that $|X \cap A|$ is the same for all $A \in \mathcal{F}$. Prove that $X = \varnothing$.

(10) Let $(V, \mathcal{F})$ be a symmetric (v, k, λ)-design and let $A \in \mathcal{F}$ be a fixed block. Let X be a subset of V such that $|X \cap B|$ is the same for all $B \in \mathcal{F} \setminus \{A\}$. Prove that $X = \varnothing$ or $X = V$ or $X = A$ or $X = V \setminus A$.

(11) Let $(V, \mathcal{F})$ be a Ryser design and let $A \in \mathcal{F}$ be a fixed block. Let X be a subset of V such that $|X \cap B|$ is the same for all $B \in \mathcal{F} \setminus \{A\}$. Prove that $X = \varnothing$ or $X = A$ or $X \supseteq V \setminus A$. Give an example of a Ryser design, a block A, and a subset $X \supset V \setminus A$, $X \neq V \setminus A$ which satisfy the given conditions.

(12) Let $\mathcal{F}$ be an equidistant family of subsets of a v-set V. Let X be a subset of V. Prove that the family $\mathcal{F} \triangle X = \{A \triangle X : A \in \mathcal{F}\}$ is also equidistant.

(13) Show that the family of subsets introduced in Example 1.4.5 is equidistant.

(14) A *regular n-simplex* is a set S of $n+1$ points of the n-dimensional real vector space $\mathbb{R}^n$ such that the (Euclidean) distance between any two points of S is the same. Prove that the following two statements are equivalent:
(a) the set of vertices of an n-dimensional cube contains a regular n-simplex;
(b) there exists a Hadamard matrix of order $n+1$.

(15) Let $\mathcal{F}$ be an equidistant family of subsets of a v-set V. Suppose that $|\mathcal{F}| = v$. Define linear polynomials f_A, $A \in \mathcal{F}$, as in the proof of Theorem 1.4.3. Prove that if $v \geq 3$, then the set $\{f_A : A \in \mathcal{F}\} \cup \{1\}$ is linearly independent. Is this true for $v = 2$?

(16) For a positive integer n, let $k = 1 + \lfloor \frac{2^{n+1}}{n+1} \rfloor$ if there exists a Hadamard matrix of order $n+1$ and $k = 1 + \lfloor \frac{2^{n+1}}{n} \rfloor$ otherwise. Prove that among any k vertices of an n-dimensional cube there are three distinct vertices of an equilateral triangle.

(17) Let V be a set of cardinality v and $\mathcal{F}$ a family of subsets of V such that $|A \cap B|$ takes at most s values for distinct $A, B \in \mathcal{F}$. Prove that $|\mathcal{F}| \leq \sum_{i=0}^{s} \binom{v}{i}$.

(18) Let p be a prime and let $V = \{1, 2, \ldots, 4p\}$. Let $\mathcal{F}$ be a family of subsets of V such that $|A| = 2p$ for all $A \in \mathcal{F}$ and $|A \cap B| \neq p$ for all $A, B \in \mathcal{F}$. Prove that $|\mathcal{F}| \leq 2\binom{4p-1}{p-1}$.
Hint: with each $A \in \mathcal{F}$, associate a multilinear polynomial f_A^* where $f_A = (\sum_{i \in A} x_i)^{p-1} - 1$ over the field of residue classes modulo p.

(19) Let X be a set of strings $(x_1, x_2, \ldots, x_v)$ of length v of elements of the set $\{0, 1, 2\}$. Suppose that for any distinct $(x_1, x_2, \ldots, x_v), (y_1, y_2, \ldots, y_v) \in X$, there is an index j such that $x_j - y_j \equiv 1 \pmod 3$. Prove that $|X| \leq 2^v$.

Notes

The topic of combinatorics of finite sets is also referred to as extremal set theory. See Bollobás (1986) and Anderson (1987) for an exposition of many famous results and methods in this area.

The technique of estimating the size of a given family of subsets of a finite set using suitable polynomials in a vector space is well known. This approach has been used, for example, by Koornwinder (1976), Delsarte, Goethals, and Seidel (1977), and more recently by Alon, Babai, and Suzuki (1991), Blokhuis (1993), Godsil (1993), Snevily (1994), and Ionin and M. S. Shrikhande (1996a) among others.

Nonuniform Fisher's Inequality was first proved in Majumdar (1953). It is a generalization of Fisher's Inequality for 2-designs considered in Section 2.3. Another proof is in Babai (1987). The proof given in Section 1.1. is adapted from Ionin and M. S. Shrikhande (1996a). The First Ray-Chaudhuri–Wilson Inequality is contained in the seminal paper by Ray-Chaudhuri and Wilson (1975). The proof given in Section 1.2. is due to Alon, Babai, and Suzuki (1991). The last paper also contains nonuniform versions of this inequality.

The Ryser–Woodall Theorem was independently proven by Ryser (1968) and Woodall (1970). The proof of this result given in Section 3 is due to Ionin and M. S. Shrikhande (1996a). The term *Ryser design* is taken from Stanton (1997). Ryser (1968) calls these structures λ*-designs* and Woodall (1970) uses the term λ*-linked designs* for a more general structure. We prefer to call these objects Ryser designs to avoid confusion with common usage of such terms as 2-design, t-design, etc. in design theory.

Theorem 1.4.6 was proven in Delsarte (1973b). Our proof follows that of Ionin and M. S. Shrikhande (1995b). Equidistant families of sets were also studied by Bose and S. S. Shrikhande (1959a) and Semakov and Zinoviev (1968).

For Exercise 17, see Alon, Babai and Suzuki (1991). The result of Exercise 18 is due to Frankl and Wilson (1981). For the polynomial proof of this result and for Exercise (19), see Blokhuis (1993).

2
Introduction to designs

Points and lines in Euclidean plane represent the oldest example of an *incidence structure*. Generally, an incidence structure can be described by two abstract sets (called the *point set* and the *block set*) and a binary relation between points and blocks. Imposing certain regularity conditions on a finite incidence structure leads to the concept of combinatorial designs that includes 2-*designs*, *symmetric designs*, and *graphs*.

2.1. Incidence structures

One of the most general notions in the theory of combinatorial designs is that of an *incidence structure*. It involves two finite sets and a binary relation between their elements.

Definition 2.1.1. A (finite) *incidence structure* is a triple $\mathbf{D} = (X, \mathcal{B}, I)$ where X and $\mathcal{B}$ are nonempty finite sets and $I \subseteq X \times \mathcal{B}$. The sets X and $\mathcal{B}$ are called the *point set* and the *block set* of $\mathbf{D}$, respectively, and their elements are called *points* and *blocks*. The set I is called the *incidence relation*. If $(x, B) \in I$, we will say that point x and block B are incident and that (x, B) is a *flag*.

The number of points incident with a block B is called the *size* or the *cardinality* of B and denoted by $|B|$. If $|B| = |X|$, the block B is said to be *complete*. The number of blocks incident with a point x is called the *replication number* of x and denoted by $r(x)$. For distinct points x and y, $\lambda(x, y)$ denotes the number of blocks incident with both x and y. An *incidence matrix* of $\mathbf{D}$ is a (0, 1)-matrix whose rows are indexed by the points of $\mathbf{D}$, columns are indexed by the blocks of $\mathbf{D}$, and the (x, B)-entry is equal to 1 if and only if $(x, B) \in I$.

Remark 2.1.2. When we have to actually form an incidence matrix of an incidence structure $\mathbf{D} = (X, \mathcal{B}, I)$ with v points and b blocks, we need to order the sets X and $\mathcal{B}$. To indicate the chosen ordering, we will write $X = \{x_1, x_2, \ldots, x_v\}$ and $\mathcal{B} = \{B_1, B_2, \ldots, B_b\}$ and refer to the $(0, 1)$-matrix $N = [n_{ij}]$ with $n_{ij} = 1$ if and only if $(x_i, B_j) \in I$ as the *corresponding incidence matrix* of $\mathbf{D}$.

If N is an incidence matrix of $\mathbf{D}$, then $|B|$ is the sum of the entries of the column of N indexed by B, $r(x)$ is the sum of the entries of the row of N indexed by x, and $\lambda(x, y)$ is the inner product of the rows of N indexed by x and y.

Definition 2.1.3. If an incidence structure $(X, \mathcal{B}, I)$ is such that $\mathcal{B}$ is a set of subsets of X, and $(x, B) \in I$ if and only if $x \in B$, then it will be denoted as $(X, \mathcal{B})$.

For any incidence structure $\mathbf{D} = (X, \mathcal{B}, I)$, we will associate with each block B the set of points incident with B. We will denote this set by the same letter B. With this notation, one should be aware that distinct blocks may have the same set of incident points. Nevertheless, it is convenient to use the set theory notation. For instance, if A and B are blocks of an incidence structure, then $A \cap B$ denotes the set of points incident with both A and B. In the same manner, we will interpret the union $A \cup B$, the difference $A \setminus B$, the symmetric difference $A \triangle B = (A \cup B) \setminus (A \cap B)$, etc. We will often use $x \in B$ or $B \ni x$ instead of $(x, B) \in I$. If Y is a set of points and B is a block, then $Y \subseteq B$ means that every point of Y is incident with B and $B \subseteq Y$ means that every point that is incident with B is in Y.

For an incidence structure $\mathbf{D} = (X, \mathcal{B}, I)$, counting flags in two ways yields the equation

$$\sum_{x \in X} r(x) = \sum_{B \in \mathcal{B}} |B|. \tag{2.1}$$

Fixing a point x and counting in two ways flags (y, B) with $y \neq x$ and $x, y \in B$, we obtain another basic equation

$$\sum_{\substack{y \in X \\ y \neq x}} \lambda(x, y) = \sum_{\substack{B \in \mathcal{B} \\ B \ni x}} |B| - r(x). \tag{2.2}$$

The notion of a *substructure* of an incidence structure can be defined in a natural way.

Definition 2.1.4. Let $\mathbf{D} = (X, \mathcal{B}, I)$ be an incidence structure. Let X_0 be a nonempty subset of X and $\mathcal{B}_0$ a nonempty subset of $\mathcal{B}$. The incidence structure $\mathbf{D}(X_0, \mathcal{B}_0) = (X_0, \mathcal{B}_0, I \cap (X_0 \times \mathcal{B}_0))$ is said to be a *substructure* of $\mathbf{D}$. If $\mathcal{B}_0 = \mathcal{B}$, we will write $\mathbf{D}(X_0)$ instead of $\mathbf{D}(X_0, \mathcal{B})$.

If N is an incidence matrix of $\mathbf{D}$, then the submatrix of N formed by the rows with indices from X_0 and columns with indices from $\mathcal{B}_0$ is an incidence matrix of $\mathbf{D}(X_0, \mathcal{B}_0)$.

The following two kinds of substructures are of special interest.

Definition 2.1.5. Let $\mathbf{D} = (X, \mathcal{B}, I)$ be an incidence structure and let Y be a proper subset of X. Let $\mathcal{B}^Y = \{B \in \mathcal{B}\colon Y \not\subseteq B\}$ and $\mathcal{B}_Y = \{B \in \mathcal{B}\colon B \not\subseteq Y\}$. If $\mathcal{B}^Y \neq \varnothing$, then the substructure $\mathbf{D}^Y = \mathbf{D}(X \setminus Y, \mathcal{B}^Y)$ is called a *residual substructure* of $\mathbf{D}$. If $\mathcal{B}_Y \neq \varnothing$, then the substructure $\mathbf{D}_Y = \mathbf{D}(Y, \mathcal{B}_Y)$ is called a *derived substructure* of $\mathbf{D}$. If Y is the set of all points incident with a block B, then we write $\mathbf{D}^B$ and $\mathbf{D}_B$ instead of $\mathbf{D}^Y$ and $\mathbf{D}_Y$ and call these substructures *block-residual* and *block-derived*, respectively. If x is a point, then we put $\mathbf{D}^x = \mathbf{D}^{\{x\}}$ and $\mathbf{D}_x = \mathbf{D}_{X\setminus\{x\}}$ and call these substructures *point-residual* and *point-derived*, respectively.

The next proposition characterizing incidence matrices of residual and derived substructures is immediate. In this proposition we denote by J the all-one matrix of an appropriate size. The following is a list of notations that will be used throughout this book without further explanation.

I	the identity matrix
J	the all-one matrix
I_n	the identity matrix of order n
J_n	the all-one matrix of order n
$J_{m,n}$	the $m \times n$ all-one matrix
O	the zero matrix
$A^\top$	the transpose of matrix A
$\mathbf{a}, \mathbf{b}, \mathbf{x}$	column vectors
$\mathbf{0}$	the zero column vector
$\mathbf{j}$	the all-one column vector

Proposition 2.1.6. *Let $\mathbf{D} = (X, \mathcal{B}, I)$ be an incidence structure and let Y be a proper subset of X. A matrix M^Y is an incidence matrix of $\mathbf{D}^Y$ if and only if there is an incidence matrix M of $\mathbf{D}$ that can be represented as a block matrix*

$$M = \begin{bmatrix} M^Y \\ P \end{bmatrix} \text{ or } M = \begin{bmatrix} M^Y & Q \\ P & J \end{bmatrix}.$$

A matrix N_Y is an incidence matrix of $\mathbf{D}_Y$ if and only if there is an incidence matrix N of $\mathbf{D}$ that can be represented as a block matrix

$$N = \begin{bmatrix} R \\ N_Y \end{bmatrix} \text{ or } N = \begin{bmatrix} R & O \\ N_Y & S \end{bmatrix}.$$

From a given incidence structure **D**, we can define the *s-fold multiple* of **D** by repeating every block s times, the *complementary* structure by replacing every block by its complement, and the *dual* incidence structure by interchanging points and blocks.

Definition 2.1.7. Let $\mathbf{D} = (X, \mathcal{B}, I)$ be an incidence structure and s a positive integer. Let $\mathcal{B} = \{B_1, B_2, \ldots, B_b\}$. The *s-fold multiple of* **D** is the incidence structure $s \times \mathbf{D} = (X, s \times \mathcal{B}, I_s)$, where $s \times \mathcal{B} = \{B_{ij} : 1 \le i \le b, 1 \le j \le s\}$ and $(x, B_{ij}) \in I_s$ if and only if $(x, B_i) \in I$.

Definition 2.1.8. Let $\mathbf{D} = (X, \mathcal{B}, I)$ be an incidence structure. The *complementary incidence structure* is $\mathbf{D}' = (X, \mathcal{B}, I')$ where $(x, B) \in I'$ if and only if $(x, B) \notin I$.

Definition 2.1.9. Let $\mathbf{D} = (X, \mathcal{B}, I)$ be an incidence structure. The *dual incidence structure* is $\mathbf{D}^\top = (\mathcal{B}, X, I^*)$ where $(B, x) \in I^*$ if and only if $(x, B) \in I$.

If N is an incidence matrix of **D**, then $N^\top$ is an incidence matrix of $\mathbf{D}^\top$ and $J - N$ is an incidence matrix of $\mathbf{D}'$.

The same incidence structure may be described in several ways. In order to make this concept precise, we define *isomorphism* between incidence structures.

Definition 2.1.10. Incidence structures $\mathbf{D}_1 = (X_1, \mathcal{B}_1, I_1)$ and $\mathbf{D}_2 = (X_2, \mathcal{B}_2, I_2)$ are called *isomorphic* if there exists a pair of bijections $f : X_1 \to X_2$ and $g : \mathcal{B}_1 \to \mathcal{B}_2$ such that $(x, B) \in I_1$ if and only if $(f(x), g(B)) \in I_2$.

If an incidence structure admits a symmetric incidence matrix, it is isomorphic to its dual. Such an incidence structure is called *self-dual*.

Definition 2.1.11. An incidence structure **D** is called *self-dual* if **D** and $\mathbf{D}^\top$ are isomorphic incident structures.

The following example of isomorphic incidence structures is an immediate corollary of Proposition 2.1.6.

Proposition 2.1.12. *Let* $\mathbf{D} = (X, \mathcal{B}, I)$ *be an incidence structure and let* $\mathbf{D}'$ *be the complementary incidence structure. Let* Y *be a proper subset of* X. *If the residual substructure* $\mathbf{D}^Y$ *of* **D** *is defined, then the complementary structure* $(\mathbf{D}^Y)'$ *is isomorphic to the derived substructure* $(\mathbf{D}')_{X \setminus Y}$ *of* $\mathbf{D}'$. *If the derived substructure* $\mathbf{D}_Y$ *of* **D** *is defined, then the complementary structure* $(\mathbf{D}_Y)'$ *is isomorphic to the residual substructure* $(\mathbf{D}')^{X \setminus Y}$ *of* $\mathbf{D}'$.

Two (0, 1)-matrices N_1 and N_2 are incidence matrices of isomorphic incidence structures if and only if there exist permutation matrices P and Q such that $PN_1 = N_2 Q$.

Proposition 2.1.13. *Let N_1 and N_2 be $v \times b$ incidence matrices of isomorphic incidence structures $\mathbf{D}_1 = (X_1, \mathcal{B}_1, I_1)$ and $\mathbf{D}_2 = (X_2, \mathcal{B}_2, I_2)$ and let bijections $f: X_1 \to X_2$ and $g: \mathcal{B}_1 \to \mathcal{B}_2$ be such that $(x, B) \in I_1$ if and only if $(f(x), g(B)) \in I_2$. For $k = 1$ and 2, for $i = 1, 2, \ldots, v$, and for $j = 1, 2, \ldots, b$, let x_i^k and B_j^k be the point and the block of X_k corresponding to the ith row and to the jth column of N_k, respectively. Let $(0, 1)$-matrices $P = [p_{ij}]$ of order v and $Q = [q_{ij}]$ of order b be defined by:*

$$p_{ij} = 1 \text{ if and only if } x_i^2 = f(x_j^1),$$
$$q_{ij} = 1 \text{ if and only if } B_i^2 = g(B_j^1).$$

Then $PN_1 = N_2Q$.

Proof. For $k = 1, 2$, let $N_k = [n_{ij}^{(k)}]$. For $i = 1, 2, \ldots, v$ and $j = 1, 2, \ldots, b$, the (i, j)-entry of PN_1 is equal to $n_{sj}^{(1)}$ with $x_i^2 = f(x_s^1)$, so it is equal to 1 if and only if $(x_i^2, g(B_j^1)) \in I_2$. Similarly, the (i, j)-entry of N_2Q is equal to $n_{it}^{(2)}$ with $B_t^2 = g(B_j^1)$, so it is equal to 1 if and only if $(x_i^2, g(B_j^1)) \in I_2$. Therefore, $PN_1 = N_2Q$. □

Remark 2.1.14. Note that the matrices P and Q defined in Proposition 2.1.13 are *permutation matrices*, that is, $(0, 1)$-matrices with exactly one entry equal to 1 in each row and each column.

Remark 2.1.15. The converse of Proposition 2.1.13 is also true. See Exercise 3

2.2. Graphs

The basic concepts of graph theory are used in many areas of combinatorics. A *graph* is determined by a set of points called *vertices* and a set of 2-subsets of the set of vertices called *edges*. All graphs under consideration are without multiple edges. Therefore, as incidence structures, they do not have repeated blocks.

Definition 2.2.1. A *graph* is a pair $\Gamma = (V, E)$ where V is a nonempty finite set (of *vertices*) and E is a set of 2-subsets of V (*edges*). If $\{x, y\}$ is an edge, then vertices x and y are said to be *adjacent*. The cardinality of V is called the *order* of Γ. For each vertex $x \in V$, $\Gamma(x)$ denotes the set of all vertices y such that $\{x, y\}$ is an edge. The cardinality of $\Gamma(x)$ is called the *degree* or *valency* of x. If all vertices of a graph are of the same degree k, then the graph is said to be *regular of degree k*.

Example 2.2.2. For $n \geq 3$, the graph C_n with vertices $x_1, x_2, \ldots, x_n$ and edges $\{x_i, x_{i+1}\}$, for $i = 1, \ldots, n-1$, and $\{x_n, x_1\}$ is called a *cycle of length* n. It is regular of degree 2.

Definition 2.2.3. A graph $\Gamma = (V, E)$ is called a *null graph* if $E = \varnothing$. A graph $\Gamma = (V, E)$ is called a *complete graph* if E is the set of all 2-subsets of V. The complete graph of order n is denoted by K_n. A graph $\Gamma = (V, E)$ is called *bipartite* if there is a partition of the vertex set V into two nonempty subsets such that no two vertices from the same partition set form an edge. A regular bipartite graph of degree 1 is called a *ladder graph*. A graph $\Gamma' = (V', E')$ is called a *subgraph* of a graph $\Gamma = (V, E)$ if $V' \subseteq V$ and $E' \subseteq E$. The subgraph Γ' is called an *induced subgraph* if E' is the set of all elements of E that are contained in V'. An induced subgraph Γ' of a graph Γ is called a *clique* if Γ' is a complete graph. An induced subgraph Γ' of a graph Γ is called a *coclique* if Γ' is a null graph. The set of vertices of a clique or a coclique is usually referred to by the same name.

With any incidence structure we associate a bipartite graph called the *Levi graph* of the structure.

Definition 2.2.4. Let $\mathbf{D} = (X, I, \mathcal{B})$ be an incidence structure with disjoint sets X and $\mathcal{B}$. The *Levi graph* of $\mathbf{D}$ is the graph with the vertex set $X \cup \mathcal{B}$ and all edges $\{x, B\}$ such that $(x, B) \in I$.

A graph $\Gamma = (V, E)$ can be regarded as a partition of the set of all 2-subsets of V into two sets: the set E of edges and the set of *non-edges*. Replacing the former set by the latter yields *the complement of the graph*.

Definition 2.2.5. The *complement* of a graph $\Gamma = (V, E)$ is the graph $\Gamma' = (V, E')$ where E' is the set of all 2-subsets of V that are not edges of Γ.

The next definition introduces some basic notions of graph theory.

Definition 2.2.6. A *walk* from a vertex x to a vertex y of a graph $\Gamma = (V, E)$ is a sequence $(x_0, x_1, \ldots, x_n)$ of vertices such that $x_0 = x$, $x_n = y$, and $\{x_{i-1}, x_i\}$ is an edge for $i = 1, 2, \ldots, n$. The number n is the *length* of the walk. The binary relation on V, given by $x \sim y$ if and only if $x = y$ or there is a walk from x to y, is an equivalence relation. If $V_1, V_2, \ldots, V_m$ are the equivalence classes, then the graphs $\Gamma_i = (V_i, E_i)$ where $E_i = \{e \in E : e \subseteq V_i\}$ are called *connected components of* Γ. A graph with only one connected component is called a *connected graph*.

We leave proof of the following proposition as an exercise.

Proposition 2.2.7. *If Γ' is the complement of a graph Γ, then at least one of these graphs is connected.*

Graphs with disjoint vertex sets can be combined into a larger graph.

Definition 2.2.8. Let $\Gamma_1 = (V_1, E_1)$ and $\Gamma_2 = (V_2, E_2)$ be graphs with $V_1 \cap V_2 = \varnothing$. The graph $\Gamma = (V_1 \cup V_2, E_1 \cup E_2)$ is called the *disjoint union* of the graphs Γ_1 and Γ_2. For positive integers m and n, the disjoint union of m copies of K_n is denoted by $m \cdot K_n$; its complement is called a *complete multipartite graph* and denoted $K_{m,n}$.

A graph can be represented via its *adjacency matrix*.

Definition 2.2.9. If $V = \{x_1, x_2, \ldots, x_v\}$ is the vertex set of a graph Γ, then the *corresponding adjacency matrix* of Γ is the $v \times v$ matrix whose (i, j) entry is equal to 1 if $\{x_i, x_j\}$ is an edge of Γ, and is equal to 0 otherwise.

A (0, 1)-matrix is an adjacency matrix of a graph if and only if it is symmetric and has zero diagonal. The following proposition can be proved by straightforward induction.

Proposition 2.2.10. *Let Γ be a graph with the vertex set $V = \{x_1, x_2, \ldots, x_v\}$ and let A be the corresponding adjacency matrix. For any positive integer k, A^k is the matrix whose (i, j) entry is equal to the number of walks of length k from vertex x_i to vertex x_j.*

If A is an adjacency matrix of a graph Γ on v vertices and J is the all-one matrix of order v, then the (i, j)-entry of AJ is the valency of x_i and the (i, j)-entry of JA is the valency of x_j. Therefore, Γ is regular if and only if $AJ = JA$. It is regular of degree k if and only if $AJ = kJ$.

If A and B are adjacency matrices of a graph Γ, then one can be obtained from the other by a suitable permutation of vertices of Γ, that is, there exists a permutation matrix P such that $B = P^\top AP$. Since permutation matrices are orthogonal, the matrices A and B have the same characteristic polynomial $\chi(\Gamma)$, which therefore can be called the *characteristic polynomial of the graph* Γ. If A is an adjacency matrix of Γ, then $\chi(\Gamma)(t) = \det(tI - A)$. The roots of $\chi(\Gamma)$ are *the eigenvalues of* Γ. The *spectrum of* Γ is the multiset of its eigenvalues taken with their respective multiplicities. Note that since adjacency matrices of graphs are symmetric matrices with zeros on the diagonal, the spectrum of any graph consists of real numbers whose sum is equal to 0. If a graph Γ has m connected components $\Gamma_1, \Gamma_2, \ldots, \Gamma_m$, then $\chi(\Gamma) = \chi(\Gamma_1)\chi(\Gamma_2)\cdots\chi(\Gamma_m)$.

Example 2.2.11. By Lemma 2.3.6, $\chi(K_n)(t) = (t - n + 1)(t + 1)^{n-1}$ and $\chi(m \cdot K_n)(t) = ((t - n + 1)(t + 1)^{n-1})^m$.

If A is an adjacency of a graph Γ, then s is an eigenvalue of Γ if and only if there exists a nonzero (column) vector $\mathbf{x}$ such that $A\mathbf{x} = s\mathbf{x}$. The vector $\mathbf{x}$ is called an *eigenvector of A corresponding to s*. All eigenvectors of A corresponding to s together with the zero vector $\mathbf{0}$ form the *eigenspace of A corresponding to s*.

The spectrum of a graph may provide useful information about the graph. For instance, the largest eigenvalue of a regular graph is the degree of the graph. In the proof of this and other results involving eigenvalues of graphs, we will use the following three results on symmetric matrices the first two of which can be found in standard linear algebra texts.

Proposition 2.2.12. *If A is a real symmetric matrix, then the dimension of the eigenspace of A corresponding to a given eigenvalue is equal to the multiplicity of this eigenvalue. If $\mathbf{x}$ and $\mathbf{y}$ are eigenvectors of A corresponding to two different eigenvalues, then $\mathbf{x}^\top \mathbf{y} = [0]$.*

Proposition 2.2.13. *If $A_1, A_2, \ldots, A_m$ are real symmetric matrices, any two of which commute, then there exists an orthogonal matrix C such that all matrices $C^\top A_i C$ $(i = 1, 2, \ldots, m)$ are diagonal matrices.*

Proposition 2.2.14. *For any matrix N, every nonzero eigenvalue of $NN^\top$ is also an eigenvalue of $N^\top N$ with the same multiplicity.*

Proof. Let s be a nonzero eigenvalue of NN^T, i.e., $NN^T\mathbf{x} = s\mathbf{x}$ for some nonzero vector $\mathbf{x}$. Then

$$N^\top \mathbf{x} \neq \mathbf{0} \quad \text{and} \quad (N^\top N)(N^\top \mathbf{x}) = s(N^\top \mathbf{x})$$

so s is an eigenvalue of $N^T N$ with the nonzero eigenvector $N^\top \mathbf{x}$. The multiplicity of an eigenvalue of a symmetric matrix is equal to the dimension of the corresponding eigenspace. Let $\mathbf{x}_1, \mathbf{x}_2, \ldots, \mathbf{x}_m$ be linearly independent eigenvectors corresponding to an eigenvalue $s \neq 0$ of NN^T. Then the corresponding eigenvectors $N^\top \mathbf{x}_1, N^\top \mathbf{x}_2, \ldots, N^\top \mathbf{x}_m$ of $N^\top N$ are also linearly independent. Indeed, if $\sum_{i=1}^m \alpha_i N^\top \mathbf{x}_i = \mathbf{0}$, then $\sum_{i=1}^m \alpha_i NN^T \mathbf{x}_i = \mathbf{0}$, so $\sum_{i=1}^m \alpha_i s\mathbf{x}_i = \mathbf{0}$, $\sum_{i=1}^m \alpha_i \mathbf{x}_i = \mathbf{0}$, and all α_i are equal to 0. Thus, each nonzero eigenvalue of NN^T is an eigenvalue of $N^T N$ with at least the same multiplicity. By interchanging N and N^T, we complete the proof. □

Corollary 2.2.15. *If N is a $v \times b$ matrix with $v \leq b$, then the spectrum of $N^\top N$ can be obtained by adjoining $b - v$ zeros to the spectrum of $NN^\top$.*

If Γ is a regular graph of degree k and A is an adjacency matrix of Γ, then $AJ = kJ$, so k is an eigenvalue of Γ with an eigenvector $\mathbf{j}$. Proposition 2.2.12 implies that if $\mathbf{x}$ is an eigenvector of Γ corresponding to an eigenvalue other than k, then $J\mathbf{x} = \mathbf{0}$.

The following proposition gives a relation between eigenvalues of a regular graph and of its complement.

Proposition 2.2.16. *Let Γ be a regular graph of order v and degree k and let s be an eigenvalue of Γ other than k. Then $-s-1$ is an eigenvalue of the complementary graph $\overline{\Gamma}$ and the multiplicity of s in Γ does not exceed the multiplicity of $-s-1$ in $\overline{\Gamma}$. Furthermore, these multiplicities are the same if and only if $s \neq k - v$.*

Proof. Let A be an adjacency matrix of Γ and let $A\mathbf{x} = s\mathbf{x}$. Then $J - A - I$ is an adjacency matrix of $\overline{\Gamma}$ and $(J - A - I)\mathbf{x} = (-s-1)\mathbf{x}$. Thus, $-s-1$ is an eigenvalue of $\overline{\Gamma}$. Furthermore, the eigenspace U of A corresponding to s is contained in the eigenspace $\overline{U}$ of $J - A - I$ corresponding to the eigenvalue $-s-1$ of $\overline{\Gamma}$. Therefore, the multiplicity of s in Γ does not exceed the multiplicity of $-s-1$ in $\overline{\Gamma}$.

If $s = k - v$, then $-s-1$ is the degree of $\overline{\Gamma}$, so $\mathbf{j} \in \overline{U}$ and $\dim(\overline{U}) > \dim(U)$. If $s \neq k - v$, then $-s-1$ is an eigenvalue of $\overline{\Gamma}$ other than the degree of $\overline{\Gamma}$ and therefore, by the first part of the proof, $\dim(\overline{U}) \leq \dim(U)$, so the multiplicities of s and $-s-1$ are the same. □

The degree of a regular graph is its largest eigenvalue.

Proposition 2.2.17. *If Γ is a regular graph of degree k with m connected components, then k is an eigenvalue of Γ of multiplicity m. If s is any eigenvalue of Γ, then $|s| \leq k$.*

Proof. First assume that $m = 1$, i.e., that Γ is a connected regular graph of degree k with the vertex set $\{x_1, x_2, \ldots, x_v\}$. Let A be the corresponding adjacency matrix of Γ. Then $AJ = kJ$ and therefore k is an eigenvalue of A with the all-one eigenvector $\mathbf{j}$.

Let $\mathbf{x} = [\alpha_1, \alpha_2, \ldots, \alpha_v]^\top$ be any nonzero vector such that $A\mathbf{x} = k\mathbf{x}$. Then (for $j = 1, 2, \ldots, v$) $k\alpha_j$ is the sum of all α_i such that x_i is adjacent to x_j. Let α_m be an entry of $\mathbf{x}$ with the largest absolute value. Then $\alpha_i = \alpha_m$ for all i such that x_i is adjacent to x_m. Since Γ is connected, this implies that all components of $\mathbf{x}$ are equal. Therefore, the eigenspace of A corresponding to k is one-dimensional and k is a simple eigenvalue of Γ.

Let s be any eigenvalue of Γ. Let $\mathbf{y}$ be an eigenvector corresponding to s and let β_m be a component of $\mathbf{y}$ with the largest absolute value. Since $A\mathbf{y} = s\mathbf{y}$, we obtain that $s\beta_m$ is the sum of k components of $\mathbf{y}$ and therefore $|s||\beta_m| \leq k|\beta_m|$, which implies $|s| \leq k$.

Suppose now that Γ has $m > 1$ connected components $\Gamma_1, \Gamma_2, \ldots, \Gamma_m$. Then each Γ_i is a connected graph of degree k. Therefore, k is a simple root of each polynomial $\chi(\Gamma_i)$, $i = 1, 2, \ldots, m$, and so it is a root of multiplicity m of $\chi(\Gamma)$. If s is another eigenvalue of Γ, then s is an eigenvalue of at least one Γ_i and therefore $|s| \leq k$. □

The following theorem gives some information on other eigenvalues of a regular graph.

Theorem 2.2.18. *Let A be an adjacency matrix of a connected regular graph of order v and degree k and let p be a polynomial with real coefficients. Then $p(A) = J$ if and only if $p(k) = v$ and $p(s) = 0$ for all eigenvalues s of Γ, other than k.*

Proof. Since $AJ = JA = kJ$, matrices A and J commute. Therefore, there exists an orthogonal matrix C such that $C^\top AC = D$ and $C^\top JC = E$ are diagonal matrices. Since the matrix J of order v has a simple eigenvalue v and an eigenvalue 0 of multiplicity $v - 1$, we assume without loss of generality that the $(1, 1)$-entry of E is v and all other entries are zeros.

Let $\mathbf{x} = C^\top \mathbf{j}$, so $C\mathbf{x} = \mathbf{j}$. Then $E\mathbf{x} = v\mathbf{x}$, which implies that $\mathbf{x} = [x_1, 0, \ldots, 0]^\top$. Since $D\mathbf{x} = k\mathbf{x}$, we obtain that the $(1, 1)$-entry of D is k.

Let p be a polynomial over the reals. Then $p(D) = C^\top p(A)C$. If $p(A) = J$, then $p(D) = E$, so $p(k) = v$ and $p(s) = 0$ for all eigenvalues s of Γ other than k.

Conversely, if $p(s) = 0$ for all these eigenvalues and $p(k) = v$, then $p(D) = E$, which implies $p(A) = J$. □

The next two propositions characterize graphs with one eigenvalue and regular graphs with two eigenvalues.

Proposition 2.2.19. *The only graphs with one eigenvalue are null graphs.*

Proof. If a graph Γ on v vertices with an adjacency matrix A has only one eigenvalue s, then $A\mathbf{x} = s\mathbf{x}$ for all vectors $x \in \mathbb{Q}^v$. In particular $A\mathbf{j} = s\mathbf{j}$, which implies that Γ is a regular graph of degree s. Now Proposition 2.2.17 implies that Γ has v connected components and therefore it is a null graph. □

Proposition 2.2.20. *A regular graph has two eigenvalues if and only if it is a K_n or a $m \cdot K_n$.*

Proof. As Example 2.2.11 shows, all graphs K_n and $m \cdot K_n$ have two eigenvalues.

Let Γ be a connected regular graph of order v and degree k with two eigenvalues, k and s. Let A be an adjacency matrix of Γ. By Proposition 2.2.17, k is a simple eigenvalue and then s is an eigenvalue of multiplicity $v - 1$. Therefore, we have $k + (v - 1)s = 0$. Let $p(t) = (s - t)/s$. Then $p(k) = v$ and $p(s) = 0$, and Theorem 2.2.18 implies that $p(A) = J$. Therefore, $A = s(I - J)$. Since A is a (0, 1)-matrix, we have $s = -1$ and $A = J - I$. Thus, $\Gamma = K_v$.

If Γ is a regular graph of order v with two eigenvalues, having $m > 1$ connected components, then each component is a complete graph. Therefore, $\Gamma = m \cdot K_{v/m}$. □

2.3. Basic properties of (v, b, r, k, λ)-designs

We will now impose certain regularity conditions on incidence structures.

Definition 2.3.1. A (v, b, r, k, λ)-design is an incidence structure $\mathbf{D} = (X, \mathcal{B}, I)$ satisfying the following conditions: (i) $|X| = v$; (ii) $|\mathcal{B}| = b$; (iii) $r(x) = r$ for all $x \in X$; (iv) $|B| = k$ for all $B \in \mathcal{B}$; (v) $\lambda(x, y) = \lambda$ for all distinct $x, y \in X$; (vi) if $I = \varnothing$ or $I = X \times \mathcal{B}$, then $v = b$.

Remark 2.3.2. Parameters v and b of a (v, b, r, k, λ)-design are positive integers; parameters r and k are nonnegative integers; if $v > 1$, then λ is a nonnegative integer; if $v = 1$, then λ is irrelevant. An incidence matrix of a (v, b, r, k, λ)-design is a $v \times b$ matrix with constant row sum r, constant column sum k, and constant inner product λ of distinct rows. If it is the all-zero or all-one matrix, then (vi) implies that it is a square matrix. The designs with incidence matrices O and J have parameters $(v, v, 0, 0, 0)$ and (v, v, v, v, v), respectively. We will call these designs *trivial*. If $v = 1$, then condition (vi) of Definition 2.3.1 implies that $b = 1$.

We now give several examples of (v, b, r, k, λ)-designs.

Example 2.3.3. Let $v \geq k \geq 2$ and let $\mathbf{D} = (X, \mathcal{B})$, where X is a set of cardinality v and $\mathcal{B}$ is the set of all k-subsets of X. Then $\mathbf{D}$ is a $(v, \binom{v}{k}, \binom{v-1}{k-1}, k, \binom{v-2}{k-2})$-design. Such a design is called *complete*.

Example 2.3.4. Let $X = \{1, 2, 3, 4, 5, 6\}$ and $\mathcal{B} = \{\{1, 2, 3\}, \{1, 2, 4\}, \{1, 3, 5\}, \{1, 4, 6\}, \{1, 5, 6\}, \{2, 3, 6\}, \{2, 4, 5\}, \{2, 5, 6\}, \{3, 4, 5\}, \{3, 4, 6\}\}$. Then $\mathbf{D} = (X, \mathcal{B})$ is a $(6, 10, 5, 3, 2)$-design.

Incidence structures introduced in Examples 1.3.1 and 1.3.3 are in fact a $(7, 7, 3, 3, 1)$-design and a $(16, 16, 6, 6, 2)$-design, respectively.

If N is an incidence matrix of a (v, b, r, k, λ)-design, then it is a $v \times b$ matrix and properties (iii) – (v) can be expressed in the form of matrix equations:

$$NJ = rJ,\ JN = kJ,\ NN^\top = (r - \lambda)I + \lambda J. \tag{2.3}$$

The complement and s-fold multiple of a (v, b, r, k, λ)-design are a $(v, b, b - r, v - k, b - 2r + \lambda)$ and a $(v, sb, sr, k, s\lambda)$-design, respectively.

Definition 2.3.5. The *order* of a (v, b, r, k, λ)-design with $v > 1$ is the non-negative integer $r - \lambda$.

Observe that a design and its complement have the same order.

If N is an incidence matrix of a (v, b, r, k, λ)-design, then the matrix $NN^\top$ is of the form $xI + yJ$. It is useful to know the determinant of such matrices.

Lemma 2.3.6. *For any real numbers x and y, $\det(xI_n + yJ_n) = (x + ny)x^{n-1}$.*

Proof. Let $A = xI_n + yJ_n$. We add to the first row of A all other rows to make all entries in the first row equal to $x + ny$. Factoring $x + ny$ out and then subtracting y times the first row from every other row yields a matrix with zeros below the diagonal and with the first diagonal entry equal to 1 and the other $n - 1$ diagonal entries equal to x. Therefore, $\det(xI_n + yJ_n) = (x + ny)x^{n-1}$. □

For a (v, b, r, k, λ)-design, equations (2.1) and (2.2) imply immediately the following result.

Proposition 2.3.7. *If $\mathbf{D} = (X, \mathcal{B}, I)$ is a (v, b, r, k, λ)-design, then*

$$vr = bk \tag{2.4}$$

and

$$\lambda(v - 1) = r(k - 1). \tag{2.5}$$

The following proposition introduces a simple but very useful counting technique known as *variance counting*.

Proposition 2.3.8. *Let* $\mathbf{D} = (X, \mathcal{B})$ *be a* (v, b, r, k, λ)*-design and let* $A \in \mathcal{B}$*. For* $i = 0, 1, \ldots, k$*, let* n_i *denote the number of blocks* $B \in \mathcal{B} \setminus \{A\}$ *such that* $|A \cap B| = i$*. Then*

$$\sum_{i=0}^{k} n_i = b - 1, \tag{2.6}$$

$$\sum_{i=0}^{k} i n_i = k(r - 1), \tag{2.7}$$

and

$$\sum_{i=0}^{k} i(i-1)n_i = k(k-1)(\lambda - 1). \tag{2.8}$$

Proof. Eq. (2.6) is obvious. Counting in two ways pairs (x, B) with $B \in \mathcal{B} \setminus \{A\}$ and $x \in A \cap B$ yields (2.7). Counting in two ways triples (x, y, B) with $B \in \mathcal{B} \setminus \{A\}$, $x \neq y$, and $x, y \in A \cap B$ yields (2.8). □

Property (vi) of Definition 2.3.1 allows us to avoid exceptions in the following classical result.

Theorem 2.3.9 (Fisher's Inequality). *For any* (v, b, r, k, λ)*-design, the number of points does not exceed the number of blocks, i.e.,* $v \leq b$*.*

Proof. Let $\mathbf{D} = (X, \mathcal{B}, I)$ be a (v, b, r, k, λ)-design. For each $x \in X$, let $\mathcal{B}_x$ denote the set of all blocks $B \in \mathcal{B}$ incident with x. If $\mathcal{B}_x = \mathcal{B}_y$ for distinct points $x, y \in X$, then $\lambda = r$ and (2.5) implies that either $r = 0$ or $v = k$. Then $I = \varnothing$ or $I = X \times \mathcal{B}$, and therefore $v = b$. Thus, we may assume that $\mathcal{B}_x \neq \mathcal{B}_y$ for any distinct points $x, y \in X$. Condition (v) of Definition 2.3.1 implies that $|\mathcal{B}_x \cap \mathcal{B}_y| = \lambda$ for any distinct $x, y \in X$. If $\lambda = 0$ and $r \neq 0$, then (2.5) implies that $k = 1$, so sets $\mathcal{B}_x$ are distinct singletons, and then $v \leq b$. If $\lambda > 0$, then Non-Uniform Fisher's Inequality applied to the family $\{\mathcal{B}_x : x \in X\}$ of subsets of $\mathcal{B}$ yields $v \leq b$. □

Remark 2.3.10. Another proof of Fisher's Inequality is proposed in Exercise 26.

Remark 2.3.11. Equations (2.4) and (2.5) and Fisher's Inequality are not sufficient for the existence of a (v, b, r, k, λ)-design. For instance, there is no $(22, 22, 7, 7, 2)$-design (see Remark 2.4.11) or a $(15, 21, 7, 5, 2)$-design (Corollary 8.2.21). However, for $k \leq 5$, these conditions are sufficient with the only exception of the parameter set $(15, 21, 7, 5, 2)$. The smallest unresolved parameter set for (v, b, r, k, λ)-designs is $(46, 69, 9, 6, 1)$.

Equations (2.4) and (2.5) indicate that some of the conditions of Definition 2.3.1 may imply the other conditions. The following three propositions confirm it.

Proposition 2.3.12. *Let* $\mathbf{D} = (X, \mathcal{B}, I)$ *be an incidence structure satisfying conditions* (i), (iv), (v), *and* (vi) *of Definition 2.3.1. If* $k \geq 2$, *then* $\mathbf{D}$ *is a* (v, b, r, k, λ)*-design with* $r = \lambda(v-1)/(k-1)$ *and* $b = vr/k$.

Proof. For the incidence structure $\mathbf{D}$, equation (2.2) reads $\lambda(v-1) = r(x)(k-1)$. Therefore, $r(x) = r = \lambda(v-1)/(k-1)$ is the same for all $x \in X$, so $\mathbf{D}$ is a (v, b, r, k, λ)-design, and then (2.1) implies that $b = vr/k$. □

Proposition 2.3.13. *Let* $\mathbf{D} = (X, \mathcal{B}, I)$ *be an incidence structure satisfying conditions* (i), (ii), (iii), (v), *and* (vi) *of Definition 2.3.1. Suppose further that there exists a real number k satisfying equations* (2.4) *and* (2.5). *Then* $\mathbf{D}$ *is a* (v, b, r, k, λ)*-design.*

Proof. For the incidence structure $\mathbf{D}$, equations (2.2) and (2.5) imply that

$$\sum_{B \ni x} |B| = \lambda(v-1) + r = rk.$$

Since $\sum_{B \in \mathcal{B}} |B|^2 = \sum_{x \in X} \sum_{B \ni x} |B|$, equation (2.4) implies that

$$\sum_{B \in \mathcal{B}} |B|^2 = vrk = bk^2.$$

Since $\sum_{B \in \mathcal{B}} |B| = vr = bk$, we obtain that

$$\sum_{B \in \mathcal{B}} (|B| - k)^2 = bk^2 - 2bk^2 + bk^2 = 0,$$

and $|B| = k$ for all $B \in \mathcal{B}$. Therefore, $\mathbf{D}$ is a (v, b, r, k, λ)-design. □

Proposition 2.3.14. *Let* $\mathbf{D} = (X, \mathcal{B}, I)$ *be an incidence structure satisfying conditions* (i) – (iv) *and* (vi) *of Definition 2.3.1. Suppose further that there exists a nonnegative integer* λ *such that* $(v-1)\lambda = r(k-1)$ *and* (i) *any two points of* $\mathbf{D}$ *are incident with at most* λ *blocks or* (ii) *any two points of* $\mathbf{D}$ *are incident with at least* λ *blocks. Then* $\mathbf{D}$ *is a* (v, b, r, k, λ)*-design.*

Proof. Fixing a point $x \in X$ and counting flags (y, B) where x is incident with B yields either $(v-1)\lambda \geq r(k-1)$ or $(v-1)\lambda \leq r(k-1)$, respectively. Since, in fact, $(v-1)\lambda = r(k-1)$, we obtain that in either case there are exactly λ blocks containing $\{x, y\}$. Therefore, $\mathbf{D}$ is a (v, b, r, k, λ)-design. □

Proposition 2.3.12 allows us to give the following definition.

Definition 2.3.15. An incidence structure **D** satisfying conditions (i) – (v) of Definition 2.3.1 is called a 2-(v, k, λ) design if $k \geq 2$.

Remark 2.3.16. A more general notion of a t-(v, k, λ) design is considered in Section 6.1.

Remark 2.3.17. Since two points of a block are contained in at least one block, we have $\lambda \geq 1$ for any 2-(v, k, λ) design.

2.4. Symmetric designs

Symmetric designs, the main subject of this book, were described informally in Chapter 1. We will now give a formal definition.

Definition 2.4.1. A *symmetric* (v, k, λ)*-design* is a (v, v, k, k, λ)-design.

Clearly, the complement of a symmetric (v, k, λ)-design is a symmetric $(v, v-k, v-2k+\lambda)$-design.

Proposition 2.3.7 yields the following basic relation for symmetric designs.

Proposition 2.4.2. *For any symmetric* (v, k, λ)*-design,*

$$\lambda(v-1) = k(k-1). \tag{2.9}$$

The Fano Plane (Example 1.3.1) is a symmetric (7, 3, 1)-design. Trivial designs (with incidence matrices O and J) are symmetric designs with parameters $(v, 0, 0)$ and (v, v, v), respectively. The block set of a symmetric $(v, 1, 0)$-design consists of all singletons of a v-set, and the block set of a symmetric $(v, v-1, v-2)$-design consists of all $(v-1)$-subsets of a v-set. Example 1.3.3 describes a symmetric (16, 6, 2)-design.

Example 2.4.3. Let a 6×6 array L contain each of the digits 1, 2, 3, 4, 5, and 6 in each row and in each column. (Such an array is called a *Latin square of order* 6.) Let $L(i, j)$ be the (i, j)-entry of L. Define the point set X to consist of the ordered pairs (i, j) with $i, j = 1, 2, 3, 4, 5, 6$. For each $x = (i, j) \in X$, define B_x to be the set of points (l, m), other than x, such that $l = i$ or $m = j$ or $L(l, m) = L(i, j)$. Let $\mathcal{B} = \{B_x : x \in X\}$. Then $\mathbf{D} = (X, \mathcal{B})$ is a symmetric (36, 15, 6)-design.

Example 2.4.4. Let $n \geq 2$ be an integer and let $\mathcal{P}$ be the set of all nonempty subsets of the set $\{1, 2, \ldots, n\}$. Consider the incidence structure $\mathbf{D} = (\mathcal{P}, \mathcal{P}, I)$

with $(X, Y) \in I$ if and only if the cardinality of the intersection $X \cap Y$ is even. Then **D** is a symmetric $(2^n - 1, 2^{n-1} - 1, 2^{n-2} - 1)$-design.

Incidence matrices of a (v, b, r, k, λ)-design satisfy the three equations (2.3). For symmetric designs, one equation suffices, as is shown by the following theorem.

Theorem 2.4.5. *A (0, 1)-matrix N of order v is an incidence matrix of a symmetric (v, k, λ)-design if and only if*

$$NN^\top = (k - \lambda)I + \lambda J, \tag{2.10}$$

where I is the identity matrix and J is the all-one matrix of order v.

Proof. If N is an incidence matrix of a symmetric (v, k, λ)-design, then (2.10) follows from (2.3).

Suppose N is a (0, 1)-matrix of order v satisfying (2.10). If $N = O$ or $N = J$, then (v, k, λ) are the parameters of a trivial symmetric design. Assume that $N \neq O$ and $N \neq J$. Then $v > 1$. Observe that the diagonal entries k and off-diagonal entries λ of $NN^\top$ represent the row sum and the inner product of two distinct rows of N, respectively. Therefore, $k > \lambda \geq 0$. By Lemma 2.3.6,

$$\det(NN^\top) = (\det N)^2 = (k + \lambda(v - 1))(k - \lambda)^{v-1}.$$

Therefore, N is nonsingular. Since the row sum of N is k, we have $NJ = kJ$, which implies $N^{-1}J = \frac{1}{k}J$. Therefore, multiplying (2.10) on the left by N^{-1} and on the right by N yields

$$N^\top N = (k - \lambda)I + \frac{\lambda}{k}JN.$$

Comparing (j, j)-entries on both sides of this equation yields

$$c_j = k - \lambda + \frac{\lambda}{k}c_j,$$

where c_j is the sum of the entries in the jth column of N. Therefore, $c_j = k$ for $j = 1, 2, \ldots, v$, and N is an incidence matrix of a symmetric (v, k, λ)-design. □

Remark 2.4.6. The proof of the above theorem shows in fact that if a (0, 1)-matrix N of order v satisfies (2.10), then

$$N^\top N = (k - \lambda)I + \lambda J,$$

i.e., the dual of a symmetric (v, k, λ)-design is a symmetric (v, k, λ)-design. This implies that any two distinct blocks of a symmetric (v, k, λ)-design meet in λ points. This also implies the following proposition.

Remark 2.4.7. If a symmetric (v, k, λ)-design $\mathbf{D}$ admits a symmetric incidence matrix, then, of course, the dual design $\mathbf{D}^\top$ is isomorphic to $\mathbf{D}$, i.e., $\mathbf{D}$ is self-dual. However, the converse is not true: there exists a self-dual symmetric $(25, 9, 3)$-design that does not admit a symmetric incidence matrix.

Proposition 2.4.8. *An incidence structure having v points and v blocks, constant block size k, and constant intersection size λ between any two distinct blocks is a symmetric (v, k, λ)-design.*

The next proposition gives another sufficient condition for an incidence structure to be a symmetric design.

Proposition 2.4.9. *Let λ and μ be positive integers and let $\mathbf{D} = (X, \mathcal{B}, I)$ be an incidence structure satisfying the following conditions*:

(i) $r(x) < |\mathcal{B}|$ *for all* $x \in X$;
(ii) $|B| < |X|$ *for all* $B \in \mathcal{B}$;
(iii) $\lambda(x, y) = \lambda$ *for any distinct* $x, y \in X$;
(iv) $|A \cap B| = \mu$ *for any distinct* $A, B \in \mathcal{B}$.

Then $\mathbf{D}$ *is either a symmetric design or a pencil.*

Proof. If $\mathbf{D}$ has distinct blocks A and B such that the set of points incident with A is the same as the set of points incident with B, then $|A| = |B| = \mu$ and, for any block C, every point incident with A is incident with C. However, this is not the case due to (i). Similarly, distinct points of $\mathbf{D}$ are incident with distinct sets of blocks. Therefore, we can consider the block set of $\mathbf{D}$ as a set of subsets of X and the block set of $\mathbf{D}^\top$ as a set of subsets of $\mathcal{B}$. Non-uniform Fisher's Inequality then implies that $|X| = |\mathcal{B}|$.

Suppose first that $\lambda > 1$. Let $A \in \mathcal{B}$ and $x \in A$. Counting in two ways flags (y, B) of $\mathbf{D}$ with $y \neq x$, $B \neq A$, $y \in A$, and $x \in B$ yields $(|A| - 1)(\lambda - 1) = (r(x) - 1)(\mu - 1)$. Therefore, $|A|$ is the same for all blocks A containing a given point x. Since any two blocks of $\mathbf{D}$ have a common point, all blocks have the same cardinality and $\mathbf{D}$ is a symmetric design. If $\mu > 1$, then, for similar reasons, $\mathbf{D}^\top$ is a symmetric design and so is $\mathbf{D}$.

Suppose now that $\lambda = \mu = 1$. If all blocks of $\mathbf{D}$ have the same cardinality or all points of $\mathbf{D}$ have the same replication number, then $\mathbf{D}$ is a symmetric design. Otherwise, by the Ryser–Woodall Theorem, applied to both $\mathbf{D}$ and $\mathbf{D}^\top$, the set X can be partitioned into nonempty subsets X_1 and X_2, and $\mathcal{B}$ can be partitioned into nonempty subsets $\mathcal{B}_1$ and $\mathcal{B}_2$ so that, for $i = 1$ and 2, all points of X_i have the same replication number r_i and all blocks of $\mathcal{B}_i$ have the same cardinality k_i. Let $A \in \mathcal{B}$ and $x \in X \setminus A$. Counting in two ways flags (y, B) of $\mathbf{D}$ with $y \in A$ and $x \in B$ yields $|A| = r(x)$. This means that every block A contains either X_1 or X_2 and, for each i, all blocks of $\mathcal{B}_i$ contain the same set X_j. Without loss of generality, we assume that the blocks of $\mathcal{B}_1$ contain X_1 and the blocks of $\mathcal{B}_2$ contain X_2. If $|\mathcal{B}_i| \geq 2$, then $|X_i| = 1$; similarly, if $|X_i| \geq 2$, then $|\mathcal{B}_i| = 1$. Therefore, we may assume that $|\mathcal{B}_1| = |X_2| = 1$. Let $\mathcal{B}_1 = \{A\}$ and $X_2 = \{x\}$. Then $A = X_1$ and therefore, every block of $\mathcal{B}_2$ contains x and one point of X_1. Thus, $\mathbf{D}$ is a pencil. □

If N is an incidence matrix of a symmetric (v, k, λ)-design, then $\det(NN^\top) = (k + \lambda(v-1))(k-\lambda)^{v-1} = k^2(k-\lambda)^{v-1}$. On the other hand, $\det(NN^\top) = (\det N)^2$ must be a perfect square. This gives the following necessary condition for the parameters of a symmetric design.

Proposition 2.4.10. *If (v, k, λ) are the parameters of a symmetric design and v is even, then $k - \lambda$ is a perfect square.*

Remark 2.4.11. This proposition shows that the necessary condition (2.9) for the parameters of a symmetric design is not sufficient. For instance, a symmetric $(22, 7, 2)$-design cannot exist even though its parameters satisfy (2.9). We now have two restrictions on the parameters of a symmetric (v, k, λ)-design with v even:

$$\lambda(v-1) = k(k-1), \quad k - \lambda \text{ is a perfect square.}$$

It is not known whether these conditions are sufficient for existence of a symmetric (v, k, λ)-design. The smallest unresolved parameter set is $(154, 18, 2)$.

In the next section, we will prove the Bruck–Ryser–Chowla Theorem that gives a necessary condition for the parameters of a symmetric (v, k, λ)-design with v odd.

Equation (2.9) implies bounds on the number of points of a symmetric design of a given order.

Proposition 2.4.12. *Let $\mathbf{D}$ be a symmetric (v, k, λ)-design of order $n = k - \lambda \geq 2$. Then*

$$4n - 1 \leq v \leq n^2 + n + 1.$$

Proof. Since $\mathbf{D}$ and its complement $\mathbf{D}'$ have the same order, we can assume without loss of generality that $v \geq 2k$. Equation (2.9) implies that λ and $v - 2n - \lambda = v - 2k + \lambda$ are the roots of the quadratic equation

$$x^2 - (v - 2n)x + n(n - 1) = 0. \tag{2.11}$$

Since the discriminant of this equation is nonnegative, we have

$$(v - 2n)^2 \geq 4n(n - 1) = (2n - 1)^2 - 1.$$

Since $(2n - 1)^2 - 1$ is not a perfect square for $n \geq 2$, we have $v - 2n \geq 2n - 1$, so $v \geq 4n - 1$.

Since the left-hand side of (2.11) is positive at $x = 0$ and since the roots of this equation are integers, it is nonnegative at $x = 1$. This implies that $v \leq n^2 + n + 1$. □

Symmetric designs meeting the bounds of Proposition 2.4.12 are projective planes and Hadamard 2-designs which will be considered in Chapters 3 and 4, respectively.

Given a symmetric design $\mathbf{D}$ with a fixed block, one can obtain the following two 2-designs as substructures of $\mathbf{D}$.

Definition 2.4.13. Let $\mathbf{D} = (X, \mathcal{B}, I)$ be a nontrivial symmetric design and let B be a block of $\mathbf{D}$. The substructures $\mathbf{D}^B$ and $\mathbf{D}_B$ are called a *residual design of* $\mathbf{D}$ and a *derived design of* $\mathbf{D}$, respectively.

The blocks of $\mathbf{D}^B$ and $\mathbf{D}_B$ can be regarded as sets $A \setminus B$ and $A \cap B$, respectively, where A is a block of $\mathbf{D}$ other than B. If N is an incidence matrix of $\mathbf{D}$ such that the last column of N corresponds to the block B, then

$$N = \begin{bmatrix} S & \mathbf{0} \\ T & \mathbf{j} \end{bmatrix}$$

where S is an incidence matrix of the residual design $\mathbf{D}^B$ and T is an incidence matrix of the derived design $\mathbf{D}_B$.

Remark 2.4.14. The residual and derived designs of a symmetric design with respect to the same block do not determine this symmetric design uniquely: there exist symmetric (25, 9, 3)-designs $\mathbf{D}$ and $\mathbf{E}$ and blocks A of $\mathbf{D}$ and B of $\mathbf{E}$ such that the residual designs $\mathbf{D}^A$ and $\mathbf{E}^B$ are isomorphic and the derived designs $\mathbf{D}_A$ and $\mathbf{E}_B$ are isomorphic, yet the designs $\mathbf{D}$ and $\mathbf{E}$ are not isomorphic.

The following proposition is straightforward.

Proposition 2.4.15. *Let* **D** *be a nontrivial symmetric* (v, k, λ)*-design with* $v > k \geq 2$ *and let* B *be a block of* **D**. *Then* $\mathbf{D}^B$ *is a* $(v-k, v-1, k, k-\lambda, \lambda)$*-design and* $\mathbf{D}_B$ *is a* $(k, v-1, k-1, \lambda, \lambda-1)$*-design.*

Proposition 2.1.12 immediately implies the following result.

Proposition 2.4.16. *Let* $\mathbf{D} = (X, \mathcal{B})$ *be a symmetric* (v, k, λ)*-design with* $v > k \geq 2$ *and let* $\mathbf{D}'$ *be the complementary design. Then, for any block* B *of* **D**, *the designs* $\mathbf{D}^B$ *and* $\mathbf{D}'_{X \setminus B}$ *are isomorphic as well as the designs* $\mathbf{D}_B$ *and* $(\mathbf{D}')^{X \setminus B}$.

Observe that if a (v, b, r, k, λ)-design is a residual of a symmetric design **D**, then $r = k + \lambda$ and **D** is a symmetric $(v + r, r, \lambda)$-design.

Definition 2.4.17. Any (v, b, r, k, λ)-design **D** with $r = k + \lambda$ is called a *quasi-residual design*. If **D** is a residual of a symmetric $(v + r, r, \lambda)$-design, then it is said to be *embeddable*. Otherwise, **D** is said to be *non-embeddable*.

Example 2.4.18 (Bhattacharya's Example). The following incidence structure $\mathbf{D} = (X, \mathcal{B})$ is a $(16, 24, 9, 6, 3)$-design, so it is quasi-residual. Let $X = \{a, b, c, \ldots, o, p\}$ and let $\mathcal{B}$ be the following family of 6-subsets of X:

abcdef	*abcdgh*	*abijlm*	*acjklo*	*adimnp*	*aegjno*	*aegkmp*	*afhikn*
afhlop	*bcijkp*	*bdlmno*	*befiop*	*behkmo*	*bfgkln*	*bghjnp*	*cdknop*
cefjmn	*cehiln*	*cfglmp*	*cghimo*	*degikl*	*dehjlp*	*dfgijo*	*dfhjkm*

This design has blocks that meet in four points, for instance, the first two blocks. Therefore, **D** cannot be a residual of a symmetric $(25, 9, 3)$-design, i.e., **D** is a non-embeddable quasi-residual design.

Two symmetric designs with the same parameters do not have to be isomorphic (see Theorem 2.4.21). Sometimes, one can prove that two symmetric designs are not isomorphic by comparing the ranks of their incidence matrices over a finite field.

Definition 2.4.19. Let **D** be a symmetric (v, k, λ)-design and let N be an incidence matrix of **D**. For any prime p, the *p-rank* of **D** is the rank of N regarded as a matrix over the field $GF(p)$ of residue classes modulo p. The p-rank of **D** is denoted as $rank_p(\mathbf{D})$.

Remark 2.4.20. Proposition 2.1.13 immediately implies that the p-rank of a symmetric design **D** is independent of the choice of an incidence matrix of the design.

The following theorem can be obtained using the 2-ranks. We leave its proof as an exercise.

Theorem 2.4.21. *There are exactly three nonisomorphic symmetric* (16, 6, 2)-*designs. Their* 2-*ranks are* 6, 7, *and* 8.

Another application of 2-ranks is given in Section 3.7 (Theorems 3.7.14 and 3.7.16.).

2.5. The Bruck–Ryser–Chowla Theorem

In this section we obtain a necessary condition on the parameters of a symmetric (v, k, λ)-design with v odd. We first develop some classical number-theoretical results related to the *Legendre symbol*. We then define the *Hilbert symbols* whose calculation uses the Legendre symbol. The Hilbert symbols are used to define the *Hasse invariants* for symmetric matrices over the integers.

Definition 2.5.1. For any odd prime p and for any integer $a \not\equiv 0 \pmod{p}$, the *Legendre symbol* $\left(\frac{a}{p}\right)$ is defined to be equal to 1 if there exists an integer x such that $a \equiv x^2 \pmod{p}$; $\left(\frac{a}{p}\right) = -1$ otherwise.

The following properties of the Legendre symbol can be found in standard number theory texts.

Theorem 2.5.2. *Let p and q be distinct odd primes and let a and b be integers not divisible by p. Then*

(i) *if* $a \equiv b \pmod{p}$, *then* $\left(\frac{a}{p}\right) = \left(\frac{b}{p}\right)$;

(ii) $\left(\frac{ab}{p}\right) = \left(\frac{a}{p}\right)\left(\frac{b}{p}\right)$;

(iii) $\left(\frac{-1}{p}\right) = (-1)^{(p-1)/2}$;

(iv) $\left(\frac{2}{p}\right) = (-1)^{(p^2-1)/8}$;

(v) $\left(\frac{q}{p}\right)\left(\frac{p}{q}\right) = (-1)^{(p-1)(q-1)/4}$.

Remark 2.5.3. Property (v) of Theorem 2.5.2 is the celebrated *Quadratic Reciprocity Law*.

Properties (i) and (ii) of Theorem 2.5.2 almost uniquely define the Legendre symbol, as the next proposition shows.

Proposition 2.5.4. *Let p be an odd prime and let a function L from the set of all integers not divisible by p to the set $\{-1, 1\}$ have the following properties:*

(i) *if* $a \equiv b \pmod{p}$, *then* $L(a) = L(b)$;
(ii) $L(ab) = L(a)L(b)$ *for all* a *and* b.

Then either $L(a) = 1$ *for all* a *or* $L(a) = \left(\frac{a}{p}\right)$ *for all* a.

Proof. Property (i) allows us to regard L as a function from the multiplicative group G of residue classes mod p to the group $\{-1, 1\}$ of order 2. Property (ii) implies that this function is a homomorphism. The kernel of this homomorphism is either the entire group G or a subgroup of index 2. In the former case, $L(a) = 1$ for all $a \in G$. In the latter case, since $L(a^2) = 1$ for all $a \in G$, the kernel is the subgroup of all squares. Therefore, in this case, $L(a) = \left(\frac{a}{p}\right)$ for all $a \in G$. □

The next theorem will allow us to define the *Hilbert symbols*.

Theorem 2.5.5. *For any odd prime* p, *there exists a unique function* $(a, b) \longrightarrow (a, b)_p$ *from* $\mathbb{Z}^* \times \mathbb{Z}^*$ *to* $\{-1, 1\}$ *that satisfies the following conditions:*

(H1) $(a, b)_p = (b, a)_p$, *for any* $a, b \in \mathbb{Z}^*$;
(H2) $(ab, c)_p = (a, c)_p(b, c)_p$, *for any* $a, b, c \in \mathbb{Z}^*$;
(H3) $(a, b)_p = 1$, *for any integers* $a, b \not\equiv 0 \pmod{p}$;
(H4) *if* $a \not\equiv 0 \pmod{p}$, *then* $(a, p)_p = \left(\frac{a}{p}\right)$;
(H5) $(-p, p)_p = 1$.

Proof. Let a function $(a, b) \longrightarrow (a, b)_p$ from $\mathbb{Z}^* \times \mathbb{Z}^*$ to $\{-1, 1\}$ satisfy conditions (H1) – (H5). Then $(p, p)_p = \left(\frac{-1}{p}\right)$ and therefore, for any nonnegative integers s and t, $(p^s, p^t)_p = \left(\frac{-1}{p}\right)^{st}$. Let $a, b \in \mathbb{Z}^*$ and let $a = p^s a_0$ and $b = p^t b_0$ where s and t are nonnegative integers and a_0 and b_0 are integers not divisible by p. Then

$$(a, b)_p = \left(\frac{-1}{p}\right)^{st} \left(\frac{a_0}{p}\right)^{t} \left(\frac{b_0}{p}\right)^{s}. \tag{2.12}$$

Conversely, if we define a function $(a, b) \longrightarrow (a, b)_p$ from $\mathbb{Z}^* \times \mathbb{Z}^*$ to $\{-1, 1\}$ by (2.12), then it is straightforward to verify that it satisfies (H1) – (H5). □

Definition 2.5.6. The functions $(a, b) \longrightarrow (a, b)_p$ from $\mathbb{Z}^* \times \mathbb{Z}^*$ to $\{-1, 1\}$ defined, for odd primes p, by (2.12) are called the *Hilbert symbols*.

The next proposition gives further properties of Hilbert symbols.

Proposition 2.5.7. *The Hilbert symbol* $(a, b)_p$ *satisfies the following properties for any nonzero integers* a *and* b *and odd prime* p:

(H6) $(a^2, b)_p = 1$;
(H7) *if* $a + b$ *is a square, then* $(a, b)_p = 1$;
(H8) $(a, -a)_p = 1$;
(H9) *if* $a + b \neq 0$, *then* $(a, b)_p = (a + b, -ab)_p$.

Proof. (H6) follows immediately from (H2).
(H7) If $a \not\equiv 0 \pmod p$ and $b \not\equiv 0 \pmod p$, then $(a, b)_p = 1$ by (H3).

Suppose that $a \not\equiv 0 \pmod p$ and $b \equiv 0 \pmod p$. Let $a + b = x^2$ and $b = p^t b_0$ where $b_0 \not\equiv 0 \pmod p$. Then $a \equiv x^2 \pmod p$, so, by (H2), (H3), and (H6), we obtain:

$$(a, b)_p = (a, b_0)_p (a, p)_p^t = (x^2, p)_p^t = 1.$$

Suppose that $a \equiv b \equiv 0 \pmod p$. Let $a = p^s a_0, b = p^t b_0$ where $a_0, b_0 \not\equiv 0 \pmod p$. Then

$$(a, b)_p = (a_0, b_0)_p (a_0, p)_p^t (b_0, p)_p^s (p, p)_p^{st}. \tag{2.13}$$

If s and t are even, then $(a, b)_p = 1$. Suppose that s is even and t is odd. Since $a + b = p^s a_0 + p^t b_0$ is a square and $s \neq t$, the smaller of the exponents s, t must be even, i.e., $s < t$. Then $a + b = p^s(a_0 + p^{t-s} b_0)$, so $a_0 + p^{t-s} b_0$ is a square. Therefore, $(a_0, p)_p = 1$ and (2.13) implies that $(a, b)_p = 1$. Suppose finally that both s and t are odd. If $s \neq t$, then the highest power of p dividing $a + b$ is odd, and $a + b$ cannot be a square. Therefore, $s = t$, and we have $a + b = p^s(a_0 + b_0)$. Since $a + b$ is a square and s is odd, $a_0 + b_0 \equiv 0 \pmod p$. Therefore, (2.13) implies that

$$\begin{aligned}(a, b)_p &= (a_0, p)_p (b_0, p)_p (p, p)_p = (a_0, p)_p (-a_0, p)_p (-1, p)_p (-p, p)_p \\ &= (a_0, p)_p^2 (-p, p)_p = 1.\end{aligned}$$

(H8) follows from (H7).
(H9) Since $a(a + b) + b(a + b) = (a + b)^2$, we apply (H7) to obtain that $(a(a + b), b(a + b))_p = 1$. Therefore,

$$\begin{aligned}(a, b)_p (a, a + b)_p (b, a + b)_p (a + b, a + b)_p &= 1, \\ (a, b)_p (ab, a + b)_p (-1, a + b)_p (-(a + b), a + b)_p &= 1, \\ (a, b)_p (-ab, a + b)_p = 1, (a, b)_p &= (-ab, a + b)_p.\end{aligned}$$

□

We next use the Hilbert symbols to define the *Hasse invariants* of symmetric matrices over the integers.

Definition 2.5.8. Let A be a symmetric matrix of order n with integral entries. For $i = 1, 2, \cdots, n$, let $D_i(A)$ be the determinant of the submatrix formed by the first i rows and the first i columns of A. Suppose that the determinants $D_1(A), D_2(A), \cdots, D_n(A)$ are not equal to zero. Let p be an odd prime. Then the product

$$c_p(A) = (-1, D_n(A))_p \prod_{i=1}^{n-1} (D_i(A), -D_{i+1}(A))_p$$

is called the *Hasse p-invariant of A*.

The following theorem is central to applications of Hasse invariants to designs. Its proof is beyond the scope of this book.

Theorem 2.5.9. *If N is a nonsingular matrix over the integers, then $c_p(NN^\top) = 1$, for every odd prime p.*

We are now ready to prove the Bruck–Ryser–Chowla Theorem, which gives a necessary condition on the parameters of a symmetric (v, k, λ)-design in case v is odd.

Theorem 2.5.10 (The Bruck–Ryser–Chowla Theorem). *If there exists a nontrivial symmetric (v, k, λ)-design with odd v, then $((-1)^{\frac{v-1}{2}}\lambda, k-\lambda)_p = 1$, for any odd prime p.*

Proof. Let N be the incidence matrix of a nontrivial symmetric (v, k, λ)-design and let $A = NN^\top$. Then $A = (k-\lambda)I + \lambda J$. For $i = 1, 2, \cdots, v$, let D_i be the determinant of the matrix formed by the first i rows and the first i columns of A. By Lemma 2.3.6, $D_i = a_i(k-\lambda)^{i-1}$ where $a_i = k + (i-1)\lambda$. Note that $a_v = k^2$, so $(-1, D_v)_p = 1$, for any odd prime p. By Theorem 2.5.9, $c_p(A) = 1$. Therefore, we have

$$\begin{aligned}
1 = c_p(A) = \prod_{i=1}^{v-1} (D_i, -D_{i+1})_p &= \prod_{i=1}^{\frac{v-1}{2}} (D_{2i-1}, -D_{2i})_p (D_{2i}, -D_{2i+1})_p \\
&= \prod_{i=1}^{\frac{v-1}{2}} (a_{2i-1}(k-\lambda)^{2i-2}, -a_{2i}(k-\lambda)^{2i-1})_p (a_{2i}(k-\lambda)^{2i-1}, \\
&\qquad - a_{2i+1}(k-\lambda)^{2i})_p \\
&= \prod_{i=1}^{\frac{v-1}{2}} (a_{2i-1}, -a_{2i}(k-\lambda))_p (a_{2i}(k-\lambda), -a_{2i+1})_p \\
&= \prod_{i=1}^{\frac{v-1}{2}} (a_{2i-1}, -a_{2i})_p (a_{2i-1}, k-\lambda)_p (a_{2i}, -a_{2i+1})_p (k-\lambda, -a_{2i+1})_p.
\end{aligned}$$

Note that $a_{2i-1} - a_{2i} = -\lambda$, and we apply (H9) to obtain that $(a_{2i-1}, -a_{2i})_p = (-\lambda, a_{2i-1}a_{2i})_p$ and $(a_{2i}, -a_{2i+1})_p = (-\lambda, a_{2i}a_{2i+1})_p$. Therefore,

$$1 = c_p(A) = \prod_{i=1}^{\frac{v-1}{2}} (-\lambda, a_{2i-1}a_{2i})_p(-\lambda, a_{2i}a_{2i+1})_p(k-\lambda, a_{2i-1}a_{2i+1})_p(k-\lambda, -1)_p$$

$$= \left((-1)^{\frac{v-1}{2}}, k-\lambda\right)_p \prod_{i=1}^{\frac{v-1}{2}} \left(-\lambda, a_{2i-1}a_{2i}^2a_{2i+1}\right)_p (k-\lambda, a_{2i-1}a_{2i+1})_p$$

$$= \left((-1)^{\frac{v-1}{2}}, k-\lambda\right)_p \left(-\lambda(k-\lambda), \prod_{i=1}^{\frac{v-1}{2}} a_{2i-1}a_{2i+1}\right)_p$$

$$= \left((-1)^{\frac{v-1}{2}}, k-\lambda\right)_p (-\lambda(k-\lambda), a_1a_v)_p = \left((-1)^{\frac{v-1}{2}}, k-\lambda\right)_p (-\lambda(k-\lambda), k)_p.$$

By (H9), $(-\lambda(k-\lambda), k)_p = (\lambda, k-\lambda)_p$, and the proof is now complete. □

Example 2.5.11. If there exists a symmetric (43, 7, 1)-design, then $(-1, 6)_p = 1$ for any odd prime p. However, $(-1, 6)_3 = (-1, 3)_3 = \left(\frac{-1}{3}\right) = -1$. Therefore, there is no symmetric (43, 7, 1)-design.

Example 2.5.12. If there exists a symmetric (29, 8, 2)-design, then $(2, 6)_3 = 1$. On the other hand, $(2, 6)_3 = (2, 3)_3 = \left(\frac{2}{3}\right) = -1$. Therefore, there is no symmetric (29, 8, 2)-design.

Remark 2.5.13. The condition of the Bruck–Ryser–Chowla Theorem is not sufficient for the existence of symmetric designs. The only known counterexample is the parameter set (111, 11, 1). It satisfies the condition of the Bruck–Ryser–Chowla Theorem (and the equation (2.9)). However, there is no symmetric (111, 11, 1)-design (Theorem 6.4.5). An unresolved parameter set for a symmetric design with the smallest number of points is (81, 16, 3).

2.6. Automorphisms of symmetric designs

In Definition 2.1.10, we introduced the notion of isomorphic incidence structures. If $\mathbf{D}_1 = (X_1, \mathcal{B}_1)$ and $\mathbf{D}_2 = (X_2, \mathcal{B}_2)$ are nontrivial symmetric designs, we can regard $\mathcal{B}_1$ and $\mathcal{B}_2$ as sets of subsets of X_1 and X_2, respectively. An isomorphism of $\mathbf{D}_1$ and $\mathbf{D}_2$ in this case can be regarded as a bijection $f : X_1 \to X_2$ such that $f(B)$ is a block of $\mathbf{D}_2$ if and only if B is a block of $\mathbf{D}_1$. It is often convenient to assume that $X_1 = X_2$; then an isomorphism of $\mathbf{D}_1$ and $\mathbf{D}_2$ can be regarded as a permutation of the point set X_1 that maps blocks of $\mathbf{D}_1$ onto blocks of $\mathbf{D}_2$.

Definition 2.6.1. Let X be a finite set and $\mathbf{D} = (X, \mathcal{B})$ a nontrivial symmetric design. Let S_X be the group of all permutations of the set X. For $\sigma \in S_X$, let $\sigma\mathbf{D} = (X, \sigma(\mathcal{B}))$ where $\sigma(\mathcal{B}) = \{\sigma B : B \in \mathcal{B}\}$. Then $\mathbf{D}$ and $\sigma\mathbf{D}$ are isomorphic symmetric designs. If $\sigma\mathbf{D} = \mathbf{D}$, i.e., $\sigma(\mathcal{B}) = \mathcal{B}$, then σ is called an *automorphism* of $\mathbf{D}$. All automorphisms of a symmetric design $\mathbf{D}$ form *the full automorphism group of* $\mathbf{D}$ denoted by Aut($\mathbf{D}$). Any subgroup of Aut($\mathbf{D}$) is called *an automorphism group of* $\mathbf{D}$. A point $x \in X$ (respectively, a block $B \in \mathcal{B}$) is called a *fixed point* (respectively, a *fixed block*) of an automorphism $\sigma \in$ Aut($\mathbf{D}$) if $\sigma x = x$ (respectively, $\sigma B = B$).

The *action of a group on a set* is one of the basic notions of group theory.

Definition 2.6.2. Let X be a set and G a group. An *action of G on X* is a homomorphism from G to the group S_X of all permutations of the set X. Let f be a fixed action of G on X. Then, for $\sigma \in G$ and $x \in X$, we denote by $\sigma(x)$ or σx the element $f(\sigma)(x)$ of X. For $x \in X$, the subgroup $G_x = \{\sigma \in G : \sigma x = x\}$ is called the *stabilizer of x in G*. The action of G on X is said to be *faithful* if any two distinct elements σ and τ of G act differently on X, that is, there is $x \in X$ such that $\sigma x \neq \tau x$.

Example 2.6.3. Any group G acts on itself by the *left multiplication:* $\sigma(\tau) = \sigma\tau$.

An action of a group G on a set X induces a partition of X into *G-orbits*.

Definition 2.6.4. Let a group G act on a set X. For $x \in X$, the set $\{\rho x : \rho \in G\}$ is called the *G-orbit of x*.

Clearly, G-orbits of elements x and y of X are either disjoint or identical, so G-orbits on X form a partition of the set X. The cardinality of each G-orbit must divide the order of G, as the following theorem implies. Its proof can be found in standard group theory texts (e.g., Humphreys (1996)).

Theorem 2.6.5 (The Orbit-Stabilizer Theorem). *Let a finite group G act on a set X. For $x \in X$, let G_x be the stabilizer of x in G. Then the cardinality of the G-orbit of x is equal to the index of G_x in G.*

If all elements of a set X form one orbit under an action of a group G, the action is *sharply transitive*.

Definition 2.6.6. An action of a group G on a set X is said to be *sharply transitive* if for any $x, y \in X$ there is a unique $\sigma \in G$ such that $\sigma x = y$.

The following proposition is straightforward.

Proposition 2.6.7. *Let a group G act on a finite set X. The following statements are equivalent*:

(i) *the action of G on X is sharply transitive*;
(ii) $|G| = |X|$ *and there is only one G-orbit on X.*

Remark 2.6.8. A sharply transitive automorphism group of a symmetric design is also called a *regular automorphism group*.

If G is an automorphism group of a nontrivial symmetric design $\mathbf{D} = (X, \mathcal{B})$, then G acts on both X and $\mathcal{B}$. We will prove two useful results comparing G-orbits on X and G-orbits on $\mathcal{B}$.

Proposition 2.6.9. *Let $\mathbf{D} = (X, \mathcal{B})$ be a nontrivial symmetric design and let $\sigma \in \mathrm{Aut}(\mathbf{D})$. Then the number of fixed points of σ is equal to the number of fixed blocks of σ.*

Proof. Let N be an incidence matrix of $\mathbf{D}$, let $|X| = |\mathcal{B}| = v$, and let (for $i = 1, 2, \ldots, v$) x_i and B_i be the point and the block of $\mathbf{D}$ corresponding to the ith row and the ith column of N, respectively. The automorphism σ can be regarded as a pair of bijections $X \to X$ and $\mathcal{B} \to \mathcal{B}$. Let $P = [p_{ij}]$ and $Q = [q_{ij}]$ be the corresponding permutation matrices from Proposition 2.1.13; then $PN = NQ$. Furthermore, $\sigma x_i = x_i$ if and only if $p_{ii} = 1$ and $\sigma B_i = B_i$ if and only if $q_{ii} = 1$. Therefore, the number of fixed points of σ is equal to the trace of P and the number of fixed blocks of σ is equal to the trace of Q. Since N is a nonsingular matrix and $Q = N^{-1}PN$, these traces are equal, so σ has as many fixed points as fixed blocks. □

Remark 2.6.10. The above proof shows that the result is true for any incidence structure with a nonsingular square incidence matrix.

We will now show that the number of point orbits of an automorphism group of a symmetric design is equal to the number of block orbits. The proof of this result relies on the basic results on group actions often called the *Burnside Lemma*. Its proof can be found in standard group theory texts (e.g., Humphreys (1996)).

Proposition 2.6.11 (The Burnside Lemma). *Let a finite group G act on a finite set X. For any $\sigma \in G$, let $f(\sigma)$ be the number of fixed points of σ, i.e., the cardinality of the set $\{x \in X : \sigma x = x\}$. Then the number of G-orbits on X is equal to*

$$\frac{1}{|G|}\sum_{\sigma \in G} f(\sigma).$$

The following theorem is an immediate corollary of Proposition 2.6.9 and the Burnside Lemma.

Theorem 2.6.12 (The Orbit Theorem). *If G is an automorphism group of a symmetric design* **D**, *then the number of G-orbits on the point set of* **D** *is equal to the number of G-orbits on the block set.*

Corollary 2.6.13. *The action of an automorphism group of a symmetric design is sharply transitive on the point set of the design if and only if it is sharply transitive on the block set.*

The following proposition places further restrictions on possible actions of an automorphism group of a symmetric design on the point set and the block set of the design.

Proposition 2.6.14. *Let G be an automorphism group of a symmetric (v, k, λ)-design* $\mathbf{D} = (X, \mathcal{B})$. *Let $X_1, X_2, \ldots, X_m$ be all distinct G-orbits on X and let $\mathcal{B}_1, \mathcal{B}_2, \ldots, \mathcal{B}_m$ be all distinct G-orbits on $\mathcal{B}$. Then, for $i, j = 1, 2, \ldots, m$, there exist integers r_{ij} and k_{ij} such that every point of X_i is contained in exactly r_{ij} blocks from $\mathcal{B}_j$ and every block of $\mathcal{B}_j$ contains exactly k_{ij} points of X_i. Furthermore, the integers r_{ij} and k_{ij} satisfy the following equations:*

$$\sum_{j=1}^{m} r_{ij} = \sum_{i=1}^{m} k_{ij} = k, \tag{2.14}$$

$$r_{ij}|X_i| = k_{ij}|\mathcal{B}_j|, \tag{2.15}$$

$$\sum_{j=1}^{m} r_{ij}k_{ij} = \lambda(|X_i| - 1) + k, \tag{2.16}$$

$$\sum_{i=1}^{m} r_{ij}k_{ij} = \lambda(|\mathcal{B}_j| - 1) + k, \tag{2.17}$$

$$\textit{for } i \neq h, \ \sum_{j=1}^{m} r_{ij}k_{hj} = \lambda|X_h|, \tag{2.18}$$

$$\textit{for } j \neq h, \ \sum_{i=1}^{m} r_{ij}k_{ih} = \lambda|\mathcal{B}_h|. \tag{2.19}$$

Proof. If $x, y \in X_i$, then $y = \sigma x$ for some $\sigma \in G$. Then, for any $B \in \mathcal{B}_j$, $x \in B$ if and only if $y \in \sigma B$. Therefore, the number of blocks of $\mathcal{B}_j$ containing x is equal to the number of blocks of $\mathcal{B}_j$ containing y. Existence of the integers k_{ij} is similar.

Equations (2.14) are immediate. Equations (2.15) are obtained by counting in two ways flags (x, B) with $x \in X_i$ and $B \in \mathcal{B}_j$. To obtain (2.16), fix $x \in X_i$

and count in two ways flags (y, B) where $y \in X_i$, $B \ni x$, and $y \neq x$:

$$\lambda(|X_i| - 1) = \sum_{j=1}^{m} r_{ij}(k_{ij} - 1) = \sum_{j=1}^{m} r_{ij}k_{ij} - k.$$

If we select y from X_h rather than X_i, we obtain (2.18). The proof of (2.17) and (2.19) is similar. □

Given an automorphism group G of a symmetric design, one can find all blocks of the design if one block of each G-orbit is known.

Definition 2.6.15. A set of blocks of a symmetric design is called a set of *base blocks* with respect to an automorphism group G of the design if it contains exactly one block from each G-orbit on the block set.

If the parameters of a symmetric design are relatively small, then, given an automorphism group of the design, the restrictions imposed by the Orbit Theorem and Propositions 2.6.9 and 2.6.14 make the number of choices for a possible set of base blocks manageable. This leads to the following strategy for constructing a symmetric (v, k, λ)-design: choose a suitable automorphism group, apply these restrictions (and the basic properties of symmetric designs) to obtain a reasonable number of possibilities for base blocks, and then try these possibilities to either find a desired symmetric design or prove non-existence of the design with this automorphism group.

In the next two sections we will illustrate this strategy by constructing symmetric designs with parameters (41, 16, 6) and (79, 13, 2).

2.7. A symmetric (41, 16, 6)-design

In this section we construct a symmetric (41, 16, 6)-design $\mathbf{D} = (X, \mathcal{B})$ that admits an automorphism group

$$G = \langle \rho, \tau, \sigma \,|\, \rho^5 = \tau^2 = \sigma^3 = 1, \rho\tau = \tau\rho^{-1}, \rho\sigma = \sigma\rho, \tau\sigma = \sigma\tau \rangle$$

acting on the point set X in such a way that ρ has a unique fixed point and $\tau Y = Y$ for every $\langle \rho \rangle$-orbit Y on X.

Note that the group G is the direct product of a dihedral group of order 10 and a cyclic group of order 3.

We will denote by ∞ the unique fixed point of ρ. By Proposition 2.6.9, there is a unique block $B_\infty \in \mathcal{B}$ fixed by ρ. Then $\tau(\infty) = \infty$ and $\sigma(\infty) = \sigma\rho(\infty) = \rho\sigma(\infty)$, so $\sigma(\infty) = \infty$. Similarly, $\tau(B_\infty) = \sigma(B_\infty) = B_\infty$.

Since $|\langle\rho\rangle| = 5$, the Orbit-Stabilizer Theorem implies that each of the sets $X \setminus \{\infty\}$ and $\mathcal{B} \setminus \{B_\infty\}$ is partitioned into eight $\langle\rho\rangle$-orbits of cardinality 5. Let them be $X_1, X_2, \ldots, X_8$ and $\mathcal{B}_1, \mathcal{B}_2, \ldots, \mathcal{B}_8$, respectively.

Since $\tau X_i = X_i$ and $\tau^2 = 1$, the Orbit-Stabilizer Theorem implies that τ fixes at least one point of X_i. If, for $x \in X_i$, $\tau x = x$ and $\tau\rho^k x = x$ for an integer k, then $\rho^{-k}\tau x = x$, $\rho^{-k}x = x$, and then $k \equiv 0 \pmod 5$. Therefore, τ fixes a unique point of each $\langle\rho\rangle$-orbit on X. Let $x_i \in X_i$ be such that $\tau x_i = x_i$. Then

$$X = \{\infty\} \cup \{\rho^m x_i : 1 \le i \le 8, 0 \le m \le 4\},$$

and we have, for any integer m,

$$\tau\rho^m x_i = \rho^{-m} x_i. \tag{2.20}$$

Since τ fixes eight points, other than ∞, by Proposition 2.6.9, it fixes eight blocks other than B_∞, one block from each $\mathcal{B}_j$ which we will denote by B_j. We then have, for integer m,

$$\tau\rho^m B_j = \rho^{-m} B_j. \tag{2.21}$$

Since $\sigma X_i = \sigma\rho X_i = \rho\sigma X_i$, we obtain that σX_i is a $\langle\rho\rangle$-orbit on X, so σ permutes the sets $X_1, X_2, \ldots, X_8$. Similarly, σ permutes $\mathcal{B}_1, \mathcal{B}_2, \ldots, \mathcal{B}_8$. If $\sigma x_i = \rho^k x_h$, then $\sigma x_i = \sigma\tau x_i = \tau\sigma x_i = \tau\rho^k x_h = \rho^{-k} x_h$. Thus, $k \equiv 0 \pmod 5$, so $\sigma x_i = x_h$. Therefore, if $\sigma X_i = X_h$, then, for any integer k,

$$\sigma\rho^m x_i = \rho^m x_h. \tag{2.22}$$

Similarly, if $\sigma(\mathcal{B}_j) = \mathcal{B}_h$, then, for any integer m,

$$\sigma\rho^m B_j = \rho^m B_h. \tag{2.23}$$

Thus, for each $\langle\rho\rangle$-orbit on X or $\mathcal{B}$, either σ fixes every element of the orbit or it maps the entire orbit onto another $\langle\rho\rangle$-orbit according to (2.22) or (2.23).

Let $Y = \{y \in X : \sigma y = y\}$. Since each $\langle\sigma\rangle$-orbit on X is of cardinality 1 or 3, we obtain that $|Y| \equiv 41 \equiv 2 \pmod 3$. Since σ fixes ∞ and either all or none of the points of each $\langle\rho\rangle$-orbit on X, we obtain that $|Y| \equiv 1 \pmod 5$. Therefore, $|Y| \equiv 11 \pmod{15}$, i.e., $|Y|$ is 11 or 26. We claim that $|Y| = 11$. Suppose $|Y| = 26$. Then, by Proposition 2.6.9, the set $\mathcal{C}$ of fixed blocks of σ is of cardinality 26. If $B \in \mathcal{B}$ and $y \in B \cap Y$, then $y = \sigma y \in (\sigma B) \cap Y$, so $B \cap Y \subseteq B \cap \sigma B$. Therefore, if $B \notin \mathcal{C}$, then $|B \cap Y| \le 6$, i.e., $|B \cap (X \setminus Y)| \ge 10$. Similarly, if $x \in X$ and $C \in \mathcal{C}$ is a block containing x, then $\sigma x \in C$. Therefore, if $x \in X \setminus Y$, then there are at most six blocks $C \in \mathcal{C}$ that contain x. Let us fix $x_0 \in X \setminus Y$ and count in two ways pairs (x, B) where $x \in X \setminus Y$, $x \ne x_0$, $B \in \mathcal{B} \setminus \mathcal{C}$, and $x_0, x \in B$. Choosing x first, we obtain at most $14 \cdot 6 = 84$ such pairs. If we choose a block $B \in \mathcal{B} \setminus \mathcal{C}$ containing x_0 first, we obtain at least $10 \cdot 9 = 90$ such pairs. This contradiction proves that $|Y| = 11$.

Since $\rho B_\infty = B_\infty$ and $|B_\infty| = 16$, we obtain that B_∞ contains ∞ and three $\langle\rho\rangle$-orbits on X. Without loss of generality, we assume that $B_\infty = \{\infty\} \cup X_1 \cup X_2 \cup X_3$. Similarly, ∞ is contained in B_∞ and all blocks from three orbits on $\mathcal{B}$. We assume that these orbits are $\mathcal{B}_1$, $\mathcal{B}_2$, and $\mathcal{B}_3$. Since $\sigma B_\infty = B_\infty$, we have $\sigma(X_1 \cup X_2 \cup X_3) = X_1 \cup X_2 \cup X_3$. Since each $\langle\sigma\rangle$-orbit on the set $\{X_1, X_2, \ldots, X_8\}$ is of cardinality 1 or 3 and since σ fixes only two elements of this set, we obtain that σ cyclically permutes X_1, X_2, and X_3. Therefore, we assume without loss of generality, that σ acts on the set $\{X_1, X_2, \ldots, X_8\}$ as the permutation $(X_1X_2X_3)(X_4X_5X_6)(X_7)(X_8)$. Let $Y_1 = X_1 \cup X_2 \cup X_3$, $Y_2 = X_4 \cup X_5 \cup X_6$, and $Y_3 = X_7 \cup X_8$.

Similarly, we assume that σ acts on the set $\{\mathcal{B}_1, \mathcal{B}_2, \ldots, \mathcal{B}_8\}$ as the permutation $(\mathcal{B}_1\mathcal{B}_2\,\mathcal{B}_3)(\mathcal{B}_4\mathcal{B}_5\mathcal{B}_6)(\mathcal{B}_7)(\mathcal{B}_8)$. We have now described the action of ρ, τ, and σ on both X and $\mathcal{B}$.

For $i, j = 1, 2, \ldots, 8$, let r_{ij} be the number of blocks $B \in \mathcal{B}_j$ that contain x_i and let $k_{ij} = |B_j \cap X_i|$. Since $|X_i| = |\mathcal{B}_j|$, (2.15) implies that $r_{ij} = k_{ij}$. Form the matrix $R = [r_{ij}]$, $i, j = 1, 2, \ldots, 8$. Our next goal is to determine this matrix. Note that the action of σ on X and $\mathcal{B}$ implies the following relations: (i) for $i = 7$ and $i = 8$, $r_{i1} = r_{i2} = r_{i3}$, $r_{i4} = r_{i5} = r_{i6}$, $r_{1i} = r_{2i} = r_{3i}$, and $r_{4i} = r_{5i} = r_{6i}$; (ii) each of the four 3×3 submatrices $[r_{ij}]$ with both i and j in $\{1, 2, 3\}$ or in $\{4, 5, 6\}$ must be circulant.

The entries of R must satisfy the following equations, which are obtained from (2.14), (2.16), and (2.18):

$$\sum_{j=1}^{8} r_{ij} = \begin{cases} 15 & \text{if } 1 \le i \le 3, \\ 16 & \text{if } 4 \le i \le 8, \end{cases} \tag{2.24}$$

$$\sum_{j=1}^{8} r_{ij}^2 = \begin{cases} 35 & \text{if } 1 \le i \le 3, \\ 40 & \text{if } 4 \le i \le 8, \end{cases} \tag{2.25}$$

and, for $i \ne h$,

$$\sum_{j=1}^{8} r_{ij}r_{hj} = \begin{cases} 25 & \text{if } i, h = 1, 2, 3, \\ 30 & \text{otherwise.} \end{cases} \tag{2.26}$$

Let $x \in X$, $x \ne \infty$. Since there are exactly six blocks containing ∞ and x, we obtain that

$$\sum_{j=1}^{3} r_{ij} = \begin{cases} 5 & \text{if } 1 \le i \le 3, \\ 6 & \text{if } 4 \le i \le 8. \end{cases} \tag{2.27}$$

Equations (2.24) – (2.27) remain true if we replace all r_{ij} by r_{ji} (and r_{ih} by r_{hi}).

Counting in two ways flags (x, B) with $x \in X_1 \cup X_2 \cup X_3$ and $B \in \mathcal{B}_7 \cup \mathcal{B}_8$ yields

$$\sum_{i=1}^{3}(r_{i7} + r_{i8}) = 12.$$

Since $r_{17} = r_{27} = r_{37}$ and $r_{18} = r_{28} = r_{38}$, $|B_7 \cap (X_1 \cup X_2 \cup X_3)| = |B_7 \cap B_\infty| = 6$, and $|B_8 \cap (X_1 \cup X_2 \cup X_3)| = 6$, we obtain that $r_{ij} = 2$ for $i = 1, 2, 3$ and $j = 7, 8$. Similarly, $r_{ij} = 2$ for $i = 7, 8$ and $j = 1, 2, 3$.

We have, for $i = 1, 2, 3$,

$$\sum_{j=4}^{6} r_{ij} = \sum_{j=4}^{6} r_{ji} = 6.$$

Thus, we have the following equations for $i = 1, 2, 3$;

$$\sum_{j=1}^{3} r_{ij} = 5, \quad \sum_{j=4}^{6} r_{ij} = 6, \quad \sum_{j=1}^{6} r_{ij}^2 = 27.$$

These equations yield the following possibilities:

$$\{r_{i1}, r_{i2}, r_{i3}\} = \{2, 2, 1\}, \ \{r_{i4}, r_{i5}, r_{i6}\} = \{4, 1, 1\}$$

and

$$\{r_{i1}, r_{i2}, r_{i3}\} = \{2, 2, 1\}, \ \{r_{i4}, r_{i5}, r_{i6}\} = \{3, 3, 0\}.$$

We will assume that $r_{12} = r_{13} = 2$ and $r_{11} = 1$. We will also assume that $r_{14} = 0$, $r_{15} = r_{16} = 3$, $r_{41} = 4$, and $r_{42} = r_{43} = 1$. This determines all r_{ij} and r_{ji} for $1 \le i \le 3$ and $1 \le j \le 6$.

Equations (2.24)–(2.26) now imply the following equations for r_{4j}, $4 \le j \le 8$:

$$\sum_{j=4}^{8} r_{4j}^2 = 22,$$

$$\sum_{j=4}^{8} r_{4j} = 10,$$

$$3(r_{45} + r_{46}) + 2(r_{47} + r_{48}) = 22,$$

$$3(r_{45} + r_{47}) + 2(r_{47} + r_{48}) = 19.$$

We let $r_{46} = a$ and then obtain $r_{45} = a$, $r_{44} = a - 1$, $r_{47} + r_{48} = 11 - 3a$, and $r_{47}^2 + r_{48}^2 = 21 - 3a^2 + 2a$. Therefore, $1 \le a \le 3$, and only $a = 2$ yields integer solutions: $r_{44} = 1$, $r_{45} = r_{46} = 2$, and $\{r_{47}, r_{48}\} = \{2, 3\}$. We choose $r_{47} = 2$ and $r_{48} = 3$. Similarly, we obtain $r_{74} = 2$ and $r_{84} = 3$. This determines all r_{ij} and r_{ji} for $4 \le i \le 6$ and $4 \le j \le 8$. Then it is straightforward to determine that $r_{77} = 4$, $r_{88} = 1$, and $r_{78} = r_{87} = 0$. Thus, we have obtained the following matrix R:

$$R = \begin{bmatrix} 1 & 2 & 2 & 0 & 3 & 3 & 2 & 2 \\ 2 & 1 & 2 & 3 & 0 & 3 & 2 & 2 \\ 2 & 2 & 1 & 3 & 3 & 0 & 2 & 2 \\ \\ 4 & 1 & 1 & 2 & 2 & 1 & 2 & 3 \\ 1 & 4 & 1 & 1 & 2 & 2 & 2 & 3 \\ 1 & 1 & 4 & 2 & 1 & 2 & 2 & 3 \\ \\ 2 & 2 & 2 & 2 & 2 & 2 & 4 & 0 \\ 2 & 2 & 2 & 3 & 3 & 3 & 0 & 1 \end{bmatrix}.$$

We will assume that the point set X and the block set $\mathcal{B}$ are ordered so that, for $i = 1, 2, \ldots, 7$, points of X_i precede points of X_{i+1} and blocks of $\mathcal{B}_i$ precede blocks of $\mathcal{B}_{i+1}$. We will further assume that each X_i and each $\mathcal{B}_i$ are ordered as follows: $X_i = \{x_i, \rho x_i, \ldots, \rho^4 x_i\}$ and $\mathcal{B}_i = \{B_i, \rho B_i, \ldots, \rho^4 B_i\}$. Let the point ∞ and the block B_∞ precede all other points and blocks, respectively.

With this ordering, we have to replace every entry r_{ij} of R by a (0, 1)-matrix M_{ij} of order 5 so that the following conditions be satisfied:

if $r_{ij} = 0$, then $M_{ij} = O$;
if $r_{ij} = 1$, then $M_{ij} = I$;
if $r_{ij} = 2$, then $M_{ij} = K$ or $M_{ij} = L = J - I - K$, where

$$K = \begin{bmatrix} 0 & 1 & 0 & 0 & 1 \\ 1 & 0 & 1 & 0 & 0 \\ 0 & 1 & 0 & 1 & 0 \\ 0 & 0 & 1 & 0 & 1 \\ 1 & 0 & 0 & 1 & 0 \end{bmatrix};$$

if $r_{ij} = 3$, then $M_{ij} = \overline{K} = J - K$ or $M_{ij} = \overline{L} = J - L$;
if $r_{ij} = 4$, then $M_{ij} = \overline{I} = J - I$.

The action of σ implies further conditions: for $i = 1, 4$ and $j = 1, 4$, $M_{ij} = M_{i+1,j+1} = M_{i+2,j+2}$, $M_{i,j+1} = M_{i+1,j+2} = M_{i+2,j}$, $M_{i,j+2} = M_{i+1,j} = M_{i+2,j+1}$; for $i = 7, 8$ and $j = 1, 4$, $M_{ij} = M_{i,j+1} = M_{i,j+2}$ and $M_{ji} = M_{j+1,i} = M_{j+2,i}$.

Thus, some of the matrices M_{ij} have been determined, others (corresponding to $r_{ij} = 2$ or 3) are yet to be determined. For this we use the intersections of blocks B_j, $1 \le j \le 8$.

If $M_{18} \neq M_{17}$, then $Y_1 \cap B_7 \cap B_8 = \varnothing$. Since also $Y_3 \cap B_7 \cap B_8 = \varnothing$, we must have $|Y_2 \cap B_7 \cap B_8| = 6$. This implies $M_{48} \neq \overline{M_{47}}$. If $M_{18} = M_{17}$, then $|Y_1 \cap B_7 \cap B_8| = 6$, and therefore $Y_2 \cap B_7 \cap B_8 = \varnothing$, so $M_{48} = \overline{M_{47}}$. Similarly, either $M_{81} \neq M_{71}$ and $M_{84} \neq \overline{M_{74}}$ or $M_{81} = M_{71}$ and $M_{84} = \overline{M_{74}}$. We will choose $M_{17} = M_{81} = M_{74} = K$, $M_{18} = M_{47} = M_{71} = L$, $M_{48} = \overline{K}$, and $M_{84} = \overline{L}$.

We have $|Y_2 \cap B_1 \cap B_7| = 2$ and $|Y_3 \cap B_1 \cap B_7| = 2$. Therefore, we must have $|Y_1 \cap B_1 \cap B_7| = 2$. This implies that one of the matrices M_{21} and M_{31} is equal to K and the other is equal to L. We choose $M_{21} = L$ and $M_{31} = K$.

We have $|Y_3 \cap B_1 \cap B_4| = 2$. Since each of the matrices M_{54} and M_{64} is equal to K or L, we obtain that $Y_2 \cap B_1 \cap B_4 = \varnothing$. Therefore, we must have $|Y_1 \cap B_1 \cap B_4| = 4$. This implies $M_{24} = \overline{K}$ and $M_{34} = \overline{L}$. The remaining yet undetermined matrices M_{ij} are those with $i, j = 4, 5, 6$. We have $|Y_1 \cap B_4 \cap B_7| = 2$ and $|Y_3 \cap B_4 \cap B_7| = 2$. Therefore, $|Y_2 \cap B_4 \cap B_7| = 2$. This implies that one of the matrices M_{54} and M_{64} is K and the other is L. We will choose $M_{54} = L$. This leads to the following matrix M:

$$M = \begin{bmatrix} I & K & L & O & \overline{L} & \overline{K} & K & L \\ L & I & K & \overline{K} & O & \overline{L} & K & L \\ K & L & I & \overline{L} & \overline{K} & O & K & L \\ \\ \overline{I} & I & I & I & K & L & L & \overline{K} \\ I & \overline{I} & I & L & I & K & L & \overline{K} \\ I & I & \overline{I} & K & L & I & L & \overline{K} \\ \\ L & L & L & K & K & K & \overline{I} & O \\ K & K & K & \overline{L} & \overline{L} & \overline{L} & O & I \end{bmatrix}.$$

It should be now verified that the matrix M augmented by the row and the column corresponding to ∞ and B_∞ is an incidence matrix of a symmetric (41, 16, 6)-design. This verification is simplified by the fact that all matrices M_{ij} are symmetric matrices that commute with each other.

For $j, h = 1, 2, \ldots, 8$, let $S_{jh} = \sum_{i=1}^{8} M_{ij} M_{ih}$. We have to show that, for $j, h = 1, 2, 3$,

$$S_{jh} = \begin{cases} 10I + 5J & \text{if } j = h, \\ 5J & \text{if } \neq j \neq h, \end{cases}$$

and, for other pairs (j, h),

$$S_{jh} = \begin{cases} 10I + 6J & \text{if } j = h, \\ 6J & \text{if } \neq j \neq h. \end{cases}$$

It suffices to verify these relations for $(j, h) = (1, 1), (1, 2), (1, 4), (1, 7), (1, 8)$, $(4, 4)$, $(4, 5)$, $(4, 7)$, $(4, 8)$, $(7, 7)$, $(7, 8)$, and $(8, 8)$ and then extend it to the remaining pairs (j, h) by the automorphism σ. We will leave this verification to the reader.

Thus we have proved the following theorem.

Theorem 2.7.1. *There exists a symmetric* $(41, 16, 6)$*-design with an automorphism group isomorphic to the direct product of a dihedral group of order* 10 *and a cyclic group of order* 3.

2.8. A symmetric (79, 13, 2)-design

The next symmetric design we will construct is the largest known biplane, that is, a symmetric design with $\lambda = 2$. It is a symmetric $(79, 13, 2)$-design $\mathbf{D} = (X, \mathcal{B})$ with an automorphism group

$$G = \langle \rho, \sigma, \tau \,|\, \rho^{11} = \sigma^5 = \tau^2 = 1, \sigma\rho = \rho^4\sigma, \tau\rho = \rho^{-1}\tau, \sigma\tau = \tau\sigma \rangle, \tag{2.28}$$

acting on X in such a way that

(i) ρ has exactly two fixed points, a_1 and a_2;
(ii) $\tau a_1 = a_2$;
(iii) if Y is a $\langle\rho\rangle$-orbit on X and $|Y| > 1$, then $\tau Y = Y$;
(iv) there is a $\langle\rho\rangle$-orbit Y on X such that $\sigma Y \neq Y$.

Let $\mathbf{D}$ be such a design. Since ρ has no fixed point, except a_1 and a_2, the set $X \setminus \{a_1, a_2\}$ is partitioned into seven $\langle\rho\rangle$-orbits of cardinality 11. Let them be $X_1, X_2, \ldots, X_7$. Since for any $x \in X$ and for any integer k, $\sigma\rho^k x = \rho^{4k}\sigma x$, we obtain that the image of the $\langle\rho\rangle$-orbit containing x is the $\langle\rho\rangle$-orbit containing σx. Therefore, the group $\langle\sigma\rangle$ acts on the set of nine $\langle\rho\rangle$-orbits on X. Since $|\langle\sigma\rangle| = 5$, condition (iv) implies that σ fixes four $\langle\rho\rangle$-orbits on X and cyclically permutes the other five $\langle\rho\rangle$-orbits. Since $\{a_1\}$ and $\{a_2\}$ are the only $\langle\rho\rangle$-orbits of cardinality 1, we obtain that, for $i = 1$ and 2, $\sigma a_i \in \{a_1, a_2\}$. Therefore, $\sigma^2 a_i = a_i$ and then $\sigma a_i = \sigma^6 a_i = a_i$, so $\langle\rho\rangle$-orbits $\{a_1\}$ and $\{a_2\}$ are fixed by σ. There are two more $\langle\rho\rangle$-orbits on X fixed (as sets) by σ. We let them be X_1 and X_2 and assume that σ acts on the set of $\langle\rho\rangle$-orbits on X as the cycle $(X_3X_4X_5X_6X_7)$.

Since $|X_1| = |X_2| = 11$, σ fixes at least one point in each of these sets. If $\sigma x = x$ for $x \in X_i$, then, since $\sigma\rho^k x = \rho^{4k}x$, σ fixes no other point of X_i. For $i = 1, 2$, let x_{i0} be the point of X_i fixed by σ. For any integer k, let $x_{ik} = \rho^k x_{i0}$.

Then $\sigma x_{ik} = x_{i,4k}$, and therefore, σ acts on each X_i $(i = 1, 2)$ as the permutation

$$(x_{i1}x_{i4}x_{i5}x_{i9}x_{i3})(x_{i2}x_{i8}x_{i,10}x_{i7}x_{i6}). \tag{2.29}$$

By Proposition 2.6.9, ρ fixes two blocks, which we denote by A_1 and A_2. Since $|A_1| = |A_2| = 13$ and $|\langle\rho\rangle| = 11$, ρ must fix at least two points in each of these blocks. Therefore, $a_1, a_2 \in A_1$ and $a_1, a_2 \in A_2$. The set $\mathcal{B} \setminus \{A_1, A_2\}$ is partitioned into seven $\langle\rho\rangle$-orbits, $\mathcal{B}_1, \mathcal{B}_2, \ldots, \mathcal{B}_7$, of cardinality 11. As before, σ acts as a permutation on the set of nine $\langle\rho\rangle$-orbits on $\mathcal{B}$, $\sigma A_1 = A_1$, and $\sigma A_2 = A_2$. Each of the sets $A_1 \setminus \{a_1, a_2\}$ and $A_2 \setminus \{a_1, a_2\}$ is fixed by ρ and therefore is a $\langle\rho\rangle$-orbit. Since each of these $\langle\rho\rangle$-orbits is fixed by σ, we assume without loss of generality that $A_i = \{a_1, a_2\} \cup X_i$ for $i = 1, 2$.

The action of σ on the set of nine $\langle\rho\rangle$-orbits on $\mathcal{B}$ must have at least four fixed orbits and each of these orbits must have at least one fixed block of σ. Since σ fixes exactly four points of **D**, Proposition 2.6.9 implies that σ fixes exactly four blocks, two of which are A_1 and A_2. We assume without loss of generality that $\sigma(\mathcal{B}_i) = \mathcal{B}_i$ for $i = 1$ and 2 and that σ acts on the set of $\langle\rho\rangle$-orbits on $\mathcal{B}$ as the cycle $(\mathcal{B}_3\mathcal{B}_4\mathcal{B}_5\mathcal{B}_6\mathcal{B}_7)$. For $i = 1, 2$, the blocks containing a_i, other than A_1 and A_2, form a $\langle\rho\rangle$-orbit on $\mathcal{B}$. Since $\sigma(\mathcal{B}_i) = \mathcal{B}_i$ and $\sigma a_i = a_i$ for $i = 1$ and 2, we obtain that these $\langle\rho\rangle$-orbits are $\mathcal{B}_1$ and $\mathcal{B}_2$. We assume without loss of generality that a_i is contained in all blocks of $\mathcal{B}_i$ for $i = 1, 2$.

For $i = 1, 2$, let B_{i0} be the block of $\mathcal{B}_i$ fixed by σ. For any integer k, let $B_{ik} = \rho^k B_{i0}$. Then the action of σ on each $\mathcal{B}_i$ $(i = 1, 2)$ can be given by permutation (2.29) with each x_{ik} replaced by B_{ik}.

For $i, j = 1, 2, \ldots, 7$, let r_{ij} and k_{ij} have the same meaning as in Proposition 2.6.14. Since $|X_i| = |\mathcal{B}_j|$, we obtain that $r_{ij} = k_{ij}$. If $B \in \mathcal{B}_1 \cup \mathcal{B}_2$, then $|B \cap \{a_1, a_2\}| = 1$, and therefore, for $i = 1, 2$, $|B \cap X_i| = |B \cap A_i| - 1 = 1$. If $B \in \mathcal{B}_j$ with $3 \le j \le 7$, then $|B \cap X_i| = 2$. Thus, $r_{ij} = 1$ for $i, j = 1, 2$ and $r_{ij} = 2$ for $i = 1, 2$ and $3 \le j \le 7$. Similarly, $r_{ij} = 2$ for $j = 1, 2$ and $3 \le i \le 7$. Since B_{10} and B_{20} are fixed blocks of σ and since $\sigma X_i = X_i$ for $i = 1$ and 2, we obtain that $x_{i0} \in B_{i0}$ for $i = 1$ and 2.

Proposition 2.6.14 now yields the following equations for $3 \le i \le 7$:

$$\sum_{j=3}^{7} r_{ij} = \sum_{j=3}^{7} r_{ji} = 9, \tag{2.30}$$

$$\sum_{j=3}^{7} r_{ij}^2 = \sum_{j=3}^{7} r_{ji}^2 = 25 \tag{2.31}$$

and, for $3 \le i < h \le 7$,

$$\sum_{j=3}^{7} r_{ij}r_{hj} = \sum_{j=3}^{7} r_{ji}r_{jh} = 14. \tag{2.32}$$

Equations (2.30) and (2.31) yield a unique and the same solution for each of the multisets $\{r_{ij} : 3 \le j \le 7\}$ and $\{r_{ji} : 3 \le j \le 7\}$ for $i = 3, 4, 5, 6, 7$, namely, $\{0, 1, 2, 2, 4\}$. The action of σ on the sets of $\langle \rho \rangle$-orbits on X and on $\mathcal{B}$ implies that the submatrix $R' = [r_{ij}]$ with $i, j \in \{3, 4, 5, 6, 7\}$ is circulant. Equations (2.32) are satisfied if we let $r_{33} = 1, r_{43} = 4, r_{53} = r_{73} = 2$, and $r_{63} = 0$. Then the action of σ yields the following matrix $R = [r_{ij}]$:

$$R = \begin{bmatrix} 1 & 1 & 2 & 2 & 2 & 2 & 2 \\ 1 & 1 & 2 & 2 & 2 & 2 & 2 \\ 2 & 2 & 1 & 2 & 0 & 2 & 4 \\ 2 & 2 & 4 & 1 & 2 & 0 & 2 \\ 2 & 2 & 2 & 4 & 1 & 2 & 0 \\ 2 & 2 & 0 & 2 & 4 & 1 & 2 \\ 2 & 2 & 2 & 0 & 2 & 4 & 1 \end{bmatrix}.$$

We will now turn our attention to the automorphism τ. Since $\tau X_i = X_i$ for $i = 1, 2, \ldots, 7$, τ fixes at least one point of each X_i. If $\tau x = x$ for $x \in X_i$, then, for any integer k, $\tau \rho^k x = \rho^{-k} x$, so τ fixes no other point of X_i. For $i = 1, 2$, $\tau x_{i0} = \tau \sigma x_{i0} = \sigma \tau x_{i0}$, which implies that $\tau x_{i0} = x_{i0}$. For $i = 3, 4, \ldots, 7$, let x_{i0} be the only point of X_i that is fixed by τ. For any integer k, let $x_{ik} = \rho^k x_{i0}$. Since $\sigma x_{i0} = \sigma \tau x_{i0} = \tau \sigma x_{i0}$, we obtain that $\sigma x_{i0} = x_{j0}$ whenever $\sigma X_i = X_j$. Then $\sigma x_{ik} = \sigma \rho^k x_{i0} = \rho^{4k} \sigma x_{i0} = x_{j,4k}$. Therefore, σ acts on the set $X_3 \cup X_4 \cup X_5 \cup X_6 \cup X_7$ as the permutation

$$\prod_{k=0}^{10} (x_{3,k} x_{4,4k} x_{5,5k} x_{6,9k} x_{7,3k}).$$

Since $\tau x_{ik} = \tau \rho^k x_{i0} = \rho^{-k} x_{i0} = x_{i,-k}$, we obtain that

$$\tau = (a_1 a_2) \prod_{i=1}^{7} (x_{i1} x_{i,10})(x_{i2} x_{i9})(x_{i3} x_{i8})(x_{i4} x_{i7})(x_{i5} x_{i6}).$$

The automorphism τ has seven fixed points and therefore seven fixed blocks. If $\tau B = B$ for $B \in \mathcal{B}_j$, we obtain, as before, that τ fixes no other block of $\mathcal{B}_j$. Since $\tau X_i = X_i$, $\tau a_1 = a_2$, and $\tau a_2 = a_1$, τ fixes A_1 and A_2. If $B \in \mathcal{B}_1$, then $a_1 \in B$. Therefore, $a_2 \in \tau B$ and $\tau B \in \mathcal{B}_2$. Thus, τ fixes no block of $\mathcal{B}_1 \cup \mathcal{B}_2$ and therefore it fixes one block in each $\mathcal{B}_j$ with $3 \le j \le 7$. Let this block be B_{j0} and let $B_{jk} = \rho^k B_{j0}$ for any integer k. Then, for $3 \le j \le 7$, $\tau B_{jk} = B_{j,-k}$. As before, we derive that the action of σ on $\mathcal{B}_3 \cup \mathcal{B}_4 \cup \mathcal{B}_5 \cup \mathcal{B}_6 \cup \mathcal{B}_7$ is given by the same permutation as the action of σ on $X_3 \cup X_4 \cup X_5 \cup X_6 \cup X_7$ (with all x_{ik} replaced by B_{ik}).

Since $x_{10} \in B_{10}$ and $\tau x_{10} = x_{10}$, we obtain that $x_{10} \in \tau B_{10}$. Therefore, $\tau B_{10} = B_{20}$, and then, for any integer k, $\tau B_{1k} = B_{2,-k}$ and $\tau B_{2k} = B_{1,-k}$. Thus, on the set $\mathcal{B}$, τ acts as the permutation

$$\prod_{k=0}^{10}(B_{1k}B_{2,-k})\prod_{j=3}^{7}(B_{j1}B_{j,10})(B_{j2}B_{j9})(B_{j3}B_{j8})(B_{j4}B_{j7})(B_{j5}B_{j6}).$$

For $3 \le i \le 7$, $r_{i1} = r_{i2} = r_{1i} = r_{2i} = 2$.

In order to obtain an incidence matrix of a symmetric (79, 13, 2)-design, we have to replace each r_{ij} in R by a circulant (0, 1)-matrix M_{ij} of order 11 with row and column sum r_{ij} and then to adjoin the resulting matrix M of order 77 by two rows and two columns corresponding to a_1, a_2, A_1, and A_2. Each $r_{ij} = 0$ must be replaced by the zero matrix and each $r_{ij} = 1$ by the identity matrix. To describe the other blocks M_{ij}, we denote by $P = [p_{ij}]$, $i, j = 0, 1, \ldots, 10$, the permutation matrix of order 11 given by

$$p_{ij} = \begin{cases} 1 & \text{if } j \equiv i+1 \pmod{11}, \\ 0, & \text{otherwise.} \end{cases}$$

Observe that, for any integer k, $P^k = [p_{ij}^{(k)}]$ where

$$p_{ij}^{(k)} = \begin{cases} 1 & \text{if } j \equiv i+k \pmod{11}, \\ 0, & \text{otherwise.} \end{cases}$$

In particular, $P^{11} = I$. Note that $(P^k)^\top = P^{-k}$ and $\sum_{k=0}^{10} P^k = J$.

We also let $Q_k = P^k + P^{-k}$ for any integer k. Then $Q_0 = 2I$, $Q_{-k} = Q_k$, and, for any integers m and n, $Q_m Q_n = Q_{m+n} + Q_{m-n}$. Observe also that, for $k \not\equiv 0 \pmod{11}$, $\sum_{i=1}^{5} Q_{ki} = J - I$.

Since $r_{31} = 2$, we have $M_{31} = P^m + P^n$ with $m \not\equiv n \pmod{11}$. The action of τ then implies that $M_{32} = P^{-m} + P^{-n}$.

For $3 \le i \le 7$ and $j = 1, 2$, $X_i \cap B_{j0} = \sigma^{i-3}(X_3 \cap B_{j0})$. Therefore, for $3 \le i \le 7$,

$$M_{i1} = P^{4^{i-3}m} + P^{4^{i-3}n}$$

and

$$M_{i2} = P^{-4^{i-3}m} + P^{-4^{i-3}n}.$$

Therefore, for $j = 1, 2$,

$$\sum_{i=1}^{7} M_{ij}^\top M_{ij} = 2I + \sum_{i=0}^{4} \left(P^{4^i m} + P^{4^i n}\right)\left(P^{-4^i m} + P^{-4^i n}\right)$$

$$= 12I + \sum_{i=0}^{4} Q_{4^i(m-n)} = 12I + \sum_{k=0}^{4} Q_{k(m-n)} = 11I + J.$$

Furthermore,

$$\sum_{i=1}^{7} M_{i2}^{\top} M_{i1} = 2I + \sum_{i=0}^{4} \left(P^{4^i m} + P^{4^i n}\right)^2$$

$$= 2I + \sum_{i=0}^{4} P^{2m\cdot 4^i} + \sum_{i=0}^{4} P^{2n\cdot 4^i} + 2\sum_{i=0}^{4} P^{2(m+n)\cdot 4^i}.$$

If we select m and n to be quadratic non-residues $\pmod{11}$ such that $m+n$ is also a quadratic non-residue $\pmod{11}$, then each of the sets $\{2m\cdot 4^i : 0 \le i \le 4\}$ and $\{2n \cdot 4^i : 0 \le i \le 4\}$ consists of all quadratic residues $\pmod{11}$, while the set $\{(m+n)\cdot 4^i : 0 \le i \le 4\}$ consists of all quadratic non-residues $\pmod{11}$. Therefore, we obtain in this case that

$$\sum_{i=1}^{7} M_{i2}^{\top} M_{i1} = 2I + 2(J - I) = 2J.$$

We will choose $m = -1$ and $n = -4$.

The action of τ implies that each of the matrices M_{13}, M_{23}, M_{53}, and M_{73} is of the form Q_s with $s \not\equiv 0 \pmod{11}$ and M_{43} is of the form $Q_s + Q_t$ with $s, t \not\equiv 0 \pmod{11}$ and $s \not\equiv t \pmod{11}$. As these matrices are chosen, the remaining matrices M_{ij} are uniquely determined by the action of σ. We will choose the following matrix $M = [M_{ij}]$:

$$M = \begin{bmatrix} I & I & Q_2 & Q_3 & Q_1 & Q_4 & Q_5 \\ I & I & Q_5 & Q_2 & Q_3 & Q_1 & Q_4 \\ P^{-1}+P^{-4} & P+P^4 & I & Q_5 & O & Q_1 & Q_5+Q_1 \\ P^{-4}+P^{-5} & P^4+P^5 & Q_2+Q_4 & I & Q_2 & O & Q_4 \\ P^{-5}+P^2 & P^5+P^{-2} & Q_5 & Q_3+Q_5 & I & Q_3 & O \\ P^2+P^{-3} & P^{-2}+P^3 & O & Q_2 & Q_1+Q_2 & I & Q_1 \\ P^{-3}+P^{-1} & P^3+P & Q_4 & O & Q_3 & Q_4+Q_3 & I \end{bmatrix}.$$

Let N be the matrix obtained from M by adjoining the two rows and two columns corresponding to a_1, a_2, A_1, and A_2. In order to prove that N is an incidence matrix of a symmetric (79, 13, 2)-design, it suffices, due to the action of σ, to verify that

$$\sum_{j=1}^{7} M_{3j} M_{3j}^{\top} = 11I + 2J$$

and, for $i = 1, 2, 4, 5, 6$,

$$\sum_{j=1}^{7} M_{3j} M_{ij}^{\top} = 2J.$$

We leave this verification to the reader.

Thus, we have proved

Theorem 2.8.1. *There exists a symmetric* (79, 13, 2)*-design with the automorphism group* (2.28) *of order* 110.

We will state without proof several results of a similar nature, giving sporadic examples of symmetric designs. The similarity of the results does not necessarily imply the similarity of constructions. Each time one has to make a right choice of an automorphism group, of its action on the point set of a design, and an appropriate choice of base blocks. The choices are usually numerous, and it could be very difficult and time consuming to make the right ones. An extensive computer search may be necessary.

Theorem 2.8.2. *There exists a symmetric* (49, 16, 5)*-design with an automorphism group of order* 15.

Theorem 2.8.3. *There exists a symmetric* (70, 24, 8)*-design with an automorphism group of order* 42.

Theorem 2.8.4. *There exists a symmetric* (71, 21, 6)*-design with an automorphism group of order* 21.

Theorem 2.8.5. *There exists a symmetric* (78, 22, 6)*-design with an automorphism group of order* 39.

Theorem 2.8.6. *There exists a symmetric* (78, 22, 6)*-design with an automorphism group of order* 168.

Theorem 2.8.7. *There exists a symmetric* (105, 40, 15)*-design with an automorphism group of order* 100.

Theorem 2.8.8. *There exists a symmetric* (189, 48, 12)*-design with an automorphism group of order* 42.

Exercises

(1) Give an example of an incidence structure which satisfies conditions (i), (ii), (iii), (v), and (vi) of Definition 2.3.1 but does not satisfy (iv).

(2) Give an example of an incidence structure that has as many points as blocks but is not self-dual.

(3) Let N_1 and N_2 be incidence matrices of incidence structures $\mathbf{D}_1$ and $\mathbf{D}_2$. Prove that if there exist permutation matrices P and Q such that $PN_1 = N_2Q$, then the structures $\mathbf{D}_1$ and $\mathbf{D}_2$ are isomorphic.

(4) Let A be a square (0, 1)-matrix and let $N = \begin{bmatrix} A & A^\top \\ A^\top & A \end{bmatrix}$. Prove that the incidence structure with incidence matrix N is self-dual.

(5) Give an example of a self-dual incidence structure that does not admit a symmetric incidence matrix.

(6) Let N be a square matrix with nonnegative integers as entries such that every row and column of N has the same sum k. The G. Birkhoff Theorem states that then N is the sum of k permutation matrices. Apply this theorem to prove the following result: if $\mathbf{D} = (X, \mathcal{B}, I)$ is an incidence structure with $|X| = |\mathcal{B}|$, with constant block size $k \geq 1$, and constant replication number, then $\mathbf{D}$ has an incidence matrix with all diagonal entries equal to 1.

(7) Prove Proposition 2.2.7.

(8) Prove Proposition 2.2.10.

(9) Prove that a graph Γ is bipartite if and only if it has no cycle of odd length.

(10) Let $\chi(\Gamma)(t) = t^n + a_1 t^{n-1} + a_2 t^{n-2} + a_3 t^{n-3} + \cdots + a_n$ be the characteristic polynomial of a graph $\Gamma = (V, E)$. Prove:

(1) $a_1 = 0$;

(2) $-a_2$ is the number of edges of Γ;

(3) $-\frac{1}{2}a_3$ is the number of triangles in Γ, that is, 3-subsets $\{x, y, z\}$ of V such that $\{x, y\}, \{y, z\}, \{z, x\} \in E$.

(11) Prove that a graph is bipartite if and only if its characteristic polynomial has no terms of odd degree.

(12) Find the spectrum of the Levi graph of a symmetric (v, k, λ)-design.

(13) Prove that if there exists a 2-(v, k, λ) design and a 2-(v, k, μ) design, then there exists a 2-$(v, k, \lambda + \mu)$ design.

(14) Let X be the set of all elements of a finite field F of order q and let Y be a k-subset of X, $2 \leq k \leq q - 1$. Let $\mathcal{B}$ be the set of all distinct subsets of X of the form $aY + b = \{ay + b\colon y \in Y\}$ where $a, b \in F$, $a \neq 0$. Prove that there exists a divisor n of $k(k-1)$ such that $(X, \mathcal{B})$ is a 2-$(q, k, k(k-1)/n)$.

(15) (*Mann's Inequality*). Under the conditions of Proposition 2.3.8, suppose further that block A is repeated s times, i.e., there are exactly s blocks (including A) which are incident with the same set of points as A.

(a) Prove that $\sum_{i=0}^{k-1} n_i = b - s$, $\sum_{i=0}^{k-1} i n_i = k(r - s)$, and $\sum_{i=0}^{k-1} i(i-1)n_i = k(k-1)(\lambda - s)$.

(b) Prove that $b \geq sv$ and $r \geq sk$ and the equalities $b = sv$ and $r = sk$ hold if and only if each of the remaining $b - s$ blocks meets A in exactly $k(k-1)/(v-1)$ points.

(16) Let $\Gamma = (V, E)$ be a regular graph of degree d. Let $V = \{1, 2, \ldots, v\}$ and let k, λ, and μ be nonnegative integers, $2 \leq k \leq v - 1$, $\lambda \neq \mu$. A (v, k, λ, μ)-design on Γ is an incidence structure $\mathbf{D} = (V, \mathcal{B})$ where $\mathcal{B}$ is a set of k-subsets of V satisfying the following conditions: (i) if $i, j \in V$ and $\{i, j\} \in E$, then there are exactly λ blocks containing $\{i, j\}$; (ii) if $i, j \in V$, $i \neq j$, and $\{i, j\} \notin E$, then there are exactly μ blocks containing $\{i, j\}$. Let $b = |\mathcal{B}|$.

(a) Prove that all points of $\mathbf{D}$ have the same replication number r satisfying equations $s\lambda + (v - d - 1)\mu = r(k-1)$ and $vr = bk$.

(b) With each block B, we associate a variable x_B. For $i \in V$, let $x_B(i) = 1$ if $i \in B$ and $x_B(i) = 0$ if $i \notin B$. For $i = 1, 2, \ldots, v$, let $f_i = \sum_{B \ni i} x_B - \mu$ be linear polynomials in the variables x_B. Let $P = span\{f_1, f_2, \ldots, f_v\}$. Prove that if dim $P \leq v - 1$, then $s = (r - \mu)/(\mu - \lambda)$ is an eigenvalue of Γ.

(c) Prove that if $s = (r - \mu)/(\mu - \lambda)$ is a simple eigenvalue of Γ, then dim $P \geq v - 1$.

(d) Prove that if $s = (r - \mu)/(\mu - \lambda)$ is a simple eigenvalue of Γ or s is not an eigenvalue of Γ, then $b \geq v$.

(e) Let $\Gamma = (V, E)$ be the disjoint union of v complete graphs K_n and let $\mathbf{C} = (W, \mathcal{A})$ be a (v, b, r, k, μ)-design whose points are the connected components of Γ. For every block A of $\mathbf{C}$, let $B_A = \bigcup_{K \in A} K$. Let $\mathcal{B} = \{B_A : A \in \mathcal{A}\}$. Show that $\mathbf{D} = (V, \mathcal{B})$ is an (nv, mk, r, μ)-design on Γ with b blocks. Observe that, for sufficiently large n, we have $b < nv$.

(17) Find all values of v, r, k and λ for which there exists a $(v, 6, r, k, \lambda)$-design.

(18) Find all values of v, r, k and λ for which there exists a $(v, 7, r, k, \lambda)$-design.

(19) Find all values of v, r, k and λ for which there exists a $(v, 8, r, k, \lambda)$-design.

(20) Show that, for every positive integer $v \geq 2$, there is a unique (up to isomorphism) 2-$(v, 2, 1)$ design.

(21) Show that any 2-$(v, 2, \lambda)$ design is isomorphic to the λ-fold multiple of a 2-$(v, 2, 1)$ design.

(22) Construct a 2-(7, 3, 3) design without repeated blocks.

(23) Construct a 2-(7, 3, 3) design which has a block repeated thrice, a block repeated twice, and a non-repeated block.

(24) Let **D** be a 2-(6, 3, 2) design.

(a) Prove that **D** has no repeated block.

(b) Prove that no two blocks of **D** are disjoint.

(25) Prove that there is a unique (up to isomorphism) 2-(6, 3, 2) design.

(26) Show that if N is an incidence matrix of a nontrivial (v, b, r, k, λ)-design, then the matrix $NN^\top$ is nonsingular. Compare the ranks of N and $NN^\top$ and obtain another proof of Fisher's Inequality.

(27) Let X be the set of all ordered pairs (i, j) with $i, j \in \{1, 2, \ldots, n\}$. Define an incidence structure $\mathbf{D} = (X, X, I)$ where $((i, j), (k, l)) \in I$ if and only if $i = k$, $j \neq l$ or $i \neq k$, $j = l$. Prove that **D** is a symmetric design if and only if $n = 4$.

(28) Let X be the set of all ordered pairs (i, j) with $i, j \in \{1, 2, \ldots, n\}$ and let L be a *Latin square of order* n, i.e., an $n \times n$ array such that, for $m = 1, 2, \ldots, n$, each row and each column of L contains a unique entry equal to m. let $L(i, j)$ denote the (i, j)-entry of L. Define an incidence structure $\mathbf{D} = (X, X, I)$ where $((i, j), (k, l)) \in I$ if and only if $i = k$, $j \neq l$ or $i \neq k$, $j = l$ or $i \neq k$, $j \neq l$, $L(i, j) = L(k, l)$. Prove that **D** is a symmetric design if and only if $n = 6$.

(29) Verify that the design of Example 2.4.4 is a symmetric design.

(30) Let B_1, B_2, and B_3 be three distinct blocks of a symmetric (v, k, λ)-design. Prove that $|B_1 \cap B_2 \cap B_3| \leq v - 3(k - \lambda)$.

(31) Let B_1, B_2, and B_3 be three distinct blocks of a (v, b, r, k, λ)-design. Prove that if this design is a residual of a symmetric design, then $|B_1 \cap B_2| + |B_1 \cap B_3| + |B_2 \cap B_3| \leq r$.

(32) Show that there is a unique (up to isomorphism) symmetric (7, 3, 1)-design.

(33) Show that there is a unique (up to isomorphism) symmetric (13, 4, 1)-design.

(34) Prove Theorem 2.4.21.

(35) Find an isomorphism between symmetric (16, 6, 2)-designs with incidence matrices represented as block matrices

$$N_1 = \begin{bmatrix} J-I & I & I & I \\ I & J-I & I & I \\ I & I & J-I & I \\ I & I & I & J-I \end{bmatrix} \quad \text{and} \quad N_2 = \begin{bmatrix} O & K & L & M \\ K & O & M & L \\ L & M & O & K \\ M & L & K & O \end{bmatrix}$$

where

$$K = \begin{bmatrix} 1 & 1 & 0 & 0 \\ 1 & 1 & 0 & 0 \\ 0 & 0 & 1 & 1 \\ 0 & 0 & 1 & 1 \end{bmatrix}, \quad L = \begin{bmatrix} 1 & 0 & 1 & 0 \\ 0 & 1 & 0 & 1 \\ 1 & 0 & 1 & 0 \\ 0 & 1 & 0 & 1 \end{bmatrix}, \quad M = \begin{bmatrix} 1 & 0 & 0 & 1 \\ 0 & 1 & 1 & 1 \\ 0 & 1 & 1 & 1 \\ 1 & 0 & 0 & 1 \end{bmatrix}.$$

(36) Let $\mathbf{D} = (X, \mathcal{B})$ be a symmetric (v, k, λ)-design with $2 \leq k \leq v - 2$. Prove that there are 3-subsets Y and Z of the point set X such that the number of blocks containing Y is not equal to the number of blocks containing Z.

(37) Let $n = k - \lambda$ be the order of a symmetric (v, k, λ)-design with $v \geq 2k$. Prove:
(a) if $v = 4n - 1$, then $k = 2n - 1$ and $\lambda = n - 1$;
(b) if $v = n^2 + n + 1$ and $n \geq 1$, then $k = n + 1$ and $\lambda = 1$;
(c) if $v = 4n$, then $k = 2n - \sqrt{n}$, and $\lambda = n - \sqrt{n}$ (so n is a square).

(38) Let N be an incidence matrix of an incidence structure $\mathbf{D}$. Replacing by -1 every 0-entry in N yields a (± 1) *incidence matrix of* $\mathbf{D}$. For $i = 1, 2$, let P_i be the (± 1) incidence matrix of a symmetric (v_i, k_i, λ_i)-design. Suppose that the matrix $P = P_1 \otimes P_2$ is the (± 1) incidence matrix of a symmetric design. Prove that $v_i = 4(k_i - \lambda_i)$ for $i = 1, 2$.

(39) Prove that there are no symmetric designs with parameters (291, 116, 46) and (1597, 133, 11).

(40) Show that there is a unique (up to isomorphism) symmetric (16, 6, 2)-design that admits a cyclic automorphism group G such that $|G| = 8$ and each non-identity element of G has no fixed point.

(41) Show that there is no symmetric (16, 6, 2)-design admitting an automorphism of order 5.

Notes

A combinatorial design is an arrangement of elements of a (finite) set into subsets so that the subsets satisfy certain regularity conditions. Problems leading to combinatorial designs go back as far as to Euler (1782) and Kirkman (1847). Euler's famous 36 Officers Problem is discussed in Section 3.3. Kirkman's School Girl Design is considered in Example 5.3.7. The notion of designs in geometrical context occurs, for example, in the papers of Plücker (1839) and Steiner (1853).

In the twentieth century an impetus for the development of design theory came from statistics, specifically from the area of design of experiments. Some of the pioneering classic works are Fisher and Yates (1934), Yates (1936), and Bose (1939). Fisher (1949) and Fisher and Yates (1963) are classical references in this area. For combinatorial aspects of design of experiments, we refer to the well known books by Raghavarao (1971) and by Street and Street (1987).

If not every k-subset of a point set of a 2-(v, k, λ) design is a block, the design is often called a *balanced incomplete block design* (*BIBD*). This term was coined by Bose (1939), though the concept of BIBD had appeared in earlier papers of Yates (1935, 1936). In the latter paper, the notation (v, b, r, k, λ) for the parameters was first laid down. The notation for the first three parameters comes from the use of the terms *varieties*, *blocks* and *replication number*, respectively, in agricultural experiments. The term *symmetrical BIBD* (*SBIBD*) was used by Bose for a BIBD having the same number of points and blocks. Another term for symmetric designs that is used in the literature is *square designs*.

In the seminal paper, Bose (1939) first laid down some systematic methods for constructing BIBDs. In this paper he introduced the use of groups, Galois fields, and finite geometries in the constructions of designs. In this context, the following anecdote concerning Bose might be appropriate. In order to get an estimation of crop yields research workers generally went to the agricultural fields. Bose's approach was more abstract. Colleagues of Bose used to joke that when everybody went to the agricultural field, Professor Bose went to the Galois field. See Gropp (1991) for an interesting account of the origins of design theory as mathematical subject and also of the influence of some of early contributors such as Bose.

The first proof that $b \geq v$ in a 2-design was given by Fisher (1940) by using variance counting. (See Exercise 15.) Another proof was given by Bose (1942) which uses an incidence matrix of the design. (See Exercise 26.) Bose's paper (1939) contains the first proof that any two distinct blocks of a symmetric (v, k, λ)-design intersect in λ points. Bose's proof uses variance counting. We mention Stinson (1982) for further applications and generalizations of the variance method in combinatorial designs. Chowla and Ryser (1950) discuss different conditions under which an arrangement of v elements into v sets forms a symmetric design. Hanani (1975) showed that the basic parameter relations for 2-designs with block size $k \leq 5$ are sufficient (with one exception) for the existence of a design (see Remark 2.3.11).

Proposition 2.3.8 is due independently to Schützenberger (1949) and S. S. Shrikhande (1950). The Bruck–Ryser–Chowla Theorem was proved for symmetric $(v, k, 1)$-designs in Bruck and Ryser (1949). For the general symmetric (v, k, λ)-designs the theorem was proved independently in Chowla and Ryser (1950) and in S. S. Shrikhande (1950). In the former paper the necessary condition for the existence of a nontrivial symmetric (v, k, λ)-design with v odd is that the equation

$$x^2 = (k - \lambda)y^2 + (-1)^{(v-1)/2}\lambda z^2$$

has a nonzero solution in integers x, y, z. The latter paper uses the Hilbert symbols and Hasse invariants. The equivalence of these two forms of the Bruck–Ryser–Chowla Theorem was shown in S. S. Shrikhande and Raghavarao (1964). It is based on the Hasse–Minkowski theory of quadratic forms (see Jones (1950)).

The concept of residual and derived designs of a symmetric design was introduced in Bose (1939) where he refers to these designs as those obtained from *block section* and *block intersection*, respectively. Example 2.4.18 is due to Bhattacharya (1944b). Some sufficient conditions for embeddability of quasi-residual designs are given in Chapter 8. Nonembeddable quasi-residual designs is the topic of Chapter 13.

Dembowski (1968) defines the notions of internal and external structures with respect to a point, which are equivalent to point-derived and point-residual substructures, and internal and external structures with respect to a block, which are equivalent to block-derived and block-residual substructures. Definition 2.1.5 generalizes Dembowski's definitions.

An example of a self-dual symmetric design not admitting a symmetric incidence matrix (see Remark 2.4.7) is due to Denniston (1982). This paper gives a complete enumeration of symmetric (25, 9, 3)-designs and contains an example of nonisomorphic symmetric designs with isomorphic residual and derived designs (see Remark 2.4.14).

Design theory is by now recognized as a well established branch of combinatorial mathematics. For further information on 2-designs, we refer to Ryser (1963), Dembowski (1968), Raghavarao (1971), Street and Wallis (1977), Hughes and Piper (1985), Hall (1986), Tonchev (1988), Cameron and van Lint (1991), van Lint and Wilson (1993). For the most comprehensive modern treatment, see Beth, Jungnickel, and Lenz (1999). Brouwer (1995) gives a broad survey of the theory of block designs. A very useful reference book is Colbourn and Dinitz (1996).

The Orbit Theorem is due to Brauer (1941). It was rediscovered by Hughes (1957a), Parker (1957), and Dembowski (1958). Our construction of a symmetric (41, 16, 6)-design follows van Trung (1982a). Another construction of a symmetric (41, 16, 6)-design is given in Bridges, Hall and Hayden (1981). A symmetric (79, 13, 2)-design is due to Aschbacher (1971). Theorems 2.8.2, 2.8.3, 2.8.4, 2.8.5, 2.8.6, 2.8.7, and 2.8.8 are due to Brouwer and Wilbrink (1984), Janko and van Trung (1984) Janko and van Trung (1985a), Janko and van Trung (1985b), Tonchev (1987b), Janko (1999), and Janko (1997), respectively. For an explicit description of many of these designs, see van Trung (1996).

Graph theory has numerous applications in combinatorics as well as in other branches of mathematics. For a modern introductory treatment of the subject see Bollobás (1998). For relations between properties of graphs and their eigenvalues and for further use of algebraic techniques in the study of graphs, see Biggs (1993) and Godsil and Royle (2001). We follow the former book in the proof of Proposition 2.2.17. Levi graphs were introduced in Levi (1942). Theorem 2.2.18 is due to Hoffman (1963). For Exercise 10, see Biggs (1993). Exercise (16) follows Ionin and M. S. Shrikhande (2002). We will discuss further relations between graphs and designs in Chapters 7 and 8.

3
Vector spaces over finite fields

Prototypes of many combinatorial designs come from finite *projective geometries* and finite *affine geometries*. Vector spaces over finite fields provide a natural setting for describing these geometries. Among the numerous incidence structures that can be constructed using affine and projective geometries are infinite families of symmetric designs, *nets* and *Latin squares*. Subspaces of a vector space over a finite field can be regarded as *linear codes* that will be used in later chapters for constructing other combinatorial structures, such as *Witt designs* and *balanced generalized weighing matrices*.

3.1. Finite fields

In this section we recall a few basic results on finite fields which will be used throughout this book.

For any prime p, the residue classes modulo p with the usual addition and multiplication form a finite field $GF(p)$ of order p. These fields are called *prime fields*. Any finite field F of characteristic p contains $GF(p)$ as a subfield. The field F then can be regarded as a finite-dimensional vector space over $GF(p)$, and therefore, $|F| = p^n$ where n is the dimension of this vector space. Conversely, for any prime power $q = p^n$, there is a unique (up to isomorphism) finite field of order q. This field is denoted by $GF(q)$ and is often called the *Galois field* of order q. In general, the field $GF(q)$ is isomorphic to (a unique) subfield of the field $GF(r)$ if and only if r is a power of q. If this is the case, the field $GF(r)$ is said to be an *extension* of $GF(q)$. Recall that, for any subfield F of a field K and any $\alpha \in K$, there is the smallest subfield of K containing F and α. It is denoted by $F(\alpha)$. If K is a finite field, then $F(\alpha) = \{f(\alpha)\colon f \text{ is a polynomial over } F\}$.

The additive group $GF(q)^+$ of $GF(q)$ is an *elementary abelian group*, i.e., the direct product of cyclic groups of prime order. The multiplicative group $GF(q)^*$ of $GF(q)$ is a cyclic group of order $q - 1$. Any generator of this group is called a *primitive element* of $GF(q)$. If $q = p^n$ for p a prime, then the automorphism group of $GF(q)$ is the cyclic group generated by the *Frobenius automorphism* $x \mapsto x^p$.

These and other basic results on finite fields can be found in any standard abstract algebra text. We will now introduce the notion of *quadratic character* that will be used in later sections.

Definition 3.1.1. Let q be an odd prime power. The *quadratic character* on the field $GF(q)$ is a function η from $GF(q)$ to the set $\{0, 1, -1\}$ of integers defined by

$$\eta(x) = \begin{cases} 0 & \text{if } x = 0, \\ 1 & \text{if } x \text{ is a nonzero square,} \\ -1 & \text{if } x \text{ is a nonsquare.} \end{cases}$$

Remark 3.1.2. If q is a prime, then the quadratic character restricted to $GF(q)^*$ is essentially the Legendre symbol.

If a is a primitive element of $GF(q)$ with q odd, then $x = a^n$ is a square if and only if n is even. This implies the following properties of the quadratic character.

Proposition 3.1.3. *Let q be an odd prime power and let η be the quadratic character on $GF(q)$. Then:*

(i) *η is multiplicative, that is, $\eta(xy) = \eta(x)\eta(y)$ for all $x, y \in GF(q)$;*
(ii) $\sum_{x \in GF(q)} \eta(x) = 0$;
(iii) *$\eta(-1) = 1$ if and only if $q \equiv 1 \pmod 4$.*

We will also need the following property of the quadratic character.

Lemma 3.1.4. *Let q be an odd prime power and let η be the quadratic character on $GF(q)$. Then, for any $a \in GF(q)^*$, there are exactly $(q-3)/2$ elements $x \in GF(q)$ such that $\eta(x + a) = \eta(x)$.*

Proof. Let $a \in GF(q)^*$. Then

$$\sum_{x \in GF(q)} \eta(x+a)\eta(x) = \sum_{x \in GF(q)^*} \eta(x+a)\eta(x) = \sum_{x \in GF(q)^*} \frac{\eta(x+a)}{\eta(x)}$$
$$= \sum_{x \in GF(q)^*} \eta\left(1 + \frac{a}{x}\right) = \sum_{y \in GF(q) \setminus \{1\}} \eta(y),$$

so

$$\sum_{x \in GF(q)} \eta(x+a)\eta(x) = -1. \qquad (3.1)$$

Therefore, among $q-2$ nonzero products $\eta(x+a)\eta(x)$, the number of (-1)s is one more than the number of $1s$. This implies the statement of the lemma. □

3.2. Affine planes and nets

Euclidean plane geometry studies the incidence structure formed by points and lines in a plane. In this structure, there is a unique line through any two distinct points and, for any point not on a given line, there is a unique line on the point that is parallel to (i.e., disjoint from) the given line. A nondegenerate incidence structure with these properties is called an *affine plane*.

Definition 3.2.1. An *affine plane* is a pair $(X, \mathcal{L})$, where X is a non-empty set of elements called *points* and $\mathcal{L}$ is a family of subsets of X called *lines*, that satisfy the following axioms:

(A1) Any two distinct points lie on a unique line.
(A2) For any line L and any point $x \notin L$, there is a unique line M that contains x and is disjoint from L.
(A3) There exists a *triangle*, i.e., a set of three points not on a common line.

Note that (A1) allows us to denote as $\overline{xy}$ the line containing distinct points x and y.

Example 3.2.2. Let X be a 2-dimensional vector space over a field F. We will consider elements of X as ordered pairs (x, y) where $x, y \in F$. For any $m, b \in F$, we will call the set $\{(x, y) \in X : y = mx + b\}$ a line with the slope m. For any $a \in F$, we will call the set $\{(x, y) \in X : x = a\}$ a line with infinite slope. If $\mathcal{L}$ is the set of all lines, then $(X, \mathcal{L})$ is an affine plane.

We will denote this affine plane as $AG(2, F)$. If F is the finite field of q elements, we will use $AG(2, q)$ rather than $AG(2, F)$.

Definition 3.2.3. Lines L and M in an affine plane $\mathbf{A} = (X, \mathcal{L})$ are called *parallel* ($L \parallel M$) if $L = M$ or $L \cap M = \varnothing$. This relation on the set of lines of an affine plane is called the *parallelism*.

Remark 3.2.4. It is easy to see that in Example 3.2.2, two lines are parallel if and only if they have the same slope.

Proposition 3.2.5. *The parallelism is an equivalence relation on the set of lines in an affine plane.*

Proof. Reflexivity and symmetry of this relation are obvious. Suppose that $K \parallel L$ and $L \parallel M$. If $K = L$ or $L = M$ or $K = M$, then, obviously, $K \parallel M$. If K, L, and M are three distinct lines, then $K \cap M = \varnothing$, since otherwise we would have had two lines, K and M, through the same point which are parallel to the same line L. □

Definition 3.2.6. Equivalence classes of lines in an affine plane with respect to the parallelism are called *parallel classes*.

Thus, in an affine plane, each point lies on one line from each parallel class. Axiom (A3) implies that any affine plane has at least three parallel classes.

We now introduce the notion of a *net*, which generalizes that of an affine plane.

Definition 3.2.7. A *net* is a pair $(X, \mathcal{L})$ where X is a non-empty set of elements called *points* and $\mathcal{L}$ is a family of subsets of X called *lines*, that satisfy the following axioms:

(N1) Any two distinct points lie on at most one line.
(N2) For any line L and any point $x \notin L$, there is a unique line M which contains x and is disjoint from L.
(N3) There exist three distinct lines, no two of which are disjoint.

Example 3.2.8. Let $\mathbf{A} = (X, \mathcal{L})$ be an affine plane with s parallel classes. Let $\mathcal{C}_1, \mathcal{C}_2, \ldots, \mathcal{C}_r$ be r distinct parallel classes of $\mathbf{A}$ where $3 \le r \le s$. Let $\mathcal{L}_0 = \mathcal{C}_1 \cup \mathcal{C}_2 \cup \ldots \cup \mathcal{C}_r$. Then $\mathbf{N} = (X, \mathcal{L}_0)$ is a net. Not all nets can be obtained in this manner. (See Remark 3.2.21).

Axiom (N1) immediately implies that two distinct lines of a net have at most one common point. If two lines have exactly one common point, we say that they *intersect*. Otherwise, if two lines coincide or are disjoint, we call them, as in the case of affine planes, *parallel*. The above proof of Proposition 3.2.5 applies to nets, so the set of lines of a net is the union of disjoint *parallel classes*, and each point lies on exactly one line of each parallel class. Axiom (N3) implies that any net has at least three parallel classes.

We will now turn our attention to *finite nets*, i.e., nets with finite point sets.

Theorem 3.2.9. *Let $\mathbf{N} = (X, \mathcal{L})$ be a net with finitely many points and $r \ge 3$ parallel classes. Then any point in $\mathbf{N}$ lies on exactly r lines and there exists an integer $n \ge r - 1$ such that any line of $\mathbf{N}$ consists of exactly n points, each parallel class consists of exactly n lines, $|X| = n^2$, and $|\mathcal{L}| = rn$.*

Proof. Since each point of **N** lies on exactly one line from each parallel class, there are exactly r lines through any point. Let $L \in \mathcal{L}$ and let $\mathcal{C}$ be a parallel class that does not contain L. Since each point of L lies on exactly one line from $\mathcal{C}$ and L intersects every line from $\mathcal{C}$, we obtain that $|L| = |\mathcal{C}|$. Since there are at least three parallel classes, all parallel classes are of the same cardinality. We denote this cardinality by n and then every line consists of exactly n points. Since the union of the n pairwise disjoint lines of a parallel class is the entire set X, we obtain that $|X| = n^2$. Counting in two ways flags of **N** yields $|\mathcal{L}| = rn$. Finally if L is a line and p is a point not on L, then exactly $r - 1$ lines through p intersect L and therefore $n \geq r - 1$. □

Definition 3.2.10. The number of points on line of a finite net is called the *order* of the net and the number of lines through a point is called the *degree* of the net. A net of order n and degree r is called an *(n, r)-net.*

Observe that $n = r - 1$ for an (n, r)-net means that there is a line through any two points of the net, i.e., the net is an affine plane. Thus, an affine plane of order n is an $(n, n + 1)$-net, and we have the following result.

Corollary 3.2.11. *For any finite affine plane* **A** *there is a positive integer* $n \geq 2$ *such that every line of* **A** *consists of exactly n points, every point lies on exactly $n + 1$ lines, and* **A** *has exactly n^2 points, $n^2 + n$ lines, and $n + 1$ parallel classes.*

Example 3.2.2 implies the next theorem.

Theorem 3.2.12. *For any prime power q, there exists an affine plane of order q.*

A finite affine plane is also a 2-design.

Proposition 3.2.13. *An affine plane of order n is a 2-$(n^2, n, 1)$ design and, conversely, for $n \geq 2$, any 2-$(n^2, n, 1)$ design is an affine plane of order n.*

Proof. Two distinct points of an affine plane lie on a unique line and two distinct points of a 2-$(n^2, n, 1)$ design are incident with a unique block. To complete the proof, observe that the line size and the number of lines through a point for an $(n, n + 1)$-net are the same as the block size and the replication number of a 2-$(n^2, n, 1)$ design. □

In Chapter 2, we used a Latin squares of order 6 to give an example of a symmetric design. Finite nets are closely related to the so-called *mutually orthogonal Latin squares*.

Definition 3.2.14. A *Latin square* of order n is an $n \times n$ array with entries $1, 2, \ldots, n$, having the property that each entry occurs exactly once in each row and in each column. For $i, j = 1, 2, \ldots, n$, we will denote by $A(i, j)$ the (i, j)-entry of a Latin square A of order n. Two Latin squares A and B of order n are said to be *orthogonal* if, for any $k, l \in \{1, 2, \ldots, n\}$, there are unique values of i and j such that $A(i, j) = k$ and $B(i, j) = l$. A set $\{A_1, A_2, \ldots, A_s\}$ of Latin squares of order n is called a set of *mutually orthogonal Latin squares* (MOLS) of order n if any two distinct squares in the set are orthogonal.

Remark 3.2.15. If A is a Latin square of order n and σ is a permutation of the set $\{1, 2, \ldots, n\}$, then we denote by σA the Latin square whose (i, j)-entry is equal to $\sigma(A(i, j))$ for all $i, j = 1, 2, \ldots, n$. If A and B are orthogonal Latin squares of order n and σ and τ are permutations of the set $\{1, 2, \ldots, n\}$, then Latin squares σA and τB are orthogonal.

Proposition 3.2.16. *There are at most $n - 1$ MOLS of order n.*

Proof. Let $\{A_1, A_2, \ldots, A_s\}$ be a set of MOLS of order n. By applying proper permutations to these squares, we can make $A_k(1, 1) = 1$ for $k = 1, 2, \ldots, s$. Now each square has $n - 1$ further entries equal to 1, none occurring in the first row or the first column. By orthogonality, these ones cannot occur in the same position in two different squares. Since there are only $(n - 1)^2$ available positions for these ones, there cannot be more than $n - 1$ squares. □

The following theorem implies that the existence of $n - 1$ MOLS of order n is equivalent to the existence of an affine plane of order n.

Theorem 3.2.17. *For positive integers $r \geq 3$ and $n \geq r - 1$, there exist $r - 2$ MOLS of order n if and only if there is an (n, r)-net.*

Proof. 1. Suppose there exists a set $\{A_1, A_2, \ldots, A_{r-2}\}$ of MOLS of order n. We will now build an (n, r)-net. Define the points of the net to be all ordered pairs (i, j) where $i, j \in \{1, 2, \ldots, n\}$. Define the following three types of lines: horizontal lines $H_j = \{(x, j) \colon 1 \leq x \leq n\}$ for $j = 1, 2, \ldots, n$, vertical lines $V_i = \{(i, y) \colon 1 \leq y \leq n\}$ for $i = 1, 2, \ldots n$, and oblique lines $L_{ij} = \{(x, y) \colon A_i(x, y) = j\}$ for $i = 1, 2, \ldots, r - 2$ and $j = 1, 2, \ldots, n$.

Let X be the set of points and $\mathcal{L}$ the set of lines. We claim that $\mathbf{N} = (X, \mathcal{L})$ is an (n, r)-net.

The definition of Latin squares implies that no two points of an oblique line have the same first coordinate or the same second coordinate. Therefore, for $i \neq h$, H_j is the only line through points (i, j) and (h, j) and, for $j \neq h$, V_i is the only line through points (i, j) and (i, h). If $x \neq u$ and $y \neq v$, then, due to the orthogonality of the given Latin squares, there is at most one square A_i with

$A_i(x, y) = A_i(u, v)$ and therefore at most one line that contains both (x, y) and (u, v). Thus, **N** satisfies (N1).

In order to verify (N2), observe that two distinct horizontal lines are disjoint as well as two distinct vertical lines, while every horizontal line meets every vertical line. Given $i \in \{1, 2, \ldots, r-2\}$ and $j, h \in \{1, 2, \ldots, n\}$, there is a unique $x \in \{1, 2, \ldots, n\}$ such that $A_i(x, h) = j$, which means that lines L_{ij} and H_h intersect. Thus, every oblique line meets every horizontal line. Similarly, every oblique line meets every vertical line. For $j \neq h$, oblique lines L_{ij} and L_{ih} are disjoint, while for distinct $i, k \in \{1, 2, \ldots, r-2\}$, the orthogonality of A_i and A_k implies that lines L_{ij} and L_{kl} intersect. Therefore, given line H_j and point $(i, h) \notin H_j$, H_h is the only line through (i, h) that is disjoint from H_j; given line V_i and point $(h, j) \notin V_i$, V_h is the only line through (h, j) that is disjoint from V_i; given line L_{ij} and point (x, y) such that $A_i(x, y) = h \neq j$, L_{ih} is the only line through (x, y) that is disjoint from L_{ij}. Thus **N** satisfies (N2).

Clearly, **N** satisfies (N3). Since it is a net with n points on each line and r lines through each point, it is an (n, r)-net.

2. Suppose there exists an (n, r)-net **N**. It has n^2 points and r parallel classes of cardinality n. We select two parallel classes, $\mathcal{H} = \{H_1, \ldots, H_n\}$ and $\mathcal{V} = \{V_1, \ldots, V_n\}$, and call their elements horizontal and vertical lines, respectively. Now any point lies on a unique horizontal line H_j and a unique vertical line V_i; we give this point coordinates (i, j).

Let $\{\mathcal{C}_1, \mathcal{C}_2, \ldots, \mathcal{C}_{r-2}\}$ be the other parallel classes and let $\mathcal{C}_i = \{L_{i1}, L_{i2}, \ldots, L_{in}\}$. For $i = 1, 2, \ldots, r-2$, define an array A_i so that, for $x, y, j = 1, 2, \ldots, n$, $A_i(x, y) = j$ if and only if $(x, y) \in L_{ij}$.

If $A_i(x, y) = A_i(x, z) = j$ for $y \neq z$, then $(x, y), (x, z) \in L_{ij}$, which is not the case, because L_{ij} is not a vertical line. Similarly, $A_i(x, y) \neq A_i(z, y)$ whenever $x \neq z$. Therefore, A_i is a Latin square.

Let $i, h \in \{1, 2, \ldots, r-2\}, i \neq h$. For $x, y, j, l \in \{1, 2, \ldots, n\}$, $A_i(x, y) = j$ and $A_h(x, y) = l$ if and only if $(x, y) \in L_{ij} \cap L_{hl}$. Since lines L_{ij} and L_{hl} are not parallel, they intersect in a unique point. Therefore, the squares A_i and A_h are orthogonal, and we have obtained $r-2$ MOLS of order n. □

Corollary 3.2.18. *For $n \geq 3$, there exist $n-1$ MOLS of order n if and only if there is an affine plane of order n.*

If $r = 3$, then, as the proof of Theorem 3.2.17 shows, one Latin square provides an $(n, 3)$-net.

Corollary 3.2.19. *For any $n \geq 2$, there exists an $(n, 3)$-net.*

Corollary 3.2.20. *For any prime power q and any r such that $3 \leq r \leq q+1$, there exists a (q, r)-net.*

Remark 3.2.21. In the following two sections we will give two different proofs to the fact that there is no affine plane of order 6 (See Remarks 3.3.7 and 3.4.9). Therefore, a (6, 3)-net cannot be obtained by deleting some parallel classes from an affine plane.

In the next section we will discuss the existence of $(n, 4)$-nets.

Remark 3.2.22. The multiplication table of a finite group is clearly a Latin square. If the group is abelian, then the Latin square is symmetric. Thus, symmetric Latin squares of order n exist for all n. Later, we will also need symmetric Latin squares with constant diagonal. If $n > 1$ is the order of such a Latin square, then n cannot be odd (Exercise 14). The next result deals with the case of even n.

Lemma 3.2.23. *For any even n, there is a symmetric Latin square L of order n with constant diagonal, i.e., for all i and j, $L(i, j) = L(j, i)$ and $L(i, i) = n$.*

Proof. Let n be an even positive integer. Define a Latin square A of order $n-1$ by: $A(i, j) = r$ if and only if $i + j \equiv r \pmod{n-1}$ and $1 \leq r \leq n-1$. Then A is symmetric and, since $n-1$ is odd, no two diagonal entries of A are the same. Now define a Latin square L of order n as follows:

$$L(i, j) = \begin{cases} A(i, j) & \text{if } i \neq j, i \neq n, j \neq n, \\ A(i, i) & \text{if } i \neq n, j = n, \\ A(j, j) & \text{if } i = n, j \neq n, \\ n & \text{if } i = j. \end{cases}$$

Then L is the required Latin square. □

3.3. The 36 officers problem

In 1779, Leonhard Euler proposed the following problem.

Thirty-six officers of six different ranks and taken from six different regiments, one of each rank and each regiment, are to be arranged, if possible, in a solid square formation of six by six, so that each row and each column contains one and only one officer of each rank and one and only one officer from each regiment.

This problem is equivalent to the existence of two orthogonal Latin squares of order six. Digits 1, 2, 3, 4, 5, and 6 in one Latin square denote the six different ranks, and the same digits denote the six different regiments in the other Latin square. Theorem 3.2.17 implies that this problem is equivalent to the existence of a $(6, 4)$-net. We will see that an $(n, 4)$-net exists whenever n is odd or is a multiple of 4. Euler conjectured that there is no $(n, 4)$-net with $n \equiv 2 \pmod 4$. Though this conjecture is true for $n = 6$ (see Theorem 3.3.6 below), it is false for all other $n \equiv 2 \pmod 4$.

We begin with nets of order $n \not\equiv 2 \pmod 4$. Corollary 3.2.20 implies that a $(q, 4)$ net exists for any prime power $q \geq 3$. If $n \not\equiv 2 \pmod 4$ (and $n \neq 1$), then n can be represented as a product of such prime powers, and Theorem 3.3.3 below implies that for any such n, there exists an $(n, 4)$-net. In order to state this theorem, we introduce the Kronecker product of two Latin squares.

Definition 3.3.1. Let A and B be Latin squares of order m and n, respectively. Then the Kronecker product of A and B is an array $A \otimes B$ of order mn whose (i, j) entry is defined as follows: if $i = (t-1)m + s$ and $j = (v-1)m + u$ with $s, u \in \{1, 2, \ldots, m\}$ and $t, v \in \{1, 2, \ldots, n\}$, then $(A \otimes B)(i, j) = (B(t, v) - 1)m + A(s, u)$.

Proposition 3.3.2. *If A and B are Latin squares of order m and n, respectively, then $A \otimes B$ is a Latin square of order mn. If A and A' are orthogonal Latin squares of order m and B and B' are orthogonal Latin squares of order n, then $A \otimes B$ and $A' \otimes B'$ are orthogonal Latin squares of order mn.*

Proof. Let A and B be Latin squares of order m and n, respectively. Let $x, y \in \{1, 2, \ldots, mn\}$. We have unique representations $x = (b-1)m + a$ and $y = (d-1)m + c$ with $a, c \in \{1, 2, \ldots, m\}$ and $b, d \in \{1, 2, \ldots, n\}$. Then $(A \otimes B)(i, j) = x$ for $i = (t-1)m + s$ and $j = (v-1)m + u$ if and only if $A(s, u) = a$ and $B(t, v) = c$. Therefore, the equation $(A \otimes B)(i, j) = x$ has a unique solution j for each i and a unique solution i for each j. This proves that $A \otimes B$ is a Latin square. Similarly, if A and A' are orthogonal Latin squares of order m and B and B' are orthogonal Latin squares of order n, then the system of equations $(A \otimes B)(i, j) = x$, $(A' \otimes B')(i, j) = y$ has a unique solution (i, j), i.e., Latin squares $A \otimes B$ and $A' \otimes B'$ are orthogonal. □

Theorem 3.2.17 and Proposition 3.3.2 immediately imply the following result.

Theorem 3.3.3. *If there exist an (m, r)-net and an (n, r)-net, then there exists an (mn, r)-net.*

Corollary 3.3.4. *If $n \not\equiv 2 \pmod 4$ and $n \geq 3$, then there exists an $(n, 4)$-net.*

The case $n \equiv 2 \pmod 4$ is significantly more complicated. In this case, if $n \neq 2$ or 6, then, contrary to Euler's conjecture, there exists a pair of orthogonal Latin squares of order n, i.e., an $(n, 4)$-net. The proof of the following theorem is beyond the scope of this book.

Theorem 3.3.5 (The Bose–Shrikhande–Parker Theorem). *If $n \equiv 2 \pmod 4$ and $n \neq 2$ or 6, then there exists an $(n, 4)$-net.*

The case $n = 2$ is obvious. We will now consider $n = 6$.

Theorem 3.3.6. *There is no $(6, 4)$-net.*

Proof. Suppose there exists a $(6, 4)$-net $(P, \mathcal{L})$ with $P = \{p_1, p_2, \ldots, p_{36}\}$, $\mathcal{L} = \{l_1, l_2, \ldots, l_{24}\}$, and four parallel classes Π_1, Π_2, Π_3, and Π_4. Let $X = \{1, 2, \ldots, 24\}$. For $j = 1, 2, 3, 4$, let $B_j = \{i \in X : l_i \in \Pi_j\}$; for $j = 5, 6, \ldots, 40$, let $B_j = \{i \in X : p_{j-4} \in l_i\}$. Consider the incidence structure $\mathbf{D} = (X, \mathcal{B})$ where $\mathcal{B} = \{B_1, B_2, \ldots, B_{40}\}$. Observe that (i) any element of X is contained in exactly 7 blocks, (ii) any 2-subset of X is contained in a unique block, (iii) the blocks B_1, B_2, B_3, and B_4 of cardinality 6 partition X, (iv) the cardinality of each block B_j with $5 \leq j \leq 40$ is equal to 4, and (v) $|B_j \cap B_k| = 1$ whenever $5 \leq j \leq 40$ and $1 \leq k \leq 4$.

Let $M = [m_{ij}]$ be the 24×40 matrix over the field $GF(2)$ with $m_{ij} = 1$ if and only if $i \in B_j$. Observe that $MM^\top = J$.

Claim 1. The rank of M (over $GF(2)$) does not exceed 20.

To prove this claim, consider the vector space S over $GF(2)$ of solutions to the system of homogeneous linear equations with matrix M, i.e., $S = \{\mathbf{x} : M\mathbf{x} = \mathbf{0}\}$. Let $\{R_1, R_2, \ldots, R_s\}$ be a linearly independent set of rows of M. Since $MM^\top = J$, we have $R_1^\top + R_i^\top \in S$ for $i = 2, 3, \ldots, s$. Since the set $\{R_1^\top + R_2^\top, R_1^\top + R_3^\top, \ldots, R_1^\top + R_s^\top\}$ is linearly independent, we conclude that $\dim(S) \geq rank(M) - 1$. Since, on the other hand, $\dim(S) = 40 - rank(M)$, we obtain that $rank(M) \leq 20$.

Let V denote the 24-dimensional vector space over $GF(2)$. We will regard every subset A of X as a vector $A = [a_1, a_2, \ldots, a_{24}]^\top \in V$ with $a_i = 1$ if and only if $i \in A$. If A and B are subsets of X, regarded as vectors, then $A + B = (A \cup B) + (A \cap B)$ (and, as a set, $A + B$ is the symmetric difference of A and B). Let U be the subspace of V formed by all solutions to the system of homogeneous linear equations with matrix $M^\top$. Then Claim 1 implies that $\dim(U) = 24 - rank(M^\top) \geq 4$. Observe that a subset A of X is an element of U if and only if $|A \cap B_j|$ is even for $j = 1, 2, \ldots, 40$. Therefore, $U_0 = \{\varnothing, X\} \cup \{B_j \cup B_k : 1 \leq j < k \leq 4\}$ is a 3-dimensional subspace of U with a basis $\{B_1 \cup B_2, B_1 \cup B_3, B_1 \cup B_4\}$. Since $\dim(U) \geq 4$, the set $U \setminus U_0$ is not empty. If $Y \in U \setminus U_0$, then $X \setminus Y \in U \setminus U_0$, so there is a subset Y of X such that $Y \in U \setminus U_0$ and $|Y| \leq 12$.

Let Y be such a subset and let b_m, for $m = 0, 2, 4, 6$, denote the number of blocks B_j such that $|Y \cap B_j| = m$. Then

$$b_0 + b_2 + b_4 + b_6 = 40. \tag{3.2}$$

Counting in two ways pairs (j, i) with $i \in Y \cap B_j$ and triples (j, i, h) with $h, i \in Y \cap B_j$ and $h \neq i$ yields two more equations:

$$2b_2 + 4b_4 + 6b_6 = 7|Y|, \tag{3.3}$$

$$2b_2 + 12b_4 + 30b_6 = |Y|(|Y| - 1). \tag{3.4}$$

Eqs. (3.3) and (3.4) imply

$$b_4 + 3b_6 = \frac{|Y|(|Y| - 8)}{8}, \tag{3.5}$$

and therefore $|Y| \geq 8$ and $|Y| \equiv 0 \pmod 4$. Thus, we have proved

Claim 2. If $Y \in U \setminus U_0$ and $|Y| \leq 12$, then $|Y| = 8$ or 12.

Suppose now that there exists $Y \in U \setminus U_0$ with $|Y| = 8$. We assume without loss of generality that $Y = \{17, 18, 19, 20, 21, 22, 23, 24\}$ is such a subset of X. Equations (3.5), (3.3), and (3.2) imply that $b_4 = b_6 = 0$, $b_2 = 28$, and $b_0 = 12$. Since $\{B_1, B_2, B_3, B_4\}$ is a partition of X into 6-subsets and since $b_4 = b_6 = 0$, we obtain that $|Y \cap B_j| = 2$ for $j = 1, 2, 3, 4$. Therefore, we assume without loss of generality that $B_1 = \{1, 2, 3, 4, 17, 18\}$, $B_2 = \{5, 6, 7, 8, 19, 20\}$, $B_3 = \{9, 10, 11, 12, 21, 22\}$, and $B_4 = \{13, 14, 15, 16, 23, 24\}$. Form the graph Γ with the vertex set $\{1, 2, 3, \ldots, 16\}$ and with all 2-sets of the form $B_j \setminus Y$ as edges. Note that if $|B_j \setminus Y| = 2$, then $|B_j| = 4$. Since $b_2 = 28$, there are 24 blocks B_j of cardinality 4 such that $|Y \cap B_j| = 2$, so the graph Γ has 16 vertices and 24 edges.

Claim 3. The graph Γ is regular of degree 3, and each vertex i of Γ has exactly one adjacent vertex in each of the blocks B_1, B_2, B_3, and B_4, except the one that contains i.

It suffices to prove this claim for vertex 1. Since no block other than B_1 contains two vertices of B_1, no vertex of B_1 is adjacent to 1. Suppose two distinct vertices from the same set B_j, $2 \leq j \leq 4$, are adjacent to 1. Without loss of generality we assume that vertices 5 and 6 are adjacent to 1. Then the two blocks of size 4, e.g., A_1 and A_2, that contain edges

$\{1,5\}$ and $\{1,6\}$, respectively, contain no other vertex of Γ. Therefore, $(A_1 \cup A_2) \cap (B_3 \cup B_4) = \{21, 22, 23, 24\}$. Consider the block that contains vertices $1 \in B_1$ and $19 \in B_2$. Since this block shares 1 with A_1, A_2, and B_1 and shares 19 with B_2, it contains no element of Y, except 19, a contradiction. Therefore, vertex 1 has at most one adjacent vertex in each of the blocks B_2, B_3, and B_4. Applying the same reasoning to each vertex of Γ shows that the degree of every vertex does not exceed 3. Since Γ has 16 vertices and $\frac{16 \cdot 3}{2} = 24$ edges, the degree of every vertex must be equal to 3, and Claim 3 is proven.

Claim 4. The graph Γ is triangle-free, i.e., it has no cycle of length 3.

To prove this claim, we assume that Γ has a triangle. Since the vertices of this triangle lie in distinct blocks B_j with $1 \le j \le 4$, we assume without loss of generality that vertices 1, 5, and 9 form a triangle. Let A_1 and A_2 be the blocks that contain edges $\{1,5\}$ and $\{1,9\}$, respectively. Then $|A_1 \setminus Y| = |A_2 \setminus Y| = 2$, and we assume without loss of generality that $A_1 = \{1, 5, 21, 23\}$ and $A_2 = \{1, 9, 19, 24\}$. Let A_3 be the block that contains edge $\{5, 9\}$. Then $|A_3 \setminus Y| = 2$. Since $5 \in A_3 \cap B_2$ and $9 \in A_3 \cap B_3$, we obtain that $19, 20, 21, 22 \notin A_3$. Since $5 \in A_1 \cap A_3$ and $9 \in A_2 \cap A_3$, we obtain that $23, 24 \notin A_3$. But then $A_3 = \{5, 9, 17, 18\}$ and therefore $|A_3 \cap B_1| = 2$. This contradiction proves Claim 4.

Claim 5. If i_1, i_2, and i_3 are the three neighbors of the same vertex of Γ, then there is no block that contains the set $\{i_1, i_2, i_3\}$.

To prove this claim, we assume without loss of generality that there is a block B_j with $5 \le j \le 40$ that contains the three neighbors of vertex 1. Then no edge of Γ is contained in B_j. Claim 3 allows us to assume without loss of generality that these three neighbors are 5, 9, and 13. Since $\{1,5\}$ is an edge, we have $1 \notin B_j$, so we assume that $B_j = \{2, 5, 9, 13\}$. Let A_1, A_2, and A_3 be the blocks of $\mathbf{D}$ that contain 2-subsets $\{1,6\}$, $\{1,7\}$, and $\{1,8\}$, respectively. Since $6, 7, 8 \in B_2$ and the vertices 6, 7, and 8 are not adjacent to 1, the blocks A_1, A_2, and A_3 are distinct and disjoint from Y. Therefore, we assume without loss of generality that $A_1 = \{1, 6, 10, 14\}$, $A_2 = \{1, 7, 11, 15\}$, and $A_3 = \{1, 8, 12, 16\}$. By Claim 3, vertex 2 has exactly one adjacent vertex in B_2. Since 1 is the only neighbor of 5 in B_1, vertices 2 and 5 are not adjacent. Without loss of generality we assume that 8 is the vertex adjacent to 2 in B_2. Let A_4 and A_5 be the blocks that contain 2-subsets $\{2,6\}$ and $\{2,7\}$, respectively. Since these 2-subsets are not edges, the blocks A_4 and A_5 are disjoint from Y. Since $6, 10, 14 \in A_1$ and $7, 11, 15 \in A_2$, we obtain that $10, 14 \notin A_4$ and $11, 15 \notin A_5$. Since $B_j = \{2, 5, 9, 13\}$ and $2 \in A_4 \cap A_5$, we obtain that $5, 9, 13 \notin A_4 \cup A_5$. If $12 \notin A_4 \cup A_5$, then $A_4 \cap B_3 = \{11\}$ and $A_5 \cap B_3 = \{10\}$. Then $15 \notin A_4$

and $14 \notin A_5$, which implies that $2, 16 \in A_4 \cap A_5$, a contradiction. Therefore, $12 \in A_4 \cup A_5$, and we assume without loss of generality that $12 \in A_4$. This implies that $10 \in A_5$ and $16 \notin A_4$. Then $14 \in A_4$ and $16 \in A_5$. Since the vertices adjacent to 2 are not contained in $A_4 \cup A_5$, we obtain that these vertices are 8, 11, and 15. Since Γ is triangle-free, there are three distinct blocks, A_6, A_7, and A_8, that are disjoint from Y and contain 2-subsets $\{8, 11\}$, $\{8, 15\}$, and $\{11, 15\}$, respectively. None of these blocks contains 1 (because $1, 11, 15 \in A_2$) or 2 (because $\{2, 8\}$, $\{2, 11\}$, $\{2, 15\}$ are edges), and therefore at least one of them contain 3 and at least one of them contains 4. Therefore, one of these blocks is disjoint from B_1, a contradiction. This proves Claim 5.

Claim 6. If distinct vertices h and i are contained in the same block of cardinality 6, then there is a unique block that contains i and exactly two vertices adjacent to h.

To prove this claim, we assume that $h, i \in B_1$ and, for $j = 2, 3, 4$, let k_j be the vertex adjacent to h in B_j. By Claim 5, 2-subsets $\{k_2, k_3\}$, $\{k_2, k_4\}$, and $\{k_3, k_4\}$ are contained in three distinct blocks A_1, A_2, and A_3, respectively. Since Γ is triangle-free, these 2-subsets are not edges and therefore, the blocks A_1, A_2, and A_3 are disjoint from Y. Since 2-subsets $\{h, k_j\}$ are edges, we obtain that the blocks A_1, A_2, and A_3 do not contain h. Since these three blocks must meet B_1 in three distinct points, exactly one of them contains i. This proves Claim 6.

For any distinct $h, i \in B_1$, we denote by $T(h, i)$ the 3-subset of X that is contained in a block and consists of i and two vertices adjacent to h.

Claim 7. For distinct $h, i, k \in B_1$ and for $j = 2, 3, 4$, if $T(h, i) \cap B_j = \varnothing$, then $T(k, i) \cap B_j = \varnothing$.

To prove this claim we assume that $T(2, 1) \cap B_4 = \varnothing$. We also assume that, for $i = 1, 2, 3, 4$, the vertices adjacent to i are $i + 4$, $i + 8$, and $i + 12$. Then $T(2, 1) = \{1, 6, 10\}$. We shall prove that $T(3, 1) \cap B_4 = \varnothing$.

Let A_2 be the block containing $T(2, 1)$ and A_3 the block containing $T(3, 1)$. Since $\{1, 13\}$ is an edge, $13 \notin A_2 \cup A_3$. Since vertices 6, 10, and 14 are adjacent to 2, Claim 5 implies that $14 \notin A_2$. Therefore, $15 \in A_2$ or $16 \in A_2$. If $15 \in A_2$, then $15 \notin A_3$, and therefore $T(3, 1) \cap B_4 = \varnothing$. If $16 \in A_2$, then $16 \notin T(4, 1)$, so $T(4, 1) = \{1, 8, 12\}$. Since $\{1, 5\}$ is an edge, $\{1, 6\} \subset T(2, 1)$, and $\{1, 8\} \subset T(4, 1)$, we obtain that $7 \in A_3$. By similar reasons, $11 \in A_3$. Then, by Claim 5, $15 \notin A_3$ and therefore again $T(3, 1)$ is disjoint from B_4.

We again assume that, for $i = 1, 2, 3, 4$, the vertices adjacent to i are $i + 4$, $i + 8$, and $i + 12$. Claim 7 allows us to assume that $T(2, 1) = \{1, 6, 10\}$,

$T(3, 1) = \{1, 7, 11\}$, and $T(4, 1) = \{1, 8, 12\}$. Let A be the block that contains $\{5, 9\}$. Since Γ is triangle-free, this 2-subset is not an edge and therefore $A \cap Y = \varnothing$. Since $\{1, 5\}$ is an edge, $1 \notin A$. Therefore, $A \cap \{2, 3, 4\} \neq \varnothing$, and we assume without loss of generality that $2 \in A$. Then $T(1, 2) = \{2, 5, 9\}$, so $T(1, 2) \cap B_4 = \varnothing$. Claim 7 implies that $T(3, 2) \cap B_4 = \varnothing$, i.e., $T(3, 2) = \{2, 7, 11\}$. Thus we obtained that $|T(3, 1) \cap T(3, 2)| = 2$, a contradiction. This contradictions rules out sets Y with $|Y| = 8$, i.e., we have proved the following

Claim 8. If $Y \in U \setminus U_0$, then $|Y| = 12$.

Let $Y \in U \setminus U_0$ and let $|Y| = 12$. For $i = 1, 2, 3, 4$, let $a_i = |Y \cap B_i|$. Without loss of generality, we assume that $a_1 \leq a_2 \leq a_3 \leq a_4$. Since $Y \notin U_0$, we have $Y \neq B_3 \cup B_4$, so $a_3 < 6$. Since a_1, a_2, a_3, and a_4 are even and add up to 12, we obtain that $a_1 + a_2 \leq 4$. Let $Y' = (B_3 \cup B_4) + Y$. Then $Y' \in U \setminus U_0$ and $|Y'| = 2(a_1 + a_2) < 12$. This contradicts Claim 8, and the proof is now complete. □

Remark 3.3.7. Theorem 3.3.6 implies that there is no affine plane of order 6. A simpler proof of this fact will be obtained from the Bruck–Ryser Theorem in the next section (see Remark 3.4.9).

3.4. Projective planes

Finite affine planes give us a family of 2-(v, k, λ) designs with $\lambda = 1$. As we will see in this section, symmetric (v, k, λ)-designs with $\lambda = 1$ are equivalent to another famous geometric structure known as finite projective planes.

Definition 3.4.1. A *projective plane* is a pair $(X, \mathcal{L})$ where X is a non-empty set of elements called *points* and $\mathcal{L}$ is a family of subsets of X called *lines*, that satisfy the following axioms:

(P1) Any two distinct points lie on a unique line.
(P2) Any two lines have a non-empty intersection.
(P3) There exists a *quadrangle*, i.e., a set of four points, no three of which lie on a common line.

The unique line containing distinct points x and y is denoted by $\overline{xy}$.

The following theorem describes the standard procedure which produces a projective plane from any affine plane.

Theorem 3.4.2. *Let $\mathbf{A} = (X, \mathcal{L})$ be an affine plane. Let Π be the set of parallel classes in $\mathbf{A}$. Put $X' = X \cup \Pi$. For each line L in $\mathbf{A}$, put $L' = L \cup \{\pi\}$ where*

π is the parallel class containing L. Finally, put

$$\mathcal{L}' = \{L' : L \in \mathcal{L}\} \cup \{\Pi\}.$$

Then $\mathbf{P} = (X', \mathcal{L}')$ *is a projective plane.*

Proof. It is convenient to call Π the infinite line and its points infinite points.

If p and q are distinct non-infinite points, then (A1) implies that there is a unique line through p and q. Since any point in $\mathbf{A}$ lies on exactly one line of each parallel class, there is exactly one line in $\mathbf{P}$ through a non-infinite point and an infinite point. Since Π is the only line through two infinite points, (P1) is satisfied.

If two lines in $\mathbf{A}$ intersect, then the corresponding lines in $\mathbf{P}$ intersect. If two lines in $\mathbf{A}$ are parallel, then they belong to the same parallel class, so the corresponding lines in $\mathbf{P}$ contain the same infinite point. Thus, (P2) is satisfied.

By (A3), there is a triangle $\{p, q, r\}$ in $\mathbf{A}$. Let $s \neq r$ be a point on the line through r that is parallel to $\overline{pq}$. Then $\{p, q, r, s\}$ is a quadrangle, so (P3) is satisfied. □

The converse result is also true. We leave its proof as an exercise.

Theorem 3.4.3. *Let* $\mathbf{P} = (X, \mathcal{L})$ *be a projective plane and let L be a line of* $\mathbf{P}$. *Let $X' = X \setminus L$ and $\mathcal{L}' = \mathcal{L} \setminus \{L\}$. Then* $\mathbf{A} = (X', \mathcal{L}')$ *is an affine plane.*

Corollary 3.4.4. *Let* $\mathbf{P} = (X, \mathcal{L})$ *be a projective plane. If X is a finite set, then there exists a positive integer $n \geq 2$ such that*

(i) *any line in* $\mathbf{P}$ *consists of exactly $n + 1$ points;*
(ii) *any point in* $\mathbf{P}$ *lies on exactly $n + 1$ lines;*
(iii) $|X| = |\mathcal{L}| = n^2 + n + 1$.

Definition 3.4.5. A projective plane $\mathbf{P}$ is said to be of *order n* if each line of $\mathbf{P}$ has cardinality $n + 1$.

The following theorem is straightforward.

Theorem 3.4.6. *A projective plane of order n is a symmetric $(n^2 + n + 1, n + 1, 1)$-design, and conversely, any symmetric $(n^2 + n + 1, n + 1, 1)$-design with $n \geq 2$ is a projective plane of order n.*

Remark 3.4.7. It was shown in Theorem 2.4.12 that $v \leq n^2 + n + 1$ for any symmetric design of order n on v points. Projective planes meet this bound.

Note that any symmetric $(v, k, 1)$-design is in fact a symmetric $(n^2 + n + 1, n + 1, 1)$-design with $n = k - 1$.

The Bruck–Ryser–Chowla Theorem applied to projective planes imposes a restriction on the order of a projective plane.

Theorem 3.4.8 (The Bruck–Ryser Theorem). *Let n be a positive integer congruent to 1 or 2 $\pmod 4$. If there exists a prime $p \equiv 3 \pmod 4$ such that the highest power of p dividing n is odd, then there is no projective plane of order n.*

Proof. Let p be a prime, $p \equiv 3 \pmod 4$, and let $n = mp^s$ where m is an integer not divisible by p and s is odd. If there exists a symmetric $(n^2+n+1, n+1, 1)$-design, then $(-1)^{(n^2+n)/2} = -1$ and, by the Bruck–Ryser–Chowla Theorem,

$$1 = (-1, n)_p = (-1, p)_p = \left(\frac{-1}{p}\right) = (-1)^{(p-1)/2} = -1,$$

a contradiction. □

Remark 3.4.9. The Bruck–Ryser Theorem rules out infinitely many orders for projective planes: there are no projective planes of order 6, 14, 21, 22, 30, etc. Theorem 3.4.2 implies that there are no affine planes of the same orders. Non-existence of projective planes of order 10 (see Theorem 6.4.5) shows that the condition of the Bruck–Ryser Theorem is not sufficient for the existence of projective planes. The smallest unresolved order for projective (and affine) planes is 12.

Projective planes of order n are equivalent to symmetric $(n^2+n+1, n+1, 1)$-designs. An automorphism of such a symmetric design is called a *collineation* of the corresponding projective plane.

Definition 3.4.10. A *collineation* of a projective plane $\mathbf{P} = (X, \mathcal{L})$ is a bijection $\alpha : X \to X$ such that $\alpha(L)$ is a line for every line $L \in \mathcal{L}$. The group of all collineations of $\mathbf{P}$ is called the *full collineation group of* $\mathbf{P}$ and is denoted by Aut($\mathbf{P}$).

Collineations that fix all lines through a particular point or all points on a particular line are of special interest.

Definition 3.4.11. Let α be a collineation of a projective plane $\mathbf{P} = (X, \mathcal{L})$. A point $c \in X$ is called a *center of* α if $\alpha(L) = L$ for every line $L \in \mathcal{L}$ containing c. A line $A \in \mathcal{L}$ is called an *axis of* α if $\alpha(x) = x$ for every point $x \in A$.

Remark 3.4.12. If α is a collineation of a projective plane $\mathbf{P}$, then α can be regarded as a collineation of the dual projective plane $\mathbf{P}^\top$. If c is a center of α (as a collineation $\mathbf{P}$), then c serves as an axis of α (as a collineation $\mathbf{P}^\top$).

Proposition 3.4.13. *A non-identity collineation of a finite projective plane has at most one center and at most one axis, and it has a center if and only if it has an axis.*

Proof. Let α be a collineation of a projective plane $\mathbf{P} = (X, \mathcal{L})$. Suppose that α has distinct centers b and c and let $L = \overline{bc}$. Let x be a point of $\mathbf{P}$ such that $x \notin L$. Since $\alpha(\overline{bx}) = \overline{bx}$ and $\alpha(\overline{cx}) = \overline{cx}$, we obtain that $\alpha(x) = x$, so $\alpha(x) = x$ for all $x \in X \setminus L$. Let $y \in L$ and let K be a line through y, other than L. Since $\alpha(x) = x$ for all $x \in K \setminus \{y\}$, we obtain that $\alpha(K) = K$ and then $\alpha(y) = y$, i.e., α is the identity. Therefore, any non-identity collineation has at most one center. Applying this result to $\mathbf{P}^\top$, we obtain that any non-identity collineation has at most one axis.

Suppose now that α has a center c. Let n be the order of $\mathbf{P}$. Since α fixes every line through c, it fixes at least $n + 1$ lines. Proposition 2.6.9 then implies that α fixes at least $n + 1$ points. If all these points lie on a line through c, then this line is an axis of α. Suppose α fixes points x and y such that $\overline{cx}$ and $\overline{cy}$ are distinct lines. Then $\alpha(\overline{xy}) = \overline{xy}$. Let $z \in \overline{xy}$. Since $\alpha(\overline{cz}) = \overline{cz}$ and $\alpha(\overline{xy}) = \overline{xy}$, we obtain that $\alpha(z) = z$. Thus $\overline{xy}$ is an axis of α. Therefore, every collineation with a unique center has a unique axis. Applying this statement to $\mathbf{P}^\top$ implies that every collineation with a unique axis has a unique center. □

Definition 3.4.14. A collineation α of a projective plane $\mathbf{P} = (X, \mathcal{L})$ that has a center (and an axis) is called a *central collineation*. For $c \in X$ and $A \in \mathcal{L}$, any central collineation with center c and axis A is called a (c, A)-*central collineation* or a (c, A)-*perspectivity*. A (c, A)-perspectivity is called a (c, A)-*elation* or a (c, A)-*homology* if $c \in A$ or $c \notin A$, respectively.

Example 3.4.15. Let $\mathbf{P}$ be the projective plane obtained by adjoining infinite points and the infinite line to the affine plane $AG(2, q)$. Let $\alpha \in GF(q)$. For each point (x, y) of $AG(2, q)$, let $t_\alpha(x, y) = (x + \alpha, y)$. Then, for any line L of $AG(2, q)$, $t_\alpha(L)$ is a line parallel to L and $t_\alpha(L) = L$ if and only if L is a horizontal line $y = a$ with $a \in GF(q)$. Letting $t_\alpha(\Pi) = \Pi$ for every parallel class of $AG(2, q)$ extends t_α to a collineation of the projective plane $\mathbf{P}$. If Π_0 is the parallel class containing the line $y = 0$, then $t_\alpha(\Pi_0) = \Pi_0$, so t_α is an elation with Π_0 as the center and the infinite line as the axis.

Let $\beta \in GF(q)^*$. For each point (x, y) of $AG(2, q)$, let $h_\beta(x, y) = (\beta x, \beta y)$. Then, for any line L of $AG(2, q)$, $h_\beta(L)$ is a line parallel to L. Letting $h_\beta(\Pi) = \Pi$ for every parallel class Π of $AG(2, q)$ extends h_β to a collineation of $\mathbf{P}$. Since $h_\beta(L) = L$ for every line L containing $(0, 0)$ and $h_\beta(0, 0) = (0, 0)$, we obtain that h_β is a homology with $(0, 0)$ as the center and the infinite line as the axis.

Proposition 3.4.16. *If α is a (c, A)-perspectivity of a projective plane* **P**, *other than the identity, then α has no fixed point, except the center c and the points of the axis A.*

Proof. Let α be a (c, A)-perspectivity of a projective plane $\mathbf{P} = (X, \mathcal{L})$ and let $\alpha(x) = x$, for some $x \in X$ such that $x \neq c$ and $x \notin A$. Since $x \notin A$, every line through L has at least two fixed points of α, and therefore $\alpha(L) = L$ for every line L containing x. Since $\alpha(K) = K$ for every line containing c, we obtain that $\alpha(y) = y$ for all $y \in X \setminus \overline{cx}$. Then $\alpha(L) = L$ for all $L \in \mathcal{L}$ and therefore α is the identity collineation. □

For a given point c and a line A of a projective plane **P**, all (c, A)-perspectivities form a subgroup of the group Aut(**P**). The next theorem places a restriction on the order of such a subgroup.

Proposition 3.4.17. *Let* **P** *be a projective plane of order q. Let c be a point of* **P** *and A a line of* **P**. *If $c \in A$, then the order of the group of all (c, A)-elations divides q. If $c \notin A$, then the order of the group of all (c, A)-homologies divides $q - 1$.*

Proof. Let G be the group of all (c, A)-perspectivities. Let L be a line of **P**, other than A, that contains c and let $Y = L \setminus (A \cup \{c\})$. Then $\alpha(y) \in Y$ for all $y \in Y$ and $\alpha \in G$. Proposition 3.4.16 and the Orbit-Stabilizer Theorem imply that the cardinality of every G-orbit on Y equals the order of G. Therefore, the order of G divides the cardinality of Y which equals q if α is an elation and equals $q - 1$ if α is a homology. □

3.5. Affine geometries over finite fields

Vector spaces over the real numbers lead to the classical affine and projective geometries. In a similar manner, vector spaces over finite fields will lead us to finite geometries.

We will denote by $V(n, q)$ the n-dimensional vector space over the field GF(q). Obviously, $|V(n, q)| = q^n$. In order to count subspaces of $V(n, q)$, we shall use the notion of *Gaussian coefficients*.

Definition 3.5.1. For a prime power q and nonnegative integers n and d, the *Gaussian coefficient* $\begin{bmatrix} n \\ d \end{bmatrix}_q$ is defined to be the number of d-dimensional subspaces of $V(n, q)$.

Proposition 3.5.2. *For a prime power q and positive integers n and $d \leq n$,*

$$\begin{bmatrix} n \\ d \end{bmatrix}_q = \frac{(q^n - 1)(q^{n-1} - 1) \cdots (q^{n-d+1} - 1)}{(q^d - 1)(q^{d-1} - 1) \cdots (q - 1)}.$$

Proof. Let $N(n,d)$ denote the number of linearly independent d-tuples $(\mathbf{x}_1, \mathbf{x}_2, \dots, \mathbf{x}_d)$ of vectors in $V(n,q)$. Counting in two ways pairs $(U, (\mathbf{x}_1, \mathbf{x}_2, \dots, \mathbf{x}_d))$, where U is a d-dimensional subspace of $V(n,q)$ and $(\mathbf{x}_1, \mathbf{x}_2, \dots, \mathbf{x}_d)$ is a basis of U, we obtain that $\left[{n \atop d}\right]_q N(d,d) = N(n,d)$, so

$$\begin{bmatrix} n \\ d \end{bmatrix}_q = \frac{N(n,d)}{N(d,d)}.$$

In order to evaluate $N(n,d)$, note that any non-zero vector can be selected as $\mathbf{x}_1$, and, as soon as linearly independent vectors $\mathbf{x}_i$ for $i \le s$ are selected, $\mathbf{x}_{s+1}$ can be any non-zero vector which is not in the s-dimensional subspace generated by $\mathbf{x}_1, \dots, \mathbf{x}_s$. Therefore,

$$\begin{aligned} N(n,d) &= (q^n - 1)(q^n - q) \cdots (q^n - q^{d-1}) \\ &= q^{d(d-1)/2}(q^n - 1)(q^{n-1} - 1) \cdots (q^{n-d+1} - 1). \end{aligned}$$

Then

$$N(d,d) = q^{d(d-1)/2}(q^d - 1)(q^{d-1} - 1) \cdots (q - 1),$$

and Proposition 3.5.2 follows. □

Corollary 3.5.3.

$$\begin{bmatrix} n \\ d \end{bmatrix}_q = \begin{bmatrix} n \\ n-d \end{bmatrix}_q.$$

Corollary 3.5.4. *Let $d \le m \le n$ be nonnegative integers. Let q be a prime power and let U be a d-dimensional subspace of $V(n,q)$. Then the number of m-dimensional subspaces of $V(n,q)$ that contain U is $\left[{n-d \atop m-d}\right]_q$.*

Proof. The natural homomorphism $V(n,q) \to V(n,q)/U$ establishes a one-to-one correspondence between m-dimensional subspaces of $V(n,q)$ that contain U and $(m-d)$-dimensional subspaces of the $(n-d)$-dimensional space $V(n,q)/U$. □

Definition 3.5.5. Let U be a d-dimensional subspace of $V(n,q)$, $0 \le d \le n-1$, and let $\mathbf{x} \in V(n,q)$. The coset $U + \mathbf{x}$ is called a *d-flat*.

Proposition 3.5.2 and Corollary 3.5.4 imply

Proposition 3.5.6. *For a prime power q and integers n and d, $0 \le d \le n-1$, the number of d-flats in $V(n,q)$ is $q^{n-d}\left[{n \atop d}\right]_q$. If $d \le m \le n-1$, then the number of m-flats that contain a fixed d-flat is $\left[{n-d \atop m-d}\right]_q$.*

We will now define the n-dimensional affine geometry over the finite field $GF(q)$.

Definition 3.5.7. The set of all flats in $V(n, q)$ is called the *n-dimensional affine geometry* over $GF(q)$ and is denoted by $AG(n, q)$. We will call 0-flats *points*, 1-flats *lines*, 2-flats *planes*, and $(n-1)$-flats *hyperplanes*.

Remark 3.5.8. If M is an $m \times n$ matrix over $GF(q)$ of rank r, then the set of all vectors $\mathbf{x} \in V(n, q)$ satisfying the equation $M\mathbf{x} = \mathbf{0}$ is a subspace of $V(n, q)$ of dimension $n - r$. In particular, any $(n-1)$-dimensional subspace of $V(n, q)$ can be described as the set of all vectors $\mathbf{x} = [x_1\ x_2 \ldots x_n]^\top$ satisfying the equation $a_1x_1 + a_2x_2 + \cdots + a_nx_n = 0$, for some nonzero vector $\mathbf{a} = [a_1\ a_2 \ldots a_n]^\top \in V(n, q)$. Any $(n-1)$-flat can be given by an equation $a_1x_1 + a_2x_2 + \cdots + a_nx_n = b$ for some $b \in GF(q)$.

The following proposition is immediate.

Proposition 3.5.9. *For $1 \le d \le n-1$, the incidence structure $AG_d(n, q)$ formed by the points and the d-flats of $AG(n, q)$ is a (v, b, r, k, λ)-design with $v = q^n$, $b = q^{n-d}\begin{bmatrix} n \\ d \end{bmatrix}_q$, $r = \begin{bmatrix} n \\ d \end{bmatrix}_q$, $k = q^d$, and $\lambda = \begin{bmatrix} n-1 \\ d-1 \end{bmatrix}_q$.*

We will often use the following special case:

Proposition 3.5.10. *Let q be a prime power. For $n \ge 1$, the incidence structure formed by the points and hyperplanes of $AG(n, q)$ is the design $AG_{n-1}(n, q)$ with parameters*

$$\left(q^n, \frac{q(q^n-1)}{q-1}, \frac{q^n-1}{q-1}, q^{n-1}, \frac{q^{n-1}-1}{q-1}\right).$$

The notion of parallelism that was introduced in Section 3.2. for affine planes immediately implies that lines $U_1 + \mathbf{x}$ and $U_2 + \mathbf{y}$ of $AG(2, q)$ are parallel if and only if $U_1 = U_2$. This allows us to introduce the parallelism of d-flats in $AG(n, q)$.

Definition 3.5.11. Let U be a d-dimensional subspace of $V(n, q)$, $0 \le d \le n-1$, and let $\mathbf{x}, \mathbf{y} \in V(n, q)$. The d-flats $U + \mathbf{x}$ and $U + \mathbf{y}$ are called *parallel*.

Obviously, parallelism is an equivalence relation on the set of all d-flats. Two parallel d-flats are disjoint or coincide. For hyperplanes, the converse is also true.

Proposition 3.5.12. *Two distinct hyperplanes are parallel if and only if they are disjoint.*

This immediately implies the following proposition, known as Playfair's Axiom.

Proposition 3.5.13. *If $\mathbf{x}$ is a point and H is a hyperplane in $A(n, q)$, then there exists a unique hyperplane through $\mathbf{x}$ that is parallel to H.*

We leave proofs of the last three propositions as exercises.

A vector space over a finite field can be obtained as an extension of this field. This allows us to give a convenient description of all hyperplanes. We begin with a lemma.

Lemma 3.5.14. *Let q be a prime power and d a positive integer. Let H be a hyperplane in the field $GF(q^d)$ regarded as a vector space over $GF(q)$ and let $\alpha \in GF(q^d)^*$. If $\alpha H = H$, then $\alpha \in GF(q)$.*

Proof. Suppose $\alpha H = H$. Then $\alpha^2 H = \alpha H = H, \alpha^3 H = \alpha^2 H = H, \dots,$ $\alpha^n H = H$ for all integers n. Let f be a polynomial over $GF(q)$ such that $f(\alpha) \neq 0$. Then $f(\alpha)H = H$. Let $F = GF(q)(\alpha)$. Then $FH = H$, and therefore H can be regarded as a vector space over F. Since F is an extension of $GF(q)$, we have $|F| = q^s$ for some positive integer s. Since F is a subfield of $GF(q^d)$, s must divide d. On the other hand, if m is the dimension of H over F, then $q^{d-1} = |H| = |F|^m = q^{sm}$, so s divides $d - 1$. Therefore, $s = 1$, i.e., $F = GF(q)$ and $\alpha \in GF(q)$. □

Corollary 3.5.15. *Let q be a prime power and d a positive integer. Let H be a hyperplane in the field $GF(q^d)$ regarded as a vector space over $GF(q)$ and let $\alpha, \beta \in GF(q^d)^*$. Then $\alpha H = \beta H$ if and only if $\alpha\beta^{-1} \in GF(q)$.*

If H is a hyperplane in the field $GF(q^d)$ regarded as a vector space over $GF(q)$, then so is αH for any $\alpha \in GF(q^d)^*$. The next proposition shows that all hyperplanes in $GF(q^d)$ can be obtained in this way.

Proposition 3.5.16. *Let q be a prime power and d a positive integer. Let H be a hyperplane in the field $GF(q^d)$ regarded as a vector space over $GF(q)$. Then every hyperplane in this vector space can be represented as αH with $\alpha \in GF(q^d)^*$.*

Proof. Corollary 3.5.15 implies that the number of distinct hyperplanes of the form αH is equal to the number of cosets of $GF(q)^*$ in $GF(q^d)^*$. Since this number is $(q^d - 1)/(q - 1)$ and it is equal to the total number of hyperplanes in a d-dimensional vector space over $GF(q)$, every hyperplane must be of the form αH. □

3.6. Projective geometries over finite fields

In Section 3.3., we saw how a projective plane can be constructed by adding "points at infinity" to an affine plane. It is possible to obtain higher dimensional projective spaces from the corresponding affine spaces by a similar approach.

However, we will first present another standard description of projective spaces and later show that these two approaches are in fact equivalent.

In order to define the n-dimensional projective geometry over $GF(q)$, we start with the $(n+1)$-dimensional vector space $V(n+1,q)$. We define an equivalence relation on the set of nonzero vectors of $V(n+1,q)$ by declaring vectors $\mathbf{a}$ and $\mathbf{b}$ equivalent if and only if there is a nonzero element $\alpha \in GF(q)$ such that $\mathbf{b} = \alpha\mathbf{a}$. We will call each equivalence class a *projective point* and the set of all projective points the *n-dimensional projective space* over $GF(q)$. Thus, a projective point is the set of all nonzero elements of a one-dimensional subspace of $V(n+1,q)$.

If U is a subspace of $V(n+1,q)$ and x is a projective point, then either $x \subseteq U$ or $x \cap U = \varnothing$. If $\dim U = d+1$, where $-1 \leq d \leq n$, we will call the set of all projective points $x \subseteq U$ a *d-dimensional subspace* of the n-dimensional projective space. The set of all subspaces of the n-dimensional projective space over $GF(q)$ is called the *n-dimensional projective geometry over $GF(q)$* and is denoted by $PG(n,q)$. Projective points are precisely the 0-dimensional subspaces. We will call 1-dimensional, 2-dimensional, and $(n-1)$-dimensional projective spaces (*projective*) *lines*, (*projective*) *planes*, and (*projective*) *hyperplanes*, respectively. Clearly, there is a unique projective line through any two distinct projective points. Note that the empty set is the subspace of dimension -1.

The following result characterizes subspaces of the space $PG(n,q)$.

Proposition 3.6.1. *A set X of projective points is a subspace of $PG(n,q)$ if and only if for any two distinct points $x, y \in X$, the projective line through x and y is contained in X.*

Proof. Let X be a set of projective points and let U be the union of all one-dimensional subspaces of $V(n+1,q)$ that represent points of X.

If X is a subspace of $PG(n,q)$, then U is a subspace of $V(n+1,q)$. Therefore, if x and y are one-dimensional subspaces of $V(n+1,q)$ that represent distinct points of X, then $x+y$ is a two-dimensional subspace of $V(n+1,q)$ that represents a line through these points. Since $x+y \subseteq U$, this line is contained in X.

Conversely, suppose X is a subset of $PG(n,q)$ that contains a line through any two of its points. Then U is subset of $V(n+1,q)$ that contains any linear combination of any two of its points. Thus, U is a subspace of $V(n+1,q)$ and therefore, X is a subspace of $PG(n,q)$. □

Proposition 3.5.2 and Corollary 3.5.4 imply the following result.

Proposition 3.6.2. *For $-1 \le d \le n$, the number of d-dimensional subspaces of $PG(n,q)$ is $\begin{bmatrix} n+1 \\ d+1 \end{bmatrix}_q$. For $-1 \le d \le m \le n$, the number m-dimensional subspaces of $PG(n,q)$ that contain a given d-dimensional subspace is $\begin{bmatrix} n-d \\ m-d \end{bmatrix}_q$.*

Corollary 3.6.3. *Projective geometry $PG(n,q)$ contains $(q^{n+1}-1)/(q-1)$ points and the same number of hyperplanes.*

The following definition extends the analogy between vector spaces and projective spaces.

Definition 3.6.4. Let X be a set of points of $PG(n,q)$. The *span of X* denoted by $\langle X \rangle$ is the intersection of all subspaces of $PG(n,q)$ that contain X.

This definition allows us to obtain an analog of the dimension formula for vector spaces. We leave proof of the following proposition as an exercise.

Proposition 3.6.5. *Let U and W be subspaces of $PG(n,q)$. Then*

$$\dim(\langle U \cup W \rangle) = \dim(U) + \dim(W) - \dim(U \cap W).$$

We can now introduce an important family of 2-designs.

Proposition 3.6.6. *Let n be a positive integer and q a prime power. For $0 \le d < n$, the incidence structure $PG_d(n,q)$ formed by points and d-dimensional subspaces of $PG(n,q)$ is a (v,b,r,k,λ)-design with $v = \begin{bmatrix} n+1 \\ 1 \end{bmatrix}_q$, $b = \begin{bmatrix} n+1 \\ d+1 \end{bmatrix}_q$, $r = \begin{bmatrix} n \\ d \end{bmatrix}_q$, $k = \begin{bmatrix} d+1 \\ 1 \end{bmatrix}_q$, and $\lambda = \begin{bmatrix} n-1 \\ d-1 \end{bmatrix}_q$.*

In the case $s = n-1$ this design is symmetric.

Corollary 3.6.7. *Let n be a positive integer and q a prime power. The design $PG_{n-1}(n,q)$ is a symmetric*

$$\left(\frac{q^{n+1}-1}{q-1}, \frac{q^n-1}{q-1}, \frac{q^{n-1}-1}{q-1} \right)\text{-design.}$$

Symmetric designs with these parameters form the so-called *natural series* of symmetric designs. The next proposition gives a convenient description of points and hyperplanes of the design $PG_{n-1}(n,q)$.

Proposition 3.6.8. *Let n be a positive integer and q a prime power. Let α be a primitive element of the field $GF(q^{n+1})$. Let $v = (q^{n+1}-1)/(q-1)$. Then $\{\langle 1 \rangle, \langle \alpha \rangle, \langle \alpha^2 \rangle, \ldots, \langle \alpha^{v-1} \rangle\}$ is the point set of $PG_{n-1}(n,q)$. Let H be an n-dimensional subspace of $GF(q^{n+1})$ regarded as a vector space over $GF(q)$. Then $\{H, \alpha H, \alpha^2 H, \ldots, \alpha^{v-1} H\}$ is the block set of $PG_{n-1}(n,q)$.*

Proof. Since $|GF(q^{n+1})^*| = q^{n+1} - 1$ and $|GF(q)^*| = q - 1$, v is the smallest positive integer, for which $\alpha^v \in GF(q)$. Therefore, $1, \alpha, \dots, \alpha^{v-1}$ generate distinct one-dimensional subspaces of $GF(q^{n+1})$ and thus represent distinct points of $PG_{n-1}(n, q)$. By Corollary 3.5.15, $\{H, \alpha H, \alpha^2 H, \dots, \alpha^{v-1} H\}$ are distinct hyperplanes. Thus, we have found v distinct points and v distinct blocks of $PG_{n-1}(n, q)$, i.e., the point set and the block set of this design. □

We will now prove that the designs $PG_{n-1}(n, q)$ are self-dual.

Proposition 3.6.9. *For any positive integer n and any prime power q, the design $PG_{n-1}(n, q)$ admits a symmetric incidence matrix and therefore is self-dual.*

Proof. Let n be a positive integer and q a prime power and let α be a primitive element of the field $GF(q^{n+1})$. Let $v = (q^{n+1} - 1)/(q - 1)$. We apply Proposition 3.6.8 and order the point set X and the block set $\mathcal{B}$ of $PG_{n-1}(n, q)$ as follows: $X = \{\langle 1\rangle, \langle\alpha\rangle, \langle\alpha^2\rangle, \dots, \langle\alpha^{v-1}\rangle\}$, $\mathcal{B} = \{\alpha^{v-1}H, \alpha^{v-2}H, \dots, \alpha H, H\}$. Let $N = [n_{ij}]$ be the corresponding incidence matrix of $PG_{n-1}(n, q)$. Then

$$n_{ij} = 1 \Leftrightarrow \alpha^{i-1} \in \alpha^{v-j} H \Leftrightarrow \alpha^{j-1} \in \alpha^{v-i} H \Leftrightarrow n_{ji} = 1,$$

i.e., N is a symmetric matrix. □

In Section 3.2, we saw the relationship between affine and projective planes. A similar relationship holds for affine and projective geometries of higher dimension. We will formulate the corresponding result in the language of symmetric designs.

Proposition 3.6.10. *Let q be a prime power and n a positive integer. If B is a block of the design $\mathbf{D} = PG_{n-1}(n, q)$, then the residual design $\mathbf{D}^B$ is isomorphic to $AG_{n-1}(n, q)$ and, for $n \geq 2$, the derived design $\mathbf{D}_B$ is isomorphic to the q-fold multiple of $PG_{n-2}(n - 1, q)$.*

We leave proof of this proposition as an exercise.

The set of points of $PG(n, q)$ admits a trivial partition into zero-dimensional subspaces (singletons). This partition is the simplest example of the important concept of a *spread of subspaces*.

Definition 3.6.11. A *spread of s-spaces* of $PG(n, q)$ is a partition of the set of points of $PG(n, q)$ into s-dimensional subspaces.

We will show that $PG(n, q)$ admits a spread of s-spaces if and only if $s + 1$ divides $n + 1$.

Lemma 3.6.12. *Let a, m, n be positive integers. If $a \neq 1$, then* $\gcd(a^m - 1, a^n - 1) = a^d - 1$ *where* $d = \gcd(m, n)$.

Proof. For $m \geq n$, $a^m - 1 = a^{m-n}(a^n - 1) + (a^{m-n} - 1)$. Therefore, $\gcd(a^m - 1, a^n - 1) = \gcd(a^n - 1, a^{m-n} - 1)$. Since also $\gcd(m, n) = \gcd(n, m - n)$, the proof can be carried on by induction on $\min(m, n)$. □

Theorem 3.6.13. *The following statements are equivalent*:

(i) *There exists a spread of s-spaces of $PG(n, q)$.*
(ii) $s + 1$ *divides* $n + 1$.

Proof. (i) ⇒ (ii). If there exists a spread of s-spaces of $PG(n, q)$, then the number of points of an s-dimensional subspace divides the number of points of $PG(n, q)$, i.e., $q^{s+1} - 1$ divides $q^{n+1} - 1$. Then, by Lemma 3.6.12, $s + 1$ divides $n + 1$.

(ii) ⇒ (i). Suppose $s + 1$ divides $n + 1$. Then there exists a tower of fields

$$GF(q) \subset GF(q^{s+1}) \subset GF(q^{n+1}).$$

Put $v = (q^{s+1} - 1)/(q - 1)$ and $w = (q^{n+1} - 1)/(q^{s+1} - 1)$ and let α be a primitive element of $GF(q^{n+1})$, i.e., a generator of the multiplicative group of this field. Then $\beta = \alpha^w$ is a primitive element of $GF(q^{s+1})$ and β^v is a primitive element of $GF(q)$.

We will regard $GF(q^{n+1})$ as the vector space $V(n + 1, q)$. Observe that $\alpha^m \notin GF(q)$ for $1 \leq m \leq vw - 1$. Therefore, $1, \alpha, \alpha^2, \ldots, \alpha^{vw-1}$ are representatives of all one-dimensional subspaces of $V(n + 1, q)$, i.e., the points of $PG(n, q)$. For $i = 0, 1, \ldots, w - 1$ and $j = 0, 1, \ldots, v - 1$, let x_{ij} denote the point of $PG(n, q)$ corresponding to the one-dimensional subspace $\alpha^i \beta^j$ of $V(n + 1, q)$.

For $i = 0, 1, \ldots, w - 1$, define $U_i = \{x_{ij} : 0 \leq j \leq v - 1\}$ and $V_i = \{a\alpha^i \beta^j : a \in GF(q), 0 \leq j \leq v - 1\}$. We claim that the set $\{U_0, U_1, \ldots, U_{w-1}\}$ is a spread of s-spaces of $PG(n, q)$.

Observe that V_0 and $GF(q^{s+1})$ consist of the same elements, so V_0 is an $(s + 1)$-dimensional subspace of $V(n + 1, q)$. Since $V_i = \alpha^i V_0$, each V_i, $i = 0, 1, \ldots, w - 1$, is an $(s + 1)$-dimensional subspace of $V(n + 1, q)$ too. Since each $(s + 1)$-dimensional subspace contains v one-dimensional subspaces, we obtain that the elements of U_i represent all the one-dimensional subspaces of V_i. Therefore, each U_i is an s-dimensional subspace of $PG(n, q)$. Since every point of $PG(n, q)$ is in one of these subspaces, they form a required spread. □

In Section 3.4. we defined the notion of collineations of projective planes. It can be extended to projective geometries of any dimension.

Definition 3.6.14. Let X be the set of all points of the projective geometry $PG(n, q)$. A bijection $\alpha\colon X \to X$ is called a *collineation of* $PG(n, q)$ if $\alpha(L)$ is a line for every line L.

Example 3.6.15. Let V be the $(n+1)$-dimensional vector space over $GF(q)$. Let $\alpha\colon V \to V$ be a non-singular linear transformation. Then $\alpha(\langle \mathbf{x} \rangle) = \langle \alpha(\mathbf{x}) \rangle$ for any $\mathbf{x} \in V$, and therefore, α can be regarded as a bijection $X \to X$, where X is the set of all points of $PG(n, q)$. Clearly, this is a collineation of $PG(n, q)$.

Any collineation of $PG(n, q)$ is an automorphism of the symmetric design $PG_{n-1}(n, q)$. The next theorem describes the full automorphism group of this design.

Definition 3.6.16. Let V be the n-dimensional vector space over $GF(q)$. A mapping $\alpha\colon V \to V$ is called *semilinear mapping of* V if (i) $\alpha(x + y) = \alpha(x) + \alpha(y)$ for all $x, y \in V$ and (ii) there exists an automorphism σ of $GF(q)$ such that $\alpha(ax) = (\sigma(a))\alpha(x)$ for all $x \in V$ and all $a \in GF(q)$. The group of all semilinear mappings of V is denoted by $\Gamma L(n, q)$.

Remark 3.6.17. With each $a \in GF(q)^*$, we associate the semilinear mapping $\mathbf{x} \mapsto a\mathbf{x}$ of V. This allows us to regard $GF(q)^*$ as a subgroup of $\Gamma L(n, q)$. This subgroup is normal.

Proof of the following theorem is beyond the scope of this book.

Theorem 3.6.18 (Fundamental Theorem of Projective Geometry). *The full automorphism group of $PG_{n-1}(n, q)$ is isomorphic to the group $\Gamma L(n + 1, q)/GF(q)^*$.*

We leave proof of the following result to the reader. (See Exercise 48 to this Chapter.)

Corollary 3.6.19. *If p is a prime and $q = p^d$, then the order of the full automorphism group of $PG_{n-1}(n, q)$ is*

$$dq^{n(n+1)/2} \cdot \prod_{i=2}^{n+1} (q^i - 1).$$

The next theorem describes groups of perspectivities of projective planes $PG(2, q)$.

Theorem 3.6.20. *Let q be a prime power and let c be a point and A a line of the projective plane $PG(2, q)$. Let G be the group of all (c, A)-perspectivities of $PG(2, q)$. If $c \in A$, then G is isomorphic to the additive group of $GF(q)$. If $c \notin A$, then G is isomorphic to the multiplicative group of $GF(q)$.*

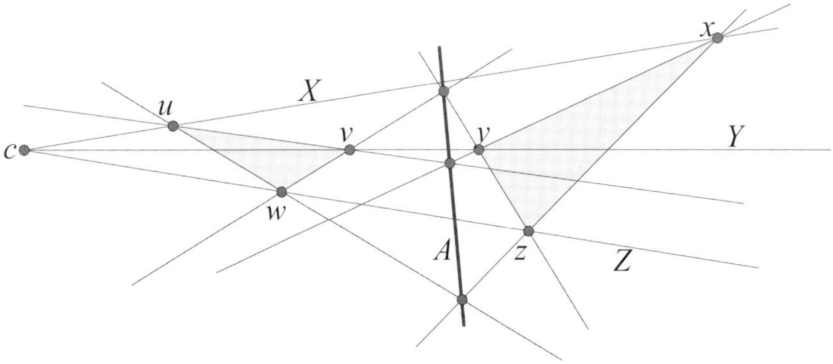

Figure 3.1 Desargues Theorem.

Proof. Let V be the 3-dimensional vector space over $GF(q)$. Let $c = \langle \mathbf{e}_1 \rangle$. If $c \notin A$, then, for $A = \langle \mathbf{e}_2, \mathbf{e}_3 \rangle$, the vectors $\mathbf{e}_1$, $\mathbf{e}_2$, and $\mathbf{e}_3$ form a basis of V. If $c \in A$, then let $A = \langle \mathbf{e}_1, \mathbf{e}_2 \rangle$ and choose $\mathbf{e}_3$ so that $\mathbf{e}_1$, $\mathbf{e}_2$, and $\mathbf{e}_3$ form a basis of V.

Suppose first that $c \in A$. Let $a \in GF(q)$. For $x = x_1\mathbf{e}_1 + x_2\mathbf{e}_2 + x_3\mathbf{e}_3 \in V$, let $T_a(x) = (x_1 + ax_3)\mathbf{e}_1 + x_2\mathbf{e}_2 + x_3\mathbf{e}_3$. Then T_a is a non-singular linear transformation of V. Let t_a be the corresponding collineation of $PG(2, q)$. It can be checked that t_a is a (c, A)-elation. Since $t_a t_b = t_{a+b}$, we obtain a group of (c, A)-elations isomorphic to the additive group of $GF(q)$. Proposition 3.4.17 implies that it is the group of all (c, A)-elations.

Suppose now that $c \notin A$. Let $a \in GF(q)^*$. For $x = x_1\mathbf{e}_1 + x_2\mathbf{e}_2 + x_3\mathbf{e}_3 \in V$, let $M_a(x) = x_1\mathbf{e}_1 + ax_2\mathbf{e}_2 + ax_3\mathbf{e}_3$. Then M_a is a non-singular linear transformation of V. Let m_a be the corresponding collineation of $PG(2, q)$. Then m_a is a (c, A)-homology. Since $m_a m_b = m_{ab}$, we obtain a group of (c, A)-homologies isomorphic to the multiplicative group of $GF(q)$. Proposition 3.4.17 implies that it is the group of all (c, A)-homologies. □

In Section 10.5 we will construct projective planes that admit non-cyclic groups of (c, A)-homologies. Theorem 3.6.20 implies that such projective planes are not isomorphic to projective planes $PG(2, q)$.

Definition 3.6.21. Projective planes $PG(2, q)$ are called *desarguesian*. All other projective planes are called *non-desarguesian*.

Remark 3.6.22. All desarguesian projective planes satisfy the *Desargues Theorem* (Fig. 3.1) stated in Exercise 45. Conversely, every projective plane satisfying the Desargues Theorem is desarguesian. The proof of the last result is beyond the scope of this book.

In Section 3.2, we gave an axiomatic description of projective planes. The following classical theorem of projective geometry introduces an axiomatic description of projective spaces of higher dimension. The proof of this theorem is beyond the scope of this book.

Theorem 3.6.23 (The Veblen–Young Theorem). *Let $\mathbf{P} = (X, \mathcal{L})$ be a finite incidence structure satisfying the following properties:*

(VY1) *For any two distinct points $x, y \in X$, there is a unique block (called a line and denoted by $\overline{xy}$) that contains x and y.*
(VY2) *If x, y, z, w are four distinct points such that $\overline{xy} \cap \overline{zw} \neq \emptyset$, then $\overline{xz} \cap \overline{yw} \neq \emptyset$.*
(VY3) *Every line is incident with at least three points.*
(VY4) *There are two disjoint lines.*

Then there exists an integer $n \geq 3$ and a prime power q such that $\mathbf{P}$ is isomorphic to the design $PG_1(n, q)$ of points and lines of $PG(n, q)$.

3.7. Combinatorial characterization of $PG_{n-1}(n, q)$

As we will see in this section (Theorem 3.7.10), there are symmetric designs with the same parameters as $PG_{n-1}(n, q)$ which are not isomorphic to $PG_{n-1}(n, q)$. However, as the Veblen–Young Theorem shows, certain geometric properties of an incidence structure may uniquely determine this structure. The famous Dembowski–Wagner Theorem (Theorem 3.7.13) shows that there are geometric properties of designs $PG_{n-1}(n, q)$ that characterize them among symmetric designs.

We begin by introducing the notion of a *line* for arbitrary 2-designs.

Definition 3.7.1. For distinct points x and y of a 2-(v, k, λ) design $\mathbf{D}$, the *line* $\overline{xy}$ is the intersection of all blocks of $\mathbf{D}$ that contain both x and y.

Proposition 3.7.2. *Let $\mathbf{D}$ be a 2-(v, k, λ) design. Then every line of $\mathbf{D}$ is contained in exactly λ blocks. For distinct points x and y of $\mathbf{D}$, the line $\overline{xy}$ is the only line of $\mathbf{D}$ that contains both x and y.*

Proof. The line $\overline{xy}$ is contained in a block B of $\mathbf{D}$ if and only if $x, y \in B$. Therefore, there are exactly λ blocks containing any given line. If $x, y \in \overline{zw}$ where z and w are distinct points of $\mathbf{D}$, then every block containing z and w contains x and y. Since $\mathbf{D}$ is a 2-design the number of blocks containing z and w is the same as the number of blocks containing x and y. Therefore, $\overline{xy} = \overline{zw}$. □

The next proposition gives an upper bound on the size of a line of a symmetric design.

Proposition 3.7.3. *Let L be a line of a nontrivial symmetric (v, k, λ)-design* **D** *with $\lambda \geq 1$. Then $|L| \leq 1 + (k-1)/\lambda$.*

Proof. The line L is contained in λ blocks of **D** and meets each of the remaining $v - \lambda$ blocks in at most one point. Therefore, counting in two ways flags (x, B) with $x \in L$ yields $|L|k \leq \lambda|L| + v - \lambda$. Therefore,

$$|L| \leq \frac{v-\lambda}{k-\lambda},$$

$$|L| - 1 \leq \frac{v-k}{k-\lambda} = \frac{(v-1)\lambda - (k-1)\lambda}{(k-\lambda)\lambda} = \frac{k-1}{\lambda},$$

giving the required bound. □

We leave it as an exercise to show that lines in $PG_{n-1}(n, q)$ are precisely one-dimensional projective subspaces and lines in $AG_{n-1}(n, q)$ are 1-flats. In particular, the size of every line in $PG_{n-1}(n, q)$ attains the upper bound of Proposition 3.7.3.

Proposition 3.7.4. *The size of every line in $PG_{n-1}(n, q)$ is $q + 1$ and the size of every line in $AG_{n-1}(n, q)$ is q.*

Another geometric notion that can be defined for any 2-design is that of a *plane*.

Definition 3.7.5. Let **D** be a 2-(v, k, λ). A set of three points of **D** that do not lie on the same line is called a *triangle*. If $\{x, y, z\}$ is a triangle in **D**, then *plane* $\overline{xyz}$ is the intersection of all blocks that contain $\{x, y, z\}$. If no block contains $\{x, y, z\}$, then $\overline{xyz}$ is the entire point set of **D**.

In a 2-design, a triangle is not necessarily contained in a unique plane. For instance, let $\mathbf{D} = (X, \mathcal{B})$ be a symmetric $(v, k, 2)$-design with $k \geq 3$. Any line of such a design consists of two points and therefore, any three points form a triangle. If three points belong to a block, then it is the only block that contains these points. Therefore, every block is a plane. If points x, y, and z do not belong to the same block, then $\overline{xyz} = X$. Therefore, X is a plane and any three points of a block B lie in two distinct planes, B and X. This example also shows that different planes of a 2-design may not lie in the same number of blocks. The plane X in this example lies in 0 blocks, while each plane, which is a block, lies in one block.

Definition 3.7.6. A nontrivial 2-(v, k, λ) design $\mathbf{D}$ is said to be *smooth* if there is a nonnegative integer ρ such that every plane of $\mathbf{D}$ is contained in exactly ρ blocks.

The following proposition is straightforward.

Proposition 3.7.7. *All designs $PG_d(n, q)$ and $AG_d(n, q)$ with $1 \le d < n$ are smooth.*

In smooth symmetric designs, the upper bound for the line size given in Proposition 3.7.3 is attained.

Proposition 3.7.8. *Let $\mathbf{D}$ be a smooth symmetric (v, k, λ)-design with $\lambda \ge 1$ and $v \ge k + 1$. Then every line of $\mathbf{D}$ has exactly $1 + (k - 1)/\lambda$ points.*

Proof. Let every plane of $\mathbf{D}$ be contained in exactly ρ blocks. Let L be a line of $\mathbf{D}$ and let x and y be distinct points on L. Counting in two ways flags (z, B) with $z \ne x$, $z \ne y$, and $B \supseteq L$ yields $(|L| - 2)\lambda + (v - |L|)\rho = \lambda(k - 2)$. Since $\lambda > \rho$, this equation implies that all lines of $\mathbf{D}$ have the same cardinality which we denote by m.

Fix a point x of $\mathbf{D}$ and let $\mathcal{L}_0$ be the set of all lines of $\mathbf{D}$ containing x and $\mathcal{B}_0$ the set of all blocks of $\mathbf{D}$ containing x. Consider the incidence structure $\mathbf{D}_0 = (\mathcal{L}_0, \mathcal{B}_0, I)$ with $(L, B) \in I$ if and only if $L \subseteq B$. We claim that $\mathbf{D}_0$ is a symmetric design. Let $B \in \mathcal{B}_0$. Since the set $\{L \setminus \{x\} \colon L \in \mathcal{L}_0, L \subseteq B\}$ partitions $B \setminus \{x\}$ into $(m - 1)$-subsets, we obtain that every block of $\mathbf{D}_0$ is incident with exactly $k_0 = (k - 1)/(m - 1)$ lines $L \in \mathcal{L}_0$. Let $L_1, L_2 \in \mathcal{L}_0$, $L_1 \ne L_2$. Let $y_1 \in L_1 \setminus \{x\}$ and $y_2 \in L_2 \setminus \{x\}$. Then a block B of $\mathbf{D}_0$ is incident with both L_1 and L_2 if and only if $\overline{xy_1y_2} \subseteq B$. Therefore, there are exactly ρ such blocks. Let $B_1, B_2 \in \mathcal{B}_0$, $B_1 \ne B_2$. Since the set $\{L \setminus \{x\} \colon L \in \mathcal{L}_0, L \subseteq B_1 \cap B_2\}$ partitions $(B_1 \cap B_2) \setminus \{x\}$ into $(m - 1)$-subsets, we obtain that there are exactly $\mu = (\lambda - 1)/(m - 1)$ lines $L \in \mathcal{L}_0$ incident with B_1 and B_2. Since $|\mathcal{B}_0| = k$, Proposition 2.4.9 implies that $\mathbf{D}_0$ is a symmetric (k, k_0, ρ)-design with any two distinct blocks meeting in μ points. Therefore, $\rho = \mu = (\lambda - 1)/(m - 1)$. By (2.9), $(k - 1)\rho = k_0(k_0 - 1)$. These equations imply that

$$\frac{\lambda - 1}{m - 1} = \frac{k - m}{(m - 1)^2}.$$

Solving this equation for m gives $m = 1 + (k - 1)/\lambda$. □

As the following theorem shows, designs $PG_{n-1}(n, q)$ generally are not determined by their parameters. We begin with a lemma.

Lemma 3.7.9. *Let q be a prime power and n a positive integer. It is possible to find $n + 1$ hyperplanes of $PG(n, q)$ whose intersection is empty.*

Proof. Note that for any nonempty set of points of $PG(n, q)$, there is a hyperplane not containing this set. Let H_1 and H_2 be two distinct hyperplanes of $PG(n, q)$. If $n = 1$, then $H_1 \cap H_2 = \varnothing$. If $n \geq 2$, we will try to choose, for each $j \geq 2$, a hyperplane H_{j+1} so that $H_{j+1} \not\supset H_1 \cap H_2 \cap \ldots \cap H_j$. If $j \leq n$ and $H_1, H_2, \ldots, H_j$ have been chosen to satisfy this condition, then $\dim(H_1 \cap H_2 \cap \ldots \cap H_j) = n - j \geq 0$, so $H_1 \cap H_2 \cap \ldots \cap H_j \neq \varnothing$, and therefore a required H_{j+1} can be chosen. With this choice, $\dim(H_1 \cap H_2 \cap \ldots \cap H_{n+1}) = -1$, i.e., $H_1 \cap H_2 \cap \ldots \cap H_{n+1} = \varnothing$. □

Theorem 3.7.10. *For any prime power q and any integer $n \geq 3$, there exists a symmetric $((q^{n+1} - 1)/(q - 1), (q^n - 1)/(q - 1), (q^{n-1} - 1)/(q - 1))$-design that is not isomorphic to $PG_{n-1}(n, q)$.*

Proof. Let q be a prime power and $n \geq 3$ an integer. Let $AG_{n-1}(n, q) = (X, \mathcal{A})$ and $PG_{n-2}(n - 1, q) = (Y, \mathcal{B})$. We assume that the point sets X and Y are disjoint. Let $\Pi_1, \Pi_2, \ldots, \Pi_r$, $r = (q^n - 1)/(q - 1)$, be all distinct parallel classes of $AG_{n-1}(n, q)$ and let $\mathcal{B} = \{H_1, H_2, \ldots, H_r\}$.

Consider the incidence structure $\mathbf{D} = (X \cup Y, \mathcal{C})$ where

$$\mathcal{C} = \bigcup_{i=1}^{r} \{A \cup H_i : A \in \Pi_i\} \cup \{Y\}.$$

We claim that $\mathbf{D}$ is a symmetric $((q^{n+1} - 1)/(q - 1), (q^n - 1)/(q - 1), (q^{n-1} - 1)/(q - 1))$-design.

We have $|X \cup Y| = q^n + (q^n - 1)/(q - 1) = (q^{n+1} - 1)/(q - 1)$, $|\mathcal{C}| = qr + 1 = (q^{n+1} - 1)/(q - 1)$, and, for $A \in \Pi_i$, $|A \cup H_i| = q^{n-1} + (q^{n-1} - 1)/(q - 1) = (q^n - 1)/(q - 1)$.

For $i, j = 1, 2, \ldots, r$ and $A_1 \in \Pi_i$, $A_2 \in \Pi_j$,

$$|(A_1 \cup H_i) \cap (A_2 \cup H_j)| = \begin{cases} |H_i| = \frac{q^{n-1}-1}{q-1} & \text{if } i = j \text{ and } A_1 \neq A_2, \\ |A_1 \cap A_2| + |H_i \cap H_j| & \text{if } i \neq j. \end{cases}$$

Since $|A_1 \cap A_2| + |H_i \cap H_j| = q^{n-2} + \frac{q^{n-2}-1}{q-1} = \frac{q^{n-1}-1}{q-1}$ and, for $A \in \Pi_i$, $|(A \cup H_i) \cap Y| = |H_i| = (q^{n-1} - 1)/(q - 1)$, $\mathbf{D}$ is a symmetric design with the required parameters.

We will now show that the block set $\mathcal{B}$ can be suitably ordered so that $\mathbf{D}$ is not isomorphic to $PG_{n-1}(n, q)$.

Let $x, y \in X$, $x \neq y$, and let $\overline{xy}$ be the line through x and y in $\mathbf{D}$. By Proposition 3.7.4, $|\overline{xy} \cap X| = q$. There are $\lambda = (q^{n-1} - 1)/(q - 1)$ blocks of $AG_{n-1}(n, q)$ that contain $\overline{xy}$. We may assume that these blocks belong to parallel classes $\Pi_1, \Pi_2, \ldots, \Pi_\lambda$. We will now apply Lemma 3.7.9 and assume that $H_1 \cap H_2 \cap \ldots \cap H_n = \varnothing$. Since $\lambda \geq n$ for $n \geq 3$, we have $H_1 \cap H_2 \cap \ldots \cap$

$H_\lambda = \varnothing$. Therefore, $\overline{xy} \cap Y = \varnothing$ and then $|\overline{xy}| = q$. Proposition 3.7.4 now implies that **D** is not isomorphic to $PG_{n-1}(n, q)$. □

Remark 3.7.11. Theorem 3.7.10 does not consider the case $n = 2$. In fact, for infinitely many values of q, designs $PG_1(2, q)$ are not determined by their parameters. However, projective planes of order $q \leq 8$ are determined by their parameters. The next proposition considers the case $q = 4$.

Proposition 3.7.12. *Any symmetric* $(21, 5, 1)$*-design is isomorphic to* $PG_1(2, 4)$.

Proof. Let **D** be a symmetric $(21, 5, 1)$-design and let B be a block of **D**. Then the residual design $\mathbf{D}^B = (X, \mathcal{A})$ consists of 16 points and 20 blocks. It suffices to show that the design $\mathbf{D}^B$ is uniquely determined.

We declare two blocks of $\mathbf{D}^B$ equivalent if they meet B at the same point. Then $\mathcal{A}$ is partitioned into 5 equivalence classes of cardinality 4. Each point of $\mathbf{D}^B$ is contained in one block of each equivalence class. Let $H = \{A_1, A_2, A_3, A_4\}$ and $V = \{B_1, B_2, B_3, B_4\}$ be two of these classes. For $i, j = 1, 2, 3, 4$, we denote by (ij) the intersection point of blocks A_i and B_j. Permuting sets H and V if necessary, we may assume that there is a block $C_1 \notin H \cup V$ that is incident with points (ii), $i = 1, 2, 3, 4$.

Blocks A_1, B_1, and C_1 are three blocks through (11). The other possible two blocks through (11) are $L_1 = \{(11), (23), (34), (42)\}$ and $L_1' = \{(11), (24), (32), (43)\}$. Similarly, we obtain blocks $L_2 = \{(22), (13), (34), (42)\}$ and $L_2' = \{(22), (14), (31), (43)\}$ through (22), blocks $L_3 = \{(33), (12), (24), (41)\}$ and $L_3' = \{(33), (14), (21), (42)\}$ through (33), and blocks $L_4 = \{(44), (12), (23), (31)\}$ and $L_4' = \{(44), (13), (21), (32)\}$ through (44). The remaining three blocks have to be disjoint from C_1. The only possible sets of four points that are disjoint from C_1 and meet each of the other 16 blocks at one point are $C_2 = \{(12), (21), (34), (43)\}$, $C_3 = \{(13), (31), (24), (42)\}$, and $C_4 = \{(14), (41), (23), (32)\}$.

Thus the design $\mathbf{D}^B$ is uniquely determined (up to an isomorphism). □

We can now give a combinatorial characterization of the designs $PG_{n-1}(n, q)$ with $n \geq 3$.

Theorem 3.7.13 (The Dembowski–Wagner Theorem). *Let* **D** *be a symmetric* (v, k, λ)*-design with* $\lambda > 1$ *and* $k > \lambda + 1$. *If*

(i) *every line of* **D** *meets every block or*
(ii) *every line of* **D** *has exactly* $1 + (k - 1)/\lambda$ *points or*
(iii) *every triangle of* **D** *is contained in exactly* $k(\lambda - 1)/(v - 1)$ *blocks, or*
(iv) **D** *is smooth,*

then there exist a prime power q and an integer $n \geq 3$ such that **D** *is isomorphic to $PG_{n-1}(n, q)$.*

Proof. First we shall show that each of the conditions (i), (ii), and (iii) implies the other two.

(i) $\Leftrightarrow$ (ii). Let L be a line of **D**. Let $|L| = \sigma$ and let τ be the number of blocks that meet L but do not contain L. Counting in two ways flags (x, B) where $x \in L$ and $B \not\supseteq L$, yields $\sigma(k - \lambda) = \tau$. Therefore, $\sigma = 1 + (k - 1)/\lambda$ if and only if $\tau = v - \lambda$, i.e., L meets every block.

(ii) $\Rightarrow$ (iii). Suppose every line has exactly $1 + (k - 1)/\lambda$ points and therefore meets every block. Let $\{x, y, z\}$ be a triangle and let $L = \overline{yz}$. Then the blocks containing $\{x, y, z\}$ are precisely the blocks containing x and L. Suppose there are ρ such blocks. Counting in two ways flags (w, B) with $w \in L$, $x \in B$, and $L \not\subseteq B$ yields $(1 + (k - 1)/\lambda)(\lambda - \rho) = k - \rho$. This implies $\rho = k(\lambda - 1)/(v - 1)$.

(iii) $\Rightarrow$ (ii). Suppose every triangle is contained in exactly $\rho = k(\lambda - 1)/(v - 1)$ blocks. Let L be a line. Fix distinct points $x, y \in L$. Counting in two ways flags (z, B) with $z \neq x$, $z \neq y$, and $L \subseteq B$ yields $(|L| - 2)\lambda + (v - |L|)\rho = \lambda(k - 2)$. This implies $|L| = 1 + (k - 1)/\lambda$,

By Proposition 3.7.8, (iv) $\Rightarrow$ (ii). Therefore, we may assume that **D** is a symmetric (v, k, λ)-design satisfying (i), (ii), and (iii).

Claim. If π is a plane, B is a block, and $|B \cap \pi| \geq 2$, then either $B \supseteq \pi$ or $B \cap \pi$ is a line.

To prove this claim, assume that $x, y \in B \cap \pi$ are distinct points. Then $\overline{xy} \subseteq B \cap \pi$. If there is a point z such that $z \in B \cap \pi$ and $z \notin \overline{xy}$, then B is one of the ρ blocks that contain triangle $\{x, y, z\}$. If π is the intersection of the ρ blocks that contain triangle $\{s, t, u\}$, then, since x, y, and z are contained in these ρ blocks, B is one of them. Therefore, $B \supseteq \pi$.

We are now ready to verify that the points and lines of **D** satisfy the conditions of the Veblen–Young Theorem.

Condition (VY1) is satisfied by Proposition 3.7.2.

To verify (VY2), assume that x, y, z, w, and t are five distinct points such that $\overline{xy} \cap \overline{zw} = \{t\}$. There are exactly λ blocks that contain $\overline{xz}$ and exactly $\rho = k(\lambda - 1)/(v - 1)$ blocks that contain triangle $\{x, z, t\}$. Since $\rho < \lambda$, there is a block B that contains $\overline{xz}$ and does not contain t. Let π be the intersection of all blocks that contain triangle $\{x, z, t\}$. Then π is a plane and, since $\overline{tx}, \overline{tz} \subseteq \pi$, we obtain that $\overline{yw} \subseteq \pi$. By the above claim, $B \cap \pi = \overline{xz}$. By (i), $\overline{yw} \cap B \neq \varnothing$. Let $s \in \overline{yw} \cap B \neq \varnothing$. Then $s \in \pi$ and therefore, $s \in \overline{xz}$. Thus, $\overline{xz} \cap \overline{yw} \neq \varnothing$.

If (VY3) is not satisfied, then every line consists of two points, so $1 + (k - 1)/\lambda = 2$, $k = \lambda + 1$, contrary to the hypothesis.

To verify (VY4), consider a line L. Since $\lambda < k$, there is a block B that does not contain L. By (i), L meets B at a unique point x. Since each line has $1 + (k - 1)/\lambda < k$ points, there is a line M such that $M \subseteq B$ and $x \notin M$. Then $L \cap M = \varnothing$.

Let X be the set of points and $\mathcal{L}$ the set of lines of **D**. The Veblen–Young Theorem implies that the incidence structure $(X, \mathcal{L})$ is isomorphic to $PG_1(n, q)$ where q is a prime power and $n \geq 3$. Therefore, $v = (q^{n+1} - 1)/(q - 1)$. Since every line of $PG_1(n, q)$ has $q + 1$ points, we obtain that $(k - 1)/\lambda = q$. The relation $\lambda(v - 1) = k(k - 1)$ then implies that $k = (q^n - 1)/(q - 1)$. By Proposition 3.6.1, the blocks of **D** are subspaces of $PG(n, q)$. Since $k = (q^n - 1)/(q - 1)$, they are $(n - 1)$-dimensional subspaces. Therefore, **D** is isomorphic to $PG_{n-1}(n, q)$. □

We will now show that the rank over $GF(2)$ can be used to characterize the designs $PG_{d-1}(d, 2)$.

Theorem 3.7.14. *Let d be a positive integer and let* **D** *be a symmetric $(2^{d+1} - 1, 2^d - 1, 2^{d-1} - 1)$-design. Let* **D**′ *be the complement of* **D**. *Then*

$$rank_2(\mathbf{D}) = 1 + rank_2(\mathbf{D}') \geq d + 2.$$

Proof. Let V be the $(2^{d+1} - 1)$-dimensional vector space over $GF(2)$. Let N be an incidence matrix of **D** and let Y be the set of all columns of N regarded as elements of V. Since the row sum of N is odd, we obtain that the sum of all elements of Y is the all-one vector $\mathbf{j}$. Let $Y' = \{\mathbf{y} + \mathbf{j} \colon \mathbf{y} \in Y\}$. Then Y' is the set of all columns of the incidence matrix $N' = J - N$ of $\mathbf{D}'$. Let U and U' be the subspaces of V generated by Y and Y', respectively. Since $\mathbf{j} \in U$, we obtain that $Y' \subseteq U$ and therefore $U' \subseteq U$.

For any $\mathbf{x} \in V$, let the weight of $\mathbf{x}$, denoted as $wt(\mathbf{x})$, be the number of nonzero components of $\mathbf{x}$. Observe that $wt(\mathbf{x} + \mathbf{y}) \equiv wt(\mathbf{x}) + wt(\mathbf{y}) \pmod 2$ for all $\mathbf{x}, \mathbf{y} \in V$. Since the column sum of N' is even, all elements of Y' have even weight and therefore all elements of U' have even weight. Since U has elements of odd weight (for instance, all elements of Y), we have $U' \neq U$. We claim that every element of U of even weight is in U'. Let $\mathbf{x} \in U$ and let $wt(\mathbf{x})$ be even. If $\mathbf{x} = \mathbf{0}$, then $\mathbf{x} \in U'$. If $\mathbf{x} \neq \mathbf{0}$, then $\mathbf{x} = \mathbf{y}_1 + \mathbf{y}_2 + \cdots + \mathbf{y}_m$, for some $\mathbf{y}_1, \mathbf{y}_2, \ldots, \mathbf{y}_m \in Y$. Since $wt(\mathbf{x})$ is even and $wt(\mathbf{y}_i)$ is odd for $i = 1, 2, \ldots, m$, we obtain that m is even. But then $\mathbf{x} = (\mathbf{y}_1 + \mathbf{j}) + (\mathbf{y}_2 + \mathbf{j}) + \cdots + (\mathbf{y}_m + \mathbf{j})$, and therefore, $\mathbf{x} \in U'$. Since the sum of any two vectors of odd weight is a vector of even weight, we conclude that U', as a subgroup of the additive group U, has index 2. Therefore, $|U| = 2|U'|$, and then $\dim U = 1 + \dim U'$. Since

$|U'| \geq |Y'| = 2^{d+1} - 1 > 2^d$, we obtain that $\dim U' \geq d + 1$. This implies that $rank_2(\mathbf{D}) = 1 + rank_2(\mathbf{D}') \geq d + 2$. □

The next theorem characterizes symmetric $(2^{d+1} - 1, 2^d - 1, 2^{d-1} - 1)$-designs of 2-rank $d + 2$. We begin with a lemma.

Lemma 3.7.15. *Let d be a positive integer and let $B_1, B_2, \ldots, B_m$ be blocks of a design $\mathbf{D}$ isomorphic to the complement of $PG_{d-1}(d, 2)$. Then the symmetric difference $B_1 \triangle B_2 \triangle \cdots \triangle B_m$ is either a block of $\mathbf{D}$ or the empty set.*

Proof. Induction on m. First let $m = 2$. If $B_1 = B_2$, then $B_1 \triangle B_2 = \varnothing$. Suppose $B_1 \neq B_2$ and let W be the $(d + 1)$-dimensional vector space over $GF(2)$. Recall that every d-dimensional subspace of W can be described as the set of vectors $\mathbf{x} = [x_0\ x_1 \ldots x_d]^\top \in W$ satisfying an equation of the form $a_0x_0 + a_1x_1 + \ldots + a_dx_d = 0$, where $\mathbf{a} = [a_0\ a_1 \ldots a_d]^\top$ is a nonzero element of W. Since blocks B_1 and B_2 are the complements of distinct d-dimensional subspaces of W, they can be described by equations $a_0x_0 + a_1x_1 + \ldots + a_dx_d = 1$ and $b_0x_0 + b_1x_1 + \ldots + b_dx_d = 1$, respectively, with distinct nonzero vectors $\mathbf{a}$ and $\mathbf{b} = [b_0\ b_1 \ldots b_d]^\top$. Since $\mathbf{a} + \mathbf{b} \neq \mathbf{0}$, the equation $(a_0 + b_0)x_0 + (a_1 + b_1)x_1 + \cdots + (a_d + b_d)x_d = 1$ gives a block C of $\mathbf{D}$. Observe now that, for any $\mathbf{x} = [x_0\ x_1 \ldots x_d]^\top \in W$, $\mathbf{x} \in C$ if and only if $\mathbf{x} \in B_1 \triangle B_2$. Therefore, $B_1 \triangle B_2 = C$ is a block of $\mathbf{D}$.

Let $m \geq 3$ and let $C = B_1 \triangle B_2 \triangle \cdots \triangle B_{m-1}$ be either a block of $\mathbf{D}$ or the empty set. Then $C \triangle B_m$ is either a block of $\mathbf{D}$ or the empty set. □

Theorem 3.7.16. *Let d be a positive integer and let $\mathbf{D}$ be a symmetric $(2^{d+1} - 1, 2^d - 1, 2^{d-1} - 1)$-design. Then $rank_2(\mathbf{D}) = d + 2$ if and only if $\mathbf{D}$ is isomorphic to $PG_{d-1}(d, 2)$.*

Proof. Let $\mathbf{D}'$ be the complement of $\mathbf{D}$. By Theorem 3.7.14, $rank_2(\mathbf{D}) = 1 + rank_2(\mathbf{D}')$. As in the proof of Theorem 3.7.14, we denote by V the $(2^{d+1} - 1)$-dimensional vector space over $GF(2)$, by N' an incidence matrix of $\mathbf{D}'$, by Y' the set of all columns of N' regarded as elements of V, and by U' the subspace of V generated by Y'. We will also denote by W the $(d + 1)$-dimensional vector space over $GF(2)$.

(i) Suppose $\mathbf{D}$ is isomorphic to $PG_{d-1}(d, 2)$. It suffices to show that $rank_2(\mathbf{D}') = d + 1$. Since $|Y' \cup \{\mathbf{0}\}| = 2^{d+1}$, we have to show that $Y' \cup \{\mathbf{0}\} = U'$, and therefore, it suffices to show that the set $Y' \cup \{\mathbf{0}\}$ is closed under addition, i.e., that the sum of any two distinct elements of Y' is in Y'.

Let $\mathbf{y}, \mathbf{z} \in Y'$, $\mathbf{y} \neq \mathbf{z}$, and let A and B be the corresponding blocks of $\mathbf{D}'$. Then $\mathbf{y} + \mathbf{z}$, regarded as a set of points of $\mathbf{D}$, is the symmetric difference $A \triangle B$. By Lemma 3.7.15, $C = A \triangle B$ is a block of $\mathbf{D}'$. Therefore, $\mathbf{y} + \mathbf{z} \in Y'$.

(ii) Suppose now that $\mathbf{D}$ is a symmetric $(2^{d+1}-1, 2^d-1, 2^{d-1}-1)$-design with $rank_2(\mathbf{D}) = d+2$. Then $rank_2(\mathbf{D}') = d+1$. Let $S = \{C_0, C_1, \ldots, C_d\}$ be a set of $d+1$ linearly independent (over $GF(2)$) columns of N'. For $i = 0, 1, \ldots, d$, let B_i be the block of $\mathbf{D}'$ corresponding to the column C_i. Let $\mathcal{B} = \{B_0, B_1, \ldots, B_d\}$. Then every block B of $\mathbf{D}'$ that is not in $\mathcal{B}$ admits a unique representation as the symmetric difference of two or more elements of $\mathcal{B}$. We now define a map φ from the point set of $\mathbf{D}'$ to W as follows: if p is a point of $\mathbf{D}'$, then $\varphi(p) = [a_0\, a_1 \ldots a_d]^\top$ with $a_i = 1$ if and only if $p \in B_i$. Since every one-dimensional vector space over $GF(2)$ consists of $\mathbf{0}$ and a unique nonzero vector, we will identify the set of all nonzero elements of W with the set of all points of $PG(d, 2)$ and show that φ is an isomorphism between $\mathbf{D}'$ and the complement of the design $PG_{d-1}(d, 2)$ of points and hyperplanes of $PG(d, 2)$.

If $\varphi(p) = \mathbf{0}$ for some point p of $\mathbf{D}'$, then $p \notin B_i$, for $i = 0, 1, \ldots, d$. This implies that no block B of $\mathbf{D}'$ contains p, a contradiction. Therefore, φ is a map from the point set of $\mathbf{D}'$ to the point set of $PG_{d-1}(d, 2)$. Suppose $\varphi(p) = \varphi(q)$ for some points p and q of $\mathbf{D}'$. Then the set of blocks of $\mathcal{B}$ that contain p is the same as the set of blocks of $\mathcal{B}$ that contain q. If a block B of $\mathbf{D}'$ is the symmetric difference of m distinct blocks of $\mathcal{B}$, then B is the set of all points that are contained in odd number of these m blocks. Therefore, for every block B of $\mathbf{D}'$, $p \in B$ if and only if $q \in B$, so $p = q$. Thus, φ is a bijection.

To complete the proof, we have to show that $\varphi(B)$ is the complement of a hyperplane of $PG(d, 2)$ for every block B of $\mathbf{D}'$. If $B = B_i \in \mathcal{B}$, then $\varphi(B)$ is the complement of the hyperplane given by the equation $x_i = 0$. If $B = B_{i_1} \triangle B_{i_2} \triangle \cdots \triangle B_{i_m}$ is the symmetric difference of m blocks of $\mathcal{B}$, then $\varphi(B) = \varphi(B_{i_1}) \triangle \varphi(B_{i_2}) \triangle \cdots \triangle \varphi(B_{i_m})$. Since $\varphi(B) \neq \emptyset$, Lemma 3.7.15 implies that $\varphi(B)$ is a block of $\mathbf{D}'$. □

We can now show that the property of the complement of $PG_{d-1}(d, 2)$ given by Lemma 3.7.15 in fact characterizes these designs.

Proposition 3.7.17. *Let d be a positive integer and let $\mathbf{D}$ be a symmetric $(2^{d+1}-1, 2^d, 2^{d-1})$-design. If, for any blocks $B_1, B_2, \ldots, B_m$ of $\mathbf{D}$, the set $B_1 \triangle B_2 \triangle \cdots \triangle B_m$ is either a block of $\mathbf{D}$ or the empty set, then $\mathbf{D}$ is isomorphic to the complement of $PG_{d-1}(d, 2)$.*

Proof. Let $r = rank_2(\mathbf{D})$. By Theorem 3.7.14, $r \geq d+1$. Let N be an incidence matrix of $\mathbf{D}$ and let $\mathcal{B}$ be a set of r blocks of $\mathbf{D}$ corresponding to linearly independent columns of N. Then the symmetric difference of all blocks of any nonempty subset of $\mathcal{B}$ is a block of $\mathbf{D}$, and all these symmetric differences are distinct blocks. This gives us $2^r - 1$ distinct blocks of $\mathbf{D}$. Therefore,

$2^r - 1 \le 2^{d+1} - 1$, which implies $r \le d + 1$. Thus, $r = d + 1$, and then Theorem 3.7.16 implies that **D** is isomorphic to the complement of $PG_{d-1}(d, 2)$. □

3.8. Two infinite families of symmetric designs

In this section, we apply vector spaces over finite fields to construct two infinite families of symmetric designs.

We begin by introducing a special order on a finite abelian group.

Lemma 3.8.1. *Given a finite abelian group of order n, it is possible to order its elements $x_1, x_2, \ldots, x_n$ so that $x_i + x_{n+1-i}$ is the same for $i = 1, 2, \ldots, n$.*

Proof. Let $G = \{x_1, x_2, \ldots, x_n\}$ be an abelian group of order n. For each $a \in G$, let $H(a) = \{x \in G : 2x = a\}$. Since the sets $H(a)$ are pairwise disjoint, either all of them are singletons or at least one of them is empty. Fix $a \in G$ such that $|H(a)| \le 1$ and partition the set $G \setminus H(a)$ into 2-subsets $\{b_i, c_i\}$ such that $b_i + c_i = a$. For $1 \le i \le \frac{n}{2}$, put $x_i = b_i$ and $x_{n+1-i} = c_i$. If $H(a) \ne \varnothing$, then n is odd, and we let $x_{(n+1)/2}$ be the only element of $H(a)$. □

We will call the order on G described in Lemma 3.8.1 *symmetric*. Throughout this section, we will always assume that a finite abelian group G is equipped with a symmetric order, and $G = \{x_1, x_2, \ldots, x_n\}$ means that $x_i + x_{n+1-i}$ is the same for $i = 1, 2, \ldots, n$.

With any subset A of a finite abelian group $G = \{x_1, x_2, \ldots, x_n\}$ we associate a (0, 1)-matrix $M(A) = [m_{ij}(A)]$ of order n where

$$m_{ij}(A) = \begin{cases} 1 & \text{if } x_{n+1-j} - x_i \in A, \\ 0 & \text{if } x_{n+1-j} - x_i \notin A, \end{cases}$$

and a (0, 1)-matrix $N(A) = [n_{ij}]$ of order n where

$$n_{ij}(A) = \begin{cases} 1 & \text{if } x_j - x_i \in A, \\ 0 & \text{if } x_j - x_i \notin A. \end{cases}$$

The definition of symmetric order implies that matrices $M(A)$ are symmetric. If $A = -A$, then the matrix $N(A)$ is symmetric. If $A \cap (-A) = \varnothing$, then $N(A) + N(A)^\top$ is a (0, 1)-matrix. The following lemma is immediate.

Lemma 3.8.2. *If A and B are subsets of a finite abelian group $G = \{x_1, x_2, \ldots, x_n\}$, then* (i) $M(A)M(B)^\top = N(A)N(B)^\top$ *and for* $l, m =$

$1, 2, \ldots, n$, *the* (l, m)-*entry of the matrix* $M(A)M(B)^\top$ *is equal to* $|(A + x_l) \cap (B + x_m)|$ *and* (ii) $M(A)J = N(A)J = |A|J$.

Let q be a prime power, d a positive integer, and V the $(d+1)$-dimensional vector space over the field $GF(q)$. The space V contains $r = (q^{d+1} - 1)/(q - 1)$ hyperplanes, which we denote by $H_1, H_2, \ldots, H_r$. We will regard $V = \{x_1, x_2, \ldots, x_{q^{d+1}}\}$ as an abelian group equipped with a symmetric order. The next two theorems introduce infinite families of symmetric designs.

Theorem 3.8.3. *Let* q *be a prime power,* d *a positive integer, and* $r = (q^{d+1} - 1)/(q - 1)$. *Let* V *be the* $(d+1)$-*dimensional vector space over* $GF(q)$ *and let* $\{H_1, H_2, \ldots, H_r, H_{r+1}\}$ *be the set consisting of all* d-*dimensional subspaces of* V *and the empty set. Let* H_s *be the empty set,* $1 \le s \le r + 1$. *Let* $L = [L(i, j)]$ *be a Latin square of order* $r + 1$. *For* $i, j = 1, 2, \ldots, r + 1$, *let* F_{ij} *be the empty set if* $L(i, j) = s$ *and let* F_{ij} *be a hyperplane parallel to* $H_{L(i,j)}$ *otherwise. Then block matrices* $M = [M(F_{ij})]$ *and* $N = [N(F_{ij})]$ $(i, j = 1, 2, \ldots, r + 1)$ *are incidence matrices of symmetric designs with parameters*

$$((r+1)q^{d+1}, rq^d, (r-1)q^{d-1}). \tag{3.6}$$

Proof. For $i, h = 1, 2, \ldots, r + 1$, let $S_{ih} = \sum_{j=1}^{r+1} M(F_{ij})M(F_{hj})^\top$. Lemma 3.8.2 implies that, for $l, m = 1, 2, \ldots, q^{d+1}$, the (l, m)-entry of S_{ih} is equal to $\sum_{j=1}^{r+1} |(F_{ij} + x_l) \cap (F_{hj} + x_m)|$.

If $L(i, j) = k \ne s$, then d-flats $F_{ij} + x_l$ and $F_{ij} + x_m$ are either equal or disjoint depending on whether $x_l - x_m$ is or is not in H_k. Therefore, the (l, m)-entry of S_{ii} is equal to rq^d if $l = m$, and it is equal to $q^d(q^d - 1)/(q - 1)$ if $l \ne m$.

If $i \ne h$, then either $F_{ij} + x_l$ and $F_{hj} + x_m$ are nonparallel d-flats, which meet in q^{d-1} points, or one of these sets is empty. Hence, the (l, m)-entry of S_{ih} is equal to $(r-1)q^{d-1} = q^d(q^d - 1)/(q - 1)$. Therefore, M is an incidence matrix of a symmetric design with the required parameters. So is N, because, by Lemma 3.8.2, $MM^\top = NN^\top$. □

Remark 3.8.4. One can see that a certain flexibility is built in the statement of the above (and the next) theorem. For instance, one may replace any hyperplane by a parallel hyperplane or choose a specific value of the parameter s. We will use this flexibility in later applications of these theorems.

Theorem 3.8.5. *Let* d *be a positive integer and let* $r = (3^{d+1} - 1)/2$. *Let* V *be the* $(d+1)$-*dimensional vector space over* $GF(3)$ *and let* $H_1, H_2, \ldots, H_r$ *be all* d-*dimensional subspaces of* V. *Fix* $s \in \{1, 2, \ldots, r\}$. *Let* $L = [L(i, j)]$

be a Latin square of order r. For $i, j = 1, 2, \ldots, r$, let F_{ij} be a hyperplane parallel to $H_{L(i,j)}$ if $L(i, j) \neq s$ and let F_{ij} be the complement of a hyperplane parallel to H_s if $L(i, j) = s$. Then block matrices $M = [M(F_{ij})]$ and $N = [N(F_{ij})]$ $(i, j = 1, 2, \ldots, r)$ are incidence matrices of symmetric designs with parameters

$$(r \cdot 3^{d+1}, (r+1) \cdot 3^d, (r+2) \cdot 3^{d-1}). \tag{3.7}$$

Proof. For $i, h = 1, 2, \ldots, r$, let $S_{ih} = \sum_{j=1}^{r} M(F_{ij})M(F_{hj})^\top$. For $l, m = 1, 2, \ldots, q^{d+1}$, the (l, m)-entry of S_{ih} is equal to $\sum_{j=1}^{r+1} |(F_{ij} + x_l) \cap (F_{hj} + x_m)|$.

If $L(i, j) = k \neq s$, then d-flats $F_{ij} + x_l$ and $F_{ij} + x_m$ are either equal or disjoint depending on whether $x_l - x_m$ is or is not in H_k. If $L(i, j) = s$, then the cardinality of $(F_{ij} + x_l) \cap (F_{ij} + x_m)$ is equal to $2 \cdot 3^d$ or 3^d depending on whether $x_l - x_m$ is or is not in H_s. Therefore, the (l, m)-entry of S_{ii} is equal to $(r+1) \cdot 3^d$ if $l = m$, and it is equal to $(r+2) \cdot 3^{d-1}$ if $l \neq m$.

If $i \neq h$, then the cardinality of $(F_{ij} + x_l) \cap (F_{hj} + x_m)$ is equal to 3^{d-1} if $L(i, j) \neq s$ and $L(h, j) \neq s$, and it is equal to $2 \cdot 3^{d-1}$ otherwise. Hence, the (l, m)-entry of S_{ih} is equal to $(r+2) \cdot 3^{d-1}$. Therefore, M is an incidence matrix of a symmetric design with the required parameters. By Lemma 3.8.2, so is N. □

3.9. Linear codes

In this section, we introduce basic notions of Coding Theory, which will be used later for constructing designs.

We assume that there is a set $\mathcal{A}$ of cardinality $q \geq 2$ called the *alphabet*. Any ordered n-tuple of elements of $\mathcal{A}$ is called a *word of length n over $\mathcal{A}$*.

Definition 3.9.1. The *Hamming space* $H(n, q)$ is a metric space which consists of all words of length n over the alphabet $\mathcal{A}$ of cardinality q endowed with the distance function d defined as follows: the distance $d(\mathbf{x}, \mathbf{y})$ between words $\mathbf{x}$ and $\mathbf{y}$ is the number of positions, at which $\mathbf{x}$ differs from $\mathbf{y}$.

The following proposition describes an isometry of $H(n, q)$, i.e., a map that does not change the distance between words. The proof of the proposition is immediate.

Proposition 3.9.2. *Let $\sigma_1, \sigma_2, \ldots, \sigma_n$ be permutations of the alphabet $\mathcal{A}$ of cardinality q. For any $\mathbf{x} = (x_1, x_2, \ldots, x_n) \in H(n, q)$ define $\sigma(\mathbf{x}) = (\sigma_1(x_1), \sigma_2(x_2), \ldots, \sigma_n(x_n))$. Then $d(\sigma(\mathbf{x}), \sigma(\mathbf{y})) = d(\mathbf{x}, \mathbf{y})$ for all $\mathbf{x}, \mathbf{y} \in H(n, q)$.*

A q-*ary code* is any subset of $H(n, q)$ of cardinality of at least 2.

Definition 3.9.3. A q-*ary* (n, m, d)-*code* is a subset $C \subseteq H(n, q)$ such that $|C| = m \geq 2$ and $d = \min\{d(\mathbf{x}, \mathbf{y})\colon \mathbf{x}, \mathbf{y} \in C, \mathbf{x} \neq \mathbf{y}\}$. Elements of C are called *codewords*, d is called the *minimum distance*. If $q = 2$, q-ary codes are called *binary*. If σ is an isometry described in Proposition 3.9.2, then codes C and $\sigma(C) = \{\sigma(\mathbf{x})\colon \mathbf{x} \in C\}$ are called *equivalent*.

Remark 3.9.4. For any $a \in \mathcal{A}$, any (n, m, d)-code is equivalent to a code containing the word $(a, a, \ldots, a)$.

Codes naturally arise from the practical need to transmit information over some "noisy" channel. If the distance between any two codewords is sufficiently large, it may be possible to detect and correct errors in their transmission. This idea is formalized in the following definitions.

Definition 3.9.5. For $\mathbf{x} \in H(n, q)$ and a positive integer e, the set $B_e(\mathbf{x}) = \{\mathbf{y} \in H(n, q)\colon d(\mathbf{x}, \mathbf{y}) \leq e\}$ is called the *ball of radius e centered at* $\mathbf{x}$.

If balls of radius e centered at the codewords are pairwise disjoint, then any word that differs from a codeword in at most e positions uniquely determines this codeword. This observation motivates the next definition.

Definition 3.9.6. A q-ary (n, m, d)-code C is called e-*error-correcting* if $B_e(\mathbf{x}) \cap B_e(\mathbf{y}) = \varnothing$ for any distinct $\mathbf{x}, \mathbf{y} \in C$.

Proposition 3.9.7. *An* (n, m, d)-*code is e-error-correcting if and only if* $d \geq 2e + 1$.

Proof. 1. Let C be an e-error-correcting (n, m, d)-code. Suppose $d \leq 2e$. Let $d(\mathbf{x}, \mathbf{y}) = d$, $\mathbf{x}, \mathbf{y} \in C$. Then there exists a sequence of words $\mathbf{x}_0 = \mathbf{x}, \mathbf{x}_1, \ldots, \mathbf{x}_d = \mathbf{y}$ such that $d(\mathbf{x}_{i-1}, \mathbf{x}_i) = 1$ for $i = 1, \ldots, d$. Let f be a nonnegative integer such that $d - e \leq f \leq e$. Then $\mathbf{x}_f \in B_e(\mathbf{x}) \cap B_e(\mathbf{y})$, a contradiction.

2. Suppose that $d \geq 2e + 1$. If there exists a word $\mathbf{z} \in B_e(\mathbf{x}) \cap B_e(\mathbf{y})$, then $d(\mathbf{x}, \mathbf{y}) \leq d(\mathbf{x}, \mathbf{z}) + d(\mathbf{z}, \mathbf{y}) \leq 2e < d$, a contradiction. Therefore, $B_e(\mathbf{x}) \cap B_e(\mathbf{y}) = \varnothing$. □

If the balls of radius e centered at the codewords of an e-error-correcting code C cover the entire Hamming space $H(n, q)$, the code C is called *perfect*.

Definition 3.9.8. An e-error-correcting code $C \subseteq H(n, q)$ is called *perfect* if

$$\bigcup_{\mathbf{x} \in C} B_e(\mathbf{x}) = H(n, q).$$

Example 3.9.9. A q-ary *repetition code* of length n is the set of all words of the form $(a, a, \dots, a)$ in $H(n, q)$. If $n = 2e + 1$, then a binary repetition code is perfect e-error-correcting.

The following theorem establishes an upper bound on the size of a q-ary e-error-correcting code.

Theorem 3.9.10 (The Hamming Bound Theorem). *Let C be a q-ary e-error-correcting (n, m, d)-code. Then*

$$m \le \frac{q^n}{\sum_{i=0}^{e} \binom{n}{i}(q-1)^i}.$$

The equality holds if and only if C is perfect.

Proof. For any $\mathbf{x} \in H(n, q)$,

$$|B_e(\mathbf{x})| = \sum_{i=0}^{e} |\{\mathbf{y} \in H(n, q) \colon d(\mathbf{x}, \mathbf{y}) = i\}| = \sum_{i=0}^{e} \binom{n}{i}(q-1)^i.$$

Since C is e-error-correcting,

$$q^n = |H(n, q)| \ge \sum_{\mathbf{x} \in C} |B_e(\mathbf{x})| = m \sum_{i=0}^{e} \binom{n}{i}(q-1)^i.$$

The equality holds if and only if

$$H(n, q) = \bigcup_{\mathbf{x} \in C} B_e(\mathbf{x}),$$

i.e., C is perfect. □

The most important class of codes is *linear codes*.

Definition 3.9.11. Let q be a prime power and let $\mathbf{e}_1, \mathbf{e}_2, \dots, \mathbf{e}_n$ be a basis of the n-dimensional vector space $V(n, q)$ over the field $GF(q)$. With each element $\mathbf{x} = x_1\mathbf{e}_1 + x_2\mathbf{e}_2 + \cdots + x_n\mathbf{e}_n$ of $V(n, q)$, we associate the word $\mathbf{x} = (x_1, x_2, \dots, x_n)$ over $GF(q)$ regarded as the alphabet. We will identify every element of $V(n, q)$ with the corresponding word and the entire space $V(n, q)$ with the Hamming space $H(n, q)$ over the alphabet $GF(q)$. For $k \ge 1$, any k-dimensional subspace of $V(n, q)$ is called a *q-ary* (linear) *$[n, k]$-code.*

Remark 3.9.12. If C is a q-ary linear $[n, k]$-code, then $|C| = q^k$.

If $\mathbf{x}$ and $\mathbf{y}$ are codewords in a linear code C, then $d(\mathbf{x}, \mathbf{y}) = d(\mathbf{x} - \mathbf{y}, \mathbf{0})$.

Definition 3.9.13. For any $\mathbf{x} \in V(n, q)$, the *weight* of $\mathbf{x}$ is $wt(\mathbf{x}) = d(\mathbf{x}, \mathbf{0})$.

The following proposition is immediate.

Proposition 3.9.14. *For any linear code C,*

$$\min\{wt(\mathbf{x})\colon \mathbf{x} \in C, \mathbf{x} \neq \mathbf{0}\} = \min\{d(\mathbf{x}, \mathbf{y})\colon \mathbf{x}, \mathbf{y} \in C, \mathbf{x} \neq \mathbf{y}\}.$$

As a subspace of a vector space, a linear code is determined by its basis.

Definition 3.9.15. A *generator matrix* of an $[n, k]$-code C is a $k \times n$ matrix whose rows form a basis of C.

Remark 3.9.16. Any $k \times n$ matrix of rank k over $GF(q)$ is a generator matrix of a q-ary linear $[n, k]$-code.

In a vector space with a fixed basis, one can naturally introduce the *inner product.*

Definition 3.9.17. Let $\mathbf{x} = (x_1, x_2, \ldots, x_n)$ and $\mathbf{y} = (y_1, y_2, \ldots, y_n)$ be codewords over $GF(q)$. Then the element

$$\langle \mathbf{x}, \mathbf{y} \rangle = \sum_{i=1}^{n} x_i y_i$$

of $GF(q)$ is called the *inner product* of $\mathbf{x}$ and $\mathbf{y}$. Words $\mathbf{x}$ and $\mathbf{y}$ are called *orthogonal* if $\langle \mathbf{x}, \mathbf{y} \rangle = 0$.

Definition 3.9.18. Let C be a q-ary linear $[n, k]$-code, $1 \leq k \leq n - 1$. The code

$$C^{\perp} = \{\mathbf{y} \in V(n, q)\colon \langle \mathbf{x}, \mathbf{y} \rangle = 0, \text{ for any } \mathbf{x} \in C\}$$

is called the *dual code* of C. If $C = C^{\perp}$, the code C is called *self-dual.* If $C \subseteq C^{\perp}$, the code C is called *self-orthogonal.*

Remark 3.9.19. The dual of an $[n, k]$-code is an $[n, n - k]$-code.

The following proposition characterizes generator matrices of C and $C^{\perp}$.

Proposition 3.9.20. *Let G be a $k \times n$ and H an $(n - k) \times n$ matrix over $GF(q)$ of rank k and $n - k$, respectively, $1 \leq k \leq n - 1$. Then $GH^{\top} = O$ if and only if there exists a q-ary $[n, k]$-code C such that G is a generator matrix of C and H is a generator matrix of $C^{\perp}$.*

Proof. 1. Suppose $GH^{\top} = O$. Let C be the $[n, k]$-code spanned by the rows of G. Since each row of H is orthogonal to each row of G, each row of H is orthogonal to every codeword from C, i.e., the rows of H belong to $C^{\perp}$. Since $\dim(C^{\perp}) = n - k$, H is a generator matrix of $C^{\perp}$.

2. Suppose G and H are generator matrices of linear codes C and $C^{\perp}$, respectively. Then the definition of the dual code implies $GH^{\top} = O$. □

Corollary 3.9.21. *For any linear code C, $(C^\perp)^\perp = C$.*

Proof. Let G and H be generator matrices of linear codes C and $C^\perp$, respectively. Then $GH^\top = O$, so each row of G belongs to $(C^\perp)^\perp$. Since $\dim((C^\perp)^\perp) = n - \dim(C^\perp) = k$ and since the k rows of G are linearly independent, the rows of G form a basis of $(C^\perp)^\perp$. Therefore, $(C^\perp)^\perp = C$. □

Definition 3.9.22. A generator matrix of $C^\perp$ is called a *parity check matrix* of C.

A parity check matrix of a code yields a lower bound for the minimum weight of the code.

Proposition 3.9.23. *A linear code has minimum weight d or greater if and only if any $d-1$ columns of its parity check matrix are linearly independent.*

Proof. Let C be a linear code with a parity check matrix H. Let $\mathbf{x} \in C$ and let $w = wt(\mathbf{x})$. If we regard $\mathbf{x}$ as a $1 \times n$ matrix, then $\mathbf{x}H^\top = \mathbf{0}$, and therefore the matrix $H^\top$ has at least w linearly dependent rows. Therefore, if any $d-1$ columns of H are linearly independent, the code C cannot have a word of weight $d-1$ or less, i.e., the minimum weight of C is d or greater.

Suppose now that H has a linearly dependent set of $d-1$ columns. Then there exists a dependence relation $\alpha_1\mathbf{y}_1 + \alpha_2\mathbf{y}_2 + \cdots + \alpha_n\mathbf{y}_n = \mathbf{0}$, where $\mathbf{y}_1, \mathbf{y}_2, \ldots, \mathbf{y}_n$ are the rows of $H^\top$, with at most $d-1$ nonzero coefficients α_i. Then the word $\mathbf{x} = (\alpha_1, \alpha_2, \ldots, \alpha_n)$ has weight less than d and belongs to C (since $\mathbf{x}H^\top = \mathbf{0}$). Therefore, the minimum weight of C is less than d. □

Definition 3.9.24. Let q be a prime power and $n \geq 2$ an integer. Let $v = \frac{q^n-1}{q-1}$. Consider the vector space $V(n, q)$ with a fixed basis as a Hamming space and let $\mathbf{x}_1, \mathbf{x}_2, \ldots, \mathbf{x}_v$ be nonzero vectors (words), one from each one-dimensional subspace of $V(n, q)$. Let H be the $n \times v$ matrix with the words $\mathbf{x}_1, \mathbf{x}_2, \ldots, \mathbf{x}_v$ as columns. Then $rank(H) = n$ and therefore H is a generator matrix of a q-ary linear $[v, n]$-code called a *simplex code* or an $S_n(q)$. The code $(S_n(q))^\perp$ is called a *Hamming code*.

Remark 3.9.25. Since any two columns of a parity check matrix of a Hamming code are linearly independent, Proposition 3.9.23 implies that Hamming codes are single-error-correcting.

The following theorem characterizes Hamming codes.

Theorem 3.9.26. *A linear single-error-correcting code is a Hamming code if and only if it is perfect.*

Proof. Let q be a prime power, $n \geq 2$ an integer, and $v = \frac{q^n-1}{q-1}$. If C is a $[v, v-n]$ Hamming code over $GF(q)$, then $|C| = q^{v-n}$ and C is a (v, q^{v-n}, d)-code. Theorem 3.9.10 implies that C is a perfect single-error-correcting code. Conversely, let C be a q-ary linear perfect single-error-correcting $[v, v-n]$-code and let H be a parity check matrix for C, so H is an $n \times v$ matrix. Theorem 3.9.10 implies that $v = \frac{q^n-1}{q-1}$. Since C is single-error-correcting, the columns of H are pairwise linearly independent, so they represent all distinct one-dimensional subspaces of $V(n, q)$. Therefore, H is a generator matrix of a simplex code, and then C is a Hamming code. □

The next theorem gives a combinatorial characterization of simplex codes.

Theorem 3.9.27. *Let q be a prime power, $n \geq 2$ an integer, and $v = \frac{q^n-1}{q-1}$. A q-ary linear $[v, n]$-code S is a simplex code if and only if $wt(\mathbf{x}) = q^{n-1}$ for every nonzero word $\mathbf{x} \in S$.*

Proof. Let a subspace S of the vector space $V(v, q)$ be a $[v, n]$-code. Let nonzero words $\mathbf{x}_1, \mathbf{x}_2, \ldots, \mathbf{x}_v$ be representatives of all distinct one-dimensional subspaces of S and let $W = [\omega_{ij}]$ be the $v \times v$ matrix with $\mathbf{x}_1, \mathbf{x}_2, \ldots, \mathbf{x}_v$ as consecutive rows. Without loss of generality, we assume that the first n rows of W form a generator matrix H of S. Let the words $\mathbf{y}_1, \mathbf{y}_2, \ldots, \mathbf{y}_v$ be the columns of W and let Y be the subspace of $V(v, q)$ spanned by $\{\mathbf{y}_1, \mathbf{y}_2, \ldots, \mathbf{y}_v\}$. Then $\dim(Y) = rank(W) = \dim(S) = n$. For $i = 1, 2, \ldots, v$, let U_i be the hyperplane of $V(v, q)$ consisting of all words with the ith component equal to 0.

1. Suppose S is a simplex code. Then the columns of H represent all distinct one-dimensional subspaces of an n-dimensional vector space over $GF(q)$, and therefore no two of them are proportional. This in turn implies that no two columns of W are proportional and therefore the words $\mathbf{y}_1, \mathbf{y}_2, \ldots, \mathbf{y}_v$ represent all distinct one-dimensional subspaces of Y. If, for some i, $Y \subset U_i$, then $\mathbf{x}_i = \mathbf{0}$, which is not the case. Therefore, $\dim(Y \cap U_i) = n - 1$, and then $Y \cap U_i$ has exactly $(q^{n-1} - 1)/(q - 1)$ one-dimensional subspaces. This implies that $wt(\mathbf{x}_i) = v - (q^{n-1} - 1)/(q - 1) = q^{n-1}$. Since every nonzero word $\mathbf{x} \in S$ is of the form $\alpha_i \mathbf{x}_i$ with $\alpha_i \in GF(q)^*$ and $i \in \{1, 2, \ldots, v\}$, we obtain that $wt(\mathbf{x}) = q^{n-1}$ for every nonzero word $\mathbf{x} \in S$.

2. Suppose $wt(\mathbf{x}) = q^{n-1}$ for every nonzero word $\mathbf{x} \in S$. Then $d(\mathbf{x}, \mathbf{x}') = wt(\mathbf{x} - \mathbf{x}') = q^{n-1}$ for any distinct words $\mathbf{x}, \mathbf{x}' \in S$.

Let $i, h \in \{1, 2, \ldots, v\}$, $i \neq h$. If $Y \cap U_i = Y \cap U_h$, then the words $\mathbf{x}_i$ and $\mathbf{x}_h$ have zeros in the same $(q^{n-1} - 1)/(q - 1)$ positions. Let $\omega_{ij} \neq 0$. Then $\omega_{hj} \neq 0$ and there is $\alpha \in GF(q)^*$ such that $\omega_{hj} = \alpha\omega_{ij}$. Then

$$d(\alpha\mathbf{x}_i, \mathbf{x}_h) \leq v - \frac{q^{n-1} - 1}{q - 1} - 1 < q^{n-1}.$$

Since $\alpha\mathbf{x}_i \neq \mathbf{x}_h$, we conclude that $Y \cap U_i \neq Y \cap U_h$, and then $\dim(Y \cap U_i \cap U_h) = n - 2$. This implies that there are exactly $(q^{n-2} - 1)/(q - 1)$ indices j such that $\omega_{ij} = \omega_{hj} = 0$. Therefore, if we replace with 1 every nonzero entry of W, we obtain a $(0, 1)$-matrix N of order v with exactly q^{n-1} nonzero entries in every row and the inner product of any two distinct rows equal to

$$v - \frac{2(q^{n-1} - 1)}{q - 1} + \frac{q^{n-2} - 1}{q - 1} = q^{n-1} - q^{n-2}.$$

Therefore, N is an incidence matrix of a symmetric $(v, q^{n-1}, q^{n-1} - q^{n-2})$-design. This implies that no two columns of N are equal and therefore no two columns of W are proportional.

We will show that S is a simplex code if we verify that the columns of H represent all distinct one-dimensional subspaces of an n-dimensional vector space over $GF(q)$. Since there are exactly v such subspaces, it suffices to show that no two columns of H are proportional. Let words $\mathbf{z}_1, \mathbf{z}_2, \dots, \mathbf{z}_v$ be the columns of H. Suppose there are distinct $k, h \in \{1, 2, \dots, v\}$ and $\alpha \in GF(q)^*$ such that $\mathbf{z}_k = \alpha\mathbf{z}_h$. Then $\omega_{ik} = \alpha\omega_{ih}$ for $i = 1, 2, \dots, n$. Let $l \in \{1, 2, \dots, v\}$. Since $\{\mathbf{x}_1, \mathbf{x}_2, \dots, \mathbf{x}_n\}$ is a basis of S, we have, for some $\beta_1, \beta_2, \dots, \beta_v \in GF(q)$, $\mathbf{x}_l = \sum_{i=1}^{n} \beta_i \mathbf{x}_i$. Therefore,

$$\omega_{lk} = \sum_{i=1}^{n} \beta_i \omega_{ik} = \alpha \sum_{i=1}^{n} \beta_i \omega_{ih} = \alpha\omega_{lh}.$$

This implies $\mathbf{y}_k = \alpha\mathbf{y}_h$, which is not the case. Therefore, H is a generator matrix of a simplex code, and S is this simplex code. □

In the course of the above proof we obtained the following result.

Proposition 3.9.28. *Let S be a q-ary simplex code of dimension n and let $\mathbf{x}_1, \mathbf{x}_2, \dots, \mathbf{x}_v$ be nonzero representatives of all distinct one-dimensional subspaces of S. Let W be the $v \times v$ matrix with $\mathbf{x}_1, \mathbf{x}_2, \dots, \mathbf{x}_v$ as consecutive rows and let N be the matrix obtained by replacing every nonzero entry of W with 1. Then N is an incidence matrix of a symmetric $(v, q^{n-1}, q^{n-1} - q^{n-2})$-design.*

Simplex codes will be used for constructing balanced generalized weighing matrices in Chapter 10. Further perfect linear codes will be discussed in Chapter 6.

Exercises

(1) Prove that if F is a field of prime characteristic p, then, for all $a, b \in F$, $(a + b)^p = a^p + b^p$ and $(a - b)^{p-1} = \sum_{i=0}^{p-1} a^i b^{p-1-i}$.

(2) Let $q > 2$ be a prime power. Show that the sum of all elements of $GF(q)$ equals 0.

(3) Prove that every element of $GF(q)$ is a root of the polynomial $x^q - x$.

(4) Let q and $r = q^n$ be prime powers. Let α be a primitive element of $GF(r)$.
   (a) Prove that $1, \alpha, \alpha^2, \ldots, \alpha^{n-1}$ is a basis of $GF(r)$ as a vector space over $GF(q)$.
   (b) Prove that there exists an irreducible over $GF(q)$ polynomial f of degree n such that $f(\alpha) = 0$.
   (c) Prove that $\alpha, \alpha^q, \alpha^{q^2}, \ldots, \alpha^{q^{n-1}}$ are all the roots of f.

(5) Let V be the two-dimensional vector space over $GF(3)$. Define multiplication on V so that V becomes a field. Choose a primitive element α of this field and determine, for every pair $(a, b) \neq (0, 0)$ of elements of $GF(3)$, the least positive integer n such that $a + b\alpha = \alpha^n$.

(6) Let V be the three-dimensional vector space over $GF(2)$. Define multiplication on V so that V becomes a field. Choose a primitive element α of this field and determine, for every triple $(a, b, c) \neq (0, 0, 0)$ of elements of $GF(2)$, the least positive integer n such that $a + b\alpha + c\alpha^2 = \alpha^n$.

(7) How many primitive elements does the field $GF(81)$ have?

(8) Let n be a positive integer and q a prime power. Let $a \in GF(q)$. Prove that the polynomial $x^{q^n} - x + na$ over $GF(q)$ is divisible by the polynomial $x^q - x + a$.

(9) Let q be an odd prime power and let $a \in GF(q)^*$. Prove that there are exactly $q - 1$ ordered pairs (x, y) such that $x, y \in GF(q)$ and $x^2 - y^2 = a$.

(10) Let $q \equiv 3 \pmod 4$ be a prime power and let η be the quadratic character on $GF(q)$. For each $a \in GF(q)$, let $B_a = \{x \in GF(q) \colon \eta(x + a) = 1\}$. Let X be the set of all elements of $GF(q)$ and let $\mathcal{B} = \{B_a \colon a \in GF(q)\}$. Prove that the incidence structure $(X, \mathcal{B})$ is a symmetric $(q, (q-1)/2, (q-3)/4)$-design.

(11) Let q be an odd prime power and let η be the quadratic character on $GF(q)$. Let $f(x) = x^2 + ax + b$ be a quadratic polynomial over $GF(q)$ and let $d = a^2 - 4b$. Prove that

$$\sum_{x \in GF(q)} \eta(f(x)) = \begin{cases} -1 & \text{if } d \neq 0, \\ q - 1 & \text{if } d = 0. \end{cases}$$

(12) A finite incidence structure $\mathbf{D} = (X, \mathcal{B}, I)$ is called a *transversal design* if there is a partition of the point set X into subsets called *point classes* such that (i) $\lambda(x, y) = 1$ for any points x and y from different point classes, (ii) $|B \cap P| = 1$ for any block B and any point class P, and (iii) there are at least three point classes.
   (a) Prove that an incidence structure is a transversal design if and only if the dual structure is a net.
   (b) Prove that all blocks of a transversal design have the same size.
   (c) Prove that all point classes of a transversal design have the same size.
   (d) Prove that if there exists a transversal design with a block size k and point classes of cardinality m and a transversal design with the same block size k and point classes of cardinality n, then there exists a transversal design with block size k and point classes of cardinality mn.

(e) For $k \geq 3$, let $TD(k)$ denote the set of all n such that there exists a transversal design with block size k and with point classes of cardinality n. Prove that the set $TD(4)$ contains all odd positive integers, the set $TD(5)$ contains all positive integers $n \equiv \pm 1, \pm 4, \pm 5 \pmod{12}$, and the set $TD(6)$ contains all positive integers $n \equiv \pm 1, \pm 5, \pm 7, \pm 8, \pm 11 \pmod{24}$.

(13) Let n be a positive integer, let a, b, and c be integers, and let a and b be relatively prime to n. Prove that the $n \times n$ array L such that $L(i, j) = r$ if and only if $ai + bj + c \equiv r \pmod{n}$ and $1 \leq r \leq n$ is a Latin square.

(14) Let L be a symmetric Latin square of odd order. Prove that no two diagonal entries of L are the same.

(15) Let $G = \{x_1, x_2, \ldots, x_n\}$ be a group of order n. A *Cayley table* of G is a Latin square L of order n such that $L(i, j) = k$ if and only if $x_i x_j = x_k$. Prove that the Cayley table satisfies the following *quadrangle criterion*:

$$\begin{aligned} &\textit{if } L(i,k) = L(i_1,k_1),\ L(j,k) = L(j_1,k_1),\ L(j,l) = L(j_1,l_1), \\ &\textit{then } L(i,l) = L(i_1,l_1). \end{aligned} \tag{3.8}$$

(16) Prove that every Latin square satisfying the quadrangle criterion (3.8) is a Cayley table of some group.

(17) A *transversal* of a Latin square L of order n is a set T of n ordered pairs of elements of the set $\{1, 2, \ldots, n\}$ such that $i \neq k$, $j \neq l$, and $L(i, j) \neq L(k, l)$ for all distinct $(i, j), (k, l) \in T$.
(a) Prove that a Cayley table of any group of odd order admits a transversal.
(b) Prove that a Cayley table of a group of odd order n admits n pairwise disjoint transversals.

(18) Prove: for a Latin square A of order n, there is an orthogonal Latin square B if and only if A admits n pairwise disjoint transversals. Derive from this that for any odd n there exist orthogonal Latin squares of order n.

(19) Construct four MOLS of order 5.

(20) Construct three MOLS of order 20.

(21) Let A and B be orthogonal Latin squares of order 8. Let X be the set of all ordered pairs (i, j) with $i, j \in \{1, 2, 3, 4, 5, 6, 7, 8\}$. For $(i, j) \in X$, let

$$\begin{aligned} B_{ij} &= \{(k,j) \in X : k \neq i\} \cup \{(i,l) \in X : l \neq j\} \cup \{(k,l) \in X : A(k,l) \\ &= A(i,j), (k,l) \neq (i,j)\} \cup \{(k,l) \in X : B(k,l) = B(i,j), (k,l) \neq (i,j)\}. \end{aligned}$$

Let $\mathcal{B} = \{B_{ij} : (i, j) \in X\}$. Prove that the incidence structure $(X, \mathcal{B})$ is a symmetric $(64, 28, 12)$-design.

(22) Prove that if there exists an $(2n, n)$-net, then there exists a symmetric $(4n^2, 2n^2 - n, n^2 - n)$-design.

(23) A Latin square L is said to be *self-orthogonal* if Latin squares L and $L^\top$ are orthogonal. Let $q \geq 4$ be a prime power and let $GF(q) = \{x_1, x_2, \ldots, x_q\}$. Let $a \in GF(q)$, $a \neq 0, \pm 1$. Let L be a $q \times q$ array defined as follows: $L(i, j) = k$ if and only if $x_i + ax_j = (1 + a)x_k$. Prove that L is a self-orthogonal Latin square of order q.

(24) Let $\mathbf{D} = (X, \mathcal{B})$ be a symmetric (v, k, λ)-design with $X = \{1, 2, \ldots, v\}$ and $\mathcal{B} = \{B_1, B_2, \ldots, B_v\}$. Prove that there is a Latin square L of order v such that $L(i, j) \in$

B_j for $i = 1, 2, \ldots, k$ and $j = 1, 2, \ldots, v$. Such a Latin square is called a *Youden square*.

(25) Prove Theorem 3.4.3.

(26) If we regard the Gaussian coefficient as a function of the real variable q given by Proposition 3.5.2 (with n and d fixed), then prove that

$$\lim_{q\to 1} \begin{bmatrix} n \\ d \end{bmatrix}_q = \binom{n}{d}.$$

(27) Prove:

$$\begin{bmatrix} n \\ d \end{bmatrix}_q = \begin{bmatrix} n \\ n-d \end{bmatrix}_q.$$

(28) Prove:

$$\begin{bmatrix} n+1 \\ d \end{bmatrix}_q = \begin{bmatrix} n \\ d-1 \end{bmatrix}_q + \begin{bmatrix} n \\ d \end{bmatrix}_q.$$

(29) Prove:

$$\prod_{i=0}^{n-1}(1+q^i t) = \sum_{d=0}^{n} q^{d(d-1)/2} \begin{bmatrix} n \\ d \end{bmatrix}_q t^d.$$

(30) Let $S_q(n)$ denote the total number of subspaces of n-dimensional vector space over $GF(q)$. Then $S_q(0) = 1$ and $S_q(1) = 2$. Prove that

$$S_q(n) = 2S_q(n-1) + (q^{n-1} - 1)S_q(n-2)$$

for $n \geq 2$.

(31) Prove Propositions 3.5.12 and 3.5.13.

(32) Prove Proposition 3.6.5.

(33) Prove Proposition 3.6.10.

(34) Let n, r, and μ be positive integers, $r \geq 3$. An incidence structure $\mathbf{D} = (X, \mathcal{B})$ is called an *$(n, r; \mu)$-net* if it has constant block size $n\mu$ and the block set $\mathcal{B}$ can be partitioned into r subsets (parallel classes) of size n so that each parallel class partitions X and any two blocks from different parallel classes meet in exactly μ points.

(a) Let X be the point set of $AG(d, q)$ and let $\mathcal{B}$ be the union of $r \geq 3$ parallel classes of $(d-1)$-flats of $AG(d, q)$. Prove that $(X, \mathcal{B})$ is a $(q, r; q^{d-2})$-net.

(b) Let $\mathbf{D} = (X, \mathcal{B})$ be an $(n, r; \mu)$-net and let $x \in X$. Let $\overline{\lambda} = r(s\mu - 1)/(s^2 \mu - 1)$. Prove the following identities:

$$\sum_{y \in X\setminus\{x\}} \lambda(x, y) = r(s\mu - 1);$$

$$\sum_{y \in X\setminus\{x\}} \lambda(x, y)(\lambda(x, y) - 1) = r(r-1)(\mu - 1);$$

$$\sum_{y \in X\setminus\{x\}} (\overline{\lambda} - \lambda(x, y))^2 = (s^2\mu - 1)\overline{\lambda}^2 - 2\overline{\lambda}r(s\mu - 1) + r(r-1)(\mu - 1) + r(s\mu - 1).$$

(c) Let $\mathbf{D} = (X, \mathcal{B})$ be an $(n, r; \mu)$-net. Prove that $r \leq (s^2\mu - 1)/(s - 1)$ with equality if and only if $\mathbf{D}$ is a 2-design.

(35) A collineation τ of an affine plane $\mathbf{A} = (X, \mathcal{L})$ is called a *translation* if (i) $\tau(L) \parallel L$ for every line $L \in \mathcal{L}$ and (ii) either τ has no fixed points or τ is the identity.

(a) Let x and y be points of an affine plane $\mathbf{A}$. Prove that there exists at most one translation τ of $\mathbf{A}$ such that $\tau x = y$.

(b) Prove that all translation of an affine plane form a group.

(c) Prove that the group of all translations of $AG(2, q)$ is isomorphic to the additive group of $GF(q^2)$.

(d) Let τ be a nonidentity translation of an affine plane $\mathbf{A}$. Prove that the set of all lines L of $\mathbf{A}$ such that $\tau(L) = L$ is a parallel class of $\mathbf{A}$. It is called the *direction* of τ.

(e) Let σ and τ be nonidentity translations of an affine plane $\mathbf{A}$. Prove that if the directions of σ and τ are different, then $\sigma\tau = \tau\sigma$.

(f) Prove that if an affine plane $\mathbf{A}$ admits nonidentity translations with different directions, then the group of all translations of $\mathbf{A}$ is abelian.

(36) An affine plane $\mathbf{A}$ is called a *translation plane* if, for any points x and y of $\mathbf{A}$, there exists a translation τ of $\mathbf{A}$ such that $\tau x = y$. Let T be the group of all translations of a translation plane $\mathbf{A}$ and let F be the set of all homomorphisms $\alpha\colon T \to T$ (with the image of a translation τ denoted by τ^α) that preserve direction, i.e., for any $\tau \in T$ and for any line L of $\mathbf{A}$, if $\tau(L) = L$, then $\tau^\alpha(L) = L$.

(a) For each parallel class Π of $\mathbf{A}$, let $T(\Pi)$ denote the set consisting of the identity and of all the translations of A with direction Π. Prove that if Π_1 and Π_2 are distinct parallel classes, then $T(\Pi_1)T(\Pi_2) = T$.

(b) For $\alpha, \beta \in F$, define a map $\alpha + \beta\colon T \to T$ by $\tau^{\alpha+\beta} = \tau^\alpha\tau^\beta$. Prove that $\alpha + \beta \in F$.

(c) For $\alpha, \beta \in F$, define a map $\alpha\beta\colon T \to T$ by $\tau^{\alpha\beta} = (\tau^\alpha)^\beta$. Prove that $\alpha\beta \in F$.

(d) Prove that with respect to the above addition and multiplication, F is an associative ring. (In fact, F is a field. See Notes for further information and references.)

(37) Construct an incidence matrix of a symmetric (21, 5, 1)-design.

(38) Construct a spread of lines of $PG(2, 3)$.

(39) A set $\mathcal{P}$ of proper subgroups of a finite group G is called a *spread of subgroups* if (i) for any nonidentity element $x \in G$, there is a unique $A \in \mathcal{P}$ such that $x \in A$, and (ii) $AB = G$ for all distinct $A, B \in \mathcal{P}$.

(a) Let $\mathbf{A}$ be a translation plane. For each parallel class Π of $\mathbf{A}$, let $T(\Pi)$ denote the set consisting of the identity and of all the translations of A with direction Π. Prove that all the subgroups $T(\Pi)$ of the group T of all translations of $\mathbf{A}$ form a spread of subgroups of T.

(b) Let q be a prime power, d a positive integer, and V the $(2d)$-dimensional vector space over $GF(q)$. Consider the projective geometry $PG(2d - 1, q)$ formed by subspaces of V and let $\mathcal{P}$ be a spread of $(d - 1)$-spaces of this projective geometry. Then each element of $\mathcal{P}$ is a d-dimensional subspace of V. Regarding V as an abelian group and $\mathcal{P}$ as a set of subgroups of V, prove that $\mathcal{P}$ is a spread of subgroups.

(c) Let $\mathcal{P}$ be a spread of subgroups of a finite group G and let $\mathcal{L}$ be the set of all left cosets of all elements of $\mathcal{P}$. Prove that the incidence structure $\mathbf{A} = (G, \mathcal{L})$

is an affine plane. Prove that the group of all translations of **A** is isomorphic to G.

(d) Prove that if a finite group G admits a spread of subgroups, then G is abelian.

(40) Let α be a nontrivial (c, A)-perspectivity of a projective plane **P** and let L be a line of **P** such that $\alpha(L) = L$. Prove that either $L = A$ or $c \in L$.

(41) Let c be a point and A a line of a projective plane **P**. Let x be a point of **P** such that $x \neq c$ and $x \notin A$. Let y be a point of the line $\overline{cx}$ such that $y \neq c$ and $y \notin A$. Then there exists at most one (c, A)-perspectivity α of **P** such that $\alpha x = y$.

(42) Let **P** be a projective plane and let L be a line of **P**. Let **A** be the affine plane obtained by deleting the line L and all its points from **P**. Prove that any translation of **A** can be uniquely extended to an elation of **P** with L as the axis. Conversely, any elation of **P** with axis L is a translation on **A**.

(43) Let L be a line of a projective plane **P**. Prove that all elations of **P** with axis L form a group. Prove that if this group contains elations with different centers, then it is abelian.

(44) Let c be a point and A a line of a projective plane **P**. The plane **P** is said to be (c, A)-*transitive* if it satisfies the following condition:

> if x is a point of **P** such that $x \neq c$ and $x \notin A$ and y is a point of the line $\overline{cx}$ such that $y \neq c$ and $y \notin A$, then there exists a (c, A)-perspectivity α such that $\alpha x = y$.
>
> The plane **P** is said to be (c, A)-*desarguesian* if it satisfies the following condition:
>
> if X, Y, and Z are three distinct lines through point c, other than A, u and x are distinct points of $X \setminus \{c\}$, v and y are distinct points of $Y \setminus \{c\}$, and w and z are distinct points of $Z \setminus \{c\}$, such that the intersection point of lines $\overline{uv}$ and $\overline{xy}$ is on A and the intersection point of lines $\overline{vw}$ and $\overline{yz}$ is on A, then the intersection point of lines $\overline{wu}$ and $\overline{zx}$ is on A.

Prove that a projective plane is (c, A)-transitive if and only if it is (c, A)-desarguesian.

(45) Let **P** be a desarguesian projective plane. Prove that **P** is (c, A)-desarguesian for any point c and any line A. Prove that **P** satisfies the following *Desargues Theorem*:

> if X, Y, and Z are three distinct lines through a point c, u and x are distinct points of $X \setminus \{c\}$, v and y are distinct points of $Y \setminus \{c\}$, and w and z are distinct points of $Z \setminus \{c\}$, then the intersection points of lines $\overline{uv}$ and $\overline{xy}$, lines $\overline{vw}$ and $\overline{yz}$, and lines $\overline{wu}$ and $\overline{zx}$ are collinear.

In fact, desarguesian projective planes are the only projective planes that satisfy the Desargues Theorem. (See Notes for references.)

(46) Let q be a prime power. Find the order of the group $GL(n, q)$ of all nonsingular matrices of order n over $GF(q)$.

(47) Let q be a prime power. With every nonsingular matrix M of order n over $GF(q)$, we associate the semilinear mapping $\mathbf{x} \mapsto M\mathbf{x}$ of the n-dimensional vector space over $GF(q)$. Then the group $GL(n, q)$ can be regarded as a subgroup of $\Gamma L(n, q)$. Prove that $GL(n, q)$ is a normal subgroup of $\Gamma L(n, q)$ and that the

factor group $\Gamma L(n, q)/GL(n, q)$ is isomorphic to the group of all automorphisms of $GF(q)$.

(48) Prove Corollary 3.6.19.

(49) Prove that the lines of the design $PG_{n-1}(n, q)$ are precisely the lines of the projective geometry $PG(n, q)$ and the lines of the design $AG_{n-1}(n, q)$ are the lines of the affine geometry $AG(n, q)$.

(50) Let **D** be a (v, b, r, k, λ)-design with $r > \lambda \geq 1$. Let L be a line of **D**. Prove that $|L| \leq (b - \lambda)/(r - \lambda)$. This is a generalization of Proposition 3.7.3.

(51) Use Theorem 3.8.3 to construct an incidence matrix of a symmetric (45, 12, 3)-design.

(52) Use Theorem 3.8.5 to construct an incidence matrix of a symmetric (36, 15, 6)-design.

(53) Show that all symmetric designs of Theorem 3.8.3 have lines of cardinality q.

(54) Construct binary codes with parameters (6, 2, 6), (3, 8, 1), and (4, 8, 2).

(55) Prove that there is no binary (5, 3, 4)-code.

(56) Prove that there is no binary (8, 29, 3)-code.

(57) Let N be an incidence matrix of the Fano Plane. Let C be the binary code in $H(7, 2)$ consisting of the seven rows of N, the seven rows of $J - N$, of the allzero word (0, 0, 0, 0, 0, 0, 0), and of the all-one word (1, 1, 1, 1, 1, 1, 1). Verify that C is a perfect code.

(58) For positive integers q, n, d, let $A_q(n, d)$ denote the largest value of m such that there exists a q-ary (n, m, d)-code.
   (a) Prove that $A_q(n, 1) = q^n$ and $A_q(n, n) = q$.
   (b) Prove that $A_q(q, 2) = q^2$.
   (c) Prove that, for $n \geq 2$, $A_2(n, d) \leq 2A_2(n - 1, d)$.
   (d) Prove that $A_2(8, 5) = 4$.

(59) Let C be the binary linear code with generator matrix $\left[\begin{smallmatrix} 1 & 1 & 1 & 1 & 0 \\ 0 & 0 & 1 & 1 & 1 \end{smallmatrix}\right]$. List the codewords and find the minimum distance of C.

(60) Let C be the binary linear code with generator matrix

$$\begin{bmatrix} 1 & 0 & 0 & 1 & 1 & 0 & 1 \\ 0 & 1 & 0 & 1 & 0 & 1 & 1 \\ 0 & 0 & 1 & 0 & 1 & 1 & 1 \end{bmatrix}.$$

Find the minimum distance of C.

(61) Let C be the ternary linear code with generator matrix $\left[\begin{smallmatrix} 1 & 0 & 1 & 1 \\ 0 & 1 & 1 & 2 \end{smallmatrix}\right]$. Show that C is a perfect code.

(62) Let q be a prime power. Prove that there exists a linear q-ary code of length q^2 and weight 2.

(63) Let C be the binary code with generator matrix

$$\begin{bmatrix} 0 & 0 & 0 & 1 & 1 & 1 & 1 \\ 0 & 1 & 1 & 0 & 0 & 1 & 1 \\ 1 & 0 & 1 & 0 & 1 & 0 & 1 \end{bmatrix}.$$

Prove that C is selforthogonal.

(64) Let C be the binary code with generator matrix

$$\begin{bmatrix} 1 & 1 & 1 & 1 & 1 & 1 & 1 & 1 \\ 0 & 0 & 0 & 1 & 1 & 1 & 1 & 0 \\ 0 & 1 & 1 & 0 & 0 & 1 & 1 & 0 \\ 1 & 0 & 1 & 0 & 1 & 0 & 1 & 0 \end{bmatrix}.$$

Prove that C is selfdual.

(65) Show that the ternary code of Exercise 61 is selfdual.

(66) Prove that if there exists a selfdual linear $[n, k]$-code, then $n = 2k$.

Notes

The origins of projective geometry might be traced back to Euclid's *Optics*, an elementary treatise on perspective. Perspective was used in Greek and Roman paintings and later it was revived by artists and architects of the Renaissance. In the seventeenth century projective geometry was taken up by a group of French mathematicians, notable amongst them were Gerard Desargues (1591–1661) and Blaise Pascal (1623–1662). After a period of neglect, this subject was revived through the efforts of Gaspard Monge (1746–1818), L. N. Carnot (1753–1823), Charles Brianchon (1785–1864), and Jean Victor Poncelet (1788–1867). Their work was followed by many mathematicians, among them Steiner (1796–1863), von Staudt (1798–1867), and Plücker (1801–1868).

Finite projective geometries were first considered by Fano (1892), who introduced the n-dimensional projective space over $GF(p)$ for p, a prime. Points and lines of each plane in this design form the famous Fano Plane. Veblen and Bussey (1906) gave this geometry the name $PG(n, p)$ and extended it to $PG(n, q)$ for q a prime power. The study of finite projective geometries was developed into a coherent theory in the classic two volume book of Veblen and Young (1916). One of the seminal papers on projective planes is that of M. Hall (1943).

Yates (1936) introduced symmetric designs formed by points and lines of projective planes, and Bose (1939) considered the designs $PG_{n-1}(n, q)$ and $AG_{n-1}(n, q)$.

A comprehensive treatment of finite geometries can be found in books by Segre (1961), Dembowski (1968), Hirschfeld (1985, 1998), and Hirschfeld and Thas (1991). Most of the material presented in this chapter can be found also in books by M. Hall (1986), Batten (1986), and Beth, Jungnickel and Lenz (1999). See also Batten and Beutelspacher (1993) and Beutelspacher (1996).

Latin squares and orthogonal Latin squares (as Graeco-Latin squares) were introduced in Euler (1782). In this paper Euler conjectured that there is no pair of orthogonal Latin squares of order n for all $n \equiv 2 \pmod 4$. Tarry (1900) verified this conjecture for $n = 6$ by complete enumeration. The presented proof of Theorem 3.3.6 is due to Stinson (1984). Another proof can be found in Dougherty (1994). Bose and S. S. Shrikhande (1959c) generalized a result from Parker (1959) to obtain a pair of orthogonal Latin squares of order 22, thus giving a counter-example to Euler's conjecture. The methods of this paper were further refined in Bose and S. S. Shrikhande (1960) where it was shown that Euler's conjecture is false for infinitely many values of n including all $n \equiv 22 \pmod{36}$. Finally, in Bose, S. S. Shrikhande and Parker (1960), the conjecture was completely disproved, i.e., Theorem 3.3.5 was proved.

For a comprehensive treatment of Latin squares, see Denés and Keedwell (1974) and Laywine and Mullen (1998). For further results on MOLS, see Abel, Brouwer, Colbourn and Dinitz (1996) and Colbourn and Dinitz (2001). The notion of a net is due to Bruck (1951). The relation between affine planes and mutually orthogonal Latin squares (Corollary 3.2.18) was proved in Bose (1938) and generalized to nets (Theorem 3.2.17) in Bruck (1951). For further connections between nets and Latin squares, see Jungnickel (1990a). The result of Exercise 34 was obtained independently by Mavron (1972) and by Drake and Jungnickel (1978). We will return to nets in Chapter 7.

The term *desarguesian* for projective planes $PG(2, q)$ arose due to the fact that such planes are precisely the projective planes satisfying the famous Desargues Theorem given in Exercise 45. For the proof of the fact that every projective plane satisfying the Desargues Theorem is isomorphic to $PG(2, q)$ for some prime power q, see, for instance, Beutelspacher and Rosenbaum (1998). The smallest order of a nondesarguesian projective plane is 9. We will construct such a plane in Section 10.5. For further information on nondesarguesian projective planes, see de Resmini (1996). For a comprehensive source on projective planes, see Hughes and Piper (1982).

For a proof of the Fundamental Theorem of Projective Geometry, see the books by Baer (1952), Artin (1957), Segre (1961), Lenz (1965), Hughes and Piper (1982), Tsuzuku (1982), and Hirschfeld (1998).

Theorem 3.6.13 for q a prime was given by Burnside (1911) for abelian groups. It was later rediscovered several times. For a proof different from the one presented here see Hirschfeld (1998, p. 93).

A proof of the Veblen–Young Theorem can be found in Veblen and Young (1916). The properties (VY1) – (VY4) in the statement of this theorem are usually called the *Veblen–Young Axioms*.

The proof of Proposition 3.7.12 is patterned after MacInnes (1907), in which it is shown that affine planes of order 5 are uniquely determined by their parameters.

The proof of Theorem 3.7.10 is taken from Beth, Jungnickel and Lenz (1999, Theorem 12.2.2). Dembowski (1968) attributes this result to W. M. Kantor. Theorem 3.7.13 is a part of the famous Dembowski–Wagner Theorem proved in Dembowski and Wagner (1960). It was generalized in Kantor (1969b). See also Beth, Jungnickel, Lenz (1999, Chapter 12).

Theorems 3.7.14 and 3.7.16 are due to Hamada and Ohmori (1975). See Tonchev (1998) for a proof of these theorems based on coding theory. A similar statement for designs $PG_{d-1}(d, q)$ with $q \neq 2$ is a part of the *Hamada conjecture*, and it is still open. (See Tonchev (1998) for details.)

Symmetric designs with parameters (3.6) were first constructed in Wallis (1971). We will give Wallis' proof in Chapter 7 (Theorem 7.1.26). The proofs of Theorems 3.8.3 and 3.8.5 are from Ionin and Kharaghani (2003a) and are modeled after McFarland (1973) and Spence (1977), respectively.

For a comprehensive treatment of finite fields, we refer to Jungnickel (1993) and to Lidl and Niederreiter (1997).

For references on codes, see MacWilliams and Sloane (1977), Hill (1986), Tonchev (1988), Pless (1989), Assmus and Key (1992), van Lint (1992), and Colbourn and Dinitz (1996, Chapter V.1). For a more recent survey of coding theory, see Pless and Huffman (1998). For connections between codes and designs, see Tonchev (1996, 1998).

The results of Exercise 39 is due to André (1954). For a comprehensive treatment of translation planes, see Lüneburg (1980) and Biliotti, Jha and Johnson (2001).

Part (d) of Exercise 12 was proved in MacNeish (1922). The characteristic property of Cayley tables given in Exercises (15) and (17) is due to Frolov (1890). The associative ring F introduced in Exercise 36 is in fact a field. For a proof of this result and further discussion, see Artin (1957).

4
Hadamard matrices

Square matrices with entries ± 1 and with pairwise orthogonal rows were introduced by Jacques Hadamard as solutions to the problem of finding the maximum determinant of matrices with entries in the unit disk. They were later called *Hadamard matrices* and turned out to be a rich source of symmetric designs and other interesting combinatorial structures. Hadamard matrices give rise to symmetric designs known as *Hadamard 2-designs*. Hadamard matrices with constant row sum represent symmetric designs known as *Menon designs*. In later chapters, certain Hadamard matrices will be used for constructing other infinite families of symmetric designs.

4.1. Basic properties of Hadamard matrices

Hadamard matrices are square matrices with entries ± 1 and with pairwise orthogonal rows.

Definition 4.1.1. A matrix H of order n with every entry equal to 1 or -1 is called a *Hadamard matrix* if $HH^\top = nI$.

Example 4.1.2. In the following examples of Hadamard matrices $-$ denotes -1:

$$\begin{bmatrix} 1 & 1 \\ 1 & - \end{bmatrix}, \begin{bmatrix} - & 1 & 1 & 1 \\ 1 & - & 1 & 1 \\ 1 & 1 & - & 1 \\ 1 & 1 & 1 & - \end{bmatrix}.$$

Permutations of rows, permutations of columns, and multiplication of all entries of a row or a column of a Hadamard matrix by -1 yields a Hadamard matrix.

Definition 4.1.3. Two Hadamard matrices of the same order are called *equivalent* if one can be obtained from the other by a sequence of operations, each of which is a permutation of rows, or a permutation of columns, or multiplying all entries of a row or a column by -1.

Clearly, every Hadamard matrix is equivalent to a matrix with all entries in the first row and the first column equal to 1.

Definition 4.1.4. A Hadamard matrix with all entries in the first row and the first column equal to 1 is called *normalized*.

In the above example, the first matrix is normalized and the second matrix is equivalent to

$$\begin{bmatrix} 1 & 1 & 1 & 1 \\ 1 & 1 & - & - \\ 1 & - & 1 & - \\ 1 & - & - & 1 \end{bmatrix}.$$

The following proposition imposes a restriction on the order of a Hadamard matrix.

Proposition 4.1.5. *If there exists a Hadamard matrix of order n, then $n = 1$ or $n = 2$ or $n \equiv 0 \pmod 4$.*

Proof. Let H be a normalized Hadamard matrix of order $n \geq 3$. Consider the submatrix formed by the second row and the third row of H. Suppose that among its n columns there are a columns equal to $\left[\begin{smallmatrix}+\\+\end{smallmatrix}\right]$, b columns equal to $\left[\begin{smallmatrix}+\\-\end{smallmatrix}\right]$, c columns equal to $\left[\begin{smallmatrix}-\\+\end{smallmatrix}\right]$, and d columns equal to $\left[\begin{smallmatrix}-\\-\end{smallmatrix}\right]$. Then $a + b + c + d = n$. Since any two rows of H are orthogonal, we obtain the following additional equations:

$$\begin{aligned} a + b - c - d &= 0, \\ a - b + c - d &= 0, \\ a - b - c + d &= 0. \end{aligned}$$

Adding these four equations yields $4a = n$, so $n \equiv 0 \pmod 4$. □

Remark 4.1.6. The equations obtained in the proof of Proposition 4.1.5 actually yield $a = b = c = d = n/4$.

If A is a (± 1)-matrix, then $N = \frac{1}{2}(A + J)$ is a $(0, 1)$-matrix. Conversely, if N is a $(0, 1)$-matrix, then $A = 2N - J$ is a (± 1)-matrix. We will use this observation to demonstrate the equivalence of Hadamard matrices of order $4n$ and symmetric $(4n - 1, 2n - 1, n - 1)$-designs.

Proposition 4.1.7. *Let n be a positive integer and let H be a (± 1)-matrix of order $4n$ with all entries in the first row and the first column equal to 1. Let A be the matrix of order $4n-1$ obtained by removing the first row and the first column of H and let $N = \frac{1}{2}(A+J)$. Then H is a Hadamard matrix if and only if N is an incidence matrix of a symmetric $(4n-1, 2n-1, n-1)$-design.*

Proof.

$$\begin{aligned} H \text{ is a Hadamard matrix} &\iff JA = AJ = -J \text{ and } AA^\top = 4nI - J \\ \iff (2N-J)J = J(2N-J) &= -J \text{ and } (2N-J)(2N-J)^\top = 4nI - J \\ &\iff NJ = JN = (2n-1)J \text{ and } NN^\top = nI + (n-1)J \\ \iff N \text{ is an incidence matrix of a symmetric } &(4n-1, 2n-1, n-1)\text{-design.} \end{aligned}$$

□

Definition 4.1.8. Let n be a positive integer. Symmetric $(4n-1, 2n-1, n-1)$-designs are called *Hadamard 2-designs* of order n.

Symmetric designs with parameters $(4n-1, 2n-1, n-1)$ form the so-called *Hadamard series* of symmetric designs.

Remark 4.1.9. It was shown in Theorem 2.4.12 that $v \geq 4n-1$ for any symmetric design of order n on v points. Hadamard 2-designs are precisely the symmetric designs that meet this bound.

The next result gives an example of another interesting design that can be obtained from a Hadamard matrix.

Proposition 4.1.10. *Let n be a positive integer and let $H = [a_{ij}]$ be a Hadamard matrix of order $4n$ with all entries in the last row equal to 1. Let $X = \{1, 2, \ldots, 4n\}$. For $i = 1, 2, \ldots, 4n-1$, let $A_i = \{j \in X : a_{ij} = 1\}$ and $B_i = \{j \in X : a_{ij} = -1\}$. Then the incidence structure $\mathbf{D} = (X, \mathcal{B})$ where $\mathcal{B} = \{A_1, A_2, \ldots, A_{4n-1}, B_1, B_2, \ldots, B_{4n-1}\}$ is a 2-$(4n, 2n, 2n-1)$ design. Furthermore, any 3-subset of X is contained in exactly $n-1$ blocks of $\mathbf{D}$.*

Proof. Since all entries in the last row of H are equal to 1, every row, except the last, has $2n$ entries equal to 1 and $2n$ entries equal to -1, i.e., $|A_i| = |B_i| = 2n$ for $i = 1, 2, \ldots, 4n-1$.

Let $\{i, j, k\}$ be a 3-subset of X. We shall show that there are exactly $n-1$ rows of H, in which the ith, the jth, and the kth column have equal entries. Let A, B, C, and D be subsets of X defined as follows: $A = \{m : a_{mi} = a_{mj} = a_{mk}\}$, $B = \{m : a_{mi} = a_{mj} = -a_{mk}\}$, $C = \{m : a_{mi} = -a_{mj} = a_{mk}\}$, and $D = \{m : a_{mi} = -a_{mj} = -a_{mk}\}$. Let a, b, c, and d be the cardinalities of the sets

$A, B, C,$ and D, respectively. Then $a, b, c,$ and d satisfy the same four equations (with n replaced by $4n$) as in the proof of Proposition 4.1.5. These equations yield $a = b = c = d = n$. Since $4n \in A$, we obtain that the incidence structure $\mathbf{D}$ has exactly $n - 1$ blocks that contain $\{i, j, k\}$.

Let $\{i, j\}$ be a 2-subset of X and let λ be the number of blocks $B \in \mathcal{B}$ that contain $\{i, j\}$. Counting in two ways flags (k, B) where $k \in X$, $k \neq i$, $k \neq j$ and $B \in \mathcal{B}$, $B \supseteq \{i, j\}$, yields $(4n - 2)(n - 1) = \lambda(2n - 2)$. Therefore, $\lambda = 2n - 1$. □

Remark 4.1.11. The design constructed in Theorem 4.1.10 is called a *Hadamard 3-design*. The general definition of t-designs is given in Chapter 6 (Definition 6.1.6).

We conclude this section with another useful property of Hadamard matrices.

Proposition 4.1.12. *Let H be a Hadamard matrix of order n and let r_i be the sum of all entries of the ith column of H. Then $r_1^2 + r_2^2 + \cdots + r_n^2 = n^2$.*

Proof. Let $\mathbf{a_i}$ be the ith column of H. The vectors $\mathbf{b_i} = \frac{1}{\sqrt{n}}\mathbf{a_i}$, $i = 1, 2, \ldots, n$, form an orthonormal basis of the n-dimensional real vector space $\mathbb{R}^n$. Since $\mathbf{j}^\top \mathbf{b_i} = \frac{1}{\sqrt{n}} r_i$, we obtain that

$$\mathbf{j} = \frac{1}{\sqrt{m}}(r_1\mathbf{b_1} + r_2\mathbf{b_2} + \cdots + r_n\mathbf{b_n}).$$

Therefore,

$$n = \mathbf{j}^\top \mathbf{j} = \frac{1}{n}(r_1^2 + r_2^2 + \cdots + r_n^2),$$

and then $r_1^2 + r_2^2 + \cdots + r_n^2 = n^2$. □

4.2. Kronecker product constructions

One of the most famous open conjectures in combinatorics asserts that for any positive integer n there exists a Hadamard matrix of order $4n$. Though there are several methods of constructing Hadamard matrices, this conjecture is still far from being resolved.

One of the earliest recursive methods of constructing Hadamard matrices is provided by the Kronecker product operation on matrices.

Definition 4.2.1. The *Kronecker product* of an $m \times n$ matrix $A = [a_{ij}]$ and an $m' \times n'$ matrix B over a commutative ring is the $(mm') \times (nn')$ block matrix

$$A \otimes B = \begin{bmatrix} a_{11}B & a_{12}B & \dots & a_{1n}B \\ a_{21}B & a_{22}B & \dots & a_{2n}B \\ \dots & \dots & \dots & \dots \\ a_{m1}B & a_{m2}B & \dots & a_{mn}B \end{bmatrix}.$$

We will also need another product of matrices called the *Hadamard product*.

Definition 4.2.2. The *Hadamard product* of $m \times n$ matrices $A = [a_{ij}]$ and $B = [b_{ij}]$ is the $m \times n$ matrix $A \circ B = [a_{ij}b_{ij}]$.

The following properties of the Kronecker product are easily verified.

Proposition 4.2.3. *Let A, B, C, and D be matrices over a commutative ring R. Then*

(i) $(\alpha A) \otimes (\beta B) = (\alpha\beta)(A \otimes B)$ *for all* $\alpha, \beta \in R$;
(ii) *if A and B are identity matrices, then so is $A \otimes B$*;
(iii) $(A \otimes B)^\top = A^\top \otimes B^\top$;
(iv) $(A + B) \otimes C = A \otimes C + B \otimes C$ *and* $C \otimes (A + B) = C \otimes A + C \otimes B$, *whenever $A + B$ is defined*;
(v) $(AB) \otimes (CD) = (A \otimes C)(B \otimes D)$, *whenever AB and CD are defined*;
(vi) $(A \circ B) \otimes (C \circ D) = (A \otimes C) \circ (B \otimes D)$, *whenever $A \circ B$ and $C \circ D$ are defined.*

These properties immediately imply that the Kronecker product of Hadamard matrices is a Hadamard matrix and that the Kronecker product of symmetric matrices is a symmetric matrix.

Proposition 4.2.4. *If H_1 and H_2 are Hadamard matrices of orders n_1 and n_2, then $H_1 \otimes H_2$ is a Hadamard matrix of order n_1n_2. If H_1 and H_2 are symmetric matrices, then so is $H_1 \otimes H_2$.*

Starting with a Hadamard matrix of order 2, one can apply the Kronecker product construction to obtain Hadamard matrices of orders 2^n. The following construction also uses the Kronecker product, but in a more creative way.

Theorem 4.2.5. *For $i = 1, 2$, let P_i, Q_i, R_i, and S_i be (± 1)-matrices of order h_i such that matrices*

$$H_i = \begin{bmatrix} P_i & Q_i \\ R_i & S_i \end{bmatrix}$$

are Hadamard matrices. Define the matrix

$$H = \frac{1}{2}\begin{bmatrix} A & B \\ C & D \end{bmatrix},$$

where $A = (P_1 + Q_1) \otimes P_2 + (P_1 - Q_1) \otimes R_2$, $B = (P_1 + Q_1) \otimes Q_2 + (P_1 - Q_1) \otimes S_2$, $C = (R_1 + S_1) \otimes P_2 + (R_1 - S_1) \otimes R_2$, *and* $D = (R_1 + S_1) \otimes Q_2 + (R_1 - S_1) \otimes S_2$. *Then* H *is a Hadamard matrix of order* $2h_1h_2$.

Proof. Since H_1 and H_2 are Hadamard matrices, we have, for $i = 1$ and 2, $P_iP_i^\top + Q_iQ_i^\top = R_iR_i^\top + S_iS_i^\top = h_iI$ and $P_iR_i^\top + Q_iS_i^\top = R_iP_i^\top + S_iQ_i^\top = O$. Routine manipulations then yield $AA^\top + BB^\top = 2h_1h_2I$ and $AC^\top + BD^\top = O$. Therefore, H is a Hadamard matrix. □

Remark 4.2.6. The order of the matrix H in the above theorem is equal to half the product of the orders of H_1 and H_2.

In order to obtain Hadamard matrices whose orders are not powers of 2, we need other construction methods.

4.3. Conference matrices

In this section, we will use the notion of the quadratic character introduced in Section 3.1 to define *Paley matrices* which then will be used to obtain infinite families of Hadamard matrices whose orders are not powers of two.

Definition 4.3.1. Let q be an odd prime power and let $GF(q) = \{a_1, a_2, \ldots, a_q\}$. Let η be the quadratic character on $GF(q)$. The matrix $P = [p_{ij}]$ of order q with $p_{ij} = \eta(a_i - a_j)$ is called a *Paley matrix* of order q.

The diagonal entries of Paley matrices are all zeros. Proposition 3.1.3 implies that a Paley matrix of order q is symmetric if $q \equiv 1 \pmod 4$ and skew-symmetric if $q \equiv 3 \pmod 4$.

Proposition 4.3.2. *If* P *is a Paley matrix of order* q, *then* $PJ = JP = O$ *and* $PP^\top = qI - J$.

Proof. Let q be an odd prime power. Since the field $GF(q)$ has equal number of non-zero squares and nonsquares, $PJ = JP = O$. Lemma 3.1.4 implies that, for $a, b \in GF(q)$,

$$\sum_{x \in GF(q)} \eta(a-x)\eta(b-x) = \begin{cases} q-1 & \text{if } a = b, \\ -1 & \text{if } a \neq b. \end{cases}$$

Therefore, $PP^\top = qI - J$. □

Corollary 4.3.3. *Let $q \equiv 3 \pmod 4$ be a prime power and let P be the Paley matrix of order q. Then $N = \frac{1}{2}(P + J - I)$ is an incidence matrix of a symmetric $(q, (q-1)/2, (q-3)/4)$-design.*

Proposition 4.3.2 shows that the rows of Paley matrices are "almost orthogonal." We can obtain a matrix C of order $q+1$ with pairwise orthogonal rows from a Paley matrix P of order q by adjoining $R = [0, 1, 1, \ldots, 1]$ as the first row and $R^\top$ or $-R^\top$ as the first column. Therefore, we can make C a symmetric or a skew-symmetric matrix with pairwise orthogonal rows depending on whether P is symmetric or skew-symmetric.

Definition 4.3.4. An $n \times n$ matrix $C = [c_{ij}]$ with entries ± 1 and 0 is called a *conference matrix* if $c_{ii} = 0$ for $i = 1, 2, \ldots, n$ and $CC^\top = (n-1)I$. If $c_{12} = c_{13} = \cdots = c_{1n}$ and $c_{21} = c_{31} = \cdots = c_{n1}$, the matrix C is said to be *normalized*.

Thus, we have the following result.

Proposition 4.3.5. *For any prime power $q \equiv 1 \pmod 4$, there exists a normalized symmetric conference matrix of order $q+1$; for any prime power $q \equiv 3 \pmod 4$, there exists a normalized skew-symmetric conference matrix of order $q+1$.*

In fact any conference matrix can be suitably normalized so that it becomes symmetric or skew-symmetric. The following theorem will be generalized in Chapter 10 (Corollary 10.4.21).

Theorem 4.3.6. *Let $C = [c_{ij}]$ be a normalized conference matrix of order n. If $n \equiv 2 \pmod 4$ and $c_{12} = c_{21}$, then C is symmetric. If $n \equiv 0 \pmod 4$ and $c_{12} = -c_{21}$, then C is skew-symmetric.*

If $[1, a_2, \ldots, a_n]$ is a row of a normalized conference matrix, other than the first row, then $a_2 + \cdots + a_n = 0$ and therefore, among the $n-2$ nonzero terms of this sum there are $(n-2)/2$ positive and $(n-2)/2$ negative terms. Therefore, we have the following proposition.

Proposition 4.3.7. *If the order of a conference matrix is greater than 1, then it is even.*

The Hasse invariants impose further restriction on the order of a conference matrix.

Theorem 4.3.8. *If there exists a conference matrix of order* $n \equiv 2 \pmod 4$, *then, for any prime* $p \equiv 3 \pmod 4$, *the highest power of* p *dividing* $n-1$ *is even.*

Proof. Let C be a conference matrix of order $n \equiv 2 \pmod 4$. We have $CC^\top = (n-1)I$. Therefore, Theorem 2.5.9 implies that $c_p((n-1)I) = 1$ for any odd prime p, where c_p is the Hasse p-invariant. Definition 2.5.8 implies that

$$c_p((n-1)I) = (-1, (n-1)^n)_p \cdot \prod_{i=1}^{n-1} ((n-1)^i, -(n-1)^{i+1})_p.$$

Since n is even, we have $(-1, (n-1)^n)_p = 1$. We also have $((n-1)^i, -(n-1)^{i+1})_p = 1$ for every even i. If i is odd, then $((n-1)^i, -(n-1)^{i+1})_p = (n-1, -1)p$. Therefore,

$$c_p((n-1)I) = ((n-1, -1)_p)^{n/2} = (n-1, -1)_p.$$

Let $p \equiv 3 \pmod 4$ be a prime divisor of $n-1$ and let p^m be the highest power of p dividing $n-1$. If m is odd, then $(n-1, -1)_p = (p, -1)_p = \left(\frac{-1}{p}\right) = -1$. Therefore, for any prime $p \equiv 3 \pmod 4$, the highest power of p dividing $n-1$ must be even. □

Example 4.3.9. There are no conference matrices of orders 22, 34, 58, and 78.

Definition 4.3.10. Let C be a normalized conference matrix. The matrix obtained from C by removing the first row and the first column is called the *core of* C.

Proposition 4.3.11. *A* $(0, \pm 1)$*-matrix* $S = [s_{ij}]$ *of order* $n-1$ *is the core of a normalized conference matrix of order* n *if and only if the following conditions are satisfied:* (i) $s_{ii} = 0$ *for* $i = 1, 2, \ldots, n-1$; (ii) $SJ = JS = O$; *and* (iii) $SS^\top = (n-1)I - J$.

Proof. If S is the core of a normalized conference matrix, then the properties (i)–(iii) are immediate. Suppose $S = [s_{ij}]$ is a $(0, \pm 1)$-matrix of order $n-1$ satisfying these properties. Adjoining the $1 \times n$ row $R = [0\ 1\ \ldots 1]$ and the column $R^\top$ or $-R^\top$ yields a normalized conference matrix of order n. □

Corollary 4.3.12. *Every Paley matrix is the core of a conference matrix.*

The Kronecker product of conference matrices is not a conference matrix. However, the next theorem shows that the Kronecker product of the cores of conference matrices can be used to obtain the core of a larger conference matrix.

Theorem 4.3.13. *Let U and V be the cores of conference matrices of order $n+1$. Then*

$$W = U \otimes V + I_n \otimes J_n - J_n \otimes I_n$$

is the core of a conference matrix of order n^2+1.

Proof. By Proposition 4.3.11, $UJ = JU = O, VJ = JV = O$, and $UU^\top = VV^\top = nI - J$. From these equations and Proposition 4.2.3, one obtains by routine manipulations that $WJ = JW = O$ and $WW^\top = n^2 I - J$. Proposition 4.3.11 then implies that W is the core of a conference matrix of order n^2+1. □

We will now prove a stronger result.

Theorem 4.3.14. *If there exists a conference matrix of order $n+1$, then, for any positive integer m, there exists a conference matrix of order n^m+1.*

Proof. Suppose there exists a conference matrix of order $n+1$. If the statement of the theorem is true for $m = s$ and $m = t$, then, since $n^{st}+1 = (n^s)^t + 1$, it is true for $m = st$. Since the statement is true for $m = 2$ (Theorem 4.3.13), it suffices to prove the theorem for odd values of m.

From now on, let $m \geq 3$ be an odd integer and let $\mathbf{Z}_m = \{0, 1, \ldots, m-1\}$ be the additive group of residue classes modulo m. Let W be the core of a symmetric or skew-symmetric conference matrix of order $n+1$. Throughout the proof, I and J denote the identity and the all-one matrices of order n and, for any positive integer k, $I_{(k)}$ and $J_{(k)}$ denote the identity and the all-one matrices of order n^k.

Let M denote the set of all maps from $\mathbf{Z}_m$ to the set $\{I, J, W\}$. With each $f \in M$, we associate the matrix

$$\Pi_f = \bigotimes_{k=0}^{m-1} f(k)$$

of order n^m. We will now specify three elements of M, denoted as u, v, and w, and three subsets of M, denoted as A, B, and C:

$$u(k) = I, \;\; v(k) = J, \;\; w(k) = W, \quad \text{for all } k \in \mathbf{Z}_m;$$

A is the set of all $f \in M$ satisfying the following condition:

$$\text{for all } k \in \mathbf{Z}_m,\ f(k) = I \text{ if and only if } f(k+1) = J; \tag{4.1}$$

$$B = \{f \in M \setminus \{u\} \colon f(k) \in \{I, W\} \text{ for all } k \in \mathbf{Z}_m\};$$

$$C = \{f \in M \colon f(k) \in \{I, J\} \text{ for all } k \in \mathbf{Z}_m\}.$$

We will prove the theorem by showing that the matrix

$$W_m = \sum_{f \in A} \Pi_f$$

is the core of a conference matrix. We will prove this result in a series of lemmas.

Lemma 4.3.15. *For any distinct $i, j = 1, 2, \ldots, n^m$, there is a unique $f \in B$ such that the (i, j)-entry of Π_f is not equal to* 0.

Proof. Let $i, j \in \{1, 2, \ldots, n^m\}$, $i \neq j$. Integers $i - 1$ and $j - 1$ have unique representations in the base n:

$$i - 1 = \sum_{k=1}^{m} a_k n^{m-k}, \quad j - 1 = \sum_{k=1}^{m} b_k n^{m-k}$$

with all $a_k, b_k \in \{0, 1, \ldots, n-1\}$. Observe that, for any $f \in M$, the (i, j)-entry of Π_f is equal to the product $x_0 x_1 \ldots x_{m-1}$, where x_k is the $(a_k + 1, b_k + 1)$-entry of $f(k)$. Since an entry of W is not equal to 0 if and only if it is an off-diagonal entry and an entry of I is not equal to 0 if and only if it is a diagonal entry, we obtain that, for $f \in B$, the (i, j)-entry of Π_f is not equal to 0 if and only if, for all $k \in \mathbf{Z}_m$,

$$f(k) = \begin{cases} W & \text{if } a_k \neq b_k, \\ I & \text{if } a_k = b_k. \end{cases}$$

This proves the lemma. □

Since m is odd, condition (4.1) immediately implies the next lemma.

Lemma 4.3.16. *For every $f \in A$, there exists $k \in \mathbf{Z}_m$ such that $f(k) = W$.*

Lemma 4.3.17. *For any $f \in B$, there is a unique $g \in A$ such that, for all $k \in \mathbf{Z}_m$,*

$$f(k) = g(k) \text{ if and only if } g(k) \neq J. \tag{4.2}$$

Proof. Let $f \in B$. If $f(k) = W$ for all $k \in \mathbf{Z}_m$, then $g = f$ is the only element of A satisfying (4.2). Suppose there is $k \in \mathbf{Z}_m$ such that $f(k-1) = W$ and

$f(k) = I$. If $g \in A$ satisfies (4.2), then $g(k) = I$ and, for $l = 1, 2, \ldots, m-1$,

$$g(k+l) = \begin{cases} f(k+l) & \text{if } g(k+l-1) \neq I, \\ J & \text{if } g(k+l-1) = I. \end{cases}$$

These equations define a unique $g \in A$ satisfying (4.2). □

Lemma 4.3.18. *Let $f, g \in A$ and let $f \neq g$. Then there exists $k \in \mathbf{Z}_m$ such that $\{f(k), g(k)\} = \{W, J\}$.*

Proof. Since $f \neq g$, Lemma 4.3.16 allows us to assume that there exists $k \in \mathbf{Z}_m$ such that $f(k) = W$ and $g(k) \neq W$. Suppose $\{f(l), g(l)\} \neq \{W, J\}$ for all $l \in \mathbf{Z}_m$. Then $g(k) = I$ and (4.1) implies that $g(k+1) = J$ and $f(k+1) \neq J$. Therefore, $f(k+1) = I$, and then $f(k+2) = J$. Therefore, $g(k+2) \neq W$ and $g(k+2) \neq J$, i.e., $g(k+2) = I$. Continuing this reasoning, we obtain that $g(k+s)$ is equal to I for s even and to J for s odd. On the other hand, $g(k+m) = g(k) = I$ with m odd, a contradiction. □

The next lemma can be proven in a similar manner.

Lemma 4.3.19. *Let $f, g \in A$ and let $f \neq g$. Then there exists $k \in \mathbf{Z}_m$ such that $\{f(k), g(k)\} = \{W, I\}$.*

Lemma 4.3.20. *Let $f, g \in A$, $f \neq g$. Then $\Pi_f \Pi_g = \Pi_f \Pi_g^\top = O$.*

Proof. Lemma 4.3.18 implies that there exist $k, l \in \mathbf{Z}_m$ such that $\{f(k), g(k)\} = \{W, J\}$. Since W is the core of a conference matrix, we have $f(k)g(k) = f(k)g(k)^\top = O$. Therefore, $\Pi_f \Pi_g = \Pi_f \Pi_g^\top = O$. □

Lemma 4.3.21. *Let $f, g \in A$, $f \neq g$. Then $\Pi_f \circ \Pi_g = O$.*

Proof. Lemma 4.3.19 implies that there exist $k, l \in \mathbf{Z}_m$ such that $\{f(k), g(k)\} = \{W, I\}$. Since all the diagonal entries of W are zeros, we have $f(k) \circ g(k) = O$. Therefore, $\Pi_f \circ \Pi_g = O$. □

Lemma 4.3.16 immediately implies the next result.

Lemma 4.3.22. *All the diagonal entries of W_m are equal to 0.*

The next lemma deals with off-diagonal entries of W_m.

Lemma 4.3.23. *All the off-diagonal entries of W_m are equal to ± 1.*

Proof. Let $i, j \in \{1, 2, \ldots, n^m\}$, $i \neq j$. If there are distinct $f, g \in A$ such that the (i, j)-entries of both Π_f and Π_g are not equal to 0, then $\Pi_f \circ \Pi_g \neq O$, contrary to Lemma 4.3.21. Therefore, it suffices to show that there exists $g \in A$ such that the (i, j)-entry of Π_g is not equal to 0.

By Lemma 4.3.15, there is $k \in \mathbf{Z}_m$ such that the (i, j)-entry of Π_f is not equal to 0. Let $g \in A$ satisfy (4.2). Since Π_g is obtained from Π_f by replacing some factors of Π_f with J, the (i, j)-entry of Π_g is not equal to 0. □

We are now ready to complete the proof of Theorem 4.3.14.

Lemma 4.3.20 implies that

$$W_m W_m^\top = \sum_{f \in A} \Pi_f \Pi_f^\top.$$

Since $WW^\top = nI - J$, $II^\top = I$, and $JJ^\top = nJ$, we use the distributive property of the Kronecker product (Proposition 4.2.3(iv)) to express each product $\Pi_f \Pi_f^\top$ as a linear combination of matrices Π_h, $h \in C$:

$$\Pi_f \Pi_f^\top = \sum_{h \in C} \alpha_f(h) \Pi_h$$

with integral coefficients $\alpha_f(h)$. The proof will be completed if we show that

$$\sum_{f \in A} \alpha_f(h) = \begin{cases} n^m & \text{if } h = u, \\ -1 & \text{if } h = v, \\ 0 & \text{if } h \in C \setminus \{u, v\}. \end{cases} \tag{4.3}$$

Observe that $\alpha_w(u) = n^m$, $\alpha_w(v) = -1$, and $\alpha_f(u) = \alpha_f(v) = 0$ for all $f \in A \setminus \{w\}$, so only the third line of (4.3) has to be verified.

Fix $h \in C \setminus \{u, v\}$ and define subsets R, S, and T of $\mathbf{Z}_m$ as follows:

$$R = \{k \in \mathbf{Z}_m \setminus \{m-1\} \colon h(k) = I \text{ and } h(k+1) = J\};$$
$$S = \{k \in \mathbf{Z}_m \setminus R \colon h(k) = I\};$$
$$T = \{k \in \mathbf{Z}_m \colon h(k) = J \text{ and } k - 1 \notin R\}.$$

Let $r = |R|$, $s = |S|$, and $t = |T|$. Then $2r + s + t = m$.

Observe that, for $f \in A$, $\alpha_f(h) \neq 0$ if and only if the following two conditions are satisfied:

$$\text{for all } k \in R,\ f(k) = f(k+1) = W \text{ or } f(k) = I \text{ (and then } f(k+1) = J); \tag{4.4}$$

$$\text{for all } k \in S \cup T,\ f(k) = W. \tag{4.5}$$

Therefore, in order to uniquely determine a map $f \in A$ with $\alpha_f(h) \neq 0$, it suffices to choose an arbitrary subset of R to be $f^{-1}(I)$. Suppose such a subset is chosen and let $i = |f^{-1}(I)|$. Then the Kronecker product

$$\Pi_f \Pi_f^\top = \bigotimes_{k=0}^{m-1} f(k) f(k)^\top$$

has i factors $II^\top = I$ followed by $JJ^\top = nJ$ that occupy the i chosen positions in R and contribute to $\alpha_f(h)\Pi_h$ the product $n(I \otimes J)$ each, $r - i$ factors $WW^\top = nI - J$ followed by $WW^\top$ that occupy the remaining positions in R and contribute to $\alpha_f(h)\Pi_h$ the product $-n(I \otimes J)$ each, s factors $WW^\top$ that occupy the s positions in S and contribute the term nI of $\alpha_f(h)\Pi_h$ each, and t factors $WW^\top$ that occupy the t positions in T and contribute the term $-J$ of $\alpha_f(h)\Pi_h$ each. Thus, $\alpha_f(h) = n^i(-n)^{r-i} n^s (-1)^t = n^{r+s}(-1)^{r+t-i}$. Therefore,

$$\sum_{f \in A} \alpha_f(h) = (-1)^t n^{r+s} \sum_{i=0}^{r} \binom{r}{i} (-1)^{r-i} = 0,$$

and the proof is now complete. □

The next theorem obtains Hadamard matrices from symmetric and skew-symmetric conference matrices.

Theorem 4.3.24. *If C is a skew-symmetric conference matrix, then $H = C + I$ is a Hadamard matrix. If C is a symmetric conference matrix, then*

$$H = \begin{bmatrix} C + I & C - I \\ C - I & -C - I \end{bmatrix}$$

is a Hadamard matrix.

Proof. If C is a skew-symmetric conference matrix of order n, then $C^\top = -C$ and therefore $(C + I)(C + I)^\top = nI$. For any conference matrix C of order n, $(C + I)(C + I)^\top + (C - I)(C - I)^\top = 2nI$. For any symmetric matrix C, $(C + I)(C - I)^\top - (C - I)(C + I)^\top = O$. Therefore, if C is a symmetric conference matrix, we have $HH^\top = 2nI$. □

Corollary 4.3.25. *If there exists a conference matrix of order $n \equiv 0 \pmod 4$, then there exists a Hadamard matrix of order n; if there exists a conference matrix of order $n \equiv 2 \pmod 4$, then there exists a Hadamard matrix of order $2n$.*

Corollary 4.3.26. *Let q be a prime power. If $q \equiv 3 \pmod 4$, then there exists a Hadamard matrix of order $q + 1$. If $q \equiv 1 \pmod 4$, then there exists a Hadamard matrix of order $2q + 2$.*

We will return to conference matrices in Section 4.6..

4.4. Regular Hadamard matrices

It was observed in Section 4.1. that a Hadamard matrix of order $4n$ induces a symmetric $(4n-1, 2n-1, n-1)$-design, and vice versa. In this section we show that Hadamard matrices with constant row sum yield another family of symmetric designs called *Menon designs*.

Definition 4.4.1. A Hadamard matrix with constant row sum is called *regular*.

There is no regular Hadamard matrix of order 2. The second matrix of Example 4.1.2 is a regular Hadamard matrix of order 4.

Proposition 4.4.2. *The row sum of a regular Hadamard matrix of order $n \geq 4$ is even and not equal to 0. If it is equal to s, then $n = s^2$.*

Proof. Let H be a regular Hadamard matrix of order n with row sum s. Then $HH^\top = nI$, so $H^{-1} = \frac{1}{n}H^\top$. Then $HJ = sJ$ implies $J = sH^{-1}J = \frac{s}{n}H^\top J$. Since H is nonsingular, $s \neq 0$, and we have $H^\top J = \frac{n}{s}J$. Thus $\frac{n}{s}$ is the constant column sum of H. Therefore, $\frac{n}{s} = s$ and $n = s^2$. Since n is divisible by 4, s must be even. □

Remark 4.4.3. The above proof shows that if H is a regular Hadamard matrix, then so is $H^\top$.

If $2h$ is the row sum of a regular Hadamard matrix of order n, then the sum of all entries of this matrix is $2hn = \pm n\sqrt{n}$. This property gives another criterion for the regularity of Hadamard matrices.

Proposition 4.4.4. *A Hadamard matrix of order n is regular if and only if the sum of all its entries is equal to $\pm n\sqrt{n}$.*

Proof. Suppose H is a Hadamard matrix of order n with the sum of all entries equal to $\pm n\sqrt{n}$. For $i = 1, 2, \ldots, n$, let r_i be the sum of all entries of the ith row of H. Then $(r_1 + r_2 + \cdots + r_n)^2 = n^3$. On the other hand, by Proposition 4.1.12, $r_1^2 + r_2^2 + \cdots + r_n^2 = n^2$. Therefore,

$$\left(\frac{1}{n}\sum_{i=1}^{n} r_i\right)^2 = \frac{1}{n}\sum_{i=1}^{n} r_i^2,$$

which implies that $r_1 = r_2 = \cdots = r_n$, i.e., H is regular. □

Replacing all positive entries of a regular Hadamard matrix by zeros and all negative entries by ones yields an incidence matrix of a symmetric design.

Theorem 4.4.5. *Let H be a (± 1)-matrix of order $n \geq 4$ and let $N = \frac{1}{2}(J - H)$. Then H is a regular Hadamard matrix with row sum $2h$ if and only if N is an incidence matrix of a symmetric $(4h^2, 2h^2 - h, h^2 - h)$-design.*

Proof. If H is a regular Hadamard matrix with row sum $2h$, then $n = 4h^2$, $HH^\top = 4h^2 I$, and $HJ = 2hJ$. Therefore, $NN^\top = \frac{1}{4}(J - H)(J - H^\top) = h^2 I + (h^2 - h)J$. Conversely, if N is an incidence matrix of a symmetric $(4h^2, 2h^2 - h, h^2 - h)$-design, then $HH^\top = (J - 2N)(J - 2N^\top) = 4h^2 I$ and $HJ^\top = 2hJ$. □

Definition 4.4.6. Let h be a nonzero integer. A symmetric $(4h^2, 2h^2 - h, h^2 - h)$-design and the complementary symmetric $(4h^2, 2h^2 + h, h^2 + h)$-design are called *Menon designs* of order h^2.

The next proposition characterizes parameters of Menon designs.

Proposition 4.4.7. *A nontrivial symmetric (v, k, λ)-design is a Menon design if and only if $v = 4(k - \lambda)$.*

Proof. If $(v, k, \lambda) = (4h^2, 2h^2 - h, h^2 - h)$ or $(v, k, \lambda) = (4h^2, 2h^2 + h, h^2 + h)$, then $v = 4(k - \lambda)$.

Conversely, let **D** be a nontrivial symmetric (v, k, λ)-design with $v = 4(k - \lambda)$. Then v is even and, by Proposition 2.4.10, $k - \lambda = h^2$ for some integer $h \neq 0$. Then $v = 4h^2$ and, by (2.9), $(4h^2 - 1)\lambda = (h^2 + \lambda)(h^2 + \lambda - 1)$. Solving this equation for λ yields $\lambda = h^2 \pm h$, and therefore, **D** is a Menon design. □

We will now show that, with obvious exceptions, any symmetric design on $4q$ points, where q is a prime power, is a Menon design.

Theorem 4.4.8. *Let (v, k, λ) be the parameters of a symmetric design. Suppose that $2 \leq k \leq v - 2$ and $v = 4p^e$ where p is a prime. Then e is even and*

$$(v, k, \lambda) = (4h^2, 2h^2 - h, h^2 - h) \tag{4.6}$$

where $h = \pm p^{e/2}$.

Proof. Replacing, if necessary, (v, k, λ) with the parameters of the complementary design, we may assume that

$$2k < v. \tag{4.7}$$

Since v is even, Proposition 2.4.10 implies that $n = k - \lambda$ must be a square, so let $n = p^{2f} n_1^2$ where $f \geq 0$ and n_1 is not divisible by p. First suppose that

$2f \geq e$. Then (4.7) implies

$$p^{2f}n_1^2 = n < k < \frac{v}{2} = 2p^e. \tag{4.8}$$

Then $1 \leq p^{2f-e}n_1^2 < 2$, so $2f = e$ and $n_1 = 1$. Therefore, $v = 4p^{2f} = 4n$ and we have (4.6) with $h = p^{e/2}$.

Now assume that

$$2f < e. \tag{4.9}$$

We complete the proof by showing that (4.9) is impossible. From (2.9), we derive that

$$n = k^2 - \lambda v. \tag{4.10}$$

Substituting for n and v in (4.10) yields

$$p^{2f}n_1^2 = k^2 - 4p^e\lambda. \tag{4.11}$$

Hence $k = p^f k_1$ for some integer k_1. Let

$$\lambda_1 = k_1 - p^f n_1^2. \tag{4.12}$$

Then $\lambda = k - n = p^f k_1 - p^{2f}n_1^2 = p^f \lambda_1$. Substituting the above expressions for k and λ in (4.11) yields

$$4p^{e-f}\lambda_1 = k_1^2 - n_1^2 = (k_1 + n_1)(k_1 - n_1). \tag{4.13}$$

Since $(k_1 + n_1) - (k_1 - n_1) = 2n_1$ and p does not divide n_1, (4.13) implies that

$$p^{e-f} \text{ divides } k_1 + n_1 \text{ or } k_1 - n_1, \tag{4.14}$$

even if $p = 2$. Now $k_1 = k/p^f < \frac{1}{2}v/p^f = 2p^{e-f}$ and

$$n_1 = \frac{\sqrt{n}}{p^f} < \frac{\sqrt{\frac{1}{2}v}}{p^f} = \sqrt{2}p^{\frac{e}{2}-f} \leq p^{e-f}. \tag{4.15}$$

Thus

$$0 < k_1 + n_1 < 3p^{e-f}, \tag{4.16}$$

$$0 < k_1 - n_1 < 2p^{e-f}, \tag{4.17}$$

Then (4.13), (4.14), (4.16), and (4.17) imply that either

$$p^{e-f} = k_1 \pm n_1, \quad 4\lambda_1 = k_1 \mp n_1 \tag{4.18}$$

or

$$2p^{e-f} = k_1 + n_1, \quad 2\lambda_1 = k_1 - n_1. \tag{4.19}$$

Suppose that (4.18) holds. Then eliminating k_1 yields $p^{e-f} \mp n_1 = k_1 = 4\lambda_1 \pm n_1$, $p^{e-f} = 4\lambda_1 \pm 2n_1$. Therefore, p is even, i.e., $p = 2$. But p does not divide n_1, so $e - f = 1$. Then (4.9) implies that $e = 1$, so $v = 8$. However, there is no symmetric $(8, k, \lambda)$-design with $2 \le k \le 6$.

Suppose that (4.18) holds. First eliminate λ_1 using (4.12) and then eliminate the quantity $k_1 + n_1$. We obtain that $k_1 - n_1 = 2\lambda_1 = 2(k_1 - p^f n_1^2)$, $2p^f n_1^2 = k_1 + n_1 = 2p^{e-f}$, $n_1^2 = p^{e-2f}$. But p does not divide n_1, so $e = 2f$ contradicting (4.9). This completes the proof. □

The Kronecker product operation preserves regularity for Hadamard matrices.

Proposition 4.4.9. *The Kronecker product of regular Hadamard matrices is a regular Hadamard matrix.*

Proof. Let H and K be regular Hadamard matrices with row sums $2h$ and $2k$, respectively. The matrix $H \otimes K$ is a Hadamard matrix of order $16h^2k^2$. We have

$$
\begin{aligned}
(H \otimes K)J_{16h^2k^2} &= (H \otimes K)(J_{4h^2} \otimes J_{4k^2}) = (HJ_{4h^2}) \otimes (KJ_{4k^2}) \\
&= (2hJ_{4h^2}) \otimes (2kJ_{4k^2}) = 4hkJ_{16h^2k^2}.
\end{aligned}
$$

Therefore, $H \otimes K$ has constant row sum $4hk$. □

Corollary 4.4.10. *If there exist Menon designs of order h^2 and k^2, then there exists a Menon design of order $(2hk)^2$.*

We will now introduce a recursive construction of regular Hadamard matrices of order $4 \cdot 3^{2n}$ for every positive integer n.

Theorem 4.4.11. *Let*

$$
Q = \begin{pmatrix} 0 & - & 1 \\ 1 & 0 & - \\ - & 1 & 0 \end{pmatrix}, \quad A_0 = \begin{pmatrix} - & 0 & 0 & 0 \\ 0 & 1 & 0 & 0 \\ 0 & 0 & 1 & 0 \\ 0 & 0 & 0 & 1 \end{pmatrix}, \quad B_0 = \begin{pmatrix} 0 & 1 & 1 & 1 \\ 1 & 0 & - & 1 \\ 1 & 1 & 0 & - \\ 1 & - & 1 & 0 \end{pmatrix},
$$

where $-$ stands for -1. Let matrices A_n and B_n be defined recursively for $n \ge 1$ by $A_n = B_{n-1} \otimes I$ and $B_n = A_{n-1} \otimes J + B_{n-1} \otimes Q$ where I and J are the identity and the all-one matrix of order 3, respectively. Then, for each $n \ge 0$, $H_n = A_n + B_n$ and $H'_n = A_n - B_n$ are Hadamard matrices and H_{2n} is a regular Hadamard matrix. Furthermore, each row of every matrix H_n can be represented as a 1×4 block-matrix $[H_{n1}\ H_{n2}\ H_{n3}\ H_{n4}]$ where each block is a 1×3^n matrix, which in turn can be represented as a block-matrix $[X_1\ X_2\ \ldots\ X_{3^{n-1}}]$ with each block being a row of $\pm J$ or a row of $\pm(I + Q)$.

Proof. For each $n \geq 0$, let $I_{(n)}$ denote the identity matrix of order $4 \cdot 3^n$. We have $QQ^\top = 3I - J$, $A_0A_0^\top = I_{(0)}$, $B_0B_0^\top = 3I_{(0)}$, and $A_0B_0^\top + B_0A_0^\top = O$.

Straightforward induction on $n \geq 0$ shows that $A_nA_n^\top = 3^n I_{(n)}$, $B_nB_n^\top = 3^{n+1}I_{(n)}$, and $A_nB_n^\top + B_nA_n^\top = O$. Then $H_nH_n^\top = H'_nH_n'^\top = 4 \cdot 3^n I_{(n)}$, so matrices H_n and H'_n have pairwise orthogonal rows.

For $n \geq 0$, let $A_n = [a_{ij}^{(n)}]$ and $B_n = [b_{ij}^{(n)}]$. Then, for $n \geq 1$, matrices H_n and H'_n can be represented as block-matrices $[C_{ij}^{(n)}]$ and $[C_{ij}'^{(n)}]$ $(i, j = 1, 2, \ldots, 4 \cdot 3^{n-1})$ with $C_{ij}^{(n)} = a_{ij}^{(n-1)}J + b_{ij}^{(n-1)}(I + Q)$ and $C_{ij}'^{(n)} = -a_{ij}^{(n-1)}J + b_{ij}^{(n-1)}(I - Q)$. Thus, each entry of $C_{ij}^{(n)}$ as well as each entry of $C_{ij}'^{(n)}$ is equal to $\pm a_{ij}^{(n-1)} \pm b_{ij}^{(n-1)}$, and therefore it is an entry of $\pm H_{n-1}$ or $\pm H'_{n-1}$. Since H_0 and H'_0 are (± 1)-matrices, it follows by induction on n that all matrices H_n and H'_n are (± 1)-matrices. Therefore, H_n and H'_n are Hadamard matrices.

For any matrix X, let $s(X)$ denote the sum of all entries of X. We have, for $n \geq 1$, $s(H_{2n}) = 3s(B_{2n-1}) + 9s(A_{2n-1}) = 27s(H_{2n-2})$. Since $s(H_0) = 8$, we obtain by induction on n that $s(H_{2n}) = (4 \cdot 3^{2n})^{3/2}$. By Proposition 4.4.4, H_{2n} is a regular Hadamard matrix.

To complete the proof, observe that, for $n \geq 1$,

$$C_{ij}^{(n)} = \begin{pmatrix} a_{ij}^{(n-1)} + b_{ij}^{(n-1)} & a_{ij}^{(n-1)} - b_{ij}^{(n-1)} & a_{ij}^{(n-1)} + b_{ij}^{(n-1)} \\ a_{ij}^{(n-1)} + b_{ij}^{(n-1)} & a_{ij}^{(n-1)} + b_{ij}^{(n-1)} & a_{ij}^{(n-1)} - b_{ij}^{(n-1)} \\ a_{ij}^{(n-1)} - b_{ij}^{(n-1)} & a_{ij}^{(n-1)} + b_{ij}^{(n-1)} & a_{ij}^{(n-1)} + b_{ij}^{(n-1)} \end{pmatrix}.$$

If $b_{ij}^{(n-1)} = 0$, then each row of $C_{ij}^{(n)}$ has three equal entries and, since they are entries of a Hadamard matrix, it is a row of $\pm J$. If $b_{ij}^{(n-1)} \neq 0$, then each row of $C_{ij}^{(n)}$ has two equal entries and therefore it is a row of $\pm(I + Q)$. □

We now describe another family of regular Hadamard matrices that will be used in the later chapters.

Definition 4.4.12. A regular Hadamard matrix K of order n^2 is said to be *of Bush type* if it can be represented as a block matrix $K = [K_{ij}]$ where each K_{ij} is either the all-one matrix of order n or a matrix of order n with the sum of entries in every row and every column equal to 0.

The second matrix of Example 4.1.2 is a Bush-type Hadamard matrix of order 4.

Remark 4.4.13. Proposition 4.4.2 immediately implies that every row and every column of blocks of a Hadamard matrix of Bush type contains exactly one all-one block.

We will now introduce a construction of an infinite family of Bush-type Hadamard matrices. The main ingredient of this construction is given by the following definition.

Definition 4.4.14. Let H be a Hadamard matrix of order n with all entries in the last row equal to 1. Let $R_1, R_2, \ldots, R_n$ be the consecutive rows of H. Then $\mathcal{K}(H)$ denotes the sequence $(C_1, C_2, \ldots, C_n)$ of (± 1)-matrices defined by $C_i = R_i^\top R_i$, for $i = 1, 2, \ldots, n$.

The following properties of the sequence $\mathcal{K}(H)$ are straightforward.

Lemma 4.4.15. *Let H be a Hadamard matrix of order n with all entries in the last row equal to* 1. *Let $\mathcal{K}(H) = (C_1, C_2, \ldots, C_n)$. Let $i, j \in \{1, 2, \ldots, n\}$, $i \neq j$. Then:* (i) $C_i^\top = C_i$; (ii) $C_n = J$; (iii) *if $i \neq n$, then* $C_i J = O$; (iv) $C_i^2 = nC_i$; (v) $C_i C_j = O$; (vi) $\sum_{k=1}^{n} C_k = nI$.

Theorem 4.4.16. *Let H be a Hadamard matrix of order n with all entries in the last row equal to* 1 *and let $\mathcal{K}(H) = (C_1, C_2, \ldots, C_n)$. Let L be a Latin square of order n and let $K = [K_{ij}]$ be a matrix of order n^2 represented as a block matrix with $n \times n$ blocks K_{ij} satisfying the following conditions:* (i) *if $L(i, j) = n$, then $K_{ij} = C_n$;* (ii) *if $L(i, j) = k \neq n$, then $K_{ij} \in \{\pm C_k\}$. Then K is a regular Hadamard matrix of Bush type.*

Proof. Lemma 4.4.15 implies that

$$\sum_{j=1}^{n} K_{ij} K_{hj}^\top = O$$

for distinct $i, h = 1, 2, \ldots, n$. Since, for $i = 1, 2, \ldots, n$,

$$\sum_{j=1}^{n} K_{ij} K_{ij}^\top = \sum_{k=1}^{n} C_k C_k^\top = n \sum_{k=1}^{n} C_k = n^2 I$$

and

$$\sum_{j=1}^{n} K_{ij} J = \sum_{k=1}^{n} C_k J = J^2 = nJ,$$

the matrix K is a regular Hadamard matrix of Bush type. □

The order of any Hadamard matrix constructed in Theorem 4.4.16 is divisible by 16, so it leaves open a question whether there exist Hadamard matrices of Bush type of order $4n^2$ for n odd. There are examples of such matrices for $n = 3$, $n = 5$, and $n = 9$. We will state (without proof) the following two recent results.

They will be used in Chapter 11 for constructing infinite families of symmetric designs.

Theorem 4.4.17. *There exists a regular Hadamard matrix of Bush type of order* 100.

Theorem 4.4.18. *For any odd integer n, there exists a symmetric regular Hadamard matrix of Bush type of order* $4n^4$.

4.5. From Paley matrices to regular Hadamard matrices

In this section we describe a construction of a regular Hadamard matrix of order n^2 based on Paley matrices of orders $n-1$ and $n+1$.

Theorem 4.5.1. *If $n-1$ and $n+1$ are odd prime powers, then there exists a symmetric regular Hadamard matrix of order n^2.*

Proof. Let $\{n-1, n+1\} = \{p, q\}$ where $p \equiv 1 \pmod 4$ and $q \equiv -1 \pmod 4$. Observe that $p - q = 2 \cdot (-1)^{\frac{n}{2}}$. Let X and Y be Paley matrices of order p and q, respectively. Then X is symmetric, Y is skew-symmetric, $J_p X = XJ_p = O_p$, $YJ_q = J_q Y = O_q$, $XX^\top = pI_p - J_p$, and $YY^\top = qI_q - J_q$.

Let $Y = [y_{ij}]$, $i, j = 1, 2, \dots, q$. Define matrices H_{ij} of order p by

$$H_{ij} = \begin{cases} (-1)^{i+j}((-1)^{\frac{n}{2}} y_{i,q+1-j} X - I_p) & \text{if } i + j \neq q + 1, \\ J_p - (-1)^{\frac{n}{2}} I_p - I_p & \text{if } i + j = q + 1. \end{cases}$$

For $j = 1, 2, \dots, q$, let $H_{0j} = (-1)^{j-1} J_{1,p}$ and $H_{j0} = (-1)^{j-1} J_{p,1}$. Finally, let H_{00} be the matrix of order 1 with entry $(-1)^{\frac{n}{2}-1}$. For $i, j = 1, 2, \dots, q$, matrices H_{ij} are symmetric (± 1)-matrices. Since $y_{i,q+1-j} = y_{j,q+1-i}$, the block matrix $H = [H_{ij}]$, $i, j = 0, 1, \dots, q$, is symmetric. We claim that H is a regular Hadamard matrix of order n^2.

The sum of the entries of the 0th row of H is $p + (-1)^{\frac{n}{2}-1} = n$. Since for $i, j = 1, 2, \dots, q$,

$$H_{ij} J_p = \begin{cases} (-1)^{i+j-1} J_p & \text{if } i + j \neq q + 1, \\ (n-1) J_p & \text{if } i + j = q + 1, \end{cases}$$

the row sum of every row of H is n.

It remains to verify that distinct rows of H are orthogonal.

For $i, j = 1, 2, \ldots, q$,

$$\sum_{j=1}^{q} H_{ij} H_{ij}^{\top} = \sum_{j \neq q+1-i} H_{ij}^2 + H_{i,q+1-i}^2$$

$$= (q-1)XX^{\top} - 2(-1)^{\frac{n}{2}} \sum_{j=1}^{q} y_{i,q+1-j} X + (q-1)I_p + (J_p - (-1)^{\frac{n}{2}} I_p - I_p)^2$$

$$= (q-1)(pI_p - J_p) + (q-1)I_p + (J_p - (-1)^{\frac{n}{2}} I_p - I_p)^2 = n^2 I_p - J_p.$$

Since $H_{i0} H_{i0}^{\top} = J_p$, we obtain that

$$\sum_{j=0}^{q} H_{ij} H_{ij}^{\top} = n^2 I_p.$$

Let $i, k = 1, 2, \ldots, q, i \neq k$. Then we use that Y is skew-symmetric to obtain that

$$\sum_{j=1}^{q} H_{ij} H_{kj}^{\top} = \sum_{\substack{j \neq q+1-i \\ j \neq q+1-k}} H_{ij} H_{kj} + (-1)^{i+k} H_{i,q+1-i} H_{k,q+1-i} + H_{i,q+1-k} H_{k,q+1-k}$$

$$= (-1)^{i+k}(q - p + 2(-1)^{\frac{n}{2}})I_p + (-1)^{i+k-1} J_p = (-1)^{i+k-1} J_p.$$

Now since $H_{i0} H_{k0}^{\top} = (-1)^{i+k} J_p$, we obtain that

$$\sum_{i=0}^{q} H_{ij} H_{kj}^{\top} = O.$$

It can be verified in the same manner that the 0th row of H is orthogonal to every other row. □

Remark 4.5.2. Observe that all diagonal entries of H, except possibly the $(0, 0)$ entry, are equal to -1. The $(0, 0)$ entry is equal to $(-1)^{\frac{n}{2}-1}$ and therefore, if $n \equiv 0 \pmod 4$, H has constant diagonal. Later we will give other constructions of symmetric regular Hadamard matrices with constant diagonal (Theorem 5.3.16, Corollary 5.3.17, and Theorem 7.4.22).

4.6. Regular sets of (±1)-matrices

In Section 4.3., we constructed conference matrices whose cores are Paley matrices. In this section, we will use Paley matrices to construct conference matrices of a different kind.

Let $q \equiv 3 \pmod 4$ be a prime power, let G be a (multiplicatively written) elementary abelian group of order q, and let $\varphi \colon G \to GF(q)$ be a fixed isomorphism from G to the additive group of $GF(q)$. We choose an ordering of G and let $\mathcal{M}$ be the set of all matrices of order q over the integers whose rows and columns are indexed by elements of G in the chosen order. For $X \in \mathcal{M}$ and $\sigma, \tau \in G$, let $X(\sigma, \tau)$ denote the (σ, τ)-entry of X. A matrix $X \in \mathcal{M}$ is said to be *G-invariant* if $X(\sigma\alpha, \tau\alpha) = X(\sigma, \tau)$ for all $\alpha, \sigma, \tau \in G$.

We will regard matrices I and J of order q as elements of $\mathcal{M}$. Let $R, E \in \mathcal{M}$ be (0, 1)-matrices such that $R(\sigma, \tau) = 1$ if and only if $\sigma\tau = 1$ and $E(\sigma, \tau) = 1$ if and only if $\sigma = 1$. Define matrices $P, M, N \in \mathcal{M}$ as follows:

$P(\sigma, \tau) = \eta(\varphi(\tau) - \varphi(\sigma))$ where η is the quadratic character on $GF(q)$, so P is a Paley matrix; $M = I + P$; $N(\sigma, \tau) = M(1, \tau)$ for all $\sigma, \tau \in G$. Observe that matrices P and M are G-invariant.

Proposition 4.3.2 implies the following properties of the matrices P, M, and N:

$$P^\top = -P, \quad PP^\top = P^\top P = qI - J, \quad PJ = JP = O; \tag{4.20}$$

$$MM^\top = M^\top M = (q+1)I - J, \quad MJ = JM = J; \tag{4.21}$$

$$NJ = J, \quad JN = qN; \tag{4.22}$$

$$PN = O, \quad MN = N, \quad MN^\top = (q+1)E - J. \tag{4.23}$$

We define an action of G on $\mathcal{M}$ as follows: for $\alpha \in G$ and $X \in \mathcal{M}$, $(\alpha X)(\sigma, \tau) = X(\sigma, \alpha\tau)$ for all $\sigma, \tau \in G$. Thus, α acts as a permutation of columns of X. The following lemma summarizes properties of this action that will be used throughout this section.

Lemma 4.6.1. *Let $\alpha, \beta \in G$ and $X, Y \in \mathcal{M}$ and let s be an integer. Then:*

(i) $\alpha(X + Y) = \alpha X + \alpha Y$, $(\alpha X)(\alpha Y)^\top = XY^\top$;
(ii) *if* $JX = sJ$, *then* $J(\alpha X) = sJ$; *if* $XJ = sJ$, *then* $(\alpha X)J = sJ$ *and* $\sum_{\rho \in G} \rho X = sJ$;
(iii) $X(\beta Y) = \beta(XY)$; *if Y is G-invariant, then* $(\alpha X)(\beta Y) = (\alpha\beta)(XY)$;
(iv) $(\alpha R)^\top = \alpha R$;
(v) *if Y is G-invariant, then* $(\alpha Y)^\top = \alpha^{-1} Y^\top$;
(vi) *if Y is G-invariant and symmetric, then* $(\alpha R)Y = Y(\alpha R)$;
(vii) *if Y is G-invariant and skew-symmetric, then* $(\alpha R)Y = -Y(\alpha R)$;
(viii) $(\alpha R)(\beta R) = \alpha^{-1}\beta I$;
(ix) $(\alpha M)(\beta N) = (\alpha N)(\beta N) = \beta N$;
(x) $(\alpha N)(\beta N)^\top = \begin{cases} qJ & \textit{if } \alpha = \beta, \\ -J & \textit{if } \alpha \neq \beta. \end{cases}$
(xi) $\sum_{\rho \in G} (\rho N)^\top (\rho N) = (q^2 + q)I - qJ.$

Proof. Since α permutes the columns of X in the same way as the columns of Y, (i) is immediate.

(ii) If $JX = sJ$, then αX has the same constant column sum s as X, so $J(\alpha X) = sJ$. Let $XJ = sJ$. Then αX has constant row sum s and therefore $(\alpha X)J = sJ$. For $\sigma, \tau \in G$,

$$\sum_{\rho \in G} (\rho X)(\sigma, \tau) = \sum_{\rho \in G} X(\sigma, \rho\tau) = \sum_{\rho \in G} X(\sigma, \rho) = sJ.$$

(iii) For $\sigma, \tau \in G$,

$$(\alpha X)(\beta Y)(\sigma, \tau) = \sum_{\rho \in G} (\alpha X)(\sigma, \rho)(\beta Y)(\rho, \tau) = \sum_{\rho \in G} X(\sigma, \alpha\rho) Y(\rho, \beta\tau).$$

If $\alpha = 1$, we obtain that $X(\beta Y)(\sigma, \tau) = (XY)(\sigma, \beta\tau) = \beta(XY)(\sigma, \tau)$. If Y is G-invariant, then

$$\begin{aligned}(\alpha X)(\beta Y)(\sigma, \tau) &= \sum_{\rho \in G} X(\sigma, \alpha\rho) Y(\alpha\rho, \alpha\beta\tau) = (XY)(\sigma, \alpha\beta\tau) \\ &= (\alpha\beta)(XY)(\sigma, \tau).\end{aligned}$$

(iv) For $\sigma, \tau \in G$,

$$\begin{aligned}(\alpha R)^\top(\sigma, \tau) = 1 &\Leftrightarrow R(\tau, \alpha\sigma) = 1 \Leftrightarrow \tau\alpha\sigma = 1 \Leftrightarrow R(\sigma, \alpha\tau) \\ &= 1 \Leftrightarrow (\alpha R)(\sigma, \tau) = 1.\end{aligned}$$

(v) Let Y be G-invariant. For $\sigma, \tau \in G$,

$$\begin{aligned}(\alpha Y)^\top(\sigma, \tau) &= (\alpha Y)(\tau, \sigma) = Y(\tau, \alpha\sigma) = Y(\alpha^{-1}\tau, \sigma) = Y^\top(\sigma, \alpha^{-1}\tau) \\ &= (\alpha^{-1} Y^\top)(\sigma, \tau).\end{aligned}$$

(vi) and (vii) Let Y be G-invariant. For $\sigma, \tau \in G$,

$$(\alpha R)Y(\sigma, \tau) = \sum_{\rho \in G} R(\sigma, \alpha\rho) Y(\rho, \tau) = Y(\alpha^{-1}\sigma^{-1}, \tau) = Y(1, \sigma\alpha\tau);$$

$$Y(\alpha R)(\sigma, \tau) = \sum_{\rho \in G} Y(\sigma, \rho) R(\rho, \alpha\tau) = Y(\sigma, \tau^{-1}\alpha^{-1}) = Y(\alpha\tau\sigma, 1).$$

If further Y is symmetric, then $(\alpha R)Y(\sigma, \tau) = Y(\alpha R)(\sigma, \tau)$; if Y is skew-symmetric, then $(\alpha R)Y(\sigma, \tau) = -Y(\alpha R)(\sigma, \tau)$.

(viii) For $\sigma, \tau \in G$,

$$(\alpha R)(\beta R)(\sigma, \tau) = \sum_{\rho \in G} R(\sigma, \alpha\rho) R(\rho, \beta\tau),$$

so

$$(\alpha R)(\beta R)(\sigma, \tau) = \begin{cases} 1 & \text{if } \alpha\sigma = \beta\tau, \\ 0 & \text{if } \alpha\sigma \neq \beta\tau. \end{cases}$$

Therefore, $(\alpha R)(\beta R)(\sigma, \tau) = I(\sigma, \alpha^{-1}\beta\tau) = (\alpha^{-1}\beta I)(\sigma, \tau)$.
(ix) For $\sigma, \tau \in G$,

$$\begin{aligned}(\alpha M)(\beta N)(\sigma, \tau) &= \sum_{\rho\in G} M(\sigma, \alpha\rho)N(\rho, \beta\tau) = \sum_{\rho\in G} M(\alpha^{-1}\sigma, \rho)N(\rho, \beta\tau)\\ &= (MN)(\alpha^{-1}\sigma, \beta\tau) = N(\alpha^{-1}\sigma, \beta\tau)\\ &= N(\sigma, \beta\tau) = (\beta N)(\sigma, \tau);\\ (\alpha N)(\beta N)(\sigma, \tau) &= \sum_{\rho\in G} N(\sigma, \alpha\rho)N(\rho, \beta\tau) = \sum_{\rho\in G} M(1, \alpha\rho)N(\rho, \beta\tau)\\ &= \sum_{\rho\in G} M(\alpha^{-1}, \rho)N(\rho, \beta\tau) = (MN)(\alpha^{-1}, \beta\tau)\\ &= N(\alpha^{-1}, \beta\tau) = N(\sigma, \beta\tau) = (\beta N)(\sigma, \tau).\end{aligned}$$

(x) For $\sigma, \tau \in G$,

$$(\alpha N)(\beta N)^\top(\sigma, \tau) = \sum_{\rho\in G} N(\sigma, \alpha\rho)N(\tau, \beta\rho) = \sum_{\rho\in G} M(1, \alpha\rho)N^\top(\beta\rho, \tau)$$

$$= \sum_{\rho\in G} M(\beta\alpha^{-1}, \beta\rho)N^\top(\beta\rho, \tau) = (MN^\top)(\beta\alpha^{-1}, \tau) = \begin{cases} q & \text{if } \alpha = \beta,\\ -1 & \text{if } \alpha \neq \beta.\end{cases}$$

(xi) Let $K = \sum_{\rho\in G}(\rho N)^\top(\rho N)$. Then, for $\sigma, \tau \in G$,

$$\begin{aligned}K(\sigma, \tau) &= \sum_{\pi\in G}\sum_{\rho\in G}(\rho N)^\top(\sigma, \pi)(\rho N)(\pi, \tau) = \sum_{\pi\in G}\sum_{\rho\in G} N(\pi, \rho\sigma)N(\pi, \rho\tau)\\ &= \sum_{\pi\in G}\sum_{\rho\in G} M(1, \rho\sigma)M(1, \rho\tau) = q\sum_{\rho\in G} M(\rho^{-1}, \sigma)M(\rho^{-1}, \tau)\\ &= q\sum_{\rho\in G} M^\top(\sigma, \rho^{-1})M(\rho^{-1}, \tau) = q(M^\top M)(\sigma, \tau).\end{aligned}$$

Therefore, (4.21) implies that $K = qM^\top M = (q^2 + q)I - qJ$. □

Let $\mathcal{N}$ denote the set of all matrices of order q^2 that can be represented as block matrices $[X_{\sigma\tau}]$, $\sigma, \tau \in G$, indexed by elements of G (in the chosen order), with all $X_{\sigma\tau} \in \mathcal{M}$. In particular, for $X, Y \in \mathcal{M}$, the Kronecker product $X \otimes Y$ is an element of $\mathcal{N}$. We will define the following matrices $A, C \in \mathcal{N}$:

$$\begin{aligned}A &= P \otimes P + I \otimes J - J \otimes I, \quad I, J \in \mathcal{M};\\ C &= [C_{\sigma\tau}], \quad C_{\sigma\tau} = \sigma N \text{ for all } \sigma, \tau \in G.\end{aligned}$$

For each $a \in GF(q)$, we define a map $\pi_a : G \to G$ by

$$\pi_a(\sigma) = \varphi^{-1}(a\varphi(\sigma)).$$

The following properties of the maps π_a are straightforward.

Lemma 4.6.2. *If $a \neq 0$, then π_a is an automorphism of G. For all $a, b \in GF(q)$ and $\sigma \in G$, $\pi_{a+b}(\sigma) = \pi_a(\sigma)\pi_b(\sigma)$.*

For each $a \in GF(q)^*$, we define matrix $B_a \in \mathcal{N}$ by

$$B_a = \sum_{\rho \in G} (\rho R) \otimes (\pi_a(\rho) M).$$

Observe that matrices $A + I$, C, and B_a, $a \in GF(q)^*$, are (± 1)-matrices.

Lemma 4.6.3. *Let $a, b \in GF(q)^*$. Then:*

(i) $A = A^\top$, $AA^\top = q^2 I - J$;
(ii) $AB_a = -B_a A$;
(iii) $AC = -CA$;
(iv) $AJ = O$, $B_a J = CJ = qJ$;
(v) *if* $a + b \neq 0$, *then* $B_a B_b = J$;
(vi) *if* $a \neq b$, *then* $B_a B_b^\top = B_a^\top B_b = J$;
(vii) $B_a B_a^\top = B_{-a}^\top B_{-a}$;
(viii) $B_a C = C B_a = B_a^\top C = C B_a^\top = J$;
(ix) $C^2 = J$, $CC^\top = ((q^2+q)I_q - qJ_q) \otimes J_q$, $C^\top C = qJ_q \otimes ((q+1)I_q - J_q)$;
(x) *if S is the set of all nonzero squares of $GF(q)$, then*

$$\sum_{a \in S} \left(B_a B_a^\top + B_a^\top B_a\right) + CC^\top + C^\top C = q^2(q+1)I.$$

Proof. For proof of (i), see Theorem 4.3.13.

(ii) Since matrices P and M are G-invariant and $PM = MP$, Lemma 4.6.1(iii) implies that $P(\pi_a(\rho)M) = (\pi_a(\rho)M)P$. Since P is also skew-symmetric, 4.6.1(vii) implies that $P(\rho R) = -(\rho R)P$. Therefore, $(P \otimes P)B_a = -B_a(P \otimes P)$. Lemma 4.6.1(iii) and (4.21) imply that $J(\pi_a(\rho)M) = (\pi_a(\rho))(JM) = J$ and $(\pi_a(\rho)M)J = (\pi_a(\rho))(MJ) = J$. Therefore,

$$(I \otimes J)B_a = B_a(I \otimes J) = \sum_{\rho \in G} (\rho R) \otimes J_q = J_q \otimes J_q = J.$$

Similarly, $(J \otimes I)B_a = B_a(J \otimes I) = J$, so $AB_a = -B_a A$.

(iii) Since P is G-invariant and $PN = O$, Lemma 4.6.1(iii) implies that $P(\sigma N) = O$ for all $\sigma \in G$ and therefore, $(P \otimes P)C = O$. On the other hand, for $\sigma, \tau \in G$, the (σ, τ) block of $C(P \otimes P)$ is equal to $(\sigma N)(sP)$ where s is the column sum of P. By (4.20), $s = 0$, so $C(P \otimes P) = O$. Lemma 4.6.1(ii) and (4.22) imply that $(I \otimes J)C = [J(\sigma N)] = [q\sigma N] = qC$ and $(J \otimes I)C = [Y_{\sigma\tau}]$ with $Y_{\sigma\tau} = \sum_{\rho \in G}(\rho N) = J$, so $AC = (q-1)C$. We further have $C(I \otimes J) = [(\sigma N)J] = J$ and $C(J \otimes I) = [q\sigma N] = qC$, so $CA = -(q-1)C$.

(iv) Since $PJ = O$, we have $(P \otimes P)J = O$. Therefore, $AJ = (I_q \otimes J_q)(J_q \otimes J_q) - (J_q \otimes I_q)(J_q \otimes J_q) = O$. We have $CJ = [q(\sigma N)J] = qJ$ and

$$B_a J = \sum_{\rho \in G} ((\rho R)J) \otimes ((\pi_a(\rho) M)J) = qJ.$$

(v) We apply Lemma 4.6.1(iii,viii) to obtain that

$$B_a B_b = \sum_{\rho, \sigma \in G} (\rho^{-1} \sigma I) \otimes (\pi_a(\rho) \pi_b(\sigma) M^2).$$

Replacing σ with $\rho\tau$ yields

$$B_a B_b = \sum_{\tau \in G} (\tau I) \otimes \sum_{\rho \in G} (\pi_a(\rho) \pi_b(\rho) \pi_b(\tau) M^2).$$

Suppose $a + b \neq 0$. By Lemma 4.6.2, $\pi_a(\rho)\pi_b(\rho) = \pi_{a+b}(\rho)$. Since $M^2 J = J$ and since π_{a+b} is an automorphism of G, Lemma 4.6.1(ii) implies that

$$\sum_{\rho \in G} (\pi_{a+b}(\rho) \pi_b(\tau) M^2) = J,$$

and then $B_a B_b = J$.

(vi) Lemma 4.6.1(iv,v) implies that

$$B_b^\top = \sum_{\sigma \in G} (\sigma R)^\top \otimes (\pi_b(\sigma) M)^\top = \sum_{\sigma \in G} (\sigma R) \otimes (\pi_{-b}(\sigma) M^\top).$$

Therefore, we let, as in the proof of (v), $\sigma = \rho\tau$ to obtain:

$$B_a B_b^\top = \sum_{\tau \in G} (\tau I) \otimes \sum_{\rho \in G} (\pi_{a-b}(\rho) \pi_{-b}(\tau)(M M^\top)). \tag{4.24}$$

Suppose $a \neq b$. Then π_{a-b} is an automorphism of G. Since $MM^\top J = J$, Lemma 4.6.1(ii) implies that $B_a B_b^\top = J$. In a similar manner, one obtains that

$$B_a^\top B_b = \sum_{\tau \in G} (\tau I) \otimes \sum_{\rho \in G} (\pi_{b-a}(\rho) \pi_b(\tau)(M^\top M)). \tag{4.25}$$

and then $B_a^\top B_b = J$.

(vii) Since $MM^\top = M^\top M$, equations (4.24) and (4.25) imply that

$$B_a B_a^\top = B_{-a}^\top B_{-a} = q \sum_{\tau \in G} (\tau I) \otimes (\pi_a(\tau)(M M^\top)). \tag{4.26}$$

(viii) Observe that, for $\rho \in G$ and $X \in \mathcal{M}$, the (σ, τ) block of $((\rho R) \otimes X)C$ is equal to $X(\sigma^{-1}\rho^{-1}N)$. Therefore, Lemma 4.6.1(ix) imply that, for $\sigma, \tau \in G$, the (σ, τ) block of $B_a C$ is equal to

$$\sum_{\rho \in G} \sigma^{-1} \rho^{-1} N.$$

Since $NJ = J$, Lemma 4.6.1(ii) implies that this sum equals J, so $B_a C = J$. Since M is G-invariant, Lemma 4.6.1(v) shows that the same reasoning works for $B_a^\top C$.

For $\sigma, \tau \in G$, the (σ, τ) block of CB_a is equal to

$$\sum_{\rho \in G} (\sigma N)(\pi_a(\rho)M).$$

Since π_a is an automorphism of G and since $NJ = J$, we obtain that $CB_a = J$. Similarly, $CB_a^\top = J$.

(ix) For $\sigma, \tau \in G$, the (σ, τ) block of C^2 is equal to

$$\sum_{\rho \in G} (\sigma N)(\rho N) = (\sigma N)J = J,$$

so $C^2 = J$.

The (σ, τ) block of $CC^\top$ is equal to $q(\sigma N)(\tau N)^\top$, and we apply Lemma 4.6.1(x) to obtain the desired result.

Lemma 4.6.1(xi) immediately implies that the (σ, τ) block of $C^\top C$ is equal to $(q^2 + q)I - qJ$, and this gives the required formula.

(x) Let S be the set of all nonzero squares of $GF(q)$ and let $Z = \sum_{a \in S}(B_a B_a^\top + B_a^\top B_a)$. Since $q \equiv 3 \pmod 4$, we have $GF(q)^* = S \cup (-S)$. Therefore, by (4.26),

$$Z = \sum_{a \in GF(q)^*} B_a B_a^\top = q \sum_{\tau \in G} (\tau I) \otimes \sum_{a \in GF(q)^*} \pi_a(\tau)(MM^\top).$$

Suppose $\tau \neq 1$. Then, as a runs through $GF(q)^*$, $a\varphi(\tau)$ runs through the same set and $\pi_a(\tau) = \varphi^{-1}(a\varphi(\tau))$ runs through $G \setminus \{1\}$. Since $MM^\top J = J$, we obtain that

$$\sum_{a \in GF(q)^*} \pi_a(\tau)(MM^\top) = J - MM^\top = 2J - (q+1)I.$$

If $\tau = 1$, then

$$\sum_{a \in GF(q)^*} \pi_a(\tau)(MM^\top) = (q-1)MM^T = (q^2 - 1)I - (q-1)J.$$

Therefore,

$$\begin{aligned} \tfrac{1}{q} Z &= I_q \otimes ((q^2 - 1)I_q - (q-1)J_q) + (J_q - I_q) \otimes (2J_q - (q+1)I_q) \\ &= (q^2 + q)I + 2J - (q+1)(I_q \otimes J_q + J_q \otimes I_q). \end{aligned}$$

From (ix),

$$\frac{1}{q}(CC^\top + C^\top C) = (q+1)(I_q \otimes J_q + J_q \otimes I_q) - 2J,$$

and (x) follows. □

We will now construct a putative family of symmetric conference matrices.

Theorem 4.6.4. *Let $q \equiv 3 \pmod 4$ be a prime power. If there exists a symmetric conference matrix of order $q+3$, then there exists a symmetric conference matrix of order $q^2(q+2)+1$.*

Proof. Let $W = [w_{ij}]$ be the core of a normalized symmetric conference matrix of order $q+3$ and let A, C, and B_a, $a \in GF(q)^*$, be matrices of order q^2 from Lemma 4.6.3. Let $S = \{a_1, a_2, \ldots, a_{(q-1)/2}\}$ be the set of all nonzero squares of $GF(q)$. Define a block matrix $U = [U_{ij}]$ of order $q^2(q+2)$ as follows: for $i, j = 1, 2, \ldots, q+2$,

$$U_{ij} = \begin{cases} w_{ij} B_{a_k} & \text{if } i - j \equiv k \pmod{q+2},\ 1 \le k \le \frac{q-1}{2}, \\ w_{ij} B_{a_k}^\top & \text{if } i - j \equiv -k \pmod{q+2},\ 1 \le k \le \frac{q-1}{2}, \\ A & \text{if } i - j \equiv 0 \pmod{q+2}, \\ w_{ij} C & \text{if } i - j \equiv \frac{q+1}{2} \pmod{q+2}, \\ w_{ij} C^\top & \text{if } i - j \equiv -\frac{q+1}{2} \pmod{q+2}. \end{cases}$$

Since $A^\top = A$ (Lemma 4.6.3(i)), we have $U_{ji} = U_{ij}^\top$ for all i and j and therefore, U is a symmetric matrix with zeros on the diagonal and ± 1 off the diagonal. We claim that U is the core of a conference matrix.

Lemma 4.6.3(iv) implies that

$$\sum_{j=1}^{q+2} U_{ij} J = \sum_{j=1}^{q+2} w_{ij} J.$$

Since W is the core of a conference matrix, we obtain that $UJ = O$.

For $i, h = 1, 2, \ldots, q+2$, let

$$P_{ih} = \sum_{j=1}^{q+2} U_{ij} U_{hj}^\top.$$

Then

$$P_{ii} = AA^\top + CC^\top + C^\top C + \sum_{a \in S} \left(B_a B_a^\top + B_a^\top B_a \right),$$

and Lemma 4.6.3(i,x) implies that $P_{ii} = q^2(q+2)I - J$.

Suppose $i \neq h$. Lemma 4.6.3(v,vi,viii) implies that, for $j \neq i$ and $j \neq h$, $U_{ij} U_{hj}^\top = (w_{ij} w_{hj}) J$. Since W is the core of a conference matrix, we have

$\sum_{j=1}^{q+2} w_{ij} w_{hj} = -1$. Therefore,

$$P_{ih} = U_{ih}A^\top + AU_{hi}^\top + \sum_{j=1}^{q+2} (w_{ij}w_{hj})J = U_{ih}A^\top + AU_{hi}^\top - J.$$

Since $w_{ih} = w_{hi}$, Lemma 4.6.3(ii,iii) implies that $P_{ih} = -J$.

Therefore, U is the core of a symmetric conference matrix of order $q^2(q+2)+1$. □

Theorem 4.6.4 and 4.3.24 and Proposition 4.3.5 imply the following result.

Corollary 4.6.5. *If $q \equiv 3 \pmod 4$ is a prime power and $q+2$ is a prime power, then there exists a symmetric conference matrix of order $q^2(q+2)+1$ and a symmetric Hadamard matrix of order $2q^2(q+2)+2$.*

Remark 4.6.6. For $q = 3$, we obtain a Hadamard matrix of order 92. This is the smallest order of a Hadamard matrix that is not obtained directly from Theorem 4.3.24.

We will now apply the matrices considered in Lemma 4.6.3 to construct a family of regular Hadamard matrices. It is convenient to introduce the following terminology.

Definition 4.6.7. A set $\mathcal{R}$ of s square (± 1)-matrices of the same order m^2 is called a *regular s-set of matrices* if it satisfies the following conditions:

(i) $XJ = mJ$ for all $X \in \mathcal{R}$;
(ii) $XY = J$ for all $X, Y \in \mathcal{R}$;
(iii) $XY^\top = X^\top Y = J$ for all distinct $X, Y \in \mathcal{R}$;
(iv) $\sum_{X \in \mathcal{R}} (XX^\top + X^\top X) = 2sm^2 I$.

If $q \equiv 3 \pmod 4$ is a prime power and S is the set of all nonzero squares of $GF(q)$, then the matrices B_a, $a \in S$, and C of Lemma 4.6.3 form a regular set. Thus, we have the following result.

Theorem 4.6.8. *For any prime power $q \equiv 3 \pmod 4$, there exists a regular $\left(\frac{q+1}{2}\right)$-set of matrices of order q^2.*

We will now construct a family of regular Hadamard matrices based on a regular s-set of (± 1)-matrices.

Let m be a positive integer such that $q = 8m^2 - 1$ is a prime power. The two ingredients of our construction are a regular $(4m^2)$-set

$$\mathcal{R} = \{A_1, A_2, \ldots, A_{4m^2}\}$$

of (± 1)-matrices of order q^2, given by Theorem 4.6.8, and a regular Hadamard matrix H of order $n = 4m$ that generates a sequence $\mathcal{K}(H) = (C_1, C_2, \ldots, C_n)$ of (± 1)-matrices of order n (Definition 4.4.14). For $k = 1, 2, \ldots, n$, let $C_k = [c_{ij}^{(k)}]$. For the remainder of this section, we assume that the set $\mathcal{R}$ and the matrix H are fixed. We will also fix the following permutation ρ of $\mathcal{R}$:

$$\rho = (A_1 A_2 \ldots A_n)(A_{n+1} A_{n+2} \ldots A_{2n}) \cdots \big(A_{(m-1)n+1} A_{(m-1)n+2} \ldots A_{4m^2}\big).$$

Lemma 4.6.9. *For $k = 1, 2, \ldots, n = 4m$, let $L_k = [L_{ij}^{(k)}]$ be block matrices of order $q^2 n$ with the following blocks of order q^2:*

$$L_{ij}^{(k)} = \begin{cases} c_{ij}^{(k)}\big(\rho^{j-i} A_{(k-1)n+1}\big) & \text{if } 1 \le k \le m \text{ or } 2m+1 \le k \le 3m, \\ c_{ij}^{(k)}\big(\rho^{j-i} A_{(k-1)n+1}\big)^{\top} & \text{if } m+1 \le k \le 2m \text{ or } 3m+1 \le k \le 4m. \end{cases}$$

Let $s, t \in \{1, 2, \ldots, n\}$. Then:

(i) *if $s \not\equiv t \pmod{2m}$, then $L_s L_t^{\top} = O$;*
(ii) *if $s \equiv t \pmod{2m}$, then $L_s L_t^{\top} = L_t L_s^{\top}$;*
(iii) $\sum_{k=1}^{n} L_k L_k^{\top} = 16qm^3 I.$

Proof. (i) and (ii). Definition 4.6.7 implies that

$$\big(\rho^{j-i} A_{(k-1)n+1}\big)\big(\rho^{j-h} A_{(l-1)n+1}\big) = \big(\rho^{j-i} A_{(k-1)n+1}\big)^{\top}\big(\rho^{j-h} A_{(l-1)n+1}\big)^{\top} = J \tag{4.27}$$

for all $k, l, i, h, j \in \{1, 2, \ldots, n\}$ and

$$\big(\rho^{j-i} A_{(k-1)n+1}\big)\big(\rho^{j-h} A_{(l-1)n+1}\big)^{\top} = \big(\rho^{j-i} A_{(k-1)n+1}\big)^{\top}\big(\rho^{j-h} A_{(l-1)n+1}\big) = J \tag{4.28}$$

for all $k, l, i, h, j \in \{1, 2, \ldots, n\}$, except the case $i \equiv h \pmod{n}$ and $k \equiv l \pmod{m}$. Therefore, if $k \not\equiv l \pmod{m}$, we apply $C_s C_t^{\top} = O$ to obtain (i).

If $k \not\equiv l \pmod{2m}$ but $k \equiv l \pmod{m}$, then $k - l = \pm m$ or $\pm 3m$. Therefore, in this case, we do not have to deal with (4.28) at all.

Suppose $t = s + 2m$. Then, for $i \ne h$, the off-diagonal blocks of $L_s L_t^{\top}$ and $L_t L_s^{\top}$ are equal to O and, for $i = 1, 2, \ldots, n$, the ith diagonal block of both matrices is either

$$\sum_{j=1}^{n} c_{ij}^{(s)} c_{ij}^{(t)}\big(\rho^{j-i} A_{(s-1)n+1}\big)\big(\rho^{j-i} A_{(s-1)n+1}\big)^{\top}$$

or

$$\sum_{j=1}^{n} c_{ij}^{(s)} c_{ij}^{(t)}\big(\rho^{j-i} A_{(s-1)n+1}\big)^{\top}\big(\rho^{j-i} A_{(s-1)n+1}\big).$$

This proves (ii).

(iii) Let $i, h \in \{1, 2, \ldots, n\}$. If $1 \le k \le m$ or $2m + 1 \le k \le 3m$, then let

$$B_{ih}^{(k)} = \sum_{j=1}^{n} c_{ij}^{(k)} c_{hj}^{(k)} \big(\rho^{j-i} A_{(k-1)n+1}\big)\big(\rho^{j-h} A_{(k-1)n+1}\big)^{\top};$$

if $m + 1 \le k \le 2m$ or $3m + 1 \le k \le 4m$, then let

$$B_{ih}^{(k)} = \sum_{j=1}^{n} c_{ij}^{(k)} c_{hj}^{(k)} \big(\rho^{j-i} A_{(k-1)n+1}\big)^{\top}\big(\rho^{j-h} A_{(k-1)n+1}\big).$$

If $i \ne h$, then Definition 4.6.7 and Lemma 4.4.15 imply that

$$\sum_{k=1}^{n} B_{ih}^{(k)} = \sum_{k=1}^{n} c_{ij}^{(k)} c_{hj}^{(k)} J = \sum_{k=1}^{n} c_{ih}^{(k)} J = O.$$

We further have

$$B_{ii}^{(k)} = \begin{cases} \sum_{X\in\mathcal{R}} XX^{\top} & \text{if } 1 \le k \le m \text{ or } 2m+1 \le k \le 3m, \\ \sum_{X\in\mathcal{R}} X^{\top}X & \text{if } m+1 \le k \le 2m \text{ or } 3m+1 \le k \le 4m. \end{cases}$$

Therefore, Definition 4.6.7 implies that

$$\sum_{k=1}^{n} B_{ii}^{(k)} = 2m \sum_{X\in\mathcal{R}} (XX^{\top} + X^{\top}X) = 16qm^3 I.$$

This proves (iii). □

Lemma 4.6.10. *Let σ be the following permutation of the set $\{L_1, L_2, \ldots, L_n\}$ of matrices of order $4q^2m$ of Lemma 4.6.9:*

$$\sigma = (L_1 L_2 \ldots L_{2m})(L_{2m+1} L_{2m+2} \ldots L_{4m}).$$

Consider block matrices $U = [U_{ij}]$ and $V = [V_{ij}]$ of order $8q^2m^2$ with blocks $U_{ij} = [\sigma^{j-i} L_1]$ and $V_{ij} = [\sigma^{j-i} L_{2m+1}]$ of order $4q^2m$. Then:

(i) *$UJ = O$ and $VJ = aJ$, for some integer a;*
(ii) *$UU^{\top} + VV^{\top}$ is a diagonal matrix;*
(iii) *$UV^{\top} = VU^{\top}$.*

Proof. (i) Since $AJ = qJ$ for all $A \in \mathcal{R}$, $C_kJ = O$ for $k = 1, 2, \ldots, n-1$, and $C_nJ = nJ$, the proof of (i) is straightforward.

(ii) The blocks of $UU^{\top} + VV^{\top}$ are of the form

$$\sum_{j=1}^{2m} (\sigma^{j-i} L_1)(\sigma^{j-h} L_1)^{\top} + \sum_{j=2m+1}^{4m} (\sigma^{j-i} L_{2m+1})(\sigma^{j-h} L_{2m+1})^{\top}.$$

For the off-diagonal blocks, $i \neq h$, and then Lemma 4.6.9(i) implies that these blocks are zero matrices. The ith diagonal block of $UU^\top + VV^\top$ is equal to

$$\sum_{j=1}^{4m} L_j L_j^\top,$$

and it is a diagonal matrix by 4.6.9(iii).

(iii) For $i, h \in \{1, 2, \ldots, 2m\}$, the $(i, h)^{th}$ block of $UV^\top$ is equal to

$$\sum_{j=1}^{2m} (\sigma^{j-i} L_1)(\sigma^{j-h} L_{2m+1})^\top,$$

while the $(i, h)^{th}$ block of $VU^\top$ is equal to

$$\sum_{j=1}^{2m} (\sigma^{j-h} L_{2m+1})(\sigma^{j-i} L_1)^\top,$$

so Lemma 4.6.9(i,ii) implies (iii). □

Theorem 4.6.11. *Let m be an integer such that $q = 8m^2 - 1$ is a prime power. If there exists a Hadamard matrix of order $4m$, then there exists a regular Hadamard matrix of order $16q^2m^2$.*

Proof. Let U and V be the matrices of order $8q^2m^2$ given by Lemma 4.6.10 and let

$$K = \begin{bmatrix} V & U \\ -U & V \end{bmatrix}.$$

Then K is a (± 1)-matrix of order $16q^2m^2$. Lemma 4.6.10(i) implies that K has a constant row sum and Lemma 4.6.10(ii,iii) implies that $KK^\top$ is a diagonal matrix. Therefore, K is a regular Hadamard matrix. □

4.7. Binary equidistant codes

In Chapter 3, we defined the notion of a q-ary code.

Definition 4.7.1. A q-ary (v, m, d)-code is called *equidistant* if the distance between any two codewords is equal to d.

In this section we explore some connections between binary equidistant codes and Hadamard 2-designs and other symmetric designs.

If C is a binary equidistant (v, m, d)-code over the alphabet $\mathcal{A} = \{0, 1\}$, then each codeword $X = (x_1, x_2, \ldots, x_v)$ can be regarded as the subset

$\{i \in \{1, 2, \ldots, v\} : x_i = 1\}$ of the set $\{1, 2, \ldots, v\}$. Then the distance between codewords X and Y is equal to the cardinality of the set $X \triangle Y$. Thus, a binary equidistant (v, m, d)-code can be viewed as an equidistant family of subsets of the set $\{1, 2, \ldots, v\}$. Theorem 1.4.3 says that, for any binary equidistant (v, m, d)-code, $m \le v + 1$. Theorem 1.4.6 and Example 1.4.5 imply that a binary equidistant $(v, v + 1, d)$-code exists if and only if $d = (v + 1)/2$ and there exists a Hadamard matrix of order $v + 1$.

Throughout this section we will denote the set $\{1, 2, \ldots, v\}$ by $[v]$.

Definition 4.7.2. If $\mathcal{B}$ is a family of subsets of $[v]$ and $X \subseteq [v]$, then families $\mathcal{B}$ and $\mathcal{B} \triangle X = \{B \triangle X : B \in \mathcal{B}\}$ are called equivalent.

It is indeed an equivalence relation on the set of subsets of $[v]$ due to the obvious equalities: $\mathcal{B} \triangle \varnothing = \mathcal{B}$, $(\mathcal{B} \triangle X) \triangle X = \mathcal{B}$, and $(\mathcal{B} \triangle X) \triangle Y = \mathcal{B} \triangle (X \triangle Y)$. Since $(A \triangle X) \triangle (B \triangle X) = A \triangle B$, for any subsets A, B, and X, a family that is equivalent to an equidistant family is itself equidistant with the same cardinality of symmetric difference.

Example 4.7.3. Let $H = [h_{ij}]$ be a Hadamard matrix of order $v + 1$ with all entries in the last column equal to 1. Let $\mathcal{B}$ be the corresponding Hadamard family of subsets of $[v]$, i.e., $\mathcal{B} = \{B_1, B_2, \ldots, B_v, B_{v+1}\}$ where $B_i = \{j \in [v] : h_{ij} = 1\}$ (cf. Example 1.4.5). Let $X = [v] \setminus B_{v+1}$. Then the family $\mathcal{B} \triangle X$ consists of $[v]$ and all blocks of a Hadamard 2-design on the point set $[v]$.

The distance between any two members of an equidistant family of more than two sets is even.

Lemma 4.7.4. *Let $\mathcal{B}$ be an equidistant family of subsets of set $[v]$. If $|\mathcal{B}| \ge 3$, then $|A \triangle B|$ is even for any distinct sets $A, B \in \mathcal{B}$.*

Proof. Let $|A \triangle B| = d$ for distinct sets $A, B \in \mathcal{B}$. Fix X in $\mathcal{B}$ and consider an equivalent family $\mathcal{B}' = \mathcal{B} \triangle X$. Since this family contains the empty set, the other sets in the family have cardinality d. If A' and B' are two nonempty sets in $\mathcal{B}'$, then $d = |A' \triangle B'| = 2|A' \setminus B'|$, so d is even. □

We will now turn our attention to binary equidistant (v, v, d)-codes, i.e., equidistant families of v subsets of a v-set. One example of such a family is the set of blocks of a symmetric (v, k, λ)-design. In this case, $d = 2(k - \lambda)$. If $v \equiv 0 \pmod 4$, another example can be constructed using a Hadamard 3-design on v points. The block set of such a design consists of $v - 1$ pairs of complementary blocks. Selecting one block from each pair and adjoining the set $[v]$, we obtain an equidistant family of v subsets of $[v]$ with $d = v/2$. The following theorem shows that these are the only possibilities up to the equivalence introduced by Definition 4.7.2.

Theorem 4.7.5. *Let $\mathcal{B}$ be an equidistant family of v subsets of $[v]$, $v \geq 3$, and let $|A \triangle B| = d$ for any distinct $A, B \in \mathcal{B}$. If $v = 2d$, then $\mathcal{B}$ is equivalent to the family $\{[v]\} \cup \mathcal{A}$ where $\mathcal{A}$ is a family of $v - 1$ blocks of a Hadamard 3-design on the point set $[v]$ with no two blocks being complementary. If $v \neq 2d$, then $\mathcal{B}$ is equivalent to the block set of a symmetric design.*

Proof. Let P be the vector space of linear polynomials in v variables $x_1, x_2, \ldots, x_v$ with rational coefficients. As in the proof of Theorem 1.4.3, we associate with each $A \in \mathcal{B}$ the polynomial

$$f_A(x_1, x_2, \ldots, x_v) = \sum_{i \notin A} x_i - \sum_{i \in A} x_i + |A| - d. \tag{4.29}$$

Then, $f_A(X) = |A \triangle X| - d$ for any subset X of $[v]$, and therefore, for any $A, B \in \mathcal{B}$,

$$f_A(B) = \begin{cases} 0 & \text{if } B \neq A, \\ -d & \text{if } B = A. \end{cases} \tag{4.30}$$

It was shown in the proof of Theorem 1.4.3 that the set $\{f_A : A \in \mathcal{B}\}$ is linearly independent. We claim that the constant polynomial 1 is not in the span of this set. Indeed, if $1 = \sum_{A \in \mathcal{B}} \alpha_A f_A$, then applying both sides of this equality to $B \in \mathcal{B}$ yields $-\alpha_B d = 1$. Therefore, $\alpha_A = -1/d$ for all $A \in \mathcal{B}$, and we have $-d = \sum_{A \in \mathcal{B}} f_A$. We now apply both sides of this equality to the empty set and to the set $[v]$ to obtain $\sum_{A \in \mathcal{B}} (|A| - d) = -d$ and $\sum_{A \in \mathcal{B}} (v - |A| - d) = -d$. These equalities imply $v^2 = 2d(v - 1)$. Therefore, $v - 1$ divides v^2, which is not the case for $v \geq 3$.

Since $\dim P = v + 1$, the set $\{f_A : A \in \mathcal{B}\} \cup \{1\}$ is a basis of P. Expand monomials x_i, $i = 1, 2, \ldots, v$, in this basis: $x_i = \sum_{A \in \mathcal{B}} \alpha_A^{(i)} f_A + \beta_i$. Since $x_i(B)$ is equal to 1 or 0 depending on whether $i \in B$ or $i \notin B$, we obtain that

$$\alpha_B^{(i)} = \begin{cases} \frac{\beta_i - 1}{d} & \text{if } i \in B, \\ \frac{\beta_i}{d} & \text{if } i \notin B. \end{cases}$$

Therefore,

$$(\beta_i - 1) \sum_{i \in A} f_A + \beta_i \sum_{i \notin A} f_A + d\beta_i = dx_i, \tag{4.31}$$

for $i = 1, 2, \ldots, v$.

We apply both sides of (4.31) to the sets $\varnothing$, $\{i\}$, and $[v]$ to obtain

$$(\beta_i - 1)\sum_{i\in A}(|A| - d) + \beta_i \sum_{i\notin A}(|A| - d) + d\beta_i = 0,$$

$$(\beta_i - 1)\sum_{i\in A}(|A| - d - 1) + \beta_i \sum_{i\notin A}(|A| - d + 1) + d\beta_i = d,$$

$$(\beta_i - 1)\sum_{i\in A}(v - |A| - d) + \beta_i \sum_{i\notin A}(v - |A| - d) + d\beta_i = d.$$

We denote by r_i the number of sets $A \in \mathcal{B}$ that contain i. We denote also $s = \sum_{A\in\mathcal{B}} |A|$ and $s_i = \sum_{A\ni i} |A|$ for $i = 1, 2, \ldots, v$, and rewrite the above equalities as follows:

$$(s - dv + d)\beta_i + dr_i - a_i = 0, \tag{4.32}$$

$$(s - dv + d + v)\beta_i + (d + 1)r_i - a_i - 2\beta_i r_i = d, \tag{4.33}$$

$$(v^2 - dv + d - s)\beta_i - (v - d)r_i + a_i = d. \tag{4.34}$$

Adding (4.32) to (4.34) and (4.33) to (4.34) eliminates s and a_i:

$$(v^2 - 2dv + 2d)\beta_i = d - (2d - v)r_i, \tag{4.35}$$

$$(v^2 - 2dv + 2d + v - 2r_i)\beta_i = 2d - (2d - v + 1)r_i. \tag{4.36}$$

If $v^2 - 2dv + 2d = 0$, then $v = d + \sqrt{(d-1)^2 - 1}$, so $(d-1)^2 - 1$ is a square, which implies $d = 2$ and $v = 2$. Therefore, $v^2 - 2dv + 2d \neq 0$, and we divide (4.36) by (4.35) to eliminate β_i:

$$\frac{v^2 - 2dv + 2d + v - 2r_i}{v^2 - 2dv + 2d} = \frac{2d - (2d - v + 1)r_i}{d - (2d - v)r_i}.$$

This equality simplifies to

$$(2r_i - d)v^2 - (2r_i^2 + 4dr_i - 2d^2 - d)v + 2d(2r_i^2 - d) = 0. \tag{4.37}$$

We now split the proof into three cases.

Case 1: $v = 2d$.

In this case, for any distinct subsets A and B, $|A \triangle B'| = |A \triangle B| = d$ where B' is the complement of B. Therefore, if we replace some subsets from the family $\mathcal{B}$ by their complements, we obtain another equidistant family with the same parameters v and d. Let $\mathcal{B}'$ be the family obtained from $\mathcal{B}$ by replacing every element of $\mathcal{B}$ that does not contain v by its complement.

Let $\mathcal{B}_1 = \{B - \{v\} \colon B \in \mathcal{B}'\}$. Then $\mathcal{B}_1$ is an equidistant family of v subsets of the set $[v-1]$. Theorem 1.4.6 implies that $\mathcal{B}_1$ is a Hadamard family. Example 4.7.3 shows that there is a subset X of $[v-1]$ such that $\mathcal{B}_1 \triangle X = \{[v-1]\} \cup \mathcal{A}_1$

where $\mathcal{A}_1$ is the block set of a Hadamard 2-design on the point set $[v-1]$. Therefore, $\mathcal{B}'\triangle X = \{[v]\} \cup \mathcal{A}$ where $\mathcal{A} = \{A \cup [v] \colon A \in \mathcal{A}_1\}$. Then $\mathcal{A}$ is the set of all blocks of a Hadamard 3-design on the point set $[v]$ which contain point v. No two of these blocks are complementary. Since the complement of any block of a Hadamard 3-design is also a block of this design, the family $\mathcal{B}\triangle X$ is a set of blocks of a Hadamard 3-design, no two of which are complementary.

Case 2: $v \neq 2d$ and there is $i \in [v]$ such that $2r_i = d$.

Let $2r_v = d$. Then, for $i = v$, (4.36) simplifies to $vd(d/2-1) = 2d^2(d/2-1)$, which implies $d = 2$ and $r_v = 1$. Let A be the only block containing v. Consider the family $\mathcal{B}' = \mathcal{B}\triangle A'$ where A' is the complement of A. Then $[v] \in \mathcal{B}'$ and $[v]$ is the only element of $\mathcal{B}'$ containing v. Since $|B\triangle[v]| = 2$ for any $B \in \mathcal{B}'$, except $[v]$, the elements of $\mathcal{B}'$ other than $[v]$ are all $(v-2)$-subsets of the set $[v-1]$. Therefore, the family $\mathcal{B}'' = \mathcal{B}'\triangle\{v\}$, which is equivalent to $\mathcal{B}$, is the set of all $(v-1)$-subsets of $[v]$. This is the block set of a symmetric $(v, v-1, v-2)$-design.

Case 3: $v \neq 2d$ and there is no $i \in [v]$ such that $2r_i = d$.

In this case, (4.37) can be regarded as a quadratic equation for v. It yields the solutions $v = 2d$ and $v = (2r_i^2 - d)/(2r_i - d)$. Since $v \neq 2d$, we obtain the following quadratic equation for r_i: $2r_i^2 - 2vr_i + (v-1)d = 0$.

Let r and r^* be the roots of this equation and let $E = \{i \in [v] \colon r_i = r^*\}$. Consider family $\mathcal{B}' = \mathcal{B}\triangle E$. Since $r + r^* = v$, each of the points $1, 2, \ldots, v$ belongs to exactly r subsets from $\mathcal{B}'$. Therefore, (4.34) applied to the family $\mathcal{B}'$ implies that $\beta_i = \beta$ does not depend on i, and then the expansion (4.31) reads

$$(\beta - 1)\sum_{i\in A} f_A + \beta \sum_{i\notin A} f_A + d\beta = dx_i. \tag{4.38}$$

In order to show that $\mathcal{B}'$ is the block set of a symmetric design, we evaluate the coefficient of x_i and the coefficient of x_j for $j \neq i$ in the left-hand side of (4.38). By (4.29), f_A contributes 1 into the coefficient of x_j if $j \notin A$, and -1 if $j \in A$. Comparing the coefficients of x_i in (4.38) yields $-(\beta-1)r + \beta(v-r) = d$, which implies

$$(2r - v)\beta = r - d. \tag{4.39}$$

Fix distinct $i, j \in [v]$ and denote by λ the number of elements of $\mathcal{B}'$ that contain both i and j. Comparing the coefficients of x_j in (4.38) yields $2\lambda = r + (2r - v)\beta$, and then (4.39) implies that $\lambda = (2r-d)/2$ does not depend on i and j.

Thus, in the incidence structure $([v], \mathcal{B}')$ every point belongs to r blocks and every pair of distinct points belongs to λ blocks. Since the number of blocks is equal to the number of points, $\mathcal{B}'$ is the block set of a symmetric (v, r, λ)-design. □

We will now address the following question: what is the maximum value of d in an equidistant (v, v, d)-code and what are the codes that attain this maximum value?

Theorem 4.7.6. *Let C be an equidistant (v, v, d)-code with $v \geq 4$.*

(i) *If $v \equiv 0 \pmod 4$, then $d \leq v/2$. An equidistant $(v, v, v/2)$-code exists if and only if there exists a Hadamard matrix of order v.*
(ii) *If $v \equiv 1 \pmod 4$, then $d \leq (v-1)/2$. An equidistant $(v, v, (v-1)/2)$-code exists if and only if $v = 2u^2 + 2u + 1$ for some positive integer u and there exists a symmetric $(2u^2 + 2u + 1, u^2, u(u-1)/2)$-design.*
(iii) *If $v \equiv 2 \pmod 4$, then $d \leq (v-2)/2$. An equidistant $(v, v, (v-2)/2)$-code exists if and only if $v = 12u^2 + 8u + 2$ where u is such an integer that $3u^2 + 2u$ is a square and there exists a symmetric $(12u^2 + 8u + 2, 6u^2 + u, 3u^2 - u)$-design.*

Proof. With every codeword $(x_1, x_2, \ldots, x_v) \in C$, we associate a set $B = \{i \in [v] : x_i = 1\}$. Let $\mathcal{B}$ be the family of all such subsets of $[v]$.

Suppose first that $v \neq 2d$. Then, by Theorem 4.7.5, $\mathcal{B}$ is the block set of a symmetric (v, k, λ)-design and $d = 2(k - \lambda)$. We transform (2.9) into

$$2k^2 - 2vk + d(v-1) = 0. \tag{4.40}$$

Proposition 2.4.12 implies that $2d \leq v + 1$. Since d is even, we obtain that $d \leq v/2$ if $v \equiv 0 \pmod 4$, $d \leq (v-1)/2$ if $v \equiv 1 \pmod 4$, and $d \leq (v-2)/2$ if $v \equiv 2 \pmod 4$.

If $v = 2d$, then $d = v/2$ and, by Lemma 4.7.4, $v \equiv 0 \pmod 4$. Thus we have obtained the upper bounds on d claimed in the statement of the theorem.

(i) Suppose that $v \equiv 0 \pmod 4$ and there exists an equidistant $(v, v, v/2)$-code. Then Theorem 4.7.5 implies that there exists a Hadamard matrix of order v. Conversely, if H is a Hadamard matrix of order v, then we replace all -1s in H by 0s to obtain a matrix whose rows form an equidistant $(v, v, v/2)$-code.

(ii) Suppose that $v \equiv 1 \pmod 4$ and there exists an equidistant $(v, v, (v-1)/2)$-code. Then (4.40) has an integer solution k, so $v^2 - 2d(v-1) = 2v - 1$ is a square. Let $2v - 1 = (2u+1)^2$, where u is a positive integer. Then $v = 2u^2 + 2u + 1$, and (4.40) implies $k = u^2$ or $k = (u+1)^2$. If $k = u^2$, then $\lambda = k - d/2 = u(u-1)/2$, so there exists a symmetric $(2u^2 + 2u + 1, u^2, u(u-1)/2)$-design. If $k = (u+1)^2$, we obtain the complement of such a design.

Conversely, if there exists a symmetric $(2u^2+2u+1, u^2, u(u-1)/2)$-design, the rows of its incidence matrix form an equidistant $(v, v, (v-1)/2)$-code with $v = 2u^2+2u+1$.

(iii) Suppose that $v \equiv 2 \pmod 4$ and there exists an equidistant $(v, v, (v-2)/2)$-code. Then (4.40) has an integer solution k, so $v^2 - 2d(v-1) = 3v - 2 = (6u+2)^2$ where u is a non-zero integer. Therefore, $v = 12u^2 + 8u + 2$, and (4.40) yields $k = 6u^2 + u$ or $k = 6u^2 + 7u + 2$. Suppose that $k = 6u^2 + u$ (the other solution leads to the complementary design). Then $\lambda = 3u^2 - u$, so there exists a symmetric $(12u^2 + 8u + 2, 6u^2 + u, 3u^2 - u)$-design of order $k - \lambda = 3u^2 + 2u$. Proposition 2.4.10 now implies that $3u^2 + 2u$ is a square. □

Remark 4.7.7. The case $v \equiv 3 \pmod 4$ is covered by Theorem 1.4.6 and Example 1.4.5.

Remark 4.7.8. Symmetric designs of case (ii) of Theorem 4.7.6 are known to exist whenever u is an odd prime power (see Corollary 11.8.4), $u = 1$ (this is the symmetric (5, 1, 0)-design), $u = 2$ (the symmetric (7, 3, 1)-design), and $u = 4$ (Theorem 2.7.1). The only known symmetric designs satisfying case (iii) of Theorem 4.7.6 are the symmetric (6, 5, 4)-design and symmetric (66, 26, 10)-designs (Theorem 11.1.1).

Exercises

(1) Let **D** be a design of order n on $4n - 1$ points. Prove that **D** is a Hadamard 2-design.

(2) Prove that if two Hadamard 2-designs are isomorphic, then the corresponding Hadamard matrices are equivalent.

(3) Prove that, for $n \le 12$, any two Hadamard matrices of order n are equivalent.

(4) Let P be a Paley matrix of order $q \equiv 3 \pmod 4$ and let $R = [r_{ij}]$ be the matrix of order q defined by

$$r_{ij} = \begin{cases} 1 & \text{if } i + j = q + 1, \\ 0 & \text{if } i + j \ne q + 1. \end{cases}$$

Prove that PR is a symmetric matrix.

(5) Prove that the sum of all entries of a Hadamard matrix of order n does not exceed $n\sqrt{n}$.

(6) Let A be a nonsingular matrix of order n over the real numbers and let vectors $\mathbf{a}_1, \mathbf{a}_2, \dots, \mathbf{a}_n$ be the columns of A.
   (a) Prove that there exist pairwise orthogonal vectors $\mathbf{e}_1, \mathbf{e}_2, \dots, \mathbf{e}_n$ such that $\mathbf{e}_1 = \mathbf{a}_1$ and, for $2 \le i \le n$, $\mathbf{e}_i - \mathbf{a}_i$ is a linear combination of $\mathbf{a}_1, \mathbf{a}_2, \dots, \mathbf{a}_{i-1}$.
   (b) Let E be the matrix of order n with consecutive columns $\mathbf{e}_1, \mathbf{e}_2, \dots, \mathbf{e}_n$. Prove that $\det(A) = \det(E)$.

(c) For any vector $\mathbf{x} = [x_1\ x_2 \ldots x_n]^\top$ over the real numbers,

$$\|\mathbf{x}\| = \sqrt{x_1^2 + x_2^2 + \cdots + x_n^2}.$$

Prove that

$$|\det(A)| \leq \|\mathbf{a}_1\| \cdot \|\mathbf{a}_2\| \cdot \cdots \cdot \|\mathbf{a}_n\|$$

with equality if and only if $\mathbf{a}_i = \mathbf{e}_i$ for $i = 1, 2, \ldots, n$.

(d) Prove that if A is a (± 1)-matrix, then $|\det(A)| \leq n^{n/2}$ with equality if and only if A is a Hadamard matrix.

(7) Let $H = [H_1\ H_2 \ldots H_t]$ be an $(s \times st)$ matrix represented as a block matrix where each block H_i is a (± 1)-matrix of order s. Suppose $HH^\top = stI$. Prove that if H has t all-one columns, then all the remaining columns of H have zero sum.

(8) A $(0, \pm 1)$-matrix N is said to be an incidence matrix of *twin* (v, k, λ)*-designs* if there exist incidence matrices N^+ and N^- of symmetric (v, k, λ)-designs such that $N^+ + N^-$ is a $(0, 1)$-matrix and $N^+ - N^- = N$. Let $H = [H_{ij}]$ be a regular Hadamard matrix of Bush type of order $4n^2$ with blocks H_{ij} of order $2n$. Prove that the matrix N obtained by replacing every all-one block with the zero matrix of order n is an incidence matrix of Siamese twin $(4n^2, 2n^2 - n, n^2 - n)$-designs.

(9) A $(0, \pm 1)$-matrix N matrix is said to be an incidence matrix of *Siamese twin* (v, k, λ)*-designs* if there exist incidence matrices N^+ and N^- of symmetric (v, k, λ)-designs and a $(0, 1)$-matrix M such that $N^+ + N^- - M$ is a $(0, 1)$-matrix and $N^+ - N^- + M = N$. Let $H = [H_{ij}]$ be a Bush-type regular Hadamard matrix of order $4n^2$ with blocks H_{ij} of order $2n$. Prove that the matrix N obtained by replacing every all-one block with the zero matrix of order n is an incidence matrix of Siamese twin $(4n^2, 2n^2 + n, n^2 + n)$-designs.

(10) Prove that if there exists a conference matrix of order m and a Hadamard matrix of order n, then there exists a Hadamard matrix of order mn.

(11) Matrices M and N of the same order are called *amicable Hadamard matrices* if $M - I$ is a skew-symmetric conference matrix, N is a symmetric Hadamard matrix, and $MN^\top = NM^\top$.

(a) Find amicable Hadamard matrices of order 2.

(b) Let M and N be amicable Hadamard matrices of order m and let M' and N' be amicable Hadamard matrices of order n. Prove that $M'' = I_m \otimes M' + (M - I_m) \otimes N'$ and $N'' = N \otimes N'$ are amicable Hadamard matrices of order mn.

(c) Construct amicable Hadamard matrices of order 4.

(d) Construct a conference matrix of order 16.

(e) Let $q \equiv 3 \pmod 4$ be a prime power and let P be a Paley matrix of order q. Let $U = [u_{ij}]$ be the permutation matrix of order q defined by: $u_{ij} = 1$ if and only if $i + j \equiv 2 \pmod q$. Prove that matrices $M = \begin{bmatrix} 1 & \mathbf{j}^\top \\ -\mathbf{j} & P+I \end{bmatrix}$ and $N = \begin{bmatrix} 1 & \mathbf{0}^\top \\ \mathbf{0} & -U \end{bmatrix}$ are amicable Hadamard matrices of order $q + 1$.

(f) Construct amicable Hadamard matrices of order 12.

(12) Let $A = [a_{ij}]$ be a $(0, \pm 1)$-matrix of order n with pairwise orthogonal nonzero rows. For $i = 1, 2, \ldots, n$, let e_i denote the number of nonzero entries in the ith row of A. For $j = 1, 2, \ldots, n$, let $\alpha_j = \{i : a_{ij} \neq 0\}$.

(a) Prove that $e_1 e_2 \cdots e_n$ is a square.

(b) Prove that, for $j = 1, 2, \ldots, n$, $\sum_{i \in \alpha_j} \frac{1}{e_i} = 1$.

(c) Prove that, for distinct $j, k \in \{1, 2, \ldots, n\}$, $\sum_{i \in \alpha_j \cap \alpha_k} \frac{a_{ij} a_{ik}}{e_i} = 0$.

(d) Prove that if A has a column without zero entries, then A is a Hadamard matrix.

(e) Let H be a Hadamard matrix of order n with all entries in the last row equal to 1. Let $\mathcal{K}(H) = (C_1.C_2, \ldots, C_n)$ be the sequence of matrices introduced in Definition 4.4.14. Prove that, for $i = 1, 2, \ldots, n-1$, there exists a permutation matrix P_i of order n such that $C_i P_i = -C_1$.

(f) Let H and K be Hadamard matrices of order m and n, respectively.

(a) Prove that if $m = n$, then the Hadamard matrix $H \otimes K$ is equivalent to a regular Hadamard matrix of Bush type.

(b) Prove that if $m < n$, then the matrix $H \otimes K$ is equivalent to a block matrix $\left[\begin{smallmatrix} A \\ B \end{smallmatrix}\right]$ such that $\frac{1}{2}(J - A)$ is an incidence matrix of an $(s^2, st, t(s-1)/2, s(s-1)/2, t(s-2)/4)$-design, and every row sum of B equals 0.

Notes

In Hadamard (1893), the celebrated French mathematician Jacques Hadamard considered the following question: among real matrices of order n with entries from the interval $[-1, 1]$, find the matrices with the maximum absolute value of the determinant. It is not difficult to see that this maximum does not exceed $n^{\frac{n}{2}}$ and that this bound is attained by matrices of order n with entries ± 1 and pairwise orthogonal rows. (See Exercise 6.)

Hadamard matrices of order 2^n (not under this name) were known to Sylvester in at least 1857. Sylvester can also be credited with introducing the Kronecker product construction of Hadamard matrices. The modification of the Kronecker product construction described in Theorem 4.2.5 is due to Agaian and Sarukhanian (see Agaian (1985)).

Hadamard himself constructed (± 1)-matrices of order 12 and 20 with pairwise orthogonal rows. He also conjectured that there exist such matrices of any order $n \equiv 0 \pmod 4$. This conjecture of Hadamard is still open. The case $n = 428$ that had been open for many years has been settled by Kharaghani and Tayfeh-Rezaie (2005). Now the first unresolved case is $n = 668$. For history, general results, and further references on Hadamard matrices see Wallis, Street and Wallis (1972), Hedayat and Wallis (1978), Seberry and Yamada (1992), Craigen and Wallis (1993), and Craigen (1996a).

Paley proposed his constructions of Hadamard matrices in 1933 (Paley (1933)). In Todd (1933), a connection between Hadamard matrices and symmetric Hadamard designs was first established.

Conference matrices were introduced in Belevitch (1950) in connection with communication problems. Theorem 4.3.13 is proven in this paper and Theorem 4.3.14 is proven in Turyn (1971). Theorem 4.3.6 is due to Delsarte, Goethals and Seidel (1971). Their proof uses the eigenvalue technique. A simple elementary proof of a more general Theorem 10.4.20 is given in Chapter 10. Conference matrices of Theorem 4.6.4 are constructed in Mathon (1978). Our exposition is a modification of that of Seberry and

Whiteman (1988). Theorem 4.3.8 is due to Raghavarao (1971). The smallest unresolved order of a conference matrix is 66. For further references on conferences matrices, see Craigen (1996b).

The connection between regular Hadamard matrices of order $4h^2$ and symmetric $(4h^2, 2h^2 - h, h^2 - h)$-designs is due to Menon (1962a). It is conjectured that, for any integer $h \neq 0$, there exists a regular Hadamard matrix of order $4h^2$. The smallest unresolved case is $h = 45$. Theorem 4.4.8 is due to McFarland (1974). For $p = 2$, it was proven by Mann (1965a, 1965b). The proof of Theorem 4.4.11 is from Ionin and Kharaghani (2005) and is modeled after Mukhopadhay (1978).

The term *regular s-set of matrices* was introduced in Seberry and Zhang (1993). Theorem 4.6.11 is due to Behbahani and Kharaghani (2004).

Bush (1971) showed that if there exists a projective plane of order $2h$, then there exists a symmetric Hadamard matrix of order $4h^2$ of Bush type with all the diagonal blocks equal to J. Wallis (1972) showed that the existence of such a matrix is implied by the existence of $n - 1$ MOLS of order $2n$. Theorem 4.4.16, which produces an infinite family of Hadamard matrices of Bush type (not necessarily symmetric), is due to Kharaghani (1985). In the recent papers Janko (2001) and Janko, Kharaghani and Tonchev (2000a, 2000b), Hadamard matrices of Bush type having orders 36, 324, and 100, respectively, are constructed. Theorem 4.4.18 is due to Muzychuk and Xiang (2005). Goldbach and Claasen (1998) study relations between symmetric Hadamard matrices of Bush type and 3-class association schemes.

The proof of Theorem 4.5.1 is from Wallis, Street and Wallis (1972). It is a modification of the original proof by Goethals and Seidel (1970).

The results of Exercises 5 and 12 are due to Best (1977) and Christian and Shader (2004), respectively.

We will return to regular Hadamard matrices in Chapter 7.

Equidistant codes have been studied from different points of view. Semakov and Zinoviev (1968) consider equidistant codes of fixed length having maximal distance. Some of their results are presented in Section 5.5. Deza (1973) obtains an upper bound for the size of nontrivial equidistant binary codes with fixed even distance and van Lint (1973) describes the codes achieving this bound. Stinson and van Rees (1984) and van Lint (1984b) proved case (ii) of Theorem 4.7.6. The proof of Theorem 4.7.6 given here follows that of Ionin and M. S. Shrikhande (1995b).

5

Resolvable designs

An affine plane of order n has $n^2 + n$ lines, any two of which are either parallel or intersecting. The relation of parallelism on the set of lines is an equivalence relation, and so it partitions the set of lines into $n + 1$ parallel classes of cardinality n. Each point lies on exactly one line from each parallel class. The block set of the complement of an affine plane of order n can be partitioned into $n + 1$ classes so that each point is contained in exactly $n - 1$ blocks from each class. Similar partitions exist in affine geometries of higher dimension. In this chapter we study a more general notion of resolution of an incidence structure, i.e., a partition of the block set of the structure into classes so that each point is contained in a constant number of blocks from each class.

5.1. Bose's Inequality

The incidence structures on which we define the notion of resolution are *pairwise balanced designs*.

Definition 5.1.1. Let λ be a positive integer. A *pairwise balanced design (PBD) of index* λ is an incidence structure $\mathbf{D} = (X, \mathcal{B}, I)$ such that $X \neq \varnothing$, every $x \in X$ is incident with more than λ blocks, and, for any distinct $x, y \in X$, there are precisely λ blocks that are incident with both x and y. If a PBD of index λ has constant replication number r, it is called an (r, λ)*-design*.

Remark 5.1.2. By this definition, $r(x) > \lambda$ for every $x \in X$. Therefore, for any two distinct points, there is a block that contains one of them but not the other.

Definition 5.1.3. Let $\mathbf{D} = (X, \mathcal{B}, I)$ be a PBD of index λ and let $\mathcal{C}$ be a subset of $\mathcal{B}$. The set $\mathcal{C}$ is called a *resolution class* in each of the following two cases:

(i) $\mathcal{C}$ contains no complete block and there exists a positive integer $\alpha(\mathcal{C})$ such that every point is contained in precisely $\alpha(\mathcal{C})$ blocks from $\mathcal{C}$ or (ii) $\mathcal{C}$ is a singleton consisting of a complete block. In the latter case, $\alpha(\mathcal{C}) = 1$. A partition of the block set $\mathcal{B}$ into resolution classes is called a *resolution* of $\mathbf{D}$.

Resolution classes that partition the point set are called *parallel classes*.

Definition 5.1.4. A resolution class $\mathcal{C}$ of a pairwise balanced design $\mathbf{D}$ is called a *parallel class* if $\alpha(\mathcal{C}) = 1$. A partition of the block set of $\mathbf{D}$ into parallel classes is called a *parallelism* of $\mathbf{D}$. A pairwise balanced design that admits a parallelism is called *resolvable*.

Any nontrivial 2-(v, k, λ) design is a PBD of index λ. The following incidence structure $(X, \mathcal{B})$ is a resolvable PBD of index 1: X is a set of cardinality 3 and $\mathcal{B}$ consists of all singletons and all 2-subsets of X; each parallel class consists of a singleton and its complement.

Remark 5.1.5. If $\mathcal{C}$ is a resolution class of a PBD of index λ, then, by replacing each block of this class with its complement, we obtain a PBD of index $\lambda + |\mathcal{C}| - 2\alpha(\mathcal{C})$.

Proposition 5.1.6. *A PBD of index λ admits a resolution if and only if it is an (r, λ)-design.*

Proof. If a PBD of index λ admits a resolution $\mathfrak{R}$, then it has constant replication number $r = \sum_{\mathcal{C}\in\mathfrak{R}} \alpha(\mathcal{C})$. Conversely, any (r, λ)-design admits a resolution consisting of one resolution class. □

Observe that the block set of the incidence structure dual to a PBD of index λ can be regarded as a family of distinct subsets of a finite set with any two subsets meeting in exactly λ points. Therefore, Nonuniform Fisher's Inequality implies that the number of points in a PBD does not exceed the number of blocks. If a PBD has resolution classes, a stronger result known as *Bose's Inequality* holds. In order to obtain Bose's Inequality, we introduce the following notions.

Let $\mathbf{D} = (X, \mathcal{B}, I)$ be a PBD of index λ that admits a set $\mathfrak{R}$ of pairwise disjoint resolution classes. With each block $B \in \mathcal{B}$, we associate a variable x_B. With the design $\mathbf{D}$, we associate a vector space $Pol(\mathbf{D})$ of linear polynomials with rational coefficients in the variables x_B. Clearly, $\dim Pol(\mathbf{D}) = |\mathcal{B}| + 1$. We will define the action of any linear polynomial $f = \sum_{B\in\mathcal{B}} a_B x_B + c$ on the point set X by

$$f(x) = \sum_{B\ni x} a_B + c \text{ for each } x \in X. \tag{5.1}$$

For each point $x \in X$, we define a polynomial $f_x \in Pol(\mathbf{D})$ by

$$f_x = \sum_{B \ni x} x_B - \lambda. \tag{5.2}$$

Then, for any points $x, y \in X$,

$$f_x(y) = \begin{cases} 0 & \text{if } x \neq y, \\ r(x) - \lambda & \text{if } x = y. \end{cases} \tag{5.3}$$

For each resolution class $\mathcal{C} \in \mathfrak{R}$, we define a polynomial $g_{\mathcal{C}} \in Pol(\mathbf{D})$ by

$$g_{\mathcal{C}} = \sum_{B \in \mathcal{C}} x_B - \alpha(\mathcal{C}). \tag{5.4}$$

Then $g_{\mathcal{C}}(x) = 0$ for all points $x \in X$.

Proposition 5.1.7. *The set* $\{f_x : x \in X\} \cup \{g_{\mathcal{C}} : \mathcal{C} \in \mathfrak{R}\}$ *of linear polynomials is linearly independent.*

Proof. Suppose

$$\sum_{x \in X} a_x f_x + \sum_{\mathcal{C} \in \mathfrak{R}} b_{\mathcal{C}} g_{\mathcal{C}} = 0$$

for some rational coefficients $a_x, b_{\mathcal{C}}$. Applying both sides of this equation to a point $y \in X$ and using (5.2) and (5.3) yields $a_y(r(y) - \lambda) = 0$. Remark 5.1.2 implies that $a_y = 0$ for all $y \in X$, and therefore

$$\sum_{\mathcal{C} \in \mathfrak{R}} b_{\mathcal{C}} g_{\mathcal{C}} = 0.$$

If $B \in \mathcal{C}$, then the coefficient of x_B in the left-hand side of this equality is $b_{\mathcal{C}}$, which implies that $b_{\mathcal{C}} = 0$ for all $\mathcal{C} \in \mathfrak{R}$. □

This proposition immediately implies the following generalization of Fisher's Inequality.

Theorem 5.1.8 (Bose's Inequality). *Let* $\mathbf{D} = (X, \mathcal{B}, I)$ *be a PBD of index* λ *that admits a set* $\mathfrak{R}$ *of pairwise disjoint resolution classes. Then* $|\mathcal{B}| \geq |X| + |\mathfrak{R}| - 1$.

Proof. Observe that $|\{f_x : x \in X\} \cup \{g_{\mathcal{C}} : \mathcal{C} \in \mathfrak{R}\}| = |X| + |\mathfrak{R}|$ and $\dim Pol(\mathbf{D}) = |\mathcal{B}| + 1$ and apply Proposition 5.1.7. □

Remark 5.1.9. Since any (r, λ)-design admits a (nonempty) resolution, Bose's Inequality implies Fisher's Inequality for (r, λ)-designs.

For the partition of an affine plane of order n into parallel classes, $|\mathcal{B}| = n^2 + n$, $|X| = n^2$, and $|\mathfrak{R}| = n + 1$, so we have equality in Bose's Inequality.

In order to obtain a necessary and sufficient condition for equality in Bose's Inequality, we introduce the notion of an *affine resolution*.

Definition 5.1.10. Let $\mathbf{D} = (X, \mathcal{B}, I)$ be a PBD of index λ. A resolution $\mathfrak{R}$ of $\mathbf{D}$ is called *affine* if the cardinality of each block and the cardinality of the intersection of any two distinct blocks in $\mathbf{D}$ depends only on their respective resolution classes in $\mathfrak{R}$.

Two classical examples of designs admitting affine resolutions are designs $AG_{d-1}(d, q)$ and Hadamard 3-designs. Parallel classes in $AG_{d-1}(d, q)$ are sets of q parallel $(d-1)$-flats. A parallel class in a Hadamard 3-design consists of a block and its complement. Remark 5.2.3 gives another example of a PBD admitting an affine resolution.

Remark 5.1.11. Let $\mathbf{D} = (X, \mathcal{B}, I)$ be a PBD admitting an affine resolution $\mathfrak{R}$ and let $\overline{\mathbf{D}} = (X, \overline{\mathcal{B}}, \overline{I})$ be a PBD obtained by adjoining one or more complete blocks to $\mathcal{B}$. If $\overline{\mathfrak{R}}$ is obtained by adjoining the corresponding singletons to $\mathfrak{R}$, then $\overline{\mathfrak{R}}$ is an affine resolution of $\overline{\mathbf{D}}$.

Theorem 5.1.12. *If $\mathfrak{R}$ is an affine resolution of a pairwise balanced design $\mathbf{D} = (X, \mathcal{B}, I)$, then $|\mathcal{B}| = |X| + |\mathfrak{R}| - 1$.*

Proof. Let $\mathbf{D} = (X, \mathcal{B}, I)$ be a PBD admitting an affine resolution $\mathfrak{R} = \{\mathcal{C}_1, \mathcal{C}_2, \dots, \mathcal{C}_t\}$. If $\mathbf{D}$ has one or more complete blocks, then removing them decreases both $|\mathcal{B}|$ and $|X| + |\mathfrak{R}| - 1$ by the same amount. Therefore, we assume that $\mathbf{D}$ has no complete block.

Let $|B| = k_i$ for any block $B \in \mathcal{C}_i$ and $|A \cap B| = m_{ij}$ for any distinct blocks $A \in \mathcal{C}_i$ and $B \in \mathcal{C}_j$. Let S be the subspace of the vector space $Pol(\mathbf{D})$ generated by the polynomials $f_x, x \in X$, and $g_i = g_{\mathcal{C}_i}$, $1 \le i \le t$, defined by (5.2) and (5.4). By Proposition 5.1.7, these polynomials are linearly independent, so $\dim S = |X| + |\mathfrak{R}|$. Therefore, it suffices to prove that $S = Pol(\mathbf{D})$.

Consider the polynomial

$$h = \sum_{x \in X} f_x - \sum_{i=1}^{t} k_i g_i. \tag{5.5}$$

Let $h = \sum_B a_B x_B + c$. Let $A \in \mathcal{C}_j$, $1 \le j \le t$. Then (5.5) implies $a_A = |A| - k_j = 0$, so h is a constant polynomial. Since the polynomials f_x, g_i are linearly independent, $h \neq 0$, and therefore all constant polynomials are in the subspace S.

Let again $A \in \mathcal{C}_j$, $1 \le j \le t$. Consider the polynomial

$$h_A = \sum_{x \in A} f_x - \sum_{i=1}^{t} m_{ij} g_i. \tag{5.6}$$

Let $h_A = \sum_B a_B x_B + c$. Then (5.6) implies $a_A = |A| - m_{jj} = k_j - m_{jj}$. If $k_j - m_{jj} = 0$, then $|A| = |B| = |A \cap B|$ for any blocks A and B in $\mathcal{C}_j$, a contradiction, since $|\mathcal{C}_j| \geq 2$. Therefore, $a_A \neq 0$. Let B be any block other than A and let $B \in \mathcal{C}_k$. Then (5.6) implies $a_B = |A \cap B| - m_{kj} = 0$. Thus, $h_A = a_A x_A + c$. Since the constant c is in S and $a_A \neq 0$, we obtain that $x_A \in S$. Since this holds for any block A, we obtain $S = Pol(\mathbf{D})$. □

In order to prove the converse theorem, we need the following result.

Lemma 5.1.13. *Let $\mathbf{D} = (X, \mathcal{B}, I)$ be a PBD of index λ and $\mathfrak{R}$ be a set of pairwise disjoint resolution classes of $\mathbf{D}$. If $|\mathcal{B}| = |X| + |\mathfrak{R}| - 1$, then $\mathfrak{R}$ is a resolution of $\mathbf{D}$.*

Proof. Let $|\mathcal{B}| = |X| + |\mathfrak{R}| - 1$. Then the set $\{f_x : x \in X\} \cup \{g_{\mathcal{C}} : \mathcal{C} \in \mathfrak{R}\}$ is a basis of the vector space $Pol(D)$. For any block $A \in \mathcal{B}$, expand the monomial x_A in this basis:

$$x_A = \sum_{x \in X} a_x^A f_x + \sum_{\mathcal{C} \in \mathfrak{R}} b_{\mathcal{C}}^A g_{\mathcal{C}}. \tag{5.7}$$

Applying both sides of (5.7) to a point $y \in X$, we obtain:

$$a_y^A = \begin{cases} \frac{1}{r(y)-\lambda}, & \text{if } y \in A, \\ 0, & \text{if } y \notin A. \end{cases}$$

Therefore,

$$x_A = \sum_{x \in A} \frac{f_x}{r(x) - \lambda} + \sum_{\mathcal{C} \in \mathfrak{R}} b_{\mathcal{C}}^A g_{\mathcal{C}}. \tag{5.8}$$

Suppose that $\mathfrak{R}$ does not partition $\mathcal{B}$. Let B be a block which does not belong to any resolution class in $\mathfrak{R}$ and let A be a block other than B. Comparing the coefficients of x_B in both sides of (5.8), we obtain

$$0 = \sum_{x \in A \cap B} \frac{1}{r(x) - \lambda}. \tag{5.9}$$

Since $r(x) - \lambda > 0$ for any point x, (5.9) implies that $A \cap B = \varnothing$, for any block A other than B. Since Fisher's Inequality implies that $\mathfrak{R} \neq \varnothing$, we have a contradiction. □

We will use this lemma to obtain a sufficient condition for a family of sets with constant intersection size to be the block set of a symmetric design.

Proposition 5.1.14. *Let v and λ be positive integers and let $\mathcal{B}$ be a family of v subsets of a set X of cardinality v such that $|A \cap B| = \lambda$ for distinct $A, B \in \mathcal{B}$ and $|B| > \lambda$ for all $B \in \mathcal{B}$. Suppose further that there is a nonempty subset Z*

of X and an integer k such that $|B \cap Z| = k$ for all $B \in \mathcal{B}$. Then $Z = X$ and $(X, \mathcal{B})$ is a symmetric (v, k, λ)-design.

Proof. Let **D** be the incidence structure dual to $(X, \mathcal{B})$. Since $|B| > \lambda$ for all $B \in \mathcal{B}$, **D** is a PBD of index λ. Let A_x denote the block of **D** corresponding to a point $x \in X$, that is, $A_x = \{B \in \mathcal{B}: x \in B\}$. Since $|B \cap Z| = k$ for all $B \in \mathcal{B}$, every point in **D** belongs to k blocks from the set $\mathcal{C} = \{A_x : x \in Z\}$. Therefore, $\mathcal{C}$ is a resolution class in **D**. Since **D** has $v + |\mathfrak{R}| - 1$ blocks where $\mathfrak{R}$ is the singleton $\{\mathcal{C}\}$, Lemma 5.1.13 implies that $\mathfrak{R}$ is a resolution of **D**, i.e., $\{\mathcal{C}\}$ consists of all blocks of **D**. Therefore, $Z = X$, which implies that all sets in the family $\mathcal{B}$ have cardinality k. Hence, $(X, \mathcal{B})$ is a symmetric (v, k, λ)-design. □

The next theorem gives structural information on the designs attaining the equality in Bose's Inequality.

Theorem 5.1.15. *Let $\mathbf{D} = (X, \mathcal{B}, I)$ be a PBD of index λ and let $v = |X|$. Let $\mathfrak{R}$ be a resolution of **D** such that $|\mathcal{B}| = v + |\mathfrak{R}| - 1$. Let r be the replication number of **D**. Then:*

(i) *$\mathfrak{R}$ is an affine resolution of **D**;*
(ii) *for any distinct blocks A and B in a resolution class $\mathcal{C} \in \mathfrak{R}$, $|A \cap B| = k(\mathcal{C}) - r + \lambda$, where $k(\mathcal{C})$ is the cardinality of any block in $\mathcal{C}$;*
(iii) for any blocks A and B in distinct resolution classes $\mathcal{C}_1$ and $\mathcal{C}_2$, $|A \cap B| = k(\mathcal{C}_1)k(\mathcal{C}_2)/v$;
(iv) for any resolution class $\mathcal{C} \in \mathfrak{R}$, $k(\mathcal{C})|\mathcal{C}| = v\alpha(\mathcal{C})$;
(v) the incidence structure dual to any resolution class $\mathcal{C} \in \mathfrak{R}$ is a 2-$(|\mathcal{C}|, \alpha(\mathcal{C}), k(\mathcal{C}) - r + \lambda)$ design;
(vi) $\mathfrak{R}$ is the only affine resolution of **D**.

Proof. Since $|\mathcal{B}| = v + |\mathfrak{R}| - 1$, Proposition 5.1.7 implies that the set of polynomials $\{f_x : x \in X\} \cup \{g_\mathcal{C} : \mathcal{C} \in \mathfrak{R}\}$ is a basis of $Pol(\mathbf{D})$. Using the same reasoning as in the proof of Lemma 5.1.13, we derive that

$$(r - \lambda)x_A = \sum_{x \in A} f_x - \sum_{\mathcal{C} \in \mathfrak{R}} b_\mathcal{C}^A g_\mathcal{C}. \tag{5.10}$$

For any block B, let $\mathcal{C}(B)$ be the resolution class containing B. Comparing the coefficients of x_A in both sides of (5.10) yields $r - \lambda = |A| - b_{\mathcal{C}(A)}^A$, so

$$b_{\mathcal{C}(A)}^A = |A| - r + \lambda. \tag{5.11}$$

Let A and B be two distinct blocks. Comparing the coefficients of x_B in both sides of (5.10) yields $0 = |A \cap B| - b_{\mathcal{C}(B)}^A$, so

$$b_{\mathcal{C}(B)}^A = |A \cap B|. \tag{5.12}$$

If $\mathcal{C}(A) = \mathcal{C}(B)$, then (5.11) and (5.12) imply

$$|A| - r + \lambda = |A \cap B|. \tag{5.13}$$

Interchanging A and B, we obtain that $|B| - r + \lambda = |A \cap B|$, so $|A| = |B|$. Thus, we have proved that any two blocks in the same resolution class have the same cardinality. Now (5.13) immediately implies (ii), which in turn implies (v).

Let $\mathcal{C}_1$ and $\mathcal{C}_2$ be distinct resolution classes. Fix $A \in \mathcal{C}_1$ and count in two ways pairs (x, B) where $B \in \mathcal{C}_2$ and $x \in A \cap B$. Since (5.12) implies that every block $B \in \mathcal{C}_2$ meets A in the same number of points equal to $b^A_{\mathcal{C}_2}$, we obtain

$$k(\mathcal{C}_1)\alpha(\mathcal{C}_2) = |\mathcal{C}_2| \cdot |A \cap B|. \tag{5.14}$$

Counting in two ways pairs (B, x) where $B \in \mathcal{C}$, $x \in B$, we obtain $k(\mathcal{C})|C| = \alpha(\mathcal{C})v$, so (iv) holds and then (iii) follows from (5.14) and (iv). Obviously, (i) follows from (ii) and (iii).

We will now prove (vi). Suppose $\mathfrak{R}'$ is an affine resolution of **D** other than $\mathfrak{R}$.

Let blocks B_1 and B_2 belong to the same class $\mathcal{C} \in \mathfrak{R}$ and to different classes $\mathcal{C}_1$ and $\mathcal{C}_2$, respectively, in $\mathfrak{R}'$. By (ii), $|B_1 \cap B_2| = k - r + \lambda$ where $k = k(\mathcal{C}_2)$. Since $k(\mathcal{C}_1) = k(\mathcal{C}_2) = k$, (iii) implies that $|B_1 \cap B_2| = k^2/v$, so

$$k - r + \lambda = \frac{k^2}{v}. \tag{5.15}$$

Form a partition $\mathfrak{R}''$ of $\mathcal{B}$ by replacing classes $\mathcal{C}_1$ and $\mathcal{C}_2$ in $\mathcal{R}'$ with $\mathcal{C}' = \mathcal{C}_1 \cup \mathcal{C}_2$. We will show that $\mathfrak{R}''$ is also an affine resolution of **D**. Let $A, B \in \mathcal{B}$, $A \neq B$. If $A, B \in \mathcal{C}_1$ or $A, B \in \mathcal{C}_2$, then, by (ii), $|A \cap B| = |A| - r + \lambda = k - r + \lambda$. If $A \in \mathcal{C}_1$ and $B \in \mathcal{C}_2$, then, by (iii), $|A \cap B| = k^2/v$, so by (5.15), $|A \cap B| = k - r + \lambda$ for any $A \neq B$ in $\mathcal{C}'$. If $A \in \mathcal{C}'$ and $B \notin \mathcal{C}'$, then, by (iii), $|A \cap B| = k|B|/v$, so the cardinality of the intersection of any two distinct blocks in $\mathcal{B}$ depends only on their respective classes in $\mathfrak{R}''$. Thus $\mathfrak{R}''$ is an affine resolution of **D** and Theorem 5.1.12 implies that $|\mathfrak{R}''| = |\mathfrak{R}'|$ which is obviously not the case. The proof is now complete. □

Corollary 5.1.16. *If* **D** *is a (v, b, r, k, λ)-design that admits an affine resolution, then any two distinct blocks from the same resolution class meet in $k - r + \lambda$ points and any two blocks from distinct resolution classes meet in k^2/v points.*

Let $\mathbf{D} = (X, \mathcal{B}, I)$ be a PBD of index λ that admits an affine resolution $\mathfrak{R}$. Each resolution class consists of at least two blocks. The next theorem classifies such designs having a resolution class consisting of precisely two blocks.

Theorem 5.1.17. *Let $\mathbf{D} = (X, \mathcal{B}, I)$ be a PBD of index λ that admits an affine resolution $\mathfrak{R}$. If $\mathfrak{R}$ has a class of cardinality 2, then $\mathbf{D}$ is a Hadamard 3-design.*

Proof. Let $\mathcal{C}_1 = \{A_1, B_1\}$ be a resolution class of cardinality 2 in $\mathfrak{R}$. Then the blocks A_1 and B_1 complement each other. Therefore, $|A_1| = |B_1| = v/2$ where $v = |X|$, and, by Theorem 5.1.15 (ii), $|A_1 \cap B_1| = 0 = v/2 - r + \lambda$ where r is the replication number of $\mathbf{D}$. Thus we have $v = 2(r - \lambda)$. By Theorem 5.1.15(ii), $|A| \geq r - \lambda = v/2$, for any block A in $\mathbf{D}$. Consider the design $\mathbf{D}'$ obtained by replacing every block of $\mathbf{D}$ with its complement. By Remark 5.1.5, $\mathbf{D}'$ is an (r', λ')-design, and, since $\mathbf{D}'$ has the same number of resolution classes as $\mathbf{D}$, $\mathbf{D}'$ admits an affine resolution. Since $\mathbf{D}'$ has a resolution class of cardinality 2, each block in $\mathbf{D}'$ has cardinality $\geq v/2$, so each block in $\mathbf{D}$ has cardinality $\leq v/2$. Therefore, each block in $\mathbf{D}$ has cardinality $v/2$, i.e., $\mathbf{D}$ is a $2-(v, v/2, \lambda)$ design. Theorem 5.1.15 now implies that any two blocks in the same resolution class are disjoint, so each class has cardinality 2 and $|\mathfrak{R}| = |\mathcal{B}|/2$. Therefore, $|\mathcal{B}| = v + |\mathcal{B}|/2 - 1$, so $|\mathcal{B}| = 2v - 2$.

Let $\mathfrak{R} = \{\mathcal{C}_1, \mathcal{C}_2, \ldots, \mathcal{C}_{v-1}\}$ and let $\mathcal{C}_i = \{A_i, B_i\}$, for $i = 1, 2, \ldots, v-1$. Let $A_v = X$ and $B_v = \emptyset$. Let $X = \{x_1, x_2, \ldots, x_v\}$. Consider the matrix $H = [h_{ij}]$ of order v defined by

$$h_{ij} = \begin{cases} 1 & \text{if } x_i \in A_j, \\ -1 & \text{if } x_i \in B_j. \end{cases}$$

By Theorem 5.1.15(iii), $|A_i \cap A_j| = |A_i \cap B_j| = |B_i \cap A_j| = |B_i \cap B_j| = \frac{v}{4}$, for $1 \leq i < j \leq v - 1$. Since also $|A_i \cap A_v| = |B_i \cap A_v| = \frac{v}{2}$ for $1 \leq i \leq v - 1$, distinct rows of H are orthogonal. Thus, H is a Hadamard matrix with an all-one row, and then $\mathcal{B}$ is the block set of the corresponding Hadamard 3-design. □

5.2. Affine α-resolvable designs

Definition 5.2.1. Let α be a positive integer. An *affine α-resolvable pairwise balanced design* is a PBD that admits an affine resolution $\mathfrak{R}$ such that $\alpha(\mathcal{C}) = \alpha$ for all $\mathcal{C} \in \mathfrak{R}$.

We now give some necessary conditions on parameters of affine α-resolvable PBDs.

Theorem 5.2.2. *Let $\mathbf{D} = (V, \mathcal{B})$ be an affine α-resolvable PBD of index λ and $\mathfrak{R}$ be its affine resolution. If $\alpha = 1$ and $\mathbf{D}$ has no complete block, then $\mathbf{D}$ is a*

2-design. *If* **D** *is not a* 2-*design, then there exist integers* u, m, c_1, *and* c_2 *such that*

(i) $r - \lambda = m\alpha$ and $v = mu$;
(ii) any resolution class of $\mathfrak{R}$ has cardinality c_1 or c_2;
(iii) $c_1 + c_2 = u + 1$ and $c_1 c_2 = u\alpha$;
(iv) all blocks of any resolution class of cardinality c_1 have cardinality $k_1 = mc_2$ and all blocks of any resolution class of cardinality c_2 have cardinality $k_2 = mc_1$.

Proof. Let $\mathbf{D} = (V, \mathcal{B})$ be an affine α-resolvable PBD of index λ and $\mathfrak{R}$ be its affine resolution. By Theorem 5.1.15(v), for any $\mathcal{C} \in \mathfrak{R}$, the dual structure is a 2-$(|\mathcal{C}|, \alpha, k(\mathcal{C}) - r + \lambda)$ design. Since its replication number is $k(\mathcal{C})$, we apply a standard 2-design relation to obtain:

$$(|\mathcal{C}| - 1)(k(\mathcal{C}) - r + \lambda) = k(\mathcal{C})(\alpha - 1).$$

Using Theorem 5.1.15(iv), we simplify this equation to

$$(r - \lambda)|\mathcal{C}|^2 - (\alpha v + r - \lambda)|\mathcal{C}| + \alpha^2 v = 0. \tag{5.16}$$

If $\alpha = 1$, then (5.16) has solutions $|\mathcal{C}| = v/(r - \lambda)$ and $|\mathcal{C}| = 1$, so $|\mathcal{C}| = v/(r - \lambda)$ for any resolution class $\mathcal{C}$. Then Theorem 5.1.15(iv) implies that any block in **D** has cardinality $r - \lambda$ and therefore **D** is a 2-design.

Suppose that **D** is not a 2-design. Then (5.16) has two different solutions c_1 and c_2 in positive integers. If $|\mathcal{C}| = c_i, i = 1, 2$, then any block of $\mathcal{C}$ has cardinality $k_i = \alpha v/c_i$. Then $k_1 + k_2 = \alpha v(c_1 + c_2)/(c_1 c_2) = v + (r - \lambda)/\alpha$. Therefore, $m = (r - \lambda)/\alpha$ is an integer and we simplify (5.16) to

$$m|\mathcal{C}|^2 - (v + m)|\mathcal{C}| + \alpha v = 0. \tag{5.17}$$

Since $c_1 + c_2 = (v + m)/m$, m divides v, so we put $v = mu$ and simplify (5.17) to

$$|\mathcal{C}|^2 - (u + 1)|\mathcal{C}| + \alpha u = 0. \tag{5.18}$$

This implies (iii). To obtain (iv), note that by Theorem 5.1.15(iv), $k_1 = \alpha v/c_1 = \alpha m u/c_1 = m c_1 c_2/c_1 = mc_2$. The proof of the equality $k_2 = mc_1$ is analogous. □

Remark 5.2.3. For any α and for $u = 2(2\alpha - 1)$, (5.18) has solutions $|\mathcal{C}| = c_1 = 2\alpha$ and $|\mathcal{C}| = c_2 = 2\alpha - 1$. For any even m, we obtain a feasible set of parameters of an affine α-resolvable PBD of index $m\alpha$ having $v = 2m(2\alpha - 1)$ points, $2m - 1$ resolution classes, each consisting of 2α blocks of cardinality $m(2\alpha - 1)$, and one resolution class consisting of $2\alpha - 1$ blocks of cardinality

$2m\alpha$. Such a design has $4m\alpha - 1$ blocks, the replication number is equal to $2m\alpha$. Here is an example of such a design for $\alpha = 2$, $m = 2$:

Consider the incidence structure $\mathbf{D} = (X, \mathcal{B})$ where $X = \{1, 2, 3, \ldots, 12\}$ and $\mathcal{B}$ consists of the following 15 blocks:

$\{1, 2, 5, 6, 9, 10\}$	$\{1, 2, 7, 8, 9, 11\}$	$\{1, 3, 5, 7, 9, 12\}$	$\{1, 2, 3, 4, 5, 6, 7, 8\}$
$\{1, 4, 5, 8, 11, 12\}$	$\{1, 3, 6, 8, 10, 12\}$	$\{1, 4, 6, 7, 10, 11\}$	$\{1, 2, 3, 4, 9, 10, 11, 12\}$
$\{2, 3, 6, 7, 11, 12\}$	$\{2, 4, 5, 7, 10, 12\}$	$\{2, 3, 5, 8, 10, 11\}$	$\{5, 6, 7, 8, 9, 10, 11, 12\}$
$\{3, 4, 7, 8, 9, 10\}$	$\{3, 4, 5, 6, 9, 11\}$	$\{2, 4, 6, 8, 9, 12\}$	

In this structure, the blocks in each column form a resolution class, each point occurs in two blocks of each class, so $\alpha = 2$, and any two points are contained in four blocks. Since the structure has 15 blocks and $15 = 12 + 4 - 1$, $\mathbf{D}$ is an affine 2-resolvable PBD of index 4. Of course, $\mathbf{D}$ is not a 2-design, since it has two different block sizes, 6 and 8.

5.3. Resolvable 2-designs

In this section, we concentrate on α-resolvable 2-designs.

If $\mathcal{C}$ is a resolution class of an α-resolvable 2-(v, k, λ) design, then counting in two ways flags $(x, \mathcal{B})$ with $B \in \mathcal{C}$ yields $k|\mathcal{C}| = v\alpha$. Thus,

Proposition 5.3.1. *All resolution classes of an α-resolvable 2-(v, k, λ) design have the same cardinality equal to $v\alpha/k$.*

Definition 5.3.2. An α-resolvable 2-design with $\alpha = 1$ is called a *resolvable 2-design.*

All designs $AG_d(n, q)$ where q is a prime power and $1 \leq d \leq n - 1$ are resolvable. If $d = n - 1$, such a design is affine-resolvable. An incidence structure $(X, \mathcal{B})$ where X is a set of even cardinality $2k$ and $\mathcal{B}$ is the set of all k-subsets of X gives another example of a resolvable 2-design.

Proposition 5.3.3. *If there exists a resolvable 2-(v, k, λ) design, then $\lambda(v - 1) \equiv 0 \pmod{k - 1}$ and $v \equiv 0 \pmod{k}$.*

Proof. The first divisibility condition follows from the identity $\lambda(v - 1) = r(k - 1)$ where r is the replication number of the design. Since each parallel class partitions the point set, the second divisibility condition follows. □

Remark 5.3.4. Though parameters $(v, k, \lambda) = (36, 6, 1)$ satisfy the conditions of Proposition 5.3.3, there is no 2-(36, 6, 1) design, i.e., an affine plane of order 6. (See Remark 3.4.9.)

The next result shows that, for $k = 2$, conditions of Proposition 5.3.3 are sufficient.

Proposition 5.3.5. *For any even v and any $\lambda \geq 1$, there exists a resolvable 2-$(v, 2, \lambda)$ design.*

Proof. It suffices to show that there exists a resolvable $(v, 2, 1)$-design, since the λ-fold multiple of such a design is a resolvable 2-$(v, 2, \lambda)$ design.

Let v be even, let $X = \{1, 2, \ldots, v\}$, and let $\mathcal{B}$ be the set of all 2-subsets of X. Then the incidence structure $\mathbf{D} = (X, \mathcal{B})$ is a 2-$(v, 2, 1)$ design. We claim that $\mathbf{D}$ is resolvable.

Let L be a symmetric Latin square of order v with $L(i, i) = v$ for $i = 1, 2, \ldots, v$. Such a Latin square exists by Lemma 3.2.23. For $k = 1, 2, \ldots, v - 1$, let $\mathcal{C}_k = \{\{i, j\} \in \mathcal{B} \colon L(i, j) = k\}$. For any $\{i, j\} \in \mathcal{B}$, $L(i, j) = L(j, i) \neq v$, so $\mathcal{C}_k$ is well-defined. The definition of Latin square implies that each $\mathcal{C}_k$ is a parallel class of $\mathbf{D}$ and that the set $\mathfrak{R} = \{\mathcal{C}_1, \mathcal{C}_2, \ldots, \mathcal{C}_{v-1}\}$ is a parallelism of $\mathbf{D}$. □

Remark 5.3.6. Resolvable 2-$(v, 2, 1)$ designs are called *round robin tournaments*. If we interpret the points of such a design as participants of a round robin tennis tournament and a block $\{x, y\}$ as a match between players x and y, then resolution classes describe rounds of the tournament. In each round each participant plays one match.

We will now introduce a famous example of a resolvable 2-(15, 3, 1) design.

Example 5.3.7 (Kirkman's School Girl Design). Let $\mathbf{T} = (X, \mathcal{A})$ be a round robin tournament with $|X| = 8$. Let ∞ be a fixed point of $\mathbf{T}$ and $\mathfrak{R}$ the parallelism of $\mathbf{T}$. Let $\mathbf{P} = (Y, \mathcal{L})$ be the Fano Plane (with the point set Y disjoint from X) and let $\varphi \colon X \setminus \{\infty\} \to Y$ be a bijection. Let $f \colon \mathcal{L} \to Y$ be a bijection such that $f(L) \notin L$ for each $L \in \mathcal{L}$. (Such a bijection can be easily constructed for the Fano Plane and, in fact, exists for every projective plane by Theorem 7.5.9.) For distinct $a, b \in X \setminus \{\infty\}$, let $p(a, b)$ be the third point of the line of $\mathbf{P}$ that contains $\varphi(a)$ and $\varphi(b)$, so $\{\varphi(a), \varphi(b), p(a, b)\}$ is a line of $\mathbf{P}$. For each $L \in \mathcal{L}$, let $\mathcal{C}_L$ be the parallel class in $\mathfrak{R}$ that contains $\{\varphi^{-1}(f(L)), \infty\}$ and let $\mathcal{B}_L$ be the set of the following five 3-subsets of $V = X \cup Y$: L, $\{f(L), \varphi^{-1}(f(L)), \infty\}$, and the three 3-subsets of the form $\{a, b, p(a, b)\}$ with $a, b \in X \setminus \{\infty\}$ and $\{a, b\} \in \mathcal{C}_L$.

Let $\mathbf{D} = (V, \mathcal{B})$ where $\mathcal{B}$ is the union of all sets $\mathcal{B}_L$, $L \in \mathcal{L}$. Observe that the five elements of $\mathcal{B}_L$ are pairwise-disjoint and the sets $\mathcal{B}_L$ and $\mathcal{B}_M$ are disjoint whenever $L, M \in \mathcal{L}$ and $L \neq M$. It is straightforward to check that $\mathbf{D}$ is a

2-(15, 3, 1) design and $\{\mathcal{B}_L : L \in \mathcal{L}\}$ is a parallelism of $\mathbf{D}$, so $\mathbf{D}$ is a resolvable 2-(15, 3, 1) design.

Proposition 5.3.3 applied to a resolvable 2-$(v, 3, 1)$ design gives $v \equiv 3 \pmod 6$. The following theorem, whose proof is far beyond the scope of this book, shows that this condition is sufficient for the existence of such designs.

Theorem 5.3.8. *If $v \equiv 3 \pmod 6$, then there exists a resolvable 2-$(v, 3, 1)$ design.*

Round robin tournaments and resolvable 2-$(v, 3, 1)$ designs are two infinite families of resolvable *Steiner systems*, i.e., 2-$(v, k, 1)$ designs. Another family of resolvable Steiner systems is affine planes. In the next theorem, we will construct an infinite family of resolvable *unitals*, that is, 2-$(n^3 + 1, n + 1, 1)$ designs.

Theorem 5.3.9. *For any prime power q, there exists a resolvable 2-$(q^3 + 1, q + 1, 1)$ design.*

Proof. The statement is true for $q = 2$, since the required 2-design is an affine plane of order 3.

Let $q > 2$ be a prime power and let F and K denote the fields $GF(q)$ and $GF(q^2)$, respectively. We will regard F as a subfield of K. Then $\varphi(x) = x^{q+1}$ is a homomorphism from K^* onto F^*. Therefore, for each $a \in F^*$, the equation $x^{q+1} = a$ has exactly $q + 1$ solutions in K^*.

Let V be the three-dimensional vector space over K and let $\mathbf{P} = (P, \mathcal{L})$ be the projective plane $PG_1(2, q^2)$ over K. For every nonzero vector $\mathbf{a} = [\alpha_1\ \alpha_2\ \alpha_3]^\top \in V$, we denote by $p(\mathbf{a})$ the point of $\mathbf{P}$ corresponding to the one-dimensional subspace $\langle \mathbf{a} \rangle$ of V and by $l(\mathbf{a})$ the line of $\mathbf{P}$ corresponding to the one-dimensional subspace of V consisting of all vectors $\mathbf{x} = [x_1\ x_2\ x_3]^\top$ such that $\alpha_1 x_1 + \alpha_2 x_2 + \alpha_3 x_3 = 0$.

Let S be the set of all vectors $\mathbf{x} = [x_1\ x_2\ x_3]^\top \in V$ such that

$$x_1^{q+1} + x_2^{q+1} + x_3^{q+1} = 0. \tag{5.19}$$

Observe that if $\mathbf{a} \in S$, then $\langle \mathbf{a} \rangle \subseteq S$. Let $X = \{p(\mathbf{a}) : \mathbf{a} \in S\}$. Since $\mathbf{0} \in S$, we have

$$|S| = 1 + (q^2 - 1)|X|. \tag{5.20}$$

Claim 1. $|X| = q^3 + 1$.

To prove this claim, observe that the equation $a_1 + a_2 + a_3 = 0$ over F has q^2 solutions. One solution is $a_1 = a_2 = a_3 = 0$, and $3(q - 1)$ solutions have exactly one of a_1, a_2, a_3 equal to 0. Thus, this equation has exactly $q^2 - 3q + 2$

solutions with $a_1, a_2, a_3 \in F^*$. Therefore, (5.19) has one solution with $x_1 = x_2 = x_3 = 0$, $3(q-1)(q+1)^2$ solutions with exactly one of x_1, x_2, x_3 equal to 0, and exactly $(q^2 - 3q + 2)(q+1)^3$ solutions with $x_1, x_2, x_3 \in K^*$. Therefore,

$$|S| = 1 + 3(q-1)(q+1)^2 + (q^2 - 3q + 2)(q+1)^3,$$

and then (5.20) implies that $|X| = q^3 + 1$.

For any nonzero vector $\mathbf{a} = [\alpha_1\ \alpha_2\ \alpha_3]^\top \in V$, let $\mathbf{a}' = [\alpha_1^q\ \alpha_2^q\ \alpha_3^q]^\top$ and let us call the line $l(\mathbf{a}')$ the polar of the point $p(\mathbf{a})$. Observe that if $\mathbf{b} = \alpha\mathbf{a}$ with $\alpha \in K^*$, then $\mathbf{b}' = \alpha^q \mathbf{b}$ and therefore, the polar of a projective point is well defined. The definition of the polar immediately implies that the polar of a projective point x contains x if and only if $x \in X$.

Observe also that if points $p(\mathbf{a})$ and $p(\mathbf{b})$ with $\mathbf{a} = [\alpha_1\ \alpha_2\ \alpha_3]^\top$ and $\mathbf{b} = [\beta_1\ \beta_2\ \beta_3]^\top$ have the same polar, then there is $\gamma \in K^*$ such that $\alpha_i^q = \gamma\beta_i^q$ for $i = 1, 2, 3$. This would imply that $\alpha_i = \gamma^q \beta_i$ and therefore $p(\mathbf{a}) = p(\mathbf{b})$. Thus, distinct points of **P** have distinct polars.

Claim 2. Let $x, y \in P$. If the polar of x contains y, then the polar of y contains x.

To prove this claim, let $x = p(\mathbf{a})$ and $y = p(\mathbf{b})$ where $\mathbf{a} = [\alpha_1\ \alpha_2\ \alpha_3]^\top$ and $\mathbf{b} = [\beta_1\ \beta_2\ \beta_3]^\top$. If the polar of x contains y, then

$$\alpha_1^q\beta_1 + \alpha_2^q\beta_2 + \alpha_3^q\beta_3 = 0.$$

Therefore,

$$\alpha_1^{q^2}\beta_1^q + \alpha_2^{q^2}\beta_2^q + \alpha_3^{q^2}\beta_3^q = 0$$

and then

$$\alpha_1\beta_1^q + \alpha_2\beta_2^q + \alpha_3\beta_3^q = 0,$$

i.e., the polar of y contains x.

Claim 3. Let $x \in X$ and let L be the polar of x. Then $L \cap X = \{x\}$.

To prove this claim, suppose that there is $y \neq x$ such that $y \in L \cap X$. Then L is the line through x and y. By Claim 2, the polar of y contains x, so L is the polar of y. Let $x = p(\mathbf{a})$ and $y = p(\mathbf{b})$ with $\mathbf{a} = [\alpha_1\ \alpha_2\ \alpha_3]^\top$ and $\mathbf{b} = [\beta_1\ \beta_2\ \beta_3]^\top$. Then the equations

$$\alpha_1^q x_1 + \alpha_2^q x_2 + \alpha_3^q x_3 = 0$$

and

$$\beta_1^q x_1 + \beta_2^q x_2 + \beta_3^q x_3 = 0$$

give the same two-dimensional subspace of V. Therefore, there is $\gamma \in K^*$ such that $\alpha_i^q = \gamma \beta_i^q$ for $i = 1, 2, 3$. This implies $\alpha_i = \gamma^q \beta_i$, and therefore $p(\mathbf{a}) = p(\mathbf{b})$, a contradiction.

Claim 4. Every line meets X in at most $q + 1$ points.

To prove this claim, let $L = l(\mathbf{a})$ with $\mathbf{a} = [\alpha_1\ \alpha_2\ \alpha_3]^\top$. Without loss of generality, we assume that $\alpha_3 = -1$. Then the system consisting of (5.19) and the equation $\alpha_1 x_1 + \alpha_2 x_2 - x_3 = 0$ reduces to the equation

$$x_1^{q+1} + x_2^{q+1} + (\alpha_1 x_1 + \alpha_2 x_2)^{q+1} = 0.$$

This equation can be transformed into

$$F(x_1, x_2) = 0 \tag{5.21}$$

with

$$F(x_1, x_2) = \left(1 + \alpha_1^{q+1}\right)x_1^{q+1} + \left(1 + \alpha_2^{q+1}\right)x_2^{q+1} + \alpha_1^q \alpha_2 x_1^q x_2 + \alpha_2^q \alpha_1 x_2^q x_1.$$

If $1 + \alpha_1^{q+1} \neq 0$, then, for each $x_2 \in K^*$, (5.21) is satisfied by at most $q + 1$ values of x_1; if $x_2 = 0$, then (5.21) implies $x_1 = 0$ and then $x_3 = 0$. Therefore, in this case, (5.21) has at most $1 + (q^2 - 1)(q + 1)$ solutions. The case $1 + \alpha_2^{q+1} \neq 0$ is similar.

If $1 + \alpha_1^{q+1} = 1 + \alpha_2^{q+1} = 0$, then, for any $x_2 \in K^*$, (5.21) is satisfied by at most q values of x_1; if $x_2 = 0$, then (5.21) is satisfied by $q^2 - 1$ nonzero values of x_1. Therefore, in this case, (5.21) has at most $1 + (q^2 - 1)q + (q^2 - 1) = 1 + (q^2 - 1)(q + 1)$ solutions.

Thus, in any case, (5.21) has at most $1 + (q^2 - 1)(q + 1)$ solutions, which means that $|L \cap X| \leq q + 1$.

Claim 5. Let $x \in X$. Then every line containing x, other than the polar of x, meets X in $q + 1$ points.

To prove this claim, denote by m the number of lines through x that contain no other point of X. By Claim 3, $m \geq 1$. Counting in two ways flags (y, L) with $y \in X$, $y \neq x$, and $L \supset \{x, y\}$ yields, due to Claim 4, the inequality $q^3 \leq (q^2 + 1 - m)q$. Therefore, $m = 1$, the inequality is in fact an equality, and therefore, every line through x, other than the polar of x, meets X in $q + 1$ points.

Claim 6. Every line meets X in 1 or $q + 1$ points.

To prove this claim, observe that each point $x \in X$ lies on q^2 lines, other than the polar of x. Since every such line contains exactly $q+1$ points of X and $|X| = q^3+1$, the total number of lines, other than the polars of points of X, that meet X is equal to $q^2(q^3+1)/(q+1) = q^2(q^2-q+1)$. Since the polars of distinct points of X are distinct lines, the total number of lines meeting X is equal to $q^2(q^2-q+1)+(q^3+1) = q^4+q^2+1$. Therefore, every line meets X. Claims 3 and 5 now imply that every line meets X in either one or $q+1$ points.

Claim 7. For any point $z \in P \setminus X$, there are $q+1$ lines through z that meet X in one point and q^2-q lines through z that meet X in $q+1$ points.

To prove this claim, we fix $z \in P \setminus X$ and denote by m_1 and m_2 the number of lines through z that meet X in one and $q+1$ points, respectively. Counting in two ways flags (y, L), where $y \in X$ and L is a line through z yields

$$q^3+1 = m_1 + m_2(q+1).$$

Since also $m_1 + m_2 = q^2+1$, we obtain that $m_1 = q+1$ and $m_2 = q^2-q$.

Claim 8. Let $z \in P \setminus X$ and let $L_1, L_2, \ldots, L_{q+1}$ be all lines through z that meet X in one point. For $i = 1, 2, \ldots, q+1$, let $L_i \cap X = \{x_i\}$. Let L be the polar of z. Then $L \cap X = \{x_1, x_2, \ldots, x_{q+1}\}$.

Since $|L \cap X| \le q+1$, in order to prove the claim, it suffices to show that $x_i \in L$ for $i = 1, 2, \ldots, q+1$. Since x_i is the only point of X on L_i, Claim 5 implies that L_i is the polar of x_i. Now Claim 2 implies that $x_i \in L$.

Let $\mathcal{B}$ be the set of all lines of $\mathbf{P}$ that meet X in $q+1$ points. Consider the substructure $\mathbf{D} = (X, \mathcal{B})$ of $\mathbf{P}$. It has q^3+1 points and all its blocks have cardinality $q+1$. Any two points of $\mathbf{D}$ lie on a unique line of $\mathbf{P}$, and this line meets X in more than one, and so in $q+1$ points. Therefore, $\mathbf{D}$ is a 2-$(q^3+1, q+1, 1)$ design. It has $q^2(q^2-q+1)$ blocks and its replication number is q^2. To complete the proof, we have to show that $\mathbf{D}$ is resolvable.

Fix a point $x_0 \in X$ and let $L_0 = \{x_0, z_1, z_2, \ldots, z_{q^2}\}$ be the polar of x_0. For $i = 1, 2, \ldots, q^2$, let L_i be the polar of z_i and let $\mathcal{C}_i$ be the subset of $\mathcal{B}$ consisting of L_i and all blocks that, as lines of $\mathbf{P}$, contain z_i. We claim that $\mathfrak{R} = \{\mathcal{C}_1, \mathcal{C}_2, \ldots, \mathcal{C}_{q^2}\}$ is a parallelism of $\mathbf{D}$.

Let $B_1, B_2 \in \mathcal{C}_i \setminus \{L_i\}$, $B_1 \ne B_2$. Then, since $B_1 \cap B_2 = \{z_i\}$, we have $B_1 \cap B_2 \cap X = \varnothing$. Suppose $y \in B_1 \cap L_i \cap X$. Then y lies on the polar of z_i and then z_i must lie on the polar of y. Therefore, B_1 is the polar of y. However, this is not the case, because $|B_1 \cap X| = q+1$. Thus, the blocks of $\mathcal{C}_i$ are pairwise disjoint. By Claim 7, the sum of their cardinalities is $(q+1)(1+q^2-q) = q^3+1 = |X|$. Therefore, each $\mathcal{C}_i$ partitions X.

It remains to prove that $\mathfrak{R}$ partitions $\mathcal{B}$. First note that, since $L_1, L_2, \ldots, L_{q^2}$ are the polars of distinct points, they are distinct lines. Therefore, these lines are all the blocks of $\mathcal{B}$ that contain x_0. If B is a block of **D** that does not contain x_0, then B meets L_0 in one of the points z_i and then $B \in \mathcal{C}_i$. Therefore, $\mathcal{B}$ is the union of the parallel classes $\mathcal{C}_i$. By Claim 7, $q^2|\mathcal{C}_i| = q^2(q^2-q+1)$. Since this is precisely the number of blocks of **D**, we obtain that $\mathfrak{R}$ partitions $\mathcal{B}$. □

The following theorem is a useful tool for constructing α-resolvable designs.

Theorem 5.3.10. *Let M be an incidence matrix of a $(v_1, b_1, r_1, k_1, \lambda_1)$-design $\mathbf{D}_1$. Let $N = [N_1, N_2, \ldots, N_s]$ be an incidence matrix of an α-resolvable $(v_2, b_2, r_2, k_2, \lambda_2)$-design $\mathbf{D}_2$, where each N_j is a $v_2 \times v_1$ incidence matrix of a resolution class of $\mathbf{D}_2$. Then $P = [N_1M, N_2M, \ldots, N_sM]$ is an incidence matrix of an (αr_1)-resolvable (v, b, r, k, λ)-design $\mathbf{D}$ with parameters*

$$v = v_2,\ b = \frac{r_2 b_1}{\alpha},\ r = r_1 r_2 + r_2\lambda_1(\alpha - 1),\ k = k_1k_2,\ \lambda = (r_1 - \lambda_1)\lambda_2 + \lambda_1 r_2 \alpha.$$

If $\mathbf{D}_1$ is symmetric and $\mathbf{D}_2$ is affine resolvable, then $\mathbf{D}$ is affine r_1-resolvable.

Proof. Since, for $j = 1, 2, \ldots, s$, $J(N_jM) = k_2JM = k_1k_2J$, the matrix P has constant column sum $k = k_1k_2$. We further have

$$\begin{aligned} PP^\top &= \sum_{j=1}^{s}(N_jM)(N_jM)^\top = \sum_{j=1}^{s} N_j(MM^\top)N_j^\top \\ &= (r_1 - \lambda_1)\sum_{j=1}^{s} N_jN_j^\top + \lambda_1 \sum_{j=1}^{s} N_jJN_j^\top \\ &= (r_1-\lambda_1)(r_2-\lambda_2)I + ((r_1-\lambda_1)\lambda_2 + \lambda_1 r_2\alpha)J, \end{aligned}$$

so **D** is a (v, b, r, k, λ)-design. Since, for $j = 1, 2, \ldots, s$, $(N_jM)J = \alpha r_1 J$, **D** is (αr_1)-resolvable.

Suppose now that $\mathbf{D}_1$ is symmetric and $\mathbf{D}_2$ is affine resolvable. Then $\alpha = 1$, so $b = r_2b_1 = r_2v_1 = b_2$. Since $\mathbf{D}_2$ is affine-resolvable, Theorem 5.1.12 implies that $b_2 = v_2 + r_2 - 1$. Since $b = b_2$, $v = v_2$, and the resolution of **D** consists of r_2 classes, Theorem 5.1.15 implies that **D** is an affine r_1-resolvable design. □

Example 5.3.11. If $q \equiv 3 \pmod 4$ is a prime power, then let $\mathbf{D}_1$ be a symmetric 2-$(q, (q-1)/2, (q-3)/4)$-design (a Hadamard 2-design) and let $\mathbf{D}_2$ be the design $AG_{n-1}(n, q)$, that is an affine resolvable 2-$(q^n, q^{n-1}, (q^{n-1}-1)/$

$(q-1))$ design. Then $\mathbf{D}$ is a $((q^n-1)/2)$-resolvable 2-(v,k,λ) design with

$$v = q^n, \quad k = \frac{q^{n-1}(q-1)}{2}, \quad \lambda = \frac{k-1}{2}.$$

We will now give a direct construction of a family of α-resolvable designs.

Theorem 5.3.12. *Let q be a prime power and let H be a subgroup of the multiplicative group of the field $GF(q)$, $|H| = h$. Let $m = (q-1)/h$ and let $a_1, a_2, \ldots, a_m$ be representatives of distinct cosets of H in $GF(q)^*$. Let $B = H \cup \{0\}$ and let $\mathcal{B} = \{a_i B + b \colon b \in GF(q), 1 \le i \le m\}$. Then $\mathbf{D} = (GF(q), \mathcal{B})$ is an $(h+1)$-resolvable 2-$(q, h+1, h+1)$ design.*

Proof. The incidence structure $\mathbf{D}$ has q points and constant block size $h+1$. Let $x, y \in GF(q)$, $x \ne y$. We have to show that there are exactly $h+1$ pairs (i, b) such that $\{x, y\} \subseteq a_i B + b$.

The function $f(b) = (x-b)/(y-b) = 1 + (x-y)/(y-b)$ gives a bijection from the set $GF(q) \setminus \{x, y\}$ to the set $GF(q)^* \setminus \{1\}$. Therefore, there are exactly $h-1$ values of b, for which $x-b$ and $y-b$ are in the same coset of H. Thus, there are exactly $h-1$ values of b for which there exists i such that $x, y \in a_i H + b$. If $b = x$, then $y \in a_i B + b$ if and only if $y - x \in a_i H$. Similarly, if $b = y$, $x \in a_i B + b$ if and only if $x - y \in a_i H$. Thus, there are exactly $h+1$ blocks of $\mathbf{D}$ that contain $\{x, y\}$.

For $i = 1, 2, \ldots, m$, let $\mathcal{C}_i = \{a_i B + b \colon b \in GF(q)\}$. The sets $\mathcal{C}_i$ partition the block set $\mathcal{B}$ of $\mathbf{D}$. For $x \in GF(q)$ and for a fixed i, we have $x \in a_i B + b$ if and only if $x - b \in a_i B$. Since $|a_i B| = h+1$, we obtain that x is contained in exactly $h+1$ blocks of $\mathcal{C}_i$. Therefore, $\mathbf{D}$ is an $(h+1)$-resolvable 2-$(q, h+1, h+1)$ design. □

The only known examples of affine resolvable 2-designs are designs $AG_{d-1}(d, q)$ and Hadamard 3-designs. The following proposition shows that parameters of any affine resolvable 2-design can be expressed in terms of two integral parameters. This result will be used in subsequent chapters.

Proposition 5.3.13. *Let $\mathbf{D}$ be an affine-resolvable (v, b, r, k, λ)-design. Let s be the cardinality of a parallel class and μ be the cardinality of the intersection of two blocks from distinct parallel classes. Then $s-1$ divides $\mu - 1$ and*

$$v = s^2\mu, \quad b = \frac{s(s^2\mu - 1)}{s-1}, \quad r = \frac{s^2\mu - 1}{s-1}, \quad k = s\mu, \quad \lambda = \frac{s\mu - 1}{s-1}. \tag{5.22}$$

Proof. Since each parallel class partitions the point set of $\mathbf{D}$ into k-subsets, we have $v = sk$. By Theorem 5.1.15(iii), $k^2 = \mu v$. This implies $k = s\mu$ and then $v = s^2\mu$. Since **D** is affine resolvable, we have $b = v + r - 1$, so $bk = vk + rk - k$. Since, on the other hand, $bk = vr$, we obtain that

$$r = \frac{(v-1)k}{v-k} = \frac{s^2\mu - 1}{s-1},$$

and then

$$\lambda = \frac{r(k-1)}{v-1} = \frac{s\mu - 1}{s-1}.$$

Since $s\mu - 1 \equiv \mu - 1 \pmod{s-1}$, we obtain that $s - 1$ divides $\mu - 1$. □

Remark 5.3.14. If **D** is an affine-resolvable design with parameters (5.22), we will say that **D** is an $A_\mu(s)$.

We will now demonstrate a construction of a symmetric design using an affine-resolvable design.

Theorem 5.3.15. *If there exists an affine-resolvable (v, b, r, k, λ)-design, then there exists a symmetric $(v(r+1), kr, k\lambda)$-design admitting a symmetric incidence matrix.*

Proof. Let **D** be an affine-resolvable (v, b, r, k, λ)-design with the point set $X = \{x_1, x_2, \dots, x_v\}$ and parallel classes $\mathcal{C}_1, \mathcal{C}_2, \dots, \mathcal{C}_r$. For $h = 1, 2, \dots, r$, let M_h be the (0, 1)-matrix of order v with (i, j)-entry equal to 1 if and only if the points x_i and x_j are in the same block of the parallel class $\mathcal{C}_h$. Note that the matrices M_h are symmetric. Furthermore, if μ is the cardinality of the intersection of two blocks of **D** from different parallel classes, then, for $h, l = 1, 2, \dots, r$,

$$M_h M_l^\top = \begin{cases} \mu J & \text{if } l \neq h, \\ kM_h & \text{if } l = h \end{cases}$$

and, if N is an incidence matrix of **D**, then

$$NN^\top = \sum_{h=1}^{r} M_h = (r - \lambda)I + \lambda J.$$

Let M_{r+1} be the zero matrix of order v and let L be a symmetric Latin square of order $r + 1$ (see Remark 3.2.22). Define a block matrix $N = [N_{ij}]$ of order $v(r+1)$ by

$$N_{ij} = M_h \text{ if and only if } L(i, j) = h.$$

Then, for $h = 1, 2, \dots, r+1$,

$$\sum_{j=1}^{r+1} N_{hj} N_{hj}^\top = k \sum_{j=1}^{r+1} M_h = (kr - k\lambda)I + k\lambda J$$

and, for distinct $i, h \in \{1, 2, \ldots, r+1\}$,

$$\sum_{j=1}^{r+1} N_{ij} N_{hj}^{\top} = (r-1)\mu J.$$

Proposition 5.3.13 implies that $(r-1)\mu = k\lambda$, and therefore, N is an incidence matrix of a symmetric $(v(r+1), kr, k\lambda)$-design. Since each M_h is symmetric and the Latin square L is symmetric, the matrix N is symmetric. □

If an affine-resolvable design in Theorem 5.3.15 is a Hadamard 3-design, based on a Hadamard matrix of order $4n$, then the resulting symmetric design is a Menon design of order $4n^2$. In this case, we have $r = 4n-1$, so $r+1$ is even. Therefore, one can select the symmetric Latin square L with all diagonal entries equal to $r+1$ (Lemma 3.2.23). Then all diagonal blocks of the incidence matrix N are zeros. This implies the following result.

Theorem 5.3.16. *If there exists a Hadamard matrix of order $4n$, then there exists a symmetric $(16n^2, 8n^2-2n, 4n^2-2n)$-design admitting a symmetric incidence matrix with zero diagonal.*

Corollary 5.3.17. *If there exists a Hadamard matrix of order $4n$, then there exists a regular symmetric Hadamard matrix of order $16n^2$ with constant diagonal.*

Remark 5.3.18. If the affine-resolvable design in Theorem 5.3.15 is the $AG_d(d+1, q)$, then the theorem gives a symmetric incidence matrix N of a symmetric design with parameters (3.6). If $r+1$ is even, which occurs if and only if dq is even, then we apply Lemma 3.2.23 to make all diagonal entries of N equal to 0.

5.4. Embedding of resolvable designs in symmetric designs

A natural question in the study of an incidence structure is whether it is possible to embed it in a larger (and often more regular) structure. A typical example is embedding of an affine plane in a projective plane. If an incidence structure is embeddable in a more regular structure, then the embedding may provide useful information about the smaller structure or suggest a way for constructing the larger structure. In the most general setting, an incidence structure **D** is embeddable in an incidence structure **E** if **D** is isomorphic to a substructure of **E** (cf. Definition 2.1.5). However, for specific incidence structures more restrictive definitions of embeddability are often used. In Chapter 13, we will explore different ways of embedding a symmetric design in a larger symmetric design. One possibility is to assume that the smaller design is isomorphic to a *normal subdesign* of the larger symmetric design.

Definition 5.4.1. A nontrivial symmetric design $\mathbf{D} = (Y, \mathcal{A})$ is called a *normal subdesign* of a symmetric design $\mathbf{S} = (X, \mathcal{B})$ if $Y \subseteq X$ and $\mathcal{A}$ is the set of all distinct nonempty sets $B \cap Y$ with $B \in \mathcal{B}$.

In the following proposition we use the notation introduced in Definition 2.1.5.

Proposition 5.4.2. *Let $\mathbf{S} = (X, \mathcal{B})$ be a symmetric design and let Y be a proper subset of X. Let $\mathcal{A}$ be the set of all distinct nonempty sets $B \cap Y$ with $B \in \mathcal{B}$. The incidence structure $\mathbf{D} = (Y, \mathcal{A})$ is a normal subdesign of $\mathbf{S}$ if and only if the substructure $\mathbf{E} = \mathbf{S}(Y, \mathcal{B}_{X \setminus Y})$ of $\mathbf{S}$ is a multiple of a symmetric design.*

Proof. If $\mathbf{E}$ is a multiple of a symmetric design, then taking each block of $\mathbf{E}$ once yields a symmetric design, and it is the normal subdesign $\mathbf{D}$.

Conversely, suppose $\mathbf{D}$ is a symmetric (v, k, λ)-design and let it be a normal subdesign of $\mathbf{S}$. Then $\mathbf{E}$ is a (v, b, r, k, λ')-design with any two blocks meeting in either k or λ points. Fix a block A of $\mathbf{E}$. Suppose $\mathbf{E}$ has m blocks that meet A in k points and $b - m$ blocks that meet A in λ points. Counting in two ways flags (x, B) where x is a point of $\mathbf{E}$ and B is a block of $\mathbf{E}$, other than A, yields $k(r-1) = mk + (b-m)\lambda$. Since $k \neq \lambda$, this equation shows that m does not depend on A and therefore $\mathbf{E}$ is an m-fold multiple of $\mathbf{D}$. Thus, $\mathbf{E}$ is a multiple of a symmetric design. □

Observe that if $\mathbf{S} = (X, \mathcal{B}, I)$ is a nontrivial symmetric (v, k, λ)-design and Y is a nonempty subset of X, then the substructure $\mathbf{S}(Y) = (Y, \mathcal{B}, I \cap (Y \times \mathcal{B}))$ is a PBD of index λ. This motivates the following definition.

Definition 5.4.3. A pairwise balanced design $\mathbf{D}$ is said to be *embeddable in a symmetric design* $\mathbf{S} = (X, \mathcal{B})$ if there is a subset Y of X such that $\mathbf{D}$ is isomorphic to $\mathbf{S}(Y)$.

The following theorem relates normal subdesigns of symmetric designs to certain pairwise balanced designs with an affine resolution.

Theorem 5.4.4. *Let $\mathbf{S} = (X, \mathcal{B})$ be a nontrivial symmetric design and Y be a proper subset of X. Then the following statements are equivalent.*

(i) $\mathbf{S}(X \setminus Y, \mathcal{B}_Y)$ is a c-fold multiple of a symmetric design.
(ii) $A \cap Y \neq \varnothing$ for any $A \in \mathcal{B}$ and there exists a positive integer α such that $\mathbf{S}(Y)$ is a PBD that admits an affine resolution with one resolution class

of replication number α and $|X|$ resolution classes of cardinality c and replication number $c - \alpha$.

If conditions (i) and (ii) are satisfied then the unique resolution class of replication number α consists of all blocks of $\mathcal{B}$ that are disjoint from $X \setminus Y$.

Proof. (i) $\Rightarrow$ (ii). Suppose $\mathbf{S} = (X, \mathcal{B})$ is a symmetric 2-(w, r, λ) design and $\mathbf{S}(X \setminus Y, \mathcal{B}_Y)$ is a c-fold multiple of a symmetric $(t-1, m, \mu)$-design $\mathbf{T}$. Let $|Y| = v$. Then $w = v + t - 1$, $r = cm$, and $\lambda = c\mu$. Let $\{B_1, B_2, \ldots, B_{t-1}\}$ be the block-set of $\mathbf{T}$. For $i = 1, 2, \ldots, t-1$, put $\mathcal{C}_i = \{A \in \mathcal{B}\colon A \cap (X \setminus Y) = B_i\}$. Then $|\mathcal{C}_i| = c$ for $i = 1, 2, \ldots, t-1$.

Put $\mathcal{C}_t = \{A \in \mathcal{B}\colon A \cap (X \setminus Y) = \varnothing\}$. Then $|\mathcal{C}_t| = w - c(t-1)$. We claim that $\mathcal{C}_t \neq \varnothing$. Indeed, if $\mathcal{C}_t = \varnothing$, then $w = c(t-1)$ and the basic relation (2.9) applied to symmetric designs $\mathbf{S}$ and $\mathbf{T}$ would yield after routine manipulations that $c = 1$. This in turn would imply that $w = t - 1$, i.e., $Y = \varnothing$, which is not the case.

For each $y \in Y$ and $i = 1, 2, \ldots, t$, denote by $\alpha_i(y)$ the number of blocks $A \in \mathcal{C}_i$ that contain y. Fixing $i \in \{1, 2, \ldots, t-1\}$ and counting in two ways pairs (y, A) where $A \in \mathcal{C}_i$ and $y \in A \cap Y$, we obtain:

$$\sum_{y \in Y} \alpha_i(y) = c(r - m) = mc(c-1). \tag{5.23}$$

Fixing $i \in \{1, 2, \ldots, t-1\}$ and counting triples (y, A, B) where $A, B \in \mathcal{C}_i$, $A \neq B$, and $y \in A \cap B \cap Y$, we obtain:

$$\sum_{y \in Y} \alpha_i(y)(\alpha_i(y) - 1) = c(c-1)(\lambda - m). \tag{5.24}$$

Equations (5.23) and (5.24) imply:

$$\sum_{y \in Y} (\alpha_i(y))^2 = \lambda c(c-1). \tag{5.25}$$

Fixing $i, j \in \{1, 2, \ldots, t-1\}, i \neq j$, and counting triples (y, A, B) where $A \in \mathcal{C}_i$, $B \in \mathcal{C}_j$, and $y \in A \cap B \cap Y$, we obtain:

$$\sum_{y \in Y} \alpha_i(y)\alpha_j(y) = c^2(\lambda - \mu) = \lambda c(c-1). \tag{5.26}$$

Equations (5.25) and (5.26) imply:

$$\sum_{y \in Y} (\alpha_i(y) - \alpha_j(y))^2 = 0,$$

so $\alpha_i(y) = \beta(y)$ is the same for $i = 1, 2, \ldots, t-1$. Put $\alpha(y) = \alpha_t(y)$ for $y \in Y$.

Let $y \in Y$ and $x \in X \setminus Y$. Since x and y occur together in λ blocks of $\mathcal{B}$ and x occurs in m of the blocks B_i, $i = 1, 2, \ldots, t-1$, we have $m\beta(y) = \lambda$, i.e., $\beta(y) = \beta = \frac{\lambda}{m}$ does not depend on y. Therefore, neither does $\alpha(y) = \alpha = r - (t-1)\beta$.

Since $\mathcal{C}_t \neq \varnothing$, we have $\alpha \neq 0$. Observe that if $A \in \mathcal{C}_i$ with $1 \leq i \leq t-1$, then $|A \cap Y| = r - m$, and if $A \in \mathcal{C}_t$, then $A \subseteq Y$. Since $\mathcal{C}_t \neq \varnothing$, we obtain that $v \geq r > r - m$ and therefore none of the sets $\mathcal{C}_1, \mathcal{C}_2, \ldots, \mathcal{C}_{t-1}$ contains a complete block of the pairwise balanced design $\mathbf{S}(Y)$. If $\mathcal{C}_t$ contains such a block, then Y is a block of $\mathbf{S}$ and $\mathcal{C}_t$ is a singleton. Therefore, $\mathbf{S}(Y)$ has a resolution $\mathfrak{R} = \{\mathcal{C}_1, \mathcal{C}_2, \ldots, \mathcal{C}_t\}$ of cardinality t. Since $\mathbf{S}(Y)$ has v points and $v + t - 1$ blocks, Theorem 5.1.15 implies that $\mathfrak{R}$ is an affine resolution.

We have to show that $\alpha + \beta = c$. Let

$$c' = |\mathcal{C}_t| = w - (t-1)c = v - (t-1)(c-1).$$

Since $t - 1 = \frac{r-\alpha}{\beta}$, we have $c'\beta = v\beta - (r-\alpha)(c-1)$. Equation (5.23) implies that $v\beta = r(c-1)$, so

$$c'\beta = (c-1)\alpha. \tag{5.27}$$

Since $(\mathcal{C}_t, Y)$ is a 2-(c', α, λ)-design with replication number r and v blocks, we have $(c'-1)\lambda = r(\alpha - 1)$ which implies

$$(c'-1)\beta = c(\alpha - 1). \tag{5.28}$$

Equations (5.27) and (5.28) imply $\alpha + \beta = c$.

(ii) $\Rightarrow$ (i). Suppose that $A \cap Y \neq \varnothing$ for any $A \in \mathcal{B}$ and that the pairwise balanced design $\mathbf{S}(Y)$ on v points admits an affine resolution $\mathcal{R} = \{\mathcal{C}_1, \mathcal{C}_2, \ldots, \mathcal{C}_t\}$ such that $\alpha(\mathcal{C}_t) = \alpha$ and, for $i = 1, 2, \ldots, t-1$, $|\mathcal{C}_i| = c$ and $\alpha(\mathcal{C}_i) = c - \alpha$. By Theorem 5.1.15(iv), if $A \in \mathcal{B} \setminus \mathcal{C}_t$, then $|A \cap Y|c = (c-\alpha)v$, so $|A \cap (X \setminus Y)| = m$ where $m = r - \frac{v(c-\alpha)}{c}$. If $A, B \in \mathcal{C}_i$ with $1 \leq i \leq t-1$ and $A \neq B$, then, by Theorem 5.1.15 (ii), $|A \cap B \cap Y| = |A \cap Y| - r + \lambda$, which implies that $|A \cap B \cap (X \setminus Y)| = \lambda - |A \cap B \cap Y| = m$, i.e., all blocks from any class $\mathcal{C}_i$, $i = 1, 2, \ldots, t-1$, in the substructure $\mathbf{D} = \mathbf{S}(X \setminus Y, \mathcal{B}_Y)$ are incident with the same m points. If $A \in \mathcal{C}_i$ and $B \in \mathcal{C}_j$ where $1 \leq i < j \leq t-1$, then Theorem 5.1.15 (iii) implies that $|A \cap B \cap (X \setminus Y)|$ is a constant, which we denote by μ. Therefore, selecting one block A_i from each $\mathcal{C}_i$ with $i = 1, 2, \ldots, t-1$ and putting $B_i = A_i \cap (X \setminus Y)$, we obtain that $\{B_1, B_2, \ldots, B_{t-1}\}$ is the block-set of a symmetric $(t-1, m, \mu)$-design $\mathbf{T}$ with point-set $X \setminus Y$. It suffices to prove that $\mathbf{T}$ is a normal subdesign of $\mathbf{S}$, that is, $A \cap (X \setminus Y) = \varnothing$ for any $A \in \mathcal{C}_t$.

Let $A \in \mathcal{C}_t$. Theorem 5.1.15(iii, iv) implies that $|A \cap B \cap Y|$ is the same for all $B \in \mathcal{B} \setminus \mathcal{C}_t$, so the set $A \cap (X \setminus Y)$ meets every block of $\mathbf{T}$ in the same number of points. By Proposition 5.1.14, we have either $A \cap (X \setminus Y) = \varnothing$ or $A \supseteq X \setminus Y$. Suppose that $A \supseteq X \setminus Y$. Since $|B \cap Y|$ is the same for all $B \in \mathcal{C}_t$, every block $B \in \mathcal{C}_t$ must contain $X \setminus Y$. Let $\mathbf{S}'$, $\mathbf{T}'$, and, for $i = 1, 2, \ldots, t$, $\mathcal{C}_i'$ be the complements of $\mathbf{S}$, $\mathbf{T}$, and the substructure $\mathbf{S}(X, \mathcal{C}_i)$, respectively. Then $\mathbf{T}'$ is a normal subdesign of $\mathbf{S}'$ and then part (i) of this proof implies that $\alpha(\mathcal{C}_i') = c - \alpha(\mathcal{C}_t')$, for $i = 1, 2, \ldots, t-1$. On the other hand, $\alpha(\mathcal{C}_i') = \alpha$ for $i = 1, 2, \ldots, t-1$ and $\alpha(\mathcal{C}_t') = |\mathcal{C}_t| - \alpha$. Therefore, $|\mathcal{C}_t| = c$. We now consider again the design $\mathbf{S}$. Since $|\mathcal{C}_t| = c$ and the replication number of the c-fold multiple of $\mathbf{T}$ is cm, we obtain that

$$r = c(m+1). \tag{5.29}$$

Since the replication number of $\mathbf{S}(Y)$ is $(c-\alpha)(t-1) + \alpha$, we derive that

$$t - 1 = \frac{r - \alpha}{c - \alpha}. \tag{5.30}$$

Applying Theorem 5.1.15(iv) to $\mathcal{C} = \mathcal{C}_1$ and using $k(\mathcal{C}_1) + m = r$, we obtain that

$$v = \frac{(r-m)c}{c-\alpha}. \tag{5.31}$$

Applying Theorem 5.1.15(iv) to $\mathcal{C} = \mathcal{C}_t$ and using (5.30) and $k(\mathcal{C}_t) + t - 1 = r$, we obtain that

$$v = \frac{cr}{\alpha} - \frac{c(r-\alpha)}{\alpha(c-\alpha)}. \tag{5.32}$$

Using (5.29), (5.31), and (5.32), we obtain $c(2\alpha - c + 1) = \alpha$, a contradiction, because $\alpha < c$.

Therefore, $A \cap (X \setminus Y) = \varnothing$ for each $A \in \mathcal{C}_t$, and then $\mathbf{T}$ is a normal subdesign of $\mathbf{S}$. The proof is now complete. □

Corollary 5.4.5. *Let B be a block of a symmetric design $\mathbf{S}$. Then the residual design $\mathbf{S}^B$ is an affine-resolvable 2-design if and only if the derived design $\mathbf{S}_B$ is a multiple of a symmetric design.*

Proof. Let $\mathbf{S}' = (X, \mathcal{B})$ be the complement of $\mathbf{S}$ and let $A = X \setminus B$. Then $\mathbf{S}'(X \setminus A, \mathcal{B}_A) = (\mathbf{S}')^A$, and the derived design $(\mathbf{S}')_A$ can be obtained by removing the only complete block from $\mathbf{S}'(A)$. Therefore, $\mathbf{S}'(A)$ satisfies condition (ii) of Theorem 5.4.4 if and only if $(\mathbf{S}')_A$ is an affine $(c-1)$-resolvable 2-design with resolution classes of cardinality c. This in turn is true if and only if the complementary 2-design $((\mathbf{S}')_A)'$ is an affine-resolvable design. Since $((\mathbf{S}')_A)'$ is isomorphic to $\mathbf{S}^B$ and $(\mathbf{S}')^A$ is isomorphic to $\mathbf{S}_B$, the statement of the corollary follows. □

Remark 5.4.6. If $\mathbf{S}$ is $PG_{n-1}(n, q)$ with $n \geq 3$, then $\mathbf{S}^B$ is $AG_{n-1}(n, q)$ and $\mathbf{S}_B$ is a multiple of $PG_{n-2}(n-1, q)$; if $\mathbf{S}$ is a symmetric $(n^2+n+1, n+1, 1)$-

design, then $\mathbf{S}^B$ is an affine plane of order n and $\mathbf{S}_B$ is a multiple of a symmetric $(n+1, 1, 0)$-design. No other realization of the condition of Corollary 5.4.5 is known.

The Dembowski–Wagner Theorem gives several combinatorial characterizations of the designs $PG_{n-1}(n, q)$ with $n \geq 3$. Corollary 5.4.5 allows us to characterize these designs in terms of their residual and derived designs.

Theorem 5.4.7. *Let $\mathbf{S} = (X, \mathcal{B})$ be a symmetric (v, k, λ)-design with $\lambda > 1$ and $v > k + 1$. Then the following conditions are equivalent.*

(i) $\mathbf{S}$ is isomorphic to $PG_{n-1}(n, q)$ for some prime power q and some integer $n \geq 3$;
(ii) the derived design $\mathbf{S}_B$ is a multiple of symmetric design for each block $B \in \mathcal{B}$;
(iii) the residual design $\mathbf{S}^B$ is an affine-resolvable 2-(v', k', λ') design with $\lambda' > 1$ for each block $B \in \mathcal{B}$.

Proof. Conditions (ii) and (iii) are equivalent by Corollary 5.4.5. Condition (i) implies both (ii) and (iii) (cf. Remark 5.4.6).

Suppose now that $\mathbf{S}$ satisfies both (ii) and (iii). Since all designs $PG_{n-1}(n, q)$ are selfdual (Proposition 3.6.9), it suffices to show that the design $\mathbf{S}^\top$ is isomorphic to some $PG_{n-1}(n, q)$ with $n \geq 3$. By the Dembowski–Wagner Theorem, it suffices to show that the cardinality of every line in $\mathbf{S}^\top$ is $1 + (k-1)/\lambda$. If A and B are distinct blocks of $\mathbf{S}$, that is, distinct points of $\mathbf{S}^\top$, then the line $\overline{AB}$ in $\mathbf{S}^\top$ is the set of all blocks C of $\mathbf{S}$ such that $A \cap B \subset C$. Let C be a block of $\mathbf{S}$ other than A or B. Since any two distinct blocks of a symmetric design meet in the same number of points, we have

$$A \cap B \subset C \Leftrightarrow C \cap B = A \cap B \Leftrightarrow (C \setminus B) \cap (A \setminus B) = \varnothing.$$

The last equality means that $C \setminus B$ and $A \setminus B$ are distinct parallel blocks of an affine-resolvable design $\mathbf{S}^B$. Therefore, $|\overline{AB}| = 1 + c$ where c is the cardinality of a parallel class of $\mathbf{S}^B$. By Theorem 5.4.4, $\mathbf{S}_B$ is a c-fold multiple of a symmetric design. Since $\mathbf{S}_B$ is a $(k, v-1, k-1, \lambda, \lambda-1)$-design, we obtain that $c = (k-1)/\lambda$ and therefore $|\overline{AB}| = 1 + (k-1)/\lambda$. The proof is now complete. □

Theorem 5.4.4 shows that, under some conditions, a symmetric design can be split into a multiple of a smaller symmetric design and a PBD with an affine resolution. The next result gives a condition under which a symmetric design and a PBD with an affine resolution can be combined into a larger symmetric design.

Theorem 5.4.8. *Let α, λ, c, and t be positive integers with $c > \alpha$ and $t \geq 2$. Suppose there exists a PBD of index λ on v points that admits an affine resolution having t resolution classes, one of them of replication number α and the other $t-1$ of cardinality c and replication number $c-\alpha$. Suppose further that there exists a symmetric $(t-1, \frac{r}{c}, \frac{\lambda}{c})$-design, where $r = (t-1)(c-\alpha)+\alpha$. Then there exists a symmetric $(v+t-1, r, \lambda)$ design.*

Proof. Let $\mathbf{D} = (X, \mathcal{B})$ be a PBD satisfying the conditions of the theorem and let $\mathbf{T} = (X', \mathcal{B}')$ be a symmetric $(t-1, \frac{r}{c}, \frac{\lambda}{c})$-design where r is the replication number of $\mathbf{D}$. We will assume that $X \cap X' = \varnothing$. Let $\mathcal{B}' = \{B_1, B_2, \ldots, B_{t-1}\}$ and let $\mathfrak{R} = \{\mathcal{C}_1, \mathcal{C}_2, \ldots, \mathcal{C}_t\}$ be the affine resolution of $\mathbf{D}$. Let $\alpha(\mathcal{C}_t) = \alpha$ and let $|\mathcal{C}_i| = c$ and $\alpha(\mathcal{C}_i) = c - \alpha$ for $i = 1, 2, \ldots, t-1$. Put $Y = X \cup X'$. For $i = 1, 2, \ldots, t-1$, put $\mathcal{C}_i^* = \{B \cup B_i : B \in \mathcal{C}_i\}$ and put $\mathcal{A} = \left(\bigcup_{i=1}^{t-1} \mathcal{C}_i^*\right) \cup \mathcal{C}_t$. We shall show that $\mathbf{S} = (Y, \mathcal{A})$ is a symmetric 2-(w, r, λ) design where $w = v + t - 1$. Since $\mathbf{S}$ has w points and w blocks, each point is replicated r times, and any two points from X as well as any two points from X' occur together in exactly λ blocks, it suffices to show that any point $x \in X$ and any point $x' \in X'$ occur together in exactly λ blocks. Since x' occurs in all the blocks of $\frac{r}{c}$ classes $\mathcal{C}_i^*$ and x occurs in $c - \alpha$ blocks of each of these classes, x and x' occur together in $\frac{r(c-\alpha)}{c}$ blocks. On the other hand, since $\mathbf{T}$ is a symmetric design, we have $(t-2)\frac{\lambda}{c} = \frac{r}{c}(\frac{r}{c} - 1)$. Using $r = (t-1)(c-\alpha)+\alpha$, we derive that $\lambda = \frac{r(c-\alpha)}{c}$. □

Applying this theorem to the complement of an affine-resolvable design $A_\mu(s)$ yields the following result.

Corollary 5.4.9. *Let s and μ be positive integers. If there exists an affine-resolvable design $A_\mu(s)$ and a symmetric $((s^2\mu - 1)/(s-1), (s\mu - 1)/(s-1), (\mu - 1)/(s-1))$-design, then there exists a symmetric $((s^3\mu - 1)/(s-1), (s^2\mu - 1)/(s-1), (s\mu - 1)/(s-1))$-design.*

Remark 5.4.10. Let $\mathbf{D}$ be a PBD with an affine resolution $\mathfrak{R}$ satisfying condition (ii) of Theorem 5.4.4. If $c = 2\alpha$, then $\mathbf{D}$ is an α-resolvable PBD embedded in a symmetric design $\mathbf{S}$. If $\mathfrak{R}$ has one class of cardinality 2α and replication number α and the other classes of cardinality c and replication number $c - \alpha$, then the complement $\mathbf{D}'$ of $\mathbf{D}$ is an affine α-resolvable PBD embedded in the complement of $\mathbf{S}$. In the next theorem, we show that these are the only types of affine α-resolvable PBDs with $\alpha \geq 2$ that are embeddable in symmetric designs.

Theorem 5.4.11. *Let $\mathbf{D} = (X, \mathcal{B}, \mathcal{R})$ be an affine α-resolvable PBD of index λ. Suppose that $\alpha \geq 2$, $\mathbf{D}$ is not a BIBD, and $\mathbf{D}$ is embedded in a symmetric design $\mathbf{S}$. Then either*

(i) **S** *is a Hadamard 2-*$(4\alpha m-1, 2\alpha m, \alpha m)$ *design where* m *is even or* $m=1$; $|X|=(4\alpha-2)m$; $\mathcal{R}$ *consists of* $2m-1$ *classes of cardinality* 2α *and one class of cardinality* $2\alpha-1$

or

(ii) **S** *is a symmetric* $((2\alpha-1)^3n+8\alpha^2-4\alpha+1, \alpha(2\alpha-1)^2n+4\alpha^2, \alpha^2(2\alpha-1)n+\alpha(2\alpha+1))$*-design where* n *is a nonnegative integer;* $|X|=2(\alpha-1)(2\alpha-1)^2n+2(2\alpha-1)^2$; $\mathcal{R}$ *consists of one class of cardinality* 2α *and* $(2\alpha-1)^2n+4\alpha-1$ *classes of cardinality* $2\alpha-1$.

Proof. Let $|X|=v$, $|\mathcal{R}|=t$, so $|\mathcal{B}|=b=v+t-1$. Let r, m, u, c_1 and c_2 have the same meaning as in Theorem 5.2.2. Then r is the block size of **S**. Since **D** is not a BIBD, Theorem 5.2.2(iv) implies that $c_1\neq c_2$. Let $c_1>c_2$. Let $\mathbf{S}=(W,\mathcal{A})$ be a symmetric (w,r,λ)-design.

Claim 1. $w=b$.

Assume that $w=b+b_0=v+t-1+b_0$ where $b_0\geq 1$. This means that $\mathcal{A}=\mathcal{A}^*\cup\mathcal{A}_0$ where $|\mathcal{A}^*|=b$ and $A\cap X\in\mathcal{B}$ for any $A\in\mathcal{A}^*$, $|\mathcal{A}_0|=b_0$, and $A\cap X=\varnothing$ for any $A\in\mathcal{A}_0$. If $x\in X$ and $y\in W\setminus X$, then there are λ blocks in $\mathcal{A}^*$ which contain both x and y. Therefore, y is contained in at most $r-\lambda$ blocks from $\mathcal{A}_0$. Since each block of **S** has cardinality r, we obtain the inequality $rb_0\leq(r-\lambda)(t-1+b_0)$ which we transform into

$$\lambda b_0\leq(r-\lambda)(t-1). \tag{5.33}$$

On the other hand, since $A\cap X=\varnothing$ for any $A\in\mathcal{A}_0$, we have $|A|\leq|W\setminus X|$, i.e., $r\leq t-1+b_0$, so

$$b_0\geq r-t+1. \tag{5.34}$$

Inequalities (5.33) and (5.34) yield

$$r-t+1\leq\frac{(r-\lambda)(t-1)}{\lambda}.$$

Theorem 5.2.2 implies that $r-\lambda=m\alpha$. Since $r=t\alpha$, we have $\lambda=(t-m)\alpha$, so we obtain the inequality

$$t\alpha-t+1\leq\frac{m(t-1)}{t-m}. \tag{5.35}$$

By Theorem 5.2.2, a block of cardinality $k_1=mc_2$ and a block of cardinality $k_2=mc_1$ of **D** meet in $\frac{k_1k_2}{v}=\frac{mc_1c_2}{u}=\alpha m$ points. Since any two blocks of **S** meet in λ points, we have $\lambda\geq\alpha m$, which implies $t\geq 2m$. Now (5.35) implies $(\alpha-1)t<t-1$, which cannot be true, since $\alpha\geq 2$.

Thus, **S** is a symmetric 2-$(v+t-1, r, \lambda)$ design.

Claim 2. $v = (4\alpha - 2)m$, $c_1 = 2\alpha$, and $c_2 = 2\alpha - 1$.

The basic symmetric design equation implies $(v+t-2)\lambda = r(r-1)$. Substituting $v = um$, $\lambda = \alpha(t-m)$, and $r = \alpha t$, we obtain by routine manipulations the equation

$$(\alpha - 1)t^2 - (um - m - 1)t + m(um - 2) = 0. \tag{5.36}$$

The discriminant of this equation must be a square, so

$$(um - m - 1)^2 - 4m(um-2)(\alpha - 1) = d^2,$$

where d is a nonnegative integer. This equation can be transformed into

$$((u-1)^2 - 4u(\alpha - 1))m^2 - 2(u - 4\alpha + 3)m + 1 = d^2. \tag{5.37}$$

Since $c_1 + c_2 = u + 1$ and $c_1c_2 = \alpha u$, we have

$$(u-1)^2 - 4u(\alpha - 1) = (c_1 - c_2)^2 \tag{5.38}$$

and

$$u - 4\alpha + 3 = c_1 + c_2 + 2 - \frac{4c_1c_2}{c_1 + c_2 - 1} = \frac{(c_1 - c_2)^2 - 1}{c_1 + c_2 - 1} + 1,$$

so

$$u - 4\alpha + 3 = \frac{(c_1 - c_2 - 1)(c_1 - c_2 + 1)}{c_1 + c_2 - 1} + 1. \tag{5.39}$$

Using (5.38), we transform (5.37) into

$$(c_1 - c_2)^2m^2 - 2(u - 4\alpha + 3)m + 1 = d^2. \tag{5.40}$$

Since (5.39) implies $u - 4\alpha + 3 \geq 1$, we obtain that $d < (c_1 - c_2)m$, so $d \leq (c_1 - c_2)m - 1$. Therefore, $d^2 \leq (c_1 - c_2)^2m^2 - 2(c_1 - c_2)m + 1$, and (5.40) implies that $u - 4\alpha + 3 \geq c_1 - c_2$. Using (5.39), we obtain that if $c_1 - c_2 > 1$, then $c_1 - c_2 + 1 \geq c_1 + c_2 - 1$, so $c_2 = 1$, which implies $\alpha = 1$, a contradiction. Therefore, $c_1 - c_2 = 1$. Since $c_1 + c_2 = u + 1$, we obtain $c_1 = \frac{u+2}{2}$ and $c_2 = \frac{u}{2}$. Since $c_1c_2 = \alpha u$, we obtain that $u = 4\alpha - 2$, so $c_1 = 2\alpha$, $c_2 = 2\alpha - 1$, and $v = um = (4\alpha - 2)m$.

Equation (5.37) now yields $d = m - 1$ and (5.36) yields $t = 2m$ or $t = 2m + \frac{m-1}{\alpha - 1}$.

Suppose $\mathcal{R}$ consists of t_1 classes of cardinality c_1 and t_2 classes of cardinality c_2. Then $|\mathcal{B}| = c_1t_1 + c_2t_2$. Since $|\mathcal{B}| = v + |\mathfrak{R}| - 1 = v + t_1 + t_2 - 1$, we have

$$(c_1 - 1)t_1 + (c_2 - 1)t_2 = v - 1.$$

Therefore,

$$(2\alpha - 1)t_1 + (2\alpha - 2)t_2 = (4\alpha - 2)m - 1. \tag{5.41}$$

Case 1. Suppose that $t = 2m$. Then $t_1 + t_2 = t$ and (5.41) yield $t_1 = 2m - 1$ and $t_2 = 1$. Then $v + t - 1 = 4\alpha m - 1$, $r = \alpha t = 2\alpha m$, and $\lambda = \alpha(t - m) = \alpha m$, so **S** is a Hadamard 2-$(4\alpha m - 1, 2\alpha m, \alpha m)$ design. If $m > 1$, then **D** has two blocks of cardinality $k_1 = (2\alpha - 1)m$ from different classes of cardinality c_1. By Theorem 5.1.15(iii), these blocks meet in $\frac{(2\alpha-1)m}{2}$ points, so m is even.

Case 2. Suppose that $t = 2m + z$ where $z = \frac{m-1}{\alpha-1}$. Then $t_1 + t_2 = t$ and (5.41) yield $t_1 = 1$ and $t_2 = 1 + (2\alpha - 1)z$. Recall that **S** is a 2-$(v + t - 1, \alpha t, \alpha(t - m))$ design. We have $v + t - 1 = 4\alpha(\alpha - 1)z + 4\alpha + z - 1$. If $m = 1$, we have the previous case, so we assume that $m \geq 2$. Then $t_2 > 1$, so we can find two blocks of cardinality k_2 in **D** from different classes of cardinality c_2. By Theorem 5.1.15(iii), these blocks meet in $\frac{2\alpha^2 m}{2\alpha-1}$ points , so $2\alpha - 1$ divides m. Since $(\alpha - 1)z \equiv -1 \pmod{m}$, we obtain that $(\alpha - 1)z \equiv 2\alpha - 2 \pmod{2\alpha - 1}$, i.e., $z \equiv 2 \pmod{2\alpha - 1}$. Let $z = 2 + (2\alpha - 1)n$ where n is a nonnegative integer. We can now express $m = (\alpha - 1)z + 1$, $v = (4\alpha - 2)m$, $t = 2m + z$, $r = \alpha t$, and $\lambda = \alpha(t - m)$ in terms of α and n and obtain that **S** is a symmetric $((2\alpha - 1)^3 n + 8\alpha^2 - 4\alpha + 1, \alpha(2\alpha - 1)^2 n + 4\alpha^2, \alpha^2(2\alpha - 1)n + \alpha(2\alpha + 1))$-design. □

We will now construct a family of affine α-resolvable pairwise balanced designs embeddable in symmetric designs. This family will correspond to case (i) of Theorem 5.4.11. We will discuss a possible realization of case (ii) in Remark 12.3.7.

Theorem 5.4.12. *Let α and m be positive integers. If there exist Hadamard matrices of orders 4α and $2m$, then there exists an affine α-resolvable PBD of index αm on $(4\alpha - 2)m$ points whose resolution consists of $2m - 1$ classes of cardinality 2α and one class of cardinality $2\alpha - 1$.*

Proof. Let H_1 and H_2 be the matrices from the statement of Theorem 4.2.5 with $h_1 = 2h$ and $h_2 = m$. We also assume that the matrices P_1, Q_1, P_2, and Q_2 in that statement have all entries in the first row equal to 1, while matrices P_1, Q_1, R_1, P_2, and R_2 have all entries in the first column equal to 1, and all entries in the first column of S_1 are equal to -1. Let H be the Hadamard matrix of order $4\alpha m$ constructed in Theorem 4.2.5. Then H is a normalized Hadamard matrix

The set $W = \{2, 3, \ldots, 4\alpha m\}$ serves as the point-set of the Hadamard 2-$(4\alpha m - 1, 2\alpha m, \alpha m)$ design $\mathbf{S}$ with the blocks $B_i = \{j \in W : (H)_{ij} = -1\}$, $i = 2, 3, \ldots, 4\alpha m$. Let $V' = \{1, 2, \ldots, m\} \cup \{2\alpha m + 1, 2\alpha_m + 2, \ldots, 2\alpha m + m\}$ and $V = W \setminus V'$. Consider the submatrix $K = [k_{ij}]$ of H formed by the columns whose index is in V'. Then K can be represented as a block-matrix $K = [K_{ij}]$, $1 \le i \le 4\alpha$, $1 \le j \le 2$, where

$$K_{i1} = \begin{cases} P_2 & \text{for } 1 \le i \le 2\alpha, \\ R_2 & \text{for } 2\alpha + 1 \le i \le 4\alpha, \end{cases} \quad K_{i2} = \begin{cases} Q_2 & \text{for } 1 \le i \le 2\alpha, \\ S_2 & \text{for } 2\alpha + 1 \le i \le 4\alpha. \end{cases}$$

For $i = 1, 2, \ldots, 4m\alpha$, put $A_i = \{j : 2 \le j \le 2m \text{ and } k_{ij} = -1\}$. Then each set $A_{\beta m+1}$ for $\beta = 0, 1, \ldots, 2\alpha - 1$ is empty and the family

$$\{A_2, A_3, \ldots, A_m\} \cup \{A_{2m\alpha+1}, A_{2\alpha_m+2}, \ldots, A_{2m\alpha+m}\}$$

is the block-set of a Hadamard 2-$(2m - 1, m, \frac{m}{2})$ design $\mathbf{T}$, so the nonempty sets A_i, $i = 1, 2, \ldots, 4m\alpha$, form the blockset of the (2α)-fold multiple of $\mathbf{T}$. By Theorem 5.4.4, $\mathbf{D} = \mathbf{S}^\top(V)$ is affine α-resolvable. □

5.5. Resolvable 2-designs and equidistant codes

In Section 4.6 we considered equidistant binary codes and some relations between these codes and symmetric designs. In this section, we explore the connections between equidistant q-ary codes and resolvable designs.

We will consider q-ary codes over the alphabet $\mathcal{A} = \{0, 1, \ldots, q - 1\}$, $q \ge 2$. A q-ary (n, m, d)-code C consists of m codewords of length n with minimum distance d between the codewords. We will identify such a code with an $m \times n$ matrix whose rows are the codewords. We will denote this matrix by the same letter C.

Definition 5.5.1. If C is a q-ary (n, m, d)-code, then

$$\overline{d} = \frac{1}{\binom{m}{2}} \sum_{x,y \in C} d(x, y)$$

is called the *mean distance of* C.

Lemma 5.5.2. *The mean distance* $\overline{d}$ *of a* q*-ary* (n, m, d)*-code* C *satisfies the inequality*

$$\overline{d} \le \frac{mn(q - 1)}{(m - 1)q}$$

with equality if and only if every element of the alphabet occurs m/q times in each column of C.

Proof. For $i = 0, 1, \dots, q-1$ and $j = 1, 2, \dots, n$, let a_{ij} denote the number of occurrences of i in the jth column of C. Since $\sum_{i=0}^{q-1} a_{ij} = m$, we obtain:

$$\sum_{x,y\in C} d(x, y) = \sum_{j=1}^{n}\left(\binom{m}{2} - \sum_{i=0}^{q-1}\binom{a_{ij}}{2}\right)$$
$$= \frac{nm(m-1)}{2} + \frac{nm}{2} - \frac{1}{2}\sum_{j=1}^{n}\sum_{i=0}^{q-1} a_{ij}^2.$$

Since

$$\frac{1}{nq}\sum_{j=1}^{n}\sum_{i=0}^{q-1} a_{ij}^2 \geq \left(\frac{1}{nq}\sum_{j=1}^{n}\sum_{i=0}^{q-1} a_{ij}\right)^2 = \left(\frac{m}{q}\right)^2,$$

we obtain that

$$\sum_{x,y\in C} d(x, y) \leq \frac{nm^2(q-1)}{2q}.$$

This implies the desired inequality. The equality holds if and only if all a_{ij} are the same, which means that every element of the alphabet occurs m/q times in each column of C. □

Definition 5.5.3. A q-ary equidistant (n, m, d)-code is called *maximal* if

$$d = \frac{mn(q-1)}{(m-1)q}.$$

Theorem 5.5.4. *A maximal q-ary equidistant (n, m, d)-code exists if and only if there exists a resolvable $(m, nq, n, m/q, n-d)$-design.*

Proof. Let $N = [N_1\ N_2\ \dots\ N_n]$ be an incidence matrix of a resolvable $(m, nq, n, m/q,\ n-d)$-design **D** where submatrices $N_1, N_2, \dots, N_n$ correspond to distinct parallel classes of **D**. We will index the q columns of each submatrix N_j by $0, 1, \dots, q-1$ and define the matrix $C = [c_{ij}]$ over the alphabet $\mathcal{A}$ by

$$c_{ij} = l \text{ if and only if the } (i, l)\text{-entry of } N_j \text{ is equal to } 1.$$

If codewords x and y correspond to i^{th} and h^{th} rows of C $(i \neq h)$, then $n - d(x, y)$ is the number of columns of N, which have 1 in both i^{th} and h^{th} rows. Since this is the number of blocks of **D** that contain the corresponding points, we obtain that $d(x, y) = d$ for all distinct x and y, i.e., C is a q-ary equidistant (n, m, d)-code. Since each element of $\mathcal{A}$ occurs m/q times in each column of C, the code C is maximal.

Conversely, suppose now that we have a maximal q-ary equidistant (n, m, d)-code $C = [c_{ij}]$. For $j = 1, 2, \ldots, n$, define a (0, 1)-matrix N_j of size $m \times q$ (with columns indexed by $0, 1, \ldots, q-1$) whose (i, l)-entry is equal to 1 if and only if $c_{ij} = l$. Then the row sum of N_j is 1 and the matrix $N = [N_1\ N_2\ \ldots\ N_n]$ is an incidence matrix of a resolvable $(m, nq, n, m/q, n-d)$-design. □

Exercises

(1) Let $\mathbf{D} = (X, \mathcal{B})$ be a resolvable PBD of index λ on v points. Prove the following statements.
   (a) If a positive integer d is such that $|B| \equiv 1 \pmod{d}$ for every $B \in \mathcal{B}$, then $(v-1)\lambda \equiv 0 \pmod{d}$.
   (b) If a positive integer d is such that $|B|^2 \equiv |B| \pmod{d}$ for every $B \in \mathcal{B}$, then $v(v-1)\lambda \equiv 0 \pmod{d}$.

(2) Construct a resolvable PBD of index 1 on 12 points.

(3) Let d and n be positive integers, $1 \le d \le n$, and let V be the n-dimensional vector space over $GF(q)$. Let X be the set of all lines of V and $\mathcal{B}$ the set of all d-flats of V. Prove that the incidence structure $\mathbf{D} = (X, \mathcal{B}, I)$ with $(x, B) \in I$ if and only if $x \subseteq B$ is a resolvable 2-design.

(4) Show that there exists a 2-(10, 5, 4) design but there is no resolvable 2-design with these parameters.

(5) Show that there is no resolvable 2-(v, k, λ) design with $(v, k, \lambda) = (14, 7, 6)$, $(15, 5, 2)$, $(18, 9, 8)$, and $(21, 7, 3)$.

(6) Construct a resolvable 2-(28, 4, 1) design.

(7) Prove that if q and $q-1$ are prime powers, then there exists a $(q-1)$-resolvable 2-(q^3+1, q^2+q, q^2+q-1) design.

Notes

The notion of a resolvable and affine resolvable 2-design was introduced in the seminal paper by Bose (1942). In this paper, Bose's Inequality for these designs was proved, and the case of equality was characterized. The notion of a pairwise balanced design was introduced in Bose and S. S. Shrikhande (1959b). S. S. Shrikhande and Raghavarao (1964) generalized the concept of resolvability and affine resolvability of 2-designs to α-resolvability, extended Bose's Inequality to these designs, and also characterized the case of equality. Hughes and Piper (1976) obtained a similar result under the assumption that different resolution classes may have different replication numbers. Vanstone (1979) extended Bose's Inequality to (r, λ)-designs, in which every point occurs exactly once in each resolution class. Ionin and M. S. Shrikhande (1998) extended the concepts of resolvability, affine resolvability, and α-resolvability to pairwise balanced designs.

The proofs of Theorems 5.1.8, 5.1.12, 5.1.15, and 5.2.2 follow that paper. For further information on PBDs, see Mullin and Gronau (1996).

Proposition 5.1.14 for the case of symmetric design was proven in Marrero (1972). The statement and the proof presented here are given in Ionin and M. S. Shrikhande (1998).

Kirkman's schoolgirl problem was introduced and solved for 15 school girls in Kirkman (1847). Theorem 5.3.8 solved a longstanding open problem. This theorem is due to Ray-Chaudhuri and Wilson (1973).

The construction of resolvable unitals in Theorem 5.3.9 is due to Bose (1958–9).

Theorem 5.3.10 is proven in S. S. Shrikhande and Raghavarao (1963). Theorem 5.3.15 is due to Lenz and Jungnickel (1979).

Embedability of pairwise balanced designs in symmetric designs was studied in Bekker, Ionin and M. S. Shrikhande (1998). Corollary 5.4.5 was obtained in Mavron (1973). Theorem 5.4.7 is due to Kantor (see Dembowski (1968), p. 75 footnote). Corollary 5.4.9 was proved in S. S. Shrikhande (1951). The survey paper by S. S. Shrikhande (1976) contains a conjecture that the only affine resolvable 2-designs are designs $AG_{d-1}(d, q)$ and Hadamard 3-designs. This conjecture is still open.

For more results on resolvable designs we refer to Abel and Furino (1996), Furino, Miao and Yin (1996), and Beth, Jungnickel and Lenz (1999).

The inequality for the minimal distance that follows from Lemma 5.5.2 is known as the *Plotkin bound* (Plotkin (1960)).

Relations between resolvable 2-designs and equidistant codes presented in Section 5.5 are due to Semakov and Zinoviev (1968).

6
Symmetric designs and t-designs

In a 2-(v, k, λ) design, every pair of distinct points is incident with exactly λ blocks. In a more general incidence structure of a t-(v, k, λ) *design*, every subset of t points occurs in exactly λ blocks. Among the most important examples of t-designs with $t > 2$ are *Witt designs*, which will be used in this chapter for constructing interesting symmetric designs. Witt designs are closely related to other famous combinatorial objects, including *binary Golay codes*.

6.1. Basic properties of t-designs

We defined 2-(v, k, λ) designs in Chapter 2. For $\lambda \geq 1$, we will now define a more general notion of a t-(v, k, λ) design.

Definition 6.1.1. Let t, v, k, and λ be integers, $v \geq k \geq t \geq 0$. A t-(v, k, λ) *design* is an incidence structure $\mathbf{D} = (X, \mathcal{B}, I)$, satisfying the following conditions:

(i) $|X| = v$;
(ii) every block $B \in \mathcal{B}$ is incident with exactly k points;
(iii) for any t-subset Y of X, there are exactly λ blocks that are incident with every point of Y.

Remark 6.1.2. The requirement $k \geq t$ in the definition of a t-(v, k, λ) design implies $\lambda \geq 1$.

Example 6.1.3. Let $\mathcal{B}$ be the set of all k-subsets of X. If $|X| = v$, then, for $t \leq k$, $(X, \mathcal{B})$ is a t-(v, k, λ) design with $\lambda = \binom{v-t}{k-t}$.

Definition 6.1.4. A t-(v, k, λ) design on the point set X is called *trivial* if every k-subset of X is incident with at least one block.

Example 6.1.5. The design described in Proposition 4.1.10 is a 3-$(4n, 2n, n-1)$ design known as a *Hadamard 3-design*.

As the next theorem shows, every t-design with $t \geq 1$ is also an s-design for $0 \leq s \leq t$.

Theorem 6.1.6. *Let* $\mathbf{D} = (X, \mathcal{B}, I)$ *be a* t-(v, k, λ) *design. Then, for* $s = 0, 1, \ldots, t$, $\mathbf{D}$ *is an* s-(v, k, λ_s) *design with*

$$\lambda_s = \lambda \frac{\binom{v-s}{t-s}}{\binom{k-s}{t-s}}.$$

Proof. Let A be an s-subset of X, $0 \leq s \leq t$. Let λ_s be the number of blocks incident with every point of A. Counting in two ways pairs (B, Y), where Y is a t-subset containing A and B is a block incident with every point of Y, yields

$$\lambda_s \binom{k-s}{t-s} = \binom{v-s}{t-s} \lambda,$$

so λ_s is independent of the s-subset A. Therefore, $\mathbf{D}$ is an s-(v, k, λ_s) design.

□

Remark 6.1.7. Obviously, $b = \lambda_0$ is the cardinality of the block-set $\mathcal{B}$ and $r = \lambda_1$ is the replication number of $\mathbf{D}$.

Corollary 6.1.8. *If there exists a* t-(v, k, λ) *design with* $t \geq 1$, *then, for* $0 \leq s \leq t-1$, *the product* $(k-s)(k-s-1)\cdots(k-t+1)$ *divides the product* $\lambda(v-s)(v-s-1)\cdots(v-t+1)$.

This result gives necessary conditions for the parameters of a t-(v, k, λ) design. That these conditions are asymptotically sufficient is the content of the following theorem, whose proof is beyond the scope of this book.

Theorem 6.1.9. *Let integers* $v \geq k \geq t > 0$ *and* λ *be such that* $(k-s)(k-s-1)\cdots(k-t+1)$ *divides* $\lambda(v-s)(v-s-1)\cdots(v-t+1)$ *for* $s = 0, 1, \ldots, t-1$. *Then there is a positive integer* n *such that, whenever* $\lambda > n$, *there exists a* t-(v, k, λ) *design.*

We now recall the following classical result, known as the *Inclusion–Exclusion Principle*.

Proposition 6.1.10 (The Inclusion–Exclusion Principle). *Let* $X_1, X_2, \ldots, X_n$ *be finite sets. For any subset* I *of the set* $\{1, 2, \ldots, n\}$, *let* $Y_I = \bigcap_{i \in I} X_i$. *Then*

$$\left| \bigcup_{i=1}^{n} X_i \right| = \sum_{k=1}^{n} (-1)^{k-1} \sum_{|I|=k} |Y_I|.$$

The Inclusion–Exclusion Principle implies the next result.

Proposition 6.1.11. *If* $\mathbf{D} = (X, \mathcal{B}, I)$ *is a* t-(v, k, λ) *design with* $k \leq v - t$, *then the complementary incidence structure is a* t-$(v, v - k, \lambda')$ *design with*

$$\lambda' = \sum_{s=0}^{t} (-1)^s \binom{t}{s} \lambda_s.$$

Proof. Let $\{x_1, x_2, \ldots, x_t\}$ be a t-subset of X. For $i = 1, 2, \ldots, t$, let $\mathcal{B}_i$ be the set of all blocks that contain x_i. Then $\lambda' = |\mathcal{B}| - |\mathcal{B}_1 \cup \mathcal{B}_2 \cup \ldots \cup \mathcal{B}_t|$, and we apply Theorem 6.1.6 and the inclusion–exclusion principle. □

If $\mathbf{D}$ is a t-design, then the point-residual and the point-derived substructures are $(t - 1)$-designs.

Proposition 6.1.12. *Let* $D = (X, \mathcal{B}, I)$ *be a* t-(v, k, λ) *design with* $t \geq 1$ *and let* $x \in X$. *Then* $\mathbf{D}^x$ *is a* $(t - 1)$-$(v - 1, k, \lambda_{t-1} - \lambda)$ *design and* $\mathbf{D}_x$ *is a* $(t - 1)$-$(v - 1, k - 1, \lambda)$ *design.*

Proof. Let Y be a $(t - 1)$-subset of $X \setminus \{x\}$. Then $Y \cup \{x\}$ is a t-subset of X. Therefore, there are exactly λ blocks $B \in \mathcal{B}$ that contain x and contain Y, which means that $\mathbf{D}_x$ is a $(t - 1)$-$(v - 1, k - 1, \lambda)$ design. Since exactly λ_{t-1} blocks $B \in \mathcal{B}$ contain Y, exactly $\lambda_{t-1} - \lambda$ of them do not contain x. This proves that $\mathbf{D}^x$ is a $(t - 1)$-$(v - 1, k, \lambda_{t-1} - \lambda)$ design. □

Corollary 6.1.13. *A* t-(v, k, λ) *design with* $t \geq 3$ *and* $k \leq v - 2$ *cannot be a symmetric design.*

Proof. Let $\mathbf{D} = (X, \mathcal{B}, I)$ be a t-(v, k, λ) design with $t \geq 3$ and $k \leq v - 2$. Let $x \in X$. Then $\mathbf{D}_x$ is a $(t - 1)$-$(v - 1, k - 1, \lambda)$ design, so $\mathbf{D}_x$ is a 2-design. If $\mathbf{D}$ is a symmetric design, then it has exactly k blocks containing x. Therefore, $\mathbf{D}_x$ has $v - 1$ points and $k \leq v - 2$ blocks. This contradicts Fisher's Inequality. □

Example 6.1.14. If $\mathbf{D}$ is a 3-$(4n, 2n, n - 1)$ design, then, for any point x, $\mathbf{D}_x$ is a symmetric $(4n - 1, 2n - 1, n - 1)$-design and D^x is a symmetric $(4n - 1, 2n, n)$-design.

We will now introduce the notion of *intersection numbers* of an incidence structure.

Definition 6.1.15. An *intersection number* of an incidence structure is the cardinality of the intersection of two distinct blocks of this structure.

The following proposition is immediate.

Proposition 6.1.16. *Let S be the set of all intersection numbers of a t-design* $\mathbf{D}$ *with $t \geq 1$. Then each intersection number of every point-derived design of* $\mathbf{D}$ *is contained in the set $\{\alpha - 1 : \alpha \in S\}$. Conversely, every nonnegative element of this set is an intersection number of some point-derived design of* $\mathbf{D}$.

Given a t-design and disjoint subsets P and Q of its point set such that $|P| + |Q| \leq t$, the number of blocks of the design that contain P and are disjoint from Q is constant.

Theorem 6.1.17. *Let $\mathbf{D} = (X, \mathcal{B}, I)$ be a t-(v, k, λ) design and let P and Q be subsets of X of cardinalities p and q, respectively. If $p + q \leq t$ and $P \cap Q = \varnothing$, then the number of blocks that contain P and are disjoint from Q is equal to $\lambda\binom{v-p-q}{k-p}/\binom{v-t}{k-t}$.*

Proof. Let $Q = \{y_1, y_2, \ldots, y_q\}$. Let $\mathcal{B}_0$ be the set of all blocks incident with every point of P and, for $i = 1, 2, \ldots, q$, let $\mathcal{B}_i$ be the set of blocks $B \in \mathcal{B}_0$ that are incident with q_i. We have to prove that

$$|\mathcal{B}_0| - |\mathcal{B}_1 \cup \mathcal{B}_2 \cup \ldots \cup \mathcal{B}_q| = \frac{\lambda\binom{v-p-q}{k-p}}{\binom{v-t}{k-t}}. \tag{6.1}$$

Theorem 6.1.6 implies that $|\mathcal{B}_0| = \lambda\binom{v-p}{t-p}/\binom{k-p}{t-p}$. If $1 \leq i_1 < i_2 < \cdots < i_m \leq q$, then, by Theorem 6.1.6,

$$|\mathcal{B}_{i_1} \cap \mathcal{B}_{i_2} \cap \ldots \cap \mathcal{B}_{i_m}| = \frac{\lambda\binom{v-p-m}{t-p-m}}{\binom{k-p-m}{t-p-m}},$$

so by the inclusion–exclusion principle,

$$|\mathcal{B}_1 \cup \mathcal{B}_2 \cup \ldots \cup \mathcal{B}_q| = \lambda \sum_{m=1}^{q} (-1)^{m-1} \binom{q}{m} \frac{\binom{v-p-m}{t-p-m}}{\binom{k-p-m}{t-p-m}}.$$

Let $n = v - p$, $l = k - p$, and $u = t - p$. Then (6.1) is equivalent to

$$\sum_{m=0}^{q} (-1)^m \binom{q}{m} \frac{\binom{n-m}{u-m}}{\binom{l-m}{u-m}} = \frac{\binom{n-q}{l}}{\binom{n-u}{l-u}}. \tag{6.2}$$

We have

$$\frac{\binom{n-m}{u-m}}{\binom{l-m}{u-m}} = \frac{(n-m)!(l-u)!}{(n-u)!(l-m)!} \quad \text{and} \quad \frac{\binom{n-q}{l}}{\binom{n-u}{l-u}} = \frac{(n-q)!(n-l)!(l-u)!}{(n-q-l)!l!(n-u)!},$$

so (6.2) is equivalent to

$$\sum_{m=0}^{q} (-1)^m \binom{q}{m} \binom{n-m}{l-m} = \binom{n-q}{l}. \tag{6.3}$$

To prove (6.3), let $A = X \setminus P$. Then $|A| = n$. Let $\mathcal{A}_0$ be the set of all l-subsets of A that are disjoint from Q. For $m = 1, 2, \dots, q$, let $\mathcal{A}_m$ be the set of all l-subsets of A that contain y_m. Then

$$|\mathcal{A}_0| = \binom{n-q}{l} = \binom{n}{l} - |\mathcal{A}_1 \cup \mathcal{A}_2 \cup \ldots \cup \mathcal{A}_q|. \tag{6.4}$$

If $1 \le i_1 < i_2 < \cdots < i_m \le q$, then $|\mathcal{A}_{i_1} \cap \mathcal{A}_{i_2} \cap \ldots \cap \mathcal{A}_{i_m}| = \binom{n-m}{l-m}$. Applying the inclusion–exclusion principle to (6.4) yields (6.3). □

Clearly, a t-(v, k, λ) design with $k = t$ is trivial. Theorem 6.1.17 implies that t-(v, k, λ) designs with $k \ge v - t$ are also trivial.

Corollary 6.1.18. *Any t-(v, k, λ) design with $k \ge v - t$ is trivial.*

Proof. Let $\mathbf{D} = (X, \mathcal{B}, I)$ be a t-(v, k, λ) design with $k \ge v - t$. If K is a k-subset of a X, then let $P = \varnothing$ and $Q = X \setminus K$ in Theorem 6.1.17. The theorem implies that there is a block disjoint from Q and therefore equal to K. □

Thus, $t < k < v - t$ is a necessary condition for the existence of a nontrivial t-(v, k, λ) design. The next theorem shows this condition is also sufficient.

Theorem 6.1.19. *If v, k, and t are nonnegative integers such that $t < k < v - t$, then there exists a nontrivial t-design on v points with block size k.*

Proof. Let v, k, and t be nonnegative integers such that $t < k < v - t$. Let X be a set of cardinality v and let V be a vector space of dimension $\binom{v}{k}$ over the rationals. We assume that the components of every vector $\mathbf{x} \in V$ are indexed by k-subsets of X. Let M be a $\binom{v}{t} \times \binom{v}{k}$ matrix whose rows and columns are indexed by the t-subsets and k-subsets of X, respectively, and the (A, B)-entry is equal to 1 if $A \subset B$ and it is equal to 0 otherwise. Since $t < k < v - t$, we have $\binom{v}{k} > \binom{v}{t}$, so the columns of M are linearly dependent over the rationals. Therefore, there exists a nonzero vector $\mathbf{x} \in V$ such that $M\mathbf{x} = \mathbf{0}$. By multiplying the components of $\mathbf{x}$ by a common multiple of their denominators, we may assume that all these components are integers. Let $-m$ be the smallest component of $\mathbf{x}$. Then $m > 0$. If $\mathbf{j}$ denotes the all-1 vector, then $\mathbf{y} = \mathbf{x} + m\mathbf{j}$ is a vector with nonnegative integral components and at least one component equal to 0. We have

$$M\mathbf{y} = M\mathbf{x} + mM\mathbf{j} = m\binom{v-t}{k-t}\mathbf{j}. \tag{6.5}$$

For each k-subset A of X, let y_A be the corresponding component of $\mathbf{y}$. Let $\mathcal{B}$ be the set of all pairs (A, i) where A is a k-subset of X with $y_A \ne 0$ and i is an integer, $1 \le i \le y_A$. We define an incidence relation I on $X \times \mathcal{B}$ by

$(x, (A, i)) \in I$ if and only if $x \in A$. Equation (6.5) implies that the incidence structure $(X, \mathcal{B}, I)$ is a t-(v, k, λ) design with $\lambda = m\binom{v-t}{k-t}$. This design is not trivial. Indeed, if y_A is a zero entry of $\mathbf{y}$, then there is no block incident with each point of A. □

The design constructed in Theorem 6.1.19 may have repeated blocks. It is much more difficult to obtain a condition on the parameters that is sufficient for the existence of a nontrivial t-design without repeated blocks. The proof of the following theorem is beyond the scope of this book.

Theorem 6.1.20 (Teirlinck's Theorem). *If $v \geq t + 1$ and $v \equiv t \pmod{(t+1)!^{2t+1}}$, then there exists a t-$(v, t+1, (t+1)!^{2t+1})$ design without repeated blocks.*

6.2. The Second Ray-Chaudhuri–Wilson Inequality

Fisher's Inequality gives a lower bound on the number of blocks of a 2-design. We will now generalize it to t-designs with $t \geq 2$.

Theorem 6.2.1 (The Second Ray-Chaudhuri–Wilson Inequality). *Let $\mathbf{D}$ be a t-(v, k, λ) design and let s be a positive integer such that $t \geq 2s$. Then $\mathbf{D}$ has at least $\binom{v}{s}$ blocks.*

Proof. Let $\mathbf{D} = (X, \mathcal{B}, I)$. Let V be a vector space of dimension $\binom{v}{s}$ over the rationals. Let $\binom{X}{s}$ denote the set of all s-subsets of X. We assume that the components of every element of V are indexed by the elements of $\binom{X}{s}$. For each $S \in \binom{X}{s}$, let $\mathbf{e}(S) \in V$ denote the vector whose S-component is equal to 1 and all the other components are zeros. Then the set $\{\mathbf{e}(S) \colon S \in \binom{X}{s}\}$ is a basis of V. We will associate with the design $\mathbf{D}$ a (0, 1)-matrix N whose rows and columns are indexed by the elements of $\binom{X}{s}$ and $\mathcal{B}$, respectively, and an (S, B)-entry is equal to 1 if and only if $S \subset B$. For each $B \in \mathcal{B}$, we will denote by $\mathbf{c}(B)$ the column of N corresponding to B. Then $\mathbf{c}(B) = \sum_{S \subset B} \mathbf{e}(S)$. Let U be the subspace of V generated by the set $\{\mathbf{c}(B) \colon B \in \mathcal{B}\}$. Since $\dim(U) \leq |\mathcal{B}|$, it suffices to show that $\mathbf{e}(S) \in U$ for every $S \in \binom{X}{s}$.

For $S \in \binom{X}{s}$ and for $0 \leq j \leq i \leq s$, let

$$\mathbf{a}_i(S) = \sum_{|T \cap S| = i} \mathbf{e}(T), \quad \mathbf{b}_{ij}(S) = \sum_{|B \cap S| = i} \sum_{\substack{T \subseteq B \\ |T \cap S| = j}} \mathbf{e}(T) = \sum_{|T \cap S| = j} \beta_{ij}(S, T)\mathbf{e}(T).$$

For fixed $S, T \in \binom{X}{s}$ such that $|S \cap T| = j$, the coefficient $\beta_{ij}(S, T)$ is equal to the number of blocks $B \in \mathcal{B}$ such that $B \supset T$ and $|B \cap S| = i$, i.e., the number of blocks $B \in \mathcal{B}$ such that $B \supset T$ and $|B \cap (S \setminus T)| = i - j$. For any $(i-j)$-subset Y of $S \setminus T$, the sum of the cardinalities of disjoint sets $S \setminus (Y \cup T)$ and $Y \cup T$ is equal to $2s - j$ and therefore does not exceed t. Now Theorem 6.1.17 implies that $\beta_{ij}(S,T) = \beta_{ij}$ depends only on i and j rather than on the choice of sets S and T and

$$\beta_{ij} = \frac{\lambda\binom{s-j}{i-j}\binom{v-2s+j}{k-s-i+j}}{\binom{v-t}{k-t}}.$$

Note that $\beta_{ij} \neq 0$ for $0 \le j \le i \le s$ and

$$\mathbf{b}_{ij}(S) = \beta_{ij} \sum_{|T\cap S|=j} \mathbf{e}(T) = \beta_{ij}\mathbf{a}_j(S).$$

For $i = 0, 1, \ldots, s$, let

$$\mathbf{y}_i(S) = \sum_{j=0}^{i} \beta_{ij}\mathbf{a}_j(S). \tag{6.6}$$

Since

$$\mathbf{y}_i(S) = \sum_{|B\cap S|=i} \sum_{j=0}^{i} \sum_{\substack{T\subseteq B \\ |T\cap S|=j}} \mathbf{e}(T) = \sum_{|B\cap S|=i} \mathbf{c}(B),$$

we obtain that all vectors $\mathbf{y}_i(S)$ are in U. Equations (6.6) form a triangular system of linear equations with respect to the vectors $\mathbf{a}_j(S)$. Since the coefficients β_{ii} are not equal to zero, this system implies that all vectors $\mathbf{a}_j(S)$ are in U. In particular, all vectors $\mathbf{a}_s(S) = \mathbf{e}(S)$ are in U. The proof is now complete. □

Corollary 6.2.2. *If* **D** *is a* $(2s)$-(v,k,λ) *design with* s *intersection numbers and without repeated blocks, then* **D** *has exactly* $\binom{v}{s}$ *blocks.*

Proof. Let b be the number of blocks of a $(2s)$-(v,k,λ) design **D**. The Second Ray-Chaudhuri–Wilson Inequality implies $b \ge \binom{v}{s}$. If **D** has s intersection numbers and no repeated blocks, then the First Ray-Chaudhuri–Wilson Inequality implies $b \le \binom{v}{s}$. Therefore, in this case we have $b = \binom{v}{s}$. □

Corollary 6.2.3. *If* **D** *is a* $(2s)$-(v,k,λ) *design without repeated blocks, then* **D** *has at least* s *intersection numbers.*

Proof. If **D** is a $(2s)$-(v,k,λ) design with fewer than s intersection numbers and without repeated blocks, then the First Ray-Chaudhuri–Wilson Inequality implies that **D** has at most $\binom{v}{s-1}$ blocks, while the Second Ray-Chaudhuri–Wilson Inequality implies that **D** has at least $\binom{v}{s}$ blocks. Since $v \ge k \ge 2s$, this is a contradiction. □

Corollary 6.2.4. *If* **D** *is a nontrivial* $(2s+1)$-(v, k, λ) *design without repeated blocks, then* **D** *has at least* $s+1$ *intersection numbers.*

Proof. Suppose there is a $(2s+1)$-(v, k, λ) design **D** without repeated blocks having fewer than $s+1$ intersection numbers. Since **D** is a $(2s)$-design, Corollary 6.2.3 implies that **D** has s intersection numbers and then Corollary 6.2.2 implies that **D** has exactly $\binom{v}{s}$ blocks. Let $\mathbf{D}_x$ be a point-derived design of **D**. Then $\mathbf{D}_x$ is a $(2s)$-design with at most s intersection numbers and without repeated blocks. Therefore, Corollary 6.2.3 implies that $\mathbf{D}_x$ has exactly s intersection numbers. Then, by Corollary 6.2.2, $\mathbf{D}_x$ has $\binom{v-1}{s}$ blocks. Since the number of blocks of $\mathbf{D}_x$ is the replication number of **D**, we obtain that $\binom{v}{s}k = v\binom{v-1}{s}$. This implies $k = v - s$, and then Corollary 6.1.18 implies that **D** is a trivial $(2s+1)$-design. Since this is not the case, **D** has at least $s+1$ intersection numbers. □

Remark 6.2.5. Symmetric designs are precisely the 2-designs with one intersection number. In Section 6.5. we will construct all possible 4-designs with two intersection numbers. There is no 6-design with three intersection numbers and it is not known whether there exists a $(2s+1)$-design with $s+1$ intersection numbers for $s \geq 3$. (See Notes to this chapter for details.)

6.3. Hadamard 3-designs

We begin this section by recalling the connections between Hadamard matrices, Hadamard 2-designs, and Hadamard 3-designs.

Let n be a positive integer. If H is a normalized Hadamard matrix of order $4n$, then deleting the first row and the first column of H and replacing all -1s by 0s yields an incidence matrix of a symmetric $(4n-1, 2n-1, n-1)$-design called a *Hadamard 2-design*. Conversely, if N is an incidence matrix of a symmetric $(4n-1, 2n-1, n-1)$-design, then adjoining a column and a row of all 1s and replacing all 0s by -1s produces a Hadamard matrix of order $4n$.

Suppose $H = [h_{ij}]$ is a Hadamard matrix of order $4n$ with all entries in the last row equal to 1. We define the incidence structure $\mathbf{D} = (X, \mathcal{B})$ with $X = \{1, 2, \ldots, 4n\}$ and $\mathcal{B} = \{A_1, A_2, \ldots, A_{4n-1}, B_1, B_2, \ldots, B_{4n-1}\}$ where $A_i = \{j \in X : h_{ij} = 1\}$ and $B_i = \{j \in X : h_{ij} = -1\}$. As Proposition 4.1.10 shows, this incidence structure is a 3-$(4n, 2n, n-1)$ design known as a *Hadamard 3-design*. The following result shows that any 3-design with these parameters arises in this manner from a Hadamard matrix.

Proposition 6.3.1. *Every* 3-$(4n, 2n, n-1)$ *design is a Hadamard 3-design.*

Proof. Let $\mathbf{D} = (X, \mathcal{B})$ be a 3-$(4n, 2n, n-1)$ design. We claim that any two distinct blocks of $\mathbf{D}$ meet in either 0 or n points.

Let $B_1, B_2 \in \mathcal{B}$. Suppose that $B_1 \neq B_2$ and $B_1 \cap B_2 \neq \varnothing$. Let $x \in B_1 \cap B_2$. Then D_x is a symmetric $(4n-1, 2n-1, n-1)$-design. Therefore, $|(B_1 \setminus \{x\}) \cap (B_2 \setminus \{x\})| = n-1$, which implies that $|B_1 \cap B_2| = n$.

We now claim that for any block B of $\mathbf{D}$, the complement $X \setminus B$ is also a block of $\mathbf{D}$.

Let $B \in \mathcal{B}$ and let m be the number of blocks $A \in \mathcal{B}$ such that $|A \cap B| = n$. Counting in two ways pairs (x, A) where $A \in \mathcal{B}$, $A \neq B$, and $x \in A \cap B$ yields $2n(4n-2) = mn$, so $m = 8n-4$. Since the design $\mathbf{D}$ has exactly $8n-2$ blocks there is a unique block in $\mathbf{D}$ that is disjoint from B. This block is the complement of B.

We now assume that $X = \{1, 2, \ldots, 4n\}$, and $\mathcal{B} = \{A_1, A_2, \ldots, A_{4n-1}, B_1, B_2, \ldots, B_{4n-1}\}$, where the first $4n-1$ blocks contain the point $4n$ and B_i is the complement of A_i, $i = 1, 2, \ldots, 4n-1$.

Let N be the corresponding incidence matrix of the derived symmetric design $\mathbf{D}_{4n}$. Since this is a Hadamard 2-design, adjoining the last row and the last column of all 1s to N and replacing all 0s by -1s yields a Hadamard matrix $H = [h_{ij}]$ of order $4n$. Observe that, for $j = 1, 2, \ldots, 4n-1$, $A_j = \{i \in X : h_{ij} = 1\}$ and $B_j = \{i \in X : h_{ij} = -1\}$, so $\mathbf{D}$ is a Hadamard 3-design. □

If $\mathbf{D}$ is a Hadamard 3-$(2^d, 2^{d-1}, 2^{d-1}-1)$ design, then its derived designs have the parameters of $PG_{d-2}(d-1, 2)$. If a derived design of $\mathbf{D}$ is $PG_{d-2}(d-1, 2)$, then $\mathbf{D}$ is $AG_{d-1}(d, 2)$.

Theorem 6.3.2. *Let* $\mathbf{D}$ *be a Hadamard* 3-$(2^d, 2^{d-1}, 2^{d-1}-1)$ *design and let* x *be a point of* $\mathbf{D}$. *If the derived design* $\mathbf{D}_x$ *is isomorphic to* $PG_{d-2}(d-1, 2)$, *then* $\mathbf{D}$ *is isomorphic to* $AG_{d-1}(d, 2)$.

Proof. Let V be the d-dimensional vector space over $GF(2)$ and let $\mathbf{D}_x$ be the design $PG_{d-2}(d-1, 2)$ of one-dimensional and $(d-1)$-dimensional subspaces of V. We will identify the point x of $\mathbf{D}$ with the zero vector of V. Every other point of $\mathbf{D}$ is a one-dimensional subspace $\{\mathbf{0}, \mathbf{a}\}$ of V, and we will identify it with the vector $\mathbf{a}$. Thus, V becomes the point set of $\mathbf{D}$.

The blocks of $\mathbf{D}$ that contain x are precisely the $(d-1)$-dimensional subspaces of V. Every block of $\mathbf{D}$ that does not contain x is the complement of a block containing x. Therefore, the blocks of $\mathbf{D}$ not containing x are the $(d-1)$-flats of V that are not subspaces, and the block set of $\mathbf{D}$ can be identified with the set of all $(d-1)$-flats of V. Thus, $\mathbf{D}$ is isomorphic to $AG_{d-1}(d, 2)$. □

6.4. Cameron's Theorem

We have seen that a derived design of a $(t+1)$-design is a t-design. We now reverse this procedure and define the notion of *extension of a t-design*.

Definition 6.4.1. A $(t+1)$-design **E** is called an *extension of a t-design* **D** if **D** is isomorphic to a derived design of the design **E**.

Thus, to extend a t-design **D** we must adjoin a new point to every block of **D** and create new blocks that do not contain this new point so that this new structure is a $(t+1)$-design. The following simple proposition will be often used in this chapter.

Proposition 6.4.2. *If* **E** *is an extension of a* t-(v, k, λ) *design* **D**, *then* **E** *is a* $(t+1)$-$(v+1, k+1, \lambda)$ *design and the replication number of* **E** *is equal to the number of blocks of* **D**. *Furthermore, if* **D** *is a symmetric* (v, k, λ)*-design, then any two distinct blocks of* **E** *meet in* 0 *or* $\lambda + 1$ *points.*

Proof. The first two statements of the proposition are immediate. Let **D** be a symmetric (v, k, λ)-design and let A and B be distinct blocks of its extension **E**. If $x \in A \cap B$, then the blocks $A \setminus \{x\}$ and $B \setminus \{x\}$ of a symmetric (v, k, λ)-design $\mathbf{E}_x$ meet in λ points, and therefore, A and B meet in $\lambda + 1$ points. □

Corollary 6.4.3. *If a* t-(v, k, λ) *design with* b *blocks is extendable to a* $(t+1)$-*design, then* $k+1$ *divides* $b(v+1)$.

Proof. Let b' be the number of blocks of an extension of a t-(v, k, λ) design with b blocks. Then $b'(k+1) = b(v+1)$. □

We will be mostly interested in extending symmetric designs. We first consider the possibility of extending designs from two classical families: Hadamard 2-designs and projective planes.

Theorem 6.4.4. *Any Hadamard* 2*-design has a unique (up to isomorphism) extension to a* 3*-design, and this* 3*-design is a Hadamard* 3*-design.*

Proof. Let $\mathbf{D} = (X, \mathcal{A})$ be a Hadamard 2-design, i.e., a symmetric $(4n-1, 2n-1, n-1)$-design. If **E** is an extension of **D**, then **E** is a 3-$(4n, 2n, n-1)$ design, so by Proposition 6.3.1, **E** is a Hadamard 3-design. Note that the existence of a symmetric $(4n-1, 2n-1, n-1)$-design implies the existence of a Hadamard matrix of order $4n$ and therefore the existence of a 3-$(4n, 2n, n-1)$ design.

In order to extend **D** to a Hadamard 3-design, we must adjoin a new point to every block of **D** and create new blocks which are necessarily the complements of the old blocks. Therefore, the extension is unique. □

Before turning our attention to extendable projective planes, we state the following result whose only known proof involved a massive computer search.

Theorem 6.4.5. *There is no projective plane of order* 10.

We can now describe all extendable projective planes.

Theorem 6.4.6. *If a projective plane of order n is extendable to a 3-design, then $n = 2$ or 4.*

Proof. A projective plane of order n is a symmetric $(n^2+n+1, n+1, 1)$-design. If it is extendable to a 3-design, then Corollary 6.4.3 implies that $n+2$ divides $(n^2+n+1)(n^2+n+2)$, which in turn implies that $n+2$ divides 12. By Theorem 6.4.5, $n \neq 10$, so $n = 2$ or $n = 4$. □

Remark 6.4.7. The projective plane of order 2 (the Fano Plane) is a Hadamard 2-design, so it is extendable to a 3-(8, 4, 1) design. We will show in the next section (Corollary 6.5.10) that the projective plane of order 4 is extendable to a 3-(22, 6, 1) design.

The Hadamard 2-designs and projective planes of order 2 and 4 are all the known symmetric designs that can be extended to 3-designs. The following theorem gives feasible parameters of all extendable symmetric designs.

Theorem 6.4.8 (Cameron's Theorem). *If a symmetric (v, k, λ)-design* **D** *is extendable, then one of the following holds:*

(i) **D** *is a Hadamard 2-design;*
(ii) $v = (\lambda+2)(\lambda^2+4\lambda+2)$, $k = \lambda^2+3\lambda+1$;
(iii) $v = 495$, $k = 39$, $\lambda = 3$.

If a symmetric (v, k, λ)-design **D** *is twice extendable, then it is $PG_1(2, 4)$.*

Proof. Let $\mathbf{E} = (X, \mathcal{B})$ be an extension of a symmetric (v, k, λ)-design **D**. Then **E** is a 3-$(v+1, k+1, \lambda)$ design. By Proposition 6.4.2, any two distinct blocks of **E** meet in 0 or $\lambda+1$ points. Fix a block $A \in \mathcal{B}$ and let $\mathcal{B}_0 = \{B \in \mathcal{B} : A \cap B = \varnothing\}$. Fix distinct points $x, y \notin A$. Suppose that there are exactly μ blocks in $\mathcal{B}_0$ which contain both x and y. By counting in two ways pairs (z, C) where $z \in A$ and C is a block containing x, y, and z, we obtain:

$$(k+1)\lambda = (k-\mu)(\lambda+1).$$

Solving this equation for μ, we obtain:

$$\mu = \frac{k-\lambda}{\lambda+1}.$$

If $|\mathcal{B}_0| = 1$, then $\mu = 1$ and then $k = 2\lambda + 1$. Since $\lambda(v-1) = k(k-1)$, we obtain that $v = 4\lambda + 3$, so **D** is a Hadamard 2-design, which is the case (i).

Suppose $|\mathcal{B}_0| \geq 2$. Then $\mathbf{D}_0 = (X \setminus A, \mathcal{B}_0)$ is a 2-$(v-k, k+1, \mu)$ design. Therefore,

$$|\mathcal{B}_0| = \frac{(v-k)(v-k-1)(k-\lambda)}{k(k+1)(\lambda+1)}. \tag{6.7}$$

By Fisher's Inequality $|\mathcal{B}_0| \geq v-k$, and (6.7) implies

$$(v-k-1)(k-\lambda) \geq k(k-1)(\lambda+1). \tag{6.8}$$

Replacing $v-1$ by $k(k-1)/\lambda$ yields

$$k+1 \geq (\lambda+1)(\lambda+2). \tag{6.9}$$

On the other hand, Theorem 6.1.6 and the equation $\lambda(v-1) = k(k-1)$ imply that **E** has $v(v+1)/(k+1)$ blocks, v divides $k(k-1)+\lambda$, and $v+1$ divides $k(k-1)+2\lambda$. Therefore,

$$(k(k-1)+\lambda)(k(k-1)+2\lambda) \equiv 0 \pmod{k+1},$$

which implies that

$$2(\lambda+1)(\lambda+2) \equiv 0 \pmod{k+1}. \tag{6.10}$$

Now (6.9) implies that $k+1 = (\lambda+1)(\lambda+2)$ or $k+1 = 2(\lambda+1)(\lambda+2)$.

If $k+1 = (\lambda+1)(\lambda+2)$, we have (ii). Suppose $k+1 = 2(\lambda+1)(\lambda+2)$. Then $k+1 \equiv 4 \pmod{\lambda}$. From $\lambda(v-1) = k(k-1)$, we derive that $k(k-1) \equiv 0 \pmod{\lambda}$. Therefore, λ divides 6, i.e., λ is equal to $1, 2, 3$, or 6. If $\lambda = 1$, then $k = 11$ and $v = 111$, so **D** is a projective plane of order 10, which does not exist (Theorem 6.4.5). If $\lambda = 2$, then $k = 23$, $v = 254$, and b is not an integer. If $\lambda = 3$, then $k = 39$ and $v = 495$, so we have (iii). If $\lambda = 6$, then $k = 111$, $v = 2036$, and b is not an integer.

Suppose now that **D** is a twice extendable symmetric (v, k, λ)-design. Then its second extension is a 4-$(v+2, k+2, \lambda)$ design. By Theorem 6.1.6, it has

$$\frac{\lambda v(v-1)(v+1)(v+2)}{k(k-1)(k+1)(k+2)}$$

blocks. Since $\lambda(v-1) = k(k-1)$, we obtain that

$$(k+1)(k+2) \text{ divides } v(v+1)(v+2). \tag{6.11}$$

In case (i), we obtain that $2\lambda+3$ divides $2(4\lambda+3)(4\lambda+5)$. Since $2\lambda+3$ and $2(4\lambda+5)$ are relatively prime, we obtain that $2\lambda+3$ divides $4\lambda+3$, and then $2\lambda+3$ must divide 3. Therefore, Hadamard 2-designs are not twice extendable.

In case (ii), we obtain that $\lambda^2 + 3\lambda + 3$ divides $(\lambda + 1)(\lambda + 2)(\lambda^2 + 4\lambda + 2)(\lambda^2 + 5\lambda + 5)(\lambda^3 + 6\lambda^2 + 10\lambda + 6)$. This implies that $\lambda^2 + 3\lambda + 3$ divides $\lambda + 6$, so $\lambda = 1$. Therefore, $PG_1(2, 4)$ is the only twice extendable design from case (ii). Finally, the design from case (iii) does not satisfy (6.11). □

Corollary 6.4.9. *For any positive integer λ, there exist at most finitely many extendable symmetric (v, k, λ)-designs.*

Remark 6.4.10. In case (ii) of Cameron's Theorem, $\lambda = 1$ gives the unique symmetric $(21, 5, 1)$-design, that is $PG_1(2, 4)$ (Proposition 3.7.12). We will show in the next section (Corollary 6.5.10) that this design is extendable. For $\lambda = 2$, we obtain parameters $(v, k, \lambda) = (56, 11, 2)$. We will show in Theorem 6.6.1 that there exists a symmetric design with these parameters, though no extendable symmetric $(56, 11, 2)$-design is known.

6.5. Golay codes and Witt designs

In this section we will construct a 5-(24, 8, 1) design $\mathbf{W}_{24}$. Its derived design is a 4-(23, 7, 1) design $\mathbf{W}_{23}$. The derived design of $\mathbf{W}_{23}$ is a 3-(22, 6, 1) design $\mathbf{W}_{22}$ whose derived design is a symmetric $(21, 5, 1)$-design. By Proposition 3.7.12, the last design is isomorphic to $PG_1(2, 4)$. Thus, we will obtain an extension of the projective plane of order 4 to a 3-(22, 6, 1) design, which can be extended to a 4-(23, 7, 1) design, which in turn can be extended to a 5-(24, 8, 1) design. It can be shown that these three designs are uniquely determined by their parameters. The designs $\mathbf{W}_{23}$ and $\mathbf{W}_{24}$ are called *large Witt designs*. Our main tool in these constructions will be certain perfect codes known as *Golay codes*.

We begin with a general relation between binary perfect error-correcting codes and t-designs.

Proposition 6.5.1. *If there exists a binary perfect e-error correcting code of length n, then there exists an $(e + 1)$-$(n, 2e + 1, 1)$ design.*

Proof. Let C be a perfect e-error correcting (n, m, d) code over the alphabet $\{0, 1\}$. We will identify each word with a subset of the set $X = \{1, 2, \dots, n\}$, so C represents a family of subsets of X with $|Y \triangle Z| \geq 2e + 1$ for any distinct $Y, Z \in C$. By Remark 3.9.4, we can assume that C contains the empty set. Therefore, any other set in C contains at least $2e + 1$ elements. Since C is perfect, for every $Z \subseteq X$, there exists a unique set $Y \in C$ such that $|Y \triangle Z| \leq e$. Let $\mathcal{B}$ be the family of sets from C of cardinality $2e + 1$. We claim that $(X, \mathcal{B})$ is an $(e + 1)$-$(n, 2e + 1, 1)$ design. Let Z be a subset of X of cardinality $e + 1$.

The code C has a unique element B such that $|B \triangle Z| \le e$. We claim that $B \in \mathcal{B}$ and $Z \subset B$. Since $|B \triangle Z| \le e$, we have $1 \le |B| \le 2e + 1$. Therefore, $|B| = 2e + 1$, i.e., $B \in \mathcal{B}$. Now we have $|B \triangle Z| = |B| - |Z|$, which implies that $Z \subset B$. The proof is now complete. □

Remark 6.5.2. The proof of Proposition 6.5.1 shows that if C is a binary perfect e-error-correcting code containing a word $\mathbf{a} = (a, a, \ldots, a)$, then the words at distance $2e + 1$ from $\mathbf{a}$ serve as blocks of an $(e + 1)$-$(n, 2e + 1, 1)$ design. If C is a linear binary perfect e-error-correcting code, then $\mathbf{0} \in C$, so the words of weight $2e + 1$ form an $(e + 1)$-$(n, 2e + 1, 1)$ design.

We will now describe the *binary Golay codes* that will play an essential role in subsequent constructions.

Definition 6.5.3. Let $\mathbf{D}$ be a Hadamard 2-(11, 5, 2) design. The complementary design $\mathbf{D}'$ is a symmetric (11, 6, 3)-design. Let $N = [n_{ij}]$ be an incidence matrix of $\mathbf{D}'$. Let $P = [p_{ij}]$ be the matrix of order 12 defined by

$$p_{ij} = \begin{cases} n_{ij} & \text{if } i \neq 12 \text{ and } j \neq 12, \\ 1 & \text{if } i = 12,\ j \neq 12 \text{ or } i \neq 12,\ j = 12, \\ 0 & \text{if } i = j = 12. \end{cases}$$

Consider a 12×24 matrix $G = [I\ P]$ over GF(2) and let G^* be a matrix obtained by removing one column from G. The binary linear codes G_{24} and G_{23} with generator matrices G and G^* are called the *extended binary Golay code* and the *binary Golay code*, respectively.

Observe that if x and y are distinct points of the incidence structure with incidence matrix G, then $r(x) \in \{8, 12\}$ and $\lambda(x, y) \in \{4, 6\}$. Therefore, any two rows of G are orthogonal.

The following lemma will determine the minimum distance for G_{24} and G_{23}.

Lemma 6.5.4. *For $1 \le n \le 12$, the weight of the sum of any n distinct rows of G is divisible by 4 and greater than or equal to 8.*

Proof. Let V be the 24-dimensional vector space over $GF(2)$. We will treat binary words as both vectors from V and subsets of the set $\{1, 2, \ldots, 24\}$. Let $X_1, X_2, \ldots, X_{12}$ be the rows of G. We will use induction on n. The sets $X_1, X_2, \ldots, X_{11}$ have cardinality 8 and $|X_{12}| = 12$, so the statement of the Lemma is true for $n = 1$. Suppose $2 \le n \le 12$ and let the statement be true for any $n - 1$ distinct rows of G. Let $X_{i_1}, X_{i_2}, \ldots, X_{i_n}$ be n distinct rows of G and let $X = X_{i_1} + X_{i_2} + \cdots + X_{i_{n-1}}$. We have to show that $|X + X_{i_n}|$ is divisible by 4 and that $|X + X_{i_n}| \ge 8$.

We have

$$|X + X_{i_n}| = |X| + |X_{i_n}| - 2|X \cap X_{i_n}|. \tag{6.12}$$

Since any two rows of G are orthogonal, the vectors X and X_{i_n} are orthogonal which implies that $|X \cap X_{i_n}|$ is even. Since $|X|$ is divisible by 4 by the induction hypothesis and $|X_{i_n}| \in \{8, 12\}$, we obtain from (6.12) that $|X + X_{i_n}|$ is divisible by 4.

Let $a = \min\{|X|, |X_{i_n}|\}$ and $b = \max\{|X|, |X_{i_n}|\}$. Note that $X_i \cap \{1, 2, \ldots, 12\} = \{i\}$, for $i = 1, \ldots, 12$. Therefore, neither of the sets X, X_{i_n} contains the other. This implies that $|X \cap X_{i_n}| \leq a - 2$. Therefore, (6.12) implies

$$|X + X_{i_n}| \geq b - a + 4.$$

If $b > a$, then, since $b \equiv a \pmod 4$, we have $b - a \geq 4$, so $|X + X_{i_n}| \geq 8$. Suppose $b = a$. Note that $|(X \setminus X_{i_n}) \cap \{1, 2, \ldots, 12\}| = n - 1$, so $|X \cap X_{i_n}| \leq a - (n - 1)$. Therefore, if $n \geq 4$, then $|X \cap X_{i_n}| \leq a - 3$, and (6.12) implies $|X + X_{i_n}| \geq 6$. Since $|X + X_{i_n}|$ is divisible by 4, it is at least 8. Thus, we have to consider $n = 2$ and $n = 3$. Since any two blocks of a symmetric $(11, 6, 3)$-design meet in 3 points, the sum of any two rows of G has weight 8. Suppose $n = 3$. If $1 \leq i < j < 12$, then the sum of the rows i, j, and 12 of G has weight 8. Let $1 \leq i < j < k \leq 12$. The sum of the rows i, j, and k has three ones in the columns $1, 2, \ldots, 12$ and 1 in the column 24. Since the total number of ones in this sum is divisible by 4, it suffices to prove that this sum has at least one 1 in columns $13, \ldots, 23$. Without loss of generality, we assume that the ith block of the symmetric $(11, 6, 3)$-design is $\{13, 14, 15, 16, 17, 18\}$ and the jth block is $\{13, 14, 15, 19, 20, 21\}$. Each of the points 22, 23 occurs in 6 blocks and they occur together in 3 blocks. Therefore, at least one of these points occurs in the kth block. Consequently, the kth block does not contain at least one of the points $\{16, 17, 18, 19, 20, 21\}$ and the corresponding entry in the sum of the rows i, j, and k is equal to 1. This completes the proof. □

Corollary 6.5.5. *The codes G_{24} and G_{23} are a $(24, 12, 8)$-code and a $(23, 12, 7)$-code, respectively.*

Theorem 6.5.6. *The code G_{23} is perfect 3-error-correcting.*

Proof. Lemma 6.5.4 implies that the minimum weight of G_{23} is at least 7, so it is 3-error-correcting. Since $|G_{23}| = 2^{12}$ and $\binom{23}{0} + \binom{23}{1} + \binom{23}{2} + \binom{23}{3} = 2^{11}$, Theorem 3.9.10 implies that G_{23} is perfect. □

Remark 6.5.7. It can be shown that any perfect binary $(23, 12, 7)$-code is equivalent to G_{23}.

We will now return to extensions of symmetric designs and prove that the design $PG_1(2,4)$ is thrice extendable. If there are three consecutive extensions of this design, they are a 3-(22, 6, 1) design, a 4-(23, 7, 1) design and a 5-(24, 8, 1) design. Conversely, if there exists a 5-(24, 8, 1) design, then its three consecutive derived designs lead to the $PG_1(2,4)$.

Theorem 6.5.8. *Let $X = \{1, 2, \ldots, 24\}$ and let $\mathcal{B}$ be the set of codewords of weight 8 from G_{24}. Then the incidence structure $\mathbf{W_{24}} = (X, \mathcal{B})$ is a 5-(24, 8, 1) design.*

Proof. Proposition 6.5.1 and Theorem 6.5.6 imply that the design $\mathbf{W_{23}}$ formed by the words of weight 7 of G_{23} is a 4-(23, 7, 1) design.

Let Y be a 5-subset of X and let $i \in Y$. By removing the ith entry from all words of G_{24}, we obtain a G_{23}. Let A be the block of the corresponding design $\mathbf{W_{23}}$ that contains $Y \setminus \{i\}$. The word A of weight 7 has been obtained by removing the ith entry from a word B of weight 8 from G_{24}. Clearly, $Y \subset B$. Suppose B' is another block from $\mathcal{B}$ containing Y. Then $|B \cap B'| \geq 5$. Therefore, $|B + B'| \leq 6$ which contradicts Lemma 6.5.4. Thus, $\mathbf{W_{24}}$ is a 5-(24, 8, 1) design. □

Corollary 6.5.9. *If A and B are blocks of the design $\mathbf{W_{24}}$ such that $|A \cap B| = 4$, then their symmetric difference $A \triangle B$ is a block of $\mathbf{W_{24}}$.*

Proof. If A and B are blocks of $\mathbf{W_{24}}$ such that $|A \cap B| = 4$, then the codeword $A + B$ is of weight 8, and therefore, it is a block of $\mathbf{W_{24}}$ equal to the symmetric difference of blocks A and B. □

Corollary 6.5.10. *The design $PG_1(2,4)$ is thrice extendable.*

Remark 6.5.11. See Exercise 16 for an alternative construction of $\mathbf{W_{24}}$ and for a sketch of a proof that there is a unique (up to isomorphism) 5-(24, 8, 1) design. It can also be shown that the design $\mathbf{W_{23}}$ formed by the words of weight 7 of G_{23} is the unique 4-(23, 7, 1) design and that there is a unique 3-(22, 6, 1) design that we will denote by $\mathbf{W_{22}}$.

For each of the designs $\mathbf{W_{24}}$, $\mathbf{W_{23}}$, and $\mathbf{W_{22}}$, let λ_i denote the number of blocks that contain any given set of i points. Proposition 6.1.6 yields the following

Proposition 6.5.12.

(i) *For* $\mathbf{W_{24}}$, $\lambda_5 = 1$, $\lambda_4 = 5$, $\lambda_3 = 21$, $\lambda_2 = 77$, $\lambda_1 = 253$, *and* $\lambda_0 = 759$;
(ii) *for* $\mathbf{W_{23}}$, $\lambda_4 = 1$, $\lambda_3 = 5$, $\lambda_2 = 21$, $\lambda_1 = 77$, *and* $\lambda_0 = 253$;
(iii) *for* $\mathbf{W_{22}}$, $\lambda_3 = 1$, $\lambda_2 = 5$, $\lambda_1 = 21$, *and* $\lambda_0 = 77$.

A vector space over a finite field has a relatively large number of subspaces in comparison to a small number of possible sizes of the intersections of subspaces. This is one of the reasons why vector spaces over finite fields can be applied to constructing symmetric designs and other regular incidence structures. We will show now that the designs $\mathbf{W}_{22}$, $\mathbf{W}_{23}$, and $\mathbf{W}_{24}$ have a relatively large number of blocks with just a few sizes for the intersections of blocks. In the next section we will apply the designs $\mathbf{W}_{24}$ and $\mathbf{W}_{22}$ to construct symmetric designs with parameters $(176, 50, 14)$ and $(56, 11, 2)$.

Since 1 is the only intersection number of the symmetric $(21, 5, 1)$-design, we apply Proposition 6.1.16 consecutively to $\mathbf{W}_{22}$, $\mathbf{W}_{23}$, and $\mathbf{W}_{24}$ and obtain the following theorem.

Theorem 6.5.13. (i) *The intersection numbers of* $\mathbf{W}_{22}$ *are* 0 *and* 2*; given a block* A *of* $\mathbf{W}_{22}$*, there are* 16 *blocks that are disjoint from* A *and* 60 *blocks that meet* A *in two points;*

(ii) *the intersection numbers of* $\mathbf{W}_{23}$ *are* 1 *and* 3*; given a block* A *of* $\mathbf{W}_{23}$*, there are* 112 *blocks that meet* A *in one point and* 140 *blocks that meet* A *in three points;*

(iii) *the intersection numbers of* $\mathbf{W}_{24}$ *are* 0, 2, *and* 4*; given a block* A *of* $\mathbf{W}_{24}$*, there are* 30 *blocks that are disjoint from* A*,* 448 *blocks that meet* A *in two points, and* 280 *blocks that meet* A *in four points.*

Proof. (i) Since 1 is the only intersection number of the symmetric $(21, 5, 1)$-design, Proposition 6.1.16 implies that 2 is the only nonzero intersection number of $\mathbf{W}_{22}$. As a 3-design, this design cannot be symmetric, so it must have more than one intersection number. Therefore, its intersection numbers are 0 and 2. Let A be a block of $\mathbf{W}_{22}$. For $i = 0, 2$, let m_i be the number of blocks that meet A in exactly i points. Proposition 6.5.12 implies that $m_0 + m_2 = 76$. Counting in two ways pairs (B, x) where B is a block of $\mathbf{W}_{22}$ other than A and $x \in A \cap B$ yields another equation: $2m_2 = 120$. Therefore, $m_2 = 60$ and $m_0 = 16$.

(ii) Since every point-derived design of $\mathbf{W}_{23}$ is a 3-$(22, 6, 1)$ design, Proposition 6.1.16 implies that the possible intersection numbers of $\mathbf{W}_{23}$ are 0, 1, and 3. Fix a block A of $\mathbf{W}_{23}$ and let, for $i = 0, 1, 3$, m_i be the number of blocks that meet A in exactly i points. Then $m_0 + m_1 + m_3 = 252$. Counting in two ways pairs (B, X) where B is a block of $\mathbf{W}_{23}$ other than A, $X \subseteq A \cap B$, and $|X|$ is equal to 1 or 2 yields two more equations: $m_1 + 3m_3 = 532$ and $3m_3 = 420$. These equations give $m_3 = 140$, $m_1 = 112$, and $m_0 = 0$.

The proof of (iii) is similar to (ii). □

6.6. Symmetric designs with parameters (56, 11, 2) and (176, 50, 14)

In this section we will use the block structure of designs $\mathbf{W}_{22}$ and $\mathbf{W}_{24}$ to construct a symmetric (56, 11, 2)-design and a symmetric (176, 50, 14)-design, respectively.

Throughout this section, λ_i, $0 \leq i \leq 5$, have the same meaning as in Proposition 6.5.12.

Theorem 6.6.1. *Let a be a fixed point of $\mathbf{W}_{22}$ and let $\mathcal{A}$ be the set of all blocks of $\mathbf{W}_{22}$ that do not contain a. Let $\mathbf{D}$ be the incidence structure defined by $\mathbf{D} = (\mathcal{A}, \mathcal{A}, I)$ with $(A, B) \in I$ if and only if $|A \cap B| = 2$. Then $\mathbf{D}$ is a symmetric* (56, 45, 36)*-design, and therefore its complement is a symmetric* (56, 11, 2)*-design.*

Proof. Since $|\mathcal{A}| = \lambda_0 - \lambda_1 = 56$, $\mathbf{D}$ has 56 points and 56 blocks. Let $A \in \mathcal{A}$. For $i = 0, 2$, let n_i denote the number of blocks $B \in \mathcal{A}$ such that $|A \cap B| = i$. Then $n_0 + n_2 = 55$. Counting in two ways pairs (B, x) with $B \in \mathcal{A}$, $B \neq A$, and $x \in A \cap B$ yields $2n_2 = 6(\lambda_2 - \lambda_1 - 1) = 90$. Therefore, $n_2 = 45$, so 45 is both the block size and the replication number of $\mathbf{D}$.

Let $A_1, A_2 \in \mathcal{A}$, $A_1 \neq A_2$. Let $\mathcal{A}_0 = \mathcal{A} \setminus \{A_1, A_2\}$. We have to show that there are exactly 36 blocks $B \in \mathcal{A}_0$ such that $|B \cap A_1| = |B \cap A_2| = 2$.

Case 1: $A_1 \cap A_2 = \varnothing$.

Note that, for $B \in \mathcal{A}_0$, $|B \cap A_1| = |B \cap A_2| = 2$ if and only if $|B \cap A_1| + |B \cap A_2| = 4$. If n is the number of such blocks B, then counting in two ways triples (A, x, y) with $A \in \mathcal{A}_0$, $x \in A_1 \cap A$, and $y \in A_2 \cap A$ yields $4n = 36(\lambda_2 - \lambda_3) = 144$, so $n = 36$.

Case 2: $|A_1 \cap A_2| = 2$.

Let $S = A_1 \triangle A_2$ and $T = A_1 \cap A_2$. For $i = 0, 1, 2$, let t_i denote the number of blocks $B \in \mathcal{A}_0$ such that $|B \cap T| = i$. Then $t_0 + t_1 + t_2 = 54$. Counting in two ways pairs (B, X) with $B \in \mathcal{A}_0$, $X \subseteq B \cap T$ and $|X| = 1$ and 2 yields two more equations: $t_1 + 2t_2 = 2(\lambda_1 - \lambda_2 - 2) = 28$ and $t_2 = \lambda_2 - \lambda_3 - 2 = 2$. Thus, $t_2 = 2$, $t_1 = 24$, and $t_0 = 28$.

For $i = 2, 4$, let s_i denote the number of blocks $B \in \mathcal{A}_0$ such that $|B \cap S| = i$. Counting in two ways pairs (B, x) with $B \in \mathcal{A}_0$ and $x \in B \cap S$ yields the equation $2s_2 + 4s_4 = 8(\lambda_1 - \lambda_2 - 1) = 120$. To get another equation, note that $s_2 + 6s_4$ is the number of pairs (B, X) with $B \in \mathcal{A}_0$, $X \subseteq B \cap S$, and $|X| = 2$. On the other hand, there are exactly $28(\lambda_2 - \lambda_3) = 112$ pairs (X, A) with $X \subset S$, $|X| = 2$, and $A \in \mathcal{A}$. Of these 112 pairs, six are of the form (X, A_1) and six are of the form (X, A_2). Therefore, we obtain that $s_2 + 6s_4 = 100$.

Thus, $s_2 = 40$ and $s_4 = 10$. Therefore, the number of blocks $B \in \mathcal{A}_0$ such that $|B \cap A_1| = |B \cap A_2| = 2$, that is, $t_1 + t_2 + s_4$, is equal to 36. □

The next construction uses blocks of $\mathbf{W_{24}}$ that contain exactly one of two fixed points.

Theorem 6.6.2. *Let a and b be distinct points of $\mathbf{W_{24}}$. Let $\mathcal{A}$ be the set of all blocks of $\mathbf{W_{24}}$ that contain a and do not contain b, and let $\mathcal{B}$ be the set of all blocks of $\mathbf{W_{24}}$ that contain b and do not contain a. Let $\mathbf{H}$ denote the incidence structure $(\mathcal{A}, \mathcal{B}, I)$ with $(A, B) \in I$ if and only if $|A \cap B| = 2$. Then $\mathbf{H}$ is a symmetric $(176, 126, 90)$-design, and therefore its complement is a symmetric $(176, 50, 14)$-design.*

We obtain a proof of this theorem through a sequence of lemmas.

Lemma 6.6.3. *The sets $\mathcal{A}$ and $\mathcal{B}$ are of cardinality 176, and each element of $\mathcal{A}$ is incident with exactly 126 elements of $\mathcal{B}$. Furthermore, for $A \in \mathcal{A}$, there are exactly 15 blocks $B \in \mathcal{B}$ that are disjoint from A.*

Proof. The cardinality of $\mathcal{A}$, as well as the cardinality of $\mathcal{B}$, is equal to $\lambda_1 - \lambda_2 = 176$. Let $A \in \mathcal{A}$. For $i = 0, 2, 4$, let n_i denote the number of blocks $B \in \mathcal{B}$ such that $|A \cap B| = i$. Counting in two ways pairs (B, x) with $B \in \mathcal{B}$ and $x \in A \cap B$ yields the equation $2n_2 + 4n_4 = 7(\lambda_2 - \lambda_3)$. Counting in two ways triples (B, x, y) with $B \in \mathcal{B}$, $x, y \in A \cap B$, and $x \neq y$ yields the equation $2n_2 + 12n_4 = 42(\lambda_3 - \lambda_4)$. These two equations imply that $n_2 = 126$ and $n_4 = 35$. Since $n_0 + n_2 + n_4 = 176$, we obtain that $n_0 = 15$. □

Lemma 6.6.4. *If $A_1, A_2 \in \mathcal{A}$ and $|A_1 \cap A_2| = 2$, then there is no block $B \in \mathcal{B}$ such that $B \cap A_1 = B \cap A_2 = \varnothing$.*

Proof. Let $A_1, A_2 \in \mathcal{A}$ and $|A_1 \cap A_2| = 2$. Let $S = A_1 \triangle A_2$, so $|S| = 12$. For any $B \in \mathcal{B}$, the cardinality of $B \cap S$ is even and does not exceed 6. For $i = 0, 2, 4, 6$, let l_i denote the number of blocks $B \in \mathcal{B}$ such that $|B \cap S| = i$. Then

$$l_0 + l_2 + l_4 + l_6 = 176. \tag{6.13}$$

We obtain three more equations by counting in two ways pairs (B, X) with $B \in \mathcal{B}$, $X \subseteq B \cap S$, and $|X| = 1, 2,$ and 3, respectively:

$$2l_2 + 4l_4 + 6l_6 = 12(\lambda_2 - \lambda_3) = 672, \tag{6.14}$$

$$l_2 + 6l_4 + 15l_6 = 66(\lambda_3 - \lambda_4) = 1056, \tag{6.15}$$

$$4l_4 + 20l_6 = 220(\lambda_4 - \lambda_5) = 880. \tag{6.16}$$

Eqs. (6.13)–(6.16) yield $l_0 = 0$. □

Lemma 6.6.5. *If $A_1, A_2 \in \mathcal{A}$ and $A_1 \cap A_2 = \{a, c\}$, then there are exactly 18 blocks $B \in \mathcal{B}$ such that $c \in B$ and $|B \cap A_1| = |B \cap A_2| = 2$.*

Proof. Let $A_1, A_2 \in \mathcal{A}$ and $|A_1 \cap A_2| = 2$. Let $S = A_1 \triangle A_2$. For $i = 0, 2, 4, 6$, let m_i denote the number of blocks $B \in \mathcal{B}$ such that $c \in B$ and $|B \cap S| = i$. By Lemma 6.6.4, $m_0 = 0$. Similarly to Lemma 6.6.4, we obtain the following equations:

$$m_2 + m_4 + m_6 = \lambda_2 - \lambda_3 = 56,$$

$$2m_2 + 4m_4 + 6m_6 = 12(\lambda_3 - \lambda_4) = 192,$$

$$m_2 + 6m_4 + 15m_6 = 66(\lambda_4 - \lambda_5) = 264.$$

These equations yield $m_2 = 18$. □

Lemma 6.6.6. *If $A_1, A_2 \in \mathcal{A}$ and $A_1 \cap A_2 = \{a, c\}$, then there are exactly 72 blocks $B \in \mathcal{B}$ such that $c \notin B$ and $|B \cap A_1| = |B \cap A_2| = 2$.*

Proof. Let $A_1, A_2 \in \mathcal{A}$ and $|A_1 \cap A_2| = 2$. Let $S = A_1 \triangle A_2$. For $i = 0, 2, 4, 6$, let n_i denote the number of blocks $B \in \mathcal{B}$ such that $c \notin B$ and $|B \cap S| = i$. By Lemma 6.6.4, $n_0 = 0$. As before, we obtain the following equations:

$$n_2 + n_4 + n_6 = \lambda_1 - 2\lambda_2 + \lambda_3 = 120,$$

$$2n_2 + 4n_4 + 6n_6 = 12(\lambda_2 - 2\lambda_3 + \lambda_4) = 480,$$

$$n_2 + 6n_4 + 15n_6 = 66(\lambda_3 - 2\lambda_4 + \lambda_5) = 792.$$

These equations yield $n_2 = 18$, $n_4 = 84$, and $n_6 = 18$.

For $i, j = 2, 4, 6$, let k_{ij} denote the number of blocks $B \in \mathcal{B}$ such that $c \notin B$, $|B \cap A_1| = i$, and $|B \cap A_2| = j$. We have to show that $k_{22} = 72$. We have the following two equations:

$$k_{22} + k_{04} + k_{40} = n_4 = 84, \tag{6.17}$$

$$k_{24} + k_{42} = n_6 = 18. \tag{6.18}$$

Let X be a 4-subset of $A_1 \setminus A_2$. Then there is a unique block B of $\mathbf{W_{24}}$ that contains X and b. Since this block meets A_1 in four points, it does not contain a or c. Therefore, B meets A_2 in 0 or 2 points. Similarly, for any 4-subset of $A_2 \setminus A_1$, there is a unique block of $\mathbf{W_{24}}$ that does not contain a or c and meets

A_1 in 0 or 2 points. Since $|A_1 \setminus A_2| = |A_2 \setminus A_1| = 6$, we obtain that

$$k_{40} + k_{42} + k_{04} + k_{24} = 30. \tag{6.19}$$

Eqs. (6.17)–(6.19) imply that $k_{22} = 72$. □

Lemmas 6.6.5 and 6.6.6 imply

Corollary 6.6.7. *If $A_1, A_2 \in \mathcal{A}$ and $|A_1 \cap A_2| = 2$, then exactly 90 blocks of **H** are incident with both A_1 and A_2.*

Lemma 6.6.8. *Let $A_1, A_2 \in \mathcal{A}$ and let $|A_1 \cap A_2| = 4$. Let $S = A_1 \triangle A_2$ and $T = (A_1 \cap A_2) \setminus \{a\}$. There is no block $B \in \mathcal{B}$ such that $|B \cap S| = 6$. There are exactly 8 blocks of $\mathcal{B}$ that are disjoint from S, exactly 52 blocks of $\mathcal{B}$ that are disjoint from T, and exactly 84 blocks $B \in \mathcal{B}$ such that $|B \cap T| = 1$.*

Proof. Clearly, $|B \cap S| \leq 7$ and $|B \cap T| \leq 3$ for any $B \in \mathcal{B}$. Besides, $|B \cap S|$ is even. For $i = 0, 2, 4, 6$, let s_i be the number of blocks $B \in \mathcal{B}$ such that $|B \cap S| = i$. For $j = 0, 1, 2, 3$, let t_j be the number of blocks $B \in \mathcal{B}$ such that $|B \cap T| = j$. As before, we apply variance counting to obtain the following equations:

$$\begin{aligned}
s_0 + s_2 + s_4 + s_6 &= 176,\\
2s_2 + 4s_4 + 6s_6 &= 8(\lambda_2 - \lambda_3) = 448,\\
s_2 + 6s_4 + 15s_6 &= 28(\lambda_3 - \lambda_4) = 448,\\
4s_4 + 20s_6 &= 56(\lambda_4 - \lambda_5) = 224,\\
t_0 + t_1 + t_2 + t_3 &= 176,\\
t_1 + 2t_2 + 3t_3 &= 3(\lambda_2 - \lambda_3) = 168,\\
t_2 + 3t_3 &= 3(\lambda_3 - \lambda_4) = 48,\\
t_3 &= \lambda_4 - \lambda_5 = 4.
\end{aligned}$$

These equations yield $s_0 = 8$, $t_0 = 52$, and $t_1 = 84$. □

Lemma 6.6.9. *If $A_1, A_2 \in \mathcal{A}$ and $|A_1 \cap A_2| = 4$, then there are exactly 90 blocks $B \in \mathcal{B}$ such that $|B \cap A_1| = |B \cap A_2| = 2$.*

Proof. Let $A_1, A_2 \in \mathcal{A}$ and $|A_1 \cap A_2| = 4$. Let $S = A_1 \triangle A_2$ and $T = (A_1 \cap A_2) \setminus \{a\}$. We define the following six pairwise disjoint subsets of $\mathcal{B}$:

$$\begin{aligned}
\mathcal{B}_1 &= \{B \in \mathcal{B} \colon B \cap A_1 = B \cap A_2 = \varnothing\},\\
\mathcal{B}_2 &= \{B \in \mathcal{B} \colon B \cap T = \varnothing, |B \cap A_1| = |B \cap A_2| = 2\},\\
\mathcal{B}_3 &= \{B \in \mathcal{B} \colon B \cap T = \varnothing, B \notin \mathcal{B}_1, B \notin \mathcal{B}_2\},
\end{aligned}$$

$$\mathcal{B}_4 = \{B \in \mathcal{B}\colon B \cap S = \varnothing, |B \cap T| = 2\},$$
$$\mathcal{B}_5 = \{B \in \mathcal{B}\colon |B \cap S| = 2, |B \cap T| = 1\},$$
$$\mathcal{B}_6 = \{B \in \mathcal{B}\colon |B \cap S| = 4, |B \cap T| = 1\}.$$

Note that if $B \in \mathcal{B}_5$, then $|B \cap A_1| = |B \cap A_2| = 2$. If $B \in \mathcal{B}_6$, then one of the cardinalities $|B \cap A_1|$ and $|B \cap A_2|$ is equal to 1 and the other is equal to 3. Note also that Lemma 6.6.8 rules out $|B \cap S| = 6$.

For $1 \le i \le 6$, let $u_i = |\mathcal{B}_i|$. We have to show that $u_2 + u_4 + u_5 = 90$.

Lemma 6.6.3 implies that the set $\mathcal{B}$ has exactly $30 - u_1$ blocks that are disjoint from at least one of the blocks A_1 and A_2. On the other hand, all such blocks form the set $\mathcal{B}_1 \cup \mathcal{B}_3$. Therefore, $u_1 + u_3 = 30 - u_1$. Lemma 6.6.8 implies that $u_1 + u_2 + u_3 = t_0 = 52$ and $u_1 + u_4 = s_0 = 8$. Thus, $30 - u_1 = 52 - u_2$ and then $u_2 = u_1 + 22$. Therefore, $u_2 + u_4 = u_1 + u_4 + 22 = 30$. In order to complete the proof, we shall show that $u_5 = 60$. By Lemma 6.6.8, $u_5 + u_6 = t_1 = 84$, so it suffices to prove that $u_6 = 24$.

Given a 3-subset X of $A_1 \cap S$ or of $A_2 \cap S$ and an element x of T, there is a unique block B of $\mathbf{W_{24}}$ that contains T, x, and b. This block meets A_1 or A_2 in four points and therefore it does not contain a. Hence, $B \in \mathcal{B}$. By Lemma 6.6.8, $|B \cap S| \le 4$, and therefore, $B \in \mathcal{B}_6$. Thus, for any 3-subset X of $A_1 \cap S$ or of $A_2 \cap S$ and for any $x \in T$, there is a unique block of $\mathcal{B}_6$ that contains X and x. Since every block of $\mathcal{B}_6$ can be described in this manner and since $|T| = 3$ and $|A_1 \cap S| = |A_2 \cap S| = 4$, we obtain that $|\mathcal{B}_6| = 24$. □

Corollary 6.6.7 and Lemma 6.6.9 imply Theorem 6.6.2.

Exercises

(1) Prove that the parameter λ of a t-(v, k, λ) design equals the number of blocks if $t = 0$ and equals the replication number if $t = 1$.

(2) Prove that a point-residual design of an affine resolvable design $A_\mu(s)$ is a 1-$(s^2\mu - 1, s\mu, s\mu)$ design with intersection numbers 0 and μ.

(3) Let $s \ge 2$ and $\mu \ge 1$ be integers and let $\mathbf{D} = (X, \mathcal{B})$ be a 1-$(s^2\mu - 1, s\mu, s\mu)$ design with intersection numbers 0 and μ. Suppose further that $\mathbf{D}$ has the following property: if A, B, and C are blocks of $\mathbf{D}$ such that $A \cap B = \varnothing$ and $B \cap C = \varnothing$, then $A \cap C = \varnothing$. Let a 2-design $\mathbf{E} = (X \cup \{\infty\}, \mathcal{C})$ and a point ∞ of $\mathbf{E}$ be such that $\mathbf{E}^\infty$ is the 1-design $\mathbf{D}$.
   (a) Prove that the block set $\mathcal{B}$ of $\mathbf{D}$ admits a unique partition Π into subsets of cardinality $s - 1$.
   (b) For each partition class $\pi \in \Pi$, let U_π denote the union of all of blocks of π. Prove that $|X \setminus U_\pi| = s\mu - 1$.
   (c) Let B be a block of $\mathbf{E}$. Prove that $\infty \in B$ if and only if $B = (X \setminus U_\pi) \cup \{\infty\}$ for some $\pi \in \Pi$.

(d) Let A and B be blocks of $\mathbf{E}$ such that $\infty \in B \setminus A$. Prove that $|A \cap B|$ is 0 or μ.

(e) Let A and B be blocks of $\mathbf{E}$ such that $\infty \in A \cap B$. Prove that $|A \cap B| = \mu$.

(f) Prove that $\mathbf{E}$ is an $(s, r; \mu)$-net and apply Exercise 34(c) to obtain that $\mathbf{E}$ an $A_\mu(s)$.

(4) Prove that a nontrivial t-(v, k, λ) design has at most $v - k - t - 1$ subsequent extensions.

(5) Prove that $\mathbf{W}_{24}$ is not extendable to a 6-design.

(6) Starting with $\mathbf{W}_{24}$, one can obtain residual and derived 4-designs, which can be used to obtain residual and derived 3-designs, and these 3-designs yield residual and derived 2-designs. List the parameters of all these 4-, 3-, and 2-designs.

(7) Let $\mathbf{D}$ be a t-$(v, k, 1)$ design with $v > k$.

(a) Prove that there exists a set T of $t + 1$ points of $\mathbf{D}$ such that every block of $\mathbf{D}$ is not incident with at least one point of T.

(b) Let T be such a set of $t + 1$ points. Prove that there exist blocks $B_1, B_2, \ldots, B_{t+1}$ of $\mathbf{D}$ such that the sets $B_i \setminus T$ are pairwise disjoint and then derive the following *Tits Inequality*:

$$(k - t + 1)(t + 1) \leq v.$$

(c) Prove that there is no 10-(72, 16, 1) design.

(8) Prove Theorem 6.5.13(iii).

(9) Prove that if A and B are disjoint blocks of $\mathbf{W}_{24}$, then there is a unique block of $\mathbf{W}_{24}$ that is disjoint from $A \cup B$. Thus, $\mathbf{W}_{24}$ is a resolvable design. Find the total number of resolutions of $\mathbf{W}_{24}$.

(10) Let A and B be distinct block of $\mathbf{W}_{22}$ and let $|A \cap B| = s$. For $0 \leq i \leq j \leq 2$, let n_{ij}^s denote the number of blocks C of $\mathbf{W}_{22}$ such that one of the sets $C \cap (A \setminus B)$ and $C \cap (B \setminus A)$ is of cardinality i and the other is of cardinality j.

(a) Show that if $i \not\equiv j \pmod 2$, then $n_{ij}^s = 0$.

(b) Find the integers n_{ij}^s and thus show that they do not depend on the choice of blocks A and B with $|A \cap B| = s$.

(c) Solve similar problems for $\mathbf{W}_{23}$ (with $0 \leq i \leq j \leq 3$).

(d) Solve similar problems for $\mathbf{W}_{24}$ (with $0 \leq i \leq j \leq 4$).

(11) A set X of 12 points of $\mathbf{W}_{24}$ is called a *dodecad* if it can be represented as the symmetric difference of two blocks of $\mathbf{W}_{24}$. (These blocks must meet in two points.)

(a) Let X be a dodecad of $\mathbf{W}_{24}$. Prove that there are at most 66 pairs $\{A, B\}$ of blocks of $\mathbf{W}_{24}$ such that $A \triangle B = X$. (This can be proved directly or derived from Exercise (10).)

(b) Let a dodecad X of $\mathbf{W}_{24}$ be such that there are exactly 66 pairs $\{A, B\}$ of blocks of $\mathbf{W}_{24}$ with $A \triangle B = X$. Let $\mathcal{B}$ be the set of all blocks of $\mathbf{W}_{24}$ that meet X in six points. Prove that the substructure $(X, \mathcal{B})$ of $\mathbf{W}_{24}$ is a 5-(12, 6, 1) design. (This design is called the *small Witt design* and is denoted by $\mathbf{W}_{12}$. It is known that any two 5-(12, 6, 1) designs are isomorphic.)

(12) The blocks of $\mathbf{W}_{24}$ are the codewords of weight 8 of the extended Golay code G_{24}.

(a) Prove that the sum of all codewords of the extended Golay code G_{24} is the all-one word.

(b) Use Exercise 11 to prove that $\mathbf{W_{24}}$ has at least 2576 dodecads.

(c) Prove that the entire code G_{24} is comprised of the empty set, the point set of $\mathbf{W_{24}}$, the 759 blocks of $\mathbf{W_{24}}$, their 759 complements, and 2576 dodecads.

(d) Prove that $\mathbf{W_{24}}$ has exactly 2576 dodecads.

(e) Prove that, for each dodecad X, there are exactly 66 pairs $\{A, B\}$ of blocks of $\mathbf{W_{24}}$ such that $A \triangle B = X$, so each dodecad serves as the point set of a 5-(12, 6, 1) subdesign of $\mathbf{W_{24}}$.

(f) Prove that the complement of a dodecad is a dodecad.

(13) Let A be a block of $\mathbf{W_{24}}$ and let a and b be two distinct points of $\mathbf{W_{24}}$ that are not contained in A. Let $\mathcal{B}$ be the set of all blocks of $\mathbf{W_{24}}$ that contain a and b and meet A in four points. Prove that the substructure $(A, \mathcal{B})$ of $\mathbf{W_{24}}$ is a Hadamard 3-design.

(14) Let $\mathbf{D}$ be an incidence structure $(X, \mathcal{B}, I)$, where X is a set of seven pairwise disjoint 2-subsets of the point set of $\mathbf{W_{24}}$, $\mathcal{B}$ is a set of seven blocks of $\mathbf{W_{24}}$, and $(x, B) \in I$ if and only if $x \subset B$. Show that the sets X and $\mathcal{B}$ can be chosen so that $\mathbf{D}$ is isomorphic to the Fano plane.

(15) For $0 \leq i \leq 5$, let λ_i denote the number of blocks of $\mathbf{W_{24}}$ containing a given set of i points. Let a numerical triangle $T = [t_{ij}]$, $0 \leq j \leq i \leq 8$, be defined by

$$t_{ij} = \begin{cases} \lambda_{5-i} & \text{for } 0 \leq i \leq 5 \text{ and } j = i, \\ 1 & \text{for } 6 \leq i \leq 8 \text{ and } j = i, \\ t_{i-1,j} - t_{i,j+1} & \text{for } 1 \leq i \leq 8 \text{ and } 0 \leq j \leq i-1. \end{cases}$$

Let A be a block of $\mathbf{W_{24}}$, X a subset of A, and Y a subset of X. Let $|X| = i$ and $|Y| = j$. Prove that t_{ij} is precisely the number of blocks B of $\mathbf{W_{24}}$ such that $B \cap X = Y$.

(16) The aim of this exercise is to prove the uniqueness of $\mathbf{W_{24}}$. Let $\mathbf{D} = (X, \mathcal{B})$ be a 5-(24, 8, 1) design and let $Y = \{y_1, y_2, y_3\}$ be a 3-subset of X. Then $\mathbf{P} = \mathbf{D}_Y$ is a 2-(21, 5, 1) design. By Proposition 3.7.12, $\mathbf{P}$ is isomorphic to $PG_1(2, 4)$. The point set of $\mathbf{P}$ is $X \setminus Y$ and the line set is $\{B \setminus Y \colon B \in \mathcal{B}, B \supset Y\}$.

(a) An *hyperoval* in a projective plane of order q is a set of $q+2$ points that meets every line in zero or two points. Prove that $\mathbf{P}$ has exactly 168 hyperovals. Prove that a set O of six points of $\mathbf{P}$ is an hyperoval if and only if there exists $B \in \mathcal{B}$ such that $O = B \setminus Y$.

(b) A *Baer subplane* of a projective plane of order q^2 is a set of q^2+q+1 points that meets every line in one or $q+1$ points. Prove that $\mathbf{P}$ has exactly 360 Baer subplanes. Prove that a set F of seven points of $\mathbf{P}$ is a Baer subplane if and only if there exists $B \in \mathcal{B}$ such that $F = B \setminus Y$.

(c) Prove that a set B of eight points of $\mathbf{P}$ is the symmetric difference of two lines if and only if B is a block of $\mathbf{D}$ disjoint from Y.

(d) We will call two hyperovals O_1 and O_2 of $\mathbf{P}$ equivalent if they meet in even number of points. Let $O_1 = B_1 \setminus Y$ and $O_2 = B_2 \setminus Y$ where $B_1, B_2 \in \mathcal{B}$. Prove that O_1 and O_2 are equivalent if and only if $B_1 \cap Y = B_2 \cap Y$, so this is indeed an equivalence relation on the set of hyperovals of $\mathbf{P}$. For $i = 1, 2, 3$, let $\mathcal{O}_i$ be the set of all hyperovals of the form $B \setminus Y$ where B is a block of $\mathbf{D}$ not containing y_i, so $\mathcal{O}_1$, $\mathcal{O}_2$, and $\mathcal{O}_3$ are the equivalence classes. Prove that $|\mathcal{O}_i| = 56$ for $i = 1, 2, 3$.

(e) For $i = 1, 2, 3$, let $\mathcal{F}_i$ be the set of all Baer subplanes that meet every hyperoval of $\mathcal{O}_i$ in even number points. Prove that, for Baer subplanes $F_1 \in \mathcal{F}_i$ and $F_2 \in \mathcal{F}_j$, $|F_1 \cap F_2|$ is odd if and only if $i = j$. Let $F_1 = B_1 \setminus Y$ and $F_2 = B_2 \setminus Y$ where $B_1, B_2 \in \mathcal{B}$. Prove that $i = j$ if and only if $B_1 \cap Y = B_2 \cap Y$. Prove that $|\mathcal{F}_i| = 120$ for $i = 1, 2, 3$.

(f) Prove that each block B of **D** is of one of the following types: (i) $B = L \cup Y$ where L is a line of **P**; (ii) $B = O \cup (Y \setminus \{y_i\})$ where $O \in \mathcal{O}_i$; (iii) $B = F \cup \{y_i\}$ where $F \in \mathcal{F}_i$; (iv) $B = L_1 \triangle L_2$ where L_1 and L_2 are distinct lines of **P**.

This proves the uniqueness of $\mathbf{W}_{24}$ and gives an alternative construction of this design.

Notes

For general properties and recent results on t-designs see the book by Beth, Jungnickel and Lenz (1999) and the papers by Hedayat and Kageyama (1980), Kageyama and Hedayat (1983), Wilson (1984), and Kreher (1996).

The proof of Theorem 6.1.19 is taken from Cameron and van Lint (1991). A much stronger result is Theorem 6.1.9, which is due to Wilson (1973).

The existence of nontrivial t-(v, k, λ) designs with $t < k < v - t$ was a longstanding open problem until Theorem 6.1.20 was proven in Teirlinck (1987).

The Second Ray-Chaudhuri–Wilson Inequality is proved in the classic paper by Ray-Chaudhuri and Wilson (1975). It was proved for $s = 2$ and conjectured for the general case in Petrenjuk (1968). Corollary 6.2.4 is due to Cameron (1973). The nonexistence of 6-designs with three intersection numbers is due to Peterson (1977). Bannai (1977) proved that for each $s \geq 5$ there exist at most finitely many $(2s)$-designs with s intersection numbers. It is an open problem whether there exists a $(2s + 1)$-design with $s + 1$ intersection numbers for $s \geq 3$ and whether the Witt design $\mathbf{W}_{24}$ and its complement are the only 5-designs with three intersection numbers. The papers by Ionin and M. S. Shrikhande (1993, 1995a) contain sufficient conditions for the uniqueness of $\mathbf{W}_{24}$ (and its complement) among $(2s + 1)$-designs with $s + 1$ intersection numbers (for $s \geq 2$) and among 5-designs with three intersection numbers. For 4-designs with two intersection numbers, see Theorem 8.2.33 and Chapter 8 Notes.

The only known proof of Theorem 6.4.5 involves an extensive computer search. (See Lam, Thiel and Swiercz (1989) and Lam (1991).) In the earlier paper by Lam, Thiel, Swiercz and McKay (1983), it was shown (also by a computer search) that a putative projective plane of order 10 could not be extended. Theorem 6.4.6 (without excluding $n = 10$) is due to Hughes (1965).

We follow Cameron (1973a) in the proof of Cameron's Theorem. The only known symmetric designs with parameters from case (ii) of Cameron's Theorem are $PG_1(2, 4)$ and five nonisomorphic symmetric (56, 11, 2)-designs. It is shown by Key and Tonchev (1997) that none of these (56, 11, 2)-designs are extendable. It is not known if there exists a symmetric (495, 39, 3) design that appears in case (iii) of Cameron's Theorem. Corollary 6.4.9 is due to Hughes (1965). Exercise 3 deals with extension of a class of 1-designs. We follow Baartmans, M. S. Shrikhande and Tonchev (1994) in this exercise.

Proposition 6.5.1 is a special case of a result obtained in Assmus and Mattson (1967). The Golay codes were discovered by Golay (1949).

There are many different constructions of Witt designs. Witt (1938a,b) based his construction of $\mathbf{W}_{24}$ on the fact that the group of automorphisms of this design is the Mathieu group $\mathbb{M}_{24}$. Different constructions of the Witt designs can be found in van Lint (1984a), Hughes and Piper (1986), Cameron and van Lint (1991), M. S. Shrikhande and Sane (1991), and Beth, Jungnickel and Lenz (1999). Our construction follows that of Cameron and van Lint (1991). Exercise 16 presents Lüneburg's (1969) construction based on the geometry of $PG(2, 4)$. Our presentation of this construction and of the uniqueness of the 5-(24, 8, 1) design in Exercise 16 follows Cameron and van Lint (1991).

A symmetric (56, 11, 2)-design was first constructed by Hall, Lane and Wales (1970). The construction of this design given in Theorem 6.6.1 is due to Jónsson (1973). Another construction of a design with these parameters is given in Janko and van Trung (1986). A symmetric (176, 50, 14)-design was constructed in Higman (1969). The automorphism group of this design is a simple group known as the *Higman–Sims group*. Our description of this design follows that of M. Smith (1976a), though our proof is different from Smith's. For further results on this design, see Brouwer (1982) and M. Smith (1976b).

The construction of a 5-(12, 6, 1) design presented in Exercises 11 and 12 is due to Beth and Jungnickel (1981). For further results on simple groups and their relation to t-designs, see Choinard II, Jajcay and Magliveras (1996) and Beth, Jungnickel and Lenz (1999).

7

Symmetric designs and regular graphs

Incidence relations defining designs and incidence relations induced by designs can sometimes be expressed in terms of graphs. Such graphs usually have a high degree of regularity reflecting the regularity of the corresponding designs.

7.1. Strongly regular graphs

Let N be an incidence matrix of a symmetric (v, k, λ)-design. If N is symmetric with zeros on the diagonal, it serves as an adjacency matrix of a graph Γ of order v. This graph is regular of degree k, and for any distinct vertices x and y of Γ, there are exactly λ vertices which are adjacent to both x and y.

If N is a symmetric incidence matrix of a symmetric (v, k, λ)-design with ones on the diagonal, then $N - I$ serves as an adjacency matrix of a regular graph of order v and degree $k - 1$. For any distinct vertices x and y of this graph, the number of vertices that are adjacent to both x and y is equal to $\lambda - 2$ if x and y are adjacent and is equal to λ otherwise.

The graphs we have just described are special cases of *strongly regular graphs*.

Definition 7.1.1. A *strongly regular graph with parameters* (v, k, λ, μ), or an $SRG(v, k, \lambda, \mu)$, is a regular graph Γ of order v and valency k that satisfies the following three conditions:

(i) for any two adjacent vertices x and y, there are precisely λ vertices adjacent to both x and y;
(ii) for any two nonadjacent vertices x and y, there are precisely μ vertices adjacent to both x and y;
(iii) Γ is neither a complete graph, nor a null graph.

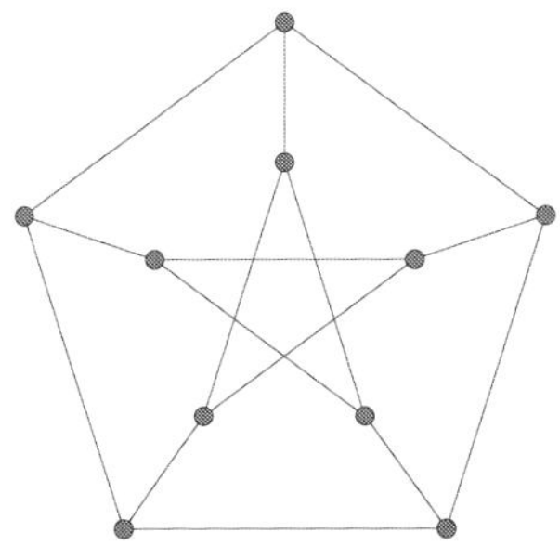

Figure 7.1 Petersen graph.

Example 7.1.2. The *Petersen graph* (Fig. 7.1) has the vertex set $\{a_i, b_i : i \in \mathbf{Z}_5\}$ of cardinality 10 and the edge set $\{\{a_i, b_i\}, \{a_i, a_{i+1}\}, \{b_i, b_{i+2}\} : i \in \mathbf{Z}_5\}$. It is an $SRG(10, 3, 0, 1)$.

Symmetric designs introduced in Examples 1.3.3 and 2.4.3 admit incidence matrices which serve as adjacency matrices of an $SRG(16, 6, 2, 2)$ and an $SRG(36, 15, 6, 6)$, respectively.

The complement of a strongly regular graph is strongly regular.

Proposition 7.1.3. *Let Γ be an SRG(v, k, λ, μ). Then its complement Γ' is an SRG$(v, v-k-1, v-2k+\mu-2, v-2k+\lambda)$.*

Corollary 7.1.4. *If there exists an SRG(v, k, λ, μ), then*

$$v \geq 2k - \mu + 2 \text{ and } v \geq 2k - \lambda. \tag{7.1}$$

Parameters of a strongly regular graph satisfy a relation similar to the basic relation for parameters of a symmetric design.

Proposition 7.1.5. *If there exists an SRG(v, k, λ, μ), then*

$$k(k - \lambda - 1) = (v - k - 1)\mu. \tag{7.2}$$

Proof. Let Γ be an SRG(v, k, λ, μ). Fix a vertex x and count in two ways the edges $\{y, z\}$ where y is adjacent to x, z is nonadjacent to x, and $z \neq x$. We have k choices for y and then $k - \lambda - 1$ choices for z or we have $v - k - 1$ choices for z and then μ choices for y. □

If an $SRG(v, k, \lambda, \mu)$ is not connected, then two vertices from different connected components have no common neighbor, so $\mu = 0$. Conversely, (7.2) implies that $\mu = 0$ if and only if $k = \lambda + 1$. Since $k = \lambda + 1$ means that the neighbors of any vertex form a complete subgraph, any $SRG(v, k, \lambda, 0)$ is a disjoint union of complete graphs, and therefore it is not connected. Thus, we have the following result.

Proposition 7.1.6. *Let Γ be an $SRG(v, k, \lambda, \mu)$. Then the following statements are equivalent:*

(i) $\mu = 0$;
(ii) $k = \lambda + 1$;
(iii) Γ *is not connected*;
(iv) Γ *is the disjoint union of v/k copies of K_k.*

We give several more examples of strongly regular graphs.

Example 7.1.7. C_n is a strongly regular graph if and only if $n = 4$ or 5. For these n, it is an $SRG(4, 2, 0, 2)$ and an $SRG(5, 2, 0, 1)$, respectively.

Example 7.1.8. A bipartite graph is strongly regular if and only if it is a complete bipartite graph $SRG(v, v/2, 0, v/2)$ with $v \geq 2$ or a ladder graph, that is, an $SRG(v, 1, 0, 0)$.

Example 7.1.9. The *triangular graph* $T(n)$ has as vertices all 2-subsets of an n-set with two distinct vertices being adjacent if and only if they are not disjoint. For $n \geq 4$, $T(n)$ is an $SRG(\frac{n(n-1)}{2}, 2(n-2), n-2, 4)$.

Example 7.1.10. The *square lattice graph* $L_2(n)$ has as vertices all ordered pairs (i, j) where $i, j \in \{1, 2, \ldots, n\}$; distinct vertices (i_1, j_1) and (i_2, j_2) are adjacent if and only if $i_1 = i_2$ or $j_1 = j_2$. For $n \geq 2$, $L_2(n)$ is an $SRG(n^2, 2(n-1), n-2, 2)$.

Example 7.1.11. The vertex set of the *Clebsch graph* is the set of all subsets of even cardinality of the set $\{1, 2, 3, 4, 5\}$, and two vertices of this graph are adjacent if and only if their symmetric difference has cardinality 4. The Clebsch graph is an $SRG(16, 5, 0, 2)$.

Example 7.1.12. Given an (n, r)-net, the *net graph* (also known as a *Latin square graph*) $L_r(n)$ has the points of the net as vertices; two distinct points form an edge if and only if there is a line through these points. If $r = n + 1$, i.e., the net is an affine plane, this graph is complete. If $r \leq n$, then $L_r(n)$ is an $SRG(n^2, r(n-1), r(r-3) + n, r(r-1))$.

Observe that, for $r \leq n$, the complement of $L_r(n)$ is an $SRG(n^2, (n-r+1)(n-1), (n-r+1)(n-r-2) + n, (n-r+1)(n-r))$. This graph has the same parameters as $L_{n-r+1}(n)$. However, an $(n, n-r+1)$-net may not exist. For instance, the complement of $L_3(6)$ has the parameters of $L_4(6)$, but there is no (6, 4)-net (Theorem 3.3.6). This observation gives rise to the following definition.

Definition 7.1.13. For positive integers n and r with $n \geq r$, a *pseudo-Latin square graph* $PL_r(n)$ is an $SRG(n^2, r(n-1), r(r-3)+n, r(r-1))$.

Remark 7.1.14. The square lattice graph $L_2(n)$ introduced in Example 7.1.10 is indeed a $PL_2(n)$.

If we formally replace n by $-n$ and r by $-r$ in the parameters of $PL_r(n)$, we obtain a set of parameters that satisfy the basic relation (7.2) between parameters of strongly regular graphs. If $n \geq r+2$, these parameters also satisfy (7.1). This motivates the following definition.

Definition 7.1.15. For positive integers n and r with $n \geq r+2$, a *negative Latin square graph* $NL_r(n)$ is an $SRG(n^2, r(n+1), r(r+3)-n, r(r+1))$.

The Clebsch graph (Example 7.1.11) is an $NL_1(4)$. We will have more examples of negative Latin square graphs in the next section.

The following proposition is straightforward.

Proposition 7.1.16. *For $1 \leq r \leq n$, the complement of a $PL_r(n)$ is a $PL_{n-r+1}(n)$; for $1 \leq r \leq n-2$, the complement of an $NL_r(n)$ is an $NL_{n-r-1}(n)$.*

Example 7.1.17. For $n \geq 3$ and for any prime power q, a *projective graph* $A_{n,2}(q)$ has the lines of $PG(n,q)$ as vertices. Two distinct lines are adjacent if and only if they intersect. Proposition 3.6.2 implies that this graph is an $SRG(v,k,\lambda,\mu)$ with

$$v = \frac{(q^{n+1}-1)(q^n-1)}{(q^2-1)(q-1)}, \quad k = \frac{q(q+1)(q^{n-1}-1)}{q-1},$$
$$\lambda = \frac{q^n-1}{q-1} + q^2 - 2, \quad \mu = (q+1)^2.$$

The defining property of strongly regular graphs can be expressed as an equation for its adjacency matrix.

Theorem 7.1.18. *Let A be an adjacency matrix of a graph Γ of order v that is neither the complete nor the null graph. Then Γ is an $SRG(v,k,\lambda,\mu)$ if and only if*

$$A^2 = (k-\mu)I + (\lambda-\mu)A + \mu J. \tag{7.3}$$

Proof. A graph Γ is an $SRG(v,k,\lambda,\mu)$ if and only if the number of walks of length 2 from a vertex x to a vertex y is equal to k if $x = y$, is equal to λ if

x and y are adjacent, and is equal to μ if $x \neq y$ and x and y are not adjacent. On the other hand, if $\{x_1, x_2, \ldots, x_v\}$ is the vertex set of Γ, then the (i, i) entry of matrix $(\lambda - \mu)A + (k - \mu)I + \mu J$ is equal to k, and the (i, j) entry of this matrix, for $i \neq j$, is equal to λ if x_i and x_j are adjacent, and is equal to μ if x_i and x_j are not adjacent. To complete the proof, we apply Theorem 2.2.10. □

Corollary 7.1.19. *Let a symmetric* $(0, 1)$*-matrix* A *with zero diagonal satisfy an equation* $A^2 = aI + bA + cJ$ *where* a, b, *and* c *are integers. If* $A \neq O$ *and* $A \neq J - I$, *then* A *is an adjacency matrix of a strongly regular graph.*

Proof. Let $\mu = c$, $\lambda = b + c$, and $k = a + c$. Let v be the order of A. Then A satisfies (7.3) and therefore it is an adjacency matrix of an $SRG(v, k, \lambda, \mu)$. □

In addition to the usual $(0, 1)$ adjacency matrices, it is sometimes convenient to represent graphs by $(0, \pm 1)$-matrices.

Definition 7.1.20. If A is an adjacency matrix of a graph Γ, then the matrix $B = J - I - 2A$ is called a *Seidel matrix* or an *S-matrix* of Γ.

Thus, the (i, j)-entry of B is -1 if the corresponding vertices are adjacent, 1 if they are distinct and nonadjacent, and 0 if $i = j$. Strongly regular graphs can be characterized in terms of their S-matrices.

Proposition 7.1.21. *Let* B *be an S-matrix of a regular graph* Γ *of order* v *and degree* k *that is neither the complete nor the null graph. Then* Γ *is an* $SRG(v, k, \lambda, \mu)$ *if and only if*

$$B^2 = (4k - 2\lambda - 2\mu - 1)I - (2\lambda - 2\mu + 2)B + (v - 4k + 2\lambda + 2\mu)J. \tag{7.4}$$

Proof. Let $A = \frac{1}{2}(J - I - B)$. Then A is an adjacency matrix of Γ. Since Γ is regular of degree k, we have $AJ = kJ$ and $BJ = (v - 2k - 1)J$. Suppose Γ is an $SRG(v, k, \lambda, \mu)$. Then A satisfies (7.3) and we derive (7.4) by routine manipulations. Conversely, if B satisfies (7.4), we replace B with $J - I - 2A$ and derive (7.3), so Γ is an $SRG(v, k, \lambda, \mu)$. □

Corollary 7.1.22. *Let a symmetric matrix* B *with zero diagonal and off-diagonal entries* ± 1 *satisfy an equation* $B^2 = aI + bB + cJ$ *where* a, b, *and* c *are integers. If* B *has a constant row sum and* $B \neq \pm(J - I)$, *then* B *is an S-matrix of a strongly regular graph.*

Strongly regular graphs with parameters $(4\mu + 1, 2\mu, \mu - 1, \mu)$ are equivalent to normalized symmetric conference matrices of order $4\mu + 2$.

Theorem 7.1.23. *For any positive integer μ, an $SRG(4\mu + 1, 2\mu, \mu - 1, \mu)$ exists if and only if there exists a symmetric conference matrix of order $4\mu + 2$.*

Proof. Let C be a symmetric matrix of order $4\mu + 2$ with all the diagonal entries equal to 0, all off-diagonal entries in the first row and the first column equal to 1, and all other off-diagonal entries equal to ± 1. Let B be the matrix obtained by deleting the first row and the first column of C. Then C is a conference matrix if and only if $B^2 = (4\mu + 1)I - J$. But this is precisely the equation (7.4) for $v = 4\mu + 1$, $k = 2\mu$, and $\lambda = \mu - 1$. □

Proposition 4.3.5 now implies

Corollary 7.1.24. *For any prime power $q \equiv 1 \pmod 4$, let $QR(q)$ be the graph with $GF(q)$ as the vertex set and with two distinct vertices adjacent if and only if their difference is a square. Then $QR(q)$ is an $SRG(q, (q - 1)/2, (q - 5)/4, (q - 1)/4)$.*

Remark 7.1.25. The graphs $QR(q)$ are called *Paley graphs*.

Recall that $A_\mu(s)$ denotes an affine resolvable 2-design with parallel classes of cardinality s and any two blocks from distinct classes intersecting in μ points. We will use such designs to construct certain strongly regular graphs.

Theorem 7.1.26. *If there exists an affine resolvable design $A_\mu(s)$ and a $(v, b, r, k, 1)$-design with $r = (s^2\mu - 1)/(s - 1)$, then there exists an*

$$SRG(vs^2\mu, (v - 1)s\mu, (k - 2)s\mu + (v - k)\mu, (v - k)\mu). \qquad (7.5)$$

Moreover, the vertex set of this graph can be partitioned into v subsets of size $s^2\mu$ in such a way that no two vertices from the same subset are adjacent.

Proof. For $i = 1, 2, \dots, v$, let $\mathbf{D}_i = (X_i, \mathcal{B}_i)$ be an $A_\mu(s)$ and let the point sets $X_1, X_2, \dots, X_v$ be pairwise disjoint. For $i = 1, 2, \dots, v$, let $\{\mathcal{C}_{ij} : 1 \le j \le r\}$ be the parallelism of $\mathbf{D}_i$. For $i = 1, 2, \dots, v$ and $j = 1, 2, \dots, r$, let $\mathcal{C}_{ij} = \{L_{ijh} : 1 \le h \le s\}$.

Let $\mathbf{D} = (X, \mathcal{B})$ be a $(v, b, r, k, 1)$-design with $r = (s^2\mu - 1)/(s - 1)$ and let $X = \{x_1, x_2, \dots, x_v\}$. For $i = 1, 2, \dots, v$, let $\{B_{ij} : 1 \le j \le r\}$ be the set of all blocks of the design $\mathbf{D}$ that contain x_i.

Let Γ be the graph on the vertex set $V = X_1 \cup X_2 \cup \dots \cup X_v$ with the following adjacency: distinct vertices α and β are adjacent if and only if there exist indices i, j, m, n, and h such that $i \ne m$, $\alpha \in L_{ijh}$, $\beta \in L_{mnh}$, and $B_{ij} = B_{mn}$.

Note that no two vertices from the same set X_i are adjacent. We will show that Γ is a strongly regular graph with parameters (7.5).

By Proposition 5.3.13, each $\mathbf{D}_i$ is an $(s^2\mu, s(s^2\mu - 1)/(s-1), r, s\mu, (s\mu - 1)/(s-1))$-design. Therefore, $|V| = vs^2\mu$.

Let $i, m \in \{1, 2, \ldots, v\}, i \neq m$. Let B be the unique block of $\mathbf{D}$ that contains both x_i and x_m. There are unique indices $j, n \in \{1, 2, \ldots, r\}$ such that $B = B_{ij} = B_{mn}$. Let $\alpha \in X_i$. There is a unique $h \in \{1, 2, \ldots, s\}$ such that $\alpha \in L_{ijh}$. Then $\beta \in X_m$ is adjacent to α if and only if $\beta \in L_{mnh}$. Since $|L_{mnh}| = s\mu$, we obtain that the degree of α is $(v-1)s\mu$.

Let indices i, m, j, and n and the block B be the same as in the previous paragraph and let $\alpha \in X_i$ and $\beta \in X_m$ be adjacent vertices of Γ. Let $\alpha \in L_{ijh}$ and $\beta \in L_{mnh}$. Let $\gamma \in X_p$ with $p \neq i$ and $p \neq m$. Suppose first that $x_p \in B$ and let $B = B_{pq}$. Then γ is adjacent to both α and β if and only if $\gamma \in L_{pqh}$. Since $|L_{pqh}| = s\mu$ and there are $k-2$ indices $p \notin \{i, j\}$ with $x_p \in B$, we obtain that there are exactly $(k-2)s\mu$ such vertices γ. Suppose now that $x_p \notin B$. Let A_1 and A_2 be the blocks of $\mathbf{D}$ that contain 2-subsets $\{x_i, x_p\}$ and $\{x_m, x_p\}$, respectively. Then $A_1 \neq B$ and $A_2 \neq B$. Since B is the only block that contains both x_i and x_m, we have $A_1 \neq A_2$. Let $B_{it} = B_{pf} = A_1$ and $B_{mu} = B_{pg} = A_2$. Then $f \neq g$. The vertex γ is adjacent to both α and β if and only if $\gamma \in L_{pfh}$ and $\gamma \in L_{pgh}$. Since $f \neq g$, nonparallel blocks L_{pfh} and L_{pgh} of $\mathbf{D}_p$ meet in μ points and therefore there are $(v-k)\mu$ choices for γ. Thus, adjacent vertices α and β of the graph Γ have exactly $(k-2)s\mu + (v-k)\mu$ common neighbors.

Let α and β be distinct vertices of Γ from the same set X_i. The design $\mathbf{D}_i$ has $\lambda = (s\mu - 1)/(s-1)$ blocks that contain both α and β, i.e., there exactly λ pairs (j, h) such that $\alpha, \beta \in L_{ijh}$. Let (j, h) be such a pair and let $B = B_{ij}$. Let $x_p \in B$, $p \neq i$, and let $\gamma \in X_p$. Then $B = B_{pf}$ for a unique f. The vertex γ is adjacent to both α and β if and only if $\gamma \in L_{pfh}$. Thus we have chosen $\lambda(k-1)s\mu$ quadruples (j, h, p, γ), which contain all vertices γ adjacent to both α and β. Changing the value of j in such a quadruple replaces the block B by a block $A \neq B$. Since $x_i \in A \cap B$, there is no other point $x_p \in A \cap B$. Therefore, the value of p in the quadruple will change, and we will not obtain the same vertex γ. Changing the value of h with a fixed j replaces the block L_{pfh} by a parallel block, so it will not lead to counting the same γ twice either. Thus, Γ has exactly $\lambda(k-1)s\mu$ vertices adjacent to both α and β. Since $v - 1 = r(k-1)$, we derive that $\lambda(k-1)s\mu = (v-k)\mu$.

To complete the proof, we have to consider the case of nonadjacent vertices $\alpha \in X_i$ and $\beta \in X_m$ with $i \neq m$. Let $\gamma \in X_p$, $p \neq i$, $p \neq m$. Let B be the block that contains both x_i and x_m. If $x_p \in B$ and γ is adjacent to both α and β, then α and β have to be adjacent, which is not the case. Let $x_p \notin B$ and let A_1 and A_2 be the blocks that contain 2-subsets $\{x_i, x_p\}$, and $\{x_m, x_p\}$, respectively. Then $A_1 \neq A_2$. Let $B_{it} = B_{pf} = A_1$ and $B_{mn} = B_{pg} = A_2$. Let $\alpha \in L_{ith}$ and $\beta \in L_{mnl}$. The vertex γ is adjacent to both α and β if and only if $\gamma \in L_{pfh} \cap L_{pgl}$. Since $f \neq g$, the blocks L_{pfh} and L_{pgl} of $\mathbf{D}_p$ are not parallel and therefore, we again have $(v-k)\mu$ choices for γ. □

7.2. Eigenvalues of strongly regular graphs

Eigenvalues of strongly regular graphs provide significant information about the graphs. For instance, Propositions 2.2.20 and 7.1.6 immediately imply

Proposition 7.2.1. *The only strongly regular graphs with two eigenvalues are the graphs $m \cdot K_n$ with m and n greater than* 1.

We begin with a theorem describing the characteristic polynomials of strongly regular graphs.

Theorem 7.2.2. *Let Γ be an $SRG(v, k, \lambda, \mu)$. Then $\chi(\Gamma)(t) = (t-k)(t-r)^f(t-s)^g$ where r and s are the two zeros of the quadratic equation*

$$\rho^2 - (\lambda - \mu)\rho - (k - \mu) = 0, \tag{7.6}$$

the multiplicities f and g are positive integers given by

$$f, g = \frac{1}{2}\left(v - 1 \pm \frac{(v-1)(\lambda-\mu) + 2k}{\sqrt{(\lambda-\mu)^2 + 4(k-\mu)}}\right). \tag{7.7}$$

and

$$\lambda = k + r + s + rs, \quad \mu = k + rs. \tag{7.8}$$

Proof. If Γ is not connected, then, by Proposition 7.2.1, $\Gamma = (v/k) \cdot K_k$. The characteristic polynomial of such a graph is given in Example 2.2.11, and equations (7.6) and (7.7) are readily verified.

From now on, assume that Γ is a connected graph.

Let A be an adjacency matrix of Γ and let $p(t) = (t^2 - (\lambda-\mu)t - (k-\mu))/\mu$. Equations (7.2) and (7.3) imply that $p(k) = v$ and $p(A) = J$, and then Theorem 2.2.18 implies that all eigenvalues of A, other than k, satisfy (7.6).

Let r and s be the solutions of (7.6), i.e.,

$$r, s = \frac{1}{2}\left(\lambda - \mu \pm \sqrt{(\lambda-\mu)^2 + 4(k-\mu)}\right).$$

Since Γ is connected, it has more than two eigenvalues. Therefore, $r \neq s$ and r, s, and k are the eigenvalues of Γ. The eigenvalue k is simple. Let f and g be the multiplicities of r and s, respectively. Since the degree of χ_A is equal to v and the trace of the matrix A is equal to 0, we have

$$v = f + g + 1, 0 = k + fr + gs. \tag{7.9}$$

Solving (7.9) for f and g and using $r - s \neq 0$, we obtain (7.7).

Equation (7.6) implies that $r + s = \lambda - \mu$ and $rs = \mu - k$, giving $\lambda = k + r + s + rs$ and $\mu = k + rs$. □

Corollary 7.2.3. *Let $\chi(\Gamma)(t) = (t - k)(t - r)^f (t - s)^g$ be the characteristic polynomial of a strongly regular graph Γ of degree k. Then $\chi_S(\Gamma)(t) = (t - v + 2k + 1)(t + 2r + 1)^f (t + 2s + 1)^g$ is the characteristic polynomial of the S-matrices of Γ.*

Proof. Let A be an adjacency matrix of Γ and let $B = J - I - 2A$ be the corresponding S-matrix. Since $AJ = JA$, there exists and orthogonal matrix C such that $C^\top AC$ and $C^\top JC$ are diagonal matrices. We can assume that $C^\top AC$ has the $(1, 1)$-entry equal to k, f diagonal entries equal to r, and g diagonal entries equal to s and that $C^\top JC$ is the matrix with the $(1, 1)$-entry equal to v and all the other entries equal to 0. Then $C^\top BC$ is a diagonal matrix with the $(1, 1)$-entry $v - 2k - 1$, f diagonal entries equal to $-2r - 1$ and g diagonal entries equal to $-2s - 1$. This gives us the characteristic polynomial of B. □

Theorem 7.2.2 implies the following necessary condition on the parameters of strongly regular graphs.

Theorem 7.2.4 (The Integrality Condition). *If there exists an SRG (v, k, λ, μ), then either*

(i) $v = 4\mu + 1$, $k = 2\mu$, *and* $\lambda = \mu - 1$ *or*
(ii) $\sqrt{(\lambda - \mu)^2 + 4(k - \mu)}$ *is an integer dividing* $(v - 1)(\lambda - \mu) + 2k$.

Proof. If $(v - 1)(\lambda - \mu) + 2k \neq 0$, then (7.7) implies (ii). Suppose that $(v - 1)(\lambda - \mu) + 2k = 0$. Then $(v - 1)(\mu - \lambda) = 2k$. Since $v > k + 1$, we obtain that $0 < \mu - \lambda < 2$, so $\lambda = \mu - 1$ and $v = 2k + 1$. Using (7.2), we obtain that $k = 2\mu$ and then $v = 4\mu + 1$. □

Remark 7.2.5. By Proposition 7.1.23, case (i) is equivalent to the existence of a conference matrix of order $4\mu + 2$.

As an application of the Integrality Condition, we will classify graphs $SRG(v, k, 0, 1)$ also known as *Moore graphs of diameter 2*.

Theorem 7.2.6. *If an $SRG(v, k, 0, 1)$ exists, then $k = 2, 3, 7$ or 57 and $v = k^2 + 1$.*

Proof. Let Γ be an $SRG(v, k, 0, 1)$. Equation (7.2) immediately implies that $v = k^2 + 1$. Then the Integrality Condition implies that either $k = 2$ or $\sqrt{4k-3}$ is an integer dividing $k(k-2)$. In either case, $4k - 3$ divides $k^2(k-2)^2$. Since any common divisor of k and $4k - 3$ must divide 3, we assume first that $k \not\equiv 0 \pmod 3$. Then $4k - 3$ and k^2 are relatively prime, so $4k - 3$ divides $(k-2)^2$. Then $4k - 3$ divides $4(k-2)^2 = 4k^2 - 16k + 16$. Since also $4k - 3$ divides $k(4k-3) = 4k^2 - 3k$, we obtain that $4k - 3$ must divide $13k - 16$. Therefore, $4k - 3$ divides $13k - 16 - 3(4k-3) = k - 7$. Since $4k - 3 > k - 7$, we obtain that $k \leq 7$. By direct trials, we determine that $k = 1, 2$ or 7. We rule out $k = 1$, since in this case $v = 2$ and Γ is a complete graph.

Suppose now that $k \equiv 0 \pmod 3$. Let $k = 3n$. Then $12n - 3$ divides $9n^2(n-2)^2$, so $4n - 1$ divides $3n^2(3n-2)^2$. Since n and $4n - 1$ are relatively prime, we obtain that $4n - 1$ must divide $3(3n-2)^2$. Therefore, $4n - 1$ divides $27n(4n-1) - 12(3n-2)^2 = 117n - 48$. Then $4n - 1$ divides $117n - 48 - 29(4n-1) = n - 19$. Obviously, $4n - 1 > n - 19$. Therefore, either $n = 19$ or $4n - 1 \leq 19 - n$, i.e., $n \leq 4$. Since $4n - 1$ does not divide $n - 19$ for $n = 2$ and $n = 3$, we have $n = 1, 3, 4$ or 19. The respective values of k are 3, 9, 12, and 57. To complete the proof, notice that for $k = 9$ or 12, $4k - 3$ is not a perfect square. □

Remark 7.2.7. Since C_5 is an $SRG(5, 2, 0, 1)$ and the Petersen graph is an $SRG(10, 3, 0, 1)$, an $SRG(k^2 + 1, k, 0, 1)$ exists for $k = 2$ and 3. No $SRG(k^2 + 1, k, 0, 1)$ for $k = 57$ is known. There is a unique $SRG(k^2 + 1, k, 0, 1)$ for $k = 7$ known as the *Hoffman–Singleton graph*. It is constructed in Exercise 18.

Proposition 7.2.1 and Theorem 7.2.2 imply that a connected strongly regular graph has exactly three eigenvalues. This property characterizes strongly regular graphs among regular connected graphs.

Theorem 7.2.8. *A regular connected graph Γ, which is not a complete or a null graph, is strongly regular if and only if it has exactly three eigenvalues.*

Proof. Suppose that Γ is a regular connected graph of degree k on v vertices, having three distinct eigenvalues, k, r, and s. Let A be an adjacency matrix of Γ. Consider the polynomial $p(t) = v(t-r)(t-s)/((k-r)(k-s))$. Then $p(k) = v$ and $p(r) = p(s) = 0$, so, by Theorem 2.2.18, $p(A) = J$, i.e., $v(A - rI)(A - sI) = (k-r)(k-s)J$. This equation can be rewritten as

$$A^2 = -rsI + (r+s)A + \frac{(k-r)(k-s)}{v}J.$$

Theorem 7.1.18 now implies that Γ is a strongly regular graph and Proposition 2.2.20 implies that Γ is connected. □

Propositions 2.2.19, 2.2.20, 7.2.1, and Theorem 7.2.8 imply the following result.

Corollary 7.2.9. *A regular graph* Γ*, which is not a complete or a null graph, is strongly regular if and only if it has at most three eigenvalues.*

The next theorem relates the smallest eigenvalue of a strongly regular graph to the maximum size of a clique and a coclique.

Theorem 7.2.10. *Let* Γ *be an* $SRG(v, k, \lambda, \mu)$ *and let* $-s$ *be the smallest eigenvalue of* Γ.

(i) *If* C *is a coclique of* Γ*, then* $|C| \leq sv/(k+s)$ *with equality if and only if every vertex* $x \notin C$ *has exactly* s *neighbors in* C.
(ii) *If* C *is a clique of* Γ*, then* $|C| \leq 1 + k/s$ *with equality if and only if every vertex* $x \notin C$ *has exactly* μ/s *neighbors in* C.

Proof. Let k, r, and $-s$ be all the eigenvalues of Γ and let $r \geq 0$. Let C be a coclique of cardinality $c > 0$. For every vertex $x \notin C$, let $e_x = |\Gamma(x) \cap C|$. Counting in two ways the number of edges $\{x, z\}$ with $z \in C$ and $x \notin C$ and the number of 2-subsets $\{y, z\}$ of $\Gamma(x)$ with $y, z \in C$ and $x \notin C$ yields equations

$$\sum_{x \notin C} e_x = ck, \quad \sum_{x \notin C} \binom{e_x}{2} = \binom{c}{2}\mu.$$

Then $e = ck/(v-c)$ is the average of all e_x, $x \notin C$, and we have

$$\begin{aligned}
0 \leq (v-c)\sum_{x \notin C}(e_x - e)^2 &= (v-c)\sum_{x \notin C}\left(2\binom{e_x}{2} + (1-2e)e_x + e^2\right) \\
&= (v-c)(c(c-1)\mu + (1-2e)ck + e^2(v-c)) \\
&= ((c-1)(v-c)\mu + (v-c-2ck)k + ck^2)c \\
&= (v(k-\mu) + (v\mu + \mu - k - k^2)c - \mu c^2)c \\
&= \left(\frac{sv}{k+s} - c\right)(r(k+s) + c\mu)c.
\end{aligned}$$

(In the last line we apply (7.8).) Hence, $c \leq sv/(k+s)$.

If $c = sv/(k+s)$, then $e_x = e$ for all $x \notin C$. Then $e = ck/(v-c) = s$, and we obtain that $e_x = s$ for all $x \notin C$.

Part (ii) can be obtained by applying the result of part (i) to the complement of Γ. □

7.3. Switching in strongly regular graphs

In this section we will explore an operation of *switching* that can be applied to any graph and is useful in constructing strongly regular graphs.

Definition 7.3.1. Let $\Gamma = (V, E)$ be a graph and let U be a proper subset of V. The operation of *switching* Γ with respect to U replaces Γ by the graph $\Gamma' = (V, E')$ where the edge set E' is defined as follows:

(i) if e is an edge of Γ, then $e \in E'$ if and only if $|e \cap U| = 0$ or 2;
(ii) if e is a 2-subset of V that is not an edge of Γ, then $e \in E'$ if and only if $|e \cap U| = 1$.

Thus, switching replaces all edges between U and the complement of U by nonedges and vice versa, and retains all the edges of Γ which lie in U or in the complement of U. Clearly, switching Γ with respect to U is the same as switching with respect to the complement of U. The following proposition is immediate.

Proposition 7.3.2. *If a graph Γ' is obtained by switching a graph Γ with respect to U, then switching the complement of Γ with respect to U yields the complement of Γ'.*

The operation of switching can be expressed in terms of both adjacency and Seidel matrices of a graph.

Proposition 7.3.3. *Let Γ be a graph and let U be a proper subset of the vertex set V of Γ. Suppose the set V is ordered so that the elements of U precede the elements of $V \setminus U$. Let the corresponding adjacency matrix A and S-matrix B of Γ be represented as block matrices*

$$A = \begin{bmatrix} A_1 & A_2 \\ A_2^\top & A_3 \end{bmatrix} \text{ and } B = \begin{bmatrix} B_1 & B_2 \\ B_2^\top & B_3 \end{bmatrix}$$

with $|U| \times |U|$ blocks A_1 and B_1. Let graph Γ' be obtained by switching Γ with respect to U. Then

$$A' = \begin{bmatrix} A_1 & J - A_2 \\ (J - A_2)^\top & A_3 \end{bmatrix} \text{ and } B' = \begin{bmatrix} B_1 & -B_2 \\ -B_2^\top & B_3 \end{bmatrix}$$

are an adjacency and an S-matrix of Γ', respectively.

There is another simple relation between S-matrices of graphs Γ and Γ' that can be obtained from each other by switching. We leave proof of the following proposition as an exercise.

Proposition 7.3.4. *Let graph Γ' be obtained by switching a graph Γ with respect to U. Let $V = \{x_1, x_2, \ldots, x_v\}$ be the vertex set of the graphs and let B and B' be the corresponding S-matrices of Γ and Γ', respectively. Then $B' = D^{-1}BD$ where $D = [d_{ij}]$ is a diagonal matrix with $d_{ii} = 1$ if $x_i \in U$ and $d_{ii} = -1$ if $x_i \notin U$.*

Corollary 7.3.5. *If a graph Γ' is obtained by switching a graph Γ, then S-matrices of Γ and Γ' have the same characteristic polynomial.*

Example 7.3.6. Switching the graph $L_2(4)$ (Example 7.1.10) with respect to the set of all vertices (i, j) with $i \in \{1, 2\}$ yields an $SRG(16, 10, 6, 6)$ that is isomorphic to the complement of the Clebsch graph (Example 7.1.11).

If Γ and Γ' are two graphs on the same vertex set, a natural question is whether one graph can be obtained by switching the other graph. An answer to this question depends on a certain incidence structure, which can be associated with any graph.

Definition 7.3.7. For any graph $\Gamma = (V, E)$, the *two-graph* of Γ is the incidence structure $\mathbf{D} = (V, \mathcal{B})$ where $\mathcal{B}$ is the set of all 3-subsets of V that contain one or three edges of Γ.

Given graphs Γ and Γ' on the same vertex set, their two-graphs determine whether Γ' can be obtained by switching Γ.

Theorem 7.3.8. *Let Γ_1 and Γ_2 be distinct graphs on the same vertex set V and let $\mathbf{D}_1 = (V, \mathcal{B}_1)$ and $\mathbf{D}_2 = (V, \mathcal{B}_2)$ be the two-graphs of Γ_1 and Γ_2, respectively. Then $\mathcal{B}_1 = \mathcal{B}_2$ if and only if Γ_2 can be obtained by switching Γ_1.*

Proof. 1. Suppose Γ_2 is obtained by switching Γ_1 with respect to a proper subset U of V. Let $T = \{x, y, z\}$ be a 3-subset of V. If T is a subset of U or of the complement of U, then Γ_1 and Γ_2 have the same number of edges contained in T. If, say, $x \in U$ and $y, z \notin U$, then each of the 2-subsets $\{x, y\}$ and $\{x, z\}$ is an edge of one and only one of the graphs Γ_1 and Γ_2, so the parity of the number of edges of each graph that are contained in T is the same. Therefore, in either case, $\mathcal{B}_1 = \mathcal{B}_2$.

2. Suppose now that Γ_1 and Γ_2 have the same two-graph $\mathbf{D} = (V, \mathcal{B})$. We define a relation $\sim$ on V as follows: $x \sim y$ if $x = y$ or $\{x, y\}$ is an edge in both Γ_1 and Γ_2 or $\{x, y\}$ is a nonedge in both Γ_1 and Γ_2. Clearly, this relation is reflexive and symmetric. It is also transitive. For instance, let x, y, and z be three distinct vertices such that $\{x, y\}$ is an edge in both Γ_1 and Γ_2 and $\{y, z\}$ is a nonedge in both Γ_1 and Γ_2. Then $\{x, z\}$ is an edge in both Γ_1 and Γ_2 if $\{x, y, z\} \notin \mathcal{B}$ and $\{x, z\}$ is a nonedge in both Γ_1 and Γ_2 if $\{x, y, z\} \in \mathcal{B}$. The

other cases are similar. In the same manner, one can verify that if $\{x, y\}$ is an edge in one and only one of the given graphs and $\{y, z\}$ is an edge in one and only one of the given graphs, then $x \sim z$. Therefore, $\sim$ is an equivalence relation on V with at most two equivalence classes. Since Γ_1 and Γ_2 are distinct graphs, there are in fact exactly two equivalence classes. Let U be an equivalence class. Then U is a proper subset of V and switching Γ_1 with respect to U yields Γ_2. □

The next proposition can be obtained by straightforward counting.

Proposition 7.3.9. *Let $\Gamma = (V, E)$ be an $SRG(v, k, \lambda, \mu)$ and let $\mathbf{D} = (V, \mathcal{B})$ be the two-graph of Γ. If e is a 2-subset of V, then the number of blocks $B \in \mathcal{B}$ containing e is equal to $v - 2k + 2\lambda$ if $e \in E$ and it is equal to $2(k - \mu)$ otherwise.*

If Γ is a complete graph or the disjoint union of two complete graphs of the same order, then the two-graph of Γ is a complete design, i.e., its block set is the set of all 3-subsets of the vertex set of Γ. The next theorem characterizes regular graphs whose two-graphs are incomplete 2-designs.

Theorem 7.3.10. *Let $\Gamma = (V, E)$ be a regular graph on v vertices and let $\mathbf{D} = (V, \mathcal{B})$ be the two-graph of Γ. Then $\mathbf{D}$ is an incomplete 2-design if and only if Γ is an $SRG(v, k, \lambda, \mu)$ with $v = 2(2k - \lambda - \mu)$ but not the disjoint union of two complete graphs of order $v/2$.*

Proof. Let Γ be an $SRG(v, k, \lambda, \mu)$ with $v = 2(2k - \lambda - \mu)$. Let $x, y \in V$, $x \neq y$. Let $B = \{x, y, z\}$ be a 3-subset of V that contains x and y. If $\{x, y\} \in E$, then $B \in \mathcal{B}$ if and only if both $\{x, z\}$ and $\{y, z\}$ are edges or both $\{x, z\}$ and $\{y, z\}$ are nonedges. Therefore, there are exactly $v - 2k + 2\lambda$ blocks $B \in \mathcal{B}$ that contain $\{x, y\}$. If $\{x, y\} \notin E$, then $B \in \mathcal{B}$ if and only if exactly one of $\{x, z\}, \{y, z\}$ is an edge, so there are $2(k - \mu)$ blocks $B \in \mathcal{B}$ that contain $\{x, y\}$. Since $v - 2k + 2\lambda = 2(k - \mu)$, we have $2(k - \mu)$ blocks containing $\{x, y\}$ in either case. Since $|B| = 3$ for all $B \in \mathcal{B}$, we obtain that $\mathbf{D}$ is a 2-$(v, 3, 2(k - \mu))$ design. If $\mu \neq 0$, then let x and y be distinct nonadjacent vertices of Γ and let z be their common neighbor. Then $\{x, y, z\} \notin \mathcal{B}$ and therefore the design $\mathbf{D}$ is incomplete. If $\mu = 0$, then, by Corollary 7.1.6, Γ is the disjoint union of $m \geq 2$ complete graphs. If $m \neq 2$, then select vertices x, y, and z from three distinct connected components of Γ. The set $\{x, y, z\}$ is not a block, so $\mathbf{D}$ is incomplete.

Conversely, suppose that Γ is a regular graph of degree k and $\mathbf{D} = (V, \mathcal{B})$ is an incomplete 2-$(v, 3, l)$ design. Let $V = \{x_1, x_2, \ldots, x_v\}$ and let $B = [b_{ij}]$ be the corresponding S-matrix of Γ. The diagonal entries of B^2 are equal to $v - 1$. An off-diagonal (i, j)-entry of B^2 is $\sum_{k=1}^{v} b_{ik}b_{kj}$. If $k \in \{i, j\}$, then $b_{ik}b_{kj} = 0$.

Let $k \notin \{i, j\}$. If $\{x_i, x_j, x_k\}$ is a block of **D**, then $b_{ik}b_{kj} = -b_{ij}$; if $\{x_i, x_j, x_k\}$ is not a block of **D**, then $b_{ik}b_{kj} = b_{ij}$. Since there are exactly l blocks that contain $\{x_i, x_j\}$, the (i, j)-entry of B^2 is equal to $(v - 2l - 2)b_{ij}$. Therefore,

$$B^2 = (v-1)I + (v - 2l - 2)B. \tag{7.10}$$

Since the design **D** is not complete, there is a set of three vertices of Γ that contains less than three edges, so Γ is not a complete graph. Since every block of **D** contains at least one edge, Γ is a not a null graph. Theorem 7.1.21 and (7.10) now imply that Γ is an $SRG(v, k, \lambda, \mu)$ with $v = 2(2k - \lambda - \mu)$. □

Corollary 7.3.11. *Let Γ be an $SRG(v, k, \lambda, \mu)$, other than the disjoint union of two complete graphs, and let a regular graph Γ' be obtained by switching Γ. The graph Γ' is strongly regular if and only if $v = 2(2k - \lambda - \mu)$. Furthermore, if Γ' is an $SRG(v, k', \lambda', \mu')$, then $v = 2(2k' - \lambda' - \mu')$.*

Proof. Let $\Gamma = (V, E)$ and let Γ' be obtained by switching Γ with respect to a proper subset U of V. Let **D** be the two-graph of Γ and Γ'.

If $v = 2(2k - \lambda - \mu)$, then Theorem 7.3.10 implies that **D** is an incomplete 2-design and therefore Γ' is an $SRG(v, k', \lambda', \mu')$ with $v = 2(2k' - \lambda' - \mu')$.

Suppose $v \neq 2(2k - \lambda - \mu)$. Then $v - 2k + 2\lambda \neq 2(k - \mu)$. Since Γ is neither a complete nor a null graph, one can find three distinct vertices x, y, and z of Γ such that $\{x, y\} \in E$, $\{x, z\} \notin E$, and $|\{y, z\} \cap U| = 1$. Proposition 7.3.9 implies that the number of blocks of **D** containing $\{x, y\}$ is not the same as the number of blocks containing $\{x, z\}$. Since either both $\{x, y\}$ and $\{x, z\}$ are edges of Γ' or both are not edges, Proposition 7.3.9 implies that Γ' is not a strongly regular graph. □

In order to apply Corollary 7.3.11 to construct strongly regular graphs, we will determine when switching a regular graph produces a regular graph.

Proposition 7.3.12. *Let $\Gamma = (V, E)$ be a regular graph of order v and degree k and let U be a proper subset of V, $|U| = u$. Let Γ' be the graph obtained by switching Γ with respect to U. The graph Γ' is regular if and only if there exist integers s and t satisfying the following conditions:*

(i) *each vertex of U has exactly s neighbors in $V \setminus U$ and each vertex of $V \setminus U$ has exactly t neighbors in U;*
(ii) *if $v = 2u$, then $s = t$;*
(iii) *if $v \neq 2u$, then $s = (v - u)/2$ and $t = u/2$.*

If these conditions are satisfied, then the degree of Γ' is equal to $k + u - 2t$.

Proof. Let A and A' be the adjacency matrices of Γ and Γ', respectively, introduced in Proposition 7.3.3. For $i = 1, 2, \ldots, u$, let s_i be the sum of the

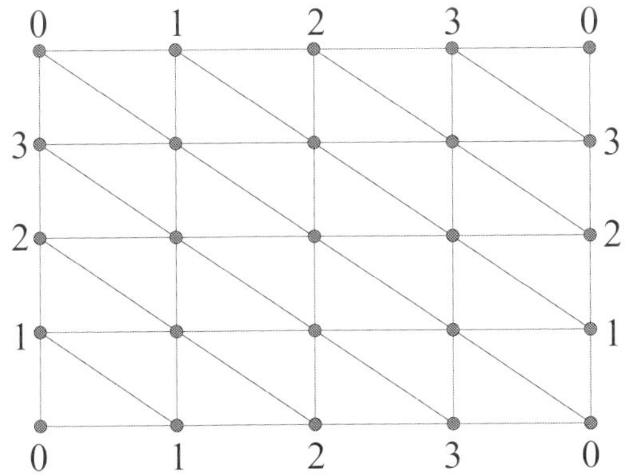

Figure 7.2 Shrikhande graph.

entries of the ith row of A_2; for $j = 1, 2, \ldots, v - u$, let t_j be the sum of the entries of the jth row of $A_2^\top$. Then the graph Γ' is regular of degree k' if and only if $(k - s_i) + (v - u - s_i) = (u - t_j) + (k - t_j) = k'$ for all i and j. Thus, Γ' is regular if and only if the matrices A_2 and $A_2^\top$ have constant row sums s and t, respectively, such that $2(s - t) = v - 2u$. Since the sum of all entries of A_2 is the same as the sum of all entries of $A_2^\top$, we obtain that s and t have also to satisfy the equation $us = (v - u)t$. The two equations for s and t are equivalent to (ii) and (iii). If these conditions are satisfied, then $k' = u + k - 2t$. □

Example 7.3.13. The *Shrikhande graph* (Fig. 7.2) is obtained by switching the graph $L_2(4)$ (Example 7.1.10) with respect to a set U of four vertices, no two of which are adjacent. Since no two vertices of U have the same first coordinate or the same second coordinate, each vertex of U has 6 neighbors that are not in U and each vertex that is not in U has 2 neighbors in U. Thus the conditions of Corollary 7.3.11 and Proposition 7.3.12 are satisfied, so the Shrikhande graph is an $SRG(16, 6, 2, 2)$. If we assume that the vertex set of both graphs is $\mathbf{Z}_4 \oplus \mathbf{Z}_4$ and let $U = \{(0, 0), (1, 1), (2, 2), (3, 3)\}$, then the vertices (a, b) and (c, d) of the Shrikhande graph are adjacent if and only if (i) $a = c$ and $b - d = \pm 1$ or (ii) $b = d$ and $a - c = \pm 1$ or (iii) $a + b = c + d$ and $a - c = \pm 1$.

Three important strongly regular graphs can be obtained by switching the graph $T(8)$ (Example 7.1.9).

Example 7.3.14. The three *Chang graphs* are the graphs obtained by switching the graph $T(8)$ with respect to the following three sets of vertices:

(i) a set of four vertices, no two of which are adjacent;

(ii) an octagon, i. e., a set $\{x_1, x_2, \ldots, x_8\}$ of eight vertices such that $\{x_i, x_j\}$ is an edge if and only if $i - j \equiv \pm 1 \pmod 8$;

(iii) the disjoint union of a triangle and an pentagon, i. e., a set $\{x_1, x_2, x_3, y_1, y_2, y_3, y_4, y_5\}$ of eight vertices such that $\{x_i, x_j\}$ is an edge for $i \neq j$, $\{y_i, y_j\}$ is an edge if and only if $i - j \equiv \pm 1 \pmod 5$, and 2-subsets $\{x_i, y_j\}$ are not edges.

The graph $T(8)$ satisfies Corollary 7.3.11. It can be verified that the Chang graphs satisfy Proposition 7.3.12, so each of these graphs is an $SRG(28, 12, 6, 4)$. These graphs will be denoted by Ch_1, Ch_2, and Ch_3, respectively.

Remark 7.3.15. It can be shown that the Shrikhande graph and the three Chang graphs do not depend on a specific choice of the set U as soon as it satisfies condition (i), (ii), or (iii) above. Though the Shrikhande graph and $L_2(4)$ have the same parameters, they are not isomorphic. Indeed, the Shrikhande graph has cycles of length 3, while $L_2(4)$ does not. It can be shown that the three Chang graphs and $T(8)$ are pairwise nonisomorphic graphs (see Exercise 13).

The problem of determining whether two incidence structures are isomorphic is a very difficult one. Only in rare cases can one find a relatively short list of numerical parameters that determine an incidence structure up to an isomorphism. A prominent exception is the Dembowski–Wagner Theorem. The next theorem describes two infinite families of strongly regular graphs that are characterized by their standard parameters. Its proof is beyond the scope of this book.

Theorem 7.3.16. *If Γ is a strongly regular graph with the same parameters as $L_2(n)$ for $n \neq 4$ or $T(n)$ for $n \neq 8$, then Γ is isomorphic to $L_2(n)$ or $T(n)$, respectively.*

Net graphs $L_r(2r)$ yield another application of Theorem 7.3.11.

Example 7.3.17. Let $\mathbf{N} = (V, \mathcal{L})$ be a $(2r, r)$-net and let $\Gamma = (V, E)$ be the corresponding net graph $L_r(2r)$. Let $\mathcal{C} = \{L_1, L_2, \ldots, L_{2r}\}$ be a parallel class of $\mathbf{N}$ and let U be the set of all vertices of Γ corresponding to the points of the set $L_1 \cup L_2 \cup \ldots \cup L_r$. Then the conditions of Proposition 7.3.12 are satisfied with $u = 2r^2$ and $s = t = r(r-1)$. The graph obtained by switching Γ with respect to U is an $NL_r(2r)$.

The next theorem gives another relation between strongly regular graphs and 2-designs with block size 3.

Theorem 7.3.18. *Let $\Gamma = (V, E)$ be a graph on v vertices and let Γ' be obtained by adjoining an isolated vertex ∞ to Γ, so $\Gamma' = (V \cup \{\infty\}, E)$ is a graph on $v + 1$ vertices. Let* **D** *be the two-graph of Γ'. Then* **D** *is an incomplete 2-design if and only if Γ is an $SRG(v, k, \lambda, \mu)$ with $k = 2\mu$. Furthermore, if* **D** *is an incomplete 2-design, then it is a 2-$(v + 1, 3, 2\mu)$ design.*

Proof. For each $x \in V$, let $k(x)$ be the valency of x in Γ. For distinct $x, y \in V$, let $\lambda(x, y)$ be the number of blocks of **D** that contain $\{x, y\}$ and let $\gamma(x, y)$ be the number of common neighbors of x and y in Γ.

Let $\{x, y\} \in E$. Then $\{x, y, \infty\}$ is a block of **D**. For $z \in V \setminus \{x, y\}$, $\{x, y, z\}$ is a block of **D** if and only if both $\{x, z\}$ and $\{y, z\}$ are edges or both $\{x, z\}$ and $\{y, z\}$ are nonedges of Γ. Therefore, $\lambda(x, y) = v - k(x) - k(y) + 2\gamma(x, y) + 1$.

Let $x, y \in V$, $x \neq y$, and $\{x, y\} \notin E$. Then $\{x, y, \infty\}$ is not a block of **D**. For $z \in V \setminus \{x, y\}$, $\{x, y, z\}$ is a block of **D** if and only if one of the 2-sets $\{x, z\}$ and $\{y, z\}$ is an edge, while the other is not. Therefore, in this case $\lambda(x, y) = k(x) + k(y) - 2\gamma(x, y)$.

Suppose now that Γ is an $SRG(v, 2\mu, \lambda, \mu)$. Then, for distinct $x, y \in V$,

$$\lambda(x, y) = \begin{cases} v - 4\mu + 2\lambda + 1 & \text{if } \{x, y\} \in E, \\ 2\mu & \text{if } \{x, y\} \notin E. \end{cases}$$

Since Γ is a graph of valency 2μ, (7.2) implies that $v - 4\mu + 2\lambda + 1 = 2\mu$. Also, for $x \in V$, the number of blocks of the form $\{\infty, x, y\}$ is equal to the valency of x, i.e., to 2μ. Therefore, **D** is a 2-$(v + 1, 3, 2\mu)$ design. If a 2-set $\{x, y\}$ is not an edge of Γ, then $\{\infty, x, y\}$ is not a block of **D**, so **D** is incomplete.

Conversely, suppose **D** is an incomplete 2-$(v + 1, 3, l)$ design. For $x \in V$, the number of blocks of **D** containing $\{\infty, x\}$ is equal to the valency of x, so Γ is a regular graph of degree l. For distinct $x, y \in V$,

$$\gamma(x, y) = \begin{cases} \dfrac{3l - v - 1}{2} & \text{if } \{x, y\} \in E, \\ \dfrac{l}{2} & \text{if } \{x, y\} \notin E. \end{cases}$$

Therefore, Γ is an $SRG(v, k, \lambda, \mu)$ with $k = 2\mu = l$. □

Example 7.3.19. The graph $SRG(5, 2, 0, 1)$ from Example 7.1.7 satisfies the condition of Theorem 7.3.18. The corresponding 2-design is a 2-(6, 3, 2) design. Strongly regular graphs $T(6)$, $L_2(3)$, and $L_{(q+1)/2}(q)$ for q a prime power also satisfy the condition of Theorem 7.3.18. The corresponding 2-designs are a 2-(16, 3, 8) design, a 2-(10, 3, 4) design, and a 2-$(q^2 + 1, 3, (q^2 - 1)/2)$ design.

Theorems 7.3.10 and 7.3.18 may allow us to obtain a strongly regular graph on v vertices from a strongly regular graph on $v+1$ vertices and vice versa. This motivates the following definition.

Definition 7.3.20. Let $\Gamma = (V, E)$ be a graph of order $v+1$ and let x be a vertex of Γ. Let U be the set of all neighbors of x. Suppose U is a proper subset of $V \setminus \{x\}$ and let Γ' be the graph obtained by switching Γ with respect to U. Let Γ_* be the graph obtained from Γ' by deleting the vertex x. The graph Γ_* is called a *descendant of* Γ and the graph Γ is called an *ascendant of* Γ_*.

We will assume that the vertex set V of the graph Γ in this definition is so ordered that x is the first vertex and every vertex of U precedes every vertex of $V \setminus U$, except x. The corresponding adjacency matrices of Γ and Γ_* can be represented as block matrices

$$A = \begin{bmatrix} 0 & \mathbf{j}^\top & \mathbf{0}^\top \\ \mathbf{j} & A_1 & A_2 \\ \mathbf{0} & A_2^\top & A_3 \end{bmatrix}, \quad A_* = \begin{bmatrix} A_1 & J - A_2 \\ (J - A_2)^\top & A_3 \end{bmatrix}. \tag{7.11}$$

The next theorem determines when a descendant of a strongly regular graph is strongly regular.

Theorem 7.3.21. *Let Γ be an $SRG(v+1, k, \lambda, \mu)$, other than the disjoint union of two complete graphs, and let Γ_* be a descendant of Γ. The graph Γ_* is strongly regular if and only if $v + 1 = 2(2k - \lambda - \mu)$. If this condition is satisfied, then Γ_* is an $SRG(v, 2(k-\mu), k+\lambda-2\mu, k-\mu)$.*

Proof. Let x and Γ' be the same as in Definition 7.3.20. Let $\mathbf{D}$ be the two-graph of Γ. By Theorem 7.3.8, $\mathbf{D}$ is the two-graph of Γ'. Let A and A_* be the adjacency matrices (7.11) of the graphs Γ and Γ_*, respectively. Then matrices A_1, A_2, $A_2^\top$, and A_3 have constant row sums λ, $k-\lambda-1$, μ, and $k-\mu$, respectively.

Suppose first that $v+1 = 2(2k-\lambda-\mu)$. Then Theorem 7.3.10 and Corollary 7.3.11 imply that $\mathbf{D}$ is an incomplete 2-design. Theorem 7.3.18 now implies that Γ_* is a strongly regular graph. Since both $(J-A_2)^\top$ and A_3 have the row sum $k-\mu$, the degree of Γ_* is $2(k-\mu)$. By Theorem 7.3.18, Γ_* is an $SRG(v, 2(k-\mu), \lambda_*, k-\mu)$. From (7.2) applied to Γ_* and the condition $v+1 = 2(2k-\lambda-\mu)$, we derive that $\lambda_* = k+\lambda-2\mu$.

Suppose now that $v+1 \neq 2(2k-\lambda-\mu)$. Then $v - 2k + 2\lambda - 1 \neq 2(k-\mu)$ and therefore the graph Γ_* is not regular. □

Example 7.3.22. The graph $T(8)$ is an $SRG(28, 12, 6, 4)$ and therefore it satisfies the condition of Theorem 7.3.10. Any descendant of $T(8)$ is an $SRG(27, 16, 10, 8)$. It can be shown that the graph $T(8)$ has a unique descendant (i.e., all its descendants are isomorphic). This descendant is called the *Schläfli graph*.

We will have another application of Theorem 7.3.21 in the next section (see the proof of Theorem 7.4.23).

We will now determine when an ascendant of a strongly regular graph is strongly regular. Given a graph $\Gamma = (V, E)$, we obtain every ascendant of Γ by adjoining an isolated vertex ∞ and then switching the graph $\Gamma' = (V \cup \{\infty\}, E)$ with respect to a proper subset of V.

Proposition 7.3.23. *Let Γ be an $SRG(v, k, \lambda, \mu)$ and let a regular graph Γ^* be an ascendant of Γ. The graph Γ^* is strongly regular if and only if $k = 2\mu$.*

Proof. Let $\Gamma = (V, E)$, $\Gamma' = (V \cup \{\infty\}, E)$ with $\infty \notin V$, and let Γ^* be obtained by switching Γ' with respect to a proper subset of V. If $k = 2\mu$, then, by Theorem 7.3.18, the two-graph of Γ^* is an incomplete 2-design. Since Γ^* has the same two-graph as Γ', Theorem 7.3.10 implies that Γ^* is strongly regular.

Conversely, suppose Γ^* is strongly regular. If Γ^* is not the disjoint union of two complete graphs, then Theorem 7.3.21 implies that $k = 2\mu$. If Γ^* is the disjoint union of two complete graphs, then Γ as a descendant of Γ^* has to be a complete graph, which is not the case. □

The next theorem determines when an ascendant of a regular graph is regular.

Theorem 7.3.24. *Let $\Gamma = (V, E)$ be a regular graph of order v and degree k, let U be a proper subset of V, $|U| = u$, and let $\Gamma' = (V, E')$ be obtained by switching Γ with respect to U. Let $\Gamma^* = (V^*, E^*)$ be the graph of order $v + 1$ with $V^* = V \cup \{\infty\}$ and $E^* = E' \cup \{(\infty, x) \colon x \in U\}$. The graph Γ^* is regular if and only if the following conditions are satisfied*:

(i) $2u^2 - (v + 2k + 1)u + vk = 0$;
(ii) *each vertex of U is adjacent in Γ to exactly $(v + k + 1)/2 - u$ vertices of $V \setminus U$;*
(iii) *each vertex of $V \setminus U$ is adjacent in Γ to exactly $k/2$ vertices of U.*

If these conditions are satisfied, then the degree of Γ^* is equal to u.

Proof. Let matrices A and A_* from (7.11) be adjacency matrices of Γ^* and Γ, respectively, with A_1 being a square matrix of order u. For $i = 1, 2, \ldots, u$, let s_i be the sum of the entries of the ith row of $J - A_2$; for $j = 1, 2, \ldots, v - u$, let t_j be the sum of the entries of the jth row of $(J - A_2)^\top$. The graph Γ^* is regular (of degree u) if and only if $1 + (k - s_i) + (v - u - s_i) = u$ for $i = 1, 2, \ldots, s$ and $(u - t_j) + (k - t_j) = u$ for $j = 1, 2, \ldots, v - u$. These equations are equivalent to conditions (ii) and (iii). Since the sum of all entries of matrices $J - A_2$ and $(J - A_2)^\top$ is the same, conditions (ii) and (iii) imply (i). □

Example 7.3.25. Let Γ be the graph whose vertex set is the field $GF(9)$ and distinct vertices x and y are adjacent if and only if $x - y$ is a square. Then Γ is the Paley graph with parameters $(9, 4, 1, 2)$. Let $U = \{0, 1, -1\}$. Then the corresponding ascendant graph of Γ is the Petersen graph.

In Chapter 8 (Theorem 8.2.26 and its corollaries), we give other examples of strongly regular ascendant graphs.

We defined a two-graph as an incidence structure induced by a graph. The next theorem gives an intrinsic characterization of two-graphs.

Theorem 7.3.26. *Let V be a finite set and $\mathcal{B}$ a set of 3-subsets of V. The incidence structure $\mathbf{D} = (V, \mathcal{B})$ is the two-graph of some graph with vertex set V if and only if every 4-subset of V contains an even number of elements of $\mathcal{B}$.*

Proof. Let $\mathbf{D} = (V, \mathcal{B})$ be the two-graph of a graph $\Gamma = (V, E)$ and let X be a 4-subset of V. Since each edge of Γ is contained in none or two 3-subsets of X and each block of $\mathbf{D}$ contains odd number of edges, the set X contains an even number of blocks.

Conversely, suppose every 4-subset of V contains an even number of blocks of the incidence structure $\mathbf{D}$. Fix an element $\infty \in V$ and define a graph $\Gamma = (V, E)$ with ∞ as an isolated vertex; distinct vertices x and y other than ∞ form an edge of Γ if and only if $\{\infty, x, y\}$ is a block of $\mathbf{D}$. We claim that $\mathbf{D}$ is the two-graph of Γ.

Let $B \in \mathcal{B}$. If $B = \{\infty, x, y\}$, then B contains one edge of Γ, so B is a block of the two-graph of Γ. Let $B = \{x, y, z\} \subseteq V \setminus \{\infty\}$. Consider 4-set $X = \{\infty, x, y, z\}$. It contains at least one block of $\mathbf{D}$. If X contains two blocks of $\mathbf{D}$, e.g. B and $\{\infty, x, y\}$, then $\{x, y\}$ is the only edge contained in B. If X contains four blocks, then B contains three edges, so in either case B is a block of the two-graph of Γ. Conversely, let A be a block of the two-graph of Γ. If $A = \{\infty, x, y\}$, then $\{x, y\}$ is an edge and therefore A is a block of $\mathbf{D}$. Let $A = \{x, y, z\} \subseteq V \setminus \{\infty\}$. If A contains only one edge, e.g. $\{x, y\}$, then $\{\infty, x, y\}$ is a block, while $\{\infty, x, z\}$ and $\{\infty, y, z\}$ are not blocks of $\mathbf{D}$. Since set $X = \{\infty, x, y, z\}$ contains an even number of blocks of $\mathbf{D}$, we have $A \in \mathcal{B}$. If A contains three edges, then $\{\infty, x, y\}$, $\{\infty, x, z\}$, and $\{\infty, y, z\}$ are blocks of $\mathbf{D}$. Then X contains four blocks of $\mathbf{D}$, so again $A \in \mathcal{B}$. □

7.4. Symmetric designs with polarities

A hyperplane in an n-dimensional projective space can be given by an equation of the form $a_0x_0 + a_1x_1 + \cdots + a_nx_n = 0$. Two such equations describe the same hyperplane if and only if the vectors of their coefficients are proportional, i.e., they define the same projective point. Thus, we have a one-to-one correspondence between the points and hyperplanes of a projective space. For any projective points x and y, point x is contained in the hyperplane corresponding to y if and only if y is contained in the hyperplane corresponding to x. This means that the design $PG_{n-1}(n, q)$ has a symmetric incidence matrix. (Another proof of this fact is given in Proposition 3.6.9.) A generalization of this relation between points and blocks to an arbitrary symmetric design leads to the notion of polarity.

Definition 7.4.1. Let $\mathbf{D} = (X, \mathcal{B})$ be a symmetric design. A bijection $\sigma : X \to \mathcal{B}$ such that for any $x, y \in X, x \in \sigma(y)$ if and only if $y \in \sigma(x)$ is called a *polarity* of $\mathbf{D}$. A point x is called an *absolute point* of a polarity σ, if $x \in \sigma(x)$.

A symmetric design admits a symmetric incidence matrix with zero diagonal if and only if it has a polarity with no absolute point. A symmetric design admits a symmetric incidence matrix with all diagonal entries equal to 1 if and only if it has a polarity with all points absolute.

As we mentioned at the beginning of Section 7.1., if N is an incidence matrix of a symmetric design, then in some cases N or $N - I$ may serve as an adjacency matrix of a strongly regular graph. The matrix N has to be symmetric and have all zeros or all ones on the diagonal.

Conversely, if $\Gamma = (V, E)$ is an $SRG(v, k, \lambda, \lambda)$, we can obtain a symmetric (v, k, λ)-design $\mathbf{D} = (V, \mathcal{B})$ where $\mathcal{B} = \{\Gamma(x) : x \in \Gamma\}$ is $\Gamma(x)$ is the set of all neighbors of x. If $\Gamma = (V, E)$ is an $SRG(v, k - 1, \lambda - 2, \lambda)$, we obtain a symmetric (v, k, λ)-design $\mathbf{D} = (V, \mathcal{B})$ where $\mathcal{B} = \{\Gamma(x) \cup \{x\} : x \in \Gamma\}$.

Thus, we have the following result.

Proposition 7.4.2. *An $SRG(v, k, \lambda, \lambda)$ exists if and only if there exists a symmetric (v, k, λ)-design that admits a polarity with no absolute point. An $SRG(v, k - 1, \lambda - 2, \lambda)$ exists if and only if there exists a symmetric (v, k, λ)-design that admits a polarity with all points absolute.*

Remark 7.4.3. If a symmetric design D has a polarity with no absolute point, then the complementary design has a polarity with all points absolute, and vice versa. Therefore, we will concentrate on symmetric designs that admit polarities with no absolute points, i.e., on strongly regular graphs with parameters (v, k, λ, λ).

Definition 7.4.4. A (v, k, λ)-*graph* is an $SRG(v, k, \lambda, \lambda)$.

Example 7.4.5. The symmetric (56, 45, 36)-design constructed in Theorem 6.6.1 admits a symmetric incidence matrix with zero diagonal, and therefore, there exists a (56, 45, 36)-graph. The complementary graph is an $SRG(56, 10, 0, 2)$ called the *Gewirtz graph*.

Theorem 5.3.16 gives an infinite family of (v, k, λ)-graphs.

Theorem 7.4.6. *If there exists a Hadamard matrix of order* $2n$*, then there exists a* $(4n^2, 2n^2 - n, n^2 - n)$*-graph.*

Strongly regular graphs with the above parameters can be also obtained from symmetric Hadamard matrices of Bush type.

Theorem 7.4.7. *Let* h *be a positive integer. The existence of a symmetric regular Hadamard matrix of Bush type of order* $4h^2$ *is equivalent to the existence of an* $(4h^2, 2h^2 - h, h^2 - h)$*-graph with the additional property that the vertex set can be partitioned into* $2h$ *cocliques of cardinality* $2h$.

Proof. 1. Let $H = [H_{ij}]$ be a symmetric regular Hadamard matrix of Bush type represented as a block matrix with $(2h) \times (2h)$ blocks H_{ij}. Applying, if necessary, the same permutation to the rows and columns of H, we may assume that $H_{ii} = J$ for $i = 1, 2, \ldots, 2h$. Then $A = \frac{1}{2}(J - H)$ is a symmetric incidence matrix of a symmetric $(4h^2, 2h^2 - h, h^2 - h)$-design (Theorem 4.4.5) with zero diagonal. Therefore, A is an adjacency of a $(4h^2, 2h^2 - h, h^2 - h)$-graph. The vertices corresponding to the rows of each H_{ii}, $i = 1, 2, \ldots, 2h$, form a coclique.

2. Let A be an adjacency of a $(4h^2, 2h^2 - h, h^2 - h)$-graph whose vertex set is partitioned into cocliques $C_1, C_2, \ldots, C_{2h}$ of cardinality $2h$. We will assume that the rows of A corresponding to the points of C_i precede the rows corresponding to the points of C_j whenever $i < j$. By Theorem 4.4.5, $H = J - 2A$ is a symmetric regular Hadamard matrix. If we represent H as a block matrix $H = [H_{ij}]$ with $(2h) \times (2h)$ blocks H_{ij}, then $H_{ii} = J$ for $i = 1, 2, \ldots, 2h$.

Observe that $-h$ is the smallest eigenvalue of Γ. Therefore, the cardinality of each coclique C_i attains the upper bound of Theorem 7.2.10(i). This theorem implies that, for $i \neq j$, every vertex of C_i has exactly h neighbors in C_j. Therefore, all row and column sums of each H_{ij} with $i \neq j$ are equal to 0. Thus, H is a matrix of Bush type. □

We apply Theorem 4.4.18 to obtain the following family of strongly regular graphs.

Corollary 7.4.8. *For any odd positive integer n, there exists a $(4n^4, 2n^4 - n^2, n^4 - n^2)$-graph.*

The restrictions on parameters of strongly regular graphs that we have obtained so far imply the following result.

Proposition 7.4.9. *If there exists a (v, k, λ)-graph, then* (i) $u = \sqrt{k - \lambda}$ *is an integer that divides both k and λ,* (ii) $v + u - 1 \equiv \lambda/u \pmod 2$, (iii) $(k - \lambda)^2 \geq k + \lambda$, *and* (iv) $k \leq \lambda^2 + \lambda$.

Proof. Let A be an adjacency matrix of a (v, k, λ)-graph. Theorem 7.2.2 implies that the eigenvalues of this graph are k and $\pm u$. The Integrality Condition implies that u is an integer dividing k. Since $u^2 = k - \lambda$, u divides λ and therefore $u \leq \lambda$, which implies (iv). Inequality (iii) follows from (7.1) and the relation $\lambda(v - 1) = k(k - 1)$. To prove (ii), observe that, by Theorem 7.2.2, one of the eigenvalues of A has multiplicity

$$f = \frac{v-1}{2} - \frac{k}{2u} = \frac{v+u-1}{2} - \frac{\lambda}{2u}.$$

Since $2f$ is even, we obtain that $v + u - 1 \equiv \lambda/u \pmod 2$. □

Corollary 7.4.10. *For any fixed λ, there are at most finitely many nonisomorphic (v, k, λ)-graphs.*

Remark 7.4.11. Marshall Hall, Jr. conjectured that, for any fixed $\lambda \geq 2$, there are at most finitely many nonisomorphic symmetric (v, k, λ)-designs. The conjecture is still open even for $\lambda = 2$.

Corollary 7.4.12. *There is no $(v, k, 1)$-graph.*

Corollary 7.4.13. *There is no (v, k, λ)-graph whose parameters are those of* (i) $PG_{d-1}(d, q)$ *or of* (ii) *a symmetric $(4n - 1, 2n - 1, n - 1)$-design.*

Proof. In both cases, either u is not an integer or u is an integer that does not divide k. □

Corollary 7.4.14. *If there exists a (v, k, λ)-graph, where (v, k, λ) are the parameters of the complement of $PG_{d-1}(d, q)$ with d even, then $d \geq 4$ and q is a power of 4. If, for a positive integer n, there exists a $(4n - 1, 2n, n)$-graph, then n is a square.*

Proof. The second statement immediately follows from Proposition 7.4.9(i). If (v, k, λ) are the parameters of the complement of $PG_{d-1}(d, q)$, then Proposition 7.4.9(i) implies that either d is odd or q is a square. If d is even and $q = p^2$

for some prime power p, then Proposition 7.4.9(ii) implies that p is even and therefore, q is a power of 4. Inequality (7.1) then implies that $d > 2$. □

Remark 7.4.15. In Chapter 10 (Corollary 10.4.17), we will prove that, for every odd d, there exists a (v, k, λ)-graph with parameters of the complement of $PG_{d-1}(d, q)$.

Some strongly regular graphs introduced in Section 7.1. are (v, k, λ)-graphs. For instance, a pseudo-Latin graph $PL_r(n)$ and a negative Latin graph $NL_r(n)$ are (v, k, λ)-graphs if and only if $n = 2r$. Thus we have the following result.

Proposition 7.4.16. *The following statements are equivalent:*

(i) *there exists a* $PL_n(2n)$;
(ii) *there exists an* $NL_n(2n)$;
(iii) *there exists a regular symmetric Hadamard matrix of order* $4n^2$ *with constant diagonal.*

Since there exists a net graph $L_3(6)$ (Corollary 3.2.19), we have the following result.

Corollary 7.4.17. *There exists a regular symmetric Hadamard matrix of order* 36 *with constant diagonal.*

Theorem 7.1.26 with $k = 2$ produces a $(vs^2\mu, (v-1)s\mu, (v-k)\mu)$-graph. This theorem requires two ingredients. One of them is a $(v, b, r, 2, 1)$-design. The blocks of this design are all 2-subsets of a v-set, so $b = v(v-1)/2$, and $r = v - 1$. Taking $A_\mu(s)$ to be the design $AG_d(d+1, q)$, we obtain the following result.

Theorem 7.4.18. *Let q be a prime power, d a positive integer, and $r = (q^{d+1} - 1)/(q - 1)$. Then there exists an $((r+1)q^{d+1}, rq^d, (r-1)q^{d-1})$-graph.*

Remark 7.4.19. The parameters of the (v, k, λ)-graphs of Theorem 7.4.18 are those of the symmetric designs constructed in Theorem 3.8.3.

Remark 7.4.20. Taking $k = 2$ and $A_\mu(s)$ to be a Hadamard 3-design in Theorem 7.1.26 yields (v, k, λ)-graphs with the same parameters as in Remark 5.3.18.

Remark 7.4.21. For $q = 2$ (and $r = 2^{d+1} - 1$), the graph described in Theorem 7.4.18 is a $PL_{2^d}(2^{d+1})$. The complementary graph is an $NL_{2^d}(2^{d+1})$.

The Kronecker product of two regular Hadamard matrices is a regular Hadamard matrix (Proposition 4.4.9). The Kronecker product of symmetric matrices is a symmetric matrix. The Kronecker product of two matrices with constant diagonals is a matrix with constant diagonal. Therefore, we combine Corollaries 7.4.17 and 5.3.17 to obtain the following result.

Theorem 7.4.22. *If there exists a Hadamard matrix of order* $4n$*, then, for any nonnegative integer* s*, there exists a regular symmetric Hadamard matrix of order* $2^{2s+4} \cdot 3^{2s} n^2$ *with constant diagonal.*

In Chapter 8 we will construct regular symmetric Hadamard matrices with constant diagonal of order 100 and 196. It will allow us to extend Theorem 7.4.22.

If we apply the necessary conditions imposed by Proposition 7.4.9 to a $(4m-1, 2m, m)$-graph, we obtain that m has to be a square.

Theorem 7.4.23. *If there exists a regular symmetric Hadamard matrix* $H = [a_{ij}]$ *of order* $4n^2$ *with* $a_{ii} = -1$ *for* $2 \leq i \leq 4n^2$*. Then there exists a* $(4n^2 - 1, 2n^2, n^2)$*-graph.*

Proof. Let $H = [a_{ij}]$ be a regular symmetric Hadamard matrix of order $4n^2$ and let $a_{ii} = -1$ for $i = 2, 3, \ldots, 4n^2$. Let $2n$ be the row sum of H.

Case 1: $a_{11} = -1$.

Then $A = \frac{1}{2}(J + H)$ is an adjacency matrix of a $(4n^2, 2n^2 - n, n^2 - n)$-graph Γ satisfying the condition of Theorem 7.3.21. Its descendant Γ_* is a $(4n^2 - 1, 2n^2, n^2)$-graph.

Case 2: $a_{11} = 1$.

We will normalize H to obtain a symmetric Hadamard matrix H' with the same diagonal entries as H in the following way: if $a_{1j} = a_{j1} = -1$, we will multiply both the jth row and the jth column of H by -1. Let K be the core of H'. Then $\frac{1}{2}(J + K)$ is a symmetric incidence matrix of a symmetric $(4n^2 - 1, 2n^2, n^2)$-design with zero diagonal, and therefore, it is an adjacency matrix of a $(4n^2 - 1, 2n^2, n^2)$-graph. □

Remark 7.4.24. Theorem 7.4.22 gives a family of regular Hadamard matrices satisfying the condition of Theorem 7.4.23. Another family of such matrices is given by Theorem 4.5.1 (see Remark 4.5.2).

We will now modify the proof of Theorem 7.1.26 to obtain another infinite family of (v, k, λ)-graphs.

Theorem 7.4.25. *For any positive integer* d*, there exists an* $(r \cdot 3^{d+1}, (r+1) \cdot 3^d, (r+2) \cdot 3^{d-1})$*-graph with* $r = (3^{d+1} - 1)/2$*.*

Proof. Let d be a positive integer and let $r = (3^{d+1} - 1)/2$. For $i = 1, 2, \ldots, r$, let $\mathbf{D}_i = (X_i, \mathcal{B}_i)$ be an $AG_d(d+1, 3)$, so each $\mathbf{D}_i$ is an affine resolvable $(3^{d+1}, 3r, r, 3^d, (r-1)/3)$-design. We will assume that the point sets $X_1, X_2, \ldots, X_r$ are pairwise disjoint and let V be their union. For $i = 1, 2, \ldots, r$, let $\{\mathcal{C}_{i1}, \mathcal{C}_{i2}, \ldots, \mathcal{C}_{ir}\}$ be the parallelism of $\mathbf{D}_i$ and let $\mathcal{C}_{ij} = \{L_{ij1}, L_{ij2}, L_{ij3}\}$.

Let Γ be the graph on the vertex set V with edges $\{\alpha, \beta\}$ defined as follows: if $\alpha \in X_i$ and $\beta \in X_j$ with $i \neq j$, then $\{\alpha, \beta\}$ is an edge of Γ if and only if there is an h such that $\alpha \in L_{ijh}$ and $\beta \in L_{jih}$; if $\alpha, \beta \in X_i$, then $\{\alpha, \beta\}$ is an edge of Γ if and only if there is an h such that $\alpha \in L_{iih}$ and $\beta \notin L_{iih}$. We claim that Γ is a (v, k, λ)-graph with the required parameters.

Let $\alpha \in X_i$. For each $j \neq i$, there are exactly 3^d vertices $\beta \in X_j$ that are adjacent to α. Since there are $2 \cdot 3^d$ vertices $\beta \in X_i$ that are adjacent to α, the valency of α is equal to $(r+1) \cdot 3^d$.

Let $\alpha \in X_i$ and $\beta \in X_j$ with $i \neq j$. Let $\alpha \in L_{iih}$ and $\beta \in L_{jil}$. Let $k \in \{1, 2, \ldots, r\}$ and let $k \neq i$ and $k \neq j$. Let $\alpha \in L_{ikf}$, $\beta \in L_{jkg}$. Then a vertex $\gamma \in X_k$ is adjacent to both α and β if and only if $\gamma \in L_{kif} \cap L_{kjg}$, so there are exactly 3^{d-1} such vertices γ. If $\gamma \in X_i$, then it is adjacent to both α and β if and only if $\gamma \notin L_{iih}$ and $\gamma \in L_{ijl}$. Thus, each of the sets X_i and X_j contains exactly $2 \cdot 3^{d-1}$ vertices adjacent to α and β, and the total number of such vertices is $(r+2) \cdot 3^{d-1}$.

Let $\alpha, \beta \in X_i$ and let $\{\alpha, \beta\}$ be an edge. Let $k \in \{1, 2, \ldots, r\}$ and let $k \neq i$. Let $\alpha \in L_{ikf}$ and $\beta \in L_{ikg}$. A vertex $\gamma \in X_k$ is adjacent to both α and β if and only if $f = g$ and $\gamma \in X_{kif}$. The design $\mathbf{D}_i$ has exactly $(r-1)/3$ blocks containing both α and β. Since α and β are adjacent, none of these blocks is in $\mathcal{C}_{ii}$. Thus, there are $(r-1)/3$ indices $k \neq i$ in $\{1, 2, \ldots, r\}$ such that $\alpha, \beta \in L_{ikf}$ for a unique $f \in \{1, 2, 3\}$, and then all vertices $\gamma \in X_{kif}$ are adjacent to both α and β. Therefore, there are exactly $(r-1) \cdot 3^{d-1}$ vertices $\gamma \notin X_i$ that are adjacent to both α and β. If $\gamma \in X_i$, $\alpha \in L_{iif}$, and $\beta \in L_{iig}$ with $f \neq g$, then γ is adjacent to both α and β if and only if $\gamma \in L_{iih}$ with $h \neq f$ and $h \neq g$. Since there are 3^d such vertices γ, the total number of vertices that are adjacent to α and β is $(r+2) \cdot 3^{d-1}$.

Finally, let $\alpha, \beta \in X_i$, $\alpha \neq \beta$, and let $\{\alpha, \beta\}$ be not an edge, i.e., $\alpha, \beta \in L_{iih}$ for some h. Then there are $(r-4) \cdot 3^{d-1}$ vertices outside X_i and $2 \cdot 3^d$ vertices in X_i that are adjacent to both α and β, so the total number of such vertices is again $(r+2) \cdot 3^{d-1}$. □

We will now consider (v, k, λ)-graphs whose parameters are those of the complements of the designs constructed in Theorems 3.8.3 and 3.8.5.

Theorem 7.4.26. *Let q be a prime power, d a positive integer, and $r = (q^{d+1}-1)/(q-1)$. An $((r+1)q^{d+1}, q^d(q^{d+1}+q-1), q^d(q^d+1)(q-1))$-graph exists if and only if qd is even.*

Proof. If a graph with the above parameters exists, then condition (ii) of Proposition 7.4.9 implies that qd is even.

Suppose now that qd is even. Then $r+1$ is even, so let L be a Latin square of order $r+1$ satisfying Lemma 3.2.23. Let $H_1, H_2, \ldots, H_r$ be all d-dimensional subspaces of the $(d+1)$-dimensional vector space V over $GF(q)$. For $i, j = 1, 2, \ldots, r+1$, let F_{ij} be the empty set if $L(i, j) = 1$ and $F_{ij} = H_{k-1}$ if $L(i, j) = k \neq 1$. Let $V = \{x_1, x_2, \ldots, x_{q^{d+1}}\}$. For $i, j = 1, 2, \ldots, r+1$, define a (0, 1)-matrix N_{ij} of order q^{d+1} with (s, t)-entry equal to 1 if and only if $x_s - x_t \in F_{ij}$.

By Theorem 3.8.3, the matrix $N = [N_{ij}]$ of order $(r+1)q^{d+1}$ is an incidence matrix of a symmetric $((r+1)q^{d+1}, rq^d, (r-1)q^{d-1})$-design. Since each H_k is a subgroup of an abelian group V, the blocks N_{ij} are symmetric matrices and, if $L(i, j) \neq 1$, all the diagonal entries of N_{ij} are equal to 1. Since $L(i, j) = L(j, i)$ and $L(i, i) \neq 1$, the matrix N is symmetric with all ones on the diagonal. Therefore, $J - N$ is an adjacency matrix of a (v, k, λ)-graph with the required parameters. □

Theorem 7.4.27. *Let d be a positive integer. A (v, k, λ)-graph with parameters*

$$(3^{d+1}(3^{d+1}-1)/2,\ 3^d(3^{d+1}-2),\ 3^d(3^{d+1}-3^d-2))$$

exists if and only if d is odd.

Proof. If a graph with the above parameters exists, then condition (ii) of Proposition 7.4.9 implies that d is odd.

Suppose now that d is odd. Then $r = (3^{d+1}-1)/2$ is even, so let L be a Latin square of order r satisfying Lemma 3.2.23. Let $r = (3^{d+1}-1)/2$ and let $H_1, H_2, \ldots, H_r$ be all d-dimensional subspaces of the $(d+1)$-dimensional vector space V over $GF(3)$. For $i, j = 1, 2, \ldots, r$, let $F_{ij} = V \setminus H_1$ if $L(i, j) = 1$ and $F_{ij} = H_k$ if $L(i, j) = k \neq 1$. Let $V = \{x_1, x_2, \ldots, x_{3^{d+1}}\}$. For $i, j = 1, 2, \ldots, r$, let (0, 1)-matrices N_{ij} be defined as in the proof of Theorem 7.4.26. Then $N = [N_{ij}]$ is a (0, 1) block matrix of order $r \cdot 3^{d+1}$. By Theorem 3.8.5, the matrix N is an incidence matrix of a symmetric $(r \cdot 3^{d+1}, (r+1) \cdot 3^d, (r+2) \cdot 3^{d-1})$-design. Moreover, N is a symmetric matrix with all ones on the diagonal. Therefore, $J - N$ is an adjacency matrix of a (v, k, λ)-graph with the required parameters. □

7.5. Symmetric designs and digraphs

A *directed graph* or *digraph* is a pair $\Gamma = (V, E)$, where V is a finite nonempty set of *vertices* and E is a set of ordered pairs (*arcs*) (x, y) with $x, y \in V$ and

$x \neq y$. If (x, y) is an arc, we will say that x *dominates* y or that y *is dominated by* x. If x dominates y or y dominates x, we will say that the vertices x and y are *adjacent*. A digraph is said to be *regular of degree k* if every vertex dominates exactly k vertices and is dominated by exactly k vertices. If $V = \{x_1, x_2, \dots, x_v\}$, then the $(0, 1)$-matrix $A = [a_{ij}]$ with $a_{ij} = 1$ if and only if x_i dominates x_j is the corresponding *adjacency matrix* of the digraph. The following proposition is immediate.

Proposition 7.5.1. *A digraph Γ with an adjacency matrix A is regular of degree k if and only if $AJ = JA = kJ$.*

Recall that a real matrix A is called *normal* if $AA^\top = A^\top A$. If A is the adjacency matrix of a digraph corresponding to the vertex set $V = \{x_1, x_2, \dots, x_v\}$, then the (i, j) entry of $AA^\top$ is equal to the number of vertices that are dominated by both x_i and x_j, while the (i, j) entry of $A^\top A$ is equal to the number of vertices that dominate both x_i and x_j. Therefore we give the following definition.

Definition 7.5.2. A digraph Γ is called *normal* if for any (not necessarily distinct) vertices x and y the number of vertices that are dominated by both x and y is equal to the number of vertices that dominate both x and y.

Thus we have the following result.

Proposition 7.5.3. *A digraph with an adjacency matrix A is normal if and only if the matrix A is normal.*

The following definition introduces a digraph analog of strongly regular graphs.

Definition 7.5.4. A digraph $\Gamma = (V, E)$ with $E \neq \varnothing$ is called a *normally regular digraph* with parameters (v, k, λ, μ) or an $NRD(v, k, \lambda, \mu)$ if it satisfies the following conditions: (i) $|V| = v$; (ii) for all $x, y \in V$, if $(x, y) \in E$, then $(y, x) \notin E$; (iii) every vertex of Γ dominates exactly k vertices; (iv) for all $x, y \in V$, if x and y are adjacent, then there are exactly λ vertices that are dominated by both x and y; (v) for all $x, y \in V$, if x and y are not adjacent and $x \neq y$, then there are exactly μ vertices that are dominated by both x and y.

If A is an adjacency matrix of Γ, then condition (ii) is satisfied if and only if $A + A^\top$ is a $(0, 1)$-matrix. The following proposition is straightforward.

Proposition 7.5.5. *A digraph Γ on v vertices with an adjacency matrix A is an $NRD(v, k, \lambda, \mu)$ if and only if $A + A^\top$ is a $(0, 1)$-matrix and*

$$AA^\top = (k - \mu)I + (\lambda - \mu)(A + A^\top) + \mu J. \tag{7.12}$$

Remark 7.5.6. If A is an adjacency matrix of an $NRD(v, k, \lambda, \mu)$, $N = A - (\lambda - \mu)I$, and $\eta = k - \mu + (\lambda - \mu)^2$, then (7.12) can be rewritten as

$$NN^\top = \eta I + \mu J. \tag{7.13}$$

Proposition 7.5.7. *A normally regular digraph is normal.*

Proof. Let Γ be an $NRD(v, k, \lambda, \mu)$ and let A, N, and η be the same as in Remark 7.5.6. Then (7.13) and Lemma 2.3.6 imply that

$$\det(NN^\top) = (\eta + v\mu)\eta^{v-1}. \tag{7.14}$$

If $\eta = 0$, then $k = \lambda = \mu = 0$ and then Γ has no arcs. Therefore, $\det N \neq 0$ and (7.13) implies $N^\top = \eta N^{-1} + \mu N^{-1}J$. Since $AJ = kJ$, we have $NJ = (k - \lambda + \mu)J$, so $N^{-1}J = (k - \lambda + \mu)^{-1}J$. Therefore,

$$N^\top = \eta N^{-1} + \frac{\mu}{k - \lambda + \mu}J,$$

$$N^\top J = \frac{\eta + v\mu}{k - \lambda + \mu}J = (N^\top J)^\top = JN,$$

so

$$N^\top N = \eta I + \frac{\mu(\eta + v\mu)}{(k - \lambda + \mu)^2}J. \tag{7.15}$$

Eq. (7.15) and Lemma 2.3.6 allow us to evaluate $\det(N^\top N)$. Since it is equal to $\det(NN^\top)$ evaluated in (7.14), we derive that $\eta + \mu v = (k - \lambda + \mu)^2$. Therefore, $N^\top N = NN^\top$ and then $AA^\top = A^\top A$, i.e., Γ is normal. □

Since matrix N satisfies (7.13), it is an incidence matrix of a symmetric design if and only if it is a (0, 1)-matrix. Therefore, N is an incidence matrix of a symmetric design if and only if $\mu = \lambda$ or $\mu = \lambda + 1$. Thus, Proposition 7.5.5 implies the following result.

Proposition 7.5.8. *Let N be an incidence matrix of a symmetric (v, k, λ)-design. Then N is an adjacency matrix of an $NRD(v, k, \lambda, \lambda)$ if and only if $N + N^\top$ is a (0, 1)-matrix. The matrix $N - I$ is an adjacency matrix of an $NRD(v, k - 1, \lambda - 1, \lambda)$ if and only if $N + N^\top - 2I$ is a (0, 1)-matrix.*

We will now give several infinite families of symmetric designs that yield adjacency matrices of normally regular digraphs. We begin with projective planes.

Theorem 7.5.9. *If there exists a projective plane of order q, then there exists an $NRD(q^2 + q + 1, q, 0, 1)$.*

Proof. Let N be an incidence matrix of a symmetric $(q^2+q+1, q+1, 1)$-design with all diagonal entries equal to 1 (see Exercise 6 of Chapter 2). If the matrix $N + N^\top - 2I$ has an entry equal to 2, then the design has two distinct blocks which meet in two points. Since this is not possible, $N - I$ is an adjacency matrix of an $NRD(q^2+q+1, q, 0, 1)$. □

We will now obtain two more infinite families of normally regular digraphs.

Theorem 7.5.10. *If h is a positive integer such that there exists a Hadamard matrix of order $2h$, then there exists an $NRD(4h^2, 2h^2-h, h^2-h, h^2-h)$.*

Proof. Let L be a Latin square of order $n = 2h$ from Lemma 3.2.23 and let H be a Hadamard matrix of order n with all entries in the last row equal to 1. Let $K = [K_{ij}]$ be the regular Hadamard matrix of Bush type from Theorem 4.4.16 with $K_{ij} = C_k$ whenever $L(i, j) = k$ and $i \le k$ and $K_{ij} = -C_k$ whenever $L(i, j) = k$ and $i > k$. Then $A = \frac{1}{2}(J - K)$ is an adjacency matrix of an $NRD(4h^2, 2h^2-h, h^2-h, h^2-h)$. □

Theorem 7.5.11. *Let q be a prime power, d a positive integer, and let $r = (q^{d+1}-1)/(q-1)$. Then there exists an $NRD((r+1)q^{d+1}, rq^d, (r-1)q^{d-1}, (r-1)q^{d-1})$.*

Proof. Let V be the $(d+1)$-dimensional vector space over $GF(q)$ and let $H_1, H_2, \ldots, H_r$ be the d-dimensional subspaces of V. For $i = 1, 2, \ldots, r$, fix an element $a_i \in V \setminus H_i$.

Suppose first that q is even. Then r is odd, so let L be a Latin square of order $r+1$ satisfying Lemma 3.2.23. Let $M = [M(F_{ij})]$ be the incidence matrix of a symmetric design from Theorem 3.8.3 with $s = r+1$, $F_{ij} = H_k$ whenever $L(i, j) = k \ne s$ and $i < j$, $F_{ij} = H_k + a_k$ whenever $L(i, j) = k \ne s$ and $i > j$, and $F_{ij} = \varnothing$ for $i = j$. Since sets F_{ij} and F_{ji} are disjoint, all matrices $M(F_{ij}) + M(F_{ji})$ are (0, 1)-matrices. Since all matrices $M(F_{ij})$ are symmetric, we obtain that $M + M^\top$ is a (0, 1)-matrix. Therefore, M is an adjacency matrix of the required digraph.

Suppose now that q is odd. Then $|V|$ is odd and therefore, for $i = 1, 2, \ldots, r$, there is a unique $b_i \in V$ such that $a_i = 2b_i$. Let a Latin square L of order $r+1$ be such that $L(i, j) \equiv i + j - 1 \pmod{r+1}$ for $i, j = 1, 2, \ldots, r+1$. Let $N = [N(F_{ij})]$ be the incidence matrix of a symmetric design from Theorem 3.8.3 with $F_{ij} = H_k + b_k$ whenever $L(i, j) = k \ne s$ and $F_{ij} = \varnothing$ whenever $L(i, j) = s$. Observe that $F_{ij} = F_{ji}$ and, since $2b_k \notin H_k$, $F_{ij} \cap (-F_{ij}) = \varnothing$. Then, for $i, j = 1, 2, \ldots, r+1$, $N_{ij} = N_{ji}$ and $N_{ij} + N_{ij}^\top$ is a (0, 1)-matrix. Therefore, N is an adjacency matrix of a required digraph. □

Tournaments represent an important class of digraphs.

Definition 7.5.12. A digraph is called a *tournament* if for any distinct vertices x and y exactly one of the pairs (x, y) and (y, x) is an arc.

The following proposition is immediate.

Proposition 7.5.13. *A digraph with an adjacency matrix A is a tournament if and only if $A + A^\top = J - I$.*

We will now characterize normally regular tournaments.

Theorem 7.5.14. *Let A be an adjacency matrix of a digraph, let $S = J - I - 2A$, and let C be the matrix obtained by adjoining the first row $R = [0, 1, 1, \ldots, 1]$ and the first column $-R^\top$ to S. The following statements are equivalent*:

(i) *A is an adjacency matrix of a normally regular tournament*;
(ii) *A is an incidence matrix of a Hadamard 2-design and $A + A^\top$ is a $(0, 1)$-matrix*;
(iii) *C is a skew-symmetric conference matrix.*

Proof. (i) $\Rightarrow$ (ii). Let A be an adjacency matrix of a normally regular tournament of degree k. Then $A + A^\top = J - I$ is a $(0, 1)$-matrix and the tournament is an $NRD(2k + 1, k, \lambda, \lambda)$ for some λ. Remark 7.5.6 implies that A is an incidence matrix of a symmetric $(2k + 1, k, \lambda)$-design, (2.9) implies that $\lambda = (k - 1)/2$, and then Proposition 4.1.7 implies that A is an incidence matrix of a Hadamard 2-design.

(ii) $\Rightarrow$ (iii). Let A be an incidence matrix of a symmetric $(4n - 1, 2n - 1, n - 1)$-design and let $A + A^\top$ be a $(0, 1)$-matrix. Since the row sum of $A + A^\top$ is equal to $2(2n - 1)$ and since it has zero diagonal, we obtain that $A + A^\top = J - I$. Then $S + S^\top = O$, i.e., S is a skew-symmetric matrix. Equations $AA^\top = nI + (n - 1)J$ and $AJ = JA^\top = (n - 1)J$ imply $SS^\top = (4n - 1)I - J$. Therefore, distinct rows of C are orthogonal, so C is a skew-symmetric conference matrix.

(iii) $\Rightarrow$ (i). Let C be a skew-symmetric conference matrix of order $4n$ (cf. Theorem 4.3.6). Then S is skew-symmetric and $SS^\top = (4n - 1)I - J$. Since the first row of C is $R = [0, 1, \ldots, 1]$, we obtain that the row sum of S is 0. Using $A = \frac{1}{2}(J - I - S)$, we obtain that $A + A^\top = J - I$ and $AA^\top = nI + (n - 1)J$, i.e., A is an adjacency matrix of a normally regular tournament. □

Corollary 7.5.15. *If Γ is a normally regular tournament of order n, then $n \equiv 3 \pmod 4$.*

Exercises

(1) Prove that the graph C_n is strongly regular if and only if $n = 4$ or $n = 5$.
(2) Prove that a regular bipartite graph of degree k on v vertices is strongly regular if and only if $k = v/2$ or $k = 1$.

(3) Prove that $-k$ is an eigenvalue of a strongly regular graph of degree k if and only if the graph is complete bipartite.
(4) Verify that the graphs of Examples 7.1.9, 7.1.10, 7.1.11, and 7.1.12 are strongly regular graphs.
(5) Find the eigenvalues of the Petersen graph.
(6) Find the eigenvalues of strongly regular graphs of Examples 7.1.2, 7.1.7, 7.1.9, 7.1.10, 7.1.11, and 7.1.12.
(7) Find the eigenvalues of the graphs $NL_r(n)$.
(8) Find the eigenvalues of Paley graphs.
(9) Find the eigenvalues of projective graphs.
(10) Prove that there is no $SRG(28, 9, 0, 4)$.
(11) Prove part (ii) of Theorem 7.2.10.
(12) Prove that there is no strongly regular graph with any two distinct vertices having exactly one common neighbor.
(13) Show that the three Chang graphs and the graph $T(8)$ are not isomorphic.
(14) Let $T = \{a, b, c, d\}$ be a tetrahedron in the projective space $PG(3, 4)$, i.e., a set of four points (vertices) that do not lie in the same plane. Let F be the set of the four faces of the tetrahedron T, i.e., the set of planes containing three vertices of T. Let G be the set of the six edges of T, i.e., the set of lines containing two vertices. Let V be the set of all points of $PG(3, 4)$ that lie in either none or exactly two faces of T and L be the set of all lines of $PG(3, 4)$ that intersect exactly two edges of T. Define the graph $\Gamma = (V, E)$ where $\{x, y\} \in E$ if and only if the line through the points x and y belongs to the set L. Prove that Γ is an $SRG(45, 12, 3, 3)$.
(15) Let Γ be the complement of the Clebsch graph and let x be a vertex of Γ. Prove that the graph $\Gamma(x)$ is isomorphic to the Petersen graph.
(16) Let $\mathbf{D} = (X, \mathcal{B})$ be the design $\mathbf{W}_{22}$. Define a graph $\Gamma = (V, E)$ of order 100 with $V = X \cup \mathcal{B} \cup \{\infty\}$. The vertex ∞ is adjacent to every vertex of X; a point x and a block B of $\mathbf{D}$ are adjacent in Γ if and only if $x \in B$; two distinct blocks are adjacent if and only if they are disjoint. Prove that Γ is an $SRG(100, 22, 0, 6)$. This graph is known as the *Higman–Sims graph*.
(17) A *partial geometry* with parameters (r, k, t) is an incidence structure $\mathbf{D} = (V, \mathcal{L})$ of points and lines satisfying the following conditions: (i) any two distinct points lie on at most one line; (ii) each line has cardinality k and each point has replication number r; (iii) for all $x \in V$ and $L \in \mathcal{L}$, if $x \notin L$, then there are exactly t lines through x that meet L.
   (a) Verify that an (n, r)-net is a partial geometry with parameters $(n, r, r - 1)$.
   (b) Verify that any $(v, b, r, k, 1)$ design is a partial geometry with parameters (r, k, k).
   (c) Let V be the vertex set of the triangular graph $T(n)$, so V is the set of all 2-subsets of $\{1, 2, \ldots, n\}$. For $i = 1, 2, \ldots, n$, let L_i be the set of all elements of V that contain i. Let $\mathcal{L} = \{L_1, L_2, \ldots, L_n\}$. Prove that the incidence structure $\mathbf{D} = (V, \mathcal{L})$ is a partial geometry with parameters $(2, n - 1, 2)$.
   (d) Given a partial geometry $\mathbf{D} = (V, \mathcal{L})$ with parameters (r, k, t), define a graph $\Gamma = (V, E)$ so that distinct vertices $x, y \in V$ form an edge of Γ if and only if there is a line $L \in \mathcal{L}$ containing x and y. Prove that Γ is a strongly regular graph and express its parameters in terms of r, k, and t.

(18) A *factor* in K_6 is a set of three pairwise disjoint edges. A *factorization* of K_6 is a partition of its edge set into five factors.

(a) Prove that there are exactly six factorizations of K_6.

(b) Prove that any two factorizations of K_6 have a unique common factor and, conversely, any factor is contained in exactly two factorizations.

(c) Let $F_1, F_2, \ldots, F_6$ be the six factorizations of K_6. For $1 \le i < j \le 6$, let $F_i \cap F_j = \{f_{ij}\}$. We will now describe the *Hoffman–Singleton graph*. The vertex set of this graph is $A \cup X \cup (A \times X) \cup \{0, \infty\}$, where $A = \{1, 2, 3, 4, 5, 6\}$ is the vertex set of K_6, $X = \{F_1, F_2, F_3, F_4, F_5, F_6\}$, and 0 and ∞ are two extra vertices, so the total number of vertices is $6 + 6 + 36 + 2 = 50$. The edges are the following 2-sets of vertices: (i) $\{a, 0\}$, where $a \in A$; (ii) $\{F, \infty\}$, where $F \in X$; (iii) $\{a, (a, F)\}$, where $a \in A$, $F \in X$; (iv) $\{F, (a, F)\}$, where $a \in A$, $F \in X$; (v) $\{(a, F_i), (b, F_j)\}$, where $1 \le i < j \le 6$ and $\{a, b\} \in f_{ij}$; (vi) $\{0, \infty\}$. Verify that the Hoffman–Singleton graph is an $SRG(50, 7, 0, 1)$.

(19) Prove that there is no (v, k, λ)-graph with parameters (15.7) of Appendix.

(20) Prove that there is no (v, k, λ)-graph with parameters (15.8) of Appendix.

(21) Prove that there is no (v, k, λ)-graph with parameters (15.9) of Appendix.

Notes

The concept of a strongly regular graph was introduced in Bose and Shimamoto (1952). Though there is as yet no monograph exclusively devoted to this subject, many results can be found in the survey papers by Hubaut (1974), Brouwer and van Lint (1984), and Brouwer (1996) and in the books by Brouwer, Cohen and Neumaier (1989), Beth, Jungnickel and Lenz (1999), and Godsil and Royle (2001).

The Petersen graph was introduced in Petersen (1898). The book by Holton and Sheehan (1993) is completely devoted to this graph. The terms *Clebsch graphs* and *Schläfli graphs* were coined in Seidel (1968). Coxeter (1963) relates them to lines on certain algebraic surfaces. Partial geometry graphs and pseudo Latin square graphs were introduced in the seminal paper by Bose (1963). Net graphs are a special case of these graphs. These were introduced in Bruck (1963). Negative Latin square graphs were defined in Mesner (1967). Adjacency matrices of graphs with entries $0, \pm 1$ were proposed in van Lint and Seidel (1966). (See also Seidel (1968) and Goethals and Seidel (1970).) We follow Bhagwandas and S. S. Shrikhande (1968) and call them *Seidel matrices*.

Theorem 7.1.26 is proved in Wallis (1971). It was the first construction of a family of symmetric designs with parameters (3.6).

The characterization of strongly regular graphs in terms of their eigenvalues (Theorem 7.2.8) occurs first, as far as we know, in S. S. Shrikhande and Bhagwandas (1965). Theorem 7.2.10 is due to Hoffman (1970). Our proof follows Brouwer, Cohen and Neumaier (1989), Proposition 1.3.2. Note that part (i) of this theorem is true for all regular graphs.

Seidel (1968) generalized strongly regular graphs to strong graphs. A nonnull and noncomplete graph of order v is called a *strong graph* if its Seidel matrix satisfies the equation

$$(A - \rho_1 I)(A - \rho_2 I) = (v - 1 + \rho_1\rho_2)J$$

with $\rho_1 \neq \rho_2$. A strong graph is strongly regular if and only if it is regular. Strong graphs with $v - 1 + \rho_1\rho_2 \neq 0$ are strongly regular.

At the problem session of the 15th British Combinatorial Conference (1995), W. Haemers asked whether there are any connected graphs with three distinct eigenvalues apart from strongly regular graphs and complete bipartite graphs. An example of such a graph is the graph obtained by switching the triangular graph $T(9)$ with respect to the set $U = \{\{1, i\}: 2 \leq i \leq 9\}$. For this and other examples, see Muzychuk and Klin (1998) and van Dam (1998).

Theorem 7.3.16 for $L_2(n)$ is due to S. S. Shrikhande (1959b). The only graph with parameters of $L_2(4)$ that is not isomorphic to $L_2(4)$ is the Shrikhande graph. Theorem 7.3.16 for $T(n)$ is due to the combined efforts of Connor (1958), S. S. Shrikhande (1959a), Chang (1960a, 1960b), and Hoffman (1960, 1961). The only graphs with parameters of $T(8)$ that are not isomorphic to $T(8)$ are the three Chang graphs. In order to find a unifying approach to these exceptional cases, Seidel (1967) developed the idea of switching. For different proofs of the $T(n)$ case, see Raghavarao (1971), Cameron and van Lint (1991), or Beth, Jungnickel and Lenz (1999). The problem of characterizing strongly regular graphs by numerical parameters was also studied for graphs other than $T(n)$ or $L_2(n)$. We mention the papers by Bose (1963), Bruck (1963), and Metsch (1991) dealing with net graphs and their generalizations. (See also Beth, Jungnickel and Lenz (1999), Chapter X.)

The notion of a two-graph was introduced by G. Higman in connection with sporadic simple groups. Taylor (1971, 1977) established the correspondence between switching classes of strong graphs and two-graphs. For further references on two-graphs, see Seidel (1976), Seidel and Taylor (1981) and Spence (1996). The proofs of theorems 7.3.8 and 7.3.10 follow Cameron and van Lint (1991).

The notions of descendant and ascendant graphs were introduced in Bose and S. S. Shrikhande (1970, 1971). We will have another application of ascendant graphs in Chapter 8. Graphs $PL_n(2n)$ and $NL_n(2n)$ were constructed in Bose and S. S. Shrikhande (1970, 1971) and Goethals and Seidel (1970). The term (v, k, λ)-*graph* can be found in Cameron and van Lint (1991). Bose and S. S. Shrikhande (1970, 1971) refer to these graphs as $G_2(d)$-*graphs*. We follow Muzychuk and Xiang (2005) in the proof of Theorem 7.4.7.

There are several versions of directed analogues of strongly regular graphs. (See Fiedler, Klin and Muzychuk (2002) and Klin, Munemasa, Muzychuk and Zieschang (2004).) Our definition of normally regular digraphs and the proof of Proposition 7.5.5 follow Jørgensen (1994). Theorem 7.5.9 is due to Fried and Sós (1975), Theorems 7.5.10 and 7.5.11 are from Ionin and Kharaghani (2003b).

Normally regular tournaments are also known as *doubly regular tournaments*. Theorem 7.5.14 was independently found by Szekeres and Szekeres (1965), Johnsen (1966), Brown and Reid (1972).

8

Block intersection structure of designs

Let V be a vector space over a finite field, let A and B be subspaces of V of a given dimension s, and let α and β be nonnegative integers. Then the number of s-dimensional subspaces of V that intersect A in an α-dimensional subspace and B in a β-dimensional subspace depends only on α, β, and $\dim(A \cap B)$, rather than on the choice of A and B. Let A and B be blocks of a Witt design and let α and β be nonnegative integers. Then the number of blocks that meet A in α points and B in β points depends only on α, β, and $|A \cap B|$, rather than on the choice of A and B.

This property is formalized (and generalized) by the notion of *association schemes*. Designs whose blocks form an association scheme can be useful in constructing other interesting combinatorial structures including symmetric designs and strongly regular graphs.

8.1. Association schemes

Let V be a finite set of cardinality v. Fix an integer s, $0 \le s \le v/2$, and let $\mathcal{P}_s$ be the set of all s-subsets of V. For $i = 0, 1, \dots, s$, define the incidence structure $\Gamma_i = (\mathcal{P}_s, \mathcal{P}_s, \mathcal{R}_i)$ with $(X, Y) \in \mathcal{R}_i$ if and only if $|X \cap Y| = s - i$. Let $\mathcal{P}_s = \{X_1, X_2, \dots, X_{\binom{v}{s}}\}$ and, for $i = 0, 1, \dots, s$, let A_i be the corresponding incidence matrix of Γ_i. The condition $s \le v/2$ ensures that, for $i = 0, 1, 2, \dots, s$, the relation $\mathcal{R}_i$ is not empty. Note several simple properties of these relations: $\mathcal{R}_0$ is the identity relation; each $\mathcal{R}_i$ is symmetric; the relations $\mathcal{R}_0, \mathcal{R}_1, \dots, \mathcal{R}_s$ form a partition of the set $\mathcal{P}_s \times \mathcal{P}_s$; for $h, i, j \in \{0, 1, \dots, s\}$, if $(X, Y) \in \mathcal{R}_h$, then the number of sets $Z \in \mathcal{P}_s$ such that $(X, Z) \in \mathcal{R}_i$ and $(Z, Y) \in \mathcal{R}_j$ depends only on h, i, and j.

These properties define *association schemes*.

Definition 8.1.1. An *s-class association scheme* consists of a finite set $\mathcal{P}$ (of points) and $s+1$ nonempty binary relations $\mathcal{R}_0, \mathcal{R}_1, \ldots, \mathcal{R}_s$ on $\mathcal{P}$ that satisfy the following axioms:

(AS1) $\mathcal{R}_0$ is the identity relation, i.e., $(X, Y) \in \mathcal{R}_0$ if and only if $X = Y$;
(AS2) each $\mathcal{R}_i$ is symmetric;
(AS3) the relations $\mathcal{R}_0, \mathcal{R}_1, \ldots, \mathcal{R}_s$ form a partition of the set $\mathcal{P} \times \mathcal{P}$;
(AS4) for every triple (h, i, j) of elements of $\{0, 1, \ldots, s\}$, there is an integer p_{ij}^h with the following property:
whenever $(X, Y) \in \mathcal{R}_h$, there are exactly p_{ij}^h elements $Z \in \mathcal{P}$ such that $(X, Z) \in \mathcal{R}_i$ and $(Z, Y) \in \mathcal{R}_j$.

The integers p_{ij}^h are called the *parameters* of the scheme. For $\mathcal{P} = \{x_1, x_2, \ldots, x_v\}$, the corresponding incidence matrix of the incidence structure $(\mathcal{P}, \mathcal{P}, \mathcal{R}_i)$ is called the ith *association matrix* of the association scheme.

The above s-class association scheme of s-subsets of a given finite set of cardinality v is called the *Johnson scheme*. Its parameters are

$$p_{ij}^h = \sum_{n=0}^{h} \binom{s-h}{n} \binom{h}{s-i-n} \binom{h}{s-j-n}.$$

We will give several more examples of association schemes.

Example 8.1.2. An n-class *Hamming scheme* consists of a Hamming space $H(n, q)$ of words of length n over an alphabet of size q and binary relations $\mathcal{R}_i$, $i = 0, 1, \ldots, n$, defined by: $(x, y) \in \mathcal{R}_i$ if and only if the Hamming distance between words x and y is equal to i.

Example 8.1.3. Let $V = V(n, q)$ be the n-dimensional vector space over $GF(q)$. Fix an integer s, $0 \le s \le n$, and let $\mathcal{V}_s$ be the set of all s-dimensional subspaces of V. For $i = 0, 1, \ldots, s$ define an incidence relation $\mathcal{R}_i$ on $\mathcal{V}_s$ by $(X, Y) \in \mathcal{R}_i$ if and only if $\dim(X \cap Y) = s - i$. We obtain an s-class association scheme, called the *q-analogue of the Johnson scheme*.

Example 8.1.4. Let $\mathbf{D} = (X, \mathcal{B})$ be a symmetric (v, k, λ)-design. A 3-class association scheme $\Gamma(\mathbf{D})$ consists of the set $X \cup \mathcal{B}$ of cardinality $2v$, the identity relation $\mathcal{R}_0$, and the following relations $\mathcal{R}_1$, $\mathcal{R}_2$, and $\mathcal{R}_3$ on $X \cup \mathcal{B}$: $(x, y) \in \mathcal{R}_1$ if and only if $x \ne y$ and $x, y \in X$ or $x, y \in \mathcal{B}$; $(x, y) \in \mathcal{R}_2$ if and only if the set $\{x, y\}$ consists of a point and a block that are incident in $\mathbf{D}$; $(x, y) \in \mathcal{R}_3$ if and only if the set $\{x, y\}$ consists of a point and a block that are not incident in $\mathbf{D}$.

Example 8.1.5. Let q be a prime power and let s be a divisor of $q - 1$. Let C_1 be the subgroup of index s of the multiplicative group of the field $GF(q)$ and let

$C_1, C_2, \ldots, C_s$ be the cosets of C_1. Let $C_0 = \{0\}$. Assume that $-1 \in C_1$. The points of an s-class *cyclotomic scheme* are the elements of $GF(q)^*$. We define relations $\mathcal{R}_i$, $i = 0, 1, 2, \ldots, s$, by $(x, y) \in \mathcal{R}_i$ if and only if $x - y \in C_i$.

The axioms (AS1)–(AS4) for s-class association schemes can be expressed equivalently in terms of association matrices A_i: $A_0 = I$; $A_0, A_1, \ldots, A_s$ are symmetric (0, 1)-matrices of the same order; $A_0 + A_1 + \cdots + A_s = J$; $A_i A_j = \sum_{h=0}^{s} p_{ij}^h A_h$.

Remark 8.1.6. The last property implies that $A_i A_j = A_j A_i$ for association matrices A_i and A_j. It further implies that the set of all linear combinations of matrices $A_0, A_1, \ldots, A_s$ is an algebra of matrices. It is called the *Bose–Mesner algebra* of the given association scheme.

Not much can be said about 1-class association schemes. Association matrices of such a scheme are I and $J - I$. We will now show that 2-class association schemes are equivalent to strongly regular graphs.

Proposition 8.1.7. *Let $A_0 = I$, A_1, and A_2 be symmetric* (0, 1)*-matrices of order v such that $A_0 + A_1 + A_2 = J$. Then $\{A_0, A_1, A_2\}$ is the set of association matrices of a 2-class association scheme if and only if A_1 and A_2 are adjacency matrices of complementary strongly regular graphs.*

Proof. If A_0, A_1, and A_2 are the three association matrices of a 2-class association scheme, then we apply (AS3) and (AS4) to obtain:

$$\begin{aligned} A_1^2 &= p_{11}^0 I + p_{11}^1 A_1 + p_{11}^2 (J - I - A_1) \\ &= (p_{11}^0 - p_{11}^2) I + (p_{11}^1 - p_{11}^2) A_1 + p_{11}^2 J. \end{aligned} \tag{8.1}$$

Since $A_1 \neq 0$ and $A_1 \neq J - I$, Theorem 7.1.18 implies that A_1 and similarly A_2 are adjacency matrices of strongly regular graphs. Since $A_1 + A_2 = J - I$, these graphs are complementary.

Conversely, if A_1 and A_2 are adjacency matrices of complementary strongly regular graphs, then $I + A_1 + A_2 = J$, so $J \in \langle I, A_1, A_2 \rangle$ and $A_1 A_2 = A_2 A_1$. Theorem 7.1.18 then implies that $A_1^2, A_2^2 \in \langle I, A_1, A_2 \rangle$. Since $A_1 A_2 = A_1(J - I - A_1)$, we obtain that $\langle I, A_1, A_2 \rangle$ is an algebra. Therefore, A_0, A_1, and A_2 are the three association matrices of a 2-class association scheme. □

Remark 8.1.8. Equation (8.1) implies the following relation between the parameters of an $SRG(v, k, \lambda, \mu)$ and the parameters p_{ij}^h of the corresponding 2-class association scheme: $k = p_{11}^0$, $\lambda = p_{11}^1$, and $\mu = p_{11}^2$. The complementary graph is an $SRG(v, p_{22}^0, p_{22}^2, p_{22}^1)$.

8.2. Quasi-symmetric designs

The main theme of this chapter is exploring the structure formed by blocks of a t-design with respect to the intersection sizes of the blocks. For any t-(v, k, λ) design $\mathbf{D} = (X, \mathcal{B})$ with s intersection numbers $\alpha_1, \alpha_2, \ldots, \alpha_s$, we can define binary relations $\mathcal{R}_i$, $1 \leq i \leq s$, on the block set $\mathcal{B}$ by $(A, B) \in \mathcal{R}_i$ if and only if $|A \cap B| = \alpha_i$. We will denote by $\mathcal{R}_0$ the identity relation on $\mathcal{B}$. The set $\mathcal{B}$ and relations $\mathcal{R}_i$ satisfy axioms (AS1) – (AS3). We will be interested in designs that also satisfy (AS4).

Definition 8.2.1. A t-(v, k, λ) design $\mathbf{D} = (X, \mathcal{B})$ with s intersection numbers $\alpha_1, \alpha_2, \ldots, \alpha_s$ is called *block schematic* if its blocks form an s-class association scheme with respect to the above relations $\mathcal{R}_i$, $0 \leq i \leq s$.

If $\mathbf{D}$ is a symmetric design, then it has only one intersection number, and therefore its blocks form a 1-class association scheme. We will now consider 2-designs with two intersection numbers.

Definition 8.2.2. A t-(v, k, λ) design with $t \geq 2$ and with exactly two intersection numbers is called a *quasi-symmetric design*.

Remark 8.2.3. Corollaries 6.2.2 and 6.2.3 imply that $2 \leq t \leq 4$, for any quasi-symmetric t-design.

Example 8.2.4. For $s \geq 2$, the s-fold multiple of a nontrivial symmetric (v, k, λ)-design is a quasi-symmetric 2-$(v, k, s\lambda)$ design with intersection numbers k and λ.

The converse is also true.

Proposition 8.2.5. *If a quasi-symmetric 2-(v, k, μ) design has intersection numbers k and λ, then it is a multiple of a symmetric (v, k, λ)-design.*

Proof. Let $\mathbf{D}$ be a quasi-symmetric 2-(v, k, μ) design with intersection numbers k and λ. Fix a block A of $\mathbf{D}$ and denote by s the number of blocks that meet A in k points. Counting in two ways flags (x, B) with $B \neq A$ and $x \in A$ yields the equation $k(r-1) = sk + (b-s-1)\lambda$ (where b is the number of blocks of $\mathbf{D}$). Therefore, s does not depend of A and then $\mathbf{D}$ is an s-fold multiple of a 2-design with a unique intersection number. Thus, $\mathbf{D}$ is an s-fold multiple of a symmetric design. □

Example 8.2.6. A nonsymmetric 2-$(v, k, 1)$ design is a quasi-symmetric design with intersection numbers 0 and 1.

Example 8.2.7. By Corollary 5.1.16, any (nonsymmetric) affine α-resolvable 2-design is quasi-symmetric.

Example 8.2.8. The design $\mathbf{W}_{23}$ is a quasi-symmetric design with intersection numbers 1 and 3. The design $\mathbf{W}_{22}$ is quasi-symmetric with intersection numbers 0 and 2.

Example 8.2.9. If $\mathbf{D}$ is a quasi-symmetric t-design with $t \geq 3$, then each of its point derived and point residual designs is either symmetric or quasi-symmetric (cf. Proposition 6.1.16). For instance, if x and y are distinct points of the Witt design $\mathbf{W}_{23}$, then $\mathbf{W}_{23}^x$ is a quasi-symmetric 3-(22, 7, 4) design and $(\mathbf{W}_{24}^x)^y$ is a quasi-symmetric 2-(21, 7, 12) design. If x is a point of $\mathbf{W}_{22}$, then $\mathbf{W}_{22}^x$ is a quasi-symmetric 2-(21, 6, 4) design.

Example 8.2.10. Proposition 6.4.2 implies that an extension of a symmetric (v, k, λ)-design is a quasi-symmetric 3-design with intersection numbers 0 and $\lambda + 1$.

In fact, any quasi-symmetric 3-design with an intersection number 0 is an extension of a symmetric design.

Proposition 8.2.11. *Let $v > k > \lambda \geq 1$ be integers. If $\mathbf{D}$ is a quasi-symmetric 3-(v, k, λ) design with an intersection number 0, then the other intersection number of $\mathbf{D}$ is $\lambda + 1$ and every derived design of $\mathbf{D}$ is a symmetric $(v - 1, k - 1, \lambda)$-design.*

Proof. By Proposition 6.1.6, any two points of $\mathbf{D}$ are contained in $\lambda(v - 2)/(k - 2) > \lambda$ blocks. Therefore, the nonzero intersection number β of $\mathbf{D}$ is at least 2. This implies that no two blocks of a derived design are disjoint. Therefore, any derived design is a 2-design with one intersection number, i.e., a symmetric $(v - 1, k - 1, \lambda)$-design. Since the intersection number of this symmetric design is λ, we have $\beta = \lambda + 1$. □

The parameters and intersection numbers of a quasi-symmetric design satisfy the following equation.

Proposition 8.2.12. *If $\mathbf{D}$ is a quasi-symmetric (v, b, r, k, λ)-design with intersection numbers α and β, then*

$$k(r - 1)(\alpha + \beta - 1) + \alpha\beta(1 - b) = k(k - 1)(\lambda - 1). \tag{8.2}$$

Proof. Fix a block A of $\mathbf{D}$ and let a be the number of blocks that meet A in α points. Counting in two ways pairs (x, B), where B is a block of $\mathbf{D}$, other than A, and $x \in A \cap B$, yields $a\alpha + (b - 1 - a)\beta = k(r - 1)$. Counting in two ways triples (x, y, B), where B is a block of $\mathbf{D}$, other than A, $x, y \in A \cap B$, and $x \neq y$, yields $a\alpha(\alpha - 1) + (b - 1 - a)\beta(\beta - 1) = k(k - 1)(\lambda - 1)$. Eliminating a from these two equations gives (8.2). □

We will now prove that quasi-symmetric designs are block-schematic. We will need the following definition.

Definition 8.2.13. Let $\mathbf{D} = (X, \mathcal{B})$ be a quasi-symmetric design with intersection numbers α and $\beta, \alpha < \beta$. The graph $\Gamma = (\mathcal{B}, E)$ with $\{A, B\} \in E$ if and only if $A \neq B$ and $|A \cap B| = \beta$ is called the *block graph of* $\mathbf{D}$.

Proposition 8.1.7 implies that in order to prove that a quasi-symmetric design is block-schematic, it suffices to show that its block graph is strongly regular.

Theorem 8.2.14. *Every quasi-symmetric design is block-schematic, i.e., its block graph is strongly regular.*

Proof. Let $\mathbf{D} = (X, \mathcal{B})$ be a quasi-symmetric 2-(v, k, λ) design with b blocks and replication number r. Let $\alpha < \beta$ be the intersection numbers of $\mathbf{D}$. Let N be an incidence matrix of $\mathbf{D}$. Then $NN^\top = (r - \lambda)I + \lambda J$, so $NN^\top$ has eigenvalues $r + \lambda(v - 1) = rk$ of multiplicity 1 (corresponding to the all-one eigenvector) and $r - \lambda$ of multiplicity $v - 1$.

Let $\mathcal{B} = \{B_1, \ldots, B_b\}$. Let $(0, 1)$-matrices $X = [x_{ij}]$ and $Y = [y_{ij}]$ of order b be defined as follows: $x_{ij} = 1$ if and only if $i \neq j$ and $|B_i \cap B_j| = \alpha$ and $y_{ij} = 1$ if and only if $i \neq j$ and $|B_i \cap B_j| = \beta$. Then Y is an adjacency matrix of the block graph Γ of $\mathbf{D}$ and X is an adjacency matrix of the complementary graph Γ'.

We have $I + X + Y = J$ and

$$N^\top N = kI + \alpha X + \beta Y = (k - \alpha)I + (\beta - \alpha)Y + \alpha J. \tag{8.3}$$

Multiplying both sides of this equation by the all-one vector $\mathbf{j}$ of size b yields

$$kr\mathbf{j} = (k - \alpha)\mathbf{j} + (\beta - \alpha)Y\mathbf{j} + \alpha b\mathbf{j},$$

which implies $Y\mathbf{j} = a\mathbf{j}$, where

$$a = \frac{(r - 1)k - \alpha b + \alpha}{\beta - \alpha}. \tag{8.4}$$

Since Y is a symmetric matrix, we have now $JY = YJ = aJ$, so there exists an orthogonal matrix C which diagonalizes simultaneously Y and J. Then (8.3) implies that $C^\top N^\top NC$ is a diagonal matrix with eigenvalues of $N^\top N$ on the diagonal. By Proposition 2.2.14, matrices $NN^\top$ and $N^\top N$ have the same nonzero eigenvalues with the same multiplicities. Therefore, we can assume that the first diagonal entry of $C^\top N^\top NC$ is rk, the next $v-1$ diagonal entries are equal to $k-\lambda$, and the remaining $b-v$ diagonal entries are equal to 0. From the equation

$$C^\top N^\top NC = (k-\alpha)I + (\beta-\alpha)C^\top YC + \alpha C^\top JC,$$

we derive that the matrix $C^\top YC$ (and therefore Y) has the eigenvalues a of multiplicity 1, $\frac{r-\lambda-k+\alpha}{\beta-\alpha}$ of multiplicity $v-1$, and $\frac{-(k-\alpha)}{\beta-\alpha}$ of multiplicity $b-v$.

Thus, the adjacency matrix Y of Γ has three distinct eigenvalues and constant row sum. Since our proof did not use that $\alpha<\beta$, the same is true for X. Therefore, Γ and Γ' are regular graphs with three eigenvalues. By Proposition 2.2.7, at least one of these graphs is connected, and then Proposition 8.1.7 implies that both graphs are strongly regular. □

Remark 8.2.15. In the course of the above proof, we have expressed the eigenvalues

$$\frac{r-\lambda-k+\alpha}{\beta-\alpha} \quad \text{and} \quad \frac{-(k-\alpha)}{\beta-\alpha}$$

of the block graph of a quasi-symmetric design in terms of parameters and intersection numbers of the design. The multiplicities of these eigenvalues are $v-1$ and $b-v$, respectively. Theorem 7.2.2 then allows us to express the parameters of the block graph in terms of the parameters and intersection numbers of the design.

Since the eigenvalues of a strongly regular graph are integers, we obtain

Corollary 8.2.16. *If α and β are the intersection numbers of a quasi-symmetric (v,b,r,k,λ)-design, then $\alpha-\beta$ divides both $k-\beta$ and $r-\lambda$.*

Any non-symmetric 2-(v,k,λ) design with $\lambda=1$ is quasi-symmetric. The next result shows that any quasi-residual 2-(v,k,λ) design with $\lambda=2$ is quasi-symmetric.

Proposition 8.2.17. *Any quasi-residual 2-$(v,k,2)$ design has two intersection numbers,* 1 *and* 2.

Proof. Let **D** be a quasi-residual $(v,b,r,k,2)$-design. Then $r=k+2$, $v=k(k+1)/2$, and $b=(k+1)(k+2)/2$. Fix a block A of **D** and, for $i=0,1,\dots,k$, let n_i denote the number of blocks B, other than A, such that $|A\cap B|=i$. Then variance counting (Proposition 2.3.8) yields equations $\sum_{i=0}^k n_i = -1+(k+1)(k+2)/2$, $\sum_{i=0}^k in_i = k(k+1)$, and $\sum_{i=0}^k i(i-1)n_i = k(k-1)$. These equations imply that $\sum_{i=0}^k (i-1)(i-2)n_i = 0$. Therefore, $n_i=0$ for all i, except 1 and 2. Since **D** is not a symmetric design, it has at least two intersection numbers, and therefore both 1 and 2 are intersection numbers of **D**. □

Any quasi-residual 2-(v, k, λ) design with $\lambda = 1$ has k^2 points and therefore it is an affine plane (Proposition 3.2.13). Theorem 3.4.2 implies that such a design is a residual design of a symmetric $(k^2 + k + 1, k + 1, 1)$-design.

Definition 8.2.18. A quasi-residual 2-design is called *embeddable* if it is a residual design of a symmetric design.

Thus, any quasi-residual 2-$(v, k, 1)$ design is embeddable. We will now prove that any quasi-residual 2-$(v, k, 2)$ design is embeddable. We will need the following technical lemma, whose proof we omit.

Lemma 8.2.19 (Connor's Lemma). *There is no* 2-(21, 6, 2) *design.*

Theorem 8.2.20 (The Hall–Connor Theorem). *Any quasi-residual* 2-$(v, k, 2)$ *design is embeddable in a symmetric*

$$\left(1 + \frac{(k+1)(k+2)}{2}, k+2, 2\right) - \textit{design}. \tag{8.5}$$

Proof. Let $\mathbf{D} = (X, \mathcal{B})$ be a quasi-residual 2-$(v, k, 2)$ design, i.e., a $(k(k+1)/2, (k+1)(k+2)/2, k+2, k, 2)$-design. By Proposition 8.2.17, **D** is a quasi-symmetric design with intersection numbers 1 and 2. Let $\Gamma = (V, E)$ be the complement of the block graph of **D**. Using Remark 8.2.15 and Proposition 7.1.3, we find that Γ has the same parameters as the triangular graph $T(k+2)$. If $k = 6$, then **D** is a 2-(21, 6, 2)-design. By Connor's Lemma, there is no 2-design with these parameters.

Therefore, $k \neq 6$. Then Theorem 7.3.16 implies that Γ is isomorphic to $T(k+2)$, so we have a one-to-one correspondence φ between the blocks of **D** and the 2-subsets of the set $Y = \{1, 2, \ldots, k+2\}$. We will assume that X and Y are disjoint and consider the incidence structure $\mathbf{E} = (X \cup Y, \mathcal{C})$ where $\mathcal{C} = \{B \cup \varphi(B) \colon B \in \mathcal{B}\} \cup \{Y\}$. We have $|\mathcal{C}| = |\mathcal{B}| + 1 = |X| + k + 2 = |\cup Y|$. If two blocks of **D** meet in one point, then the corresponding 2-subsets of Y meet in one point, so the corresponding blocks of **E** meet in two points. If two blocks of **D** meet in two points, then the corresponding 2-subsets of Y are disjoint, so the corresponding blocks of **E** meet in two points again. Finally, $|(B \cup \varphi(B)) \cap Y| = 2$, so any two blocks of **E** meet in two points. Therefore, **E** is a symmetric design, and $\mathbf{D} = \mathbf{E}^Y$. □

Corollary 8.2.21. *If k is congruent* 0 *or* 1 (mod 4) *and k is not a square, then there is no* 2-$(k(k+1)/2, k, 2)$ *design.*

Proof. If there exists a 2-$(k(k+1)/2, k, 2)$ design, then, by the Hall–Connor Theorem, it is embeddable in a symmetric design with parameters (8.5). However, if k is congruent 0 or 1 (mod 4), then the number of points of such a symmetric design is even and therefore, by Proposition 2.4.10, k has to be a square. □

Example 2.4.18 shows that quasi-residual 2-(v, k, λ) designs with $\lambda > 2$ do not have to be embeddable. The following theorem gives a sufficient condition for such designs to be embeddable.

Theorem 8.2.22 (The Bose–Shrikhande–Singhi Theorem). *Let a function $g(\lambda)$ for integers $\lambda \geq 3$ be defined by*

$$g(3) = 76,$$
$$g(\lambda) = \frac{1}{2}(\lambda - 1)(\lambda^4 - 2\lambda^2 + \lambda + 2) \text{ for } 4 \leq \lambda \leq 9,$$
$$g(\lambda) = \frac{(\lambda-1)^2(\lambda-2)(\lambda^2-3\lambda+3)}{2} - \lambda + 1 + \frac{(\lambda-1)(\lambda-2)}{2}\sqrt{(\lambda-1)^2(\lambda^2-3\lambda+3)^2 + 4(\lambda-1)} \text{ for } \lambda > 9.$$

Then any quasi-residual 2-(v, k, λ) *design with* $k > g(\lambda)$ *is embeddable.*

The proof of this theorem is beyond the scope of this book. Note that no example of a quasi-residual 2-(v, k, λ) design with $k > g(\lambda)$ is known.

Theorem 8.2.23. *A quasi-symmetric* 2-(v, k, λ) *design with* $b = 2v - 2$ *blocks is either a Hadamard* 3-*design or a* 2-(6, 3, 2) *design.*

Proof. Let $\mathbf{D} = (V, \mathcal{B})$ be a quasi-symmetric 2-(v, k, λ) design with $b = 2v - 2$. Then (2.3.7) reads: $2(v-1)k = vr$. Therefore, $v - 1$ divides r. Since $r < b = 2(v-1)$, we obtain that $r = v - 1$, and then $v = 2k$, so $b = 4k - 2$.

Let the block graph Γ of $\mathbf{D}$ be an SRG(b, a, c, d). Replacing Γ, if necessary, by the complementary graph, we may assume that the degree a of Γ dos not exceed $(b-1)/2$. Fix a block A of $\mathbf{D}$. Let there be a blocks meeting A in β points and $4k - a - 3$ blocks meeting A in α points, so α and β are the intersection numbers of $\mathbf{D}$. Counting in two ways pairs (B, x), where B is a block, $B \neq A$, and $x \in A \cap B$, we obtain

$$a\beta + (4k - a - 3)\alpha = 2k(k-1). \tag{8.6}$$

Let A be the adjacency matrix of Γ. It has an eigenvalue a of multiplicity 1 and eigenvalues ρ and σ of multiplicities $v - 1$ and $b - v = v - 2$, respectively. (Remark 8.2.15.) By Proposition 2.2.17, $|\rho| \leq a$ and $|\sigma| \leq a$. Consider matrices A, A^2, and A^3. They have diagonal entries $0, a, ac$, eigenvalues a, a^2, a^3

of multiplicity 1, eigenvalues ρ, ρ^2, ρ^3 of multiplicity $v-1$, and eigenvalues $\sigma, \sigma^2, \sigma^3$ of multiplicity $v-2$, respectively. Therefore,

$$\operatorname{tr}(A) = a + (v-1)\rho + (v-2)\sigma = 0, \tag{8.7}$$
$$\operatorname{tr}(A^2) = a^2 + (v-1)\rho^2 + (v-2)\sigma^2 = 2(v-1)a, \tag{8.8}$$
$$\operatorname{tr}(A^3) = a^3 + (v-1)\rho^3 + (v-2)\sigma^3 = 2(v-1)ac. \tag{8.9}$$

Equation (8.7) implies that $a \equiv \sigma \pmod{v-1}$. Since $a \leq \frac{b-1}{2} = \frac{2v-3}{2}$, we have $a \leq v-2$. Since $|\sigma| \leq a$, we obtain that $\sigma = a$ or $\sigma = a - v + 1$.

Suppose first that $\sigma = a$. Then (8.7) yields $\rho = -a$ and (8.8) yields $a = 1$, i.e., Γ is a ladder graph. Substituting $a = 1$ in (8.6), we obtain that $\beta + 4(k-1)\alpha = 2k(k-1)$. Therefore, $2(k-1)$ divides β. Since $\beta < k$, we have $\beta = 0$. Now (8.6) yields $\alpha = \frac{k}{2}$.

Thus, $\mathcal{B} = \{A_2, \cdots, A_v; B_2, \cdots, B_v\}$ where A_i and B_i are disjoint for $i = 2, \cdots, v$, and otherwise, any two blocks meet in $\frac{k}{2}$ points. Let $V = \{1, 2, \cdots, 2k\}$. Let $H = [h_{ij}]$ be a matrix of order $2k$ with $h_{ij} = 1$ if $i = 1$ or $j \in A_i$, -1 otherwise. Then H is a Hadamard matrix with all ones in the first row, and **D** is a Hadamard 3-design.

Suppose now that $\sigma = a - v + 1$. Then (8.7) yields $a = v - 2 - \rho$. From (8.8), we obtain $v = 2(\rho^2 + \rho + 1)$, and then (8.9) simplifies to $-(\rho+1)^2 = (2\rho^2 + 2\rho + 1)c$. Therefore, $2\rho^2 + 2\rho + 1$ divides $(\rho+1)^2$. Since both integers are nonnegative and $2\rho^2 + 2\rho + 1 \geq (\rho+1)^2$, we conclude that $(\rho+1)^2 = 0$, i.e., $\rho = -1$. Then $v = 6$, so $b = 10$, $k = 3$, and (2.3.7) yields $r = 5$ and $\lambda = 2$. Thus, **D** is a 2-(6, 3, 2) design. □

Remark 8.2.24. It can be shown that there is a unique 2-(6, 3, 2) design (Exercise 25 of Chapter 2).

We will now apply block graphs of certain quasi-symmetric designs to obtain (v, k, λ)-graphs.

Theorem 8.2.25. *Let* **D** *be a* 2-$(2k^2 - k, k, 1)$ *design. The block graph of* **D** *is a* $(4k^2 - 1, 2k^2, k^2)$*-graph having a clique of cardinality* $2k+1$.

Proof. The design **D** is a $(2k^2 - k, 4k^2 - 1, 2k+1, k, 1)$-design. It is a quasi-symmetric design with intersection numbers 0 and 1. Let Γ be the block graph of **D**. The eigenvalues of Γ, other than the valency, are $\pm k$ (Remark 8.2.15). By Theorem 7.2.2, Γ is a $(4k^2 - 1, 2k^2, k^2)$-graph. If x is a point of **D**, then any two of the $2k+1$ blocks containing x represent adjacent vertices of Γ, and therefore these $2k+1$ blocks form a clique. □

The next theorem introduces a putative family of 2-designs satisfying the condition of Theorem 8.2.25.

Theorem 8.2.26. *Let k be a positive integer and G be an abelian group of order $2k^2 - k$. Suppose G has k-subsets A and B and a subgroup H of order k such that*

$$G = \{x - y : x, y \in A\} \cup \{x - y : x, y \in B\} \cup H. \tag{8.10}$$

Let $U = \{A + z : z \in G\}$, $V = \{B + z : z \in G\}$, and $W = \{H + z : z \in G\}$. Then

(i) *$\mathbf{D} = (G, U \cup V \cup W)$ is a 2-$(2k^2 - k, k, 1)$ design and therefore there exists a $(4k^2 - 1, 2k^2, k^2)$-graph Γ;*
(ii) *Γ has an ascendant, which is a $(4k^2, 2k^2 - k, k^2 - k)$-graph;*
(iii) *the complement of Γ has an ascendant, which is a $(4k^2, 2k^2 + k, k^2 + k)$-graph.*

Proof. Observe that each element of $U \cup V \cup W$ is a k-subset of G. Observe further that there are at most $2k(k - 1)$ distinct nonzero differences $x - y$ where $x, y \in A$ or $x, y \in B$. Since $|G \setminus \{0\}| = |H \setminus \{0\}| + 2k(k - 1)$, condition (8.10) implies that every nonzero element of G has a unique representation as $x - y$ where $x, y \in A$, or $x, y \in B$, or $x, y \in H$.

Let $a, b \in G, a \neq b$. Let $a - b = x - y$ where $x, y \in C$ and $C \in \{A, B, H\}$. Then $a, b \in C + (b - y)$. Conversely, if $a, b \in C + z$ for some $z \in G$, then $a - b = x - y$ with $x, y \in C$. Therefore, there is a unique block of $\mathbf{D}$ that contains $\{a, b\}$, i.e., $\mathbf{D}$ is a $(2k^2 - k, 4k^2, 2k + 1, k, 1)$-design. Let Γ be the block graph of $\mathbf{D}$. By Theorem 8.2.25, Γ is a $(4k^2 - 1, 2k^2, k^2)$-graph.

Observe that $|W|$ is the index of H in G, i.e., $2k - 1$. Since $|U| \leq |G|$ and $|V| \leq |G|$ and since $\mathbf{D}$ has $4k^2 - 1 = 2|G| + (2k - 1)$ blocks, we obtain that $|U| = |V| = 2k^2 - k$. Since every element of G is contained in exactly k blocks from U and since no two blocks of $\mathbf{D}$ meet in more than one point, we obtain that every vertex of $V \cup W$ has exactly k^2 neighbors in U and every vertex of U has exactly $k(k - 1)$ neighbors in U. Therefore, every vertex of U has exactly $k(k + 1)$ neighbors in $V \cup W$.

Thus, the graph Γ and subset U of its vertex set satisfy the condition of Theorem 7.3.24. By this theorem the corresponding ascendant graph Γ^* is regular of degree $u = 2k^2 - k$, and then Proposition 7.3.23 implies that Γ^* is strongly regular. Let Γ^* be an $SRG(4k^2, u, \lambda, \mu)$. Then, by Theorem 7.3.21, Γ is an $SRG(4k^2 - 1, 2(u - \mu), u + \lambda - 2\mu, u - \mu)$. Therefore, $\lambda = \mu = k^2 - k$, i.e., Γ^* is a $(4k^2, 2k^2 - k, k^2 - k)$-graph.

The complement Γ' of Γ is an $SRG(4k^2 - 1, 2k^2 - 2, k^2 - 3, k^2 - 1)$. Then Theorem 7.3.24 and Proposition 7.3.23 imply that the ascendant of Γ' with respect to the subset $V \cup W$ of its vertex set is a $(4k^2, 2k^2 + k, k^2 + k)$-graph. □

We will now give three examples of realization of the conditions of Theorem 8.2.26:

(i) for $k = 3$, let $G = \mathbf{Z}_{15}$, $A = \{0, 1, 4\}$, $B = \{0, 2, 8\}$, and $H = \langle 5 \rangle$;
(ii) for $k = 5$, let $G = \mathbf{Z}_3^2 \times \mathbf{Z}_5$, $A = \{(0, 1, 0), (0, 2, 0), (1, 0, 2), (2, 0, 2), (0, 0, 1)\}$, $B = \{(2, 1, 0), (1, 2, 0), (2, 2, 2), (1, 1, 2), (0, 0, 1)\}$, and $H = \langle (0, 0, 1) \rangle$;
(iii) for $k = 7$, let $G = \mathbf{Z}_{91}$, $A = \{0, 10, 27, 28, 31, 43, 50\}$, $B = \{0, 11, 20, 25, 49, 55, 57\}$, and $H = \langle 13 \rangle$.

Thus, we have the following result.

Corollary 8.2.27. *For $k = 3$, $k = 5$, and $k = 7$, there exist (v, k, λ)-graphs with the parameters $(4k^2, 2k^2 - k, k^2 - k)$ and $(4k^2, 2k^2 + k, k^2 + k)$.*

Corollary 8.2.28. *There exist regular Hadamard matrices with constant diagonal of orders* 36, 100, *and* 196.

We can now extend Theorem 7.4.22

Theorem 8.2.29. *If there exists a Hadamard matrix of order $4n$, then, for any nonnegative integers s, t, and u, there exists a regular symmetric Hadamard matrix of order $2^{2s+2t+2u+2} \cdot 3^{2s} \cdot 5^{2t} \cdot 7^{2u} \cdot n^2$ with constant diagonal.*

Remark 8.2.30. A $(2k^2 - k, 4k^2 - 1, 2k + 1, k, 1)$-design is also known to exist for $k = 2, 4, 6$, and 8.

We saw (Propositions 6.4.2 and 8.2.11) that quasi-symmetric 3-designs with an intersection number 0 are precisely the extensions of symmetric designs. Therefore, the Cameron Theorem immediately allows us to classify such 3-designs.

Theorem 8.2.31. *If* **D** *is a quasi-symmetric* 3-(v, k, λ) *design with an intersection number* 0, *then one of the following holds*:

(i) $v = 4(\lambda + 1)$ *and* $k = 2(\lambda + 1)$;
(ii) $v = (\lambda + 1)(\lambda^2 + 5\lambda + 5)$ *and* $k = (\lambda + 1)(\lambda + 2)$;
(iii) $v = 496$, $k = 40$, *and* $\lambda = 3$.

The following two theorems classify quasi-symmetric 3-designs with the smaller intersection number 1, and also quasi-symmetric 4-designs. We do not give their proofs.

Theorem 8.2.32. *If* **D** *is a quasi-symmetric* 3-(v, k, λ) *design with the smaller intersection number* 1, *then* **D** *is the Witt* 4-(23, 7, 1) *design or its residual* 3-(22, 7, 4) *design.*

Theorem 8.2.33. *If* **D** *is a quasi-symmetric* 4-(v, k, λ) *design, then it is the Witt* 4-(23, 7, 1) *design or its complement.*

8.3. Multiples of symmetric designs

If **D** is a nontrivial symmetric (v, k, λ)-design, then, for any positive integer $s \geq 2$, $s \times$ **D** is a quasi-symmetric $(v, sv, sk, k, s\lambda)$-design (with intersection numbers k and λ). Is it true that any quasi-symmetric $(v, sv, sk, k, s\lambda)$-design is a multiple of a symmetric (v, k, λ)-design? The answer to this question is negative, as the following example shows.

Example 8.3.1. Consider the Witt 4-(23, 7, 1) design $\mathbf{W_{23}}$. It is a (23, 253, 77, 7, 21)-design. Let **D** be a point-residual design of $\mathbf{W_{23}}$. Then **D** is a (22, 176, 56, 7, 16)-design, so it has parameters of an 8-fold multiple of a symmetric (22, 7, 2)-designs. However, there is no symmetric design with these parameters (Remark 2.4.11).

We will now show that, if k and $(s-1)\lambda$ are relatively prime, then any quasi-symmetric $(v, sv, sk, k, s\lambda)$-design is a multiple of a symmetric (v, k, λ)-design. We begin with the following lemma.

Lemma 8.3.2. *Let* $v > k > \lambda \geq 1$ *and* $s \geq 2$ *be integers and let* **D** *be a quasi-symmetric* $(v, sv, sk, k, s\lambda)$*-design with intersection numbers* α *and* β. *Let* $P(z) = (z-\alpha)(z-\beta)$. *Then*

$$k(sk-1)(\alpha+\beta-1) - (sv-1)\alpha\beta = k(k-1)(s\lambda-1), \tag{8.11}$$

$$(s-1)\lambda P(k) = -sk(k-1)P(\lambda), \tag{8.12}$$

$$0 \leq -P(\lambda) < \lambda. \tag{8.13}$$

Proof. We obtain (8.11) by applying (8.2) to the design **D**. The parameters of **D** satisfy the basic relation $(v-1)s\lambda = sk(k-1)$, which implies $\lambda(v-1) = k(k-1)$. Substituting $v = 1 + k(k-1)/\lambda$ in (8.11) yields after routine manipulations the equation (8.12). Since $P(k) \geq 0$, (8.12) implies $P(\lambda) \leq 0$. We further have $P(k) \leq k(k-1)$ and $s - 1 < s$. Therefore, if $P(k) \neq 0$, then (8.12) implies $\lambda > -P(\lambda)$. If $P(k) = 0$, then $P(\lambda) = 0$, so again $\lambda > -P(\lambda)$. □

Theorem 8.3.3. *Let* $v > k > \lambda \geq 1$ *and* $s \geq 2$ *be integers and let* **D** *be a quasi-symmetric* $(v, sv, sk, k, s\lambda)$*-design with intersection numbers* α *and* β. *If* k *and* $(s-1)\lambda$ *are relatively prime, then* **D** *is an* s*-fold multiple of a symmetric* (v, k, λ)*-design.*

Proof. If one of the intersection numbers is k, then (8.12) implies that $P(\lambda) = 0$, so the other intersection number is λ. Then (8.4) implies that, for any block B of $\mathbf{D}$ there are exactly $s-1$ blocks that meet B in k points. This means that $\mathbf{D}$ is an s-fold multiple of a symmetric (v, k, λ)-design.

From now on, we assume that $\alpha < \beta < k$ and that k and $(s-1)\lambda$ are relatively prime. Since $\lambda(v-1) = k(k-1)$ and k and λ are relatively prime, λ divides $k-1$. Let $k-1 = m\lambda$. Reducing (8.12) modulo k yields $(s-1)\lambda\alpha\beta \equiv 0 \pmod{k}$. Since k and $(s-1)\lambda$ are relatively prime, k divides $\alpha\beta$. Let $\alpha\beta = nk$. Then $\alpha\beta = mn\lambda + n$.

We derive from (8.13):

$$0 < -\lambda^2 + (\alpha+\beta)\lambda - mn\lambda - n < \lambda,$$
$$\lambda + mn + \frac{n}{\lambda} < \alpha + \beta < \lambda + mn + \frac{n}{\lambda} + 1. \tag{8.14}$$

Since $P(\lambda) < 0$, we have $\alpha < \lambda < \beta$. Then $\alpha\beta < \lambda k$ and therefore $n < \lambda$. Thus, $0 < n/\lambda < 1$, and then (8.14) implies that $\alpha + \beta = \lambda + mn + 1$. We now have

$$P(\lambda) = \lambda^2 - (\alpha+\beta)\lambda + (mn\lambda + n) = n - \lambda$$

and

$$P(k) = k^2 - (\alpha+\beta)k + nk = k(k - \alpha - \beta + n) = k(k - \lambda - mn - 1 + n).$$

We rewrite (8.12) as

$$s(\lambda P(k) + k(k-1)P(\lambda)) = \lambda P(k).$$

This equation implies that

$$\lambda P(k) + k(k-1)P(\lambda) > 0,$$
$$\lambda k(k - \lambda - mn - 1 + n) + k(k-1)(n-\lambda) > 0,$$

which simplifies to $n > \lambda$. This contradicts the previously obtained inequality $n < \lambda$, and the proof is now complete. □

The next theorem considers quasi-symmetric designs with parameters of a multiple of a projective plane.

Theorem 8.3.4. *Let $v > k > \lambda \geq 1$ be integers and let $\mathbf{D}$ be a quasi-symmetric $(v, (k+1)v, (k+1)k, k, (k+1)\lambda)$ design with the smaller intersection number equal to 1. Then either $\mathbf{D}$ is a $(k+1)$-fold multiple of a projective plane of order $k-1$ or a 2-(22, 7, 16) design.*

Proof. If $\mathbf{D}$ is a $(k+1)$-fold multiple of a symmetric (v, k, λ)-design, then the smaller intersection number of $\mathbf{D}$ is λ, so $\lambda = 1$ and $\mathbf{D}$ is a $(k+1)$-fold multiple of a projective plane of order $k-1$.

From now on, we assume that **D** is not a $(k+1)$-fold multiple of a symmetric (v, k, λ)-design. Let $\alpha = 1$ and β be the intersection numbers of **D**, $1 < \beta < k$. Substituting $s = k+1$ and $\alpha = 1$ in (8.12) yields after routine manipulations

$$\lambda\beta(k+2) - (k+1)\beta = (k+1)\lambda^2 - \lambda. \tag{8.15}$$

This equation implies that λ divides $(k+1)\beta$. Since $\lambda(v-1) = k(k-1)$, λ divides $k(k-1)$. Since $k(k-1)\beta = (k-2)(k+1)\beta + 2\beta$, we obtain that

$$\lambda \text{ divides } 2\beta. \tag{8.16}$$

Equation (8.15) also implies that

$$k+1 \text{ divides } \lambda(\beta+1). \tag{8.17}$$

Let $g = gcd(\lambda, k+1)$. We now claim that $g = 2$. Note that since $k+1 > \beta+1$, (8.17) implies that $g > 1$. Let p be an odd prime dividing g and let p^s be the highest power of p dividing λ. Then (8.16) implies that p^s divides β, and then reducing (8.15) modulo p^{s+1} yields $\lambda \equiv 0 \pmod{p^{s+1}}$, a contradiction. Thus, no odd prime divides g. Finally, assume that 4 divides g and let 2^s be the highest power of 2 that divides λ. Then 4 divides $k+1$, (8.16) implies that 2^{s-1} divides β, and then reducing (8.15) modulo 2^{s+1} yields $\lambda \equiv 0 \pmod{2^{s+1}}$, a contradiction.

Thus $gcd(\lambda, k+1) = 2$. We derive from (8.15) that

$$\frac{\lambda}{2}\beta(k+2) - \frac{k+1}{2}\beta \equiv \frac{k+1}{2}\lambda^2 - \frac{\lambda}{2} \left(\text{mod}\frac{k+1}{2}\right),$$
$$\frac{\lambda}{2}\beta \equiv -\frac{\lambda}{2} \left(\text{mod}\frac{k+1}{2}\right).$$

Since $\lambda/2$ and $(k+1)/2$ are relatively prime, we obtain that $(k+1)/2$ divides $\beta+1$. Therefore, $k+1 \leq 2(\beta+1)$. Since $k+1 > \beta+1$, we obtain that $k = 2\beta+1$. Now (8.15) can be rewritten as

$$\lambda\beta(2\beta+3) + \lambda = 2(\beta+1)(\beta+\lambda^2),$$

which simplifies to $\lambda(2\lambda-1) = 2\beta(\lambda-1)$. By (8.16), we obtain that $\lambda - 1$ divides $2\lambda - 1$ and therefore, $\lambda = 2$. This implies that $\beta = 3, k = 7$, and $v = 22$, so **D** is a 2-(22, 7, 16) design. □

Remark 8.3.5. It is possible to prove that any quasi-symmetric 2-(22, 7, 16) design with the smaller intersection number 1 is isomorphic to the design from Example 8.3.1.

We will now consider quasi-symmetric designs with parameters of multiples of symmetric $(v, k, 1)$- and $(v, k, 2)$-designs.

Theorem 8.3.6. *Any quasi-symmetric (v, sv, sk, k, s)-design is isomorphic to the s-fold multiple of a symmetric $(v, k, 1)$-design.*

Proof. Let **D** be a quasi-symmetric (v, sv, sk, k, s)-design with intersection numbers α and β, $\alpha < \beta$. Inequalities (8.13) imply that $\alpha \leq 1$ and $\beta \geq 2$. If $\alpha = 1$, then (8.12) implies $\beta = k$, and then **D** is the s-fold multiple of a symmetric $(v, k, 1)$-design.

Suppose $\alpha = 0$. Then (8.11) and $\beta \geq 2$ imply $sk - 1 \leq (k-1)(s-1)$, so $k + s \leq 2$, a contradiction. □

Theorem 8.3.7. *Let s be a positive integer and let **D** be a quasi-symmetric $(v, sv, sk, k, 2s)$-design. Then one of the following three possibilities holds:*

(i) **D** *is isomorphic to the s-fold multiple of a symmetric $(v, k, 2)$-design;*
(ii) **D** *is a* 2-(22, 7, 16) *design and the intersection numbers of* **D** *are* 1 *and* 3;
(iii) **D** *is a* 2-(37, 9, 8) *design and the intersection numbers of* **D** *are* 1 *and* 3.

Proof. Let $\alpha < \beta$ be the intersection numbers of **D**. As in the proof of Theorem 8.3.6, we obtain that $\alpha \leq 2$, $\beta \geq 3$, and, if $\alpha = 2$, then **D** is an s-fold multiple of a symmetric $(v, k, 2)$-design.

Suppose $\alpha = 0$. Then (8.11) and $\beta \geq 2$ imply $2(sk - 1) \leq (k-1)(2s-1)$, so $k + 2s \leq 3$, a contradiction.

Suppose $\alpha = 1$. Since also $v = 1 + k(k-1)/2$, equation (8.11) can be rewritten as $(sk + 2s - 2)\beta = 4sk - 2k$. This implies $\beta < 4$. Therefore, $\beta = 3$ and we have $(s-2)k = 6(s-1)$. Thus $s - 2$ divides 6, i.e., the possible values of s are 3, 4, 5, or 8.

If $s = 3$, then $k = 12$ and $v = 12$, which is not possible.

If $s = 4$, then $k = 9$, $v = 37$, and we have case (iii).

If $s = 5$, then $k = 8$ and $v = 29$, so **D** is a $(29, 145, 40, 8, 10)$-design with intersection numbers 1 and 3. Fix a block A of **D** and a point $x \in A$. Suppose there are m blocks that contain x and meet A in three points. Counting in two ways pairs (B, y) where B is a block that contains x and meets A in three points, $y \in A \cap B$, and $y \neq x$, yields $2m = 7 \cdot 9$, a contradiction.

If $s = 8$, then $k = 7$, $v = 22$, and we have case (ii). □

Remark 8.3.8. For case (ii) of Theorem 8.3.7, see Remark 8.3.5. No design satisfying case (iii) is known.

8.4. Quasi-3 symmetric designs

Let **D** be a symmetric (v, k, λ)-design with $2 \le k \le v - 2$. The size of the intersection of three distinct blocks of **D** cannot be constant, because otherwise the dual design $\mathbf{D}^\top$ would have been a 3-design. However, this is not possible by Corollary 6.1.13. This motivates the following notion.

Definition 8.4.1. A symmetric design **D** is said to be *quasi*-3 if there exist integers α and β called *triple intersection numbers* such that $|A \cap B \cap C| \in \{\alpha, \beta\}$ for any three distinct blocks A, B, and C of **D**.

Remark 8.4.2. A design described by Definition 8.4.1 is sometimes called *quasi-3 for blocks* and then the dual design is called *quasi-3 for points*.

We will now give several examples of families of quasi-3 symmetric designs.

Example 8.4.3. If **D** is a symmetric (v, k, λ)-design with $\lambda \le 2$, then any three distinct blocks of **D** have at most one common point, so **D** is quasi-3. Conversely, if **D** is a quasi-3 symmetric (v, k, λ)-design with triple intersection numbers 0 and 1, then $\lambda \le 2$.

Example 8.4.4. For $d \ge 2$, the design $PG_{d-1}(d, q)$ is quasi-3, because the intersection of three distinct d-dimensional subspaces of the $(d + 1)$-dimensional vector space over $GF(q)$ is a subspace of dimension $d - 1$ or $d - 2$.

The following proposition is a straightforward application of the Inclusion–Exclusion principle.

Proposition 8.4.5. *The complement of a quasi-3 symmetric (v, k, λ)-design with the triple intersection numbers α and β is a quasi-3 symmetric design with the triple intersection numbers $v - 3k + 3\lambda - \alpha$ and $v - 3k + 3\lambda - \beta$.*

The next result is straightforward.

Proposition 8.4.6. *A nontrivial symmetric design* **D** *is quasi-3 if and only if every derived design of* **D** *is quasi-symmetric.*

The next definition introduces a putative family of quasi-3 symmetric designs.

Definition 8.4.7. A symmetric design **D** is said to have the *symmetric difference property* (or that **D** is an *SDP-design*) if the symmetric difference of any three blocks of **D** is either a block of **D** or the complement of a block of **D**.

Proposition 8.4.8. *Any SDP-design is quasi-3.*

Proof. Let $\mathbf{D} = (X, \mathcal{B})$ be a (v, k, λ) *SDP*-design and let A, B, and C be three distinct blocks of $\mathbf{D}$. Then

$$|A \triangle B \triangle C| = 3k - 6\lambda + 2|A \cap B \cap C|. \tag{8.18}$$

If $A \triangle B \triangle C$ is a block of $\mathbf{D}$, then $|A \cap B \cap C| = 3\lambda - k$. If $A \triangle B \triangle C$ is the complement of a block of $\mathbf{D}$, then $|A \cap B \cap C| = v/2 + 3\lambda + 2k$. Therefore, $\mathbf{D}$ is quasi-3. □

The inclusion–exclusion principle immediately implies the following result.

Proposition 8.4.9. *The complement of an SDP-design is an SDP-design.*

The next theorem gives an infinite family of SDP-designs.

Theorem 8.4.10. *Let H_1 be a regular Hadamard matrix of order 4. For $n \geq 2$, define matrices H_n recursively by $H_n = H_1 \otimes H_{n-1}$. Then, for all $n \geq 1$, $\frac{1}{2}(J + H_n)$ is an incidence matrix of an SDP-design.*

Proof. It suffices to prove the following statement: if C is the Hadamard product of three columns of H_n, then either C or $-C$ is a column of H_n. Since any regular Hadamard matrix of order 4 is equivalent to the matrix of Example 4.1.2, this statement is true for $n = 1$. Let $n \geq 2$ and let statement be true for H_{n-1}. Let $\{C_1, C_2, \ldots, C_{2^{2n-2}}\}$ be the set of all columns of H_{n-1}. For $i = 1, 2, 3, 4$, let $[\varepsilon_{i1}\ \varepsilon_{i2}\ \varepsilon_{i3}\ \varepsilon_{i4}]^\top$ be the ith column of H_1. Then the set $\{D_{il} \colon 1 \leq i \leq 4, 1 \leq l \leq 2^{2n-2}\}$ where $D_{il} = [\varepsilon_{i1}C_l^\top\ \varepsilon_{i2}C_l^\top\ \varepsilon_{i3}C_l^\top\ \varepsilon_{i4}C_l^\top]^\top$ is the set of all columns of H_n. Let $D = D_{hk} \circ D_{il} \circ D_{jm}$ be the Hadamard product of three columns of H_n. Let $\Delta = [\delta_1\ \delta_2\ \delta_3\ \delta_4]^\top$ be the Hadamard product of the hth, i^{th}, and j^{th} columns of H_1 and let $C = C_k \circ C_l \circ C_m$. Then $D = [\delta_1 C^\top\ \delta_2 C^\top\ \delta_3 C^\top\ \delta_4 C^\top]^\top$. By the induction hypothesis, either C or $-C$ is a column of H_{n-1}. Since either Δ or $-\Delta$ is a column of H_1, we obtain that either D or $-D$ is a column of H_n. □

Corollary 8.4.11. *For any positive integer m, there exist SDP-designs with parameters $(2^{2m}, 2^{2m-1} \pm 2^{m-1}, 2^{2m-2} \pm 2^{m-1})$.*

The next theorem shows that these parameters are the only possible parameters of nontrivial SDP-designs on more than two points.

Theorem 8.4.12. *Let $\mathbf{D}$ be a nontrivial (v, k, λ) SDP-design with $v \geq 3$. Then there exists a positive integer m such that $v = 2^{2m}$, $k = 2^{2m-1} \pm 2^{m-1}$, and $\lambda = 2^{2m-2} \pm 2^{m-1}$. Furthermore, the dual design $\mathbf{D}^\top$ is an SDP-design.*

Proof. Since the complement of an SDP-design is an SDP-design, we assume that $v \geq 2k$.

Let $\mathbf{D} = (X, \mathcal{B})$ and let $X = \{x_1, x_2, \ldots, x_v\}$. We will identify every subset Y of X with the vector $Y = [\alpha_1\, \alpha_2 \ldots \alpha_v]^\top$ over the field $GF(2)$ where $\alpha_i = 1$ if and only if $x_i \in Y$. Note that, for any subsets Y_1 and Y_2 of X, the vector $Y_1 + Y_2$, as a subset of X, is the symmetric difference $Y_1 \triangle Y_2$. With this representation, the symmetric difference property of $\mathbf{D}$ can be stated as follows:

$$\text{for any distinct } B_1, B_2, B_3 \in \mathcal{B}, \text{ either } B_1 + B_2 + B_3 \in \mathcal{B}$$
$$\text{or} \quad X + B_1 + B_2 + B_3 \in \mathcal{B}.$$

If the latter case is never realized, then, by (8.18), D has only one triple intersection number. Since this is not possible, there exist distinct $B_1, B_2, B_3 \in \mathcal{B}$ such that $A = X + B_1 + B_2 + B_3 \in \mathcal{B}$. Then $X + A + B_1 = B_2 + B_3$ and therefore, $v - |A + B_1| = |B_2 + B_3|$. This implies that $v = 4(k - \lambda)$ and therefore, by Proposition 4.4.7, there exists an integer $h \neq 0$ such that $v = 4h^2$, $k = 2h^2 - h$, and $\lambda = h^2 - h$. Since $v \geq 2k$, we have $h > 0$. If $h = 1$, then $\mathbf{D}$ has the required parameters with $m = 1$, so from now on let $h \geq 2$.

Let $\mathcal{H} = \{B + C\colon B, C \in \mathcal{B}, B \neq C\} \cup \{X + B + C\colon B, C \in \mathcal{B}, B \neq C\}$ and let $\mathbf{A}$ denote the incidence structure $(X, \mathcal{H})$. Our goal is to show that $\mathbf{A}$ is isomorphic to $AG_d(d + 1, 2)$ for some $d \geq 3$.

We will fix a block A of $\mathbf{D}$ and note that if $H = B + C$ for distinct blocks $B, C \in \mathcal{B} \setminus \{A\}$, then either $H + A$ or $X + H + A$ is a block of $\mathbf{D}$. Since $H = A + (H + A) = X + A + (X + H + A)$, we obtain that

$$\mathcal{H} = \{A + B\colon B \in \mathcal{B} \setminus \{A\}\} \cup \{X + A + C\colon C \in \mathcal{B} \setminus \{A\}\}.$$

If $A + B = X + A + C$ for some $B, C \in \mathcal{B} \setminus \{A\}$, then $B = X + C$ and therefore $B \cap C = \varnothing$, a contradiction. Thus, every $H \in \mathcal{H}$ has a unique representation as $A + B$ or $X + A + B$ with $B \in \mathcal{B} \setminus \{A\}$. This implies that $|\mathcal{H}| = 2v - 2$.

Note that, for $B \in \mathcal{B} \setminus \{A\}$, $|A + B| = |X + A + B| = 2(k - \lambda) = 2h^2$. Let $x, y \in X$, $x \neq y$. If $x, y \in A$, then there are exactly $v - 2k + \lambda = h^2 + h$ blocks $B \in \mathcal{B} \setminus \{A\}$, such that $x, y \in A + B$, and $\lambda - 1 = h^2 - h - 1$ blocks $C \in \mathcal{B} \setminus \{A\}$, such that $x, y \in X + A + C$; if $x, y \notin A$, then there are exactly $h^2 - h$ blocks $B \in \mathcal{B} \setminus \{A\}$, such that $x, y \in A + B$, and $h^2 + h - 1$ blocks $C \in \mathcal{B} \setminus \{A\}$, such that $x, y \in X + A + C$; if $x \in A$ and $y \notin A$, then there are exactly $k - \lambda = h^2$ blocks $B \in \mathcal{B} \setminus \{A\}$, such that $x, y \in A + B$, and $k - \lambda - 1 = h^2 - 1$ blocks $C \in \mathcal{B} \setminus \{A\}$, such that $x, y \in X + A + C$. Thus, in every case, there are exactly $2h^2 - 1$ blocks of $\mathbf{A}$ that contain $\{x, y\}$. Therefore, $\mathbf{A}$ is a 2-$(4h^2, 2h^2, 2h^2 - 1)$-design with $8h^2 - 2$ blocks.

If $H_1 = A + B$ and $H_2 = X + A + B$ with $B \in \mathcal{B} \setminus \{A\}$, then $H_1 \cap H_2 = \varnothing$. We claim that otherwise the intersection of two distinct blocks of $\mathbf{A}$ has cardinality h^2. Let $B_1, B_2 \in \mathcal{B} \setminus \{A\}$, $B_1 \neq B_2$. If $H_1 = A + B_1$ and $H_2 = A + B_2$ or $H_1 = X + A + B_1$ and $H_2 = X + A + B_2$, then $|H_1 \cap H_2| = \frac{1}{2}(4h^2 - |H_1 + H_2|) = \frac{1}{2}(4h^2 - |B_1 + B_2|) = h^2$.

If $H_1 = A + B_1$ and $H_2 = A + X + B_2$, then $|H_1 \cap H_2| = \frac{1}{2}(4h^2 - |X + B_1 + B_2|) = h^2$. Thus, $\mathbf{A}$ is a quasi-symmetric design with v points and $2v - 2$ blocks. By Theorem 8.2.23, $\mathbf{A}$ is a Hadamard 3-design.

Let $x \in X$. Then the point-derived design $\mathbf{A}_x$ is a symmetric $(4h^2 - 1, 2h^2 - 1, h^2 - 1)$-design. Since, by assumption, $h \geq 2$, we have $h^2 - 1 > 1$. We claim that $\mathbf{A}_x$ is isomorphic to $PG_{d-1}(d, 2)$ for some $d \geq 3$. By the Dembowski–Wagner Theorem and Proposition 3.7.3, it suffices to show that every line of $(\mathbf{A}_x)^\top$ consists of at least three points. Thus, we have to show that, for any distinct blocks H_1 and H_2 of $\mathbf{A}_x$, there is a block H of $\mathbf{A}_x$, other than H_1 or H_2 that contains $H_1 \cap H_2$. We will show that $H = X + H_1 + H_2$ is such a block. Clearly, $x \in H$ and $H_1 \cap H_2 \subset H$, so we have to verify that H is a block of $\mathbf{A}$. If $H_1 = A + B_1$ and $H_2 = A + B_2$ or $H_1 = X + A + B_1$ and $H_2 = X + A + B_2$ with distinct $B_1, B_2 \in \mathcal{B} \setminus \{A\}$, then $H = X + B_1 + B_2$, so H is a block of $\mathbf{A}$. If $H_1 = A + B_1$ and $H_2 = X + A + B_2$, then again $H = X + B_1 + B_2$ is a block of $\mathbf{A}$.

Thus, $\mathbf{A}_x$ is isomorphic to $PG_{d-1}(d, 2)$ for some $d \geq 3$ and then, by Theorem 6.3.2, $\mathbf{A}$ is isomorphic to $AG_d(d + 1, 2)$. Therefore, $v = 4h^2 = 2^{d+1}$, $d = 2m - 1$ is odd, and then $k = 2^{2m-1} - 2^{m-1}$ and $\lambda = 2^{2m-2} - 2^{m-1}$.

We now prove that $\mathbf{D}^\top$ is an SDP-design. Since $\mathbf{A}$ is isomorphic to $AG_{2m-1}(2m, 2)$ with $m \geq 2$, we will regard blocks of $\mathbf{A}$ as hyperplanes of the $(2m)$-dimensional vector space V over $GF(2)$. For each $x \in X$, let $\mathcal{B}(x)$ be the set of all blocks of $\mathbf{D}$ containing x (so $\mathcal{B}(x)$ is a block of $\mathbf{D}^\top$). The design $\mathbf{D}^\top$ is an SDP-design if and only if, for any 3-subset $\{x, y, z\}$ of X, there exists $w \in X$ such that the set of all blocks of $\mathbf{D}$ that meet $\{x, y, z\}$ in one or three points coincides with either $\mathcal{B}(w)$ or $\mathcal{B} \setminus \mathcal{B}(w)$.

Let x, y, and z be three distinct points of $\mathbf{D}$. Since each line of V consists of two points, the points x, y, and z of V are not collinear, and therefore, the set $\{x, y, z\}$ is contained in a unique plane $\pi = \{x, y, z, w\}$. Note that every hyperplane of V meets π in two or four points. Suppose there is a block $A \in \mathcal{B}(w)$ that meets $\{x, y, z\}$ in an odd number of points. If B is any other block of $\mathbf{D}$ that meets $\{x, y, z\}$ in an odd number of points, then the hyperplane $A + B$ meets $\{x, y, z\}$ in an even number of points. Therefore, $w \notin A + B$, i.e., $w \in B$. If C is a block of $\mathbf{D}$ that meets $\{x, y, z\}$ in an even number of points, then the hyperplane $A + C$ meets $\{x, y, z\}$ in an odd number of points and therefore $w \in A + C$, i.e., $w \notin C$. Thus, $\mathcal{B}(w)$ is the set of all blocks of $\mathbf{D}$

that meet $\{x, y, z\}$ in one or three points. If there exists a block $A \in \mathcal{B}(w)$ that meets $\{x, y, z\}$ in an even number of points, then, in a similar manner, one can show that $\mathcal{B} \setminus \mathcal{B}(w)$ is the set of all blocks of $\mathbf{D}$ that meet $\{x, y, z\}$ in one or three points. $\square$

If α is a triple intersection number of a symmetric (v, k, λ)-design, then $0 \leq \alpha \leq \lambda$. We will now characterize quasi-3 symmetric (v, k, λ)-designs having a triple intersection number equal to λ or 0.

Theorem 8.4.13. *Let* $\mathbf{D}$ *be a quasi-3 symmetric* (v, k, λ)*-design with* $\lambda \geq 3$.

(i) *If* λ *is a triple intersection number of* $\mathbf{D}$, *then* $\mathbf{D}$ *is isomorphic to* $PG_{d-1}(d, q)$ *for some* d *and* q.
(ii) *If* 0 *is a triple intersection number of* $\mathbf{D}$, *then either* $\mathbf{D}$ *is isomorphic to* $PG_{d-1}(d, 2)$ *for some* d *or* $v = \beta(\beta^3 + 5\beta^2 + 6\beta - 1)$, $k = \beta(\beta^2 + 3\beta + 1)$, *and* $\lambda = \beta(\beta + 1)$ *where* β *is the nonzero triple intersection number of* $\mathbf{D}$. *Furthermore, the dual design* $\mathbf{D}^\top$ *is quasi-3.*

Proof. Let $\mathbf{D} = (X, \mathcal{B})$.

(i) Let λ be a triple intersection number of $\mathbf{D}$. Let A and B be distinct blocks of $\mathbf{D}$. Observe that the line $\overline{AB}$ of the dual design $\mathbf{D}^\top$ is the set of all blocks of $\mathbf{D}$ that contain $A \cap B$. Therefore, if three distinct blocks A, B, and C of $\mathbf{D}$ are not collinear, as points of $\mathbf{D}^\top$, then $|A \cap B \cap C| \neq |A \cap B| = \lambda$, i.e., $|A \cap B \cap C| = \alpha$ where α is the smaller triple intersection number of $\mathbf{D}$. Thus, the design $\mathbf{D}^\top$ satisfies condition (iv) of the Dembowski–Wagner Theorem. Therefore, $\mathbf{D}^\top$ is $PG_{d-1}(d, q)$ for some d and q, and so is $\mathbf{D}$.

(ii) Let 0 and β be the triple intersection numbers of $\mathbf{D}$. Since $\lambda \geq 3$, we have $\beta \geq 2$. Let $x \in X$ and let $\mathcal{B}_1 = \{B \in \mathcal{B} \colon x \in B\}$. Consider the substructure $\mathbf{D}_1 = (\mathcal{B}_1, X \setminus \{x\})$ of $\mathbf{D}^\top$. Since 0 is a triple intersection number of $\mathbf{D}$, any three distinct blocks of $\mathbf{D}$ containing x meet in β points. Therefore, $\mathbf{D}_1$ is a 3-$(k, \lambda, \beta - 1)$ design. Let $\mathbf{D}_2$ be a point-derived design of $\mathbf{D}_1$. Then $\mathbf{D}_2$ has $k - 1$ blocks, and therefore it is a symmetric $(k - 1, \lambda - 1, \beta - 1)$-design. By the Cameron Theorem, one of the following is true:

(1) $k = 4\beta$ and $\lambda = 2\beta$;
(2) $k = \beta(\beta^2 + 3\beta + 1)$ and $\lambda = \beta(\beta + 1)$;
(3) $k = 496$ and $\lambda = 40$.

In case (2), $v = \beta(\beta^3 + 5\beta^2 + 6\beta - 1)$. In case (3), we obtain $v = 6138$. Therefore, v is even, while $k - \lambda$ is not a square, which contradicts Proposition 2.4.10.

In case (1), $\mathbf{D}$ is a symmetric $(8\beta - 1, 4\beta, 2\beta)$-design, so the complementary design $\mathbf{D}'$ is a Hadamard 2-$(8\beta - 1, 4\beta - 1, 2\beta - 1)$ design with triple intersection numbers $\beta - 1$ and $2\beta - 1$. By (i), **D** is the design $PG_{d-1}(d, q)$ for some d and q. Then $2\beta = k - \lambda = q^{d-1}$, so q is even. If $q \neq 2$, then $v = 1 + q + \cdots + q^d \equiv 1 \pmod 4$. On the other hand, $8\beta - 1 \equiv -1 \pmod 4$, and therefore $q = 2$, i.e., **D** is $PG_{d-1}(d, 2)$.

We now show that $\mathbf{D}^\top$ is quasi-3. Let $\{x, y, z\}$ be a 3-subset of X. It will follow that $\mathbf{D}^\top$ is quasi-3 if we show that the number of blocks of **D** containing $\{x, y, z\}$ is either 0 or β. Suppose there is a block $A \in \mathcal{B}$ containing $\{x, y, z\}$. Consider the above design $\mathbf{D}_1$ and let $\mathbf{D}_2$ be the derived design of $\mathbf{D}_1$ with respect to the point A of $\mathbf{D}_1$. Since $\mathbf{D}_2$ is a symmetric $(k-1, \lambda-1, \beta-1)$-design, there are exactly $\beta - 1$ blocks $B \in \mathcal{B}_1 \setminus \{A\}$ that contain y and z. These $\beta - 1$ blocks and the block A are precisely the blocks of **D** that contain $\{x, y, z\}$, so **D** has exactly β blocks containing $\{x, y, z\}$. □

In both cases of Theorem 8.4.13, the dual of the quasi-3 symmetric design is also a quasi-3 symmetric design. However, this is not always the case.

Example 8.4.14. Let

$$X_1 = \begin{bmatrix} 0 & 1 & 1 & 0 \\ 1 & 0 & 0 & 1 \\ 1 & 0 & 0 & 1 \\ 0 & 1 & 1 & 0 \end{bmatrix}, \quad X_2 = \begin{bmatrix} 0 & 0 & 1 & 1 \\ 0 & 0 & 1 & 1 \\ 1 & 1 & 0 & 0 \\ 1 & 1 & 0 & 0 \end{bmatrix}, \quad X_3 = \begin{bmatrix} 0 & 1 & 0 & 1 \\ 1 & 0 & 1 & 0 \\ 0 & 1 & 0 & 1 \\ 1 & 0 & 1 & 0 \end{bmatrix}$$

and let $\overline{X_i} = J - X_i$ for $i = 1, 2, 3$. Next define, for $i = 1, 2, 3$, block matrices

$$A_i = \begin{bmatrix} X_i & X_i \\ X_i & X_i \end{bmatrix} \text{ and } \overline{A_i} = \begin{bmatrix} X_i & \overline{X_i} \\ \overline{X_i} & X_i \end{bmatrix}.$$

We further define block matrices

$$B = \begin{bmatrix} A_1 & A_2 \\ A_2 & A_1 \end{bmatrix} \text{ and } C = \begin{bmatrix} A_3 & O \\ O & A_3 \end{bmatrix}$$

and correspondingly

$$\overline{B} = \begin{bmatrix} \overline{A_1} & \overline{A_2} \\ \overline{A_2} & \overline{A_1} \end{bmatrix} \text{ and } \overline{C} = \begin{bmatrix} \overline{A_3} & O \\ O & \overline{A_3} \end{bmatrix}.$$

One can verify that the block matrix

$$M = \begin{bmatrix} \overline{B} & \overline{C} & B & C \\ \overline{C} & \overline{B} & C & B \\ B & C & \overline{B} & \overline{C} \\ C & B & \overline{C} & \overline{B} \end{bmatrix}$$

is an incidence matrix of an SDP-design with parameters (64, 28, 12). We will now modify M to obtain a symmetric quasi-3 design, whose dual is not quasi-3.

Let

$$P = \begin{bmatrix} X_1 & \overline{X_1} \\ X_1 & \overline{X_1} \end{bmatrix} \text{ and } Q = \begin{bmatrix} X_1 & X_1 \\ \overline{X_1} & \overline{X_1} \end{bmatrix}$$

and then we replace B and $\overline{B}$ with

$$B_1 = \begin{bmatrix} Q & X_2 \\ X_2 & X_1 \end{bmatrix} \text{ and } \overline{B_1} = \begin{bmatrix} P & \overline{X_2} \\ \overline{X_2} & \overline{X_1} \end{bmatrix}.$$

Then

$$N = \begin{bmatrix} \overline{B_1} & \overline{C} & B_1 & C \\ \overline{C} & \overline{B} & C & B \\ B & C & \overline{B} & \overline{C} \\ C & B & \overline{C} & \overline{B} \end{bmatrix}$$

is an incidence matrix of a symmetric quasi-3 design with parameters (64, 28, 12), whose dual design can be shown to be not quasi-3.

The next theorem characterizes quasi-3 Hadamard 2-designs.

Theorem 8.4.15. *Let* $\mathbf{D}$ *be a quasi-3 symmetric* $(4n-1, 2n-1, n-1)$ *design with* $n \geq 3$. *Then* $\mathbf{D}$ *is isomorphic to* $PG_{d-1}(d, 2)$ *for some* d.

Proof. Let $\mathbf{D} = (X, \mathcal{B})$. We will show that every line of $\mathbf{D}^\top$ has at least three points. Then Proposition 3.7.3 and the Dembowski–Wagner Theorem would imply that $\mathbf{D}^\top$ is isomorphic to $PG_{d-1}(d, 2)$ for some d, and then so is $\mathbf{D}$.

Let A and B be distinct blocks of $\mathbf{D}$. Then the cardinality of the line $\overline{AB}$ of $\mathbf{D}^\top$ is greater than 2 if and only if there exists a block C of $\mathbf{D}$, other than A or B, such that $A \cap B \subset C$.

The complementary design $\mathbf{D}'$ is a quasi-3 symmetric $(4n-1, 2n, n)$-design. Let $A' = X \setminus A$ and $B' = X \setminus B$. The derived design $(\mathbf{D}')_{A'}$ is a quasi-symmetric 2-design with $2n$ points and $4n-2$ blocks. By Theorem 8.2.23, it is a Hadamard 3-design. Since the complement of a block of Hadamard 3-design is also a block, we obtain that there exists a block C' of $\mathbf{D}'$ such that $C' \cap A'$ is disjoint from $B' \cap A'$. Since $|C' \cap B'| = |B' \setminus A'| = |C' \setminus A'| = n$, we obtain that $C' \cap B' = B' \setminus A'$, and therefore $C' \subset A' \cup B'$. Let $C = X \setminus C'$. Then C is a block of $\mathbf{D}$, and $A \cap B \subset C$. The proof is now complete. □

8.5. Block schematic designs with three intersection numbers

In the previous sections we investigated quasi-symmetric designs, which are precisely block schematic designs with two association classes. In this section we will consider possible restrictions on designs with three intersection numbers that would lead to 3-class association schemes.

The classical example of a block schematic design with three association classes is the Witt design $\mathbf{W}_{24}$. Recall that it is a 5-(24, 8, 1) design with intersection numbers 0, 2, and 4. For $0 \le i \le 5$, let λ_i denote the number of blocks of $\mathbf{W}_{24}$ that contain a fixed i-subset of the point set. We begin with a lemma.

Lemma 8.5.1. *Let* $\mathbf{W}_{24} = (X, \mathcal{B})$, $L = \{0, 2, 4\}$, *and let* U *be the set of ordered pairs* (s, t) *of nonnegative integers with* $s + t \le 3$. *Then there exists a unique function* $f : L \times U \to \mathbb{Z}$ *with the following property: if* $\alpha \in L$ *and* $(s, t) \in U$, *then, for any* $A \in \mathcal{B}$, *any* s*-subset* S *of* X *disjoint from* A, *and any* t*-subset* T *of* A, *there are exactly* $f(\alpha, s, t)$ *blocks* $B \in \mathcal{B}$ *that contain* $S \cup T$ *and meet* A *in exactly* α *points.*

Proof. Let $(s, t) \in U$ and $A \in \mathcal{B}$. Let S be an s-subset of X disjoint from A and let T be a t-subset of A. Let $\mathcal{C}$ be the set of all blocks $B \in \mathcal{B} \setminus \{A\}$ that contain $S \cup T$. For $\alpha \in L$, let n_α be the number of blocks $B \in \mathcal{C}$ that meet A in α points. Then

$$n_0 + n_2 + n_4 = \begin{cases} \lambda_{s+t} & \text{if } s \neq 0, \\ \lambda_{s+t} - 1 & \text{if } s = 0. \end{cases} \tag{8.19}$$

If $t \le 2$, then counting in two ways pairs (B, z) with $B \in \mathcal{C}$ and $z \in (A \cap B) \setminus T$ yields

$$(2 - t)n_2 + (4 - t)n_4 = \begin{cases} (8 - t)\lambda_{s+t+1} & \text{if } s \neq 0, \\ (8 - t)(\lambda_{s+t+1} - 1) & \text{if } s = 0. \end{cases} \tag{8.20}$$

If $t = 0$, then counting in two ways pairs (B, Z) where $B \in \mathcal{C}$ and Z is a 2-subset of $A \cap B$ yields

$$n_2 + 6n_4 = \begin{cases} 28\lambda_{s+2} & \text{if } s \neq 0, \\ 28(\lambda_{s+2} - 1) & \text{if } s = 0. \end{cases} \tag{8.21}$$

If $t = 0$ and $s \le 3$, then equations (8.19) – (8.21) yield a unique solution (n_0, n_2, n_4). If $t = 1$ or $t = 2$ and $s + t \le 3$, then $n_0 = 0$, and equations (8.19) and (8.20) yield a unique pair (n_2, n_4). Finally, if $t = 3$, then $n_0 = n_2 = 0$, and then $n_4 = b/4$. Thus, in any case, the numbers n_0, n_2, and n_4 depend only on α, s, and t. □

Theorem 8.5.2. *The design* $\mathbf{W}_{24}$ *is block-schematic.*

Proof. Let $\mathcal{B}$ be the block set of $\mathbf{W}_{24}$. Let $\mathcal{R}_0$ denote the identity relation on $\mathcal{B}$ and, for $i = 1, 2, 3$, let $\mathcal{R}_i$ denote the relation on $\mathcal{B}$ defined by $(A, B) \in \mathcal{R}_i$ if and only if $|A \cap B| = 2i - 2$.

Fix blocks A_1 and A_2 of $\mathbf{W}_{24}$. For $i, j = 0, 1, 2, 3$, let p_{ij} denote the number of blocks $B \in \mathcal{B}$ such that $(A_1, B) \in \mathcal{R}_i$ and $(A_2, B) \in \mathcal{R}_j$. We have to show that parameters p_{ij} depend only on i, j, and $\alpha = |A_1 \cap A_2|$ rather than on the choice of A_1 and A_2.

Case 1: $A_1 = A_2$.

Then $p_{ij} = 0$ whenever $i \neq j$, and $p_{00} = 1$. Theorem 6.5.13(iii) implies that $p_{11} = 30$, $p_{22} = 448$, and $p_{33} = 280$.

Case 2: $A_1 \neq A_2$.

Then $p_{00} = 0$ and, for $j = 1, 2, 3$,

$$p_{0j} = p_{j0} = \begin{cases} 0 & \text{if } 2j - 2 \neq \alpha, \\ 1 & \text{if } 2j - 2 = \alpha. \end{cases}$$

For $i, j = 1, 2, 3$ and $t = 1, 2, 3, 4$, let $\pi_{ij}(t)$ denote the number of blocks $B \in \mathcal{B}$ that meet A_1 in $2i - 2$ points, A_2 in $2j - 2$ points, and $A_1 \cap A_2$ in t points. Then

$$p_{ij} = \sum_{t=0}^{\alpha} \pi_{ij}(t),$$

so it suffices to show that integers $\pi_{ij}(t)$ depend only on i, j, t, and α.

Let s and t be nonnegative integers such that $s + t \leq 3$ and let f be the function provided by Lemma 8.5.1. For $i, j = 1, 2, 3$, let

$$g_i(s, t) = \begin{cases} f(2i - 2, s, t) & \text{if } 2i - 2 \neq \alpha, \\ f(2i - 2, s, t) - 1 & \text{if } 2i - 2 = \alpha. \end{cases}$$

Fix $i \in \{1, 2, 3\}$ and nonnegative integers s and t with $s + t \leq 3$ and count in two ways triples (T, S, B) where T is a t-subset of $A_1 \cap A_2$, S is an s-subset of $A_2 \setminus A_1$, and B is a block, other than A_2, that meets A_1 in $2i - 2$ points and contains $S \cup T$:

$$\binom{\alpha}{t}\binom{8 - \alpha}{s} g_i(s, t) = \sum_{j=1}^{3} \pi_{ij}(t) \binom{2j - 2 - t}{s}. \tag{8.22}$$

Note that $\binom{2j - 2 - t}{s} = 0$ if and only if $j < (s + t + 2)/2$. Therefore, for $0 \leq t \leq 3$, letting $s = 3 - t$ in (8.22) uniquely determines $\pi_{i3}(t)$. Besides,

$\pi_{i3}(4) = 0$ for $i = 1, 2$ and $\pi_{33}(4) = 0$ if $\alpha \neq 4$. If $\alpha = 4$, then $\pi_{33}(4)$ counts the blocks that contain the 4-set $A_1 \cap A_2$, and, by Theorem 6.5.12(iii), there are three such blocks, besides A_1 and A_2. Therefore, $\pi_{33}(4) = 3$. Thus, integers $\pi_{i3}(t)$ depend only on i, j, t, and α.

Now, for $0 \leq t \leq 2$, letting $s = 2 - t$ in (8.22) uniquely determines $\pi_{i2}(t)$. Besides, $\pi_{i2}(3) = \pi_{i2}(4) = 0$, so integers $\pi_{i2}(t)$ depend only on i, j, t, and α.

If we put $s = t = 0$ in (8.22), we uniquely determine $\pi_{i1}(0)$. Since $\pi_{i1}(t) = 0$ for $t \geq 1$, we have now shown that all integers $\pi_{ij}(t)$ are uniquely determined by i, j, t, and α. This completes the proof. □

The following theorem generalizes both Theorem 8.2.14 and Theorem 8.5.2. Its proof is beyond the scope of this book.

Theorem 8.5.3. (The Cameron–Delsarte Theorem). *For $s \geq 2$, any $(2s-2)$-(v, k, λ) design with s intersection numbers is block schematic.*

Our next goal is to introduce a putative family of block schematic 2-designs with three intersection numbers.

We begin with the following theorem that gives a lower bound for the intersection numbers of a 2-design. Recall that the order of a (v, b, r, k, λ)-design is $n = r - \lambda$.

Theorem 8.5.4. *Let B_1 and B_2 be distinct blocks of a 2-(v, k, λ)-design* **D** *of order n. Then $|B_1 \cap B_2| \geq k - n$. If $|B_1 \cap B_2| = k - n$, then, for any block C of* **D**, *other than B_1 or B_2, $|C \cap B_1| = |C \cap B_2|$.*

Proof. Let $\beta = |B_1 \cap B_2|$ and let $\mathcal{C} = \mathcal{B} \setminus \{B_1, B_2\}$. For $i = 1, 2$ and for any block $C \in \mathcal{C}$, let $f_i(C) = |C \cap (B_i \setminus B_{3-i})|$. Counting in two ways pairs (C, x) with $C \in \mathcal{C}$ and $x \in C \cap (B_i \setminus B_{3-i})$ yields

$$\sum_{C \in \mathcal{C}} f_i(C) = (k - \beta)(r - 1). \tag{8.23}$$

Counting in two ways pairs (C, Y) where $C \in \mathcal{C}$ and Y is a 2-subset of $C \cap (B_i \setminus B_{3-i})$ yields

$$\sum_{C \in \mathcal{C}} f_i(C)(f_i(C) - 1) = (k - \beta)(k - \beta - 1)(\lambda - 1). \tag{8.24}$$

Finally, counting in two ways triples (C, x, y) where $C \in \mathcal{C}$, $x \in B_1 \setminus B_2$, and $y \in B_2 \setminus B_1$ yields

$$\sum_{C \in \mathcal{C}} f_1(C) f_2(C) = (k - \beta)^2(\lambda - 1). \tag{8.25}$$

Equations (8.23)–(8.25) imply

$$\sum_{C \in \mathcal{C}} (f_1(C) - f_2(C))^2 = 2(k - \beta)(\beta + n - k). \tag{8.26}$$

If $\beta = k - n$, then (8.26) implies that $f_1(C) = f_2(C)$ for any $C \in \mathcal{C}$. If $\beta \neq k$, then (8.26) implies that $\beta \geq k - n$. If $\beta = k$, then again $\beta \geq k - n$. For $i = 1, 2$, we have $|C \cap B_i| = f_i(C) + |C \cap B_1 \cap B_2|$. Therefore, $|C \cap B_1| = |C \cap B_2|$. □

If **D** is a quasi-residual 2-(v, k, λ) design, then its order equals k, and we obtain the following result.

Corollary 8.5.5. *If a quasi-residual design* **D** *has disjoint blocks* B_1 *and* B_2, *then, for any block* C *of* **D**, *other than* B_1 *or* B_2, $|C \cap B_1| = |C \cap B_2|$.

We will now concentrate on 2-(v, k, λ) designs with an intersection number $k - n$. If $\mathbf{D} = (X, \mathcal{B})$ is such a design, then we define an equivalence relation on $\mathcal{B}$ as follows: $B_1 \sim B_2$ if and only if $B_1 = B_2$ or $|B_1 \cap B_2| = k - n$. Theorem 8.5.4 implies that it is indeed an equivalence relation. The corresponding partition of the block set of **D** will be called the *M-partition of* **D**.

Definition 8.5.6. The M-partition of a 2-(v, k, λ) design is said to be *regular* if all partition classes are of the same size.

Example 8.5.7. Proposition 5.3.1 implies that any affine α-resolvable design admits a regular M-partition.

The next proposition provides a putative family of 2-designs with three intersection numbers admitting a regular M-partition.

Proposition 8.5.8. *Any* 2-(v, k, λ) *design of order* n *with three intersection numbers, one of which is* $k - n$, *admits a regular* M*-partition.*

Proof. Let **D** be a 2-(v, k, λ) design of order n with three intersection numbers, $\alpha_1 = k - n$, α_2, and α_3. Let A be a block of **D** and, for $i = 1, 2, 3$, let n_i be the number of blocks other than A, that meet A in α_i points. Applying variance counting to this design (Proposition 2.3.8) gives a linear system of three equations in n_1, n_2, n_3 with a nonzero determinant. Since every M-partition class is of size $n_1 + 1$, the M-partition is regular. □

The order of a quasi-residual design is equal to the block size of the design. Therefore, a quasi-residual design admits the M-partition if and only if it has an intersection number 0. The following result determines all intersection numbers of a quasi-residual 2-(v, k, λ) design admitting a regular M-partition with the class size $\lambda - 1$.

Proposition 8.5.9. *Let* $\mathbf{D} = (X, \mathcal{B})$ *be a quasi-residual* 2-(v, k, λ) *design. Suppose the block set* $\mathcal{B}$ *admits a partition with all partition classes of the same size* $\lambda - 1$ *and any two blocks from the same class being disjoint. Then the partition is the* M*-partition of* **D** *and, if* A *and* B *are blocks from different* *partition classes, they intersect in* $\lambda - 1$ *or* λ *points. Furthermore, for any block* A *of* $\mathbf{D}$*, there are exactly* $k\lambda$ *blocks meeting* A *in* $\lambda - 1$ *points and* $v + k - k\lambda$ *blocks meeting* A *in* λ *points.*

Proof. Fix a block A of $\mathbf{D}$. For $i = 0, 1, 2, \ldots, k$, let m_i denote the number of blocks $B \in \mathcal{B}$, other than A, that meet A in i points. Then $m_0 \geq \lambda - 2$. Since the replication number of $\mathbf{D}$ is $k + \lambda$ and the number of blocks is $v + k + \lambda - 1$, variance counting (Proposition 2.3.8) yields

$$\sum_{i=1}^{k} m_i \leq v + k,$$

$$\sum_{i=1}^{k} i m_i = k(k + \lambda - 1),$$

$$\sum_{i=1}^{k} i(i-1)m_i = k(k-1)(\lambda - 1).$$

Therefore,

$$\sum_{i=1}^{k} (i - \lambda)(i - \lambda + 1)m_i \leq 0.$$

Since the product of two consecutive integers is nonnegative, all terms in the last sum are nonnegative. Therefore $m_0 = \lambda - 2$, which implies that the given partition is the M-partition of $\mathbf{D}$, and all terms in the last sum are equal to zero. Since $(i - \lambda)(i - \lambda + 1) > 0$ for all i, except $i = \lambda$ and $i = \lambda - 1$, we obtain that $m_i = 0$, except for these two values of i. Solving the variance equations, we obtain that $m_{\lambda-1} = k\lambda$ and $m_\lambda = v + k - k\lambda$. □

With each 2-(v, k, λ) design of order n having three intersection numbers, one of which is $k - n$, we will associate a graph in the following way.

Definition 8.5.10. Let $\mathbf{D}$ be a 2-(v, k, λ) design of order n with three intersection numbers, $k - n$, α_1, and α_2, and let $\alpha_1 < \alpha_2$. Suppose $\mathbf{D}$ has an M-partition $\mathfrak{M}$. Let Γ be the graph with vertex set $\mathfrak{M}$ and with distinct vertices $\mathcal{M}_1$ and $\mathcal{M}_2$ adjacent if and only if any block from $\mathcal{M}_1$ meets any block from $\mathcal{M}_2$ in α_2 points. The graph Γ is called the *class graph of* $\mathbf{D}$.

Remark 8.5.11. Theorem 8.5.4 implies that the class graph is well defined.

Lemma 8.5.12. *If* $\mathbf{D}$ *is a* 2-(v, k, λ) *design of order* n *with three intersection numbers, one of which is* $k - n$*, then the class graph of* $\mathbf{D}$ *is regular.*

Proof. By Proposition 8.5.8, the M-partition of $\mathbf{D}$ is regular. Let m be the cardinality of each partition class. Let α_i $(i = 1, 2, 3)$ be the intersection numbers of $\mathbf{D}$. Let $\alpha_3 = k - n$ and $\alpha_1 < \alpha_2$. Variance counting (Proposition 2.3.8) now implies that for any block A of $\mathbf{D}$, the number of blocks meeting A in α_2 points does not depend on A. Dividing this number by m gives the degree of Γ. □

The next theorem shows that class graphs are in fact strongly regular.

Theorem 8.5.13. *If* $\mathbf{D}$ *is a* 2-(v, k, λ) *design of order* n *with three intersection numbers, one of which is* $k - n$*, then the class graph of* $\mathbf{D}$ *is strongly regular.*

Proof. Let $\mathbf{D}$ be a 2-(v, k, λ) design of order n with three intersection numbers, $k - n$, α_1, and α_2, and let $\alpha_1 < \alpha_2$. Let $\mathfrak{M}$ be the M-partition of $\mathbf{D}$ and let N be an incidence matrix of $\mathbf{D}$ such that all blocks from each M-partition class correspond to consecutive columns of N. Let Γ be the class graph of $\mathbf{D}$ and A an adjacency matrix of Γ. Let $c = |\mathfrak{M}|$. By Proposition 8.5.8, the partition $\mathfrak{M}$ is regular. Let m be the cardinality of each partition class.

Since $NN^\top = nI + \lambda J$, the matrix $NN^\top$ has two eigenvalues. Proposition 2.2.15 implies that the matrix $N^\top N - nI$ has three eigenvalues. Observe that $N^\top N - nI = B \otimes J_m$ where $B = (k - n - \alpha_1)I + (\alpha_2 - \alpha_1)A + \alpha_1 J_c$.

Let s be an eigenvalue of B and $\mathbf{x} = [x_1\ x_2 \ldots x_c]^\top$ a corresponding eigenvector. Let $\mathbf{y} = \mathbf{x} \otimes J_{m,1}$, so each component of $\mathbf{x}$ is repeated m times in $\mathbf{y}$. Then $(B \otimes J_m)\mathbf{y} = ms\mathbf{y}$. Therefore, if s is an eigenvalue of B, then ms is an eigenvalue of $N^\top N - nI$. This implies that B has at most three eigenvalues. If d is the degree of Γ, then $k - n - \alpha_1 + (\alpha_2 - \alpha_1)d + \alpha_1 c$ is the eigenvalue of B corresponding to the eigenvector $\mathbf{j}$. Since B has at most two more eigenvalues, A has at most two eigenvalues other than d. Since Γ is neither a null nor a complete graph, Proposition 7.2.9 and Theorem 7.2.8 imply that Γ is strongly regular. □

Remark 8.5.14. A more careful analysis of the relationship between matrices $N^\top N - nI$, B, and A in the above proof would yield expressions for the eigenvalues of the class graph Γ of a (v, b, r, k, λ)-design in terms of the parameters and intersection numbers of the design and the cardinality c of the M-partition:

$$\theta_0 = \frac{\lambda vc - b(k - n - \alpha_1 + \alpha_1 c)}{b(\alpha_2 - \alpha_1)}, \qquad \theta_1 = \frac{\alpha_1 - k + n}{\alpha_2 - \alpha_1},$$

$$\theta_2 = \frac{b(\alpha_1 - k + n) - cn}{b(\alpha_2 - \alpha_1)}.$$

The multiplicities of these eigenvalues are $1, c - b + v - 1$, and $b - v$, respectively.

Theorem 8.5.13 immediately implies the following result.

Corollary 8.5.15. *If* **D** *is a 2-(v, k, λ) design of order n with three intersection numbers, one of which is $k - n$, then* **D** *is block-schematic.*

In the next section, we will consider another approach to constructing block-schematic designs with three intersection numbers.

8.6. Designs with a nearly affine decomposition

In this section, we will consider another type of block-schematic designs with three intersection numbers.

Definition 8.6.1. Let $\mathbf{D} = (X, \mathcal{B})$ be a (v, b, r, k, λ)-design having intersection numbers ρ_1, ρ_2, and ρ_3. Suppose the block set $\mathcal{B}$ is partitioned as $\mathcal{B} = \mathcal{B}_1 \cup \mathcal{B}_2 \cup \cdots \cup \mathcal{B}_s$, so that for distinct blocks $A, B \in \mathcal{B}$,

$$|A \cap B| = \begin{cases} \rho_3 & \text{if } A \in \mathcal{B}_i, B \in \mathcal{B}_j, i \neq j, \\ \rho_1 \text{ or } \rho_2 & \text{if } A, B \in \mathcal{B}_i, i = 1, 2, \ldots, s. \end{cases}$$

Assume further that $\rho_1 > \rho_2$ and that both ρ_1 and ρ_2 are realized in at least one $\mathcal{B}_i$. Then we shall say that **D** admits a *nearly affine* (ρ_1, ρ_2, ρ_3)-*decomposition*.

Theorem 8.6.2. *Let* $\mathbf{D} = (X, \mathcal{B})$ *be a* (v, b, r, k, λ)*-design. Suppose that* **D** *admits a nearly affine* (ρ_1, ρ_2, ρ_3)*-decomposition* $\mathcal{B} = \mathcal{B}_1 \cup \mathcal{B}_2 \cup \cdots \cup \mathcal{B}_s$. *For* $i = 1, 2, \ldots, s$, *let* Γ_i *be the graph with vertex set* $\mathcal{B}_i$ *and with distinct vertices* $A, B \in \mathcal{B}_i$ *adjacent if and only if* $|A \cap B| = \rho_1$. *Then the graphs* Γ_i, $i = 1, 2, \ldots, s$, *are strongly regular graphs with the same parameters.*

Proof. Let $b_i = |\mathcal{B}_i|$ $(i = 1, 2, \ldots, s)$ and let $A \in \mathcal{B}_i$ be a fixed block. Suppose that $\mathcal{B}_i$ contains m_j blocks meeting A in ρ_j points, $j = 1, 2$. Let $m_3 = b - b_i$. Then the following relations can be obtained by two-way counting:

$$m_1 + m_2 + m_3 = b - 1. \tag{8.27}$$

$$\rho_1 m_1 + \rho_2 m_2 + \rho_3 m_3 = k(r - 1). \tag{8.28}$$

$$\rho_1(\rho_1 - 1)m_1 + \rho_2(\rho_2 - 1)m_2 + \rho_3(\rho_3 - 1)m_3 = k(k - 1)(\lambda - 1). \tag{8.29}$$

Equations (8.27), (8.28), and (8.29) in m_1, m_2, and m_3 form a linear system with a nonsingular coefficient matrix and therefore yield a unique solution for m_1, m_2 and m_3, which is independent of A and of i. Since in at least one $\mathcal{B}_i$, both m_1 and m_2 are positive, this implies that in all $\mathcal{B}_i$ both m_1 and m_2 are positive. Observe that each graph Γ_i is regular of valency $a_i = m_1$. Therefore, if X_i is an adjacency matrix of Γ_i, then $X_i J_{b_i} = J_{b_i} X_i = m_1 J_{b_i}$.

Let M_i be an incidence matrix of the substructure $(X, \mathcal{B}_i)$ of **D**. Then,

$$M_i^\top M_i = (k - \rho_2)I_{b_i} + (\rho_1 - \rho_2)A_i + \rho_2 J_{b_i}. \tag{8.30}$$

Let M be an incidence matrix of **D**. Then, since $MM^\top = (r - \lambda)I_v + \lambda J_v$, the eigenvalues of $MM^\top$ are kr and $r - \lambda$. Note that the eigenvalue kr corresponds to the all-one vector. Let $B_i = M_i^\top M_i - \rho_3 J_{b_i}$. Then,

$$M^\top M = \rho_3 J_b + diag(B_1, B_2, \ldots, B_s), \tag{8.31}$$

where $diag(B_1, B_2, \ldots, B_s)$ is the block diagonal matrix with blocks B_i along the diagonal. Let $Spec(B_i)$ denote the set of distinct eigenvalues of B_i. By Proposition 2.2.14, $MM^\top$ and $M^\top M$ have the same nonzero eigenvalues with the same multiplicities. Since $(M^\top M)J_b = J_b(M^\top M) = kr J_b$, we obtain from (8.31) that, for $i = 1, 2, \ldots, s$, $Spec(B_i) \subseteq \{kr - \rho_3 b, r - \lambda, 0\}$. Equation (8.30) now gives

$$A_i = \frac{1}{(\rho_1 - \rho_2)} M_i^\top M_i - \frac{(k - \rho_2)}{(\rho_1 - \rho_2)} I_{b_i} - \frac{\rho_2}{(\rho_1 - \rho_2)} J_{b_i}.$$

Since $M_i^\top M_i = B_i + \rho_3 J B_i$, we obtain the following relation:

$$A_i = \frac{1}{(\rho_1 - \rho_2)} B_i - \frac{(k - \rho_2)}{(\rho_1 - \rho_2)} I_{b_i} + \frac{(\rho_3 - \rho_2)}{(\rho_1 - \rho_2)} J_{b_i}. \tag{8.32}$$

Therefore,

$$Spec(A_i) \subseteq \left\{ \frac{kr - \rho_3 b - k + \rho_2 + (\rho_3 - \rho_2)b_i}{(\rho_1 - \rho_2)}, \frac{r - \lambda - k + \rho_2}{(\rho_1 - \rho_2)}, \frac{-(k - \rho_2)}{(\rho_1 - \rho_2)} \right\}.$$

Theorems 7.2.8 and 7.2.2 imply that Γ_i is an $SRG(b_i, a_i, c_i, d_i)$ with

$$a_i = \frac{kr - \rho_3 b - k + \rho_2 + (\rho_3 - \rho_2)b_i}{(\rho_1 - \rho_2)}, \tag{8.33}$$

$$c_i = a_i + \frac{(r - \lambda - 2k + 2\rho_2)}{(\rho_1 - \rho_2)} - \frac{(r - \lambda - k + \rho_2)(k - \rho_2)}{(\rho_1 - \rho_2)^2}, \tag{8.34}$$

$$d_i = a_i - \frac{(r - \lambda - k + \rho_2)(k - \rho_2)}{(\rho_1 - \rho_2)^2}. \tag{8.35}$$

Since $a_i = m_1$ and $b_i = m_1 + m_2 + 1$, do not depend on i, the strongly regular graphs Γ_i all have the same parameters ($i = 1, 2, \ldots, s$). This completes the proof. □

Corollary 8.6.3. *If a 2-design* **D** *admits a nearly affine decomposition* $\mathcal{B} = \mathcal{B}_1 \cup \mathcal{B}_2 \cup \cdots \cup \mathcal{B}_s$, *then the sets* $\mathcal{B}_1, \mathcal{B}_2, \ldots, \mathcal{B}_s$ *have the same cardinality.*

Proposition 8.6.4. *If a 2-design* **D** *admits a nearly affine* (ρ_1, ρ_2, ρ_3)*-decomposition with distinct* ρ_1, ρ_2, *and* ρ_3, *then* **D** *is block-schematic.*

Proof. Let (n, a, c, d) be the parameters of the graphs Γ_i in Theorem 2.2. Then the association scheme parameters p^i_{jh} for $i, j, h \in \{0, 1, 2\}$ are the same as the 2-class association scheme represented by the strongly regular graph Γ_i. It is straightforward to verify that $p^1_{33} = p^2_{33} = b - n$, $p^3_{03} = 1$, $p^3_{13} = a$, $p^3_{23} = n - a - 1$, $p^3_{33} = b - 2n$, and the remaining $p^i_{jh} = 0$. This shows that **D** is block schematic. □

We will now obtain an upper bound on the parameter ρ_3 of a 2-design with nearly affine (ρ_1, ρ_2, ρ_3)-decomposition.

Proposition 8.6.5. *Suppose a* (v, b, r, k, λ)*-design* $\mathbf{D} = (X, \mathcal{B})$ *admits a nearly affine* (ρ_1, ρ_2, ρ_3)*-decomposition* $\mathcal{B} = \mathcal{B}_1 \cup \mathcal{B}_2 \cup \cdots \cup \mathcal{B}_s$ *Let* $m = |\mathcal{B}_i|$ *and let* a *be the valency of the graphs* Γ_i, $(i = 1, 2, \ldots, s)$. *Then*

$$\rho_3 m \le (k - \rho_2) + (\rho_1 - \rho_2)a + \rho_2 m,$$

with equality if and only if $\{\mathcal{B}_1, \mathcal{B}_2, \ldots, \mathcal{B}_s\}$ *is a resolution of* **D**.

Proof. Let M be an incidence matrix of **D** written as $M = [M_1, M_2, \ldots, M_s]$, where M_i is an incidence matrix of $(X, \mathcal{B}_i)$, $i = 1, 2, \ldots, s$. Then, for $i = 1, 2, \ldots, s$,

$$M_i^\top M_i = (k - \rho_2)I_m + (\rho_1 - \rho_2)A_i + \rho_2 J_m,$$

where A_i is an adjacency matrix of Γ_i. For distinct i and j, $M_i^\top M_j = \rho_3 J_m$.

For $i = 1, 2, \ldots, s$, let $\mathbf{x}_i = M_i\mathbf{j}$. Then $\mathbf{x}_i^\top \mathbf{j} = \mathbf{j}^\top M_i \mathbf{j} = \mathbf{j}^\top(k\mathbf{j}) = km$. Besides, for $i \ne l$, $\mathbf{x}_i^\top \mathbf{x}_l = (M_i\mathbf{j})^\top(M_l\mathbf{j}) = \rho_3 m^2$, and $\mathbf{x}_i^\top \mathbf{x}_i = (M_i\mathbf{j})^\top(M_i\mathbf{j}) = (k - \rho_2)m + (\rho_1 - \rho_2)am + \rho_2 m^2 = \mathbf{x}_l^\top \mathbf{x}_l$. By the Cauchy–Schwarz Inequality, $(\mathbf{x}_i^\top \mathbf{x}_l)^2 \le (\mathbf{x}_i^\top \mathbf{x}_i)(\mathbf{x}_l^\top \mathbf{x}_l)$ with equality if and only if $\mathbf{x}_i = c\mathbf{x}_l$, for some scalar c. This gives $\rho_3 m \le (k - \rho_2) + (\rho_1 - \rho_2)a + \rho_2 m$, with equality if and only if $\mathbf{x}_i = c\mathbf{x}_l$ for some scalar c. Since $\mathbf{x}_i^\top \mathbf{j} = \mathbf{x}_l^\top \mathbf{j}$, we have (in the case of equality) $c = 1$. Thus, in this case, $\mathbf{x}_1 = \mathbf{x}_2 = \ldots = \mathbf{x}_s$. Since $\sum_{i=1}^{s} \mathbf{x}_i = r\mathbf{j}$, we obtain that $\mathbf{x}_i = (r/s)\mathbf{j}$, for $i = 1, 2, \ldots, s$. Thus $\rho_3 m = (k - \rho_2) + (\rho_1 - \rho_2)a + \rho_2 m$ if and only if $\{\mathcal{B}_1, \mathcal{B}_2, \ldots, \mathcal{B}_s\}$ is a resolution of **D**. The proof is now complete. □

We next give a sufficient condition under which the extremal case of the above proposition is realized.

Theorem 8.6.6. *Let* $\mathbf{D}_1$ *be a quasi-symmetric* $(v_1, b_1, r_1, k_1, \lambda_1)$*-design with intersection numbers* ρ_1 *and* ρ_2 $(\rho_1 > \rho_2)$*. Let* $\mathbf{D}_2$ *be an affine resolvable* $(v_2, b_2, r_2, k_2, \lambda_2)$*-design with* $b_2 = r_2 v_1$*. Then there exists an* r_1*-resolvable* $(v_2, b_1 r_2, r_1 r_2, k_1 k_2, \lambda)$*-design* $\mathbf{D}$ *with* $\lambda = r_1\lambda_2 + \lambda_1(r_2 - \lambda_2)$*. Furthermore,* $\mathbf{D}$ *admits a nearly affine* (ρ_1, ρ_2, ρ_3)*-decomposition with* $\rho_3 = \frac{k^2}{v}$*.*

Proof. The existence of the required design $\mathbf{D}$ follows from Theorem 5.3.10, and it only remains to show that $\mathbf{D}$ has the desired intersection numbers. Let M be an incidence matrix of $\mathbf{D}_1$. Then $MM^\top = (r_1\lambda_1)I_{v_1} + \lambda_1 J_{v_1}$. Since $\mathbf{D}_1$ is quasi-symmetric with intersection numbers ρ_1 and ρ_2, letting A denote the adjacency matrix of the block graph of $\mathbf{D}_1$, we obtain $M^\top M = (k_1\rho_2)I_{b_1} + (\rho_1 - \rho_2)A + \rho_2 J_{b_1}$.

Let $N = [N_1\ N_2 \ldots N_{r_2}]$ be an incidence matrix of $\mathbf{D}_2$, where each N_i is the incidence matrix of a resolution class. Then Theorem 5.3.10 implies that, $P = [N_1 M\ N_2 M \ldots N_{r_2} M]$ is an incidence matrix of an r_1-resolvable 2-design $\mathbf{D}$ having the desired parameters. Now to prove that any two distinct blocks of $\mathbf{D}$ intersect in either $k_2\rho_1$, $k_2\rho_2$, or $\rho_3 = \frac{k^2}{v}$ points, we show that P has these column inner products. We have

$$P^\top P = [N_1 M\ N_2 M, \ldots, N_{r_2} M]^\top [N_1 M\ N_2 M \ldots N_{r_2} M].$$

Observe that $(N_i M)^\top (N_j M) = (M^\top N_i^\top)(N_j M) = M^\top (N_i^\top N_j) M$. If $i = j$, then using that $\mathbf{D}_1$ is quasi-symmetric, we get $(N_i M)^\top (N_i M) = (k - k_2\rho_2)I_{b_1} + (k_2\rho_1 - k_2\rho_2)A + k_2\rho_2 J_{b_1}$.

Next for $i \neq j$, $(N_i M)^\top (N_j M) = \frac{k^2}{v_2} J_{b_1} = \frac{k^2}{v} J_{b_1}$. This proves that $\mathbf{D}$ has at most three intersection numbers $k_2\rho_1, k_2\rho_2$, or $\rho_3 = \frac{k^2}{v}$ and admits a nearly affine decomposition. □

Corollary 8.6.7. *Let* q *be a prime power. Suppose there exists a* 2-$(q, l, 1)$*-design. Then, for any integer* $d \geq 2$ *there exists a* (v, b, r, k, λ)*-design with*

$$v = q^d,\ b = \frac{q(q^d - 1)}{l(l-1)},\ r = \frac{q^d - 1}{l - 1},\ k = lq^{d-1},\ \lambda = \frac{lq^{d-1} - 1}{l - 1} \tag{8.36}$$

admitting a nearly affine $(q^{d-1}, 0, l^2 q^{d-2})$*-decomposition.*

Proof. In the previous theorem, take $\mathbf{D}_1$ to be a 2-$(q, l, 1)$ design. Then $\mathbf{D}_1$ has parameters $v_1 = q, b_1 = \frac{q(q-1)}{l(l-1)}, r_1 = \frac{q-1}{l-1}, k_1 = l, \lambda_1 = 1$ with intersection numbers $\rho_1 = 1, \rho_2 = 0$. Take $\mathbf{D}_2$ to be the design $AG_{d-1}(d, q)$. Then $\mathbf{D}_2$ is an affine resolvable design with parameters $v_2 = q^d, b_2 = \frac{q(q^d-1)}{q-1}, r_2 = \frac{q^d-1}{q-1}, k_2 = q^{d-1}, \lambda_2 = \frac{q^{d-1}-1}{q-1}$. Theorem 5.3.10 yields an $r_1 = \frac{q-1}{l-1}$-resolvable design with parameters (8.36). Also by Theorem 8.6.6, **D** has intersection numbers $k_2\rho_1 = q^{d-1}$, $k_2\rho_2 = 0$, and $\frac{k^2}{v} = l^2 q^{d-2}$ and admits a nearly affine decomposition with respect to these intersection numbers. □

8.7. A symmetric (71, 15, 3)-design

In this section we first construct a quasi-residual 2-(56, 12, 3) design **D** admitting a regular M-partition with class size 2. Then we construct a 2-(15, 3, 1) design **E** whose block graph is isomorphic to the complement of the class graph of **D**. Finally, we combine the design **D** and the two-fold multiple of the design **E** into a symmetric (71, 15, 3)-design.

The polynomial $p(x) = x^3 + x + 1$ is irreducible over the field $GF(2)$. We will adjoin a root θ of this polynomial to $GF(2)$ to obtain the field $F = GF(2)(\theta)$ of cardinality 8. Thus, $F = \{a + b\theta + c\theta^2 : a, b, c \in GF(2)\}$ with $\theta^3 = 1 + \theta$. This implies $\theta^4 = \theta + \theta^2, \theta^5 = 1 + \theta + \theta^2, \theta^6 = 1 + \theta^2$, and $\theta^7 = 1$.

Let X be the set of all ordered pairs (a, b) with $a, b \in F$ and $a \neq b$. Let $\mathcal{B}$ be the set of all 4-subsets of F. For any $B \in \mathcal{B}$, we denote by $s(B)$ the sum of the four elements of B. We have $|X| = 56$ and $|\mathcal{B}| = 70$. We will consider an incidence structure $\mathbf{D} = (X, \mathcal{B}, I)$ with the incidence relation I to be defined later.

Let L be the group of nonsingular linear maps $f : F \to F$, i.e., the maps of the form $f(x) = cx + d$ with $c, d \in F$ and $c \neq 0$. The group L is of order 56. If $c = 1$ and $d \neq 0$, then $f(f(x)) = x$, i.e., the order of $f(x) = x + d$, as an element of L, is 2; if $c \neq 1$, then the order of $f(x) = cx + d$ is 7. The group L acts on both the point set X and the block set $\mathcal{B}$ of **D**. It is sharply transitive on X, i.e., for any $(a_1, b_1), (a_2, b_2) \in X$ there is a unique $f \in L$ with $f(a_1) = a_2$ and $f(b_1) = b_2$. If $f(x) = cx + d \in L$, then, for $B \in \mathcal{B}$, $s(f(B)) = cs(B)$. The action of L on $\mathcal{B}$ is described in the following lemma.

Lemma 8.7.1. *For $A, B \in \mathcal{B}$, if $s(A) \neq 0$ and $s(B) \neq 0$, then there is a unique $f \in L$ such that $f(A) = B$; if $s(A) = s(B) = 0$, then there are exactly four functions $f \in L$ with $f(A) = B$.*

Proof. Let $A, B \in \mathcal{B}$. The set of all functions $f \in L$ such that $f(A) = B$ is a right coset of the group $H = \{g \in L : g(B) = B\}$ (the stabilizer of B), so it suffices to determine $|H|$. Since, for distinct $f, g \in L$, the equation $f(x) = g(x)$ has at most one solution, different elements of H act as different permutations of the set B. Therefore, we may regard H as a subgroup of the symmetric group S_4. This implies that H has no element of order 7. If $g(x) = x + d \in H$, $d \neq 0$, then $B = \{a, a+d, b, b+d\}$ and therefore $s(B) = 0$ and d is the sum of two elements of B. Therefore, if $s(B) \neq 0$, then the group H is trivial. If $s(B) = 0$, then $B = \{a, b, c, a+b+c\}$ and there are exactly three nonidentity elements in H: $g_1(x) = x + a + b$, $g_2(x) = x + a + c$, and $g_3(x) = x + b + c$, i.e., $|H| = 4$. □

Corollary 8.7.2. *The sets* $\mathcal{B}_0 = \{B \in \mathcal{B} \colon s(B) = 0\}$ *and* $\mathcal{B}_1 = \{B \in \mathcal{B} \colon s(B) \neq 0\}$ *are the two orbits of* $\mathcal{B}$ *with respect to the action of* L. *Furthermore,* $|\mathcal{B}_0| = 14$ *and* $|\mathcal{B}_1| = 56$.

Proof. Since the cardinality of the stabilizer of B is 4 if $B \in \mathcal{B}_0$ and it is 1 if $B \in \mathcal{B}_1$, each block of $\mathcal{B}_0$ lies in an orbit of cardinality 14 and each block of $\mathcal{B}_1$ lies in an orbit of cardinality 56. Since $14 + 56 = |\mathcal{B}|$, there is only one orbit of each size. □

The automorphism group of the field F is the cyclic group of order 3 generated by the Frobenius automorphism $\sigma(x) = x^2$. For any block B, we let $B^{(1)} = B$, $B^{(2)} = \sigma(B)$, and $B^{(4)} = (B^{(2)})^{(2)}$. Then $(B^{(4)})^{(2)} = B$. For $f(x) = cx + d \in L$, let $f^{(1)} = f$, $f^{(2)}(x) = c^2 x + d^2$, and $f^{(4)} = (f^{(2)})^{(2)}$. Then $(f^{(4)})^{(2)} = f$. Note that $\sigma(B^{(2)}) = (\sigma(B))^{(2)}$ and $f^{(2)}(B^{(2)}) = (f(B))^{(2)}$.

We are now ready to define the incidence relation I.

Let $A_0 = \{0, 1, \theta, \theta^3\}$, $A_1 = \{0, 1, \theta, \theta^2\}$, $A_2 = \{0, 1, \theta^2, \theta^3\}$, $A_3 = \{\theta, \theta^2, \theta^3, \theta^4\}$, $A_4 = \{\theta^2, \theta^3, \theta^4, \theta^5\}$. We declare the point $(0, 1) \in X$ incident with the 15 blocks that form the set

$$\mathcal{A} = \{A_i^{(j)} \colon i = 0, 1, 2, 3, 4;\ j = 1, 2, 4\}.$$

Observe that $A_0, A_0^{(2)}, A_0^{(4)} \in \mathcal{B}_0$ and that every other element of $\mathcal{A}$ is in $\mathcal{B}_1$. For any point $(a, b) \in X$, we declare (a, b) incident with the 15 blocks that form the set $\{f(A) \colon A \in \mathcal{A}\}$ where f is a unique element of L such that $f(0) = a$ and $f(1) = b$.

Lemma 8.7.3. *Each block* $B \in \mathcal{B}$ *is incident with exactly 12 points* $(a, b) \in X$.

Proof. The number of points a block B is incident with is equal to the number of pairs (f, A) where $f \in L$, $A \in \mathcal{A}$, and $f(A) = B$. The set $\mathcal{A}$ has three elements of $\mathcal{B}_0$ and 12 elements of $\mathcal{B}_1$. If $B \in \mathcal{B}_0$, then $f(A) = B$ implies that $A \in \mathcal{B}_0$, and Lemma 8.7.1 implies that there are exactly 12 pairs (f, A), where $f \in L$, $A \in \mathcal{A} \cap \mathcal{B}_0$, and $f(A) = B$. If $B \in \mathcal{B}_1$, then $f(A) = B$ implies $A \in \mathcal{B}_1$, and we again apply Lemma 8.7.1 to obtain that B is incident with exactly 12 points. □

Lemma 8.7.4. *Let $(a, b) \in X$ and let B and $\overline{B}$ be complementary 4-subsets of F. Then*

(i) *if the 2-set $\{a, b\}$ is contained in B, then (a, b) is incident with one and only one of the blocks B and $\overline{B}$;*
(ii) *if (a, b) is incident with B, then $\{a, b\} \subset B$ or $\{a, b\} \subset \overline{B}$;*
(iii) *the blocks B and $\overline{B}$ are disjoint, i.e., there is no point incident with both B and $\overline{B}$.*

Proof. (i) The set F has fifteen 4-subsets that contain the 2-set $\{a, b\}$. Their complements are the fifteen 4-subsets that are disjoint from $\{a, b\}$. Observe that each element of $\mathcal{A}$ either contains $\{0, 1\}$ or is disjoint from this 2-set. Since $|\mathcal{A}| = 15$, we conclude that if a block A contains $\{0, 1\}$, then either A or $\overline{A}$ is incident with $(0, 1)$. Let $f \in L$ be such that $f(a) = 0$ and $f(b) = 1$. If $\{a, b\} \subset B$, then $\{0, 1\} \subset f(B)$. If $(0, 1)$ is incident with $f(B)$, then (a, b) is incident with B. If $(0, 1)$ is not incident with $f(B)$, then $(0, 1)$ is incident with $f(\overline{B})$ and then (a, b) is incident with $\overline{B}$.

(ii) and (iii). There are 24 points $(x, y) \in X$ such that $\{x, y\} \subset B$ or $\{x, y\} \subset \overline{B}$. By (i), each of these points is incident with B or $\overline{B}$. Lemma 8.7.3 now implies that blocks B and $\overline{B}$ are disjoint and contain no other points. Therefore, (ii) and (iii) follow. □

Theorem 8.7.5. *The incidence structure $\mathbf{D} = (X, \mathcal{B}, I)$ is a quasi-residual 2-(56, 12, 3) design with three intersection numbers, 0, 2, and 3. Furthermore, $\mathbf{D}$ admits a regular M-partition with class size 2.*

Proof. The structure $\mathbf{D}$ has 56 points and 70 blocks of size 12 (Lemma 8.7.3). Let (a_1, b_1) and (a_2, b_2) be distinct points of $\mathbf{D}$ and let $f_1 \in L$ be such that $f_1(a_1) = 0$ and $f_1(b_1) = 1$. Let $f_1(a_2) = a_0$ and $f_1(b_2) = b_0$. A block $B \in \mathcal{B}$ is incident with both (a_1, b_1) and (a_2, b_2) if and only if $A = f_1(B) \in \mathcal{A}$ and A is incident with (a_0, b_0). Let $f_0 \in L$ be such that $f_0(a_0) = 0$ and $f_0(b_0) = 1$. Then A is incident with (a_0, b_0) if and only if $f_0(A) \in \mathcal{A}$. Therefore, it suffices to show that any nonidentity element $g \in L$ satisfies the following property:

$$\text{there are exactly three blocks } A \in \mathcal{A} \text{ such that } g(A) \in \mathcal{A}. \tag{8.37}$$

Note that $g(A) \in \mathcal{A}$ if and only if $g^{(2)}(A^{(2)}) \in \mathcal{A}$. Therefore, if g satisfies (8.37), then so does $g^{(2)}$ (and $g^{(4)}$). Also, if g satisfies (8.37), then so does g^{-1}. If $g(x) = cx + d$, then the leading coefficients of $g^{(2)}$, $g^{(4)}$, g^{-1}, $(g^{-1})^{(2)}$, and $(g^{-1})^{(4)}$ are c^2, c^4, $c^{-1} = c^6$, c^5, and c^3, respectively. The free terms of $g^{(2)}$ and $g^{(4)}$ are d^2 and d^4. Therefore, it suffices to verify $(*)$ for the following 11 functions: $g_0(x) = x + 1$, $g_1(x) = x + \theta$, $g_2(x) = x + \theta^3$, and, for each $d \in F$, for the function $h_d(x) = \theta x + d$. This can be done by direct inspection of the set $\mathcal{A}$.

Thus, **D** is a 2-(56, 12, 3) design. Since its replication number is 15, **D** is quasi-residual. For any block B of **D**, the block $\overline{B}$, formed by the complementary subset of F, is disjoint from B (Lemma 8.7.4(iii)). Therefore, the block set of **D** admits a partition into classes $\{B, \overline{B}\}$ of disjoint blocks. Theorem 8.5.9 now implies that the intersection numbers of **D** are 0, 2, and 3 and the partition of $\mathcal{B}$ into sets $\{B, \overline{B}\}$ is the M-partition. □

For the remainder of the section, we will call the 2-(56, 12, 3) design of Theorem 8.7.5 the *BH-design*.

Lemma 8.7.6. *The blocks A and B of the BH-design **D** meet in two points if and only if the cardinality of the intersection of 4-subsets A and B of F is 2.*

Proof. Let A and B be blocks of **D**. We may assume that $A \neq B$ and, due to Lemma 8.7.4(iii) that $A \cup B \neq F$. Let $\overline{A} = F \setminus A$ and $\overline{B} = F \setminus B$. Then $\{A, \overline{A}\}$ and $\{B, \overline{B}\}$ are two classes of the M-partition of **D**. Suppose blocks A and B meet in α points. By Theorem 8.5.4, blocks A and $\overline{B}$ meet in α points as well as blocks $\overline{A}$ and B and blocks $\overline{A}$ and $\overline{B}$. Since A and $\overline{A}$ are disjoint blocks as well as B and $\overline{B}$, there are exactly 4α points that are incident with one of the blocks A and $\overline{A}$ and one of the blocks B and $\overline{B}$. Lemma 8.7.4 now implies that the set F has exactly 2α subsets of cardinality 2, each of which is contained in one of the sets $A, \overline{A}$ and in one of the sets $B, \overline{B}$.

If the sets A and B meet in two points, then so do the sets A and $\overline{B}$, $\overline{A}$ and B, and $\overline{A}$ and $\overline{B}$. Therefore, in this case, $2\alpha = 4$ and $\alpha = 2$. If the sets A and B meet in one point, then the sets $\overline{A}$ and $\overline{B}$ meet in one point, while the sets A and $\overline{B}$ as well as the sets $\overline{A}$ and B meet in three points. In this case, $2\alpha = 6$ and $\alpha = 3$. Similarly, if the sets A and B meet in three points, we obtain $\alpha = 3$. Thus, $\alpha = 2$ if and only if the cardinality of the subset $A \cap B$ of F is 2. □

Let Γ be the class graph of the BH-design **D**. Each vertex of Γ (and of the complementary graph $\overline{\Gamma}$) is an unordered pair $\{A, \overline{A}\}$ of complementary 4-subsets of F. Lemma 8.7.6 implies that distinct vertices $\{A, \overline{A}\}$ and $\{B, \overline{B}\}$ of $\overline{\Gamma}$ are adjacent if and only if $|A \cap B| = 2$ (and then $|A \cap \overline{B}| = |\overline{A} \cap B| = |\overline{A} \cap \overline{B}| = 2$). Note that this description of $\overline{\Gamma}$ does not make any use of the field structure of F. In fact, F can be any set of cardinality 8.

In order to construct a quasi-symmetric 2-(15, 3, 1) design **E** with the block graph $\overline{\Gamma}$, we now assume that F is a fixed block of the Witt design $\mathbf{W_{24}}$. Let a

be a fixed point of $\mathbf{W}_{24}$ that does not lie in the block F. Let the point set Y of $\mathbf{E}$ be the set of all points of $\mathbf{W}_{24}$, other than a, that do not lie in F. Then $|Y| = 15$. Let the block set of $\mathbf{E}$ be the vertex set V of $\overline{\Gamma}$. We define a point $y \in Y$ and a block $\{A, \overline{A}\} \in V$ to be incident if and only if there is a block G of $\mathbf{W}_{24}$ such that $a \in G$, $y \in G$, and $F \cap G$ is equal to A or $\overline{A}$.

Theorem 8.7.7. *The design* $\mathbf{E}$ *is a* 2-(15, 3, 1) *design whose block graph is* $\overline{\Gamma}$.

Proof. Let $\{A, \overline{A}\} \in V$. Since $\mathbf{W}_{24}$ is a 5-(24, 8, 1) design, it has a unique block G that contains $A \cup \{a\}$. By Corollary 6.5.9, $\overline{G} = G \triangle F$ is a block of $\mathbf{W}_{24}$, and it is the unique block that contains $\overline{A} \cup \{a\}$. Since $G \cap Y = \overline{G} \cap Y$, the 3-set $G \cap Y$ is the set of all points of $\mathbf{E}$ that are incident with the block $\{A, \overline{A}\}$. Thus, the block size of $\mathbf{E}$ is 3. Let y and z be distinct points of $\mathbf{E}$. For $i = 0, 2, 4$, let m_i denote the number of blocks of $\mathbf{W}_{24}$ that contain a, y, and z and meet F in i points. Counting in two ways pairs (H, Z) where Z is a 1- or 2-subset of F and H is a block of $\mathbf{W}_{24}$ that contains a, y, z, and Z, yields the following two equations:

$$2m_2 + 4m_4 = 8 \cdot 5 = 40,$$
$$m_2 + 6m_4 = 28.$$

These equations give $m_4 = 2$. If G is a block of $\mathbf{W}_{24}$ that contains a, y, and z and meets F in four points, then $\overline{G}$ is the other such block. Therefore, for $A = G \cap F$, the block $\{A, \overline{A}\}$ is the only block of E that is incident with both y and z. Thus $\mathbf{E}$ is a 2-(15, 3, 1) design. It is quasi-symmetric with intersection numbers 0 and 1. The block graph of $\mathbf{E}$ has vertex set V. We have to show that distinct vertices $\{A, \overline{A}\}$ and $\{B, \overline{B}\}$ are adjacent in $\overline{\Gamma}$ if and only if they are not disjoint as blocks of $\mathbf{E}$.

If $\{A, \overline{A}\}$ and $\{B, \overline{B}\}$ are adjacent vertices of $\overline{\Gamma}$, then $|A \cap B| = 2$. Therefore, if G is the block of $\mathbf{W}_{24}$ that contains $A \cup \{a\}$ and H is the block of $\mathbf{W}_{24}$ that contains $B \cup \{a\}$, then $G \cap H$ contains two points of F and point a and therefore exactly one point of Y. Conversely, suppose distinct blocks $\{A, \overline{A}\}$ and $\{B, \overline{B}\}$ of $\mathbf{E}$ are incident with a point $y \in Y$, and let G and H be the blocks of $\mathbf{W}_{24}$ that contain $A \cup \{a, y\}$ and $B \cup \{a, y\}$, respectively. Since no two blocks of $\mathbf{E}$ meet in more than one point, we obtain that $G \cap H \cap Y = \{y\}$. Since $a, y \in G \cap H$ and $A \cap B \neq \varnothing$, we obtain that $|G \cap H| > 2$. Therefore, $|G \cap H| = 4$ and then $|A \cap B| = 2$, i.e., $\{A, \overline{A}\}$ and $\{B, \overline{B}\}$ are adjacent vertices of $\overline{\Gamma}$. The proof is now complete. □

We will combine the BH-design $\mathbf{D}$ and the design $\mathbf{E}$ from the above theorem into a symmetric (71, 15, 3)-design.

Theorem 8.7.8. *Let* **D** *be the BH-design and let* Γ *be its class graph. Let* **E** *be a* 2-(15, 3, 1) *design whose block graph is* $\overline{\Gamma}$. *Then there exists a symmetric* (71, 15, 3)*-design* **S** *and a block* Y *of* **S** *such that* $\mathbf{S}^Y$ *is isomorphic to* **D** *and* $\mathbf{S}_Y$ *is isomorphic to the two-fold multiple of* **E**.

Proof. Let $\mathbf{D} = (X, \mathcal{A})$ and $\mathbf{E} = (Y, \mathcal{B})$. We assume that the point sets X and Y are disjoint. Let $\mathfrak{M}$ be the M-partition of **D**. Since the block graph of **E** is $\overline{\Gamma}$, there exists a bijection $\varphi : \mathfrak{M} \to \mathcal{B}$ such that $\varphi(\mathcal{M}_1)$ and $\varphi(\mathcal{M}_2)$ are disjoint blocks of **E** if and only if each block of $\mathcal{M}_1$ meets each block of $\mathcal{M}_2$ in three points. The blocks of **S** are all sets $A \cup \varphi(\mathcal{M})$ where $\mathcal{M} \in \mathfrak{M}$ and $A \in \mathcal{M}$ and the set Y.

The incidence structure **S** has 71 points and 71 blocks. All blocks of **S** are of cardinality 15, so it suffices to prove that any two distinct blocks of **S** meet in three points.

Let A_1 and A_2 be distinct blocks of **D** and let $\mathcal{M}_1$ and $\mathcal{M}_2$ be their respective M-partition classes.

If $\mathcal{M}_1 = \mathcal{M}_2$, then $|A_1 \cap A_2| = 0$ and $|\varphi(\mathcal{M}_1) \cap \varphi(\mathcal{M}_2)| = 3$.

If $\mathcal{M}_1$ and $\mathcal{M}_2$ represent adjacent vertices of Γ, then $|A \cap B| = 3$ and $|\varphi(\mathcal{M}_1) \cap \varphi(\mathcal{M}_2)| = 0$.

If $\mathcal{M}_1$ and $\mathcal{M}_2$ represent distinct nonadjacent vertices of Γ, then $|A \cap B| = 2$ and $|\varphi(\mathcal{M}_1) \cap \varphi(\mathcal{M}_2)| = 1$.

Therefore, in each case $|(A_1 \cup \varphi(\mathcal{M}_1)) \cap (A_2 \cup \varphi(\mathcal{M}_2))| = 3$.

Finally, since sets X and Y are disjoint, the intersection of Y and any other block of **S** is a block of **E** and therefore, the cardinality of the intersection is 3.

Since each M-partition class of **D** consists of two blocks, the derived design $\mathbf{S}_Y$ is the two-fold multiple of **E**. The residual design $\mathbf{S}^Y$ is **D**. □

The above construction can be generalized to a necessary and sufficient condition for a quasi-residual design with a regular M-partition to be embeddable.

Theorem 8.7.9. *Let* **D** *be a quasi-residual* 2-(v, k, λ) *design that admits a regular* M*-partition with class size* m. *Let* Γ *be the class graph of* **D**. *The design* **D** *is embeddable if and only if there exists a quasi-symmetric* 2-$(k + \lambda, \lambda, (\lambda - 1)/m)$ *design whose block graph is isomorphic to the complement* $\overline{\Gamma}$ *of* Γ.

We will give only a sketch of a proof of this result. It is straightforward to verify that if **D** is embeddable in a symmetric $(v + k + \lambda, k + \lambda, \lambda)$-design, then the corresponding derived design is an m-fold multiple of a quasi-symmetric 2-$(k + \lambda, \lambda, (\lambda - 1)/m)$ design whose block graph is isomorphic $\overline{\Gamma}$. The crucial part in proving the converse is to show that if α_1, α_2, and $k - n$ are the intersection numbers of **D**, then $\lambda - \alpha_1$ and $\lambda - \alpha_2$ are the intersection numbers of **E**. It can be done by expressing the eigenvalues of $\overline{\Gamma}$ in terms of each set of intersection numbers and comparing these expressions.

Exercises

(1) Find the parameters of the Hamming scheme.
(2) Find the parameters of the q-analog of the Johnson scheme.
(3) Verify that the cyclotomic scheme is an association scheme and find its parameters.
(4) Show that the design $PG_{d-2}(d, q)$ is quasi-symmetric and find the parameters of its block graph.
(5) Show that, for $k = 2$, there is no group G satisfying the conditions of Theorem 8.2.26. However, the graphs obtained in the conclusion of this theorem exist.
(6) Prove that if a quasi-symmetric 2-(v, k, λ) design has intersection numbers 0 and 1, then $\lambda = 1$.
(7) Let $\mathbf{D} = (X, \mathcal{B})$ be a quasi-symmetric (v, b, r, k, λ)-design with intersection numbers 0 and α. For each block B of **D** let $\mathcal{C}(B)$ be the set of all blocks of **D** that are disjoint from B and let $\mathbf{E}(B)$ be the substructure $\mathbf{E}(B) = (X \setminus B, \mathcal{C}(B))$ of **D**.
    (a) Prove that $\mathbf{E}(B)$ is a 1-$(v - k, k, r - k\lambda/\alpha)$ design.
    (b) Suppose $\mathbf{E}(B)$ is a 2-$(v - k, k, \mu)$ design. Prove that

$$\lambda - \mu = \frac{k\lambda(\alpha - 1)}{\alpha(k - 1)} = \frac{k\lambda(k - \alpha - 1)}{\alpha(v - k - 1)}$$

    and then derive that $r = v - 1$.
    (c) Prove that **D** is an extension of a symmetric design if and only if $\mathbf{E}(B)$ is a 2-design for every block B of **D**. *Hint*: one part of this result is contained in the proof of Cameron's Theorem.

Notes

The notion of association schemes was first introduced in Bose and Shimamoto (1952), though the concept had already appeared in Bose and Nair (1939) in the context of partially balanced designs. The equivalence of 2-class association schemes and strongly regular graphs was established in Bose (1963). The Bose–Mesner algebra was discovered independently in Bose (1955) and Mesner (1956) and then studied in the joint paper by Bose and Mesner (1959). In Delsarte (1973a) association schemes were used to unify certain aspects of design theory and coding theory. For further results and references on association schemes, see MacWilliams and Sloane (1977), Bannai and Ito (1984), Cameron and van Lint (1991), Godsil (1993, 1996), Brouwer and Haemers (1995), van Lint and Wilson (2001), Bailey (2003).

The concept of quasi-symmetric designs goes back to S. S. Shrikhande (1952). Theorem 8.2.14 is due to S. S. Shrikhande and Bhagwandas (1965) and to Goethals and Seidel (1970). The term *quasi-symmetric design* was proposed in Stanton and Kalbfleisch (1968). Cameron's Theorem characterizes quasi-symmetric 3-designs that can be obtained as extensions of symmetric designs. Another characterization of such 3-designs is given in Exercise 7 which follows Baartmans and M. S. Shrikhande (1985). The question of which strongly regular graphs can serve as block graphs of quasi-symmetric designs appears to be a very difficult problem. Some negative results are known. For instance, there is no quasi-symmetric design whose block graph is $L_2(n)$ or its complement (Goethals and Seidel (1970)). For a comprehensive reference on quasi-symmetric designs, see the monograph by M. S. Shrikhande and Sane (1991). For the latest table of known quasi-symmetric designs on 70 or fewer points, see M. S. Shrikhande (1996).

The Hall–Connor Theorem was first proven in Hall and Connor (1954). The presented proof is due to S. S. Shrikhande (1960). For Connor's Lemma, see Connor (1952). Symmetric $(1 + k(k-1)/2, k, 2)$-designs are called *biplanes*. In the previous chapters we constructed biplanes with block size k for $k = 3, 4, 5, 6$, and 11. In Chapter 9 we will describe biplanes with $k = 9$ and $k = 13$. Proposition 2.4.10 rules out all biplanes with $k \equiv 2$ or 3 (mod 4) such that $k - 2$ is not a square. In particular, it rules out biplanes with block size $k = 7, 10, 14$, and 15. The Bruck–Ryser–Chowla Theorem rules out infinitely many other biplanes including those with $k = 8$ and 12. The first undecided case is $k = 16$. For further discussion of biplanes see Cameron (1973, 1976). Theorem 8.2.22 is due to Singhi and S. S. Shrikhande (1973) for $\lambda = 3$ and to Bose, S. S. Shrikhande and Singhi (1976) for $\lambda > 3$.

For the existence results referred to in Remark 8.2.30, see Mathon and Rosa (1996).

The proof of Theorem 8.2.26 and the subsequent construction of regular Hadamard matrices of orders 100 and 196 is based on ideas proposed in Bose and S. S. Shrikhande (1971) and in Goethals and Seidel (1970).

Theorem 8.2.32 was first proved in Calderbank and Morton (1990). A shorter proof is in Pawale and Sane (1991). Theorem 8.2.33 is due to combined efforts of Ito (1975, 1978), Enomoto, Ito and Noda (1979), and Bremner (1979). It was conjectured in Sane and M.S. Shrikhande (1987) that there is no nontrivial quasi-symmetric 3-(v, k, λ) design with $v \leq 2k$ and both intersection numbers greater than 1. This conjecture is still open.

Theorem 8.3.3 was conjectured in Jungnickel and Tonchev (1991a). It was partially proven in Ionin and M. S. Shrikhande (1994b). A complete proof was obtained in Sane (2001). Theorem 8.3.6 is due to Jungnickel and Tonchev (1991a). Theorem 8.3.7 is due to Sane (2000).

The notion of quasi-3 symmetric designs was introduced (as quasi-3 for points) in Cameron (1973c) where Theorems 8.4.13 and 8.4.15 were proven. In the paper by Kantor (1975), the SDP-designs were introduced and Theorems 8.4.12 and 8.4.10 were proven. No other examples of symmetric quasi-3 designs, except those introduced in Section 8.4., are known. The smallest open parameter set is (144, 66, 30). In the paper by Cameron (1973c), the following question was raised: does the dual of a quasi-3 symmetric design have to be quasi-3? Example 8.4.14, which is due to Bracken (2004), settles this question negatively. For other results and problems on quasi-3 symmetric designs, see Broughton and McGuire (2003). Jungnickel and Tonchev (1992) considered quasi-symmetric design with the symmetric difference property. For connections between quasi-3 symmetric designs and certain spin models, see Guo and Huang (2001), Bannai and Sawano (2002), and Bracken and McGuire (2002, 2003). For connections between quasi-3 symmetric designs and codes, see Dillon and Schatz (1987), Jungnickel and Tonchev (1991b, 1992), McGuire and Ward (1998), and Broughton and McGuire (1999).

The term *block schematic designs* should be attributed to Cameron (1975). The Cameron–Delsarte Theorem is proven independently in Cameron (1973) and Delsarte (1973). The paper Cameron (1975) gives a necessary condition for a 3-$(v, k, 1)$ design to be block schematic.

Theorem 8.5.4 is due to Majumdar (1953). We introduced the term M-partition for partitions generated by this theorem to reflect both Majumdar's contribution and the term *maximal decomposition* for these partitions proposed in Beker and Haemers (1980). We will use the term *decomposition of symmetric designs* in subsequent chapters in a different sense. Beker and Haemers (1980) proved Theorem 8.5.13. Proposition 8.5.9 is due to Singhi and S.S. Shrikhande (1974).

The results of Section 8.6. are due to Ionin and M. S. Shrikhande (2000).

We follow Beker and Haemers (1980) in the construction of a symmetric (71, 15, 3)-design in Section 8.7. The description of a 2-(15, 3, 1) design in Theorem 8.7.7 is due to Bussemaker and Seidel (1970). Another description of the same design is given in Conwell (1910).

9
Difference sets

If the action of an automorphism group of a symmetric design on the block set is known, then the design can be constructed by finding one (base) block from each block orbit and then applying the automorphism group to obtain the remaining blocks. If a symmetric design admits an automorphism group such that all blocks of the design form a single orbit, then the group itself can be regarded as the point set of the design. The base block becomes a subset of the group and such subsets are called *difference sets*. The designs obtained from difference sets admit *group invariant* incidence matrices.

Group rings are a natural setting for investigating difference sets. The notion of a *group of symmetries* of a subset of a group ring will be crucial to constructing symmetric designs in subsequent chapters.

9.1. Group invariant matrices and group rings

A *group invariant matrix* is a matrix of order v whose columns can be obtained from the first column by applying all elements of a certain permutation group of order v to the entries of the first column. If we assume that the rows and columns of the matrix are indexed by the elements of the group, then this description leads us to the following definition.

Definition 9.1.1. Let G be a group of order v. A matrix M of order v over a ring R is said to be G-*invariant* if there exists a bijection $\varphi\colon G \to \{1, 2, \ldots, v\}$ such that, for all $x, y, z \in G$, the $(\varphi(x), \varphi(y))$-entry of M is equal to the $(\varphi(xz), \varphi(yz))$-entry. If $\varphi(x) = i$ and $\varphi(y) = j$, we will denote by $M(x, y)$ the (i, j)-entry of M. We will say that the matrix $M = [M(x, y)]$ is indexed by the elements of G.

The bijection φ can be regarded as an ordering of the set of elements of G, and vice versa.

Example 9.1.2. Let $G = \{1, x, x^2, \ldots, x^{v-1}\}$ be a cyclic group of order v and let $\varphi(x^k) = k + 1$ for $k = 0, 1, \ldots, v - 1$. The corresponding G-invariant matrix $C = [c_{ij}]$ satisfies the following property: $c_{i+1,j+1} = c_{ij}$ and $c_{i+1,1} = c_{iv}$ for $i, j = 1, 2, \ldots, v - 1$. Such matrices are called *circulant*.

Remark 9.1.3. If G is a cyclic group of order v with a fixed generator x, we will always assume that the elements of G are ordered in the following natural way: $G = \{1, x, x^2, \ldots, x^{v-1}\}$.

Example 9.1.4. Let $G = \langle x, y | x^4 = y^4 = 1, xy = yx\rangle$ be the direct product of two cyclic groups of order 4. We will assume that an element $x^a y^b$ of G precedes $x^c y^d$ $(a, b, c, d \in \{0, 1, 2, 3\})$ if and only if $b < d$ or $b = d$ and $a < c$. The following $(0, 1)$-matrix N of order 16 is G-invariant:

$$N = \begin{bmatrix}
0&0&0&0 & 1&1&0&0 & 1&0&1&0 & 1&0&0&1 \\
0&0&0&0 & 0&1&1&0 & 0&1&0&1 & 1&1&0&0 \\
0&0&0&0 & 0&0&1&1 & 1&0&1&0 & 0&1&1&0 \\
0&0&0&0 & 1&0&0&1 & 0&1&0&1 & 0&0&1&1 \\[4pt]
1&0&0&1 & 0&0&0&0 & 1&1&0&0 & 1&0&1&0 \\
1&1&0&0 & 0&0&0&0 & 0&1&1&0 & 0&1&0&1 \\
0&1&1&0 & 0&0&0&0 & 0&0&1&1 & 1&0&1&0 \\
0&0&1&1 & 0&0&0&0 & 1&0&0&1 & 0&1&0&1 \\[4pt]
1&0&1&0 & 1&0&0&1 & 0&0&0&0 & 1&1&0&0 \\
0&1&0&1 & 1&1&0&0 & 0&0&0&0 & 0&1&1&0 \\
1&0&1&0 & 0&1&1&0 & 0&0&0&0 & 0&0&1&1 \\
0&1&0&1 & 0&0&1&1 & 0&0&0&0 & 1&0&0&1 \\[4pt]
1&1&0&0 & 1&0&1&0 & 1&0&0&1 & 0&0&0&0 \\
0&1&1&0 & 0&1&0&1 & 1&1&0&0 & 0&0&0&0 \\
0&0&1&1 & 1&0&1&0 & 0&1&1&0 & 0&0&0&0 \\
1&0&0&1 & 0&1&0&1 & 0&0&1&1 & 0&0&0&0
\end{bmatrix}$$

This matrix can be represented as a 4×4 block matrix with blocks of order 4, and this block matrix is circulant. In this case, we say that the matrix N is *block circulant.*

If an incidence structure $\mathbf{D} = (X, \mathcal{B}, I)$ admits a group invariant incidence matrix, then $|X| = |\mathcal{B}|$. The next proposition characterizes incidence structures with group invariant incidence matrices.

Proposition 9.1.5. *Let G be a group of order v and let $\mathbf{D} = (X, \mathcal{B}, I)$ be an incidence structure with $|X| = |\mathcal{B}| = v$. Then $\mathbf{D}$ admits a G-invariant incidence matrix if and only if $\mathbf{D}$ has a sharply transitive automorphism group isomorphic to G.*

Proof. 1. Suppose $\mathbf{D}$ admits a G-invariant incidence matrix N. For any $\tau \in G$, let x_τ and B_τ be, respectively, the point and block of $\mathbf{D}$ corresponding to the row and column of N indexed by τ. For each $\sigma \in G$, define a bijection $f_\sigma : X \to X$

by $f_\sigma(x_\tau) = x_{\tau\sigma^{-1}}$. Since N is G-invariant, we have, for all $\rho, \sigma, \tau \in G, x_{\tau\sigma^{-1}} \in B_{\rho\sigma^{-1}}$ if and only if $x_\tau \in B_\rho$. Thus, $f_\sigma(B_\rho) = B_{\rho\sigma^{-1}}$ and therefore, f_σ is an automorphism of **D**. Since, for all $\rho, \sigma, \tau \in G$, $f_{\rho\sigma}(x_\tau) = f_\rho(f_\sigma(x_\tau))$, the set $\{f_\sigma : \sigma \in G\}$ is a sharply transitive group of automorphisms of **D** isomorphic to G.

2. Suppose the group G acts on X as a sharply transitive group of automorphisms of **D**. Fix $x_0 \in X$ and $B_0 \in \mathcal{B}$ and let $x_\sigma = \sigma^{-1}(x_0)$ and $B_\sigma = \sigma^{-1}(B_0)$ for each $\sigma \in G$. Then, for all $\rho, \sigma, \tau \in G$,

$$x_{\sigma\tau} \in B_{\rho\tau} \Leftrightarrow \tau^{-1}(x_\sigma) \in \tau^{-1}(B_\rho) \Leftrightarrow x_\sigma \in B_\rho,$$

and therefore the corresponding incidence matrix of **D** is G-invariant. □

The Kronecker product of group invariant matrices is group invariant.

Proposition 9.1.6. *Let G_1 and G_2 be finite groups. If A is a G_1-invariant matrix and B is a G_2-invariant matrix, then $A \otimes B$ is a $(G_1 \times G_2)$-invariant matrix.*

Proof. Every row R of $A \otimes B$ is obtained by multiplying elements of a row R_1 of A by elements of a row R_2 of B. If $x_1 \in G_1$ is the index of R_1 and $x_2 \in G_2$ is the index of R_2, then the element (x_1, x_2) of $G_1 \times G_2$ can be regarded as the index of R. In a similar way, we index each column of $A \otimes B$ by an element of $G_1 \times G_2$. This indexation has the following property:

$$(A \otimes B)((x_1, x_2), (y_1, y_2)) = A(x_1, y_1)B(x_2, y_2).$$

Therefore, we have, for all $(x_1, x_2), (y_1, y_2), (z_1, z_2) \in G_1 \times G_2$,

$$\begin{aligned}
&(A \otimes B)((x_1, x_2)(z_1, z_2), (y_1, y_2)(z_1, z_2)) \\
&\quad = (A \otimes B)((x_1z_1, x_2z_2), (y_1z_1, y_2z_2)) \\
&\quad = A(x_1z_1, y_1z_1)B(x_2z_2, y_2z_2) = A(x_1, y_1)B(x_2, y_2) \\
&\quad = (A \otimes B)((x_1, x_2), (y_1, y_2)).
\end{aligned}$$

Therefore, $A \otimes B$ is $(G_1 \times G_2)$-invariant □

All G-invariant matrices over a ring R form a ring.

Proposition 9.1.7. *Let G be a group of order v and let R be a ring. The set of all G-invariant matrices over R is a subring of the ring of matrices of order v over R.*

Proof. Let A and B be G-invariant matrices over R and let $x, y, z \in G$. Then $(A-B)(xz, yz) = A(xz, yz) - B(xz, yz) = A(x, y) - B(x, y) = (A - B)(x, y)$ and

$$(AB)(xz, yz) = \sum_{u \in G} A(xz, u)B(u, yz) = \sum_{u \in G} A(x, uz^{-1})B(uz^{-1}, y) = (AB)(x, y),$$

so $A - B$ and AB are G-invariant. □

A group invariant matrix is completely determined by its first row or column. A convenient algebraic interpretation of this property can be obtained with the use of *group rings*.

Definition 9.1.8. Let G be a finite group and let R be a ring with unity. Then RG is the ring (called the *group ring of G over R*) whose elements are formal sums $\sum_{x \in G} a_x x$ with $a_x \in R$. The operations of addition and multiplication on RG are defined as follows:

$$\sum_{x \in G} a_x x + \sum_{x \in G} b_x x = \sum_{x \in G} (a_x + b_x)x,$$

$$\sum_{x \in G} a_x x \cdot \sum_{y \in G} b_y y = \sum_{z \in G} \sum_{xy = z} (a_x b_y)z.$$

Remark 9.1.9. We will identify each $a \in R$ with $\sum_{x \in G} a_x x$ where $a_e = a$ for the identity element e of G and $a_x = 0$ for all $x \neq e$. We will identify each element y of G with $\sum_{x \in G} a_x x$ where $a_x = 1$ if $x = y$ and $a_x = 0$ otherwise. Then R becomes a subring of RG and G a subgroup of the multiplicative semigroup of RG. The unity of R and the identity element of G are identified with the unity of RG, and we will denote this common unity by 1. Furthermore, for any subset X of G, we will denote (by abuse of notation) by the same letter X the element $\sum_{x \in X} x$ of RG. Conversely, if $\alpha = \sum_{x \in G} a_x x$ is an element of RG such that $a_x \in \{0, 1\}$ for all $x \in G$, we will denote by the same letter α the subset $\{x \in G : a_x = 1\}$ of G. With this notation G denotes not only the group G but also the sum of its elements in the group ring RG.

The first column of a G-invariant matrix M over the ring R determines an element of the group ring RG which we call the *support of M*.

Definition 9.1.10. Let G be a finite group and let M be a G-invariant matrix over a ring R with unity. The *support of M* is the element *supp*$(M) = \sum_{x \in G} M(x, 1)x$ of the group ring RG.

Example 9.1.11. Let $G = \{1, x, x^2, x^3, x^4\}$ be a cyclic group of order 5 and let $S = \langle\sigma\rangle$ be a cyclic group of order 3. Let

$$M = \begin{bmatrix} 0 & 1 & \sigma & \sigma & 1 \\ 1 & 0 & 1 & \sigma & \sigma \\ \sigma & 1 & 0 & 1 & \sigma \\ \sigma & \sigma & 1 & 0 & 1 \\ 1 & \sigma & \sigma & 1 & 0 \end{bmatrix}.$$

Then M is a G-invariant matrix over the group ring $\mathbb{Z}S$ and $supp(M) = x + \sigma x^2 + \sigma x^3 + x^4$.

The next proposition shows that the map $M \mapsto supp(M)$ is a homomorphism from the ring of G-invariant matrices over R to the group ring RG.

Proposition 9.1.12. *Let G be a finite group and let A and B be G-invariant matrices over a ring R with unity. Then* (i) $supp(A + B) = supp(A) + supp(B)$ *and* (ii) $supp(AB) = supp(A)supp(B)$.

Proof. (i) is immediate.

(ii) We have

$$supp(A)supp(B) = \sum_{x\in G} A(x, 1)x \cdot \sum_{y\in G} B(y, 1)y = \sum_{z\in G}\sum_{y\in G} A(zy^{-1}, 1)B(y, 1)z$$

$$= \sum_{z\in G}\sum_{y\in G} A(z, y)B(y, 1)z = \sum_{z\in G}(AB)(z, 1)z = supp(AB).$$

□

Let us return to the matrix M of Example 9.1.11. Though it is a matrix over the group ring $\mathbb{Z}S$, its entries belong to the subset $S \cup \{0\}$ of this ring.

Definition 9.1.13. Let G be a finite group. A $(0, G)$-matrix is any matrix over the group ring $\mathbb{Z}G$ with all entries from the subset $G \cup \{0\}$ of this ring.

The sum and the product of two $(0, G)$ matrices (whenever they are defined) are matrices with every entry of the form $\sum_{x\in G} a_x x$ where all a_x are nonnegative integers.

Remark 9.1.14. If the group $S = \{1\}$ is trivial, then $(0, S)$-matrices become the usual $(0, 1)$-matrices.

Remark 9.1.15. Let G and S be finite groups and let M be a G-invariant $(0, S)$-matrix. Then $supp(M)$ is an element of the group ring $(\mathbb{Z}S)G$. One can regard this ring as the group ring of the direct product $S \times G$ over the integers, and then $supp(M)$ can be regarded as a subset of $S \times G$. For instance, for the matrix M of Example 9.1.11, $supp(M)$ can be identified with the following subset of $S \times G$: $\{(1, x), (\sigma, x^2), (\sigma, x^3), (1, x^4)\}$.

Proposition 9.1.16. *Let G and S be finite groups. A subset D of $S \times G$ is the support of a G-invariant $(0, S)$-matrix if and only if $\alpha\beta^{-1} \notin S$ for all distinct $\alpha, \beta \in D$.*

Proof. 1. Let M be a G-invariant $(0, S)$-matrix and let $D = supp(M)$ be regarded as a subset of $S \times G$. Then $D = \{M(x, 1)x : x \in G, M(x, 1) \neq 0\}$. If $\alpha = M(x, 1)x$ and $\beta = M(y, 1)y$ are distinct elements of D, then $xy^{-1} \neq 1$ and therefore $\alpha\beta^{-1} \notin S$.

2. Suppose D is a subset of $S \times G$ such that $\alpha\beta^{-1} \notin S$ for all distinct $\alpha, \beta \in D$. Define the following $(0, S)$ matrix $M = [M(x, y)]$ indexed by elements of G:

$$M(x, y) = \begin{cases} \sigma & \text{if } \sigma \in S \text{ and } \sigma xy^{-1} \in D, \\ 0 & \text{if } S(xy^{-1}) \cap D = \varnothing. \end{cases} \tag{9.1}$$

Then M is G-invariant and $supp(M) = D$. □

Definition 9.1.17. Let G and S be finite groups and let D be a subset of $S \times G$ such that $\alpha\beta^{-1} \notin S$ for all distinct $\alpha, \beta \in D$. The matrix M defined by (9.1) is called the development of D.

If the group S is trivial, i.e., M is a $(0, 1)$-matrix, then we will also call the incidence structure with M as an incidence matrix the *development of D* and denote it by $dev(D)$. The set D in this case can be regarded as a subset of G and the group G itself can be taken as the point set of $dev(D)$. Then the block set of $dev(D)$ is $\{Dx : x \in G\}$.

Definition 9.1.18. Let D be a subset of a finite group G. The incidence structure $dev(D) = (G, \mathcal{B})$ where $\mathcal{B} = \{Dx : x \in G\}$ is called the *development of D*.

We will now seek a description of subsets of finite groups whose development is a symmetric design.

Definition 9.1.19. Let G be a finite group of order v. A k-subset D of G is called a (v, k, λ)-*difference set* if the multiset

$$\{xy^{-1} : x, y \in D, x \neq y\}$$

contains exactly λ copies of every nonidentity element of G. Difference sets in abelian groups are called *abelian difference sets* and difference sets in cyclic groups are called *cyclic difference sets*.

Proposition 9.1.20. *A k-subset D of a group G of order v is a (v, k, λ)-difference set if and only if the incidence structure $dev(D)$ is a symmetric (v, k, λ)-design.*

Proof. Let D be a k-subset of a group G of order v and let $\mathbf{D} = dev(D)$. Then every block of $\mathbf{D}$ is a k-subset of G. Let $a, b \in G$, $a \neq b$, and let $c = ab^{-1}$. For $x \in G$, we have $\{a, b\} \subseteq Dx$ if and only if $y = ax^{-1}$ and $z = bx^{-1}$ are elements of D. Since $c = (ax^{-1})(bx^{-1})^{-1}$, we obtain that the number of blocks Dx containing $\{a, b\}$ is equal to the number of ordered pairs (y, z) of elements of D such that $yz^{-1} = c$. Therefore, D is a (v, k, λ)-difference set if and only if $\mathbf{D}$ is a symmetric (v, k, λ)-design. □

Proposition 9.1.5 now implies the following result.

Corollary 9.1.21. *A symmetric design* $\mathbf{D}$ *can be obtained as the development of a difference set in a group G if and only if* $\mathbf{D}$ *has a sharply transitive automorphism group isomorphic to G.*

Example 9.1.22. Let $G = \langle x \rangle$ be a cyclic group of order 7 and let $D = \{1, x, x^3\}$. Then D is a $(7, 3, 1)$-difference set and $dev(D)$ is the Fano Plane. As we will see in Section 9.2., every symmetric design $PG_{d-1}(d, q)$ can be obtained as the development of a cyclic difference set known as a *Singer difference set*.

Example 9.1.23. The incidence structure given by the G-invariant matrix N from Example 9.1.4 is the development of the following subset $D = supp(N)$ of G:

$$D = \{y, xy, y^2, x^2y^2, y^3, x^3y^3\}.$$

One can verify that D is a $(16, 6, 2)$-difference set and therefore N is an incidence matrix of a symmetric $(16, 6, 2)$-design. The difference set D belongs to the family of *McFarland difference sets*, which will be constructed in Section 9.5.

Example 9.1.24. Let $q \equiv 3 \pmod 4$ be a prime power. Let G be the additive group of the field $GF(q)$ and let D be the set of nonzero squares of the field. Corollary 4.3.3 and Proposition 9.1.20 imply that D is a $(q, (q-1)/2, (q-3)/4)$-difference set. It belongs to the family of *Paley–Hadamard difference sets* which will be considered in Section 9.2.

Definition 9.1.25. A sharply transitive automorphism group of a symmetric design is called a *Singer group* of the design.

If G is a finite group, then every subset of G can be regarded as an element of the group ring $\mathbb{Z}G$.

In order to characterize difference sets in terms of the ring $\mathbb{Z}G$, we will define the following operation of *conjugation*.

Definition 9.1.26. For any $\alpha = \sum_{x \in G} a_x x \in RG$ of the group ring of a finite group G, the *conjugate* element $\overline{\alpha}$ is defined by $\overline{\alpha} = \sum_{x \in G} a_x x^{-1}$.

Proposition 9.1.27. *Let R be a commutative ring with unity and let G be a finite group. Then, for all $\alpha, \beta \in RG$, $\overline{\alpha + \beta} = \overline{\alpha} + \overline{\beta}$ and $\overline{\alpha\beta} = \overline{\beta}\,\overline{\alpha}$.*

Proof. The first equality is immediate. Let $\alpha = \sum_{x \in G} a_x x$ and $\beta = \sum_{y \in G} b_y y$. Then

$$\begin{aligned}\overline{\beta}\,\overline{\alpha} &= \sum_{y \in G} b_y y^{-1} \cdot \sum_{x \in G} a_x x^{-1} = \sum_{w \in G} \sum_{(xy)^{-1} = w} b_y a_x w \\ &= \sum_{w \in G} \sum_{xy = w^{-1}} b_y a_x w = \sum_{z \in G} \sum_{xy = z} b_y a_x z^{-1} = \overline{\alpha\beta}.\end{aligned}$$

□

We will now characterize difference sets as elements of the group ring RG.

Proposition 9.1.28. *A subset D of a group G of order v is a (v, k, λ)-difference set if and only if it satisfies the following equation in the group ring RG:*

$$D\overline{D} = k - \lambda + \lambda G. \tag{9.2}$$

Proof. For any subset D of G, we have

$$D\overline{D} = \sum_{z \in G} \sum_{\substack{x, y \in D \\ xy^{-1} = z}} z = \sum_{z \in G} a_z z,$$

where a_z is the number of pairs (x, y) of elements of D such that $xy^{-1} = z$. Therefore, D is a (v, k, λ)-difference set if and only if

$$a_z = \begin{cases} \lambda & \text{for } z \neq 1, \\ k & \text{for } z = 1, \end{cases}$$

i.e., $D\overline{D} = k + \lambda(G - 1) = k - \lambda + \lambda G$. □

In case the group G is abelian, the basic equation (9.2) can be rewritten in an equivalent form using character theory for abelian groups.

Definition 9.1.29. Let G be a finite abelian group. A *character* χ of G (or a *character of the group ring* $\mathbb{Z}G$) is a ring homomorphism from $\mathbb{Z}G$ to the field $\mathbb{C}$ of complex numbers such that $\chi(1) = 1$. If $\chi(x) = 1$ for all $x \in G$, the character χ is called *principal*. All other characters are called *nonprincipal*. If $\chi(x) = 1$ for all elements x of a subgroup U of G, the character χ is said to be *principal on* U.

Remark 9.1.30. If χ is a character of $\mathbb{Z}G$, then $\chi(a) = a$ for all $a \in \mathbb{Z}$, and therefore,

$$\chi\left(\sum_{x\in G} a_x x\right) = \sum_{x\in G} a_x \chi(x).$$

If $|G| = v$, then, for each $x \in G$, $\chi(x)$ is a vth root of unity. Since the inverse of a complex root of unity is the conjugate complex number, we obtain that for any $\alpha \in \mathbb{Z}G$, the complex numbers $\chi(\alpha)$ and $\chi(\overline{\alpha})$ are conjugate.

We will now state several basic properties of characters. We leave proofs of these properties as exercises.

Proposition 9.1.31. *Let G be a finite abelian group and U a subgroup of G. The group of all characters of G, which are principal on U, is isomorphic to the group of all characters of the factor group G/U. The character χ' of G/U corresponding to a character χ of G is given by $\chi'(Ux) = \chi(x)$.*

Proposition 9.1.32. *Let G be a finite abelian group. If χ is a nonprincipal character of $\mathbb{Z}G$, then $\chi(G) = 0$ (with G regarded as an element of $\mathbb{Z}G$).*

Proposition 9.1.33. *Let G be a finite abelian group of order v. There are exactly v distinct characters of the ring $\mathbb{Z}G$. If $\chi_0, \chi_1, \ldots, \chi_{v-1}$ are all the distinct characters, then, for any $x \in G \setminus \{1\}$, $\sum_{i=0}^{v-1} \chi_i(x) = 0$.*

Corollary 9.1.34. *Let G be a finite abelian group of order v. If $\chi_0, \chi_1, \ldots, \chi_{v-1}$ are all the distinct characters of $\mathbb{Z}G$, then, for any integers a, a_x,*

$$\sum_{i=0}^{v-1} \chi_i\left(a + \sum_{x\in G\setminus\{1\}} a_x x\right) = va.$$

The following theorem gives a useful tool for proving identities in a group ring.

Theorem 9.1.35. *Let G be a finite abelian group and let $\alpha, \beta \in \mathbb{Z}G$. If $\chi(\alpha) = \chi(\beta)$ for every character χ of $\mathbb{Z}G$, then $\alpha = \beta$.*

Proof. Let $\alpha = \sum_{x\in G} a_x x$ and $\beta = \sum_{x\in G} b_x x$. Let $\chi_0, \chi_1, \ldots, \chi_{v-1}$ be all the distinct characters of $\mathbb{Z}G$. Suppose $\chi_i(\alpha) = \chi_i(\beta)$ for $i = 0, 1, \ldots, v-1$.

Let $y \in G$. Proposition 9.1.33 implies that

$$\sum_{i=0}^{v-1} \chi_i(\alpha y^{-1}) = va_y \text{ and } \sum_{i=0}^{v-1} \chi_i(\beta y^{-1}) = vb_y.$$

Since $\chi_i(\alpha y^{-1} - \beta y^{-1}) = \chi_i(\alpha - \beta)\chi_i(y^{-1}) = 0$ for $i = 0, 1, \ldots, v-1$, we obtain that $a_y = b_y$ for all $y \in G$, i.e., $\alpha = \beta$. □

Corollary 9.1.36. *Let G be a finite abelian group and let α be an element of $\mathbb{Z}G$ such that $\chi(\alpha) = 0$ for every nonprincipal character χ of G. Then there is an integer t such that $\alpha = tG$.*

Proof. Let $\beta = |G|\alpha$. Let χ_0 be the principal character of G and let $s = \chi_0(\beta)/|G|$. Then s is an integer and $\chi(\beta) = \chi(sG)$ for every character χ of G. By Theorem 9.1.35, $\beta = sG$. Thus, $|G|\alpha = sG$, and therefore, $\alpha = tG$ where $t = s/|G|$ is an integer. □

We will now give another characterization of difference sets in abelian groups.

Theorem 9.1.37. *Let G be a finite abelian group of order v. A subset D of G is a (v, k, λ)-difference set if and only if $|\chi(D)| = \sqrt{k-\lambda}$ for every nonprincipal character χ of $\mathbb{Z}G$ and $\chi_0(D) = k$ for the principal character χ_0 of $\mathbb{Z}G$.*

Proof. Let $\chi_0, \chi_1, \ldots, \chi_{v-1}$ be all the distinct characters of $\mathbb{Z}G$ and let χ_0 denote the principal character.

If D is a (v, k, λ)-difference set in G, then $\chi_0(D) = |D| = k$ and (9.2) and Proposition 9.1.32 imply that for $i = 1, 2, \ldots, v-1$, $\chi_i(D)\overline{\chi_i(D)} = k - \lambda$, so $|\chi_i(D)| = \sqrt{k-\lambda}$.

Conversely, suppose D is a subset of G such that $|\chi_i(D)| = \sqrt{k-\lambda}$ for $i = 1, 2, \ldots, v-1$ and $\chi_0(D) = k$. Then $\chi_i(D\overline{D}) = k - \lambda = \chi_i(k - \lambda + \lambda G)$ for $i = 1, 2, \ldots, v-1$ and $|D| = k$. Let $D = \{x_1, x_2, \ldots, x_k\}$. Then

$$D\overline{D} = (x_1 + x_2 + \cdots + x_k)(x_1^{-1} + x_2^{-1} + \cdots + x_k^{-1}) = k + \sum_{x \in G\setminus\{1\}} a_x x$$

for some integers a_x. By Proposition 9.1.33,

$$\sum_{i=0}^{v-1} \chi_i(D\overline{D}) = vk.$$

On the other hand, since $k - \lambda + \lambda G = k + \lambda(G - 1)$, Proposition 9.1.33 implies that

$$\sum_{i=0}^{v-1} \chi_i(k - \lambda + \lambda G) = vk.$$

Since $\chi_i(D\overline{D}) = \chi_i(k - \lambda + \lambda G)$ for $i = 1, 2, \ldots, v - 1$, we obtain that $\chi_0(D\overline{D}) = \chi_0(k - \lambda + \lambda G)$. Theorem 9.1.35 now implies that $D\overline{D} = k - \lambda + \lambda G$, and then Proposition 9.1.28 implies that D is a (v, k, λ)-difference set. □

9.2. Singer and Paley–Hadamard difference sets

The classical designs $PG_{d-1}(d, q)$ have a cyclic Singer group.

Theorem 9.2.1. *For every positive integer d and prime power q, the design $PG_{d-1}(d, q)$ has a cyclic Singer group.*

Proof. Let d be a positive integer, q a prime power, and $v = (q^{d+1} - 1)/(q - 1)$. Let $\mathbf{D} = (X, \mathcal{B})$ be the design $PG_{d-1}(d, q)$. Let α be a primitive element of the field $GF(q^{d+1})$. Then α^v is a primitive element of $GF(q)$. Let $G = GF(q^{d+1})^*/GF(q)^*$ and let $\pi : GF(q^{d+1})^* \to G$ be the natural homomorphism. Then G is the cyclic group of order v generated by $x = \pi(\alpha)$. Observe that each coset x^s is the set of all nonzero elements of the one-dimensional vector space over $GF(q)$ spanned by α^s. Therefore, we can identify the point set X of $\mathbf{D}$ with the group G. With each $\sigma \in G$, we will identify the map $\sigma : X \to X$ given by $\sigma(x^s) = \sigma x^s$ for $s = 0, 1, \ldots, v - 1$. If H is a block of $\mathbf{D}$, then Corollary 3.5.15 implies that $\sigma(H)$ is a block of $\mathbf{D}$. Thus, the group G can be regarded as a group of automorphisms of $\mathbf{D}$. The action of G on $\mathbf{D}$ is sharply transitive, and therefore, by Corollary 9.1.21, G is a Singer group of $\mathbf{D}$. □

Definition 9.2.2. A difference set D is called a *Singer difference set* if $dev(D)$ is isomorphic to $PG_{d-1}(d, q)$ for some d and q. Difference sets with parameters $(4n - 1, 2n - 1, n - 1)$ are called *Paley–Hadamard difference sets of order n*.

Example 9.1.24 shows that a Paley–Hadamard difference set of order n exists whenever $4n - 1$ is a prime power. The next theorem gives another family of Paley–Hadamard difference sets.

Theorem 9.2.3. *Let q and $q + 2$ be odd prime powers. Let η be the quadratic character of the field $GF(q)$ and ζ the quadratic character of the field $GF(q + 2)$. Let G be the direct product of the additive groups of these fields. Let a subset D of G consist of all pairs (x, y) with nonzero components x and y such that $\eta(x) = \zeta(y)$ and of all pairs (x, y) with $y = 0$. Then D is a Paley–Hadamard difference set of order $(q + 1)^2/4$.*

Proof. Let $R = GF(q) \times GF(q+2)$. Then R is a ring and G is the additive group of R. Let $S = \{(x, 0) \in R : x \in GF(q)\}$, $T = \{(0, y) \in R : y \in GF(q+2)\}$, $U = \{(x, y) \in R : x \neq 0, y \neq 0\}$, and $V = \{(x, y) \in U : \eta(x) = \zeta(y)\}$. Then $D = S \cup V$.

We have $|G| = q(q+2) = 4n - 1$ where $n = (q+1)^2/4$. Since q and $q + 2$ are odd prime powers and $q \not\equiv q + 2 \pmod 4$, we have $\eta(-1) \neq \zeta(-1)$. Therefore, if $(x, y) \in V$, then $(-x, -y) \in U \setminus V$, and vice versa. This implies $V \cap (-V) = \varnothing$ and $U = V \cup (-V)$. Therefore, $|D| = |S| + \frac{1}{2}|U| = 2n - 1$.

For each nonzero element (a, b) of R, let $\Delta(a, b)$ be the set of all quadruples (x_1, y_1, x_2, y_2) with $(x_1, y_1), (x_2, y_2) \in D$ such that $x_1 - x_2 = a$ and $y_1 - y_2 = b$. We have to show that $|\Delta(a, b)| = n - 1$.

Observe that, for $a \neq 0$, the set $\Delta(a, 0)$ consists of all quadruples $(x + a, y, x, y)$ satisfying one of the following two conditions: (a) $y = 0$; (b) $y \neq 0$ and $\eta(x) = \eta(x + a) = \zeta(y)$. There are q quadruples satisfying (a) and, by Lemma 3.1.4, there are $(q - 3)(q + 1)/4$ quadruples satisfying (b). Therefore, $|\Delta(a, 0)| = n - 1$.

For $b \neq 0$, the set $\Delta(0, b)$ consists of quadruples $(x, y + b, x, y)$ satisfying one of the following three conditions: (a) $\eta(x) = \eta(y + b) = \zeta(y)$; (b) $y = 0$ and $\eta(x) = \zeta(b)$; (c) $y = -b$ and $\eta(x) = \zeta(-b)$. By Lemma 3.1.4, there are $(q - 1)^2/4$ quadruples satisfying (a). Since there are $(q - 1)/2$ quadruples satisfying (b) and $(q - 1)/2$ quadruples satisfying (c), we obtain that $|\Delta(0, b)| = n - 1$.

Let $(a, b) \in U$ and $(c, d) \in V$. The following three statements are equivalent: (a) $(x_1, y_1, x_2, y_2) \in \Delta(a, b)$, (b) $(cx_1, dy_1, cx_2, dy_2) \in \Delta(ac, bd)$; (c) $(-x_1, -y_1, -x_2, -y_2) \in \Delta(-a, -b)$. Therefore, $|\Delta(a, b)| = \lambda$ is the same for all $(a, b) \in U$. Thus, the set of all quadruples (x_1, y_1, x_2, y_2) with $(x_1, y_1), (x_2, y_2) \in D$ and $(x_1, y_1) \neq (x_2, y_2)$ consists of $(2n - 1)(2n - 2)$ elements, of which $(q - 1)(n - 1)$ belong to the sets $\Delta(a, 0)$ with $a \neq 0$, $(q + 1)(n - 1)$ belong to the sets $\Delta(0, b)$ with $b \neq 0$, and $(q^2 - 1)\lambda$ belong to the sets $\Delta(a, b)$ with $(a, b) \in U$. Therefore,

$$(2n - 1)(2n - 2) = (q - 1)(n - 1) + (q + 1)(n - 1) + (q^2 - 1)\lambda,$$

which implies $\lambda = n - 1$. The proof is now complete. □

Remark 9.2.4. The difference sets constructed in Theorem 9.2.3 are called *Stanton–Sprott difference sets.*

We list without proof several families of difference sets known as *cyclotomic difference sets.*

Theorem 9.2.5. *Let q be a prime power. Each of the following sets D is a difference set in the additive group of the field $GF(q)$:*

(i) $D = \{x^4 : x \in GF(q)^*\}$ *in case* $(q-1)/4$ *is an odd square;*
(ii) $D = \{x^4 : x \in GF(q)\}$ *in case* $(q-9)/4$ *is an odd square;*
(iii) $D = \{x^8 : x \in GF(q)^*\}$ *in case* $(q-1)/8$ *and* $(q-9)/64$ *are odd squares;*
(iv) $D = \{x^8 : x \in GF(q)\}$ *in case* $(q-49)/8$ *is an odd square and* $(q-441)/64$ *is an even square.*

Remark 9.2.6. It is not known whether any of the families of difference sets described in Theorem 9.2.5 is infinite. In fact, for some of them, just a few examples are known.

9.3. Symmetries in a group ring

The group ring $\mathbb{Z}G$ is naturally equipped with an inner product defined as follows: for $\alpha = \sum_{x\in G} a_x x$ and $\beta = \sum_{x\in G} b_x x$,

$$\langle \alpha, \beta \rangle = \sum_{x\in G} a_x b_x.$$

This inner product is bilinear. For any $\alpha \in \mathbb{Z}G$, let $r(\alpha) = \langle \alpha, G\rangle$ (with G regarded as an element of $\mathbb{Z}G$). Then $r(\alpha+\beta) = r(\alpha) + r(\beta)$ and $r(\alpha x) = r(x\alpha) = r(\alpha)$ for all $\alpha, \beta \in \mathbb{Z}G$ and $x \in G$. We also need the following property of the above inner product.

Lemma 9.3.1. *Let G be a finite group and H a subgroup of G. Let $x_1, x_2, \ldots, x_h$ be all right coset representatives of H in G. Let $\alpha_1, \ldots, \alpha_h$ and $\beta_1, \ldots, \beta_h$ be elements of the subring $\mathbb{Z}H$ of $\mathbb{Z}G$. Then*

$$\langle \sum_{i=1}^{h} \alpha_i x_i, \sum_{i=1}^{h} \beta_i x_i \rangle = \sum_{i=1}^{h} \langle \alpha_i, \beta_i \rangle.$$

Proof. For $i = 1, 2, \ldots, h$, let

$$\alpha_i = \sum_{x\in H} a_x^{(i)} x, \quad \beta_i = \sum_{x\in H} b_x^{(i)} x,$$

where all $a_x^{(i)}, b_x^{(i)}$ are integers. Then

$$\sum_{i=1}^{h} \alpha_i x_i = \sum_{i=1}^{h} \sum_{x\in H} a_x^{(i)} x x_i$$

and

$$\sum_{i=1}^{h} \beta_i x_i = \sum_{i=1}^{h} \sum_{x\in H} b_x^{(i)} x x_i.$$

Since $G = \{xx_i : x \in H, 1 \le i \le h\}$, we obtain that

$$\left\langle \sum_{i=1}^{h} \alpha_i x_i, \sum_{i=1}^{h} \beta_i x_i \right\rangle = \sum_{i=1}^{h} \sum_{x \in H} a_x^{(i)} b_x^{(i)} = \sum_{i=1}^{h} \langle \alpha_i, \beta_i \rangle.$$

□

Definition 9.3.2. Let G be a finite group. A subset A of $\mathbb{Z}G$ is called *uniform* if $\alpha x \in A$ for any $\alpha \in A$ and any $x \in G$.

We will now introduce the central notion of this section.

Definition 9.3.3. Let G be a finite group and let A be a nonempty subset of $\mathbb{Z}G$. A finite group S of bijections $A \to A$ is called a *group of symmetries of* A if the following conditions are satisfied:

(i) if $\alpha \in A$ and $\alpha \subseteq G$, then $\sigma(\alpha) \subseteq G$ for all $\sigma \in S$;
(ii) $\langle \sigma(\alpha), \sigma(\beta) \rangle = \langle \alpha, \beta \rangle$ and $r(\sigma(\alpha)) = r(\alpha)$, for all $\alpha, \beta \in A$ and all $\sigma \in S$;
(iii) for every $\alpha \in A$, there is an integer $t(\alpha)$ such that

$$\sum_{\sigma \in S} \sigma(\alpha) = t(\alpha)G.$$

Remark 9.3.4. If we apply r to both sides of the last equation, we obtain that $|S|r(\alpha) = t(\alpha)|G|$, so $t(\alpha) = r(\alpha)|S|/|G|$.

Example 9.3.5. For any $\sigma \in G$ and any $\alpha \in \mathbb{Z}G$, define $\sigma(\alpha) = \sigma\alpha$. Then G acts as a group of permutations on $\mathbb{Z}G$. Since conditions (i)–(iii) are satisfied, G is a group of symmetries of $\mathbb{Z}G$.

The following proposition gives an example of a cyclic group of symmetries.

Proposition 9.3.6. *Let $G = \{x_1, x_2, \ldots, x_g\}$ be a group of order g. For any $\alpha = \sum_{i=1}^{g} a_i x_i \in \mathbb{Z}G$, define*

$$\sigma(\alpha) = a_g x_1 + \sum_{i=1}^{g-1} a_i x_{i+1}.$$

The cyclic group generated by σ is a group of symmetries of $\mathbb{Z}G$

Proof. Properties (i) and (ii) of Definition 9.3.3 are immediate. Let S be the group generated by σ. If $\alpha = \sum_{x \in G} a_x x$, then

$$\sum_{\rho \in S} \rho\alpha = \sum_{x \in G} a_x G,$$

so $t(\alpha) = r(\alpha)$. □

We will say that a subset A of $\mathbb{Z}G$ admits a *symmetry of order s* if it has a cyclic group of symmetries whose order divides s. If this is the case and $\alpha \in A$, we will also say that α admits a *symmetry of order s*.

Our interest in symmetries can be explained by the following theorem that will be proven in Chapter 11 (Corollary 11.3.2).

Theorem 9.3.7. *Let D be a (v, k, λ)-difference set in a group G. Suppose that $q = k^2/(k - \lambda)$ is a prime power. If D (as an element of $\mathbb{Z}G$) admits a symmetry of order $q - 1$, then, for any positive integer m, there exists a symmetric design with parameters*

$$\left(\frac{v(q^{m+1} - 1)}{q - 1}, kq^m, \lambda q^m\right).$$

In this and the next section, we describe a recursive construction of several infinite families of abelian (v, k, λ)-difference sets. These difference sets admit a symmetry of order $q - 1$ with $q = k^2/(k - \lambda)$.

We begin with the following simple but useful result, which is readily verified.

Proposition 9.3.8. *Let G be a finite group, A a subset of $\mathbb{Z}G$, and S a group of symmetries of A. Let $A' = \{G - \alpha : \alpha \in A\}$. For any $\sigma \in S$ and $\alpha \in A$, let $\sigma'(G - \alpha) = G - \sigma(\alpha)$. Then $\sigma' : A' \to A'$ is a bijection and $S' = \{\sigma' : \sigma \in S\}$ is a group of symmetries of A' isomorphic to S. If A is uniform, then so is A'.*

For any abelian group G, we will now describe a useful direct construction of uniform subsets of $\mathbb{Z}G$, admitting a cyclic group of symmetries. We begin with the following definition.

Definition 9.3.9. Let G be a finite abelian group. For any subgroup U of G, denote by $X_U(G)$ the set of elements $\alpha \in \mathbb{Z}G$ such that $\chi(\alpha) = 0$ for every nonprincipal character χ of $\mathbb{Z}G$ which is principal on U.

Proposition 9.3.10. *Let G be a finite abelian group. For any subgroup U of G, the set $X_U(G)$ is uniform and has a symmetry of order $u = |U|$.*

Proof. The uniformity of $X_U(G)$ is immediate. Let $|G| = mu$ and let $x_1, x_2, \ldots, x_m$ be distinct coset representatives of $U = \{y_1, y_2, \ldots, y_u\}$ in G, so $G = \{x_i y_j : 1 \le i \le m, 1 \le j \le u\}$. Define a bijection $\rho : \mathbb{Z}G \to \mathbb{Z}G$ by

$$\rho\left(\sum_{i=1}^{m}\sum_{j=1}^{u} a_{ij} x_i y_j\right) = \sum_{i=1}^{m}\left(a_{i1} x_i y_u + \sum_{j=2}^{u} a_{ij} x_i y_{j-1}\right).$$

Clearly, ρ satisfies conditions (i) and (ii) of Definition 9.3.3. Let S be the cyclic group generated by ρ. Then $|S| = u$. It suffices to show that $\rho(\alpha) \in X_U(G)$

whenever $\alpha \in X_U(G)$ and that condition (iii) of Definition 9.3.3 is satisfied for $\alpha \in X_U(G)$.

If χ is a character of $\mathbb{Z}G$ which is principal on U, then, for $\alpha = \sum a_{ij}x_i y_j$, we have $\chi(\rho(\alpha)) = \chi(\alpha) = \sum a_{ij}\chi(x_i)$. Therefore, $\rho(\alpha) \in X_U(G)$ whenever $\alpha \in X_U(G)$. We further have that for any character χ of $\mathbb{Z}G$,

$$\chi\left(\sum_{\sigma\in S}\sigma(\alpha)\right) = \sum_{\sigma\in S}\chi(\sigma(\alpha)) = \sum_{i=1}^{m}\sum_{j=1}^{u} a_{ij}\chi(x_i)\chi(U).$$

If χ is nonprincipal on U, then $\chi(U) = 0$. Therefore,

$$\chi\left(\sum_{\sigma\in S}\sigma(\alpha)\right) = \begin{cases} 0 & \text{if } \chi \text{ is nonprincipal on } U, \\ u\chi(\alpha) & \text{if } \chi \text{ is principal on } U. \end{cases}$$

Thus, if $\alpha \in X_U(G)$, then $\chi(\sum_{\sigma\in S}\sigma(\alpha)) = 0$ for any nonprincipal character χ of $\mathbb{Z}G$. By Corollary 9.1.36, $\sum_{\sigma\in S}\sigma(\alpha) = t(\alpha)G$ where $t(\alpha)$ is an integer. □

The next three propositions describe recursive constructions of symmetries.

Proposition 9.3.11. *Let G be a finite group and G' a group containing G as a subgroup of index h. Let A be a subset of $\mathbb{Z}G$ having a symmetry of order s. Let $x_1, x_2, \ldots, x_h$ be distinct right coset representatives of G in G' and let*

$$A' = \left\{\sum_{i=1}^{h}\alpha_i x_i : \alpha_i \in A\right\}.$$

Then

(i) *A' has a symmetry of order sh;*
(ii) *if A is uniform, then so is A'.*

Proof. (i) Let $S = \langle\sigma\rangle$ be a cyclic group of symmetries of A with $|S|$ dividing s. For $\alpha = \sum_{i=1}^{h}\alpha_i x_i$ where $\alpha_1, \alpha_2, \ldots, \alpha_h \in A$, define

$$\tau(\alpha) = \sigma(\alpha_h)x_1 + \sum_{i=1}^{h-1}\alpha_i x_{i+1}.$$

Then τ is a bijection on A', τ^{sh} is the identity map,

$$r(\tau(\alpha)) = r(\sigma(\alpha_h)) + \sum_{i=1}^{h-1} r(\alpha_i) = \sum_{i=1}^{h} r(\alpha_i) = r(\alpha),$$

and

$$\sum_{i=1}^{sh}\tau^i(\alpha) = (aG)\sum_{j=1}^{h} x_j = aG',$$

where a is an integer.

If $\alpha \subseteq G'$, then $\alpha_i \subseteq G$ for $i = 1, 2, \ldots, h$, so $\sigma(\alpha_i) \subseteq G$ and $\tau(\alpha) \subseteq G'$. For $\alpha_i, \beta_i \in A$, we apply Lemma 9.3.1 to obtain

$$\left\langle \tau\left(\sum_{i=1}^{h} \alpha_i x_i\right), \tau\left(\sum_{i=1}^{h} \beta_i x_i\right)\right\rangle = \langle \sigma(\alpha_h)x_1, \sigma(\beta_h)x_1\rangle + \sum_{i=1}^{h-1}\langle \alpha_i x_{i+1}, \beta_i x_{i+1}\rangle$$
$$= \sum_{i=1}^{h}\langle \alpha_i, \beta_i\rangle = \left\langle \sum_{i=1}^{h} \alpha_i x_i, \sum_{i=1}^{h} \beta_i x_i\right\rangle.$$

Thus, the group $S' = \langle\tau\rangle$ is a cyclic group of symmetries of A' and $|S'|$ divides sh.

(ii) Suppose A is uniform and let $\alpha = \sum \alpha_i x_i \in A'$. For any $y \in G'$ there are $y_1, y_2, \ldots, y_h \in G$ such that $\{x_1 y, x_2 y, \ldots, x_h y\} = \{y_1 x_1, y_2 x_2, \ldots, y_h x_h\}$. Since A is uniform, we obtain that $\alpha y = \sum \alpha_i(x_i y) = \sum \beta_i x_i$ with $\beta_i \in A$, so $\alpha y \in A'$. □

Proposition 9.3.12. *Let G be a finite group and U a normal subgroup of G, $|U| = u$, $|G| = mu$. Let $x_1, x_2, \ldots, x_m$ be distinct coset representatives of U in G. Let A be a subset of $\mathbb{Z}(G/U)$ having a symmetry of order s and let*

$$A' = \left\{U\left(\sum_{i=1}^{m} a_i x_i\right) \in \mathbb{Z}G : \sum_{i=1}^{m} a_i(Ux_i) \in A\right\}.$$

Then

(i) *A' has a symmetry of order s;*
(ii) *if A is uniform, then so is A'.*

Proof. (i) Let

$$C = \left\{U\left(\sum_{i=1}^{m} a_i x_i\right) : a_i \in \mathbb{Z}\right\}.$$

Then C is a subgroup of the additive group of $\mathbb{Z}C$, and the map

$$f\left(U\left(\sum_{i=1}^{m} a_i x_i\right)\right) = \sum_{i=1}^{m} a_i(Ux_i)$$

is an isomorphism from C to the additive group of $\mathbb{Z}(G/U)$. Clearly, $\langle\alpha, \beta\rangle = u\langle f(\alpha), f(\beta)\rangle$ and $r(\alpha) = ur(f(\alpha))$ for any $\alpha, \beta \in C$. We have $A' = f^{-1}(A)$. Let $S = \langle\sigma\rangle$ be a cyclic group of symmetries of A. Define $\sigma' = f^{-1}\sigma f$. If $\alpha \in A'$ is a subset of G, then $f(\alpha) \subseteq G/U$, so $\sigma(f(\alpha)) \subseteq G/U$ and $\sigma'(\alpha) \subseteq G$. For any $\alpha, \beta \in A'$, $\langle\sigma'(\alpha), \sigma'(\beta)\rangle = \langle\alpha, \beta\rangle$ and $r(\sigma'(\alpha)) = r(\alpha)$. Let $S' = \langle\sigma'\rangle$.

Then S' is isomorphic to S and, for any $\alpha \in A'$, there is an integer t such that

$$\sum_{\tau' \in S'} \tau'(\alpha) = f^{-1}\left(\sum_{\tau \in S} \tau(f(\alpha))\right) = f^{-1}(t(f(\alpha))(G/U)) = t(f(\alpha))G,$$

i.e., S' is a group of symmetries of A'.

(ii) Let $\alpha = U(\sum a_i x_i) \in C$ and $x \in G$. Since $\{Ux_1x, Ux_2x, \ldots, Ux_mx\} = \{Ux_1, Ux_2, \ldots, Ux_m\}$, we have $\alpha x \in C$, i.e., C is uniform. Suppose A is uniform and $\alpha \in A'$. Then $f(\alpha x) = f(\alpha)(Ux) \in A$, so $\alpha x \in A'$ and A' is uniform. □

Proposition 9.3.13. *Let G be a finite abelian group and U a subgroup of G, $|U| = u$, $|G| = mu$. Let $x_1, x_2, \ldots, x_m$ be distinct coset representatives of U in G. Let A be a subset of $\mathbb{Z}(G/U)$ having a symmetry of order s, let A' be defined as in Proposition 9.3.12, and let $A'' = A' \cup X_U(G)$. Then*

(i) *A'' has a symmetry of order the least common multiple of s and u;*
(ii) *if A is uniform, then so is A''.*

Proof. (i) Let C and f be defined as in Proposition 9.3.12. Note that $A' \cap X_U(G) \subseteq \{aG : a \in \mathbb{Z}\}$. Indeed, if $\alpha \in C$ and $\chi(\alpha) = 0$ for any nonprincipal character χ of RG which is principal on U, then $\chi'(f(\alpha)) = 0$ for any nonprincipal character χ' of $\mathbb{Z}(G/U)$. By Proposition 9.1.36, $f(\alpha) = a(G/U)$ where a is an integer, and then $\alpha = aG$.

Let $S' = \langle \sigma' \rangle$ and $T = \langle \tau \rangle$ be the cyclic groups of symmetries of A' and $X_U(G)$ defined in Proposition 9.3.12 and Proposition 9.3.10, respectively. For any $\alpha \in A''$, define

$$\rho(\alpha) = \begin{cases} \sigma'(\alpha) & \text{if } \alpha \in A', \\ \tau(\alpha) & \text{if } \alpha \notin A'. \end{cases}$$

If $\alpha \in A' \cap X_U(G)$, then $\tau(\alpha) = \alpha$. Therefore, ρ is a bijection on A'' and ρ^l is the identity map where l is the least common multiple of s and u. We claim that the group $\langle \rho \rangle$ is a group of symmetries of A''. Since S' and T are groups of symmetries of A' and $X_U(G)$, respectively, it suffices to verify that $\langle \sigma'(\alpha), \tau(\beta) \rangle = \langle \alpha, \beta \rangle$, for $\alpha \in A'$ and $\beta \in X_U(G)$.

Let $U = \{y_1, y_2, \ldots, y_u\}$ and let elements $\alpha \in A'$ and $\beta \in X_U(G)$ be given by

$$\alpha = U\left(\sum_{i=1}^{m} a_i x_i\right), \quad \beta = \sum_{i=1}^{m} \sum_{j=1}^{u} b_{ij} x_i y_j.$$

Then

$$\langle \alpha, \beta \rangle = \sum_{i=1}^{m} \sum_{j=1}^{u} a_i b_{ij}.$$

Let

$$\sigma(f(\alpha)) = \sum_{i=1}^{m} a'_i(Ux_i).$$

Then, with $y_0 = y_u$,

$$\langle \sigma'(\alpha), \tau(\beta)\rangle = \left\langle U\left(\sum_{i=1}^{m} a'_i x_i\right), \sum_{i=1}^{m}\sum_{j=1}^{u} b_{ij}x_i y_{j-1}\right\rangle = \sum_{i=1}^{m}\sum_{j=1}^{u} a'_i b_{ij}.$$

Let χ be a nonprincipal character of RG which is principal on U. Then $\chi(\beta) = 0$, i.e., $\sum b_{ij}\chi(x_i) = 0$. By Proposition 9.1.31, $\chi'(\sum b_{ij}(Ux_i)) = 0$ for any nonprincipal character χ' of $\mathbb{Z}(G/U)$. This implies that $\sum b_{ij}(Ux_i) = b(G/U)$ where $b \in \mathbb{Z}$, i.e., $\sum_{j=1}^{u} b_{ij} = b$ for $i = 1, 2, \ldots, m$. Therefore,

$$\sum_{i=1}^{m}\sum_{j=1}^{u} a_i b_{ij} = b\sum_{i=1}^{m} a_i$$

and

$$\sum_{i=1}^{m}\sum_{j=1}^{u} a'_i b_{ij} = b\sum_{i=1}^{m} a'_i.$$

Since $\sum a'_i = r(\sigma(f(\alpha))$ and $\sum a_i = r(f(\alpha))$, we obtain that $\langle \sigma'(\alpha), \tau(\beta)\rangle = \langle \alpha, \beta\rangle$.

(ii) Suppose A is uniform. Then A'' is uniform as union of uniform sets. □

9.4. Building blocks and building sets

In this section we describe a putative construction of an infinite family of abelian (v, k, λ)-difference sets. These difference sets admit symmetries of order $q - 1$ where $q = k^2/(k - \lambda)$.

We begin with the notions of a *building block*, a *building set*, and a *covering extended building set*.

Definition 9.4.1. For a positive integer m, a *building block with modulus m* in a finite abelian group G is a subset B of G such that $|\chi(B)| \in \{0, m\}$ for any nonprincipal character χ of $\mathbb{Z}G$.

Definition 9.4.2. For positive integers a, m, and t, an (a, m, t) *building set* (or simply an (a, m, t)-BS) on a finite abelian group G *relative to a subgroup* U is a collection $\mathcal{B}$ of t building blocks with modulus m, each of cardinality a, such that

(i) if χ is a character of $\mathbb{Z}G$, which is nonprincipal on U, then there is a unique $B \in \mathcal{B}$ with $\chi(B) \neq 0$, and

(ii) if χ is a nonprincipal character of $\mathbb{Z}G$, which is principal on U, then $\chi(B) = 0$ for each $B \in \mathcal{B}$.

Definition 9.4.3. For positive integers a, m, and h, a *covering* $(a, m, h, +)$ *extended building set* (or a covering $(a, m, h, +)$-EBS) on a finite abelian group G is a collection $\mathcal{E}$ of h building blocks in G with modulus m, of which $h - 1$ have cardinality a and the remaining one has cardinality $a + m$, such that for each nonprincipal character χ of $\mathbb{Z}G$ there is a unique $B \in \mathcal{E}$ with $\chi(B) \neq 0$. A *covering* $(a, m, h, -)$-EBS on G is defined in the same way, with $a + m$ replaced by $a - m$.

Remark 9.4.4. We often refer to parameters of an EBS as $(a, m, h, \pm)$ to cover both types of covering extended building sets simultaneously. If $\pm$ occurs in a statement and/or in a proof of a theorem more than once, it is always assumed that $\pm$ stands for the same sign in all the occurrences.

Definition 9.4.2 and Proposition 9.3.11 immediately imply the following result.

Proposition 9.4.5. *If $\mathcal{B}$ is a building set on a finite abelian group G relative to a subgroup U, then $\mathcal{B} \subseteq X_U(G)$.*

The following theorem shows the significance of covering extended building sets.

Theorem 9.4.6. *Let $\mathcal{E} = \{B_1, B_2, \ldots, B_h\}$ be a covering $(a, m, h, \pm)$-EBS on a finite abelian group G. Let G' be an abelian group containing G as a subgroup of index h. Let $x_1, x_2, \ldots, x_h$ be distinct coset representatives of G in G'. Then*

$$D = \bigcup_{i=1}^{h} x_i B_i$$

is a $(|G|h, ah \pm m, ah \pm m - m^2)$-difference set in G'. Furthermore, if $\mathcal{E}$ admits a symmetry of order s, then D admits a symmetry of order sh.

Proof. Let $|B_1| = a \pm m$ and $|B_i| = a$ for $i = 2, \ldots, h$. We have $|G'| = |G|h$ and

$$|D| = \sum_{i=1}^{h} |B_i| = ah \pm m.$$

We will show that $|\chi(D)| = m$ for every nonprincipal character χ of $\mathbb{Z}G'$, and then Theorem 9.1.37 would imply that D is difference set with the required parameters.

Suppose first that χ is a nonprincipal character of $\mathbb{Z}G'$ that is principal on G. Let χ' be the corresponding nonprincipal character of the group ring of the factor group G'/G given by Proposition 9.1.31. Then

$$\sum_{i=1}^{h} \chi(x_i) = \sum_{i=1}^{h} \chi'(\pi x_i) = \sum_{z \in G'/G} \chi'(z) = 0.$$

Therefore,

$$\chi(D) = \sum_{i=1}^{h} \chi(x_i)|B_i| = \chi(x_1)(a \pm m) + a \sum_{i=2}^{h} \chi(x_i)$$
$$= \pm\chi(x_1)m + a \sum_{i=1}^{h} \chi(x_i) = \pm\chi(x_1)m,$$

and we have $|\chi(D)| = m$.

Suppose now that χ is a character of $\mathbb{Z}G'$ that is nonprincipal on G. The definition of a covering EBS implies that there is a unique $B_j \in \mathcal{E}$ with $\chi(B_j) \neq 0$. The definition of a building block then implies that $|\chi(B_j)| = m$. Since $\chi(D) = \chi(B_j)\chi(x_j)$, we again have $|\chi(D)| = m$.

Now let $\mathcal{E}$ admit a symmetry of order s, i.e., there is a subset A of $\mathbb{Z}G$ that admits a symmetry of order s and contains $\mathcal{E}$. Then the subset A' of $\mathbb{Z}G'$, described in Proposition 9.3.11, is uniform, contains D, and admits a symmetry of order sh. Therefore, the difference set D admits a symmetry of order sh. □

The next theorem describes a recursive construction of covering extended building sets.

Theorem 9.4.7. *Let G be a finite abelian group and let U be a subgroup of G of order u. Let $\mathcal{E} = \{B_1, B_2, \ldots, B_h\}$ be a covering $(am, m, h, \pm)$-EBS on G/U and let $\mathcal{B} = \{B_{h+1}, B_{h+2}, \ldots, B_{h+t}\}$ be a (uam, um, t)-BS on G relative to U. Let*

$$B_i' = \begin{cases} \{x \in G : xU \in B_i\} & \text{for } i = 1, 2, \ldots, h, \\ B_i & \text{for } i = h+1, h+2, \ldots, h+t. \end{cases}$$

Then $\mathcal{E}' = \{B_1', B_2', \ldots, B_{h+t}'\}$ is a covering $(uam, um, h+t, \pm)$-EBS on G.

Furthermore, if $\mathcal{E}$ admits a symmetry of order s, then $\mathcal{E}'$ admits a symmetry of order the least common multiple of s and u.

Proof. Let χ be a nonprincipal character of $\mathbb{Z}G$ that is principal on U. Let χ' be the corresponding character of $\mathbb{Z}(G/U)$ given by Proposition 9.1.31. Then there is a unique $j \in \{1, 2, \ldots, h\}$ such that $\chi'(B_j) \neq 0$ and $\chi'(B_i) = 0$ for all $i \in \{1, 2, \ldots, h\}$, except $i = j$. Since, for $1 \leq i \leq h$, B_i' is the union of $|B_i|$ cosets of U and since $\chi(U) = u$, we obtain that $\chi(B_j') \neq 0$ and $\chi(B_i') = 0$ for

$i \neq j$. For $i = h+1, h+2, \ldots, h+t$, the definition of a building set implies $\chi(B_i) = 0$. Thus, B'_j is the only element of $\mathcal{E}'$ with $\chi(B'_j) \neq 0$.

Let χ be a character of $\mathbb{Z}G$ that is nonprincipal on U. Then $\chi(U) = 0$ and therefore $\chi(B_i) = 0$ for $i = 1, 2, \ldots, h$. The definition of a building set implies that there is a unique $j \in \{h+1, h+2, \ldots, h+t\}$ with $\chi(B'_j) \neq 0$.

Thus, we have proved that $\mathcal{E}'$ is a covering $(uam, um, h+t, \pm)$-EBS on G.

Suppose $\mathcal{E}$ admits a symmetry of order s and let A be a uniform subset $\mathbb{Z}(G/U)$ having a symmetry of order s and containing $\mathcal{E}$. Let A' and A'' be the sets defined in Propositions 9.3.12 and 9.3.13, respectively. As an element of $\mathbb{Z}G$, each B'_i is the sum of cosets that form B_i. Since $B_i \in A$, this implies that $B'_1, B'_2, \ldots, B'_h \in A'$. By Proposition 9.4.5, $B_{h+1}, B_{h+2}, \ldots, B_{h+t} \in X_U(G)$. Therefore $\mathcal{E}' \subseteq A''$ and we apply Proposition 9.3.13 to complete the proof. □

We will combine Theorems 9.4.6 and 9.4.7 into the following theorem.

Theorem 9.4.8. *Let a, m_0, and h_0 be positive integers and let $\mathcal{E}_0$ be a covering $(am_0, m_0, h_0, \pm)$-EBS on a finite abelian group G_0. Suppose $\mathcal{E}_0$ admits a symmetry of order s_0. Let $\{U_d\}_{d=0}^{\infty}$ be a sequence of finite abelian groups, $|U_d| = u_d$. For each $d \geq 1$, let $G_d = G_{d-1} \times U_{d-1}$, $m_d = m_{d-1}u_{d-1}$, s_d be the least common multiple of s_{d-1} and u_{d-1}, and $\mathcal{B}_d$ be an (am_d, m_d, t_d)-BS on G_d relative to U_{d-1}. For each $d \geq 1$, let $h_d = h_{d-1} + t_d$. For each $d \geq 0$, let G'_d be an abelian group containing G_d as a subgroup of index h_d.*

Then, for each $d \geq 1$, there exists a difference set D_d in G'_d with parameters

$$(|G_d|h_d, am_dh_d \pm m_d, am_dh_d \pm m_d - m_d^2)$$

that admits a symmetry of order s_dh_d.

Proof. Theorem 9.4.7 and induction on d imply that, for each $d \geq 1$, there is a covering $(am_d, m_d, h_d, \pm)$-EBS $\mathcal{E}_d$ on G_d admitting a symmetry of order s_d. Theorem 9.4.6 now implies the existence of required difference sets D_d. □

9.5. McFarland, Spence, and Davis–Jedwab difference sets

In this section we apply the approach developed in the previous section to construct three infinite families of abelian difference sets admitting symmetries of an appropriate order. The construction is based on Theorem 9.4.8, and therefore requires a starting extended building set $\mathcal{E}_0$ and a family of building sets $\mathcal{B}_d$. All groups that we will be using are elementary abelian groups.

The following theorem provides families of building sets that will be used in the subsequent constructions. Proof of this theorem is beyond the scope of this book.

Theorem 9.5.1. *For every positive integer d and a prime power q, the following building sets exist:*

(a) *a (q^d, q^d, q^d)-BS on an elementary abelian group of order q^{d+1} relative to a subgroup of order q;*
(b) *a $(2^{2d+2}, 2^{2d+1}, 2^{2d})$-$BS$ on an elementary abelian group of order 2^{2d+4} relative to an elementary abelian group of order 4.*

We are now ready to construct three infinite families of difference sets.

Theorem 9.5.2. *Let q be a prime power and d a positive integer. Let $r_d = (q^{d+1} - 1)/(q - 1)$. Then there exists an abelian difference set with parameters*

$$((r_d + 1)q^{d+1}, r_d q^d, (r_d - 1)q^{d-1})$$

admitting a symmetry of order $q(r_d + 1)$.

Proof. For $d = 0, 1, 2, \ldots$, let G_d be an elementary abelian group of order q^{d+1} and U_d be a subgroup of G_d of order q. Let $\mathcal{E}_0 = \{\varnothing, \{1\}\}$. Then $\mathcal{E}_0$ can be regarded as a covering $(1, 1, 2, -)$-EBS on G_0. For each $d \geq 1$, let $\mathcal{B}_d$ be a (q^d, q^d, q^d)-BS on G_d relative to U_{d-1} provided by Theorem 9.5.1(a). Since $|G_0| = q$, Proposition 9.3.6 implies that $\mathcal{E}_0$ admits a symmetry of order q. Therefore, the conditions of Theorem 9.4.8 are satisfied with $a = 1$, $m_0 = 1$, $h_0 = 2$, $s_0 = q$, and all $t_d = q^d$. We then have $m_d = q^d$, $s_d = q$, and $h_d = r_d + 1$. Therefore, Theorem 9.4.8 yields a difference set with the required parameters. □

Remark 9.5.3. Difference sets constructed in Theorem 9.5.2 are known as *McFarland difference sets*. Note that symmetric designs with the same parameters were constructed in Theorems 3.8.3 and 7.4.18.

The following McFarland difference set will be used in a later construction.

Corollary 9.5.4. *There exists a $(16, 6, 2)$-difference set in an elementary abelian group of order 16 admitting a symmetry of order 8.*

Proof. Take $q = 2$ and $d = 1$ in Theorem 9.5.2. □

Theorem 9.5.5. *For every positive integer d, there exists an abelian difference set with parameters*

$$\left(\frac{3^{d+1}(3^{d+1}-1)}{2}, \frac{3^d(3^{d+1}+1)}{2}, \frac{3^d(3^d+1)}{2}\right)$$

admitting a symmetry of order $3(3^{d+1}-1)/2$.

Proof. For $d = 0, 1, 2, \ldots$, let G_d be an elementary abelian group of order 3^{d+1} and U_d be a subgroup of G_d of order 3. Let $G_0 = \{1, \rho, \rho^2\}$ and $\mathcal{E}_0 = \{\{\rho, \rho^2\}\}$. Then $\mathcal{E}_0$ can be regarded as a covering $(1, 1, 1, +)$-EBS on G_0. For each $d \geq 1$, let $\mathcal{B}_d$ be a $(3^d, 3^d, 3^d)$-BS on G_d relative to U_{d-1} provided by Theorem 9.5.1(a). The rest of the proof is similar to Theorem 9.5.2. □

Remark 9.5.6. Difference sets constructed in Theorem 9.5.5 are known as *Spence difference sets*. Symmetric designs with the same parameters were constructed in Theorem 3.8.5.

Theorem 9.5.7. *For each positive integer d, there exists an abelian difference set with parameters*

$$\left(\frac{2^{2d+4}(2^{2d+2}-1)}{3}, \frac{2^{2d+1}(2^{2d+3}+1)}{3}, \frac{2^{2d+1}(2^{2d+1}+1)}{3}\right)$$

admitting a symmetry of order $8(2^{2d+2}-1)/3$.

Proof. Let G_0 be an elementary abelian group of order 16 and let F be a McFarland $(16, 6, 2)$-difference set from Corollary 9.5.4. Then $\mathcal{E}_0 = \{F\}$ is a covering $(4, 2, 1, +)$-EBS on G_0 admitting a symmetry of order $s_0 = 8$. For $d = 0, 1, 2, \ldots$, let U_d be an elementary abelian group of order 4. Then, for each d, G_d is an elementary abelian group of order 2^{2d+4}. For $d \geq 1$, let $\mathcal{B}_d$ be a $(2^{2d+2}, 2^{2d+1}, 2^{2d})$-$BS$ on G_d relative to U_{d-1} provided by Theorem 9.5.1(b). Then the conditions of Theorem 9.4.8 are satisfied with $a = 4$, $m_d = 2^{2d+1}$, $s_d = 8$, $t_d = 2^{2d}$, and $h_d = (2^{2d+2}-1)/3$. It results in difference set with the required parameters. □

Remark 9.5.8. Difference sets constructed in Theorem 9.5.7 are known as *Davis–Jedwab difference sets*.

We will now describe (without proofs) two more infinite families of difference sets that can be constructed in a similar way. However, we do not know whether these difference sets admit symmetries that would allow us to apply Theorem 9.3.7.

Definition 9.5.9. A *Hadamard difference set* of order h^2 is a difference set with parameters $(4h^2, 2h^2 - h, h^2 - h)$ where h is a positive integer.

Theorem 9.5.10. *If h or $2h$ or $3h$ or $6h$ is a square, then there exists an abelian Hadamard difference set of order h^2.*

If we do not require a Hadamard difference set to be abelian, the following result is true.

Theorem 9.5.11. *There exists a Hadamard difference set of order* 25.

One can obtain more nonabelian Hadamard difference sets using the Kronecker product.

Theorem 9.5.12. *Let N_1 and N_2 be incidence matrices of symmetric designs obtained by developing Hadamard difference sets in groups G_1 and G_2, respectively. Let $H_1 = J - 2N_1$ and $H_2 = J - 2N_2$. Then $N = \frac{1}{2}(J - (H_1 \otimes H_2))$ is an incidence matrix of a symmetric design that can obtained by developing a Hadamard difference set in $G_1 \times G_2$.*

Proof. Since, for $i = 1, 2$, the matrix N_i is an incidence matrix of the development of a difference set, it is G_i-invariant. Therefore, the regular Hadamard matrix H_i is G_i-invariant. By Propositions 9.1.6 and 4.4.9, the matrix $H_1 \otimes H_2$ is a $(G_1 \times G_2)$-invariant regular Hadamard matrix. Therefore, N is an incidence matrix of the development of a Hadamard difference set in $G_1 \times G_2$. □

Finally, *Chen difference sets* are difference sets with parameters

$$\left(\frac{4q^{2d}(q^{2d}-1)}{q^2-1}, q^{2d-1}\left(\frac{2(q^{2d}-1)}{q+1}+1\right), \frac{q^{2d-1}(q-1)(q^{2d-1}+1)}{q+1}\right), \tag{9.3}$$

where q is a prime power.

Theorem 9.5.13. *If q is a prime power such that q or $2q$ or $3q$ is a square, then for any positive integer d there exists a difference set with parameters* (9.3).

9.6. Relative difference sets

A *relative difference set* is a generalization of a difference set. We will use relative difference sets in the next chapter.

Definition 9.6.1. Let G be a group of order mn and N a normal subgroup of G of order n. A k-subset R of G, where $0 < k < mn$, is called a *relative*

(m, n, k, λ)-*difference set* (or an (m, n, k, λ)-RDS) in G relative to N if the multiset $\{xy^{-1}: x, y \in R\}$ contains no nonidentity element of N and contains exactly λ copies of every element of the set $G \setminus N$.

Example 9.6.2. The subset $R = \{1, 2, 4, 8\}$ of the group $\mathbf{Z}_{15}$ of residue classes modulo 15 is a $(5, 3, 4, 1)$-RDS relative to the subgroup of order 5.

Remark 9.6.3. If N is the trivial subgroup of G, then a relative difference set in G relative to N is a difference set in G.

Example 9.6.4. Let G be the direct product of a cyclic group of order 13 and the symmetric group S_3. Let σ be a generator of the cyclic group and let $a = (12)$ and $b = (123)$ be elements of S_3. Then the set $R = \{\sigma a, \sigma^2, \sigma^3 ab, \sigma^5, \sigma^6, \sigma^7 ab^2, \sigma^8 a, \sigma^9 ab^2, \sigma^{11} ab\}$ is a $(13, 6, 9, 1)$-RDS in G relative to S_3.

The following theorem gives an infinite family of relative difference sets.

Theorem 9.6.5. *Let q be an odd prime power and let G be the additive group of the field $GF(q)$. Let R be the subset of the group $G \times G$ consisting of all ordered pairs (x, x^2) with $x \in G$. Then R is a $(q, q, q, 1)$-RDS in $G \times G$ relative to $N = \{(0, x): x \in GF(q)\}$.*

Proof. First observe that, for $(x, x^2), (y, y^2) \in R$ with $x \neq y$, the difference $(x - y, x^2 - y^2)$ is not in N.

Suppose there are distinct $(x, y), (u, v) \in G \times G$ such that $(x - y, x^2 - y^2) = (u - v, u^2 - v^2)$. Then $x - y = u - v$ and $(x + y)(x - y) = (u + v)(u - v)$. Therefore, $x + y = u + v$, and then $2x = 2u$. Since q is odd, we obtain that $x = u$ and then $y = v$. Thus, the multiset of differences $\alpha - \beta$ of distinct elements of R contains no element of N and no repeated elements. Since $|R| = q$, the cardinality of this multiset is $q(q - 1)$ and it is equal to $|(G \times G) \setminus N|$. Therefore, the multiset of differences contains every element of $(G \times G) \setminus N$ once and then R is a $(q, q, q, 1)$-RDS in $G \times G$ relative to N. □

The relative difference set constructed in Theorem 9.6.5 is an example of a *splitting relative difference set.*

Definition 9.6.6. A relative difference set in the direct product $G \times N$ of two groups relative to the subgroup $\{1\} \times N$ is called a *splitting relative difference set.*

The proof of the following characterization of relative difference sets in terms of the group ring is similar to Proposition 9.1.28.

Proposition 9.6.7. *A subset R of a group G of order mn is an (m, n, k, λ)-RDS in G relative to a normal subgroup N of order n if and only if*

$$R\overline{R} = k - \lambda N + \lambda G.$$

The following result implies an important relation between relative difference sets and difference sets.

Proposition 9.6.8. *Let R be an (m, n, k, λ)-RDS in a group G relative to a normal subgroup N and let $f : G \to G'$ be a surjective group homomorphism. Let $U = \ker(f)$ and let $u = |U|$. If $U \subseteq N$, then $f(R)$ is an $(m, n/u, k, \lambda u)$-RDS in G' relative to $N' = f(N)$. In particular, if $U = N$, then $f(R)$ is an $(m, k, \lambda u)$-difference set in G'.*

Proof. Let $x, y \in R$. If $f(x)f(y)^{-1} \in N'$, then $xy^{-1} \in aN$ with $a \in U$. Since $U \subseteq N$, we have $xy^{-1} = 1$ and then $f(x)f(y)^{-1} = 1$.

Let $z' \in G' \setminus N'$ and let z be a fixed element of G such that $f(z) = z'$. Then $f(x)f(y)^{-1} = z'$ if and only if $xy^{-1} = az$ for some $a \in U$. For each $a \in U$, there are exactly λ such pairs (x, y), and therefore, there are exactly λu pairs (x, y) with $f(x)f(y)^{-1} = z'$. To complete the proof, note that distinct elements of R have distinct images in R'. Indeed, if $f(x) = f(y)$ for $x, y \in R$, then $xy^{-1} \in N$ and therefore, $x = y$. □

The next theorem gives another relation between relative difference sets and difference sets. First we need the following simple lemma.

Lemma 9.6.9. *Let H and K be subgroups of a group G. Let X be a subset of H and Y a subset of K. If, for all distinct $y_1, y_2 \in Y$, $y_1 y_2^{-1} \notin H \cap K$, then XY is a subset of G (where X, Y, and XY are regarded as elements of $\mathbb{Z}G$).*

Proof. We should verify that $x_1 y_1 \neq x_2 y_2$ for $x_1, x_2 \in X$ and $y_1, y_2 \in Y$, unless $x_1 = x_2$ and $y_1 = y_2$. If $x_1 y_1 = x_2 y_2$, then $x_1^{-1} x_2 = y_1 y_2^{-1}$, so $y_1 y_2^{-1} \in H \cap K$, which implies $y_1 = y_2$ and $x_1 = x_2$. □

Theorem 9.6.10. *Let R be an (m, n, k, λ)-RDS in a group G relative to a normal subgroup N and D an (n, l, μ)-difference set in N. If $k\mu = \lambda l^2$, then DR is an $(mn, kl, k\mu)$-difference set in G.*

Proof. Lemma 9.6.9 (with $H = N$ and $K = G$) implies that DR is a subset of G.

We have $R\overline{R} = k - \lambda N + \lambda G$ and $D\overline{D} = l - \mu + \mu N$. Therefore,

$$(DR)\overline{DR} = D(R\overline{R})\overline{D} = D(k - \lambda N + \lambda G)\overline{D} = kD\overline{D} - \lambda DN\overline{D} + \lambda DG\overline{D}.$$

Since D and $\overline{D}$ are subsets of N and G, we have $DN\overline{D} = l^2 N$ and $DG\overline{D} = l^2 G$. Therefore,

$$(DR)\overline{DR} = k(l - \mu + \mu N) - \lambda l^2 N + \lambda l^2 G.$$

If $k\mu = \lambda l^2$, then $(DR)\overline{DR} = (kl - k\mu) + k\mu G$, and Proposition 9.1.28 implies that DR is an $(mn, kl, k\mu)$-difference set in G. □

We will be using mostly the so-called *relative difference sets with classical parameters*.

Definition 9.6.11. Let q be a prime power and d a positive integer. A relative difference set with parameters

$$((q^{d+1} - 1)/(q - 1), q - 1, q^d, q^{d-1}) \tag{9.4}$$

in a cyclic group is said to be a *relative difference set with classical parameters*.

The next theorem yields relative difference sets with classical parameters.

Theorem 9.6.12. *Let q be a prime power and d a positive integer. Let $f : GF(q^{d+1}) \to GF(q)$ be a nondegenerate linear map over $GF(q)$, i.e.,* (i) *$f(ax + by) = af(x) + bf(y)$ for all $x, y \in GF(q^{d+1})$ and $a, b \in GF(q)$, and* (ii) *there exists $x_0 \in GF(q^{d+1})$ such that $f(x_0) \neq 0$. Then $f^{-1}(1)$ is an RDS in $(GF(q^{d+1})^*$ relative to $(GF(q))^*$ with parameters $((q^{d+1} - 1)/(q - 1), q - 1, q^d, q^{d-1})$.*

Proof. Since f is nondegenerate, $\dim(\ker f) = d$, i.e., $H = \ker f$ is a hyperplane in $GF(q^{d+1})$ regarded as a vector space over $GF(q)$. Therefore, $R = f^{-1}(1)$ is a d-flat and $|R| = q^d$.

Suppose, for $x, y \in R, a = xy^{-1} \in GF(q)$. Then $x = ay$ and therefore, $1 = f(x) = af(y) = a$. Thus, the identity is the only element of $(GF(q))^*$ of the form xy^{-1} with $x, y \in R$.

Let $t \in (GF(q^{d+1}))^* \setminus (GF(q))^*$. We have to verify that there are exactly q^{d-1} pairs (x, y) of elements of R such that $xy^{-1} = t$. Define a linear map $g : GF(q^{d+1}) \to GF(q)$ by $g(y) = f(ty)$. Then we have to show that there are exactly q^{d-1} elements $y \in (GF(q^{d+1})^*$ such that $f(y) = 1$ and $g(y) = 1$. Since g is a nondegenerate linear map over $GF(q)$, the set $Y = g^{-1}(1)$ is a d-flat over $GF(q)$. We have $\ker g = t^{-1} \ker f$. Since $t \notin GF(q)$, Lemma 3.5.14 implies that $\ker f \neq \ker g$, and therefore, the intersection of d-flats R and Y is a $(d - 1)$-flat. Thus, $|R \cap Y| = q^{d-1}$. The proof is now complete. □

We will now apply Theorems 9.6.10 and 9.6.12 to obtain a family of cyclic difference sets.

Example 9.6.13. Let p be a prime power, and let s, d, and m be positive integers such that $m+1=(d+1)(s+1)$. Let $q=p^{s+1}$ and let R be a cyclic $((q^{d+1}-1)/(q-1), q-1, q^d, q^{d-1})$-$RDS$ in a group G relative to a subgroup N. Let U be the unique subgroup of G of order $p-1$. Then $U \subset N$. Let $\pi\colon G \to G/U$ be the natural homomorphism. Then Proposition 9.6.8 implies that $R'=\pi(R)$ is a $((q^{d+1}-1)/(q-1), (q-1)/(p-1), q^d, q^{d-1}(p-1))$-$RDS$ in G/U relative to $N'=\pi(N)$. Let D be a cyclic $((q-1)/(p-1), p^s, p^s-p^{s-1})$-difference set in N'. (It can be, for instance, the complement of a Singer difference set.) Then, by Theorem 9.6.10, DR' is a $(p^{m+1}-1)/(p-1), p^m, p^m-p^{m-1})$-difference set in G/U. The complement D' of DR' is known as a *Gordon–Mills–Welch* or a *GMW-difference set*. It can be shown that if R is a difference set constructed in Theorem 9.6.12 and D is the complement of a Singer difference set, then the symmetric design $dev(D')$ is not isomorphic to the design $PG_{m-1}(m, p)$.

In the next theorem, we use Hadamard difference sets to obtain splitting relative difference sets.

Theorem 9.6.14. *Let D be a $(4h^2, 2h^2-h, h^2-h)$-difference set in a group G and let $N=\{1, t\}$ be a group of order 2. Let R be the following subset of the group $G \times N$:*

$$R=\{(x,1)\colon x \in D\} \cup \{(y,t)\colon y \notin D\}.$$

Then R is a $(4h^2, 2, 4h^2, 2h^2)$-RDS in $G \times N$ relative to $\{1\} \times N$.

Proof. We will identify G and N with subgroups $G \times \{1\}$ and $\{1\} \times N$ of $K=G \times N$, respectively. Then we have the following representation of R as an element of the group ring $\mathbb{Z}K$: $R=D+(G-D)(N-1)=D(2-N)+(K-G)$. Since $XN=NX$ and $XG=GX=|X|G$ for any subset X of G and $YK=KY=|Y|K$ for any subset Y of K, we obtain that $\overline{R}=\overline{D}(2-N)+(K-G)$ and $R\overline{R}=D\overline{D}(4-2N)+2(2h^2-h)(2-N)(K-G)$. Since $D\overline{D}=h^2+(2h^2-h)G$ and $(2-N)(K-G)=K-2G$, we obtain after routine manipulations that $R\overline{R}=4h^2+2h^2(K-N)$. Proposition 9.6.7 now implies that R is a $(4h^2, 2, 4h^2, 2h^2)$-RDS in K relative to N. □

We will now construct an infinite family of nonabelian relative difference sets. We first need the following lemma.

Lemma 9.6.15. *Let q be a prime power and α a primitive element of $GF(q^2)$. For $i=0,1,\ldots,q$, let $C_i=\{a\alpha^i : a \in GF(q)^*\}$. For $i, j=0,1,\ldots,q$, let M_{ij} be the multiset of all differences $u-v$ with $u \in C_i$ and $v \in C_j$. If $i \neq j$, then M_{ij} contains exactly one copy of every element of the set $GF(q^2)^* \setminus (C_i \cup C_j)$.*

If $i = j$, then M_{ij} consists of $q - 1$ copies of 0 and $q - 2$ copies of every element of C_i.

Proof. The case $i = j$ is immediate, so let $i, j \in \{0, 1, \ldots, q\}$, $i \neq j$. Let $\beta = \alpha^{q+1}$. Then β is a primitive element of $GF(q)$ and therefore, $1, \alpha, \alpha^2, \ldots, \alpha^q$ are representatives of all distinct cosets of $GF(q)^*$ in $GF(q^2)^*$. Thus, the sets $C_0, C_1, \ldots, C_q$ are these cosets. Therefore, the set $\{\alpha^i, \alpha^j\}$ is linearly independent, so it is a basis of $GF(q^2)$ over $GF(q)$. Hence, each element of $GF(q^2)^*) \setminus (C_i \cup C_j)$ has a unique representation as $a\alpha^i - b\alpha^j$ with $a, b \in GF(q)^*$. This completes the proof. □

Theorem 9.6.16. *Let q be a prime power, α a primitive element of $GF(q^2)$, and $G = \{x_1, x_2, \ldots, x_q\}$ a group of order q. For $i = 0, 1, \ldots, q$, let $C_i = \{a\alpha^i : a \in GF(q)^*\}$. Let A be the additive group of $GF(q^2)$. Let e be the identity element of G and let R be a subset of $A \times G$ defined by*

$$R = \{(0, e)\} \cup \{(a, e) \colon a \in C_0\} \cup \bigcup_{i=1}^{q} \{(a, x_i) \colon a \in C_i\}.$$

Then R is a (q^2, q, q^2, q)-RDS in $A \times G$ relative to the subgroup $N = \{(0, x) \colon x \in G\}$ of $A \times G$ that is isomorphic to G.

Proof. Observe that $|R| = q^2$ and therefore, the cardinality of the multiset $\{yz^{-1} \colon y, z \in R, y \neq z\}$ is $q^2(q^2 - 1)$. Observe also that if (a, x_i) and (b, x_j) are distinct elements of R, then $a \neq b$ and therefore $(a, x_i)(b, x_j)^{-1} \notin N$. For $c \in A \setminus \{0\}$ and $m \in \{1, 2, \ldots, q\}$, let $P(c, m)$ denote the set of all ordered pairs $((a, x_i), (b, x_j))$ of elements of R such that $(a, x_i)(b, x_j)^{-1} = (c, x_m)$, i.e., $a - b = c$ and $x_i x_j^{-1} = x_m$. If we show that $|P(c, m)| \geq q$ for all c and m, then, since

$$\sum_{c \in A \setminus \{0\}} \sum_{m=1}^{q} |P(c, m)| = q^2(q^2 - 1),$$

we obtain that $|P(c, m)| = q$ for all c and m.

Case 1: $x_m = e$ and $c \in C_0$.

By Lemma 9.6.15, there are $q - 2$ pairs (a, b) with $a, b \in C_0$ and $a - b = c$. For any such pair, we have $((a, e), (b, e)) \in P(c, m)$. Besides, $((c, e), (0, e)) \in P(c, m)$ and $((0, e), (-c, e)) \in P(c, m)$. Therefore, $|P(c, m)| \geq q$.

Case 2: $x_m = e$ and $c \in C_n$ with $1 \leq n \leq q$.

By Lemma 9.6.15, there are $q - 2$ pairs (a, b) with $a, b \in C_n$ and $a - b = c$. For any such pair, we have $((a, x_n), (b, x_n)) \in P(c, m)$. If $n \neq m$, then there is a pair (a, b) with $a \in C_m$, $b \in C_0$, and $a - b = c$ and a pair (a', b') with $a' \in C_0$, $b' \in C_m$, and $a' - b' = c$. Then $((a, x_m), (b, e)) \in P(c, m)$ and $((a', e), (b', x_m)) \in P(c, m)$, so $|P(c, m)| \geq q$. If $n = m$, then $((c, x_m), (0, e)) \in P(c, m)$ and $((0, e), (-c, x_m)) \in P(c, m)$, so again $|P(c, m)| \geq q$.

Case 3: $x_m \neq e$.

Define a permutation σ of the set $\{1, 2, \ldots, q\}$ by $x_{\sigma(i)} = x_m x_i$. Then $\sigma(i) \neq i$ for $i = 1, 2, \ldots, q$ and Lemma 9.6.15 implies that the multiset M_i of all differences $u - v$ with $u \in C_{\sigma(i)}$ and $v \in C_i$ is in fact the set $GF(q^2)^* \setminus (C_{\sigma(i)} \cup C_i)$. Therefore, the multiset $M_1 \cup M_2 \cup \ldots \cup M_q$ contains q copies of every element of C_0 and $q - 2$ copies of every element of $C_1 \cup C_2 \cup \ldots \cup C_q$. This gives us q elements of $P(c, m)$ if $c \in C_0$ and $q - 2$ elements of $P(c, m)$ if $c \notin C_0$.

Let $x_m^{-1} = x_k$. If $c \notin C_0 \cup C_m \cup C_k$, then there is a pair (a, b) with $a \in C_m$, $b \in C_0$, and $a - b = c$ and a pair (a', b') with $a' \in C_0$, $b' \in C_k$, and $a' - b' = c$. This gives two more elements of $P(c, m)$, namely $((a, x_m), (b, e))$ and $((a', e), (b', x_k))$. If $k \neq m$ and $c \in C_m$, then $((c, x_m), (0, e)) \in P(c, m)$ and there is a pair (a, b) with $a \in C_0$, $b \in C_k$, and $a - b = c$, so $((a, e), (b, x_k)) \in P(c, m)$. If $k \neq m$ and $c \in C_k$, then $(0, e), (-c, x_k) \in P(c, m)$ and there is a pair (a, b) with $a \in C_m$, $b \in C_0$, and $a - b = c$, so $((a, x_m), (b, e)) \in P(c, m)$. Finally, if $k = m$ and $c \in C_m$, then $((c, x_m), (0, e)) \in P(c, m)$ and $((0, e), (-c, x_k)) \in P(c, m)$. Thus, in every case $|P(c, m)| \geq q$. The proof is now complete. □

Corollary 9.6.17. *Let p be a prime and let $m \geq n$ be positive integers. Let A be the additive group of $GF(p^{2m})$ and G any group of order p^n. Then there exists a $(p^{2m}, p^n, p^{2m}, p^{2m-n})$-RDS in $A \times G$ relative to G.*

Proof. Let H be a group of order p^{m-n}. By Theorem 9.6.16, there exists a $(p^{2m}, p^m, p^{2m}, p^m)$-$RDS$ in $A \times (H \times G)$ relative to $H \times G$. Since $A \times G$ is a homomorphic image of $A \times (H \times G)$ under a homomorphism with kernel H, Proposition 9.6.8 implies that there exists a $(p^{2m}, p^n, p^{2m}, p^{2m-n})$-$RDS$ in $A \times G$ relative to G. □

Exercises

(1) Let G be a group of order m and H a group of order n. Let M be a matrix of order mn represented as a block matrix $M = [M_{ij}]$ with H-invariant blocks M_{ij}. Suppose further that there exists a bijection $\varphi\colon G \to \{1, 2, \ldots, m\}$ such that $M_{\varphi(x),\varphi(y)} = M_{\varphi(xz),\varphi(yz)}$ for all $x, y, z \in G$. Prove that the matrix M is $(G \times H)$-invariant.

(2) Use Exercise 35 of Chapter 2 to obtain difference sets D_1 in the group $\mathbf{Z}_2^4$ and $\mathbf{D}_2$ in the group $\mathbf{Z}_4^2$ such that the symmetric designs $dev(D_1)$ and $dev(D_2)$ are isomorphic.

(3) Prove Proposition 9.1.31.

(4) Prove Proposition 9.1.32.

(5) Let G be a finite abelian group. Then G can be represented as a direct product of finite cyclic groups.

   (a) Prove that the group ring of a cyclic group of order v has exactly v distinct characters.

   (b) Prove that the group ring of a finite abelian group of order v has exactly v distinct characters.

   (c) Define the product of characters χ_1 and χ_2 of $\mathbb{Z}G$ by: $(\chi_1\chi_2)(x) = \chi_1(x)\chi_2(x)$. Prove that all characters of $\mathbb{Z}G$ form a group G^* isomorphic to G. The group G^* is called the *character group of G*.

   (d) Prove Proposition 9.1.33.

   (e) Prove Corollary 9.1.34.

   (f) Let U be a subgroup of G and let $U^\perp = \{\chi \in G^* : \chi(x) = 1 \text{ for all } x \in U\}$. Prove that $U^\perp$ is a subgroup of G^* of order $|G|/|U|$.

(6) Let G be a finite abelian group of order v and let $\alpha = \sum_{x\in G} a_x x$ be an element of $\mathbb{Z}G$. Let $\chi_0, \chi_1, \ldots, \chi_{v-1}$ be all the characters of $\mathbb{Z}G$. Prove the *inversion formula*:

$$\text{for each } x \in G, \quad a_x = \frac{1}{v}\sum_{i=0}^{v-1} \chi_i(\alpha)\chi_i(x^{-1}).$$

(7) Let G be a finite abelian group and let χ be a character of $\mathbb{C}G$ ($\mathbb{C}$ is the field of complex numbers). Let M be a G-invariant matrix. Let $\mathbf{e}$ be a vector whose components are indexed by elements of G (in the same order as the rows and columns of M). Let $\mathbf{e}(x) = \chi(x^{-1})$ for each $x \in G$. Prove that $\mathbf{e}$ is an eigenvector of M (over $\mathbb{C}$) corresponding to the eigenvalue $\chi(\alpha)$, where $\alpha = \sum_{x\in G} M(x, 1)x$.

(8) Let H and K be groups of order 2 and let $G = H \times K$.

   (a) Find a $(2, 2, 2)$-BS in G relative to H.

   (b) Find a $(2, 2, 4, -)$-EBS on G admitting a symmetry of order 2.

   (c) Construct the difference set of Corollary 9.5.4.

(9) Let R be an (m, n, k, λ)-RDS and let $\mathbf{D} = dev(R)$. Prove that the point set of $\mathbf{D}$ can be partitioned into n-subsets $X_1, X_2, \ldots, X_m$ so that, for any distinct points $x \in X_i$ and $y \in X_j$ of $\mathbf{D}$,

$$\lambda(x, y) = \begin{cases} \lambda & \text{if } i \neq j, \\ 0 & \text{if } i = j. \end{cases}$$

Incidence structures with this property are called *group divisible designs*.

(10) Let G be a group of order mn with a subgroup N of order n. A k-subset D of G is called a *divisible difference set with parameters* (m, n, k, λ, μ) (or an (m, n, k, λ, μ)-DDS) in G relative to N if the multiset $\{xy^{-1} : x, y \in D\}$ contains exactly λ copies of every nonidentity element of N and exactly μ copies of every element of $G \setminus N$.

(a) Show that a subset D of G is an (m, n, k, λ, μ)-DDS relative to N if and only if the following equation in $\mathbb{Z}G$ is satisfied: $D\overline{D} = k - \lambda + (\lambda - \mu)N + \mu G$.

(b) Prove that if D is an (m, n, k, λ, μ)-DDS, then $\lambda(n-1) + \mu n(m-1) = k(k-1)$.

(c) Let R be an (m, n, k, λ)-RDS in G relative to N and let D be an (s, t, l, μ_1, μ_2)-DDS in N relative to a subgroup T of N (so $n = st$). Prove that if $k\mu_2 = \lambda l^2$, then the element DR of $\mathbb{Z}G$ can be regarded as a subset of G, and this subset is an $(ms, t, kl, k\mu_1, k\mu_2)$-$DDS$ in G relative to T.

(d) Construct a $(7, 7, 21, 7, 9)$-DDS.

Notes

The notion of a cyclic difference set (in relation to projective planes) first appeared in Singer (1938). For this reason, a sharply transitive automorphism group of a symmetric design is often referred to as a Singer group of the design. Bose (1939) introduced abelian difference sets as a method for constructing designs. The systematic study of difference sets began in the seminal paper by Hall (1947). The general definition of a difference set in an arbitrary finite group occurs in Bruck (1955). The theory of difference sets is now a well established area of research which uses various techniques from finite groups, theory of rings and fields, and algebraic number theory. We have discussed a relatively small part of this theory that will be used for constructing symmetric designs in subsequent chapters.

Circulant incidence matrices of symmetric designs were considered in Hall and Ryser (1951). Group invariant matrices is the natural generalization of circulant matrices. The character theory of finite abelian groups serves as a bridge between theory of difference sets and algebraic number theory. It often provides restrictions on the abelian group containing a difference set with given parameters. We mention the monographs of Mann (1965a), Lander (1983), and Schmidt (2002) and the paper of Turyn (1965).

For the most compr hensive treatment of difference sets and for further references, see the monographs Beth, Jungnickel and Lenz (1999) and Pott (1995). (See also Jungnickel (1992) and Jungnickel and Pott (1996).) For cyclic and cyclotomic difference sets, see Storer (1967), Baumert (1971), and Hall (1986).

Stanton-Sprott difference sets were discovered in Stanton and Sprott (1958). Our proof follows van Lint and Wilson (2001).

The notion of a group of symmetries of a subset of a group ring, and construction methods of symmetries were introduced in Ionin (1999b). A more general notion of a group of symmetries of a set of matrices will appear in Chapter 10. Building blocks, building sets, and extended building sets were introduced in Davis and Jedwab (1997, 1998) as a unified approach to constructing several infinite families of difference sets, including McFarland difference sets first constructed in McFarland (1973), Spence difference sets first constructed in Spence (1977), Davis–Jedwab difference sets, Hadamard difference sets, and Chen difference sets. The first infinite family of Hadamard difference sets (also known as *Menon* or *Hadamard–Menon difference sets*) was constructed in Menon (1960). Theorem 9.5.12 is due to Menon (1962b). (See Davis and Jedwab (1996) for further references on abelian Hadamard difference sets.) Chen difference sets are due to Chen (1998). Theorem 9.5.11 is due to K. Smith (1995) where a nonabelian (100, 45, 20)-difference set is constructed.

Relative difference sets seem to have been introduced in Plackett and Burman (1946). In the paper by Butson (1963), relative difference sets in cyclic groups were introduced and relative difference sets with classical parameters were constructed. Elliott and Butson (1966) gave the general definition of an RDS, constructed several families of relative difference sets in elementary abelian groups, and proved Proposition 9.6.8. (See Pott (1995) for further results and references on relative difference sets.)

Theorem 9.6.10 is proven in Pott (1995) though it was applied (in a different context) to construct difference sets in Gordon, Mills and Welch (1962). Kantor (2001) proved that symmetric designs obtained as the developments of GMW-difference sets are not isomorphic to the designs $PG_{d-1}(d, q)$.

We follow Pott (1995) in the proof of Theorem 9.6.5 and in Exercise 7. Theorem 9.6.14 is due to Jungnickel (1982b).

The relative difference sets of Theorem 9.6.16 and Corollary 9.6.17 and the relative difference set of Example 9.6.4 were discovered in de Launey (1989) and in Glynn (1978), respectively, in the form of group invariant balanced generalized weighing matrices (see Theorem 10.3.8 and Corollary 10.3.7).

For the connection between character values and the eigenvalues of group invariant matrices, described in Exercise 7, see Pott (1995), p. 24. The results of Exercise 10 are due to Ionin (2000).

10

Balanced generalized weighing matrices

Balanced generalized weighing matrices and their special case, *generalized Hadamard matrices*, are matrices over groups that generalize both incidence matrices of symmetric designs and Hadamard matrices. They are of significant interest by themselves, and are a powerful tool for constructing symmetric designs.

10.1. Basic properties of BGW-matrices

We begin with some basic properties of matrices over a group ring. In Section 9.1 we defined the operation of conjugation on the group ring $\mathbb{Z}G$. For $\alpha = \sum_{x \in G} a_x x$, we have $\overline{\alpha} = \sum_{x \in G} a_x x^{-1}$. We will now extend this notion to matrices over $\mathbb{Z}G$.

Definition 10.1.1. Let G be a finite group and let $A = [\alpha_{ij}]$ be a matrix over $\mathbb{Z}G$. Let $\overline{A} = [\overline{\alpha_{ij}}]$ and $A^* = \overline{A}^\top$. Matrices A and A^* are called *conjugate*.

Proposition 9.1.27 implies the following result.

Proposition 10.1.2. *Let G be a finite group and let A and B be matrices over $\mathbb{Q}G$. If $A + B$ is defined, then $(A + B)^* = A^* + B^*$; if AB is defined, then $(AB)^* = B^* A^*$.*

We will now define the central notion of this chapter.

Definition 10.1.3. Let G be a finite group. A *balanced generalized weighing matrix with parameters* (v, k, λ) *over* G or a $BGW(v, k, \lambda; G)$ is a $(0, G)$ matrix $W = [\omega_{ij}]$ of order v such that (i) every row of W contains exactly k nonzero entries and (ii) for any distinct $i, h \in \{1, 2, \ldots, v\}$, the multiset $\{\omega_{ij}\overline{\omega_{hj}} : 1 \leq$

$j \leq v\}$ contains exactly $\lambda/|G|$ copies of every element of G. The integer $\lambda/|G|$ is called the *index* of the matrix W.

Remark 10.1.4. We will sometimes write $BGW(v, k, \lambda)$ for $BGW(v, k, \lambda; G)$.

This definition can be expressed as an equation over the group ring.

Proposition 10.1.5. *Let G be a finite group. A $(0, G)$ matrix W of order v is a $BGW(v, k, \lambda; G)$ if and only if*

$$WW^* = \left(k - \frac{\lambda}{|G|}G\right) I + \frac{\lambda}{|G|}GJ. \tag{10.1}$$

in $\mathbb{Z}G$.

Corollary 10.1.6. *A matrix $BGW(v, k, \lambda)$ over the trivial group is an incidence matrix of a symmetric (v, k, λ)-design, and vice versa.*

We will now give two examples of balanced generalized weighing matrices over nontrivial groups.

Example 10.1.7. Let $S = \langle \sigma \rangle$ be a cyclic group of order 3. Then

$$W = \begin{bmatrix} 0 & 1 & 1 & 1 & 1 \\ 1 & 0 & 1 & \sigma & \sigma^2 \\ 1 & 1 & 0 & \sigma^2 & \sigma \\ 1 & \sigma & \sigma^2 & 0 & 1 \\ 1 & \sigma^2 & \sigma & 1 & 0 \end{bmatrix}$$

is a $BGW(5, 4, 3; S)$. The matrix M of Example 9.1.11 is another $BGW(5, 4, 3; S)$. Matrices W and M are monomially equivalent. (See Definition 10.1.14.)

Example 10.1.8. Let $G = \langle \alpha, \beta \rangle$ be the direct product of two cyclic groups of order 2. Then

$$H = \begin{bmatrix} 1 & 1 & 1 & 1 \\ 1 & \alpha & \beta & \alpha\beta \\ 1 & \beta & \alpha\beta & \alpha \\ 1 & \alpha\beta & \alpha & \beta \end{bmatrix}$$

is a $BGW(4, 4, 4; G)$.

Observe that any Hadamard matrix of order $n \geq 2$ can be regarded as a $BGW(n, n, n)$ over a group of order 2. This motivates the following definition.

Definition 10.1.9. Let G be a finite group. Any matrix $BGW(v, v, v; G)$ is called a *generalized Hadamard matrix of order* v *over* G or a $GH(G; \lambda)$ where $\lambda = v/|G|$.

Remark 10.1.10. The index of $GH(G; \lambda)$ is λ.

Thus, a Hadamard matrix of order $n \geq 2$ is a $GH(\mathbf{Z}_2, n/2)$. In general, if $H = [\omega_{ij}]$ is a $GH(G; \lambda)$, then, for distinct $i, h \in \{1, 2, \ldots, \lambda|G|\}$, the multiset $\{\omega_{ij}\overline{\omega_{hj}} \colon 1 \leq j \leq \lambda|G|\}$ contains λ copies of every element of G.

Hadamard matrices are not the only matrices with pairwise orthogonal rows that can be regarded as GH-matrices.

Proposition 10.1.11. *Let p be a prime and let C_p be the group of complex p^{th} roots of unity. Let H be a matrix of order n with all entries from C_p such that the equation $HH^* = nI$ holds over the complex numbers. Then H is a generalized Hadamard matrix over C_p.*

Proof. We have to show that if $HH^* = nI$ holds over the complex numbers, then $HH^* = nI + (n/p)C_p(J - I)$ holds in the group ring $\mathbb{Z}C_p$. This will follow if we show an equation

$$\sum_{j=1}^{n} \eta_j \overline{\zeta_j} = 0 \tag{10.2}$$

for complex p^{th} roots of unity $\eta_1, \eta_2, \ldots .\eta_n$ and $\zeta_1, \zeta_2, \ldots .\zeta_n$ implies the equation

$$\sum_{j=1}^{n} \eta_j \overline{\zeta_j} = \frac{n}{p} C_p \tag{10.3}$$

in $\mathbb{Z}C_p$.

Let $\gamma \neq 1$ be a pth root of unity. Then (10.2) can be rewritten as

$$\sum_{j=0}^{p-1} a_j \gamma^j = 0, \tag{10.4}$$

where $a_0, a_1, \ldots, a_{p-1}$ are nonnegative integers. Since $\gamma^{p-1} = -(1 + \gamma + \cdots + \gamma^{p-2})$, we rewrite (10.4) as

$$\sum_{j=0}^{p-2} (a_j - a_{p-1})\gamma^j = 0. \tag{10.5}$$

Since the polynomial $x^{p-1} + x^{p-2} + \cdots + p + 1$ is irreducible over $\mathbb{Q}$, the set $\{1, \gamma, \ldots, \gamma^{p-2}\}$ is linearly independent over $\mathbb{Q}$. Therefore, (10.5) implies that

$a_1 = a_2 = \cdots = a_{p-1}$. Thus, the multiset $\{\eta_j \overline{\zeta_j} \colon 1 \le j \le n\}$ contains every element of C_p the same number of times. This implies (10.3). □

Remark 10.1.12. The statement of Proposition 10.1.11 is not true without the assumption that p is a prime. (See Exercise 2.)

If W is a $BGW(v, k, \lambda; G)$ and W' is obtained by permuting rows or columns of W, then W' is a $BGW(v, k, \lambda; G)$. The next proposition describes two more operations preserving the parameters of BGW-matrices.

Proposition 10.1.13. *Let W be a $BGW(v, k, \lambda; G)$ and let $x \in G$. Let W' be obtained by multiplying all entries of one row of W by x on the left or by multiplying all entries of one column of W by x on the right. Then W' is a $BGW(v, k, \lambda; G)$.*

Proof. Let $W = [\omega_{ij}]$. If we fix j and replace each ω_{ij} with $\omega_{ij}x$, then $(\omega_{ij}x)(\overline{\omega_{hj}x}) = \omega_{ij}\overline{\omega_{hj}}$ and therefore the multiset $M_{ih} = \{\omega_{ij}\overline{\omega_{hj}} \colon 1 \le j \le v\}$ does not change. If we fix i and replace each ω_{ij} with $x\omega_{ij}$, then every element of the multiset M_{ih} is multiplied by x on the left. Since the multiset M_{ih} contains equal number of copies of every element of G, it does not change. Similarly, every element of M_{hi} is multiplied by $\overline{x}$ on the right and therefore this multiset does not change either. □

Definition 10.1.14. Two BGW-matrices over a group G are called *monomially equivalent* if one can be obtained from the other by applying any combination of the following operations: permuting rows, permuting columns, multiplying all entries of one row by the same element of G on the left, multiplying all entries of one column by the same element of G on the right.

Remark 10.1.15. If $G = \langle x \rangle$ is a cyclic group, then the elements of G can be thought of as monomials x^n. This explains the term "monomially equivalent".

Observe that any BGW-matrix is monomially equivalent to a *normalized* matrix, that is, a matrix having all nonzero entries in the first row and the first column equal to 1 and all zeros preceding ones.

Definition 10.1.16. Let $W = [\omega_{ij}]$ be a $BGW(v, k, \lambda)$. The matrix W is called *normalized* if $\omega_{1j} = \omega_{j1} = 1$ for $j = v - k + 1, v - k + 2, \ldots, v$.

BGW-matrices described in examples 10.1.7 and 10.1.8 belong to infinite families that are constructed in the following two propositions.

Proposition 10.1.17. *Let q be a prime power and let G be the multiplicative group of the field $GF(q) = \{a_1, a_2, \ldots, a_q\}$. Let a $(0, G)$ matrix $W = [\omega_{ij}]$ of* *order $q+1$ be defined by*

$$\omega_{ij} = \begin{cases} a_{i-1} - a_{j-1} & \text{if } i \neq 1 \text{ and } j \neq 1, \\ 0 & \text{if } i = j = 1, \\ 1 & \text{otherwise.} \end{cases}$$

Then W is a normalized $BGW(q+1, q, q-1; G)$.

Proof. Clearly, each row of W has exactly q nonzero entries. Let $i, h \in \{1, 2, \dots, q, q+1\}$, $i \neq h$. If $i \neq 1$ and $h \neq 1$, then we have, for $j \neq h$ and $j \neq 1$,

$$\omega_{ij}\overline{\omega_{hj}} = (a_{i-1} - a_{j-1})(a_{h-1} - a_{j-1})^{-1} = (a_{i-1} - a_{h-1})(a_{h-1} - a_{j-1})^{-1} + 1.$$

As j runs through the set $\{2, 3, \dots, q+1\} \setminus \{h\}$, the difference $a_{h-1} - a_{j-1}$ runs through the set G. Since $a_{i-1} - a_{h-1} \neq 0$, the product $(a_{i-1} - a_{h-1})(a_{h-1} - a_{j-1})^{-1}$ runs through G. Since $\omega_{i1}\overline{\omega_{h1}} = 1$, the multiset $\{\omega_{ij}\overline{\omega_{hj}} : 1 \leq j \leq q+1\}$ contains one copy of every element of G (and two copies of 0). We leave the case $i = 1$ or $h = 1$ to the reader. □

Observe that the above proof would be valid for a finite algebraic structure that is "almost a finite field", i.e., a structure satisfying all the axioms of a finite field, except the commutativity of multiplication and the left distributive law. Such structures are known as *right nearfields*.

Definition 10.1.18. A *right nearfield* is a set F with two binary operations, $+$ and $\circ$, such that (i) $(F, +)$ is an abelian group (with 0 as the identity element), (ii) $(F \setminus \{0\}, \circ)$ is a group, and (iii) $(x+y) \circ z = x \circ z + y \circ z$ for all $x, y, z \in F$.

Remark 10.1.19. *Left nearfields* can be defined in a similar fashion with (iii)$'$ $z \circ (x+y) = z \circ x + z \circ y$ instead of (iii). We will not consider left nearfields in this book and therefore we will use the term *nearfields* for right nearfields.

Example 10.1.20. Let q be an odd prime power, let α be a primitive element of $GF(q^2)$, and let F be the set of all elements of $GF(q^2)$. Define an operation $\circ$ on F as follows:

$$x \circ y = \begin{cases} 0 & \text{if } y = 0, \\ xy & \text{if } y = \alpha^s, s \text{ is even}, \\ x^q y & \text{if } y = \alpha^s, s \text{ is odd}. \end{cases}$$

Then $(F, +, \circ)$ is a nearfield of order q^2. The multiplicative group of this nearfield is not abelian.

We have the following result.

Proposition 10.1.21. *Let G be the multiplicative group of a finite nearfield $F = \{a_1, a_2, \ldots, a_q\}$. Let a $(0, G)$ matrix $W = [\omega_{ij}]$ of order $q + 1$ be defined by*

$$\omega_{ij} = \begin{cases} a_{i-1} - a_{j-1} & \text{if } i \neq 1 \text{ and } j \neq 1, \\ 0 & \text{if } i = j = 1, \\ 1 & \text{otherwise}. \end{cases}$$

Then W is a normalized $BGW(q + 1, q, q - 1; G)$.

This result leads to the following question: which groups arise as the multiplicative groups of nearfields? The next theorem, which we state without proof, addresses this question. We begin with the following definition.

Definition 10.1.22. The *kernel of a nearfield F* is the set of elements $a \in F$ such that $ax = xa$ for all $x \in F$.

Theorem 10.1.23. *Let F be a finite nearfield. Then the kernel of F is a field, and therefore F can be regarded as a vector space over $GF(q)$ for some prime power q. Moreover, if $n = \dim F$, then* (i) *every prime divisor of n divides $q - 1$ and* (ii) *if $q \equiv 3 \pmod 4$, then $n \not\equiv 0 \pmod 4$. Conversely, if a prime power q and a positive integer n satisfy* (i) *and* (ii)*, then there exists a nearfield F of order q^n with $GF(q)$ as the kernel and with the multiplicative group F^* generated by elements x and y such that $y^m = 1$, $x^n = y^t$, and $xyx^{-1} = y^q$ with $m = (q^n - 1)/n$ and $t = m/(q - 1)$.*

Remark 10.1.24. Besides the nearfields introduced in Theorem 10.1.23, there are seven exceptional nearfields, $N_1, N_2, \ldots, N_7$, of order p^2 with $p = 5, 11, 7, 23, 11, 29$, and 59, respectively. Their multiplicative groups can be described as groups generated by matrices of order 2 over $GF(p)$ in the following way: $N_i^* = \langle A, B_i \rangle$ for $i = 1, 3, 5$ and $N_i^* = \langle A, B_i, C_i \rangle$ for $i = 2, 4, 6, 7$ where

$$A = \begin{bmatrix} 0 & -1 \\ 1 & 0 \end{bmatrix}, \; B_1 = \begin{bmatrix} 1 & -2 \\ -1 & -2 \end{bmatrix}, \; B_2 = \begin{bmatrix} 1 & 5 \\ -5 & -2 \end{bmatrix}, \; C_2 = \begin{bmatrix} 4 & 0 \\ 0 & 4 \end{bmatrix},$$
$$B_3 = \begin{bmatrix} 1 & 3 \\ -1 & -2 \end{bmatrix}, \; B_4 = \begin{bmatrix} 1 & -6 \\ 12 & -2 \end{bmatrix}, \; C_4 = \begin{bmatrix} 2 & 0 \\ 0 & 2 \end{bmatrix}, \; B_5 = \begin{bmatrix} 2 & 4 \\ 1 & -3 \end{bmatrix},$$
$$B_6 = \begin{bmatrix} 1 & -7 \\ -12 & -2 \end{bmatrix}, \; C_6 = \begin{bmatrix} -13 & 0 \\ 0 & -13 \end{bmatrix}, \; B_7 = \begin{bmatrix} 9 & 15 \\ -10 & -10 \end{bmatrix}, \; C_7 = \begin{bmatrix} 4 & 0 \\ 0 & 4 \end{bmatrix}.$$

We will now return to BGW-matrices. Observe that any conference matrix of order $n + 1$ is a $BGW(n + 1, n, n - 1)$ over a group of order 2.

Definition 10.1.25. Any $BGW(n + 1, n, n - 1; G)$ is called a *generalized conference matrix*.

Remark 10.1.26. The above constructions imply that generalized conference matrices $BGW(n+1, n, n-1; G)$ exist whenever n is a prime power and G is the multiplicative group of the field or a nearfield of order n. Another known family of generalized conference matrices is $BGW(n+1, n, n-1; G)$, where $n = 2^m(2^{m+1} - 1)$ with $m \geq 2$ and G is a cyclic group of order $2^m - 1$. Here is an example of a $BGW(29, 28, 27; G)$ where $G = \{1, \omega, \overline{\omega}\}$ is a cyclic group of order 3:

$$W = \begin{bmatrix} 0 & 1 & \mathbf{j}^\top & \mathbf{j}^\top & \mathbf{j}^\top \\ 1 & 0 & \mathbf{j}^\top & \omega\mathbf{j}^\top & \overline{\omega}\mathbf{j}^\top \\ \mathbf{j} & \mathbf{j} & A & B & C \\ \mathbf{j} & \omega\mathbf{j} & B & D & E \\ \mathbf{j} & \overline{\omega}\mathbf{j} & C & E & F \end{bmatrix}$$

where A, B, C, D, E, and F are symmetric circulant matrices of order 9 whose first rows are $[0\ 1\ \omega\ \omega\ \overline{\omega}\ \overline{\omega}\ \omega\ \omega\ 1]$, $[\overline{\omega}\ \overline{\omega}\ 1\ \omega\ 1\ 1\ \omega\ 1\ \overline{\omega}]$, $[\omega\ \overline{\omega}\ \omega\ 1\ \overline{\omega}\ \overline{\omega}\ 1\ \omega\ \overline{\omega}]$, $[0\ \overline{\omega}\ 1\ \overline{\omega}\ \omega\ \omega\ \overline{\omega}\ 1\ \overline{\omega}]$, $[1\ \omega\ \omega\ \overline{\omega}\ 1\ 1\ \overline{\omega}\ \omega\ \omega]$, and $[0\ \omega\ \overline{\omega}\ 1\ 1\ 1\ 1\ \overline{\omega}\ \omega]$, respectively.

Since generalized Hadamard matrices $GH(G; \lambda)$ do not have zero entries, we may write the group G additively. We will employ the additive notation in the following proposition.

Proposition 10.1.27. *Let q be a prime power and let G be the additive group of the field $GF(q) = \{a_1, a_2, \ldots, a_q\}$ with $a_1 = 0$. Then the matrix $H = [a_i a_j]$ of order q is a normalized $GH(G; 1)$.*

Proof. For distinct $i, h \in \{1, 2, \ldots, q\}$, the multiset $\{a_i a_j - a_h a_j\}$ is in fact the entire group G. Since all the entries in the first row and the first column of H are equal to 0, H is a normalized $GH(G; 1)$. □

Remark 10.1.28. Almost all known matrices $GH(G; \lambda)$ have the group G isomorphic to the additive group of a finite field. For a prime power q, we will use $GH(q, \lambda)$ instead of $GH(G; \lambda)$ if G is isomorphic to the additive group of $GF(q)$.

Applying a group homomorphism to every nonzero entry of a BGW-matrix W over G yields a BGW-matrix with the same parameters over the homomorphic image of G. Proof of the following proposition is straightforward.

Proposition 10.1.29. *Let G and G' be finite groups. Let $f : \mathbb{Z}G \to \mathbb{Z}G'$ be a ring homomorphism such that $f(G) = G'$. If $W = [\omega_{ij}]$ is a $BGW(v, k, \lambda; G)$, then $W' = [f(\omega_{ij})]$ is a $BGW(v, k, \lambda; G')$.*

If G' is the trivial group, we obtain the following results.

Corollary 10.1.30. *If W is a $BGW(v, k, \lambda)$, then replacing every nonzero entry of W by 1 yields an incidence matrix of a symmetric (v, k, λ)-design.*

Corollary 10.1.31. *If $v > k$, then no two distinct rows or distinct columns of a $BGW(v, k, \lambda)$ are proportional.*

Since every column of an incidence matrix of a symmetric (v, k, λ)-design has k nonzero entries and the inner product of distinct columns is equal to λ, we obtain the following result.

Corollary 10.1.32. *If $W = [\omega_{ij}]$ is a $BGW(v, k, \lambda; G)$, then every column of W has exactly k nonzero entries and, for distinct $j, h \in \{1, 2, \ldots, v\}$,*

$$\sum_{i=1}^{v} \omega_{ij}\overline{\omega_{ih}} = \sum_{x \in G} a_x x,$$

where all a_x are nonnegative integers and $\sum_{x \in G} a_x = \lambda$.

The following modification of the definition of BGW-matrices is sometimes useful.

Proposition 10.1.33. *Let G be a finite group and let $W = [\omega_{ij}]$ be a $(0, G)$-matrix of order v. Suppose there exist positive integers k and λ such that* (i) *every column of W has exactly k nonzero entries and* (ii) *for any distinct $i, h \in \{1, 2, \ldots, v\}$, the multiset $\{\omega_{ij}\overline{\omega_{hj}} : 1 \leq j \leq v\}$ contains exactly $\lambda/|G|$ copies of every element of G. Then W is a $BGW(v, k, \lambda; G)$.*

Proof. Replacing every nonzero entry of W by 1 yields a $(0, 1)$-matrix of order v with constant column sum k and inner product of distinct rows λ, i.e., an incidence matrix of a symmetric (v, k, λ)-design. Therefore, every row of W has exactly k nonzero entries and W is a $BGW(v, k, \lambda; G)$. □

Note that if W is a BGW-matrix over a nonabelian group, then neither $W^\top$ nor $\overline{W}$ has to be a BGW-matrix (cf. Exercise 4c). However, W^* is a BGW-matrix.

Proposition 10.1.34. *Let G be a finite group. If W is a $BGW(v, k, \lambda; G)$, then W^* is a $BGW(v, k, \lambda; G)$.*

Proof. Let W be a $BGW(v, k, \lambda; G)$. For this proof, we will regard W as a matrix over $\mathbb{Q}G$ rather than $\mathbb{Z}G$. Then W satisfies (10.1) in $\mathbb{Q}G$. Let $\widetilde{G} = \{aG \in \mathbb{Q}G : a \in \mathbb{Q}\}$. Since $xG = Gx = G$ for any $x \in G$, we obtain that $\widetilde{G}$ is an ideal of the ring $\mathbb{Q}G$. Let R denote the factor-ring $\mathbb{Q}G/\widetilde{G}$ and let $\pi : \mathbb{Q}G \to R$ be the natural homomorphism. For any matrix A over $\mathbb{Q}G$, let πA denote the matrix obtained by applying π to each entry of A. Reducing (10.1) modulo $\widetilde{G}$ yields

$(\pi W)(\pi W^*) = kI$. Therefore, the matrix πW is invertible, $\pi W^* = k(\pi W)^{-1}$, and then $(\pi W^*)(\pi W) = kI$. Therefore, $W^* W = kI + A$ where A is a matrix over $\widetilde{G}$. Let $A = [a_{jh}G]$ where all a_{jh} are rational numbers. Corollary 10.1.32 implies that $a_{jj} = 0$ for $j = 1, 2, \ldots, v$, and, for distinct j and h, $va_{jh} = \lambda$. Therefore, $A = (\lambda/|G|)G(J - I)$, and then

$$W^* W = \left(k - \frac{\lambda}{|G|}G\right) I + \frac{\lambda}{|G|} GJ.$$

Proposition 10.1.5 now implies that W^* is a $BGW(v, k, \lambda; G)$. □

Corollary 10.1.35. *Let G be a finite abelian group. If W is a BGW $(v, k, \lambda; G)$, then so are the matrices $\overline{W}$ and $W^\top$.*

Proof. Let W be a $BGW(v, k, \lambda; G)$. Since G is abelian, the map $x \mapsto x^{-1}$ is an automorphism of G. By Proposition 10.1.29, $\overline{W}$ is a $BGW(v, k, \lambda; G)$. Then by Proposition 10.1.34, the matrix $(\overline{W})^* = W^\top$ is a $BGW(v, k, \lambda; G)$.

□

10.2. BGW-matrices with classical parameters

For a prime power q and a positive integer m, BGW-matrices over a cyclic group of order $q - 1$ and with the parameters of the complement of the design $PG_{m-1}(m, q)$ are often called *balanced generalized weighing matrices with classical parameters*. If W is such a matrix, it is convenient to assume that G is the multiplicative group of $GF(q)$ and each zero entry of W is the zero of $GF(q)$. Then W can be regarded as a matrix over $GF(q)$. In this case, we will say that W is a $PBGW(m, q)$.

Definition 10.2.1. Let q be a prime power and m a positive integer. A matrix W over $GF(q)$ is said to be a $PBGW(m, q)$ if it is a balanced generalized weighing matrix $BGW(v, k, \lambda; G)$ where G is the multiplicative group of $GF(q)$ and

$$(v, k, \lambda) = \left(\frac{q^{m+1} - 1}{q - 1}, q^m, q^m - q^{m-1}\right).$$

Remark 10.2.2. The matrix $BGW(q + 1, q, q - 1)$ constructed in Proposition 10.1.17 is in fact a $PBGW(1, q)$.

Since any $PBGW(m, q)$ is a matrix over a field, we can consider its rank.

Proposition 10.2.3. *Let q be a prime power and m a positive integer. If W is a $PBGW(m, q)$, then $rank(W) \geq m + 1$.*

Proof. Let W be a $PBGW(m,q)$. Let $v=(q^{m+1}-1)/(q-1)$ and let $\mathbf{x}_1,\mathbf{x}_2,\ldots,\mathbf{x}_v$ be the columns of W regarded as elements of the vector space $V(m+1,q)$. Corollary 10.1.31 implies that the vectors $\mathbf{x}_1,\mathbf{x}_2,\ldots,\mathbf{x}_v$ represent v distinct one-dimensional subspaces of $V(m+1,q)$. Since $V(m+1,q)$ has exactly v one-dimensional subspaces, these vectors represent all one-dimensional subspaces of $V(m+1,q)$. Therefore, among them there are $m+1$ linearly independent vectors, which implies that $rank(W)\geq m+1$. □

In this section we will characterize the $PBGW$-matrices of minimal rank. We begin with a construction of a matrix $PBGW(m,q)$ of rank $m+1$. This construction is based on the properties of simplex codes presented in Section 3.9.

Theorem 10.2.4. *Let q be a prime power and m a positive integer. Let $v=(q^{m+1}-1)/(q-1)$ and let $\mathbf{x}_1,\mathbf{x}_2,\ldots,\mathbf{x}_v$ be nonzero representatives of all distinct one-dimensional subspaces of a q-ary simplex code $S_{m+1}(q)$ of dimension $m+1$. Let W be the matrix of order v over $GF(q)$ with $\mathbf{x}_1,\mathbf{x}_2,\ldots,\mathbf{x}_v$ as consecutive rows. Then W is a $PBGW(m,q)$ of rank $m+1$.*

Proof. Since $\dim S_{m+1}(q)=m+1$, we obtain that $rank(W)=m+1$.

Let $W=[\omega_{ij}]$. Proposition 3.9.28 implies that every row of W has exactly q^m nonzero entries. It also implies that, for distinct $i,h\in\{1,2,\ldots,v\}$, the multiset $M_{ih}=\{\omega_{ij}\overline{\omega_{hj}}: 1\leq j\leq v\}$ has exactly q^m-q^{m-1} nonzero elements. We shall show that every nonzero element of $GF(q)$ occurs exactly q^{m-1} times in M_{ih}. Suppose there is $\beta\in GF(q)^*$ that occurs more than q^{m-1} times in M_{ih}. Then there are more than $(q^{m-1}-1)/(q-1)+q^{m-1}$ indices j such that $\omega_{ij}=\omega_{hj}\beta$. But then

$$wt(\mathbf{x}_i-\beta\mathbf{x}_h)=d(\mathbf{x}_i,\beta\mathbf{x}_h)<v-\frac{q^{m-1}-1}{q-1}-q^{m-1}=q^m,$$

a contradiction. Thus, every element of $GF(q)^*$ occurs at most q^{m-1} times in M_{ih}. Since M_{ih} has exactly $q^{m-1}(q-1)$ elements, we conclude that every element of $GF(q)^*$ occurs exactly q^{m-1} times in M_{ih}. The proof is now complete. □

Theorem 10.2.4 and Proposition 10.1.29 immediately imply the following result.

Theorem 10.2.5. *If q is a prime power, m is a positive integer, and G is a cyclic group whose order divides $q-1$, then there exists a*

$$BGW\left(\frac{q^{m+1}-1}{q-1},q^m,q^m-q^{m-1};G\right). \tag{10.6}$$

If two $PBGW$-matrices are monomially equivalent, then they have same rank. The next theorem shows that the matrices monomially equivalent to those constructed in Theorem 10.2.4 are in fact all the $PBGW$-matrices of minimal rank.

Theorem 10.2.6. *Let W be a $PBGW(m, q)$. If $rank(W) = m + 1$, then W is monomially equivalent to a matrix whose rows represent all distinct one-dimensional subspaces of a simplex code of dimension $m + 1$.*

Proof. Let $W = [\omega_{ij}]$ and let $rank(W) = m + 1$. Let $\mathbf{y}_1, \mathbf{y}_2, \ldots, \mathbf{y}_v$ be the columns of $W^\top$ and let Y be the subspace of $V(v, q)$ spanned by $\{\mathbf{y}_1, \mathbf{y}_2, \ldots, \mathbf{y}_v\}$. Then $\dim Y = rank(W) = m + 1$, so Y is a $[v, m + 1]$-code. Without loss of generality, we assume that the first $m + 1$ rows of W are linearly independent. Then the matrix H formed by these rows is a generator matrix of Y. We claim that no two columns of H are proportional. Suppose there are distinct $j, h \in \{1, 2, \ldots, v\}$ and $\alpha \in GF(q)^*$ such that $\omega_{ij} = \alpha\omega_{ih}$ for $i = 1, 2, \ldots, m + 1$. Let $k \in \{1, 2, \ldots, v\}$. Since $\{\mathbf{y}_1, \mathbf{y}_2, \ldots, \mathbf{y}_{m+1}\}$ is a basis of Y, we have $\mathbf{y}_k = \sum_{i=1}^{m+1} \beta_i \mathbf{y}_i$ for some $\beta_1, \beta_2, \ldots, \beta_{m+1} \in GF(q)$. Then

$$\omega_{kj} = \sum_{i=1}^{m+1} \beta_i \omega_{ij} = \alpha \sum_{i=1}^{m+1} \beta_i \omega_{ih} = \alpha\omega_{kh},$$

i.e., the jth and h^{th} columns of W are proportional. Since W is a BGW-matrix, this is not possible, and therefore, no two columns of H are proportional.

Let X be the subspace of $V(m + 1, q)$ spanned by the columns of H. Then $\dim X = rank(H) = m + 1$, so the columns of H represent all distinct one-dimensional subspaces of $V(m + 1, q)$. Therefore, H is a generator matrix of $S_{m+1}(q)$. Then Y is a simplex code of dimension $m + 1$. Since no two rows of W are proportional, they represent all distinct one-dimensional subspaces of Y. The proof is now complete. □

We will have other constructions of BGW-matrices with classical parameters in later sections (Example 10.3.3 and Theorem 10.5.1).

We have already constructed matrices $GH(q, 1)$ in Proposition 10.1.27. The next theorem gives a putative construction of matrices $GH(q, n)$ where n is the order of a Hadamard matrix.

Theorem 10.2.7. *Let q be an odd prime power, let $GF(q) = \{a_1, a_2, \ldots, a_q\}$, and let η be the quadratic character on $GF(q)$. Let $n \geq 2$ and let $x_1, x_2, \ldots, x_n$ and $y_1, y_2, \ldots, y_n$ be $2n$ distinct elements of $GF(q)^*$ such that the matrix $H = [\eta(x_i - y_j)]$ is a Hadamard matrix of order n. Let $W = [W_{ij}]$ be the $n \times n$ block matrix with $q \times q$ blocks W_{ij} such that, for $s, t = 1, 2, \ldots, q$, the*

(s,t)-entry of W_{ij} is equal to

$$\omega(i,j,s,t) = \frac{a_t^2 + 2a_s a_t + x_i^{-1} y_j a_s^2}{x_i - y_j}.$$

Then W, regarded as a matrix over the additive group of $GF(q)$, is a $GH(q,n)$.

Proof. For $i, j, h = 1, 2, \ldots, n$ and $s, u = 1, 2, \ldots, q$, let $M(i,j,h,s,u)$ denote the multiset $\{\omega(i,j,s,t) - \omega(h,j,u,t) \colon 1 \le t \le q\}$ of cardinality q. For $i, h = 1, 2, \ldots, n$ and $s, u = 1, 2, \ldots, q$, let the multiset $N(i,h,s,u)$ of cardinality nq be the union of the multisets $M(i,j,h,s,u)$ with $1 \le j \le n$. We shall show that, whenever $(i,s) \ne (h,u)$, the multiset $N(i,h,s,u)$ contains n copies of every element of $GF(q)$.

Case 1: $i = h$ and $s \ne u$.

We have $\omega(i,j,s,t) - \omega(i,j,u,t) = \alpha(i,j,s,u)a_t + \beta(i,j,s,u)$ with $\alpha(i,j,s,u) \ne 0$. Therefore, $M(i,j,h,s,u)$ contains one copy of every element of $GF(q)$ and then $N(i,i,s,u)$ contains n copies of every element of $GF(q)$.

Case 2: $i \ne h$.

We have

$$\omega(i,j,s,t) - \omega(h,j,u,t) = \gamma(i,j,h)(a_t + \delta(i,j,h,s,u))^2 + \varepsilon(i,h,s,u),$$

where

$$\gamma(i,j,h) = \frac{1}{x_i - y_j} - \frac{1}{x_h - y_j} = \frac{x_h - x_i}{(x_i - y_j)(x_h - y_j)} \ne 0,$$

$$\delta(i,j,h,s,u) = \frac{a_s(x_h - y_j) - a_u(x_i - y_j)}{x_h - x_i},$$

and

$$\varepsilon(i,h,s,u) = \frac{x_h a_s^2}{x_i(x_i - x_h)} + \frac{x_i a_u^2}{x_h(x_i - x_h)} - \frac{2a_s a_u}{x_i - x_h}.$$

Let S be the set of all nonzero squares of $GF(q)$ and T the set of all nonsquares. Then $M(i,j,h,s,u)$ consists of one copy of $\varepsilon(i,h,s,u)$ and two copies of every element of the set $\gamma(i,j,h)S + \varepsilon(i,h,s,u)$, which is equal to $S + \varepsilon(i,h,s,u)$ if $\eta(\gamma(i,j,h)) = 1$ and is equal to $T + \varepsilon(i,h,s,u)$ if $\eta(\gamma(i,j,h)) = -1$.

We have $\eta(\gamma(i,j,h)) = \eta(x_h - x_i)\eta(x_i - y_j)\eta(x_h - y_j)$. Since H is a Hadamard matrix, the product $\eta(x_i - y_j)\eta(x_h - y_j)$ is equal to 1 for as many values of $j \in \{1, 2, \ldots, n\}$ as it is equal to -1. Therefore, the multiset $N(i,h,s,u)$ contains n copies of every element of $S + \varepsilon(i,h,s,u)$, n copies of every element of $T + \varepsilon(i, h, s, u)$, and n copies of $\varepsilon(i, h, s, u)$, i.e., n copies of every element of $GF(q)$. □

Remark 10.2.8. Let q be an odd prime power and let $GF(q) = \{a_1, a_2, \ldots, a_q\}$ with $a_q = 0$. Let P be the corresponding Paley matrix and let P' be the matrix obtained by deleting the last row and last column of P. Then, by Theorem 10.2.7, in order to find a $GH(q, n)$, it suffices to find a Hadamard submatrix of order n of P'. In general, this could be a daunting task. However, for $n = 2$ it can be done easily.

Theorem 10.2.9. *For any odd prime power q, there exists a $GH(q, 2)$.*

Proof. Let q be an odd prime power. If $q = 3$, then the matrix

$$\begin{bmatrix} 0 & 0 & 0 & 0 & 0 & 0 \\ 1 & 2 & 0 & 2 & 0 & 1 \\ 1 & 0 & 2 & 2 & 1 & 0 \\ 0 & 2 & 2 & 0 & 1 & 1 \\ 2 & 2 & 0 & 1 & 1 & 0 \\ 2 & 0 & 2 & 1 & 0 & 1 \end{bmatrix}$$

is a $GH(3, 2)$ over the group $\mathbf{Z}_3$.

Suppose $q \geq 5$. Let $GF(q)^* = \{a_1, a_2, \ldots, a_{q-1}\}$ and let P and P' be the matrices as in Remark 10.2.8. Since the inner product of any two rows of P is equal to -1 (Proposition 4.3.2), the inner product of the ith and j^{th} row of P' is equal to $-1 - \eta(a_i)\eta(a_j)$ where η is the quadratic character on $GF(q)$. Let us choose i and j so that $\eta(a_i) \neq \eta(a_j)$. If $[\alpha_1\ \alpha_2 \ldots \alpha_{q-1}]$ and $[\beta_1\ \beta_2 \ldots \beta_{q-1}]$ are the ith and j^{th} row of P', then two of the products $\alpha_j\beta_j$ are equal to 0, $(q-3)/2$ are equal to 1, and $(q-3)/2$ are equal to -1. Since $q \geq 5$, we can find j and h such that $\alpha_j\beta_j = 1$ and $\alpha_h\beta_h = -1$. Then $\left[\begin{smallmatrix} \alpha_j & \alpha_h \\ \beta_j & \beta_h \end{smallmatrix}\right]$ is a required Hadamard submatrix of order 2 of P'. □

Remark 10.2.10. A direct construction of $GH(q, 2)$ for q odd is given in Exercise 5.

We will give without proof three more applications of Theorem 10.2.7.

Theorem 10.2.11. *For any odd prime power q, there exists a $GH(q, 4)$.*

Theorem 10.2.12. *For any prime $p \geq 23$, there exists a $GH(p, 8)$; for any odd prime power q such that $25 \leq q \leq 169$, there exists a $GH(q, 8)$.*

Theorem 10.2.13. *Suppose there exists a Hadamard matrix of order $n \geq 8$. Then, for all odd prime powers $q \geq ((n-2)2^{n-2} - n(n-1)/2 + 3)^2$, there exists a $GH(q, n)$.*

We will have more constructions of matrices $GH(q, \lambda)$ in Section 10.4.

10.3. BGW-matrices and relative difference sets

In this section we will investigate relations between balanced generalized weighing matrices and relative difference sets. The first result shows that every relative difference set gives rise to a BGW-matrix.

Theorem 10.3.1. *Let R be an (m, n, k, λ)-RDS in a group G relative to a normal subgroup N. Let $x_1, x_2, \ldots, x_m$ be representatives of all distinct cosets of N in G. Let $W = [\omega_{ij}]$ be a $(0, N)$ matrix of order m such that, for each $\alpha \in N$, $\omega_{ij} = \alpha$ if and only if $(x_i N) \cap (Rx_j) = \{x_i\alpha\}$. Then W is a BGW $(m, k, \lambda n; N)$.*

Proof. We first show that, for $i, j = 1, 2, \ldots, m$, $|(x_i N) \cap (Rx_j)| \le 1$. Indeed, if $a, b \in (x_i N) \cap (Rx_j)$, then $a = x_i\alpha = tx_j$ and $b = x_i\beta = ux_j$ with $t, u \in R$ and $\alpha, \beta \in N$. Then $t = x_i\alpha x_j^{-1}$ and $u = x_i\beta x_j^{-1}$, so $tu^{-1} = x_i(\alpha\beta^{-1})x_i^{-1}$. Since N is a normal subgroup of G, we have $tu^{-1} \in N$, which implies $t = u$ and then $a = b$. Therefore, every entry of W is uniquely defined.

Since, for $j = 1, 2, \ldots, m$,

$$k = |Rx_j| = \left|\bigcup_{i=1}^{m}(x_i N) \cap (Rx_j)\right| = \sum_{i=1}^{m} |(x_i N) \cap (Rx_j)|,$$

we obtain that every column of W has exactly k nonzero entries.

Let $i, h \in \{1, 2, \ldots, m\}$, $i \ne h$, and let $\gamma \in N$.

First suppose that $\gamma = \omega_{ij}\overline{\omega_{hj}}$ for some $j \in \{1, 2, \ldots, m\}$. There are unique $t, u \in R$ such that $tx_j = x_i\omega_{ij}$ and $ux_j = x_h\omega_{hj}$. Then $tu^{-1} = x_i\gamma x_h^{-1}$.

Conversely, suppose $x_i\gamma x_h^{-1} = tu^{-1}$ for some $t, u \in R$. There is a unique $j \in \{1, 2, \ldots, m\}$ such that $u^{-1}x_h \in x_j N$, i.e., $u^{-1}x_h = x_j\beta$ with $\beta \in N$. Then $ux_j = x_h\beta^{-1}$ and $tx_j = tu^{-1}x_h\beta^{-1} = x_i\gamma\beta^{-1}$ with $\gamma\beta^{-1} \in N$. Therefore, $\omega_{ij} = \gamma\beta^{-1}$, $\omega_{hj} = \beta^{-1}$, and $\gamma = \omega_{ij}\overline{\omega_{hj}}$.

Thus, the number of indices j such that $\gamma = \omega_{ij}\overline{\omega_{hj}}$ is equal to the number of ordered pairs (t, u) of elements of R such that $tu^{-1} = x_i\gamma x_h^{-1}$. Since $x_i\gamma x_h^{-1} \notin N$, there are exactly λ such pairs. Therefore, by Proposition 10.1.33, W is a $BGW(m, k, \lambda n; N)$. □

Example 10.3.2. Let $G = \langle a \rangle$ be a cyclic group of order 12 and let $R = \{1, a, a^2, a^4, a^9\}$. Then R is a (6, 2, 5, 2)-RDS in G relative to the subgroup $N = \{1, \omega\}$ where $\omega = a^6$. The set $\{x_i = a^{i-1} : 1 \le i \le 6\}$ is a set of representatives of all distinct cosets of N in G. The corresponding matrix

$$W = \begin{bmatrix} 1 & 1 & 1 & \omega & 1 & 0 \\ 0 & 1 & 1 & 1 & \omega & 1 \\ \omega & 0 & 1 & 1 & 1 & \omega \\ 1 & \omega & 0 & 1 & 1 & 1 \\ \omega & 1 & \omega & 0 & 1 & 1 \\ \omega & \omega & 1 & \omega & 0 & 1 \end{bmatrix}$$

is a $BGW(6, 5, 4; N)$.

Example 10.3.3. Let q be a prime power, d a positive integer, and $v = (q^{d+1} - 1)/(q - 1)$. Let G be a cyclic group of order v. Then there exists a $(v, q - 1, q^d, q^{d-1})$-RDS in G relative to the unique cyclic subgroup N of order $q - 1$ (Theorem 9.6.12). The corresponding BGW-matrix is a $PBGW(d, q)$.

The rows and columns of the matrix W constructed in Theorem 10.3.1 can be indexed in the natural way by the elements of the factor group G/N. However, the matrix W does not have to be (G/N)-invariant. In fact, the matrix from Example 10.3.2 is not group-invariant at all (see Exercise 6). The next theorem shows that group-invariant BGW-matrices are equivalent to splitting relative difference sets.

Theorem 10.3.4. *Let H and N be finite groups, $|H| = m$, $|N| = n$, and let $G = H \times N$. The following statements are equivalent:*

(i) *there exists an (m, n, k, λ)-RDS in G relative to N;*
(ii) *there exists an H-invariant $BGW(m, k, \lambda n; N)$.*

Proof. We will regard H and N as subgroups of G and write $x\alpha$ instead of (x, α) for $x \in H$ and $\alpha \in N$. Of course, $x\alpha = \alpha x$.

(i) $\Rightarrow$ (ii).

Suppose R is an (m, n, k, λ)-RDS in G relative to N. Let $H = \{x_1, x_2, \ldots, x_m\}$ and let W be the $BGW(m, k, \lambda n; N)$ constructed in Theorem 10.3.1. For $x = x_i$ and $y = x_j$, let $W(x, y)$ denote the (i, j)-entry of W. Let $x, y, z \in H$. If $W(x, y) = \alpha \in N$, then $(xN) \cap (Ry) = \{x\alpha\}$ and therefore, $(xzN) \cap (Ryz) = (Nxz) \cap (Ryz) = \{x\alpha z\} = \{xz\alpha\}$. Then $W(xz, yz) = \alpha$. If $W(x, y) = 0$, then $(xN) \cap (Ry) = \varnothing$, so $(Nxz) \cap (Ryz) = \varnothing$, and therefore, $W(xz, yz) = 0$. Thus, W is H-invariant.

(ii) $\Rightarrow$ (i).

Let W be an H-invariant $BGW(m, k, \lambda n; N)$. Consider the element

$$R = \sum_{x \in H} x W(x, 1)$$

of the group ring $\mathbb{Z}G$. Since $x \in H$ and $W(x, 1) \in N \cup \{0\}$, this sum does not contain equal nonzero terms and therefore represents a subset R of G. We have

$$R\overline{R} = \sum_{x \in H} x W(x, 1) \cdot \sum_{y \in H} y^{-1} \overline{W(y, 1)} = \sum_{x \in H} \sum_{y \in H} x y^{-1} W(x, 1) \overline{W(y, 1)}.$$

Let $xy^{-1} = z$, so $x = zy$. Then

$$\begin{aligned} R\overline{R} &= \sum_{z \in H} \sum_{y \in H} z W(zy, 1) \overline{W(y, 1)} = \sum_{y \in H} W(y, 1) \overline{W(y, 1)} \\ &\quad + \sum_{z \in H \setminus \{1\}} z \sum_{y \in H} W(zy, 1) \overline{W(y, 1)} \\ &= k + \sum_{z \in H \setminus \{1\}} z \sum_{y \in H} W(z, y^{-1}) \overline{W(1, y^{-1})} = k + \sum_{z \in H \setminus \{1\}} z(\lambda N) \\ &= k - \lambda N + \lambda \sum_{z \in H} zN = k - \lambda N + \lambda G. \end{aligned}$$

Proposition 9.6.7 now implies that R is an (m, n, k, λ)-RDS in G relative to N. □

The direct product of two cyclic groups is cyclic if and only if the orders of the groups are relatively prime. Therefore, Theorems 10.2.5 and 10.3.4 imply the following result.

Corollary 10.3.5. *Let q be a prime power, m a positive integer, and s a divisor of $q - 1$ that is relatively prime to $v = (q^{m+1} - 1)/(q - 1)$. Then there exists a circulant $BGW(v, q^m, q^{m-1})$ over a cyclic group of order s.*

Theorem 9.6.5 allows us to obtain another application of Theorem 10.3.4.

Corollary 10.3.6. *Let q be an odd prime power and let G be the additive group of the field $GF(q)$. Then there exists a G-invariant $GH(q, 1)$.*

Example 9.6.4 implies the next result.

Corollary 10.3.7. *There exists a circulant $BGW(13, 9, 6; S_3)$.*

Theorem 10.3.4 and Corollary 9.6.17 yield a family of generalized Hadamard matrices over an arbitrary p-group.

Corollary 10.3.8. *Let p be a prime and let $m \geq n$ be positive integers. For any group G of order p^n, there exists a $GH(G; p^{2m-n})$.*

Remark 10.3.9. Suppose there exists a Hadamard difference set D in a group G of order $4h^2$. Then the symmetric $(4h^2, 2h^2 - h, h^2 - h)$-design $dev(D)$ admits a G-invariant incidence matrix N. The (± 1)-matrix $J - 2N$ is a regular Hadamard matrix and therefore it is a $GH(2, 2h^2)$. Since this matrix is also G-invariant, Theorem 10.3.4 implies that there exists a splitting $(2h^2, 2, 4h^2, 2h^2)$-RDS. This is an alternative proof of Theorem 9.6.14.

We will state without proof the following result.

Theorem 10.3.10. *Let q be an even prime power and let G be the direct product of two cyclic groups of order $2(q-1)$ and 2. Then there exists a group-invariant*

$$BGW\left(\frac{q^7-1}{q-1}, q^6, q^6 - q^5; G\right).$$

The BGW-matrix of Example 10.3.2 is "almost circulant". In fact it is ω-*circulant* according to the following definition.

Definition 10.3.11. Let ω be a generator of a cyclic group G. A $(0, G)$ matrix $W = [\gamma_{ij}]$ of order m is said to be ω-*circulant* if, for $i, j \in \{1, 2, \ldots, m-1\}$, $\gamma_{i+1,j+1} = \gamma_{ij}$ and $\gamma_{i+1,1} = \omega\gamma_{im}$.

All entries of an ω-circulant matrix are determined by the first row of the matrix. The following proposition is straightforward.

Proposition 10.3.12. *Let $W = [\gamma_{ij}]$ be an ω-circulant $(0, G)$ matrix of order m. Let $[\alpha_1\ \alpha_2 \ldots \alpha_m]$ be the first row of W. Then, for $i, j = 1, 2, \ldots, m$, $\gamma_{ij} = \alpha_{j-i+1}$ if $i \leq j$, and $\gamma_{ij} = \omega\alpha_{m+j-i+1}$ if $i > j$.*

The following theorem shows that ω-circulant BGW-matrices are equivalent to cyclic relative difference sets.

Theorem 10.3.13. *Let G be a cyclic group of order mn and let $N = \langle\omega\rangle$ be the unique subgroup of G of order n. The following statements are equivalent:*

(i) *there exists an (m, n, k, λ)-RDS in G relative to N;*
(ii) *there exists an ω-circulant $BGW(m, k, \lambda n; N)$.*

Proof. Let β be a generator of G such that $\beta^m = \omega^{-1}$. Then the set $\{x_i = \beta^{i-1} : 1 \leq i \leq m\}$ is a set of representatives of all distinct cosets of N in G.

(i) $\Rightarrow$ (ii). Let R be an (m, n, k, λ)-RDS in G relative to N and let $W = [\gamma_{ij}]$ be the $BGW(m, k, \lambda n; N)$ constructed in Theorem 10.3.1. We will show that W is ω-circulant.

Let $i, j \in \{1, 2, \ldots, m-1\}$. Let $\alpha \in N$. Then

$$\gamma_{ij} = \alpha \Leftrightarrow (\beta^{i-1}N) \cap (R\beta^{j-1}) = \{\beta^{i-1}\alpha\} \Leftrightarrow (\beta^i N) \cap (R\beta^j)$$
$$= \{\beta^i \alpha\} \Leftrightarrow \gamma_{i+1,j+1} = \alpha$$

and

$$\gamma_{im} = \alpha \Leftrightarrow (\beta^{i-1}N) \cap (R\beta^{m-1}) = \{\beta^{i-1}\alpha\} \Leftrightarrow (\beta^i N) \cap (R\omega^{-1}) = \{\beta^i \alpha\}$$
$$\Leftrightarrow (\beta^i N\omega) \cap R = \{\beta^i \omega\alpha\} \Leftrightarrow (\beta^i N) \cap R = \{\beta^i \omega\alpha\} \Leftrightarrow \gamma_{i+1,1} = \omega\alpha.$$

Therefore, W is ω-circulant.

(ii) $\Rightarrow$ (i).

Let $W = [\gamma_{ij}]$ be an ω-circulant $BGW(m, k, \lambda n; N)$ and let $[\alpha_1\ \alpha_2 \ldots \alpha_m]$ be the first row of W. Consider the element

$$R = \sum_{i=0}^{m-1} \beta^{-i}\alpha_{i+1}$$

of the group ring $\mathbb{Z}G$. Since $1, \beta^{-1}, \ldots, \beta^{-(m-1)}$ represent distinct cosets of N in G, no two nonzero terms of this sum are equal. Therefore, R represents a subset of G. We will show that R is an (m, n, k, λ)-RDS in G relative to N. We have

$$\overline{R} = \sum_{i=0}^{m-1} \beta^i \overline{\alpha_{i+1}} = \sum_{j=0}^{m-1} \beta^{m-1-j} \overline{\alpha_{m-j}}.$$

In the following calculations we apply Proposition 10.3.12. Whenever the lower limit of a summation is greater than the upper limit, we assume that the sum is equal to 0.

$$R\overline{R} = \sum_{i=0}^{m-1}\sum_{j=0}^{m-1} \beta^{m-1-i-j}\alpha_{i+1}\overline{\alpha_{m-j}}$$
$$= \sum_{i=0}^{m-1}\left(\sum_{j=0}^{m-1-i} \beta^{-1-i-j}\alpha_{i+1}\overline{\omega\alpha_{m-j}} + \sum_{j=m-i}^{m-1} \beta^{m-1-i-j}\alpha_{i+1}\overline{\alpha_{m-j}}\right)$$
$$\sum_{s=0}^{m-1}\beta^{-s}\left(\sum_{i=0}^{s-1}\alpha_{i+1}\overline{\omega\alpha_{m+1+i-s}} + \sum_{i=s}^{m-1}\alpha_{i+1}\overline{\alpha_{i+1-s}}\right)$$
$$= \sum_{i=0}^{m-1}\alpha_{i+1}\overline{\alpha_{i+1}} + \sum_{s=1}^{m-1}\beta^{-s}\sum_{i=0}^{m-1}\gamma_{1,i+1}\overline{\gamma_{s+1,i+1}}$$
$$= k + \sum_{s=1}^{m-1}\beta^{-s}(\lambda N) = k - \lambda N + \lambda\sum_{s=0}^{m-1}\beta^{-s}N = k - \lambda N + \lambda G.$$

By Proposition 9.6.7, R is an (m, n, k, λ)-RDS in G relative to N. □

We state the following result without proof.

Theorem 10.3.14. *Let q be an even prime power and let $G = \langle\omega\rangle$ be a cyclic group of order $2(q-1)$. Then, for any even m, there exists an ω-circulant*

$$BGW\left(\frac{q^{m+1}-1}{q-1}, q^m, q^m - q^{m-1}; G\right).$$

10.4. Kronecker product constructions

The Kronecker product of matrices $A = [a_{ij}]$ and B is the block matrix $A \otimes B = [a_{ij}B]$. In this section, we will extend this notion to the case when each a_{ij} is a function acting on a set of matrices of the same size containing the matrix B and $a_{ij}B$ is the result of applying this function to B.

Definition 10.4.1. Let $\mathcal{M}$ be a nonempty set of matrices of the same size $m \times n$, let S a group of mappings $\mathcal{M} \to \mathcal{M}$, and let $W = [\omega_{ij}]$ be a $(0, S)$ matrix. Then, for any $X \in \mathcal{M}$, $W \otimes_S X$ is the block matrix obtained from W by replacing every $\omega_{ij} \in S$ with $\omega_{ij}X \in \mathcal{M}$ and every $\omega_{ij} = 0$ with the $m \times n$ zero matrix. The matrix $W \otimes_S X$ is called the *Kronecker product of W and X over S*.

Example 10.4.2. Let $\mathcal{M}$ be the set of all (0, 1)-matrices of order 3. For each $X = [X_1\ X_2\ X_3] \in \mathcal{M}$ (where X_1, X_2, X_3 are the columns of X), let $\sigma X = [X_3\ X_1\ X_2]$. Let S be the cyclic group of order 3 generated by σ and let M be the $BGW(5, 4, 3; S)$ from Example 9.1.11. Then

$$M \otimes_S I = \begin{bmatrix} O & I & K & K & I \\ I & O & I & K & K \\ K & I & O & I & K \\ K & K & I & O & I \\ I & K & K & I & O \end{bmatrix},$$

where $K = \left[\begin{smallmatrix} 0 & 1 & 0 \\ 0 & 0 & 1 \\ 1 & 0 & 0 \end{smallmatrix}\right]$.

The Kronecker product $W \otimes_S X$ has useful properties if we impose certain restrictions on the action of the group S on the set of matrices containing X.

Definition 10.4.3. Let G be a finite group and let $\mathcal{M}$ be a nonempty set of $(0, G)$ matrices of the same size. A group S of bijections $\mathcal{M} \to \mathcal{M}$ is called a *group of symmetries of $\mathcal{M}$* if

(i) $(\sigma X)(\sigma Y)^* = XY^*$ for all $X, Y \in \mathcal{M}$ and all $\sigma \in S$ and

(ii) for each $X \in \mathcal{M}$ there is $\alpha_S(X) \in \mathbb{Z}G$ such that

$$\sum_{\sigma \in S} \sigma X = \alpha_S(X) J.$$

The following propositions give several examples of groups of symmetries.

Proposition 10.4.4. *Let G be a finite group, n a positive integer, and $\alpha \in \mathbb{Z}G$. Let $\mathcal{M}$ be the set of all $(0, G)$ matrices X of order n such that $XJ = \alpha J$. Let S be a transitive group of permutations of the set $\{1, 2, \ldots, n\}$. For $\sigma \in S$ and $X = [X_1\ X_2 \ldots X_n] \in \mathcal{M}$, let $\sigma X = [X_{\sigma(1)}\ X_{\sigma(2)} \ldots X_{\sigma(n)}]$. Then, with respect to this action, S is a group of symmetries of $\mathcal{M}$.*

Proof. For $X, Y \in \mathcal{M}$ and $\sigma \in S$, the matrices σX and σY are obtained from X and Y by the same permutation of columns. Therefore, $(\sigma X)(\sigma Y)^* = XY^*$. For $X \in \mathcal{M}$, $\sum_{\sigma \in S} \sigma X = [Y_1\ Y_2 \ldots Y_n]$, where $Y_i = \sum_{\sigma \in S} X_{\sigma(i)}$. Since S is transitive, we obtain that

$$Y_i = \frac{|S|}{n} \sum_{i=1}^{n} X_i = \frac{|S|}{n} \alpha J.$$

Therefore, S is a group of symmetries of $\mathcal{M}$. □

Definition 10.4.5. The group of symmetries of a set $\mathcal{M}$ of matrices given by Proposition 10.4.4 is called a *group of rotations* of $\mathcal{M}$.

The next two propositions give groups of symmetries for sets of generalized Hadamard matrices and balanced generalized weighing matrices, respectively.

Proposition 10.4.6. *Let G be a finite group and λ a positive integer. Let $\mathcal{M}$ be the set of all matrices $GH(G; \lambda)$. Let S be a finite group and let f be a homomorphism from S onto G. For each $\sigma \in S$ and $X \in \mathcal{M}$, let $\sigma X = X f(\sigma)$ be the matrix obtained by multiplying every entry of X by $f(\sigma)$ on the right. If $\mathcal{M} \neq \varnothing$, then S is a group of symmetries of $\mathcal{M}$.*

Proof. Clearly, each $\sigma \in S$ acts as a bijection $\mathcal{M} \to \mathcal{M}$. For $X, Y \in \mathcal{M}$, $(\sigma X)(\sigma Y)^* = (X f(\sigma))(Y f(\sigma))^* = XY^*$. Since also

$$\sum_{\sigma \in S} \sigma X = X \cdot \sum_{\sigma \in S} f(\sigma) = \frac{|S|}{|G|} XG = \frac{|S|}{|G|} GJ,$$

the group S satisfies Definition 10.4.3. □

Proposition 10.4.7. *Let $G = \langle \omega \rangle$ be a finite cyclic group and let $v \geq k \geq \lambda$ be positive integers. Let $\mathcal{M}$ be the set of all matrices $BGW(v, k, \lambda; G)$. Let a bijection $\rho \colon \mathcal{M} \to \mathcal{M}$ be defined as follows: if $X = [X_1\ X_2 \ldots X_v] \in \mathcal{M}$,* *then $\rho X = [X_v\omega\ X_1\ X_2 \ldots X_{v-1}]$. Let S be the cyclic group generated by ρ. If $\mathcal{M} \neq \varnothing$, then S is a group of symmetries of $\mathcal{M}$.*

Proof. Let $X = [x_{ij}] \in \mathcal{M}$ and $Y = [y_{ij}] \in \mathcal{M}$. Then the (s,t)-entry of XY^* is $\sum_{j=1}^{v} x_{sj}\overline{y_{tj}}$ and the (s,t)-entry of $(\rho X)(\rho Y)^*$ is

$$x_{sv}\omega\overline{y_{tv}\omega} + \sum_{j=2}^{v} x_{s,j-1}\overline{y_{t,j-1}} = \sum_{j=1}^{v} x_{sj}\overline{y_{tj}}.$$

Therefore, for any integer k, $(\rho^k X)(\rho^k Y)^* = XY^*$.

We have further that $\rho^v X = X\omega$ for all $X \in \mathcal{M}$. Therefore, $|S| = v|G|$ and

$$\sum_{i=0}^{v|G|-1} \rho^i X = \sum_{i=0}^{v-1}\sum_{j=0}^{|G|-1} (\rho^i X)\omega^j = \sum_{i=0}^{v-1}(\rho^i X)G = kGJ.$$

Therefore, S satisfies Definition 10.4.3. □

Remark 10.4.8. For the groups of symmetries described in Propositions 10.4.6 and 10.4.7, the coefficients $\alpha_S(X)$ required by Definition 10.4.3 are of the form tG with $t \in \mathbb{Z}$.

The next theorem gives a sufficient condition for a Kronecker product of matrices over a group to be a BGW-matrix.

Theorem 10.4.9. *Let G be a finite group and let $\mathcal{M}$ be a nonempty set of matrices, each of which is a $BGW(v,k,\lambda;G)$ with the same parameters v, k, and λ. Let S be a group of symmetries of $\mathcal{M}$ such that, for each $X \in \mathcal{M}$, $\alpha_S(X) = t(X)G$ with $t(X) \in \mathbb{Z}$. Let W be a $BGW(w,l,\mu;S)$ such that $k^2\mu = v\lambda l$. Then, for any $X \in \mathcal{M}$, the matrix $\overline{W} \otimes_S X$ is a $BGW(vw, kl, \lambda l; G)$.*

Proof. Let $W = [\omega_{ij}]$ and let $X = [x_{ij}] \in \mathcal{M}$. Then, by Proposition 10.1.5,

$$XX^* = \left(k - \frac{\lambda}{|G|}G\right)I + \frac{\lambda}{|G|}GJ.$$

For $i, h \in \{1,2,\ldots,w\}$, let $P_{ih} = \sum_{j=1}^{w}(\overline{\omega_{ij}}X)(\overline{\omega_{hj}}X)^*$. It suffices to show that

$$P_{ih} = \begin{cases} \left(kl - \frac{\lambda l}{|G|}G\right)I + \frac{\lambda l}{|G|}GJ & \text{if } i = h,\\ \frac{\lambda l}{|G|}GJ & \text{if } i \neq h.\end{cases}$$

Let $i \in \{1,2,\ldots,w\}$. Since each row of W has exactly l nonzero entries, there are $\sigma_1, \sigma_2, \ldots, \sigma_l \in S$ such that

$$P_{ii} = \sum_{j=1}^{l}(\sigma_j X)(\sigma_j X)^*.$$

Since S is a group of symmetries of $\mathcal{M}$, we obtain

$$P_{ii} = \sum_{j=1}^{l} XX^* = lXX^* = \left(kl - \frac{\lambda l}{|G|}G\right) I + \frac{\lambda l}{|G|} GJ.$$

Let $i, h \in \{1, 2, \ldots, w\}$, $i \neq h$. By condition (ii) of Definition 10.4.3 and by the definition of BGW-matrices,

$$P_{ih} = \sum_{j=1}^{w} (\omega_{hj}\overline{\omega_{ij}}X)(\omega_{hj}\overline{\omega_{hj}}X)^* = \frac{\mu}{|S|}\left(\sum_{\sigma \in S} \sigma X\right) X^* = \frac{\mu}{|S|} t(X) GJX^*.$$

By Proposition 10.1.34, X^* is a $BGW(v, k, \lambda; G)$ and therefore, $GJX^* = J(GX^*) = kGJ$. In order to get an explicit expression for $t(X)$, we multiply both sides of the equation $\sum_{\sigma \in S} \sigma X = t(X)GJ$ by the matrix GJ:

$$\sum_{\sigma \in S} (\sigma X)GJ = t(X)v|G|GJ.$$

Since $\sigma X \in \mathcal{M}$, we have $(\sigma X)GJ = kGJ$, and then $t(X) = (k|S|)/(v|G|)$. Therefore, $P_{ih} = (\lambda l/|G|)GJ$. The proof is now complete. □

Application of Theorem 10.4.9 to GH-matrices is straightforward. Let G, $\mathcal{M}$, and S be the same as in Proposition 10.4.6. Let W be a $GH(S; \mu)$. Then the conditions of Theorem 10.4.9 are satisfied, and we obtain the following results.

Corollary 10.4.10. *If there exist matrices $GH(G; \lambda)$ and $GH(S; \mu)$ and there exists a surjective homomorphism from S onto G, then there exists a $GH(G; \lambda\mu|S|)$.*

If G and S are isomorphic groups, then we obtain the following results.

Corollary 10.4.11. *If there exist matrices $GH(G; \lambda)$ and $GH(G; \mu)$, then there exists a $GH(G; \lambda\mu|G|)$.*

Corollary 10.4.12. *For any prime power q and any positive integer m, there exists a $GH(q, q^{m-1})$.*

Proof. Proposition 10.1.27 gives a $GH(q, 1)$. Applying Corollary 10.4.10 to a $GH(q, q^{m-1})$, $GH(q, 1)$, and the identity homomorphism yields a $GH(q, q^m)$, so we apply induction on m to complete the proof. □

Remark 10.4.13. Application of Theorem 10.4.9 to BGW-matrices that are not GH-matrices requires certain restrictions on the parameters of the matrices. Let v, k, λ, $G = \langle\omega\rangle$, $S = \langle\rho\rangle$ be the same as in Proposition 10.4.7. Let q be a prime power, s a positive integer, and W a $BGW(w, l, \mu; S)$ with

$w = (q^{s+1} - 1)/(q - 1)$, $l = q^s$, and $\mu = q^s - q^{s-1}$. Then

$$k^2\mu = v\lambda l \Leftrightarrow k^2(q-1) = v\lambda q \Leftrightarrow 1 - \frac{1}{q} = \frac{k^2 - k + \lambda}{k^2} \Leftrightarrow q = \frac{k^2}{k - \lambda}.$$

Let p be a prime power, d a positive integer, and let $\mathcal{M}$ be a set of matrices $BGW(v, k, \lambda)$ with $(v, k, \lambda) = ((p^{d+1} - 1)/(p - 1), p^d, p^d - p^{d-1})$. Let $q = k^2/(k - \lambda) = p^{d+1}$. Suppose the set $\mathcal{M}$ is closed under the action of the group S and let W be a $BGW((q^{s+1} - 1)/(q - 1), q^s, q^s - q^{s-1}; S)$. Since $|S| = v|G|$ divides $q - 1$, we obtain that, for any $X \in \mathcal{M}$, the matrix $W \otimes_S X$ is a $BGW((p^{m+1} - 1)/(p - 1), p^m, p^m - p^{m-1}; G)$ where $m + 1 = (s + 1)(d + 1)$. BGW-matrices with these parameters were constructed before. However, this Kronecker product construction will allow us to obtain BGW-matrices with further properties such as symmetric and skew-symmetric BGW-matrices with zero diagonals.

It is clear what a symmetric BGW-matrix is. In order to define skew-symmetric BGW-matrices, note that if W is a $BGW(v, k, \lambda; G)$, where G is a subgroup of $GF(q)^*$ with odd q, then the equality $W^\top = -W$ implies that $-1 \in G$. Observe that if this is the case, then -1 is the only element of order 2 in G. This motivates the following definition.

Definition 10.4.14. An abelian group G with a unique element of order 2 is called a *signed group*. If G is a signed group, then a $(0, G)$-matrix W is said to be *skew-symmetric* if $W^\top = \varepsilon W$ where ε is the unique element of order 2 in G.

Remark 10.4.15. Obviously, any finite signed group is of even order. Every cyclic group of even order is a signed group. In general, a finite abelian group G of even order is a signed group if and only if its Sylow 2-subgroup is cyclic.

Theorem 10.4.16. *Let q be a prime power and let n be a divisor of $q - 1$. Let G be a cyclic group of order n. If $q(q - 1)/n$ is even, then, for any positive integer d, there exists a symmetric balanced generalized weighing matrix*

$$BGW\left(\frac{q^{2d} - 1}{q - 1}, q^{2d-1}, q^{2d-1} - q^{2d-2}; G\right) \tag{10.7}$$

with all diagonal entries equal to 0. *If n is even and $(q - 1)/n$ is odd, then there exists a skew-symmetric BGW with parameters* (10.7).

Proof. We begin with the case $d = 1$. Let $GF(q) = \{a_1, a_2, \ldots, a_q\}$, let $U_1 = [\alpha_{ij}]$ be the $BGW(q + 1, q, q - 1)$ constructed in Proposition 10.1.17, and let $U_2 = [\beta_{ij}]$ be the matrix obtained from U_1 by multiplying all entries of the first row by -1. Let $V_1 = [\alpha_{ij}^{(q-1)/n}]$ and $V_2 = [\beta_{ij}^{(q-1)/n}]$. If $(q - 1)/n$ is even, then V_1 is a symmetric $BGW(q + 1, q, q - 1)$ with zero diagonal over a cyclic group of order n. If q is even, then $GF(q)$ is a field of characteristic 2, and therefore again V_1 is symmetric. If n is even and $(q-1)/n$ is odd, then V_2 is a skew-symmetric $BGW(q+1, q, q-1)$ over a cyclic group of order n.

We now consider the general case and let $v = (q^{d+1}-1)/(q-1)$. Let ω be a generator of G and let $\mathcal{M}_c$ be the set of all ω-circulant $BGW(v, q^d, q^d - q^{d-1})$ matrices over G. Let R be the matrix of order v with all back diagonal entries equal to 1 and all other entries equal to 0. Then, for any $X \in \mathcal{M}_c$, XR is a symmetric $BGW(v, q^d, q^d - q^{d-1})$ over G. Let $\mathcal{M}_s = \{XR\colon X \in \mathcal{M}_c\}$. Let ρ be the same as in Proposition 10.4.7 with ω replaced by ω^{-1}. Then $\rho Y \in \mathcal{M}_s$ for all $Y \in \mathcal{M}_s$, so the cyclic group S of order nv generated by ρ can be regarded as a group of bijections on $\mathcal{M}_s$.

Case 1. Suppose $q(q-1)/n$ is even.

Since nv divides $q^{d+1}-1$ and $q^{d+1}(q^{d+1}-1)/(nv)$ is even, there exists a symmetric $BGW(q^{d+1}+1, q^{d+1}, q^{d+1}-1; S)$ with zero diagonal. Let W be such a matrix and let $Y \in \mathcal{M}_s$. By Remark 10.4.13, $W \otimes_S Y$ is a BGW with parameters (10.7). Since both Y and W are symmetric, so is $W \otimes_S Y$. Since the diagonal entries of W are zeros, so are the diagonal entries of $W \otimes_S Y$.

Case 2. Suppose n is even and $(q-1)/n$ is odd.

Let ω be a generator of G. Then $\omega^{n/2}$ is the unique element of order 2 in G. Since nv divides $q^{d+1}-1$, nv is even, and $(q^{d+1}-1)/(nv)$ is odd, there exists a skew-symmetric $BGW(q^{d+1}+1, q^{d+1}, q^{d+1}-1; S)$. Let $W = [\gamma_{ij}]$ be such a matrix and let $Y \in \mathcal{M}_s$. By Remark 10.4.13, $W \otimes_S Y$ is a BGW-matrix over G with parameters (10.7). The definition of the map ρ implies that $\rho^{\frac{nv}{2}}(Y) = \omega^{\frac{n}{2}} Y$. Since Y is symmetric and W is skew-symmetric, we obtain that

$$(W \otimes_S Y)^\top = W^\top \otimes_S Y^\top = (\rho^{nv/2} W) \otimes_S Y$$

$$= W \otimes_S (\rho^{nv/2}(Y)) = W \otimes_S (\omega^{n/2} Y) = \omega^{n/2}(W \otimes_S Y).$$

Therefore, the matrix $W \otimes_S Y$ is skew-symmetric. □

If the group G is trivial, then Theorem 10.4.16 yields an adjacency matrix of a (v, k, λ)-graph, and we obtain the following result.

Corollary 10.4.17. *For any prime power q and any odd positive integer d, there exists a (v, k, λ)-graph with parameters of the complement of $PG_{d-1}(d, q)$.*

The definition of BGW-matrices allows us to find the product WW^* whenever W is a $BGW(v, k, \lambda; G)$. Another matrix, for which this kind of product can be found, is the *core* of a GH-matrix or of a $BGW(q+1, q, q-1)$.

Definition 10.4.18. Let W be a normalized generalized Hadamard or conference matrix. The matrix obtained by deleting the first row and the first column of W is called the *core of* W.

The next theorem shows that the cores of normalized generalized conference matrices over abelian groups are either symmetric or skew-symmetric. We begin with a lemma.

Lemma 10.4.19. *Let π be the product of all elements of a finite abelian group G. If G is not a signed group, then $\pi = 1$. If G is a signed group, then π is the unique element of order 2 in G.*

Proof. Let E be the set of all elements of G of order 1 or 2. Then E is a subgroup of G and the set $G \setminus E$ can be partitioned into 2-subsets of the form $\{\sigma, \sigma^{-1}\}$. This implies that π is the product of all elements of E, and therefore, $\pi \in E$.

If G is a signed group, then $E = \{1, \varepsilon\}$, where ε is the unique element of order 2 in G. Therefore, $\pi = \varepsilon$.

Suppose that G is not a signed group and that $\pi \neq 1$. Then $|E| = e \geq 3$. Let $\tau \in E \setminus \{1, \pi\}$. Then the subgroup $\langle \pi, \tau \rangle$ of E is of order 4, and therefore, $e \equiv 0 \pmod 4$. The group E can be partitioned into 2-subsets of the form $\{\sigma, \pi\sigma\}$. Since the product of the elements of such a 2-subset is π, we obtain that $\pi = \pi^{e/2}$. Since $e/2$ is even, this implies that $\pi = 1$. The proof is now complete. □

Theorem 10.4.20. *Let C be the core of a normalized generalized conference matrix of index λ over a finite abelian group G. If G is a signed group and λ is odd, then C is skew-symmetric. Otherwise, C is symmetric.*

Proof. Let $C = [\omega_{ij}]$ be of order n and let $s, t \in \{1, 2, \ldots, n\}$, $s \neq t$. Let π denote the product of all elements of G. Since each row of C contains exactly λ copies of every element of G, we obtain that

$$\prod_{j \neq s,t} \omega_{sj} = \pi^{\lambda} \omega_{st}^{-1}, \quad \prod_{j \neq s,t} \omega_{tj} = \pi^{\lambda} \omega_{ts}^{-1}.$$

Since the multiset $\{\omega_{sj}\omega_{tj}^{-1} : j \neq s, j \neq t\}$ contains exactly λ copies of every nonidentity element of G, we have

$$\prod_{j \neq s,t} \omega_{sj}\omega_{tj}^{-1} = \pi^{\lambda}.$$

Therefore, $\omega_{st}^{-1}\omega_{ts} = \pi^{\lambda}$, $\omega_{ts} = \pi^{\lambda}\omega_{st}$, and we apply Lemma 10.4.19 to complete the proof. □

Since the usual $(0, \pm 1)$ conference matrix of order $n \geq 2$ can be regarded as a generalized conference matrix of index $n/2$ over the group $G = \{1, -1\}$ of order 2, we obtain the following result.

Corollary 10.4.21. *Let C be the core of a conference matrix of order n. If $n \equiv 0 \pmod 4$, then C is skew-symmetric. If $n \equiv 2 \pmod 4$, then C is symmetric.*

The next two propositions describe the cores of Hadamard and conference matrices over G in terms of the group ring $\mathbb{Z}G$. Their proofs are straightforward.

Proposition 10.4.22. *Let G be a finite group and λ a positive integer. A $(0, G)$ matrix C of order $\lambda|G| - 1$ is the core of a $GH(G; \lambda)$ if and only if $CC^* = (\lambda|G| - 1)I + (\lambda G - 1)(J - I)$ and $CJ = (\lambda G - 1)J$.*

Proposition 10.4.23. *Let G be a finite group and $q \geq 2$ an integer. A $(0, G)$ matrix C of order q is the core of a $BGW(q + 1, q, q - 1; G)$ if and only if*

$$CC^* = (q - 1)I + \left(\frac{q - 1}{|G|}G - 1\right)(J - I)$$

and

$$CJ = \frac{q - 1}{|G|}GJ.$$

Remark 10.4.24. If $\mathcal{M}$ is the set of the cores of all matrices $GH(G; \lambda)$ (for given G and λ) or of all matrices $BGW(q + 1, q, q - 1; G)$ (for given G and q), then $\mathcal{M}$ is a subset of the set of matrices of the corresponding order described in Proposition 10.4.4. Both sets are closed under the action of the group S of rotations described in that Proposition, so S serves in either case as a group of rotations. Also note that any group of order n can be regarded as a sharply transitive group of permutations of the set $\{1, 2, \ldots, n\}$. Therefore, the Kronecker product $W \otimes_S C$ is defined in the following cases:

(i) C is the core of a $GH(G; \lambda)$ and W is a $BGW(v, k, \lambda; S)$, where S is any group of order $\lambda|G| - 1$;
(ii) C is the core of a $BGW(q + 1, q, q - 1; G)$ and W is a $BGW(v, k, \lambda; S)$, where S is any group of order q.

Theorems 10.4.25 and 10.4.27 give examples of application of this Kronecker product for constructing GH- and BGW-matrices.

Theorem 10.4.25. *Let G be a finite group and λ a positive integer. Let S be a group of order $q = \lambda|G| - 1$. If there exist matrices $GH(G; \lambda)$ and $GH(S, 1)$, then there exists a matrix $GH(G; \lambda q)$.*

Proof. Let $\mathcal{M}$ be the set of the cores of all matrices $GH(G; \lambda)$ and let the group S act as a group of rotations of $\mathcal{M}$. Let W be a $GH(S; 1)$ and let $C \in \mathcal{M}$.

We claim that the block matrix

$$K = \begin{bmatrix} \overline{W} \otimes_S C & C \otimes J_{q,1} \\ C \otimes J_{1,q} & J_q \end{bmatrix}$$

is a $GH(G; \lambda q)$.

Observe that $\overline{W} \otimes_S C$ is a matrix of order q^2, $C \otimes J_{q,1}$ is a $q^2 \times q$ matrix, $C \otimes J_{1,q}$ is a $q \times q^2$ matrix, and J_q is the all-one matrix of order q, so K is a $(0, G)$ matrix of order $q^2 + q = \lambda q|G|$ without zero entries. Thus, it suffices to verify that if $\mathbf{x}$ and $\mathbf{y}$ are distinct rows of K, then $\mathbf{x}\mathbf{y}^* = \lambda q G$.

Let $W = [\omega_{ij}]$. For $i, h = 1, 2, \ldots, q$, let

$$P_{ih} = \sum_{j=1}^{q} (\overline{\omega_{ij}} C)(\overline{\omega_{hj}} C)^*.$$

Then $P_{ii} = qCC^* = q^2 I + q(\lambda G - 1)J$. For $i \neq h$, we have

$$P_{ih} = \sum_{j=1}^{q} (\omega_{hj} \overline{\omega_{ij}} C)C^* = \sum_{\sigma \in S} (\sigma C)C^* = (\lambda G - 1)JC^* = (\lambda G - 1)^2 J.$$

We will partition the set of rows of K into $q + 1$ subsets $R_1, R_2, \ldots, R_{q+1}$, each consisting of q consecutive rows.

If $\mathbf{x}$ and $\mathbf{y}$ are distinct rows from the same set R_i with $1 \leq i \leq q$, then they can be represented as $\mathbf{x} = [\mathbf{u}\ \mathbf{c}]$ and $\mathbf{y} = [\mathbf{v}\ \mathbf{c}]$, where $\mathbf{u}$ and $\mathbf{v}$ are distinct rows of $\overline{W} \otimes_S C$ and $\mathbf{c}$ is a row of C. Since $\mathbf{u}\mathbf{v}^*$ is an off-diagonal entry of P_{ii}, we obtain that $\mathbf{x}\mathbf{y}^* = q(\lambda G - 1) + q = \lambda q G$.

If $\mathbf{x} \in R_i$ and $\mathbf{y} \in R_h$ with $i, h \in \{1, 2, \ldots, q\}$ and $i \neq h$, then $\mathbf{x} = [\mathbf{u}\ \mathbf{c}]$ and $\mathbf{y} = [\mathbf{v}\ \mathbf{d}]$ where $\mathbf{u}$ and $\mathbf{v}$ are rows of $\overline{W} \otimes_S C$ and $\mathbf{c}$ and $\mathbf{d}$ are distinct rows of C. In this case, $\mathbf{u}\mathbf{v}^*$ is an entry of P_{ih}, so $\mathbf{x}\mathbf{y}^* = (\lambda G - 1)^2 + (\lambda G - 1) = \lambda^2|G|G - \lambda G = \lambda q G$.

If $\mathbf{x}$ and $\mathbf{y}$ are distinct rows of R_{q+1}, then $\mathbf{x}\mathbf{y}^* = q\mathbf{c}\mathbf{d}^* + q$ where $\mathbf{c}$ and $\mathbf{d}$ are distinct rows of C, so again $\mathbf{x}\mathbf{y}^* = \lambda q G$.

Finally, suppose $\mathbf{x} \in R_i$ with $1 \leq i \leq q$ and $\mathbf{y} \in R_{q+1}$. Then $\mathbf{x} = [\mathbf{a}_1\ \mathbf{a}_2 \ldots \mathbf{a}_q\ \mathbf{c}]$ where $\mathbf{c}$ is a row of C and each $\mathbf{a}_i$ is obtained by permuting the entries of a row of C. Besides, $\mathbf{y} = [c_1 J_{1,q}\ c_2 J_{1,q} \ldots c_q J_{1,q}\ J_{1,q}]$, where $[c_1\ c_2 \ldots c_q]$ is a row of C. Therefore,

$$\mathbf{x}\mathbf{y}^* = \sum_{j=1}^{q} (\lambda G) - 1)c_j^{-1} + (\lambda G - 1) = (\lambda G - 1)^2 + (\lambda G - 1) = \lambda q G.$$

The proof is now complete. □

Corollary 10.4.26. *If q and $q - 1$ are prime powers, then there exists a $GH(q, q - 1)$.*

Theorem 10.4.27. *Let G be a finite group and $q \geq 2$ an integer. Let S be a group of order q. If there exist matrices $BGW(q+1, q, q-1; G)$ and $GH(S; 1)$, then, for any positive integer m, there exists a $BGW((q^{m+1} - 1)/(q-1), q^m, q^m - q^{m-1}; G)$.*

Proof. Let W be a normalized $BGW(q+1, q, q-1; G)$, let C be the core of W, and let H be a $GH(S; 1)$. Let $n = (q-1)/|G|$. Then, by Proposition 10.4.23, $CJ = nGJ$ and $CC^* = qI + (nG-1)(J-I)$.

Let $\mathcal{M}$ be the set of cores of all normalized matrices $BGW(q+1, q, q-1; G)$. For any positive integer m, let H_m be a $GH(S; q^{m-1})$ (such matrices exist by Corollary 10.4.11). We will assume that S acts on $\mathcal{M}$ as a group of rotations. Let $C_0 = C$. Define recursively for $m \geq 1$ matrices C_m by

$$C_m = \begin{bmatrix} \overline{H_m} \otimes_S C \\ C_{m-1} \otimes J_{1,q} \end{bmatrix}.$$

The matrix C_m is a $(0, G)$ matrix of size $(q^{m+1} + q^m) \times q^{m+1}$. Since C_0 has exactly one zero entry in each row, it follows by a straightforward induction that C_m has exactly q^m zeros in each row.

Claim 1. $C_m J = q^m nGJ$.

We will prove this claim by induction on m. It is true for $m = 0$, so let $m \geq 1$ and let $C_{m-1}J = q^{m-1}nGJ$. We have

$$(H_m \otimes_S C)J = q^m CJ = q^m nGJ$$

and

$$(C_{m-1} \otimes J_{1,q})J = qC_{m-1}J = q^m nGJ.$$

For each m, we partition the set of rows of C_m into $q^m + q^{m-1}$ subsets $R_1^{(m)}, R_2^{(m)}, \ldots$, each consisting of q consecutive rows.

Claim 2. If $\mathbf{x} \in R_i^{(m)}$ and $\mathbf{y} \in R_h^{(m)}$, then

$$\mathbf{xy}^* = \begin{cases} q^{m+1} - q^m & \text{if } \mathbf{x} = \mathbf{y}, \\ q^m(nG-1) & \text{if } i = h \text{ and } \mathbf{x} \neq \mathbf{y}, \\ (q^m - q^{m-1})nG & \text{if } i \neq h. \end{cases}$$

Since each row of C_m has q^{m+1} entries, of which q^m are zeros, we obtain that $\mathbf{xx}^* = q^{m+1} - q^m$. We will prove the other two cases of the claim by induction on m. For $m = 0$, we have $i = h = 1$. If $\mathbf{x} \neq \mathbf{y}$, then $\mathbf{xy}^* = nG - 1$.

Let $m \geq 1$ and suppose the claim is true for rows of C_{m-1}. If $\mathbf{x}, \mathbf{y} \in R_i^{(m)}$ with $i \leq q^m$ and $\mathbf{x} \neq \mathbf{y}$, then $\mathbf{xy}^*$ is an off-diagonal entry of a matrix of the form

$$P_1 = \sum_{j=1}^{q^m} (\overline{\eta_j} C)(\overline{\eta_j} C)^*,$$

where $[\eta_1\ \eta_2 \ldots \eta_{q^m}]$ is a row of H_m. Since

$$P_1 = q^m CC^* = q^{m+1} I + q^m (nG - 1)(J - I),$$

we obtain that $\mathbf{xy}^* = q^m(nG - 1)$. If $\mathbf{x} \in R_i^{(m)}$ and $\mathbf{y} \in R_h^{(m)}$ with distinct $i, h \in \{1, 2, \ldots, q^m\}$, then $\mathbf{xy}^*$ is an entry of a matrix of the form

$$P_1 = \sum_{j=1}^{q^m} (\overline{\eta_j} C)(\overline{\zeta_j} C)^*,$$

where $[\eta_1\ \eta_2 \ldots \eta_{q^m}]$ and $[\zeta_1\ \zeta_2 \ldots \zeta_{q^m}]$ are distinct rows of H_m. Since

$$\begin{aligned} P_2 = \sum_{j=1}^{q^m} (\zeta_j \overline{\eta_j} C) C^* &= q^{m-1} \left(\sum_{\sigma \in S} \sigma C \right) C^* = q^{m-1} nGJC^* \\ &= q^{m-1} n^2 |G| GJ = q^{m-1}(q - 1) nGJ, \end{aligned}$$

we obtain that $\mathbf{xy}^* = (q^m - q^{m-1})nG$.

If $\mathbf{x} \in R_i^{(m)}$ and $\mathbf{y} \in R_h^{(m)}$ with $i > q^m$ and $h > q^m$, then $\mathbf{xy}^* = q\mathbf{uv}^*$ where $\mathbf{u}$ and $\mathbf{v}$ are rows of C_{m-1}, so we apply the induction hypothesis to obtain the asserted values of $\mathbf{xy}^*$.

Finally, let $\mathbf{x} \in R_i^{(m)}$ and $\mathbf{y} \in R_h^{(m)}$ with $i \leq q^m$ and $h > q^m$. Then $\mathbf{xy}^*$ is an entry of a matrix of the form

$$P_3 = \sum_{j=1}^{q^m} (\overline{\eta_j} C) a_j,$$

where $[\eta_1\ \eta_2 \ldots \eta_{q^m}]$ is a row of H_m and $[a_1\ a_2 \ldots a_{q^m}]$ is a row of C_{m-1}. Any row of $\overline{\eta_j} C$ is a permutation of a row of C. Therefore,

$$\mathbf{xy}^* = \sum_{j=1}^{q^m} \sum_{k=1}^{q} c_{jk} a_j,$$

where $[c_{j1}\ c_{j2} \ldots c_{jk}]$ is a permutation of a row of C for each j. Thus,

$$\mathbf{xy}^* = \sum_{j=1}^{q^m} nGa_j.$$

Since

$$Ga_j = \begin{cases} G & \text{if } a_j \neq 0, \\ 0 & \text{if } a_j = 0 \end{cases}$$

and since C_{m-1} has q^{m-1} zeros in each row, we obtain that $\mathbf{xy}^* = (q^m - q^{m-1})nG$. The case $i > q^m$ and $h \leq q^m$ is similar, so Claim 2 is now proven.

Let $W_1 = W$ and define recursively for $m \geq 2$ matrices W_m by

$$W_m = \begin{bmatrix} \mathbf{0}^\top & \mathbf{j}^\top \\ W_{m-1} \otimes J_{q,1} & C_{m-1} \end{bmatrix}.$$

We claim that W_m is a $BGW(w_m, q^m, q^m - q^{m-1}; G)$ with $w_m = (q^{m+1} - 1)/(q-1)$. This is true for $m = 1$, so let $m \geq 2$ and let W_{m-1} be a $BGW(w_{m-1}, q^{m-1}, q^{m-1} - q^{m-2}; G)$.

Observe that W_m is a matrix of order $1 + qw_{m-1} = w_{m-1} + q^m = w_m$ and each row of W_m has exactly q^m nonzero entries. Starting with the second row, we will partition the set of rows of W_m into subsets, each formed by q consecutive rows.

If $\mathbf{x}$ and $\mathbf{y}$ are distinct rows from the same subset, then the induction hypothesis and Claim 2 imply that

$$\mathbf{xy}^* = q^{m-1} + q^{m-1}(nG - 1) = q^{m-1}nG = \frac{q^m - q^{m-1}}{|G|}G.$$

If $\mathbf{x}$ and $\mathbf{y}$ are rows from different subsets, then the induction hypothesis and Claim 2 imply that

$$\mathbf{xy}^* = q^{m-1}nG + (q^m - q^{m-1})nG = q^m nG.$$

Finally, if $\mathbf{x}$ or $\mathbf{y}$ is the first row of W_m, we apply Claim 1. The proof is now complete. □

If q is a prime power, then there exists a matrix $GH(q, 1)$. Therefore, we obtain the following corollary of Theorem 10.4.27.

Corollary 10.4.28. *If, for a prime power q and a finite group G, there exists a $BGW(q+1, q, q-1; G)$, then, for any positive integer m, there exists a $BGW((q^{m+1} - 1)/(q-1), q^m, q^m - q^{m-1}; G)$.*

This gives another construction for matrices (10.6). However, if we use the more general Theorem 10.1.21, we obtain infinite families of matrices over nonabelian groups.

Theorem 10.4.29. *Let q be a prime power and let G be the multiplicative group of a nearfield of order q. Then, for any positive integer m, there exists a $BGW((q^{m+1}-1)/(q-1), q^m, q^m - q^{m-1}; G)$.*

With the next result and Corollary 10.4.28, we will obtain an infinite family of BGW-matrices over an arbitrary group of order $q+1$ where q is a prime power.

Theorem 10.4.30. *Let q be a prime power and let $GF(q^2) = \{a_1, a_2, \dots, a_{q^2}\}$ with $a_1 = 0$. Let α be a primitive element of $GF(q^2)$. For $s = 0, 1, \dots, q$, let $C_s = \{a\alpha^s : a \in GF(q)^*\}$. Let $G = \{x_0, x_1, \dots, x_q\}$ be a group of order $q+1$. Let $C = [c_{ij}]$ be a $(0, G)$ matrix of order q^2 defined as follows: for $s = 0, 1, \dots, q$, $c_{ij} = x_s$ if and only if $a_i - a_j \in C_s$. Then C is the core of a $BGW(q^2+1, q^2, q^2-1; G)$.*

Proof. Since $C_0 \cup C_1 \cup \dots \cup C_q = GF(q^2)^*$, we have $c_{ij} = 0$ if and only if $i = j$. Since $|C_s| = q - 1$ for $s = 0, 1, \dots, q$, each row of C has $q-1$ copies of every element of G, i.e., $CJ = (q-1)GJ$. Let $i, h \in \{1, 2, \dots, q^2\}$, $i \neq h$. Let $s \in \{0, 1, \dots, q\}$ and let a permutation σ of $\{0, 1, \dots, q\}$ be defined by $x_{\sigma(k)} = x_s x_k$ for $k = 0, 1, \dots, q$.

Case 1. x_s is not the identity element of G.

Then $\sigma(k) \neq k$ for $k = 0, 1, \dots, q$. By Lemma 9.6.15, for $k = 0, 1, \dots, q$, the multiset

$$M_k = \{u - v : u \in C_{\sigma(k)}, v \in C_k\}$$

is in fact the set $GF(q^2)^* \setminus (C_k \cup C_{\sigma(k)})$. Therefore, the multiset $M_0 \cup M_1 \cup \dots M_q$ contains exactly $q-1$ copies of every element of $GF(q^2)^*$.

Let $j \in \{1, 2, \dots, q^2\} \setminus \{i, h\}$. Then $c_{ij}c_{hj}^{-1} = x_s$ if and only if there is $k \in \{0, 1, \dots, q\}$ such that $a_i - a_j \in C_{\sigma(k)}$ and $a_h - a_j \in C_k$. Since $(a_i - a_j) - (a_h - a_j) = a_i - a_h$ is a fixed element of $GF(q^2)^*$, there are exactly $q-1$ indices $j \in \{1, 2, \dots, q^2\} \setminus \{i, h\}$ such that $c_{ij}c_{hj}^{-1} = x_s$.

Case 2. x_s is the identity element of G.

In this case, σ is the identity permutation. By Lemma 9.6.15, for $k = 0, 1, \dots, q$, the multiset $N_k = \{u - v : u, v \in C_k\}$ contains exactly $q-2$ copies of every element of C_k (and $q-1$ zeros). Similar to Case 1, we obtain that there are exactly $q-2$ indices $j \in \{1, 2, \dots, q^2\}$ such that $c_{ij} = c_{hj}$.

Thus, $CC^* = (q^2-1)I + ((q-1)G - 1)(J - I)$. By Proposition 10.4.23, C is the core of a $BGW(q^2+1, q^2, q^2-1; G)$. □

Theorems 10.4.27 and 10.4.30 immediately imply the next result.

Corollary 10.4.31. *For any prime power q, any group G of order $q+1$, and any positive integer m, there exists a*

$$BGW\left(\frac{q^{2m+2}-1}{q^2-1}, q^{2m}, q^{2m}-q^{2m-2}; G\right).$$

10.5. BGW-matrices and projective geometries

In this section we obtain BGW-matrices from spreads of subspaces of a projective space described in Theorem 3.6.13. We will also investigate relations between generalized conference and generalized Hadamard matrices on the one hand and homologies and elations of projective planes on the other hand.

Theorem 10.5.1. *Let q be a prime power and let n and d be distinct positive integers such that $d+1$ divides $n+1$. Then there exists a nonempty set $\mathcal{M}$ of incidence matrices of the complement of $PG_{d-1}(d,q)$, a cyclic group S of symmetries of $\mathcal{M}$, and a BGW-matrix W over S such that, for any $X \in \mathcal{M}$, the matrix $W \otimes_S X$ is an incidence matrix of the complement of $PG_{n-1}(n,q)$.*

Proof. Consider the following tower of finite fields: $F = GF(q) \subset K = GF(q^{d+1}) \subset L = GF(q^{n+1})$. Set $w = (q^{n+1}-1)/(q^{d+1}-1)$ and $v = (q^{d+1}-1)/(q-1)$. Let α be a primitive element of L. Then $\beta = \alpha^w$ is a primitive element of K and β^v is a primitive element of F. We regard L as an $(n+1)$-dimensional vector space over F and K as a $(d+1)$-dimensional subspace of L. Fix an n-dimensional subspace H_0 of L. Proposition 3.6.8 implies that

$$X = \{\langle \alpha^i \beta^j \rangle : 0 \le i \le w-1, 0 \le j \le v-1\}$$

and

$$\mathcal{B} = \{\alpha^i \beta^j H_0 : 0 \le i \le w-1, 0 \le j \le v-1\}$$

are, respectively, the set of points and the set of hyperplanes of the projective geometry $PG(n,q)$ over F. For $i = 0, 1, \ldots, w-1$, let $U_i = \{\langle \alpha^i \beta^j \rangle : 0 \le j \le v-1\}$ and $\mathcal{H}_i = \{\alpha^i \beta^j H_0 : 0 \le j \le v-1\}$. By Theorem 3.6.13, the set $\{U_0, U_1, \ldots, U_{w-1}\}$ is a spread of d-spaces of $PG(n,q)$.

Let $H \in \mathcal{B}$. For $i = 0, 1, \ldots, w-1$, either $U_i \subset H$ or $\dim(U_i \cap H) = d-1$. Let m be the number of indices i such that $U_i \subset H$. Then

$$|H| = \frac{q^n-1}{q-1} = m \cdot \frac{q^{d+1}-1}{q-1} + \left(\frac{q^{n+1}-1}{q^{d+1}-1} - m\right) \cdot \frac{q^d-1}{q-1},$$

which implies $m = (q^{n-d}-1)/(q^{d+1}-1)$.

Let $k \in \{0, 1, \ldots, w-1\}$ and let $H, H' \in \mathcal{H}_k$, $H \neq H'$. Then $H' = \beta^s H$ for some $s \in \{1, 2, \ldots, v-1\}$ and therefore, for $i = 0, 1, \ldots, w-1$, $U_i \subset H$ if and only if $U_i \subset H'$. Since $\dim(U_i \cap H \cap H') \geq d-2$ for $i = 0, 1, \ldots, w-1$, we obtain that

$$|H \cap H'| = \frac{q^{n-1}-1}{q-1} \geq m \cdot \frac{q^{d+1}-1}{q-1} + \left(\frac{q^{n+1}-1}{q^{d+1}-1} - m\right) \cdot \frac{q^{d-1}-1}{q-1} = \frac{q^{n-1}-1}{q-1}.$$

Therefore, for $i = 0, 1, \ldots, w-1$, either $U_i \subset H$ and $U_i \subset H'$ or $\dim(U_i \cap H) = \dim(U_i \cap H') = d-2$ and therefore, $U_i \cap H$ and $U_i \cap H'$ are distinct $(d-1)$-dimensional subspaces of U_i.

For $i, k = 0, 1, \ldots, w-1$, let $\mathbf{D}_{ik}$ denote the incidence structure $(U_i, \mathcal{H}_k)$. If $U_i \subset H$ for some and thus for all $H \in \mathcal{H}_k$, then $\mathbf{D}_{ik}$ is a symmetric (v, v, v)-design. Otherwise, the v blocks of $\mathbf{D}_{ik}$ are distinct $(d-1)$-dimensional subspaces of U_i. Since v is the total number of $(d-1)$-dimensional subspaces of the d-dimensional projective space U_i, we obtain that $\mathbf{D}_{ik}$ is a symmetric design isomorphic to $PG_{d-1}(d, q)$. The complement of $\mathbf{D}_{ik}$ is either a trivial design with zero incidence matrix or a design isomorphic to the complement of $PG_{d-1}(d, q)$.

From now on, we assume that each of the sets U_i and $\mathcal{H}_i$ is ordered so that, for $j = 0, 1, \ldots, v-1$, $\langle \alpha^i \beta^j \rangle$ precedes $\langle \alpha^i \beta^l \rangle$ and $\alpha^i \beta^j H_0$ precedes $\alpha^i \beta^l H_0$ if and only if $j < l$. For $i, k = 0, 1, \ldots, w-1$, let N_{ik} be the corresponding incidence matrix of $\mathbf{D}_{ik}$ and let $M_{ik} = J - N_{ik}$. Then the block matrices $N = [N_{ik}]$ and $M = [M_{ik}]$ are incidence matrices of $PG_{n-1}(n, q)$ and its complement, respectively. Let S be the factor group K^*/F^* and let $b = \beta F^*$. Then S is the cyclic group of order v generated by b. Since $\alpha^i \beta^j \in \alpha^k \beta^l H_0$ if and only if $\alpha^i \beta^{j+1} \in \alpha^k \beta^{l+1} H_0$, all matrices N_{ik} and M_{ik} are S-invariant and therefore these matrices are circulant.

For any matrix $X = [X_1\ X_2 \ldots X_v]$ of order v, let $bX = [X_v\ X_1 \ldots X_{v-1}]$. Then S acts as a group of rotations on the set of all $(0, 1)$-matrices of order v. Since $\alpha^i \beta^j \in \alpha^k \beta^l H_0$ if and only if $\alpha^{i+1} \beta^j \in \alpha^{k+1} \beta^l H_0$, we obtain that, for $i = 0, 1, \ldots, w-2$, $N_{ik} = N_{i+1,k+1}$ and therefore $M_{ik} = bM_{i+1,k+1}$. Since $\alpha^i \beta^j \in \alpha^{w-1} \beta^l H_0$ if and only if $\alpha^{i+1} \beta^j \in \beta^{l+1} H_0$, we obtain that, for $i = 0, 1, \ldots, w-2$, $N_{i+1,0} = bN_{i,w-1}$ and $M_{i+1,0} = bM_{i,w-1}$. Our first goal is to show that all nonzero matrices M_{ik} can be obtained one from another by applying suitable elements of S. It suffices to show this for matrices M_{i0}, $i = 0, 1, 2, \ldots, w-1$.

Let V be the union of all subspaces U_i that are contained in H_0. Then

$$V = \bigcap_{H \in \mathcal{H}_0} H,$$

so V is a subspace of $PG(n, q)$. Since $|V| = m(q^{d+1} - 1)/(q - 1) = (q^{n-d} - 1)/(q - 1)$, we have $\dim V = n - d - 1$.

Claim 1. Let $i, h \in \{0, 1, \ldots, w - 1\}$ be such that $i \neq h$, $U_i \not\subset H_0$, and $U_h \not\subset H_0$. Then there exists a bijection $f\colon U_i \to U_h$ such that every line $\overline{xf(x)}$, $x \in U_i$, meets V.

To prove this claim, fix $x \in U_i$ and consider the set of all lines $\overline{xy}$ with $y \in V$. Since all lines of $PG(n, q)$ are of cardinality $q + 1$, each line $\overline{xy}$ with $y \in V$ meets exactly $q - 1$ subsets U_k with $k \neq i$ such that $U_k \not\subset H_0$. Suppose lines $\overline{xy}$ and $\overline{xz}$ with distinct $y, z \in V$ meet the same subspace U_k ($k \neq i$, $U_k \not\subset H_0$) at points y' and z', respectively. Let π be the plane through the lines $\overline{xy}$ and $\overline{xz}$. Then $\pi \cap V = \overline{yz}$ and $\pi \cap U_k = \overline{y'z'}$. Since $V \cap U_k = \varnothing$, we have found two disjoint lines in a projective plane. This is not possible, and therefore the lines $\overline{xy}$, $y \in V$, meet altogether $(q - 1)|V| = q^{n-d} - 1$ distinct subspaces U_k such that $k \neq i$ and $U_k \not\subset H_0$. Since there are exactly $w - m - 1 = q^{n-d} - 1$ such subspaces, we obtain that there is a unique line $\overline{xy}$ with $y \in V$ that meets U_h. Let $U_h \cap \overline{xy} = \{f(x)\}$. This proves Claim 1.

Claim 2. Let $i, h \in \{0, 1, \ldots, w - 1\}$ and let $M_{i0} \neq O$ and $M_{h0} \neq O$. Then $M_{h0} = b^s M_{i0}$, for some integer s.

To prove this claim, we assume that $i \neq h$ and consider the bijection f from Claim 1. Let $H \in \mathcal{H}_0$. Since $V \subset H$ and all lines $\overline{xf(x)}$, $x \in U_i$, meet V, we obtain that, for any $x \in U_i$, $x \in H$ if and only if $f(x) \in H$. Therefore, if $f(\langle \alpha^i \rangle) = \langle \alpha^h \beta^l \rangle$ with $0 \leq l \leq v - 1$, then the first row of M_{i0} is equal to the $(l + 1)^{th}$ row of M_{h0}. Since both M_{i0} and M_{h0} are circulant matrices, we obtain that $M_{h0} = b^{-l} M_{i0}$.

Since $M_{i+1,k+1} = M_{ik}$ and $M_{i+1,0} = M_{i,w-1}$ for $i = 0, 1, \ldots, w - 2$, we obtain that, whenever $M_{ik} \neq O$ and $M_{hl} \neq O$, there is $\gamma \in S$ such that $M_{hl} = \gamma M_{ik}$. Fix one of the nonzero matrices M_{ik} and denote it by X. Let $\mathcal{M} = \{\gamma X : \gamma \in S\}$. Then $\mathcal{M}$ contains all nonzero matrices M_{ik}. Let $W = [\omega_{ik}]$ be the $(0, S)$ matrix of order w defined as follows: for $i, k = 0, 1, \ldots, w - 1$,

$$\omega_{ik} = \begin{cases} 0 & \text{if } M_{ik} = O, \\ \gamma & \text{if } M_{ik} = \gamma X. \end{cases}$$

Then $M = W \otimes_S X$. We shall prove that W is a $BGW(w, q^{n-d}, q^{n-d} - q^{n-2d-1}; S)$.

Since each hyperplane H of L contains exactly m subspaces U_i, each column of W contains exactly $w - m = q^{n-d}$ nonzero entries.

Claim 3. Let $j, l \in \{0, 1, \ldots, w-1\}$, $j \neq l$, and let $H \in \mathcal{H}_j$ and $H' \in \mathcal{H}_l$. Then there are exactly $t = q^{n-2d-1}(q-1)$ subspaces U_i such that $U_i \cap H = U_i \cap H' \neq U_i$.

To prove this claim, let V_j and V_l be the intersection of all hyperplanes from $\mathcal{H}_j$ and the intersection of all hyperplanes from $\mathcal{H}_l$, respectively. Let s be the number of subspaces U_k that are contained in $H \cap H'$. Since $\dim(V_j \cap H') = \dim V_j - 1 = n - d - 2$, we obtain that

$$|V_j \cap H'| = \frac{q^{n-d-1}-1}{q-1} = s \cdot \frac{q^{d+1}-1}{q-1} + (m-s) \cdot \frac{q^d - 1}{q-1},$$

which gives $s = (q^{n-2d-1} - 1)/(q-1)$.

Let t be the number of subspaces U_i such that $U_i \cap H = U_i \cap H' \neq U_i$. Then $\dim(U_i \cap H \cap H')$ is equal to d for s indices i, it is equal to $d-1$ for $2(m-s)+t$ indices i, and it is equal to $d-2$ for the remaining $w+s-2m-t$ indices i. Therefore,

$$|H \cap H'| = \frac{q^{n-1}-1}{q-1} = s \cdot \frac{q^{d+1}-1}{q-1} + (2m - 2s + t) \cdot \frac{q^d - 1}{q-1} + (w + s - 2m - t) \cdot \frac{q^{d-1}-1}{q-1},$$

which implies $t = q^{n-2d-1}(q-1)$.

Let $j, l \in \{0, 1, \ldots, w-1\}$ and let $k \in \{0, 1, \ldots, v-1\}$. By Claim 3, there are exactly t indices i such that $U_i \cap (\alpha^j H_0) = U_i \cap (\alpha^l \beta^k H_0) \neq U_i$. These are precisely the indices i such that $M_{ij} \neq O$, $M_{il} \neq O$, and the first row of M_{ij} is equal to the $(k+1)^{th}$ row of M_{lj}. Since both M_{ij} and M_{lj} are circulant matrices, we obtain that $M_{ij} = b^k M_{lj}$. Thus, there are exactly t indices i such that $M_{ij} = \omega_{ij}$ and $M_{lj} = b^k \omega_{hj}$, i.e., $\omega_{ij} \neq 0, \omega_{hj} \neq 0$, and $\omega_{ij} = b^k \omega_{hj}$. Since this is true for $k = 0, 1, \ldots, v-1$, $W^\top$ is a $BGW(w, q^{n-d}, q^{n-d} - q^{n-2d-1}; S)$. Since S is an abelian group, the matrix W is a BGW-matrix with the same parameters. □

In the next theorem, we will show that generalized Hadamard matrices of index 1 and generalized conference matrices of index 1 are equivalent to projective planes admitting certain groups of collineations.

Theorem 10.5.2. *Let E be a group of order $q \geq 3$ and G a group of order $q-1$. Then:*

(i) *the existence of a $GH(E; 1)$ is equivalent to the existence of a projective plane of order q admitting a group of (c, A)-elations isomorphic to E;*

(ii) *the existence of a* $BGW(q+1, q, q-1; G)$ *is equivalent to the existence of a projective plane of order* q *admitting a group of* (c, A)*-homologies isomorphic to* G.

Proof. We will assume that the group E is additive and the group G is multiplicative. Let H be a matrix of order $q-1$ with entries from $E \setminus \{0\}$ and let C be a matrix of order q with all off-diagonal entries from G and all the diagonal entries equal 0. We will assume that the rows and columns of $H = [H(\sigma, \tau)]$ are indexed by elements of G and the rows and columns of $C = [C(a, b)]$ are indexed by elements of E.

Let $\mathbf{P}(H) = (X, \mathcal{L}(H))$ and $\mathbf{P}(C) = (X, \mathcal{L}(C))$ be incidence structures with the same point set

$$X = (G \times E) \cup G \cup E \cup \{\theta, \infty\}$$

of cardinality $q^2 + q + 1$ and the line (block) sets

$$\begin{aligned}\mathcal{L}(H) = \{J(\sigma)\colon \sigma \in G\} \cup \{K(a)\colon a \in E\} \cup \{L_H(\sigma, a)\colon a \in E, \sigma \in G\}\\ \cup \{A_\theta, A_\infty\}\end{aligned}$$

and

$$\begin{aligned}\mathcal{L}(C) = \{J(\sigma)\colon \sigma \in G\} \cup \{K(a)\colon a \in E\} \cup \{L_C(\sigma, a)\colon a \in E, \sigma \in G\}\\ \cup \{A_\theta, A_\infty\},\end{aligned}$$

where

$$\begin{aligned} J(\sigma) &= \{(\sigma, x)\colon x \in E\} \cup \{\theta\},\\ K(a) &= \{(\tau, a)\colon \tau \in G\} \cup \{a, \infty\},\\ L_H(\sigma, a) &= \{(\tau, -H(\sigma^{-1}, \tau) + a)\colon \tau \in G\} \cup \{\sigma, a\},\\ L_C(\sigma, a) &= \{(C(x, a)\sigma^{-1}, x)\colon x \in E \setminus \{a\}\} \cup \{\sigma, a\},\\ A_\theta &= E \cup \{\theta\},\ A_\infty = G \cup \{\theta, \infty\}. \end{aligned}$$

Observe that $|\mathcal{L}(H)| = |\mathcal{L}(C)| = q^2 + q + 1$, that each line $L \in \mathcal{L}(H) \cup \mathcal{L}(C)$ is of cardinality $q+1$, and that each point $x \in X$ is contained in $q+1$ lines of $\mathcal{L}(H)$ and in $q+1$ lines of $\mathcal{L}(C)$.

After these preparations, we begin the proof.

(i) Suppose first that H is the core of a normalized $GH(E; 1)$. We claim that $\mathbf{P}(H)$ is a projective plane of order q. It suffices to show that $|L_1 \cap L_2| = 1$, for any distinct $L_1, L_2 \in \mathcal{L}(H)$. The only case in which it is not immediate is $L_1 = L_H(\rho, a)$ and $L_2 = L_H(\sigma, b)$ with distinct $\rho, \sigma \in G$ and distinct

$a, b \in E$. In this case, for $\tau \in G$ and $x \in E$, $(\tau, x) \in L_1 \cap L_2$ if and only if $x = -H(\rho^{-1}, \tau) + a = H(\sigma^{-1}, \tau) + b$. We have

$$-H(\rho^{-1}, \tau) + a = H(\sigma^{-1}, \tau) + b \Leftrightarrow H(\sigma^{-1}, \tau) - H(\rho^{-1}, \tau) = b - a.$$

Since H is the core of a $GH(E; 1)$, the last equation has a unique solution τ, so $\mathbf{P}(H)$ is indeed a projective plane of order q.

For each $a \in E$, define a map $t_a : X \to X$ as follows: $t_a(\sigma) = \sigma$, $t_a(x) = x - a$, and $t_a(\sigma, x) = (\sigma, x - a)$ for all $\sigma \in G$ and $x \in E$, $t_a(\theta) = \theta$, and $t_a(\infty) = \infty$. Then $t_a(J(\sigma)) = J(\sigma)$, $t_a(K(b)) = K(b - a)$, $t_a(L_H(\sigma, b)) = L_H(\sigma, b - a)$, $t_a(A_\theta) = A_\theta$, and $t_a(A_\infty) = A_\infty$. Therefore, t_a is a (θ, A_∞)-elation of $\mathbf{P}(H)$. Moreover, since $t_a t_b = t_{a+b}$ for all $a, b \in E$, we have found a group of (θ, A_∞)-elations isomorphic to E.

Conversely, suppose $\mathbf{P}$ is a projective plane of order q with two fixed points, θ and ∞. Let A_∞ be the line through θ and ∞ and let the group of all (θ, A_∞)-elations of $\mathbf{P}$ be isomorphic to E. For each $a \in E$, let t_a denote the corresponding (θ, A_∞)-elation. Fix a line A_θ containing θ and other than A_∞, choose a point $0 \neq \theta$ on this line, identify every $a \in E$ with $t_{-a}(0)$, and let $K(a)$ denote the line through a and ∞. Choose a one-to-one correspondence between the elements of G, the points of A_∞, other that ∞ or θ, and the lines through θ, other than A_∞ or A_θ. According to this one-to-one correspondence, we will identify every $\sigma \in G$ with a point of $A_\infty \setminus \{\theta, \infty\}$ and denote by $J(\sigma)$ the line through θ corresponding to σ. For $\sigma \in G$ and $a \in E$, let $L(\sigma, a)$ denote the line through σ and a and let (σ, a) denote the intersection point of lines $J(\sigma)$ and $K(a)$. Observe that, for all $a, b \in E$, $t_a(b) = t_a(t_{-b}(0)) = t_{a-b}(0) = b - a$. Therefore, $t_a(K(b)) = K(b - a)$ and, since $t_a(J(\sigma)) = J(\sigma)$, we obtain that $t_a(\sigma, b) = (\sigma, b - a)$.

For $\sigma, \tau \in G$ and $b \in E$, let $H(\sigma, \tau) = b$ if and only if $(\tau, -b)$ is the intersection point of lines $J(\tau)$ and $L(\sigma, 0)$. We claim that the matrix $H = [H(\sigma, \tau)]$ of order $q - 1$ is the core of a $GH(E; 1)$. Since $L(\sigma, 0) \cap K(0) = \{0\}$, we obtain that the intersection point of $J(\tau)$ and $L(\sigma, 0)$ is not $(\tau, 0)$, i.e., $H(\sigma, \tau) \neq 0$. Since $L(\sigma, 0)$ meets every line $K(-b)$ with $b \neq 0$ at some point $(\tau, -b)$, each row of H contains every element of $E \setminus \{0\}$. Suppose $H(\rho, \tau_1) - H(\sigma, \tau_1) = H(\rho, \tau_2) - H(\sigma, \tau_2)$, for some distinct $\rho, \sigma \in G$ and distinct $\tau_1, \tau_2 \in G$. For $i = 1, 2$, let $a_i = H(\rho, \tau_i)$ and $b_i = H(\sigma, \tau_i)$. Then $b_1 - a_1 = b_2 - a_2 = c$, and we have $t_c(\tau_i, -a_i) = (\tau_i, -b_i)$. Since $J(\tau_i) \cap L(\rho, 0) = (\tau_i, a_i)$ and $J(\tau_i) \cap L(\sigma, 0) = (\tau_i, b_i)$, we obtain that $t_c(L(\rho, 0)) = L(\sigma, 0)$. However, $\rho \in L(\rho, 0)$, $\rho \notin L(\sigma, 0)$, and $t_c(\rho) = \rho$, a contradiction.

Therefore, H is the core of $GH(E; 1)$. Note that $L(\sigma, a) = L_H(\sigma, a)$, and therefore $\mathbf{P} = \mathbf{P}(H)$.

(ii) Suppose first that C is the core of a normalized $BGW(q+1, q, q-1; G)$. We claim that $\mathbf{P}(C)$ is a projective plane of order q. As in the previous case, it reduces to showing that $|L_C(\rho, a) \cap L_C(\sigma, b)| = 1$ for distinct $\rho, \sigma \in G$ and distinct $a, b \in E$. For $x \in E$,

$$\begin{aligned} C(x, a)\rho^{-1} &= C(x, b)\sigma^{-1} \Leftrightarrow C(x, b)^{-1}C(x, a) \\ &= \sigma^{-1}\rho \Leftrightarrow C^*(b, x)C^*(a, x)^{-1} = \sigma^{-1}\rho. \end{aligned}$$

Proposition 10.1.34 implies that C^* is the core of a $BGW(q+1, q, q-1; G)$, and therefore the last equation has a unique solution x. Thus, $\mathbf{P}(C)$ is a projective plane.

For each $\tau \in G$, we define a map $h_\tau \colon X \to X$ as follows: $h_\tau(\sigma) = \sigma\tau^{-1}$, $h_\tau(x) = x$, and $h_\tau(\sigma, x) = (\sigma\tau^{-1}, x)$ for all $\sigma \in G$ and $x \in E$, $h_\tau(\theta) = \theta$, and $h_\tau(\infty) = \infty$. It is straightforward to verify that h_τ is an (∞, A_θ)-homology and that $h_\rho h_\sigma = h_{\rho\sigma}$ for all $\rho, \sigma \in G$, so we have obtained a group of (∞, A_θ)-homologies isomorphic to G.

Conversely, suppose $\mathbf{P}$ is a projective plane of order q with two fixed points, θ and ∞. Let A_θ be a line containing θ and not containing ∞ and let $\mathbf{P}$ admit a group of (∞, A_θ)-homologies isomorphic to G. For each $\sigma \in G$, let h_σ denote the corresponding (∞, A_θ)-homology. Let A_∞ be the line containing ∞ and θ. Fix a point 1 on this line, other than ∞ or θ, and a line $J(1)$ through θ, other than A_∞ or A_θ, and, for every $\sigma \in G$, identify σ with $h_{\sigma^{-1}}(1)$ and let $J(\sigma) = h_{\sigma^{-1}}(J(1))$. Choose a one-to-one correspondence between the elements of E and the points of A_θ, other than θ, identify every $a \in E$ with the corresponding point, and let $K(a)$ be the line through ∞ and a. For all $\sigma \in G$ and $a \in E$, let $L(\sigma, a)$ be the line through σ and a and (σ, a) the intersection point of lines $J(\sigma)$ and $K(a)$. Since $h_\tau(K(a)) = K(a)$, we obtain that $h_\tau(1, a) = (\tau^{-1}, a)$, and then $h_\tau(\sigma, a) = h_\tau h_{\sigma^{-1}}(1, a) = h_{\tau\sigma^{-1}}(1, a) = (\sigma\tau^{-1}, a)$.

For distinct $a, b \in E$ and for $\tau \in G$, let $C(a, b) = \tau$ if and only if (τ, a) is the intersection point of lines $K(a)$ and $L(1, b)$. For all $a \in E$, let $C(a, a) = 0$. We claim that the matrix $C = [C(a, b)]$ of order q is the core of a $BGW(q+1, q, q-1; G)$. By Proposition 10.1.34, it suffices to show that the matrix C^* is the core of a $BGW(q+1, q, q-1; G)$.

Since $L(1, b) \cap K(b) = \{b\}$, the line $L(1, b)$ meets every line $J(\tau)$ at some point (τ, a) with $a \neq b$. Therefore, each row of C^* contains all elements of $G \cup \{0\}$. Suppose $C^*(a, c_1)C^*(b, c_1)^{-1} = C^*(a, c_2)C^*(b, c_2)^{-1}$ for some distinct $a, b \in E$ and distinct $c_1, c_2 \in E \setminus \{a, b\}$. Then $C(c_1, a)^{-1}C(c_1, b) = C(c_2, a)^{-1}C(c_2, b) = \rho \in G$. For $i = 1, 2$, let $C(c_i, a) = \sigma_i$ and $C(c_i, b) = \tau_i$. Then $\rho^{-1}(\sigma_i, c_i) = (\tau_i, c_i)$ for $i = 1, 2$. This implies that $\rho^{-1}(L(1, a)) = L(1, b)$. However, $a \in L(1, a)$, $a \notin L(1, b)$, and $\rho^{-1}(a) = a$, a contradiction.

Therefore, C^* and C are the cores of required BGW-matrices. Note that $L(\sigma, a) = L_C(\sigma, a)$, and therefore $\mathbf{P} = \mathbf{P}(C)$. □

Remark 10.5.3. Theorem 3.6.20 implies that the group of all (c, A)-homologies of a desarguesian projective plane is cyclic. If G is the multiplicative group of a nearfield of order q (that is not a field) and W is a $BGW(q+1, q, q-1; G)$, then the projective plane corresponding to W has the noncyclic group G as a group of homologies. Therefore, this projective plane is nondesarguesian. Using the nearfields of order q^2 introduced in Example 10.1.20, we obtain a nondesarguesian projective plane of order q^2 for every odd prime power q. In particular, we obtain a nondesarguesian projective plane of order 9 which is the least possible order.

If there exists a generalized Hadamard matrix or a generalized conference matrix with group invariant core, the result of Theorem 10.5.2 can be strengthened.

Theorem 10.5.4. *Let E be a group of order $q \geq 3$ and G a group of order $q-1$. The following statements are equivalent:*

(i) *there exists a normalized $GH(E; 1)$ whose core is G-invariant;*
(ii) *there exists a normalized $BGW(q+1, q, q-1; G)$ whose core is E-invariant;*
(iii) *there exists a projective plane $\mathbf{P}$ of order q with distinct points θ and ∞ and distinct lines A_θ and A_∞ such that $\theta \in A_\theta \cap A_\infty$, $\infty \in A_\infty$, the group of all (θ, A_∞)-elations is isomorphic to E, and the group of all (∞, A_θ)-homologies is isomorphic to G.*

Proof. As in the proof of Theorem 10.5.2, we assume that the group E is additive and the group G is multiplicative. Let H be a matrix of $q-1$ with entries from $E \setminus \{0\}$ and let C be a matrix of order q with all off-diagonal entries from G and all the diagonal entries equal to 0. As in the proof of Theorem 10.5.2, we assume that the rows and columns of $H = [H(\sigma, \tau)]$ are indexed by elements of G and the rows and columns of $C = [C(a, b)]$ are indexed by elements of E. The incidence structures $\mathbf{P}(H)$ and $\mathbf{P}(C)$ are defined as in the proof of Theorem 10.5.2.

(i) $\Rightarrow$ (ii).

Suppose that H is the core of a normalized $GH(E; 1)$ and that H is G-invariant. For distinct $a, b \in E$ and for $\rho \in G$, let $C(a, b) = \rho$ if and only if $H(1, \rho) = b - a$. Let $C(a, a) = 0$ for all $a \in E$. Then the matrix $C = [C(a, b)]$ is E-invariant.

Let $\sigma, \tau \in G$, $a \in E$, and $x \in E \setminus \{a\}$. Then

$$C(x,a)\sigma^{-1} = \tau \Leftrightarrow C(x,a) = \tau\sigma \Leftrightarrow H(1,\tau\sigma) = a - x$$
$$\Leftrightarrow H(\sigma^{-1},\tau) = a - x \Leftrightarrow -H(\sigma^{-1},\tau) = x.$$

Therefore, $L_C(\sigma, a) = L_H(\sigma, a)$, and then $\mathbf{P}(H)$ and $\mathbf{P}(C)$ is the same incidence structure. Since H is the core of a $GH(E; 1)$, $\mathbf{P}(H)$ is a projective plane. For each $\tau \in G$, define $h_\tau : X \to X$ as in Theorem 10.5.2(ii). Then $\{h_\tau : \tau \in G\}$ is a group of (∞, A_θ)-homologies of $\mathbf{P}(C)$ isomorphic to G. Therefore, C is the core of a $BGW(q+1, q, q-1; G)$.

(ii) $\Rightarrow$ (i).

Suppose that $C = [C(a,b)]$ is the core of a normalized $BGW(q+1, q, q-1; G)$ and that C is E-invariant. For $\sigma, \tau \in G$ and $b \in E \setminus \{0\}$, let $H(\sigma,\tau) = b$ if and only if $C(0,b) = \tau\sigma^{-1}$. Then the matrix $H = [H(\sigma,\tau)]$ of order $q-1$ is G-invariant. We have, for all $\sigma, \tau \in G$, $a \in E$, and $x \in E \setminus \{a\}$,

$$-H(\sigma^{-1},\tau) + x = a \Leftrightarrow H(\sigma^{-1},\tau) = a - x \Leftrightarrow C(x,a)$$
$$= \tau\sigma \Leftrightarrow C(x,a)\sigma^{-1} = \tau,$$

so $L_H(\sigma, a) = L_C(\sigma, a)$. Therefore, $\mathbf{P}(H) = \mathbf{P}(C)$, so $\mathbf{P}(H)$ is a projective plane of order q. For each $a \in E$, define $t_a : X \to X$ as in Theorem 10.5.2(i). Then $\{t_a : a \in E\}$ is the group of all (θ, A_∞)-elations of $\mathbf{P}(H)$ isomorphic to E. This implies that H is the core of a $GH(E; 1)$.

Observe that in the course of the above proof, we have obtained that E is the group of all (θ, A_∞)-elations and G is the group of all (∞, A_θ)-homologies of $\mathbf{P} = \mathbf{P}(H) = \mathbf{P}(C)$. Thus, (i) and (ii) imply (iii).

(iii) $\Rightarrow$ (i).

Let $\mathbf{P}$ be a projective plane satisfying (iii). Let $a \mapsto t_a$ be an isomorphism between E and the group of all (θ, A_∞)-elations of $\mathbf{P}$. Let $\sigma \mapsto h_\sigma$ be an isomorphism between G and the group of all (∞, A_θ)-homologies of $\mathbf{P}$.

Fix a point $(\theta, 0) \in A_\theta \setminus \{\theta\}$, a point $(1, \theta) \in A_\infty \setminus \{\theta, \infty\}$, and a line $J(1)$ through θ so that $J(1) \neq A_\theta$ and $J(1) \neq A_\infty$. Let $(\theta, a) = t_{-a}(\theta, 0)$ for every $a \in E$. Let $(\sigma, \theta) = h_\sigma^{-1}(1, \theta)$ and $J(\sigma) = h_\sigma^{-1}(J(1))$ for every $\sigma \in G$. For $a \in E$ and $\sigma \in G$, let $K(a)$ be the line through ∞ and a, $L(\sigma, a)$ be the line through (σ, θ) and (θ, a), and (σ, a) be the intersection point of $J(\sigma)$ and $K(a)$. Then

$$X = (G \times E) \cup \{(\sigma, \theta) \colon \sigma \in G\} \cup \{(\theta, a) \colon a \in E\} \cup \{\theta, \infty\}$$

is the set of all points of **P**. The set

$$\mathcal{L} = \{J(\sigma)\colon \sigma \in G\} \cup \{K(a)\colon a \in E\} \cup \{L(\sigma, a)\colon \sigma \in G, a \in E\} \cup \{A_\theta, A_\infty\}$$

is the set of all lines of **P**.

Elations t_a and homologies h_σ act on **P** as in the proof of Theorem 10.5.2. Let matrices H and C be defined as follows: $H(\sigma, \tau) = b \Longleftrightarrow (\tau, -b) \in J(\tau) \cap L(\sigma, 0)$; $C(a, a) = 0$; for $a \neq b$, $C(a, b) = \tau \Longleftrightarrow (\tau, a) \in K(a) \cap L(1, b)$. Then H is the core of a $GH(E; 1)$ and C is the core of a $BGW(q+1, q, q-1; G)$. We shall show that H is G-invariant.

Let $\rho, \sigma, \tau \in G$ and let $H(\sigma, \tau) = b$. Then $(\tau, -b) \in J(\tau) \cap L(\sigma, 0)$. Therefore, $\rho^{-1}(\tau, -b) \in \rho^{-1}(J(\tau)) \cap \rho^{-1}(L(\sigma, 0))$. We have $\rho^{-1}(\tau, -b) = \rho^{-1}\tau^{-1}(1, -b) = (\tau\rho, -b)$ and $\rho^{-1}(J(\tau)) = \rho^{-1}\tau^{-1}(J(1)) = J(\tau\rho)$. Since $\rho^{-1}(\sigma, \theta) = (\sigma\rho, \theta)$ and $\rho^{-1}(\theta, 0) = (\theta, 0)$, we obtain that $\rho^{-1}(L(\sigma, 0)) = L(\sigma\rho, 0)$. Therefore, $(\tau\rho, -b) \in J(\tau\rho) \cap L(\sigma\rho, 0)$, and then $H(\sigma\rho, \tau\rho) = b = H(\sigma, \tau)$. Thus, H is G-invariant. □

One of the most famous open questions in finite geometries is the *Prime Power Conjecture* which states that the order of any finite projective plane is a prime power. There are several results that prove this conjecture for projective planes admitting certain collineation groups. The next theorem implies the Prime Power Conjecture for certain projective planes of Theorem 10.5.4.

Theorem 10.5.5. *Let E be a group of order $q \geq 3$ and G an abelian group of order $q-1$. Suppose there exists a normalized $BGW(q+1, q, q-1; G)$ with E-invariant core. Then*:

(i) *if q is even, then E is an elementary abelian group and therefore q is a power of* 2;
(ii) *if q is odd, then G is a signed group*;
(iii) *if E is an abelian group, then E is an elementary abelian group and therefore q is a prime power.*

Proof. Let C be the core of a normalized $BGW(q+1, q, q-1; G)$ and let $C = [C(x, y)]$ be E-invariant.

(i) and (ii). Suppose that G is not a signed group. Then, by Theorem 10.4.20, C is symmetric. For any $a \in E$, we have $C(1, a) = C(a, 1) = C(1, a^{-1})$. Since no two entries of the same row of C are equal, we obtain that $a = a^{-1}$. Thus, every nonidentity element of E is of order 2. This implies that E is an elementary abelian 2-group.

Thus, if q is even, then $q-1$ is odd, and, since the group G of order $q-1$ cannot be a signed group, we obtain that E is an elementary abelian group and

q is a power of 2. If q is odd, then q is not a power of 2, and therefore G must be a signed group.

(iii) Suppose both G and E are abelian groups. Let p be a prime divisor of q and let R be the group ring of E over the field $GF(p)$. Note that R is a commutative ring of characteristic p and therefore, $(\alpha+\beta)^p = \alpha^p + \beta^p$ for all $\alpha, \beta \in R$.

Since $q \equiv 0 \pmod{p}$, Proposition 10.4.23 implies the following equations over the group ring RG:

$$CC^* = -GI + (G-1)(J-I); \tag{10.8}$$

$$CJ = GJ. \tag{10.9}$$

We also have

$$GC = G(J-I); \tag{10.10}$$

$$C(GJ) = -J. \tag{10.11}$$

We will prove by induction on m that, for any positive integer m,

$$C^m(CC^*) = (-1)^{m-1}G(I-(m+2)J). \tag{10.12}$$

First use equations (10.8)–(10.11) to obtain:

$$C(CC^*) = -GC + C(GJ) - CJ = G(I-3J).$$

Let $m \geq 2$ and let $C^{m-1}(CC^*) = (-1)^{m-2}G(I-(m+1)J)$. Then

$$C^m(CC^*) = (-1)^{m-2}(GC-(m+1)C(GJ))$$
$$= (-1)^{m-2}(CJ - GI + (m+1)GJ) = (-1)^{m-1}G(I-(m+2)J),$$

and (10.12) is proved. For $m = p-1$, (10.12) implies

$$C^pC^* = C^{p-1}(CC^*) = (-1)^pG(I-J). \tag{10.13}$$

Recall that $supp(C) = \sum_{x\in E\setminus\{1\}} C(x,1)x$. Therefore,

$$supp(C^p) = \left(\sum_{x\in E\setminus\{1\}} C(x,1)x\right)^p = \sum_{x\in E\setminus\{1\}} C(x,1)^p x^p.$$

Therefore,

$$supp(C^pC^*) = \sum_{x\in E\setminus\{1\}} C(x,1)^p x^p \cdot \sum_{y\in E\setminus\{1\}} C^*(y,1)y = \sum_{z\in E} a_z z$$

with $a_z \in R$. For $z = 1$, we obtain that

$$a_1 = \sum_{y\in E\setminus\{1\}} \sum_{\substack{x\in E\setminus\{1\}\\ x^p=y^{-1}}} C(x,1)^p C^*(y,1).$$

Since (10.13) implies that $a_1 = 0$, we obtain that, for all $x, y \in E \setminus \{1\}$, $x^p \neq y^{-1}$. This means that $x^p = 1$ for all $x \in E$, i.e., E is an elementary abelian p-group. Then q is a power of p. □

Theorems 10.5.4 and 10.5.5 imply the following case of the Prime Power Conjecture.

Theorem 10.5.6. *Let* **P** *be a projective plane of order q with distinct points θ and ∞ and distinct lines A_θ and $A\infty$ satisfying the following conditions:*

(i) $\theta \in A_\theta \cap A_\infty$ *and* $\infty \in A_\infty$;
(ii) *the group E of all (θ, A_∞)-elations is an abelian group of order q;*
(iii) *the group of all (∞, A_θ)-homologies is an abelian group of order $q - 1$.*

Then the group E is an elementary abelian group, and therefore, q is a prime power.

Exercises

(1) Let H be a $GH(G; \lambda)$ where G is the group of all complex n^{th} roots of unity. Prove that if H and H^* are regarded as matrices over $\mathbb{C}$, then $HH^* = nI$.
(2) Let n be a composite positive integer and let G be the group of all complex n^{th} roots of unity. Prove that there exists a matrix $H = [\eta_{ij}]$ of order $n\lambda$ with all $\eta_{ij} \in G$ satisfying the following conditions: (i) if H and H^* are regarded as matrices over $\mathbb{C}$, then $HH^* = nI$; (ii) as a matrix over $\mathbb{Z}G$, H is not a $GH(q, \lambda)$.
(3) Prove that matrices W and M of Example 10.1.7 are monomially equivalent.
(4) The field $GF(9)$ can be described as the set $\{a + bi : a, b \in GF(3)\}$ with $i^2 = -1$ and the usual addition and multiplication.
 (a) Use Example 10.1.20 to obtain the multiplication table of the (noncommutative) nearfield F of order 9.
 (b) Identify two elements of F^* as j and k so that $F^* = \{\pm 1, \pm i, \pm j, \pm k\}$ becomes the group of quaternions.
 (c) Give an example of a $BGW(10, 9, 8; F^*)$ matrix W such that $W^\top$ and $\overline{W}$ are not BGW-matrices.
(5) Let q be an odd prime power and let a be a nonsquare element of $GF(q)$. Define matrices $A = [A(x, y)]$, $B = [B(x, y)]$, $C = [C(x, y)]$, and $D = [D(x, y)]$ of order q over $GF(q)$ with rows and columns indexed by elements of $GF(q)$ as follows: $A(x, y) = xy + x^2/4$, $B(x, y) = xy + ax^2/4$, $C(x, y) = xy - y^2 - x^2/4$, and $D(x, y) = (xy - y^2 - x^2/4)/a$. Prove that the matrix $H = \left[\begin{smallmatrix} A & B \\ C & D \end{smallmatrix}\right]$, regarded as a matrix over the additive group E of $GF(q)$, is a $GH(E; 2)$.
(6) Prove that the matrix W of Example 10.3.2 is not group-invariant.
(7) Prove that if there exists a $GH(G; \lambda)$ of order greater than 2, then either G is not a signed group or λ is even.

Notes

Weighing matrices, i.e., matrices with entries 0, ± 1 and pairwise orthogonal rows, were introduced in Yates (1935) in connection with accuracy of measurements. *Balanced weighing matrices*, i.e., weighing matrices that become incidence matrices of symmetric designs when all -1 entries are replaced with 1s, were studied in connections with combinatorial designs in Mullin and Stanton (1975a, 1975b). *Generalized weighing matrices* are matrices with pairwise orthogonal rows whose entries are zeros and complex roots of unity. These matrices were studied in Butson (1962, 1963), S. S. Shrikhande (1964), and Berman (1977, 1978). The first two authors consider generalized weighing matrices without zero entries and call them generalized Hadamard matrices. Sometimes these matrices are called BH-matrices. The notion of a BH-matrix over the group of complex n^{th} roots of unity and the notion of a GH-matrix over a cyclic group of order n are equivalent if and only if n is a prime. (See Proposition 10.1.11 and Exercise 2.)

Generalized Hadamard matrices over arbitrary groups were introduced in Drake (1979). They were also considered in Rajkundlia (1978, 1983) as *Hadamard systems*, together with *balanced weighing systems* which are equivalent to balanced generalized weighing matrices. Before that, BGW-matrices were also studied in Delsarte (1968) as *orthogonal configurations over a group*. (See also Cameron, Delsarte, and Goethals (1979).) The term *balanced generalized weighing matrices* seems to have been introduced in Seberry (1979). The seminal paper Jungnickel (1982b) reintroduces the term together with *generalized conference matrices*.

Balanced generalized weighing matrices with classical parameters were constructed in Berman (1978). Characterization of $PBGW$-matrices of minimal rank and their construction (Theorems 10.2.3 and 10.2.4) are due to Jungnickel and Tonchev (1999b, 2002).

Theorem 10.2.9 was obtained independently in Jungnickel (1979) (see Exercise 5) and Street (1979) and, for q prime, in Masuyama (1957) and Butson (1962). The more general Theorem 10.2.7 is due to de Launey and Dawson (1994). Theorem 10.2.11 is in Dawson (1985). Theorems 10.2.12 and 10.2.13 are proven in de Launey and Dawson (1992) and de Launey and Dawson (1994), respectively.

Relations between relative difference sets and balanced generalized weighing matrices were explored in Jungnickel (1982b). In particular, Theorems 10.3.1, 10.3.4, and 10.3.13 were proven in this paper. Special cases of the last theorem were obtained earlier in Delsarte, Goethals and Seidel (1971) and Berman (1978). The definition of ω-circulant matrices is given in Berman (1978). It generalizes *negacyclic matrices*, i.e., (-1)-circulant matrices, of Delsarte, Goethals and Seidel (1971). BGW-matrices over nonabelian groups of Corollaries 10.3.7 and 10.3.8 are due to Glynn (1978) and de Launey (1989), respectively. Theorem 10.3.10 is due to Leung, Ma and Schmidt (2002). Theorem 10.3.14 is due to Arasu, Dillon, Leung and Ma (2001).

The definition of the Kronecker product over a group is due to Ionin (1998b), although this concept was used earlier for constructing symmetric designs and BGW-matrices. (See, for instance, Rajkundlia (1978, 1983), de Launey (1992a), and Fanning (1995).) The definition of a group of symmetries was given in Ionin (1999b) for (0, 1)-matrices. Theorem 10.4.9 was proven for (0, 1)-matrices in Ionin (1999b) and subsequently generalized in Ionin and Kharaghani (2003a). It was applied to constructing symmetric and skew-symmetric BGW-matrices (Theorem 10.4.16) in Kharaghani (2003) and Ionin and Kharaghani (2003a, 2003b). The Kronecker product of GH-matrices over the same group was considered in Drake (1979), where Corollaries 10.4.11 and 10.4.12 were obtained.

Theorem 10.4.25 was obtained for $\lambda = 1$ in Rajkundlia (1978, 1983) and Seberry (1979). The general case is due to de Launey (1986). Theorem 10.4.27 is essentially due to Rajkundlia (1978, 1983). The present proof generalizes that of Ionin (1999a). Theorem 10.4.30 is mentioned in Gibbons and Mathon (1987b) and a more general Corollary 10.4.31 is stated in de Launey (1989).

For Theorem 10.1.23 and other references on nearfields, see Dembowski (1968), Lüneburg (1980), or Motose (2001). The monograph Pilz (1983) contains a chapter on nearfields. (See also Wähling (1987).) A nice introduction to nearfields and projective planes is given in Room and Kirkpatrick (1971). This book contains constructions of three nonisomorphic projective planes of order 9, including the Hughes plane discovered by Hughes (1957b). All finite nearfields were constructed in Dickson (1905a, 1905b). Dickson proposed a general construction for nearfields of order q^n with q and n satisfying the conditions of Theorem 10.1.23 and found the seven exceptional nearfields described in Remark 10.1.24. It was shown in the famous paper by Zassenhaus (1935/36) that the finite nearfields constructed by Dickson are all the possible finite nearfields. The BGW-matrices of Theorem 10.4.29 were discovered (with a different proof) in Mavron, McDonough and Pallikaros (2001).

Besides the families of BGW-matrices described in this chapter, there are a few sporadic examples. A $BGW(15, 7, 3; \mathbf{Z}_3)$ was constructed in Baker (1977); a $BGW(19, 9, 4; \mathbf{Z}_2)$ was constructed in de Launey and Sarvate (1983) and in Gibbons and Mathon (1987a). The latter paper enumerates all BGW-matrices of order 19 and less. A matrix $BGW(45, 12, 3; \mathbf{Z}_3)$ is due to Mathon (1987).

Balanced generalized weighing matrices are a special case of *generalized Bhaskar Rao designs*. See Bhaskar Rao (1966, 1970) and de Launey (1996) for basic definitions and main results.

For all known matrices $GH(G; \lambda)$, the group G is a p-group. If the group G is abelian, then, in all known cases, it is elementary abelian. The papers by Din and Mavron (1992) and McDonough, Mavron and Pallikaros (2000) investigate relations between GH-matrices and nets and obtain conditions on a matrix $GH(G; \lambda)$ that imply that the group G has to be elementary abelian.

Theorem 10.5.2 is due to Jungnickel (1979, 1982b). Dembowski and Piper (1967) classified projective planes of order q admitting a quasi-regular collineation group of order greater than $(q^2 + q + 1)/2$. (A group G of collineations of a projective plane is called *quasi-regular* if it induces a regular action on each G-orbit.) One of the classes in this classification consists of the projective planes satisfying the condition of Theorem 10.5.6. Cases (i) and (ii) of Theorem 10.5.4 are due to Ionin (2005). They strengthen the respective results of Ganley (1977) and Pott (1994) who obtained these results under the assumption that the group of elations is abelian. Case (iii) of Theorem 10.5.4 is due to Jungnickel and de Resmini (2002). See the survey by Ghinelli and Jungnickel (2003) for further results on the Prime Power Conjecture.

For further references on BGW-matrices and their applications, see Jungnickel (2005) and Jungnickel and Kharaghani (2004).

11

Decomposable symmetric designs

Balanced generalized weighing matrices and the Kronecker product over a group can be used to piece together small symmetric designs into a larger symmetric design. All known infinite families of symmetric designs that were not introduced in the previous chapters can be constructed in this manner.

11.1. A symmetric (66, 26, 10)-design

In this section we will use the notion of the Kronecker product over a group of symmetries to construct a symmetric (66, 26, 10)-design.

Theorem 11.1.1. *Let P be a Paley matrix of order 11 and let $N = \frac{1}{2}(J + P - I)$. Let $S = \{1, -1\}$ be the group of order 2 acting as the group of permutations of the set $\{N, N^\top\}$. Let $W = [\omega_{ij}]$ be the following $BGW(6, 5, 4; S)$:*

$$W = \begin{bmatrix} 0 & + & + & + & + & + \\ - & 0 & - & + & + & - \\ - & - & 0 & - & + & + \\ - & + & - & 0 & - & + \\ - & + & + & - & 0 & - \\ - & - & + & + & - & 0 \end{bmatrix}.$$

Then the matrix $M = W \otimes_S N + I_6 \otimes I_{11}$ is an incidence matrix of a symmetric (66, 26, 10)-design.

Proof. The matrix M can be regarded as a block matrix $M = [M_{ij}]$ of order 66 with

$$M_{ij} = \begin{cases} I_{11} & \text{if } i = j, \\ N & \text{if } \omega_{ij} = 1, \\ N^\top & \text{if } \omega_{ij} = -1. \end{cases}$$

For $1 \leq i \leq h \leq 6$, let

$$P_{ih} = \sum_{j=1}^{6} M_{ij} M_{ih}^T.$$

It suffices to show that

$$P_{ih} = \begin{cases} 16I + 10J & \text{if } i = h, \\ 10J & \text{if } i \neq h. \end{cases}$$

Since $Q = NN^T = N^T N = 3I + 2J$, we obtain that $P_{ii} = I + 5(3I + 2J) = 16I + 10J$. If $h > 1$, then $P_{1h} = 2N + 2Q + 2N^2$ and, if $h > i > 1$, then $P_{ih} = N + N^T + 2Q + N^2 + (N^T)^2$. Since $N^2 = N(J - I - N^T) = 5J - N - Q$ and $(N^T)^2 = 5J - N^T - Q$, we obtain that $P_{ih} = 10J$ whenever $i \neq h$. □

11.2. Global decomposition of symmetric designs

We defined a substructure of an incidence structure in Chapter 2 (Definition 2.1.4). If a substructure $\mathbf{D}_1$ of a symmetric design $\mathbf{D}$ is itself a symmetric design, we will say that $\mathbf{D}_1$ is a *symmetric subdesign of* $\mathbf{D}$.

Definition 11.2.1. A symmetric subdesign $\mathbf{D}_1 = (X_1, \mathcal{B}_1)$ of a symmetric (v, k, λ) design $\mathbf{D}$ is said to be *proper* if $1 < |X_1| < v$.

The incidence matrix M of the symmetric (66, 26, 10)-design constructed in the previous section is a 6×6 block-matrix. All of its 36 blocks are incidence matrices of symmetric designs with parameters either (11, 5, 2) or (11, 1, 0). This motivates the following definition.

Definition 11.2.2. A family $\{\mathbf{D}_1, \mathbf{D}_2, \ldots, \mathbf{D}_s\}$ of proper symmetric subdesigns of a symmetric design $\mathbf{D}$ is called a *global decomposition* of $\mathbf{D}$ if the sets of flags of the designs $\mathbf{D}_i$ partition the set of flags of $\mathbf{D}$. If all the designs $\mathbf{D}_i$ have the same block size, the decomposition is said to be *uniform*. If, for any two designs $\mathbf{D}_i = (X_i, \mathcal{B}_i)$ and $\mathbf{D}_j = (X_j, \mathcal{B}_j)$ in the decomposition, $X_i = X_j$ or $X_i \cap X_j = \varnothing$ and $\mathcal{B}_i = \mathcal{B}_j$ or $\mathcal{B}_i \cap \mathcal{B}_j = \varnothing$, the decomposition is called *regular*.

Remark 11.2.3. If a family of symmetric designs $\mathbf{D}_i$, $i = 1, 2, \ldots, s$, is a regular uniform global decomposition of a symmetric design $\mathbf{D}$, then all $\mathbf{D}_i$ have the same parameters.

In the language of matrices, a symmetric design is globally decomposable if and only if its incidence matrix can be split into nonoverlapping submatrices (of order greater than 1), each of which is either an incidence matrix of a symmetric design with nonzero block size or a zero matrix. A symmetric design admits a regular global decomposition if its incidence matrix is a block matrix with each nonzero block being an incidence matrix of a smaller symmetric design. The following proposition is immediate.

Proposition 11.2.4. *If a symmetric design admits a global decomposition, then so does its dual design. If a symmetric design admits a regular global decomposition, then so do both the dual and the complementary designs.*

The symmetric (66, 26, 10)-design from the previous section admits a regular (but not uniform) global decomposition. Another example is the design obtained as the development of a (16, 6, 2)-difference set in the group $G = \mathbf{Z}_2^4$. If $G = \langle a, b, c, d\rangle$, then $D = \{c, ac, d, bd, cd, abcd\}$ is a (16, 6, 2)-difference set, and the symmetric (16, 6, 2)-design $dev(D)$ admits a regular global decomposition into 40 symmetric designs, of which 32 are (2, 1, 0)-designs and eight are (2, 2, 2)-designs.

An example of an infinite family of globally decomposable symmetric designs comes from regular Hadamard matrices. If H_1 and H_2 are regular Hadamard matrices of order $4h_1^2$ and $4h_2^2$, respectively, then the Kronecker product $H = H_1 \otimes H_2$ is a regular Hadamard matrix of order $16h_1^2h_2^2$. The corresponding symmetric $(16h_1^2h_2^2, 8h_1^2h_2^2 - 2h_1h_2, 4h_1^2h_2^2 - 2h_1h_2)$-design is decomposable into symmetric designs, each of which is isomorphic to the symmetric $(4h_2^2, 2h_2^2 - h_2, h_2^2 - h_2)$-design corresponding to H_2 or to the complement of this design.

Theorem 10.5.1 shows that if n and d are distinct positive integers such that $d + 1$ divides $n + 1$, then the complement of the design $PG_{n-1}(n, q)$ admits a regular and uniform global decomposition into symmetric designs isomorphic to the complement of $PG_{d-1}(d, q)$.

In Chapter 10 (Definition 10.4.3), we defined a group of symmetries of a set of $(0, G)$ matrices for an arbitrary group G. Applying this definition to the trivial group introduces a group of symmetries of (0, 1)-matrices. Let $v > k > \lambda$ be positive integers and let $\mathcal{M}$ be a nonempty set of incidence matrices of symmetric (v, k, λ)-designs. Let S be a group of symmetries of $\mathcal{M}$ and W a $BGW(w, l, \mu; S)$. Then the Kronecker product $W \otimes_S X$ is a block matrix with each block being an incidence matrix of a symmetric (v, k, λ)-design or the zero matrix of order v. If $W \otimes_S X$ is an incidence matrix of a symmetric design $\mathbf{D}$, then its parameters are $(vw, kl, \lambda l)$ and $\mathbf{D}$ admits a regular and uniform decomposition into symmetric (v, k, λ)-designs. Theorem 10.4.9 immediately implies the following result.

Theorem 11.2.5. *Let $\mathcal{M}$ be a nonempty set of $(0, 1)$-matrices, each of which is an incidence matrix of a symmetric (v, k, λ)-design with the same v, k, and λ. Let S be a group of symmetries of $\mathcal{M}$ and let W be a $BGW(w, l, \mu; S)$ such that $k^2\mu = v\lambda l$. Then, for any $X \in \mathcal{M}$, the matrix $\overline{W} \otimes_S X$ is an incidence matrix of a symmetric $(vw, kl, \lambda l)$-design that admits a regular and uniform global decomposition into symmetric (v, k, λ)-designs.*

Under certain restrictions, we obtain the converse result.

Theorem 11.2.6. *Let a symmetric design* **D** *admit a regular and uniform global decomposition into nontrivial symmetric (v, k, λ)-designs. Let M be an incidence matrix of* **D** *represented as a block matrix $M = [M_{ij}]$, $i, j = 1, 2, \ldots, w$, where each M_{ij} is either an incidence matrix of a symmetric (v, k, λ)-design or the zero matrix of order v. Suppose further that there exists a linearly independent (over the rationals) set $\mathcal{M}$ of incidence matrices of symmetric (v, k, λ)-designs that contains all nonzero matrices M_{ij} and admits a sharply transitive group S of symmetries. Then, for any $X_0 \in \mathcal{M}$, there exists a balanced generalized weighing matrix W over S with parameters (w, l, μ) such that $k^2\mu = v\lambda l$ and $\overline{W} \otimes_S X_0 = M$.*

Proof. Let W_0 be the matrix of order w whose (i, j)-entry is equal to 0 if $M_{ij} = O$ and is equal to 1 if $M_{ij} \neq O$. Let l be the number of nonzero matrices among $M_{i1}, \ldots, M_{iw}$. Note that lk is the row sum of M, so l does not depend on i. We have, for each i,

$$\sum_{j=1}^{w} M_{ij} M_{ij}^{\top} = l((k - \lambda)I + \lambda J).$$

Therefore, M is an incidence matrix of a symmetric $(vw, kl, \lambda l)$-design.

Let $i, h \in \{1, 2, \ldots, w\}$, $i \neq h$. Let μ be the number of indices $j \in \{1, 2, \ldots, w\}$ such that $M_{ij} \neq O$ and $M_{hj} \neq O$. Fix a row m in the ith row of blocks of M and count in two ways pairs (n, t) where n is a row in the hth row of blocks of M and both the (m, t)-entry and the (n, t)-entry of M are equal to 1. We obtain $v\lambda l = \mu k^2$, so μ does not depend on i, h. Therefore, W_0 is an incidence matrix of a symmetric (w, l, μ)-design.

Fix $X_0 \in \mathcal{M}$ and define a $(0, S)$ matrix $W = [\omega_{ij}]$ of order w by

$$\omega_{ij} = \begin{cases} 0 & \text{if } M_{ij} = O, \\ \sigma^{-1} & \text{if } M_{ij} = \sigma X_0. \end{cases}$$

Then $M = \overline{W} \otimes_S X_0$. We shall show that W is a $BGW(w, l, \mu; S)$.

Let $i, h \in \{1, 2, \ldots, w\}$, $i \neq h$. Then

$$\sum_{j=1}^{w} M_{ij} M_{hj}^{\top} = \lambda l J.$$

Since there are exactly μ indices j, for which $M_{ij} \neq O$ and $M_{hj} \neq O$, we have, for some $\sigma_j, \tau_j \in S$,

$$\sum_{j=1}^{w} (\overline{\omega_{ij}} X_0)(\overline{\omega_{ih}} X_0)^{\top} = \sum_{j=1}^{\mu} (\sigma_j X_0)(\tau_j X_0)^{\top} = \lambda l J.$$

Then

$$\sum_{j=1}^{\mu} \left(\tau_j^{-1} \sigma_j X_0\right) X_0^{\top} = \lambda l J.$$

Since $J X_0^{\top} = kJ$, we have $\frac{1}{k} J = J(X_0^{\top})^{-1}$, so

$$\sum_{j=1}^{\mu} \tau_j^{-1} \sigma_j X_0 = \frac{\lambda l}{k} J.$$

On the other hand, we have

$$\sum_{\sigma \in S} \sigma X_0 = aJ,$$

where $a = k|S|/v$. Since the group S is sharply transitive, we have

$$\sum_{X \in \mathcal{M}} \frac{\lambda v l}{k^2 |S|} X = \frac{\lambda l}{k} J.$$

Since the set $\mathcal{M}$ is linearly independent, we obtain that the multiset

$$\left\{\tau_j^{-1} \sigma_j \colon 1 \leq j \leq \mu\right\}$$

contains $\dfrac{\lambda v l}{k^2 |S|} = \dfrac{\mu}{|S|}$ copies of every element of S. This implies that W is a $BGW(w, l, \mu; S)$. □

For BGW-matrices (10.6), the condition $k^2 \mu = v \lambda l$ is equivalent to $q = k^2/(k - \lambda)$ (see Remark 10.4.13). This implies that if one of the matrices (10.6) (for a given q) can be used in Theorem 11.2.5, then any of these matrices can be used in this theorem. Thus, we obtain the corollary of Theorem 11.2.5 that will be central to many constructions of symmetric designs in this chapter.

Corollary 11.2.7. *Let $\mathcal{M}$ be a nonempty set of* $(0, 1)$*-matrices, each of which is an incidence matrix of a nontrivial symmetric* (v, k, λ)*-design with the same*

v, k, and λ. Let S be a cyclic group of symmetries of $\mathcal{M}$. If $q = k^2/(k-\lambda)$ is a prime power and $|S|$ divides $q-1$, then, for any positive integer m, there exists a symmetric design with parameters

$$\left(\frac{v(q^{m+1}-1)}{q-1}, kq^m, \lambda q^m\right). \tag{11.1}$$

In order to apply Corollary 11.2.7, we need to have a "starting" symmetric (v, k, λ)-design with $q = k^2/(k-\lambda)$ a prime power, and then to find a set $\mathcal{M}$ and a cyclic group S, as required by the corollary. The simplest example of such a group S is a group of rotations acting on a set of matrices of order v (cf. Definition 10.4.5). In Theorem 10.5.1, the incidence matrix of the complement of $PG_{n-1}(n, q)$ is equal to $W \otimes_S X$ where X is an incidence matrix of the complement of $PG_{d-1}(d, q)$ and S is a group of rotations. However, as the next theorem shows, all symmetric designs that can be obtained in this way have parameters of the complement of $PG_{n-1}(n, q)$.

Theorem 11.2.8. *Let $\mathcal{M}$ be a nonempty set of incidence matrices of symmetric (v, k, λ)-designs. Suppose that $q = k^2/(k-\lambda)$ is a prime power. Let S be a group of permutations of degree v and let $|S|$ divide $q-1$. For each $X \in \mathcal{M}$ and each $\sigma \in S$, let σX be the matrix obtained by applying σ to the set of columns of X. Suppose that, for each $X \in \mathcal{M}$, there is an integer $a(X)$ such that $\sum_{\sigma \in S} \sigma X = a(X)J$. Then (v, k, λ) are the parameters of the complement of the design $PG_{d-1}(d, p^a)$ where p is a prime, d and a are positive integers, and $p^{(d+1)a} = q$. Furthermore, if W is a $BGW((q^{m+1}-1)/(q-1), q^m, q^m - q^{m-1}; S)$, then, for any $X \in \mathcal{M}$, $W \otimes_S X$ is an incidence matrix of a symmetric design whose parameters are those of the complement of $PG_{n-1}(n, p^a)$ with $n+1 = (d+1)(m+1)$.*

Proof. Let $X \in \mathcal{M}$. Since each matrix σX has row sum k and the matrix $a(X)J$ has row sum $a(X)v$, we obtain that $a(X) = k|S|/v$. Let C be a column of X and let $\mathcal{C}$ be the orbit of C with respect to the action of the group S. The submatrix Y formed by all the rows of X and the columns from $\mathcal{C}$ has constant column sum k and constant inner product λ of any two distinct columns. Since $\sum_{\sigma \in S} \sigma X = (k|S|/v)J$, the matrix Y has constant row sum $r = k|\mathcal{C}|/v$. Therefore, $Y^\top$ is an incidence matrix of a $(|\mathcal{C}|, v, k, r, \lambda)$-design. Then

$$(|\mathcal{C}|-1)\lambda = k(r-1),$$

which implies that $|\mathcal{C}| = v$.

Thus v divides $|S|$ and then v divides $q - 1 = v\lambda/(k-\lambda)$. This in turn implies that $k-\lambda$ divides λ. Let $\lambda = t(k-\lambda)$. Then $q = \lambda(t+1)^2/t$. If p is the prime divisor of q, then both $t+1$ and λ/t are powers of p. Let $t+1 = p^a$ and

$\lambda = tp^b$. Then $k = p^{a+b}$ and $v - 1 = k(k-1)/\lambda = p^a(p^{a+b} - 1)/(p^a - 1)$. Therefore, $p^a - 1$ divides $p^{a+b} - 1$ which implies, by Lemma 3.6.12, that a divides b. Let $b = (d-1)a$. Then (v, k, λ) are the parameters of the complement of $PG_{d-1}(d, p^a)$. If W is a $BGW((q^{m+1} - 1)/(q-1), q^m, q^m - q^{m-1}; S)$ and $X \in \mathcal{M}$, then (11.1) implies that $W \otimes_S X$ is an incidence of a symmetric design whose parameters are those of the complement of $PG_{n-1}(n, p^a)$ with $n + 1 = (d+1)(m+1)$. □

In Section 9.3 we constructed groups of symmetries for several families of difference sets. In the next section we will show how these symmetries can be transformed into symmetries of the corresponding symmetric designs and then use these designs as the starters in constructing parametrically new families of symmetric designs.

11.3. Six infinite families of globally decomposable symmetric designs

In Section 9.3, we defined groups of symmetries of subsets of a group ring and, in particular, symmetries of difference sets. The next theorem shows that these symmetries can serve as symmetries of the corresponding symmetric designs.

Theorem 11.3.1. *Let D be a (v, k, λ)-difference set in a group G. Let A be a uniform subset of $\mathbb{Z}G$ containing D and let S be a group of symmetries of A. Then there exists a set $\mathcal{M}$ of incidence matrices of symmetric (v, k, λ)-designs with S acting on $\mathcal{M}$ as a group of symmetries.*

Proof. Let $G = \{x_1, x_2, \ldots, x_v\}$ and let M be the $(0, 1)$-matrix of order v whose (i, j)-entry is equal to 1 if and only if $x_j \in Dx_i$. Then M is an incidence matrix of a symmetric (v, k, λ)-design (which in fact is the dual of $dev(D)$). Since A is uniform, $Dx_i \in A$ and therefore $\sigma(Dx_i)$ is a subset of G for $i = 1, 2, \ldots, v$ and any $\sigma \in S$. Let σM be the $(0, 1)$-matrix of order v whose (i, j)-entry is equal to 1 if and only if $x_j \in \sigma(Dx_i)$. Then σM is an incidence matrix of a symmetric (v, k, λ)-design. For any $\sigma, \tau \in S$, the (i, j)-entry of $(\sigma M)(\tau M)^\top$ is equal to $\langle \sigma(Dx_i), \tau(Dx_j) \rangle$.

Let $\mathcal{M} = \{\rho M : \rho \in S\}$. Let $X, Y \in \mathcal{M}$, $X = \sigma M$, $Y = \tau M$ with $\sigma, \tau \in S$. For any $\rho \in S$, the (i, j)-entries of $XY^\top$ and $(\rho X)(\rho Y)^\top$ are equal to $\langle \sigma(Dx_i), \tau(Dx_j) \rangle$ and $\langle \rho\sigma(Dx_i), \rho\tau(Dx_j) \rangle$, respectively. Therefore, $XY^\top = (\rho X)(\rho Y)^\top$.

Since $r(\sigma(Dx_i)) = r(Dx_i) = k$, we obtain that

$$\sum_{\rho \in S} \rho M = \frac{k|S|}{v} J.$$

Thus, the group S is a group of symmetries of $\mathcal{M}$. □

Corollary 11.2.7 now implies the following result.

Corollary 11.3.2. *Let D be a (v, k, λ)-difference set in a group G with $q = k^2/(k - \lambda)$ a prime power. Suppose there exists a uniform subset A of $\mathbb{Z}G$ containing D and having a cyclic group of symmetries whose order divides $q - 1$. Then, for any positive integer m, there exists a symmetric design with parameters*

$$\left(\frac{v(q^{m+1} - 1)}{q - 1}, kq^m, \lambda q^m\right)$$

which admits a regular and uniform global decomposition into symmetric (v, k, λ)-designs.

We will now apply the groups of symmetries constructed in Section 9.5 to obtain six infinite families of globally decomposable symmetric designs.

Theorem 9.5.2, which produces McFarland difference sets, yields the following symmetric designs.

Theorem 11.3.3. *Let q be a prime power and d a positive integer. If $r = (q^{d+1} - 1)(q - 1)$ is a prime power, then, for any positive integer m, there exists a symmetric design with parameters*

$$\left(\frac{q^{d+1}(r^{2m} - 1)}{r - 1}, r^{2m-1}q^d, (r - 1)r^{2m-2}q^{d-1}\right). \tag{11.2}$$

Proof. In order to apply Corollary 11.3.2, one should verify that $q(r + 1)$ divides $r^2 - 1$. This verification is straightforward. □

The difference set complementary to a McFarland difference set has parameters

$$(v, k, \lambda) = ((r + 1)q^{d+1}, q^d(q^{d+1} + q - 1), q^d(q^d + 1)(q - 1)), \tag{11.3}$$

where $q, d,$ and r are as in Theorem 11.3.3. Then $k^2/(k - \lambda) = (q^{d+1} + q - 1)^2$. Since a difference set has the same symmetry as its complement (see Proposition 9.3.8) and since $q(r + 1)$ divides $(q^{d+1} + q - 1)^2 - 1$, we obtain the following theorem.

Theorem 11.3.4. *Let q be a prime power and d a positive integer. If $p = q^{d+1} + q - 1$ is a prime power, then, for any positive integer m, there exists a symmetric design with parameters*

$$\left(\frac{q^d(p^{2m}-1)}{(q-1)(q^d+1)}, q^d p^{2m-1}, q^d(q^d+1)(q-1)p^{2m-2}\right). \tag{11.4}$$

Replacing McFarland difference sets by Spence difference sets from Theorem 9.5.5 yields the following symmetric designs.

Theorem 11.3.5. *Let d be a positive integer. If $q = \frac{3^{d+1}+1}{2}$ is a prime power, then, for any positive integer m, there exists a symmetric design with parameters*

$$\left(\frac{2\cdot 3^d(q^{2m}-1)}{3^d+1}, 3^d q^{2m-1}, \frac{3^d(3^d+1)q^{2m-2}}{2}\right). \tag{11.5}$$

Replacing Spence difference sets by their complements yields another infinite family of symmetric designs.

Theorem 11.3.6. *Let d be a positive integer. If $q = 3^{d+1} - 2$ is a prime power, then, for any positive integer m, there exists a symmetric design with parameters*

$$\left(\frac{3^d(q^{2m}-1)}{2(3^d-1)}, 3^d q^{2m-1}, 2\cdot 3^d(3^d-1)q^{2m-2}\right). \tag{11.6}$$

Davis–Jedwab difference sets (Theorem 9.5.7) and their complements yield the following two families of symmetric designs in the same manner.

Theorem 11.3.7. *Let d be a positive integer. If $q = \frac{2^{2d+3}+1}{3}$ is a prime power, then, for any positive integer m, there exists a symmetric design with parameters*

$$\left(\frac{2^{2d+3}(q^{2m}-1)}{q+1}, 2^{2d+1}q^{2m-1}, 2^{2d-1}(q+1)q^{2m-2}\right). \tag{11.7}$$

Theorem 11.3.8. *Let d be a positive integer. If $q = 2^{2d+3} - 3$ is a prime power, then, for any positive integer m, there exists a symmetric design with parameters*

$$\left(\frac{2^{2d+3}(q^{2m}-1)}{3(q-1)}, 2^{2d+1}q^{2m-1}, 3\cdot 2^{2d-1}(q-1)q^{2m-2}\right). \tag{11.8}$$

11.4. Productive Hadamard matrices

In the previous section, we constructed infinite families of globally decomposable symmetric designs, starting with designs generated by difference sets. Another rich source of starting designs is Menon designs, that is, symmetric designs equivalent to regular Hadamard matrices.

If a symmetric (v, k, λ)-design $\mathbf{D}$ is a Menon design, its parameters can be written as $(4h^2, 2h^2 - h, h^2 - h)$ where h is a nonzero integer. A $(0, 1)$-matrix N is an incidence matrix of such a design if and only if $J - 2N$ is a regular Hadamard matrix (Theorem 4.4.5). For these parameters, we have $k^2/(k - \lambda) = (2h - 1)^2$. Therefore, there exists a possibility for applying Corollary 11.2.7 if $q = (2h - 1)^2$ is a prime power and there is a set of incidence matrices of such designs admitting a cyclic group of symmetries whose order divides $q - 1 = 4h(4h - 1)$. The following definition describes a set of regular Hadamard matrices such that the corresponding set of $(0, 1)$-matrices admits such a group of symmetries.

Definition 11.4.1. Let H be a regular Hadamard matrix with row sum $2h$. We will say that H is *productive* if there exists a set $\mathcal{H}$ of regular Hadamard matrices with row sum $2h$ and a bijection $\sigma : \mathcal{H} \to \mathcal{H}$, called a *symmetry of* H, such that (i) $H \in \mathcal{H}$, (ii) σ^{4h} is the identity map, (iii) $(\sigma X)(\sigma Y)^\top = XY^\top$ for all $X, Y \in \mathcal{H}$, and (iv) $H + \sigma H + \sigma^2 H + \cdots + \sigma^{4|h|-1} H = \pm 2J$.

If H, $\mathcal{H}$, and σ satisfy Definition 11.4.1, then the cyclic group S generated by σ is a group of symmetries of the set $\mathcal{M} = \{\frac{1}{2}(J - H) \colon H \in \mathcal{H}\}$ of $(0, 1)$-matrices under the following action: $\sigma(\frac{1}{2}(J - H)) = \frac{1}{2}(J - \sigma H)$. Since $\mathcal{M}$ is a set of incidence matrices of symmetric $(4h^2, 2h^2 - h, h^2 - h)$-designs and $|S|$ divides $4h$, Corollary 11.2.7 implies the following result.

Theorem 11.4.2. *Let H be a productive regular Hadamard matrix with row sum $2h$ and let $q = (2h - 1)^2$. If q is a prime power, then, for any positive integer m, there exists a symmetric design with parameters*

$$\left(\frac{4h^2(q^{m+1} - 1)}{q - 1}, (2h^2 - h)q^m, (h^2 - h)q^m\right). \tag{11.9}$$

Hadamard matrices of Bush type were introduced in Definition 4.4.12. The next theorem shows that they are productive.

Theorem 11.4.3. *Any regular Hadamard matrix of Bush type is productive.*

Proof. Let H be a regular Hadamard matrix of Bush type with row sum $2h$, i.e., H can be represented as block matrix $H = [B_{ij}(H)]$ with $(2h) \times (2h)$ blocks H_{ij} satisfying the following conditions:

(i) for each i there is a unique j such that $B_{ij}(H) = J$ and, for $k \neq j$, $B_{ik}(H)$ has all row and column sums equal to 0;
(ii) if $B_{ij}(H) = J$ and $B_{hk}(H) = J$ with $i \neq h$, then $j \neq k$.

Let $\mathcal{H}$ be the set of all (± 1)-matrices X of order $4h^2$ that can be represented as block matrices with $(2h) \times (2h)$ blocks $B_{ij}(X)$ satisfying conditions (i) and (ii). Define a bijection $\sigma : \mathcal{H} \to \mathcal{H}$ as follows: if $\sigma(X) = X'$, then, for $i, j = 1, 2, \dots, 2h$,

$$B_{ij}(X') = \begin{cases} B_{i,j-1}(X) & \text{if } j \neq 1, \\ B_{i,2h}(X) & \text{if } j = 1 \text{ and } B_{i,2h}(X) = J, \\ -B_{i,2h}(X) & \text{if } j = 1 \text{ and } B_{i,2h}(X) \neq J. \end{cases}$$

Conditions (i) and (ii) of Definition 11.4.1 are satisfied. Let $X, Y \in \mathcal{H}$ and let $\sigma(X) = X'$ and $\sigma(Y) = Y'$. In order to obtain condition (iii) of Definition 11.4.1, it suffices to verify that

$$B_{i1}(X')(B_{i1}(Y'))^\top = B_{i,2h}(X)(B_{i,2h}(Y))^\top. \tag{11.10}$$

The verification is straightforward if both $B_{i,2h}(X)$ and $B_{i,2h}(Y)$ are equal to J or both are not equal to J. If one of these matrices is equal to J and the other is not, then both sides of (11.10) are equal to O.

The verification of condition (iv) of Definition 11.4.1 is also straightforward. □

In Theorem 4.4.11, we constructed, for every positive integer n, a regular Hadamard matrix of order $4 \cdot 3^{2n}$. We will now show that these matrices are productive.

Theorem 11.4.4. *Let*

$$Q = \begin{pmatrix} 0 & - & 1 \\ 1 & 0 & - \\ - & 1 & 0 \end{pmatrix}, \quad A_0 = \begin{pmatrix} - & 0 & 0 & 0 \\ 0 & 1 & 0 & 0 \\ 0 & 0 & 1 & 0 \\ 0 & 0 & 0 & 1 \end{pmatrix}, \quad B_0 = \begin{pmatrix} 0 & 1 & 1 & 1 \\ 1 & 0 & - & 1 \\ 1 & 1 & 0 & - \\ 1 & - & 1 & 0 \end{pmatrix},$$

where $-$ stands for -1. Let matrices A_n and B_n be defined recursively for $n \geq 1$ by $A_n = B_{n-1} \otimes I$ and $B_n = A_{n-1} \otimes J + B_{n-1} \otimes Q$ where I and J denote the identity and the all-one matrix of order 3, respectively. Then, for each $n \geq 0$, $H_n = A_n + B_n$ and $H'_n = A_n - B_n$ are Hadamard matrices and H_{2n} is a productive regular Hadamard matrix of order $4 \cdot 3^{2n}$. Furthermore, each row of every matrix H_n can be represented as a 1×4 block-matrix $[H_{n1}\ H_{n2}\ H_{n3}\ H_{n4}]$ where each block is a 1×3^n matrix, which in turn can be represented as a block-matrix $[X_1\ X_2\ \dots\ X_{3^{n-1}}]$ with each block being a row of $\pm J$ or a row of $\pm(I + Q)$.

Proof. If we remove the word "productive" from the above statement, we obtain Theorem 4.4.11. Therefore, all we have to prove is that H_{2n} is a productive matrix.

We will define recursively, for $n \geq 0$, sets $\mathcal{P}_n$ and $\mathcal{Q}_n$ of 1×3^n matrices with entries ± 1. The matrices H_{ni} $(i = 1, 2, 3, 4)$ will be contained in $\mathcal{P}_n \cup \mathcal{Q}_n$. We will also define a cyclic group of symmetries for each set $\mathcal{P}_n \cup \mathcal{Q}_n$. Finally, we will describe a cyclic group of symmetries of order $4 \cdot 3^n$ for a set of regular Hadamard matrices of order $4 \cdot 3^{2n}$ containing H_{2n}.

Let sets $\mathcal{P}_0$ and $\mathcal{Q}_0$ of 1×1 matrices be defined by $\mathcal{P}_0 = \mathcal{Q}_0 = \{[1], [-1]\}$. Define a bijection $\rho_0 \colon \mathcal{P}_0 \cup \mathcal{Q}_0 \to \mathcal{P}_0 \cup \mathcal{Q}_0$ to be the identity map.

Let $n \geq 1$ and let sets $\mathcal{P}_{n-1}$ and $\mathcal{Q}_{n-1}$ of $1 \times 3^{n-1}$ matrices and a bijection $\rho_{n-1} \colon \mathcal{P}_{n-1} \cup \mathcal{Q}_{n-1} \to \mathcal{P}_{n-1} \cup \mathcal{Q}_{n-1}$ be defined. Let $\mathcal{P}_n$ be the set of all block-matrices $[X\ X\ X]$ with $X \in \mathcal{Q}_{n-1}$. Let $\mathcal{Q}_n$ be the set of all block-matrices $[X_1\ X_2\ X_3]$, where one of the three blocks is an element of $\mathcal{P}_{n-1}$ and the other two blocks are of the form Y and $-Y$ with $Y \in \mathcal{Q}_{n-1}$. For $A = [X\ X\ X] \in \mathcal{P}_n$, let

$$\rho_n(A) = [\rho_{n-1}(X)\ \rho_{n-1}(X)\ \rho_{n-1}(X)].$$

For $B = [X_1\ X_2\ X_3] \in \mathcal{Q}_n$, let

$$\rho_n(B) = \begin{cases} [X_3\ X_1\ X_2] & \text{if } X_3 \in \mathcal{Q}_{n-1}, \\ [\rho_{n-1}(X_3)\ X_1\ X_2] & \text{if } X_3 \in \mathcal{P}_{n-1}. \end{cases}$$

Observe that all matrices of $\mathcal{P}_n \cup \mathcal{Q}_n$ are (± 1)-matrices, the sets $\mathcal{P}_n$ and $\mathcal{Q}_n$ are disjoint if $n \neq 0$, $\rho_n(\mathcal{P}_n) = \mathcal{P}_n$, and $\rho_n(\mathcal{Q}_n) = \mathcal{Q}_n$.

For $n \geq 0$ and $A \in \mathcal{P}_n \cup \mathcal{Q}_n$, let $r(A)$ be the sum of all entries of A and let

$$e(A) = \begin{cases} 3^{n/2} & \text{if } n \text{ is even}, \\ 3^{(n-1)/2} & \text{if } n \text{ is odd and } A \in \mathcal{P}_n, \\ 3^{(n+1)/2} & \text{if } n \text{ is odd and } A \in \mathcal{Q}_n. \end{cases}$$

Claim. For $n \geq 0$, the sets $\mathcal{P}_n$ and $\mathcal{Q}_n$ and the map ρ_n satisfy the following conditions:

(i) $r(A) = \begin{cases} \pm e(A) & \text{if } n \text{ is even and } A \in \mathcal{P}_n \cup \mathcal{Q}_n, \\ \pm 3e(A) & \text{if } n \text{ is odd and } A \in \mathcal{P}_n, \\ \pm \frac{1}{3}e(A) & \text{if } n \text{ is odd and } A \in \mathcal{Q}_n; \end{cases}$

(ii) for all $A, B \in \mathcal{P}_n \cup \mathcal{Q}_n$, $\rho_n(A)\rho_n(B)^\top = AB^\top$;

(iii) for all $A \in \mathcal{P}_n$ and all $B \in \mathcal{Q}_n$, $\rho_n(A)B^\top = A\rho_n(B)^\top = AB^\top$;

(iv) for all $A \in \mathcal{P}_n \cup \mathcal{Q}_n$, $\rho_n^{e(A)}(A) = A$;

(v) for all $A \in \mathcal{P}_n \cup \mathcal{Q}_n$, $\sum_{k=1}^{e(A)} \rho_n^k(A) = \pm E_n$ where E_n is the 1×3^n all-one matrix.

We will prove this claim by induction on n. All five statements are immediate for $n = 0$. Let $n \geq 1$ and let the induction hypothesis be true for $\mathcal{P}_{n-1}$, $\mathcal{Q}_{n-1}$, and ρ_{n-1}. We first verify (i), (iv), and (v).

Case 1. Let $A \in \mathcal{P}_n$, $A = [X\ X\ X]$ with $X \in \mathcal{Q}_{n-1}$.

Then, for any k, $\rho_n^k(A) = [\rho_{n-1}^k(X)\ \rho_{n-1}^k(X)\ \rho_{n-1}^k(X)]$. We have $r(A) = 3r(X)$ and $e(A) = e(X)$, and the induction hypothesis immediately implies (i) and (iv). We further have $\sum_{k=1}^{e(A)} \rho_n^k(A) = [S\ S\ S]$ where $S = \sum_{k=1}^{e(A)} \rho_{n-1}^k(X)$. Since $e(A) = e(X)$, the induction hypothesis implies that $S = \pm E_{n-1}$ and (v) follows.

Case 2. Let $A = [X_1\ X_2\ X_3] \in \mathcal{Q}_n$ where X_1, X_2, and X_3 are $1 \times 3^{n-1}$ matrices. Let $X \in \{X_1, X_2, X_3\} \cap \mathcal{P}_n$.

We have $r(A) = r(X)$ and $e(A) = 3e(X)$, so the induction hypothesis implies (i). To verify (iv), define $\rho_* : \mathcal{P}_{n-1} \cup \mathcal{Q}_{n-1} \to \mathcal{P}_{n-1} \cup \mathcal{Q}_{n-1}$ by

$$\rho_*(Y) = \begin{cases} \rho_{n-1}(Y) & \text{if } Y \in \mathcal{P}_{n-1}, \\ Y & \text{if } Y \in \mathcal{Q}_{n-1}. \end{cases}$$

Then $\rho_n(A) = [\rho_*(X_3)\ X_1\ X_2]$ and $\rho_n^{e(A)}(A) = [\rho_*^{e(A)/3}(X_1)\ \rho_*^{e(A)/3}(X_2)\ \rho_*^{e(A)/3}(X_3)]$. Since $e(X) = e(A)/3$, the induction hypothesis implies (iv).

In order to verify (v), let $\sum_{k=1}^{e(A)} \rho_n^k(A) = [S_1\ S_2\ S_3]$ where S_1, S_2, and S_3 are $1 \times 3^{n-1}$ matrices. Then each S_i is equal to $\sum_{k=1}^{e(X)} \rho_{n-1}^k(X)$. Therefore, $S_1 = S_2 = S_3 = \pm E_{n-1}$ by the induction hypothesis, and (v) follows.

We shall now verify (ii) and (iii).

Case 3. Let $A = [X\ X\ X] \in \mathcal{P}_n$ and $B = [Y_1\ Y_2\ Y_3] \in \mathcal{Q}_n$ where X, Y_1, Y_2, and Y_3 are $1 \times 3^{n-1}$ matrices, and let $Y \in \{Y_1, Y_2, Y_3\} \cap \mathcal{P}_{n-1}$.

Then $AB^\top = XY^\top$, $\rho_n(A)\rho_n(B)^\top$ is equal to $\rho_{n-1}(X)Y^\top$ or $\rho_{n-1}(X)\rho_{n-1}(Y)^\top$, $\rho_n(A)B^\top = \rho_{n-1}(X)Y^\top$, and $A\rho_n(B)^\top$ is equal to $XY^\top$ or $X\rho_{n-1}(Y)^\top$. Therefore, the induction hypothesis implies (ii) and (iii).

Case 4. Let $A, B \in \mathcal{P}_n$, $A = [X\ X\ X]$, $B = [Y\ Y\ Y]$ with $X, Y \in \mathcal{Q}_{n-1}$.

Then $\rho_n(A)\rho_n(B)^\top = 3\rho_{n-1}(X)\rho_{n-1}(Y)^\top = 3XY^\top = AB^\top$.

Case 5. Let $A, B \in \mathcal{Q}_n$, $A = [X_1\ X_2\ X_3]$, $B = [Y_1\ Y_2\ Y_3]$ with $X_i, Y_i \in \mathcal{P}_{n-1} \cup \mathcal{Q}_{n-1}$.

If $X_3, Y_3 \in \mathcal{Q}_{n-1}$, then $\rho_n(A)\rho_n(B)^\top = \sum_{i=1}^3 X_i Y_i^\top = AB^\top$.

If $X_3, Y_3 \in \mathcal{P}_{n-1}$, then

$$\rho_n(A)\rho_n(B)^\top = \rho_{n-1}(X_3)\rho_{n-1}(Y_3)^\top + \sum_{i=1}^{2} X_i Y_i^\top = \sum_{i=1}^{3} X_i Y_i^\top = AB^\top.$$

If $X_3 \in \mathcal{P}_{n-1}$ and $Y_3 \in \mathcal{Q}_{n-1}$, then

$$\rho_n(A)\rho_n(B)^\top = \rho_{n-1}(X_3)Y_3^\top + \sum_{i=1}^{2} X_i Y_i^\top = \sum_{i=1}^{3} X_i Y_i^\top = AB^\top.$$

The proof of the claim is now complete.

Conditions (ii), (iv), and (v) of the claim imply that for even n the cyclic group generated by ρ_n is a group of symmetries for the set of matrices $\mathcal{P}_n \cup \mathcal{Q}_n$ and the order of this group divides $3^{n/2}$.

Let n be a nonnegative integer and let $\mathcal{M}_n$ be the set of all regular Hadamard matrices of order $4 \cdot 3^{2n}$ that can be represented as block-matrices $[A_{ij}]$, $i = 1, 2, \ldots, 4 \cdot 3^{2n}$, $j = 1, 2, 3, 4$, with $A_{ij} \in \mathcal{P}_{2n} \cup \mathcal{Q}_{2n}$ for all i, j. Note that the matrix H_{2n} constructed in Theorem 4.4.11 is in the set $\mathcal{M}_n$. Define a bijection $\sigma_n : \mathcal{M}_n \to \mathcal{M}_n$ as follows: If $A = [A_{ij}] \in \mathcal{M}_n$, then $\sigma_n A = B = [B_{ij}]$ where, for each i,

$$B_{ij} = \begin{cases} A_{i,j-1} & \text{if } j = 2, 3, 4, \\ \rho_{2n}(A_{i4}) & \text{if } j = 1. \end{cases}$$

Let $h = \pm 3^n$ and $q = (2h-1)^2$. Let G_n be the cyclic group generated by σ_n. Then the above claim immediately implies that G_n is a group of symmetries of the set $\mathcal{M}_n$ and the order of G_n divides $4 \cdot 3^n$. Therefore, H_{2n} is a productive regular Hadamard matrix. □

Another construction of regular Hadamard matrices was given in Theorem 4.6.11. Some of these matrices are productive, as the next theorem shows. We do not provide a proof of this theorem.

Theorem 11.4.5. *Let m be a positive integer such that $q = 8m^2 - 1$ is a prime. Suppose also that there exists a Hadamard matrix of order $4m$. Let K be the regular Hadamard matrix of order $16q^2m^2$ constructed in Theorem 4.6.11. Then K is productive.*

The Kronecker product of two regular Hadamard matrices is a regular Hadamard matrix. If one of the matrices is of Bush type and the other is productive, then the Kronecker product is a productive regular Hadamard matrix.

Theorem 11.4.6. *If B is a regular Hadamard matrix of Bush type and H is a productive regular Hadamard matrix, then $B \otimes H$ is a productive regular Hadamard matrix.*

Proof. Let B be a regular Hadamard matrix of Bush type of order $4k^2$ and let H be a productive regular Hadamard matrix with row sum $2h$.

Let $\mathcal{H}$ be a set of regular Hadamard matrices with row sum $2h$, containing H, and let $\sigma : \mathcal{H} \to \mathcal{H}$ be a symmetry of H. Let $\mathcal{H}_0 = \{H, \sigma H, \sigma^2 H, \dots, \sigma^{4|h|-1} H\}$.

Let $\mathcal{Z}$ be the set of all (± 1)-matrices of order $2k$ with the sum of entries of each row and each column equal to 0 and let $\mathcal{B} = \{Z \otimes H : Z \in \mathcal{Z}\} \cup \{J \otimes K : K \in \mathcal{H}_0\}$, where J is the all-one matrix of order $2k$. Define a bijection $\tau : \mathcal{B} \to \mathcal{B}$ by

$$\begin{aligned} \tau(Z \otimes H) &= -Z \otimes H \quad &&\text{for each } Z \in \mathcal{Z}, \\ \tau(J \otimes K) &= J \otimes (\sigma K) \quad &&\text{for each } K \in \mathcal{H}_0. \end{aligned}$$

Observe that τ^{4h} is the identity map. We claim that

$$(\tau A)(\tau B)^\top = AB^\top \tag{11.11}$$

for all $A, B \in \mathcal{B}$. It is immediate if $A = Y \otimes H$ and $B = Z \otimes H$ with $Y, Z \in \mathcal{Z}$. If $A = J \otimes K$ and $B = J \otimes L$ with $K, L \in \mathcal{H}_0$, then

$$(\tau A)(\tau B)^\top = (JJ^\top) \otimes ((\sigma K)(\sigma L)^\top) = (JJ^\top) \otimes (KL^\top) = AB^\top.$$

If $A = Z \otimes H$ and $B = J \otimes K$ with $Z \in \mathcal{Z}$ and $K \in \mathcal{H}_0$, then $AB^\top = (ZJ^\top) \otimes (HK^\top) = O$ and $(\tau A)(\tau B)^\top = -(ZJ^\top) \otimes (H(\sigma K)^\top) = O$. The case $A = J \otimes K$ and $B = Z \otimes H$ is similar.

Let $\mathcal{K}$ be the set of all (± 1)-matrices X of order $16k^2h^2$ that can be represented as block matrices $X = [X_{ij}]$ so that all X_{ij} are matrices of order $8kh^2$ from $\mathcal{B}$. Since B is a matrix of Bush type, it can be represented as a block matrix $B = [B_{ij}]$ with all $B_{ij} \in \mathcal{Z} \cup \{J\}$. Therefore, $B \otimes H = [B_{ij} \otimes H] \in \mathcal{K}$. We will now define a symmetry $\pi : \mathcal{K} \to \mathcal{K}$ of $B \otimes H$.

First we define a bijection $\rho : \mathcal{K} \to \mathcal{K}$ as follows. If $X = [X_{ij}] \in \mathcal{K}$ with $X_{ij} \in \mathcal{B}$, then $\rho X = [X'_{ij}]$ where

$$X'_{ij} = \begin{cases} X_{i,j-1} & \text{for } j = 2, 3, \dots, 2k, \\ X_{i,2k} & \text{for } j = 1. \end{cases}$$

Since ρX is obtained by permuting columns of X, we have $(\rho X)(\rho Y)^\top = XY^\top$ for all $X, Y \in \mathcal{K}$. Clearly, ρ^{2k} is the identity map.

For $t = 1, 2, \dots, 2k$, we define a bijection $\tau_t : \mathcal{K} \to \mathcal{K}$ as follows. If $X = [X_{ij}] \in \mathcal{K}$ with $X_{ij} \in \mathcal{B}$, then $\tau_t X = [X'_{ij}]$ where

$$X'_{ij} = \begin{cases} X_{ij} & \text{if } j \neq t, \\ \tau X_{ij} & \text{if } j = t. \end{cases}$$

Observe that $\tau_s\tau_t = \tau_t\tau_s$ for all $s, t \in \{1, 2, \dots, 2k\}$ and $\rho\tau_t = \tau_{t+1}\rho$ (with $\tau_{2k+1} = \tau_1$). Since τ^{4h} is the identity map, so is each τ_t^{4h}. Equation (11.11) implies that $(\tau_t X)(\tau_t Y)^\top = XY^\top$ for all $X, Y \in \mathcal{K}$.

Let $\pi = \tau_1\rho$. Then $(\pi X)(\pi Y)^\top = XY^\top$ for all $X, Y \in \mathcal{K}$. We further have

$$\pi^{2k} = \tau_1\tau_2\cdots\tau_{2k}\rho^{2k} = \tau_1\tau_2\cdots\tau_{2k},$$

and therefore, $\pi^{8kh} = (\tau_1\tau_2\cdots\tau_{2k})^{4h}$ is the identity map.

Let $\mathcal{K}_0 = \{B\otimes H, \pi(B\otimes H), \pi^2(B\otimes H), \dots, \pi^{8k|h|-1}(B\otimes H)\}$ and let S be the sum of the elements of $\mathcal{K}_0$. It remains to show that $S = \pm 2J$. Let $S = [S_{ij}]$ where each block S_{ij} is of order $8kh^2$. Then

$$S_{ij} = \sum_{m=1}^{2k}\sum_{n=0}^{4|h|-1} \tau^n(B_{im}\otimes H).$$

If $B_{im} = J$, then

$$\sum_{n=0}^{4|h|-1} \tau^n(B_{im}\otimes H) = J\otimes\left(\sum_{n=0}^{4|h|-1}\sigma^n H\right) = J\otimes(\pm 2J) = \pm 2J.$$

(Letter J stands here for all-one matrices of different orders.) If $B_{im} \neq J$, then

$$\sum_{n=0}^{4|h|-1} \tau^n(B_{ij}\otimes H) = 2|h|(B_{im}\otimes H + \tau(B_{im}\otimes H)) = O.$$

Since, for each i, there is a unique block B_{im} equal to J, we obtain that $S = \pm 2J$, and the proof is now complete. □

11.5. Symmetric designs with irregular global decomposition

In this section, we will describe a family of symmetric designs, which admits a global decomposition that is neither regular nor uniform.

Theorem 11.5.1. *Let q be a positive integer. If there exists a symmetric $(q^2+q+1, q+1, 1)$-design and a $BGW(q^2+q+2, q^2+q+1, q^2+q)$ over $\mathbf{Z}_{q+1}$, then there exists a globally decomposable symmetric $(\lambda^3+\lambda+1, \lambda^2+1, \lambda)$-design with $\lambda = q+1$.*

Proof. Let $\mathcal{M}$ be the set of all permutation matrices of order $q+1$. (Of course, each of these matrices can be regarded as an incidence matrix of the symmetric $(q+1, 1, 0)$-design.) For each $X \in \mathcal{M}$, let ρX be the matrix obtained by applying the permutation $\rho = (1, 2, \dots, q+1)$ to the set of columns of X. The cyclic group S generated by ρ is a group of symmetries on $\mathcal{M}$. Let $W = [\omega_{ij}]$ be a $BGW(q^2+q+2, q^2+q+1, q^2+q; S)$ with all the diagonal entries equal to 0.

Let Y be an incidence matrix of a symmetric $(q^2+q+1, q+1, 1)$-design and let $R_1, R_2, \ldots, R_{q^2+q+1}$ be the rows of Y. Let Y_1 be the $(q+1)\times(q^2+q+1)$ zero matrix and, for $i = 2, 3, \ldots, q^2+q+2$, let Y_i be the $(q+1)\times(q^2+q+1)$ matrix with every row equal to R_{i-1}. Then $XY_i = Y_i$ for $i = 1, 2, \ldots, q^2+q+2$ and any $X \in \mathcal{M}$. Define a block-matrix $N = [N_{ij}]$ with the following blocks N_{ij}, $i, j = 1, 2, \ldots, q^2+q+3$:

$$N_{ij} = \begin{cases} J_q & \text{if } i = j = 1,\\ Y_i & \text{if } j = q^2+q+3,\ i \neq q^2+q+3,\\ Y_j^\top & \text{if } i = q^2+q+3,\ j \neq q^2+q+3,\\ I_{q^2+q+1} & \text{if } i = j = q^2+q+3,\\ \omega_{ij} I_{q+1} & \text{otherwise.} \end{cases}$$

We claim that N is an incidence matrix of a symmetric $(\lambda^3+\lambda+1, \lambda^2+1, \lambda)$-design with $\lambda = q+1$. We shall show that, for $i, h = 1, 2, \ldots, q^2+q+3$,

$$P_{ih} = \sum_{j=1}^{q^2+q+3} N_{ij} N_{hj}^\top = \begin{cases} (q^2+q+1)I + (q+1)J & \text{if } i = h,\\ (q+1)J & \text{if } i \neq h. \end{cases}$$

If i or h is equal to 1 or q^2+q+3, it is straightforward. If $i, h = 2, 3, \ldots, q^2+q+2$, then, for some $\sigma_j \in S$,

$$P_{ii} = \sum_{j=1}^{q^2+q+1} (\sigma_j I)(\sigma_j I)^\top + Y_i Y_i^\top = (q^2+q+1)I + (q+1)J$$

and, for $i \neq h$, there are $\sigma_j, \tau_j \in S$, $j = 1, 2, \ldots, q^2+q$ such that

$$\begin{aligned} P_{ih} &= \sum_{j=1}^{q^2+q} (\sigma_j I)(\tau_j I)^\top + Y_i Y_h^\top = \sum_{j=1}^{q^2+q} \tau_j^{-1}\sigma_j I + J\\ &= q\sum_{\sigma\in S} \sigma I + J = (q+1)J. \end{aligned}$$

If $h > 1$, then a similar calculation shows that $P_{1h} = (q+1)J$. Finally, let $2 \le i \le q^2+q+2$ and $h = q^2+q+3$. Then

$$P_{ih} = \sum_{j=1}^{q^2+q+2} (\omega_{ij} I) Y_j + Y_i = \sum_{j=2}^{q^2+q+2} Y_j = (q+1)J.$$

Thus, N is an incidence matrix of a symmetric design. This design is decomposable into the following symmetric designs: the $(q+1, q+1, q+1)$-design with the incidence matrix N_{11}; $(q+1, 1, 0)$-designs with the incidence matrices N_{ij}, $i, j = 1, 2, \dots, q^2+q+2$, except $i = j = 1$; the $(q^2+q+1, 1, 0)$-design with the incidence matrix N_{q^2+q+3, q^2+q+3}; $(q^2+q+1, q+1, 1)$-designs with incidence matrices M_k, $k = 1, 2, \dots, q+1$, formed by the last q^2+q+1 columns of N and the rows of N whose index is congruent $k \pmod{q+1}$ and does not exceed $(q^2+q+1)(q+1)$; and, finally, $(q^2+q+1, q+1, 1)$-designs with incidence matrices M_k^*, $k = 1, 2, \dots, q+1$, formed by the last q^2+q+1 rows of N and the columns of N whose index is congruent $k \pmod{q+1}$ and does not exceed $(q^2+q+1)(q+1)$. □

Corollary 11.5.2. *If q and q^2+q+1 are prime powers, then there exists a globally decomposable symmetric $(\lambda^3+\lambda+1, \lambda^2+1, \lambda)$-design with $\lambda = q+1$.*

Below are a $BGW(8,7,6)$ over $\mathbf{Z}_3 = \{1, \rho, \rho^2\}$ and the incidence matrix of the symmetric $(31, 10, 3)$-design obtained according to the above proof.

$$
\begin{bmatrix}
0 & 1 & 1 & 1 & 1 & 1 & 1 & 1 \\
1 & 0 & \rho & \rho^2 & 1 & \rho & \rho^2 & 1 \\
1 & \rho & 0 & 1 & \rho & 1 & \rho^2 & \rho^2 \\
1 & \rho^2 & 1 & 0 & \rho & \rho^2 & \rho & 1 \\
1 & 1 & \rho & \rho & 0 & \rho^2 & 1 & \rho^2 \\
1 & \rho & 1 & \rho^2 & \rho^2 & 0 & 1 & \rho \\
1 & \rho^2 & \rho^2 & \rho & 1 & 1 & 0 & \rho \\
1 & 1 & \rho^2 & 1 & \rho^2 & \rho & \rho & 0
\end{bmatrix},
\quad
\begin{bmatrix}
111 & 100 & 100 & 100 & 100 & 100 & 100 & 100 & 0000000 \\
111 & 010 & 010 & 010 & 010 & 010 & 010 & 010 & 0000000 \\
111 & 001 & 001 & 001 & 001 & 001 & 001 & 001 & 0000000 \\[1ex]
100 & 000 & 010 & 100 & 010 & 100 & 001 & 001 & 1101000 \\
010 & 000 & 001 & 010 & 001 & 010 & 100 & 100 & 1101000 \\
001 & 000 & 100 & 001 & 100 & 001 & 010 & 010 & 1101000 \\[1ex]
100 & 010 & 000 & 100 & 010 & 100 & 001 & 001 & 0110100 \\
010 & 001 & 000 & 010 & 001 & 010 & 100 & 100 & 0110100 \\
001 & 100 & 000 & 001 & 100 & 001 & 010 & 010 & 0110100 \\[1ex]
100 & 001 & 100 & 000 & 010 & 001 & 010 & 100 & 0011010 \\
010 & 100 & 010 & 000 & 001 & 100 & 001 & 010 & 0011010 \\
001 & 010 & 001 & 000 & 100 & 010 & 100 & 001 & 0011010 \\[1ex]
100 & 100 & 010 & 010 & 000 & 001 & 010 & 100 & 0001101 \\
010 & 010 & 001 & 001 & 000 & 100 & 001 & 010 & 0001101 \\
001 & 001 & 100 & 100 & 000 & 010 & 100 & 001 & 0001101 \\[1ex]
100 & 010 & 100 & 001 & 001 & 000 & 100 & 010 & 1000110 \\
010 & 001 & 010 & 100 & 100 & 000 & 010 & 001 & 1000110 \\
001 & 100 & 001 & 010 & 010 & 000 & 001 & 100 & 1000110 \\[1ex]
100 & 001 & 001 & 010 & 100 & 100 & 000 & 100 & 0100011 \\
010 & 100 & 100 & 001 & 010 & 010 & 000 & 010 & 0100011 \\
001 & 010 & 010 & 100 & 001 & 001 & 000 & 001 & 0100011 \\[1ex]
100 & 100 & 001 & 100 & 001 & 010 & 010 & 000 & 1010001 \\
010 & 010 & 100 & 010 & 100 & 001 & 001 & 000 & 1010001 \\
001 & 001 & 010 & 001 & 010 & 100 & 100 & 000 & 1010001 \\[1ex]
000 & 111 & 000 & 000 & 000 & 111 & 000 & 111 & 1000000 \\
000 & 111 & 111 & 000 & 000 & 000 & 111 & 000 & 0100000 \\
000 & 000 & 111 & 111 & 000 & 000 & 000 & 111 & 0010000 \\
000 & 111 & 000 & 111 & 111 & 000 & 000 & 000 & 0001000 \\
000 & 000 & 111 & 000 & 111 & 111 & 000 & 000 & 0000100 \\
000 & 000 & 000 & 111 & 000 & 111 & 111 & 000 & 0000010 \\
000 & 000 & 000 & 000 & 111 & 000 & 111 & 111 & 0000001
\end{bmatrix}.
$$

11.6. Decomposable symmetric designs and regular graphs

Incidence matrices of symmetric designs may serve as adjacency matrices of strongly regular graphs and normally regular digraphs. In Sections 7.4 and 7.5 we constructed several infinite families of such graphs. In this section, we will construct more of them using symmetric and skew-symmetric BGW-matrices with zero diagonal.

We begin with the following modification of Theorem 11.2.5.

Theorem 11.6.1. *Let $v > k > \lambda$ be positive integers and let $\mathcal{M}$ be a nonempty set of symmetric incidence matrices of symmetric (v, k, λ)-designs. Let S be a group of symmetries of $\mathcal{M}$, and let W be a symmetric BGW(w, l, μ) over S with zero diagonal. If $k^2\mu = v\lambda l$, then, for $N \in \mathcal{M}$, $W \otimes_S N$ is an adjacency matrix of a $(vw, kl, \lambda l)$-graph.*

Proof. Let $W = [\alpha_{ij}]$, $i, j = 1, 2, \dots, w$. By Theorem 11.2.5, $W \otimes_S N$ is an incidence matrix of a symmetric $(vw, kl, \lambda l)$-design. Since each matrix $\alpha_{ij} N$ is symmetric and $\alpha_{ij} = \alpha_{ji}$ for $i, j = 1, 2, \dots, w$, $W \otimes_S N$ is a symmetric matrix. Since $\alpha_{ii} = 0$ for $i = 1, 2, \dots, w$, the diagonal entries of $W \otimes_S N$ are equal to 0. Therefore, $W \otimes_S N$ is an adjacency matrix of an $SRG(vw, kl, \lambda l, \lambda l)$. □

Let q be a prime power, d a positive integer, and V the $(d+1)$-dimensional vector space over the field $GF(q)$. The space V contains $r = (q^{d+1} - 1)/(q - 1)$ d-dimensional subspaces, which we denote $H_1, H_2, \dots, H_r$. All d-flats parallel to H_i form a parallel class Π_i of cardinality q. For $i = 1, 2, \dots, r$, we fix a cyclic permutation π_i of Π_i.

Let $\mathcal{F}$ be the set consisting of all d-flats, their complements, the empty set, and the entire space V (so $|\mathcal{F}| = 2(qr + 1)$). We define a bijection $\pi : \mathcal{F} \to \mathcal{F}$ as follows: if $F \in \Pi_i$, then $\pi(F) = \pi_i(F)$ and $\pi(V \setminus F) = V \setminus \pi_i(F)$; $\pi(\emptyset) = \emptyset$; $\pi(V) = V$.

We will regard V as an abelian group equipped by a symmetric order (see Lemmas 3.8.1 and 3.8.2), so, for any subset A of V, a symmetric matrix $M(A)$ is defined.

Lemma 11.6.2. *For $A, B \in \mathcal{F}$, (i) $M(\pi A)M(\pi B)^T = M(A)M(B)^T$ and (ii) $\sum_{k=1}^{q} M(\pi^k A)$ is a constant matrix.*

Proof. Statement (i) follows immediately from Lemma 3.8.2. To prove (ii) note that if A is a d-flat, then the d-flats $\pi^k A$, $k = 1, 2, \dots, q$, partition V. Therefore, $\sum_{j=1}^{q} M(\pi^k A) = J$ and $\sum_{j=1}^{q} M(\pi^k(V \setminus A)) = (q-1)J$. Of course, $\sum_{j=1}^{q} M(\pi^k(\emptyset)) = O$ and $\sum_{j=1}^{q} M(\pi^k V) = qJ$. The proof is now complete. □

Let m be a positive integer and let $\mathcal{F}^m$ be the set of all ordered m-tuples of elements of $\mathcal{F}$. For $\mathbf{A} = (A_1, A_2, \dots, A_m) \in \mathcal{F}^m$, we define a symmetric (0, 1)-matrix $P(\mathbf{A})$ of order mq^{d+1} as a block matrix $[P_{ij}(\mathbf{A})]$, $i, j = 1, 2, \dots, m$ where

$$P_{ij}(\mathbf{A}) = \begin{cases} M(A_{i+j-1}) & \text{if } i + j \le m + 1, \\ M(\pi(A_{i+j-m-1})) & \text{if } i + j \ge m + 2. \end{cases}$$

If $\mathbf{A} = (A_1, A_2, \dots, A_m) \in \mathcal{F}^m$, we denote by $\overline{\mathbf{A}}$ the complementary m-tuple $(V \setminus A_1, V \setminus A_2, \dots, V \setminus A_m)$. Clearly, $P(\overline{\mathbf{A}}) = J - P(\mathbf{A})$.

Definition 11.6.3. A *McFarland* $(r+1)$*-tuple* is an $(r+1)$-tuple $\mathbf{A} = (A_1, A_2, \dots, A_{r+1}) \in \mathcal{F}^{r+1}$ such that one of the sets $A_1, A_2, \dots, A_{r+1}$ is empty and the other r are pairwise nonparallel d-flats. A *Spence* r*-tuple* is an r-tuple $\mathbf{A} = (A_1, A_2, \dots, A_r) \in \mathcal{F}^r$ such that one of the sets $A_1, A_2, \dots, A_r$ is the complement of a d-flat parallel to H_1 and the other r are pairwise nonparallel d-flats, which are not parallel to H_1.

Theorems 3.8.3 and 3.8.5 immediately imply the following result.

Proposition 11.6.4. *If* $\mathbf{A}$ *is a McFarland* $(r+1)$*-tuple, then* $P(\mathbf{A})$ *is a symmetric incidence matrix of a symmetric* $((r+1)q^{d+1}, rq^d, (r-1)q^{d-1})$*-design. If* $q = 3$ *and* $\mathbf{A}$ *is a Spence* r*-tuple, then* $P(\mathbf{A})$ *is a symmetric incidence matrix of a symmetric* $(3^{d+1}(3^{d+1}-1)/2, 3^d(3^{d+1}+1)/2, 3^d(3^{d-1}-1)/2)$*-design.*

Now we apply Theorem 11.6.1 to the set of matrices $\{P(\mathbf{A})\colon \mathbf{A} \in \mathcal{M}\}$ where $\mathcal{M}$ is the set of McFarland $(r+1)$-tuples or the set of their complements or the set of Spence r-tuples or the set of their complements.

We begin with defining a bijection $\sigma : \mathcal{F}^m \to \mathcal{F}^m$ as follows:

$$\sigma(A_1, A_2, \dots, A_m) = (A_2, A_3, \dots, A_m, \pi(A_1)).$$

Lemma 11.6.5. *For* $\mathbf{A}, \mathbf{B} \in \mathcal{F}^m$, (i) $P(\sigma\mathbf{A})P(\sigma\mathbf{B})^T = P(\mathbf{A})P(\mathbf{B})^T$ *and* (ii) $\sum_{k=1}^{mq} P(\sigma^k\mathbf{A})$ *is a constant matrix.*

Proof. Let $\mathbf{A} = (A_1, A_2, \dots, A_m)$ and $\mathbf{B} = (B_1, B_2, \dots, B_m)$. Let $P(\mathbf{A}) = [P_{ij}]$, $P(\mathbf{B}) = [Q_{ij}]$, $P(\sigma\mathbf{A}) = [P'_{ij}]$, and $P(\sigma\mathbf{B}) = [Q'_{ij}]$, $i, j = 1, 2, \dots, m$. Then

$$\sum_{j=1}^{m} P'_{ij} Q'^T_{kj} = \sum_{j=2}^{m} P_{ij} Q^T_{kj} + M(\pi A_i)M(\pi B_k)^T,$$

and we apply Lemma 11.6.2(i).

If $\mathbf{A} = (A_1, A_2, \ldots, A_m)$, then

$$\sum_{k=1}^{mq} P(\sigma^k \mathbf{A}) = [S_{ij}],$$

where each block S_{ij} is equal to $\sum_{j=1}^{m} \sum_{k=1}^{q} M(\pi^k A_j)$, and we apply Lemma 11.6.2(ii). □

Let $\mathcal{M}$ be the set of all matrices $P(\mathbf{A})$ where $\mathbf{A}$ is a McFarland $(r+1)$-tuple. For $M = P(\mathbf{A}) \in \mathcal{M}$, let $\sigma M = P(\sigma \mathbf{A})$. The cyclic group S of bijections $\mathcal{M} \to \mathcal{M}$ generated by σ is a group of symmetries of $\mathcal{M}$. If $(v, k, \lambda) = ((r+1)q^{d+1}, rq^d, (r-1)q^{d-1})$, then $k^2/(k-\lambda) = r^2$. We have $r^2(r^2-1)/|S| = r^2(r-1)/q$. Therefore, if r is a prime power and $r(r-1)/q$ is even, Theorem 10.4.16 yields, for any positive integer m, a symmetric $BGW((r^{4m}-1)/(r^2-1), r^{4m-2}, r^{4m-2} - r^{4m-4})$ with zero diagonal. This leads to the following result.

Theorem 11.6.6. *Let q be an odd prime power and d a positive integer. If $r = (q^{d+1}-1)/(q-1)$ is a prime power, then, for any positive integer m, there exists a*

$$\left(\frac{q^{d+1}(r^{4m}-1)}{r-1}, q^d r^{4m-1}, q^{d-1} r^{4m-2}(r-1)\right) - \textit{graph}.$$

Let $\mathcal{M}$ be the set of all matrices $P(\mathbf{A})$ where $\mathbf{A}$ is the complement of a McFarland $(r+1)$-tuple. Then again σ can be regarded as a bijection $\mathcal{M} \to \mathcal{M}$. The complement of a McFarland symmetric design has parameters $((r+1)q^{d+1}, q^d(q^{d+1}+q-1), q^d(q^d+1)(q-1))$. In this case we want $s = q^{d+1}+q-1$ to be a prime power and $s^2(q^d+1)(q-1)$ to be even.

Theorem 11.6.7. *Let q be an odd prime power and d a positive integer. If $s = q^{d+1}+q-1$ is a prime power, then, for any positive integer m, there exists a*

$$\left(\frac{q^{d+1}(s^{4m}-1)}{(q-1)(s+1)}, q^d s^{4m-1}, q^d s^{4m-2}(q^d+1)(q-1)\right) - \textit{graph}.$$

Let $\mathcal{M}$ be the set of all matrices $P(\mathbf{A})$ where $\mathbf{A}$ is a Spence r-tuple over $GF(3)$ with $r = (3^{d+1}-1)/2$. In this case, $|S| = 3r$, and we want $q = (3^{d+1}+1)/2$ to be a prime power.

Theorem 11.6.8. *Let d be a positive integer. If $q = (3^{d+1}+1)/2$ is a prime power, then, for any positive integer m, there exists a*

$$\left(\frac{2 \cdot 3^{d+1}(q^{4m}-1)}{q+1}, 3^d q^{4m-1}, \frac{3^d(3^d+1)q^{4m-2}}{2}\right) - \textit{graph}.$$

The complement of a Spence symmetric design has parameters $(3^{d+1}(3^{d+1}-1)/2, 3^d(3^{d+1}-2), 2\cdot 3^d(3^d-1))$, and we obtain

Theorem 11.6.9. *Let d be a positive integer. If $q = 3^{d+1}-2$ is a prime power, then, for any positive integer m, there exists a*

$$\left(\frac{3^{d+1}(q^{4m}-1)}{2(q-1)}, 3^d q^{4m-1}, 2\cdot 3^d(3^d-1)q^{4m-2}\right)-\textit{graph}.$$

Next we construct two infinite families of symmetric designs whose incidence matrices serve as adjacency matrices of normally regular digraphs. For these constructions, we need another modification of Theorem 11.2.5.

Theorem 11.6.10. *Let $v > k > \lambda$ be positive integers and let $\mathcal{M}$ be a nonempty set of symmetric incidence matrices of symmetric (v, k, λ)-designs. Let $S = \langle\sigma\rangle$ be a cyclic group of bijections $\mathcal{M} \to \mathcal{M}$ that satisfies the following conditions:*

(i) $|S| = 2n$ *is even;*
(ii) $(\sigma X)(\sigma Y) = XY$ *for all* $X, Y \in \mathcal{M}$;
(iii) *for each* $X \in \mathcal{M}$, $\sum_{i=1}^{2n} \sigma^i X$ *is a constant matrix;*
(iv) *for each* $X \in \mathcal{M}$, $X + \sigma^n X$ *is a* $(0, 1)$*-matrix.*

Let W be a skew-symmetric $BGW(w, l, \mu)$ over S with $k^2\mu = v\lambda l$. Then, for $X \in \mathcal{M}$, $W \otimes_S X$ is an adjacency matrix of an $NRD(vw, kl, \lambda l, \lambda l)$.

Proof. By Theorem 11.2.5, $W \otimes_S X$ is an incidence matrix of a symmetric $(vw, kl, \lambda l)$-design. Let $W = [\alpha_{ij}]$. Then $W \otimes_S X = [\alpha_{ij}X]$. Since W is skew and X is symmetric, we have $(W \otimes_S X)^\top = W^\top \otimes_S X = [\sigma^n \alpha_{ij} X]$. Therefore, $(W \otimes_S X) + (W \otimes_S X)^\top = [\alpha_{ij}X + \sigma^n(\alpha_{ij}X)]$ is a $(0, 1)$-matrix, which implies that $W \otimes_S X$ is an adjacency matrix of a $NRD(vw, kl, \lambda l, \lambda l)$. □

Let h be a positive integer and H a Hadamard matrix of order $n = 2h$ with all entries in the last row equal to 1. Let L be the Latin square of order n with $L(i, j) \equiv i + j - 1 \pmod{n}$ for $i, j = 1, 2, \ldots, n$. We will denote by $\mathcal{K}$ the set of all Hadamard matrices of order n^2 of Bush type constructed in Theorem 4.4.16. Define a map $\rho\colon \mathcal{K} \to \mathcal{K}$ as follows. If $K = [K_{ij}] \in \mathcal{K}$ where each K_{ij} is a matrix of order n, then $\rho(K) = [K'_{ij}]$ where

$$K'_{ij} = \begin{cases} K_{i,j+1} & \text{if } j \neq n,\\ -K_{i1} & \text{if } j = n \text{ and } K_{i1} \neq J,\\ J & \text{if } j = n \text{ and } K_{i1} = J.\end{cases}$$

It is readily verified that, for $K, L \in \mathcal{K}$, $(\rho K)(\rho L)^\top = KL^\top$.

Let $\mathcal{K}_s$ be the subset of $\mathcal{K}$ consisting of all matrices $K = [K_{ij}] \in \mathcal{K}$ with

$$K_{ij} = \begin{cases} K_{1,i+j-1} & \text{if } i + j \leq n + 1, \\ -K_{1,i+j-n-1} & \text{if } i + j \geq n + 2 \text{ and } K_{1,i+j-n-1} \neq J, \\ J & \text{if } i + j \geq n + 2 \text{ and } K_{1,i+j-n-1} = J. \end{cases}$$

Then all matrices $K \in \mathcal{K}_s$ are symmetric and $\rho(K) \in \mathcal{K}_s$ for all $K \in \mathcal{K}_s$.

Let $\mathcal{M} = \{\frac{1}{2}(J - K) \colon K \in \mathcal{K}_s\}$. Then $\mathcal{M}$ is a set of symmetric incidence matrices of symmetric $(4h^2, 2h^2 - h, h^2 - h)$-designs. We will define the map $\sigma \colon \mathcal{M} \to \mathcal{M}$ by $\sigma(\frac{1}{2}(J - K)) = \frac{1}{2}(J - \rho(K))$. Let S be the cyclic group generated by σ. Then $|S| = 2n$ and, for all $X \in \mathcal{M}$,

$$\sum_{k=1}^{2n} \sigma^k X = (n-1)J.$$

If $X = [X_{ij}] \in \mathcal{M}$ where each block X_{ij} is a matrix of order n, then $\sigma^n X = [X'_{ij}]$ where $X'_{ij} = J - X_{ij}$ if $X_{ij} \neq O$, and $X'_{ij} = O$ if $X_{ij} = O$. Therefore, $X + \sigma^n X$ is a (0, 1)-matrix. Also, for $X, Y \in \mathcal{M}$, $(\sigma X)(\sigma Y) = XY$. Thus, conditions (i)–(iv) of Theorem 11.6.10 are satisfied, and we obtain the following result.

Theorem 11.6.11. *Let h be a positive integer such that there exists a Hadamard matrix of order $2h$. If $q = (2h - 1)^2$ is a prime power, then, for any positive integer d, there exists an*

$$NRD\left(\frac{h(q^{2d} - 1)}{h + 1}, hq^{2d}, h(h+1)q^{2d-1}, h(h+1)q^{2d-1}\right). \tag{11.12}$$

Proof. We apply Theorem 11.6.10 to the set $\mathcal{M}$, the cyclic group $S = \langle\sigma\rangle$, and a skew balanced generalized weighing matrix supplied by Theorem 10.4.16. □

To obtain another family of normally regular digraphs, we again let q be a prime power, d a positive integer, and V the $(d + 1)$-dimensional space over $GF(q)$. Let $\mathcal{F}_0$ be the set consisting of all d-flats and the empty set. Let $r = (q^{d+1} - 1)/(q - 1)$ and let $\mathcal{F}_0^{r+1}$ denote the set of all $(r + 1)$-tuples of elements of $\mathcal{F}_0$. Let π and σ be the same as in Lemmas 11.6.2 and 11.6.5.

Since, for $1 \leq s \leq q - 1$ and for all $A \in \mathcal{F}_0$, sets A and $\pi^s A$ are disjoint, we have the following result.

Lemma 11.6.12. *Let $\mathbf{A} = (A_1, A_2, \ldots, A_{r+1}) \in \mathcal{F}_0^{r+1}$ and let $\mathbf{B} = (\pi^s A_1, \pi^s A_2, \ldots, \pi^s A_{r+1})$. If $1 \leq s \leq q - 1$, then $P(\mathbf{A}) + P(\mathbf{B})$ is a (0, 1)-matrix.*

Let now q be an even prime power. Let $\mathcal{M}$ be the set of all matrices $P(\mathbf{A})$ where $\mathbf{A}$ is a McFarland $(r+1)$-tuple over $GF(q)$. For $M = P(\mathbf{A}) \in \mathcal{M}$, let $\sigma M = P(\sigma \mathbf{A})$. Let S be the cyclic group of bijections $\mathcal{M} \to \mathcal{M}$ generated by σ. Then $|S| = (r+1)q$, so condition (i) of Theorem 11.6.10 is satisfied. Conditions (ii) and (iii) are satisfied due to Lemma 11.6.5. If $\mathbf{A} = (A_1, A_2, \dots, A_{r+1}) \in \mathcal{F}_0^{r+1}$, then, for $1 \le s \le q-1$,

$$\sigma^{(r+1)s}(\mathbf{A}) = (\pi^s A_1, \pi^s A_2, \dots, \pi^s A_{r+1}),$$

and Lemma 11.6.12 implies condition (iv) of Theorem 11.6.10. If $(v, k, \lambda) = ((r+1)q^{d+1}, rq^d, (r-1)q^{d-1})$, then $k^2/(k-\lambda) = r^2$. Observe that $(r^2 - 1)/|S| = (r-1)/q$ is odd. Therefore, if r is a prime power, we apply Theorem 10.4.16 to obtain, for any positive integer m, a skew $BGW((r^{4m}-1)/(r^2-1), r^{4m-2}, r^{4m-2} - r^{4m-4})$ with zero diagonal. Now all conditions of Theorem 11.6.10 are satisfied and we obtain the following result.

Theorem 11.6.13. *Let $q = 2^t$ and let d be a positive integer. If $r = (q^{d+1} - 1)/(q-1)$ is a prime power, then, for any positive integer m, there exists an*

$$NRD\left(\frac{q^{d+1}(r^{4m}-1)}{r-1}, q^d r^{4m-1}, q^{d-1} r^{4m-2}(r-1)\right).$$

11.7. Local decomposition of symmetric designs

If $\mathbf{D}$ is a symmetric (v, k, λ)-design with $\lambda \ge 1$ and B is a block of $\mathbf{D}$, then the residual design $\mathbf{D}^B$ is a 2-$(v-k, k-\lambda, \lambda)$-design and the derived design $\mathbf{D}_B$ is a 2-$(k, \lambda, \lambda - 1)$-design. The matrices X and Y below are the incidence matrices of a residual and a derived design of a symmetric $(25, 9, 3)$-design corresponding to the same block, which we denote by B:

$$X = \begin{bmatrix}
000000\ 101010\ 101010\ 101010 \\
000000\ 100101\ 100101\ 100101 \\
000000\ 011001\ 011001\ 011001 \\
000000\ 010110\ 010110\ 010110 \\[6pt]
101010\ 000000\ 101010\ 010101 \\
100101\ 000000\ 100101\ 011010 \\
011001\ 000000\ 011001\ 100110 \\
010110\ 000000\ 010110\ 101001 \\[6pt]
101010\ 010101\ 000000\ 101010 \\
100101\ 011010\ 000000\ 100101 \\
011001\ 100110\ 000000\ 011001 \\
010110\ 101001\ 000000\ 010110 \\[6pt]
101010\ 101010\ 010101\ 000000 \\
100101\ 100101\ 011010\ 000000 \\
011001\ 011001\ 100110\ 000000 \\
010110\ 010110\ 101001\ 000000
\end{bmatrix}, \quad
Y = \begin{bmatrix}
110000\ 110000\ 110000\ 110000 \\
001100\ 001100\ 001100\ 110000 \\
000011\ 000011\ 000011\ 110000 \\[6pt]
110000\ 000011\ 001100\ 001100 \\
001100\ 110000\ 000011\ 001100 \\
000011\ 001100\ 110000\ 001100 \\[6pt]
110000\ 001100\ 000011\ 000011 \\
001100\ 000011\ 110000\ 000011 \\
000011\ 110000\ 001100\ 000011
\end{bmatrix}.$$

The matrix X is an incidence matrix of a 2-(16,6,3) design, and it is naturally divided into 4×6 blocks X_{ij}, $i, j = 1, 2, 3, 4$, so that each X_{ij} with $i \neq j$ is an incidence matrix of a residual design of the Fano Plane. The matrix Y is an incidence matrix of a 2-(9, 3, 2) design. For $i, j = 1, 2, 3$, the submatrices Y_{ij} formed by rows $3i - 2, 3i - 1, 3i$ and columns $6i - 5$ through $6i$ of Y are incidence matrices of derived designs of the Fano Plane as well as the submatrices Y_{i4} formed by the rows $i, i + 3, i + 6$ and the columns 19 through 24. For each $j \in \{1, 2, 3, 4\}$, we form three disjoint pairs (X_{ij}, Y_{hj}) with $i \neq j$ and $h \in \{1, 2, 3\}$. For every such pair, we form a Fano Plane with four points corresponding to the rows of X_{ij}, three points corresponding to the rows of Y_{hj}, six blocks corresponding to the columns of X_{ij} and Y_{hj} and the block B. The flags of the twelve Fano Planes we have obtained cover every flag of the initial symmetric (25, 9, 3)-design once, except the flags (x, B) which are covered four times each.

This example motivates the following definition.

Definition 11.7.1. Let $\mathbf{D} = (X, \mathcal{B})$ be a symmetric design, $B \in \mathcal{B}$, and let $\{\mathbf{D}_i : 1 \leq i \leq s\}$ be a family of proper symmetric subdesigns of the design $\mathbf{D}$. Let F be the set of flags of $\mathbf{D}$ and F_i, $i = 1, 2, \ldots, s$, be the set of flags of $\mathbf{D}_i$. The family $\{\mathbf{D}_i : 1 \leq i \leq s\}$ is called a *local decomposition* of $\mathbf{D}$ if (i) $F = F_1 \cup F_2 \cup \ldots \cup F_s$ and (ii) if $(x, A) \in F_i \cap F_j$ with $i \neq j$, then $A = B$.

In this section, we will describe a method for constructing several locally decomposable symmetric designs. The incidence matrix of every design in these families will be of the form $\left[\begin{smallmatrix} X & \mathbf{0} \\ Y & \mathbf{j} \end{smallmatrix}\right]$ where X and Y are incidence matrices of a residual and a derived design.

In order to apply BGW-matrices for constructing locally decomposable symmetric designs, we will extend the notion of a group of symmetries to sets of rectangular matrices.

Definition 11.7.2. Let $\mathcal{M}$ be a nonempty set of (0, 1)-matrices of the same size $m \times n$. A group S of bijections $\mathcal{M} \to \mathcal{M}$ is called a *group of symmetries of* $\mathcal{M}$ if it satisfies the following two conditions:

(i) $(\sigma X)(\sigma Y)^\top = XY^\top$ for all $X, Y \in \mathcal{M}$ and all $\sigma \in S$;
(ii) for each $X \in \mathcal{M}$, there is an integer $t(X)$ such that $\sum_{\sigma \in S} \sigma X = t(X) J_{m,n}$.

The following theorem generalizes Theorem 11.2.5. Its proof goes along the same lines as the proof of Theorem 10.4.9.

Theorem 11.7.3. *Let $\mathcal{M}$ be a nonempty set of* (0, 1)*-matrices, each of which is an incidence matrix of a (v, b, r, k, λ)-design with the same v, b, r, k, and λ. Let S be a group of symmetries of $\mathcal{M}$ and let W be a $BGW(w, l, \mu; S)$ such that*

$kr\mu = v\lambda l$. Then, for any $X \in \mathcal{M}$, the matrix $\overline{W} \otimes_S X$ is an incidence matrix of a $(vw, bw, rl, kl, \lambda l)$-design. Furthermore, if X is an incidence matrix of a quasi-residual design, then so is $\overline{W} \otimes_S X$.

Proof. Let $W = [\omega_{ij}]$ and let $X \in \mathcal{M}$. For $i, h = 1, 2, \ldots, w$, let

$$P_{ih} = \sum_{j=1}^{w} (\overline{\omega_{ij}} X)(\overline{\omega_{hj}} X)^\top .$$

Then

$$P_{ii} = lXX^\top = (rl - \lambda l)I + \lambda l J.$$

Condition (ii) of Definition 11.7.2 implies that $|S|r = t(X)b$. Therefore, we have, for $i \neq h$,

$$P_{ih} = \sum_{j=1}^{w} (\omega_{hj}\overline{\omega_{ij}} X)(\omega_{hj}\overline{\omega_{hj}} X)^\top = \frac{\mu}{|S|}\left(\sum_{\sigma \in S} \sigma X\right) X^\top = \frac{\mu}{|S|} t(X) J X^\top = \lambda l J.$$

Therefore, $\overline{W} \otimes_S X$ is an incidence matrix of a $(vw, bw, rl, kl, \lambda l)$-design. If X is an incidence matrix of a quasi-residual design, then $r = k + \lambda$. This implies that $rl = kl + \lambda l$, and then $\overline{W} \otimes_S X$ is an incidence matrix of a quasi-residual design. □

Remark 11.7.4. If (v, b, r, k, λ) are the parameters of a quasi-residual design and $(w, l, \mu) = ((q^{m+1} - 1)/(q - 1), q^m, q^m - q^{m-1})$, then the equation $kr\mu = v\lambda l$ is equivalent to $r = q$.

The following proposition introduces a group of symmetries that will be used throughout this and the next section.

Proposition 11.7.5. *Let X be an incidence matrix of a (v, b, r, k, λ)-design and let G be a sharply transitive group of permutations of set $\{1, 2, \ldots, v\}$. For any $\rho \in G$, let ρX be the $v \times b$ matrix whose i^{th} row is equal to the $\rho(i)^{th}$ row of X, $1 \leq i \leq v$. Let $\mathcal{M} = \{\rho X : \rho \in G\}$. Then G is a group of symmetries of $\mathcal{M}$.*

Proof. For $i = 1, 2, \ldots, v$, let X_i be the ith row of X. Let $\rho, \sigma, \tau \in G$. Let $Y = \sigma X$ and $Z = \tau X$. For $i, j = 1, 2, \ldots, v$, the (i, j)-entry a_{ij} of $YZ^\top$ is equal to the inner product of $X_{\sigma(i)}$ and $X_{\tau(j)}$, while the (i, j)-entry b_{ij} of $(\rho Y)(\rho Z)^\top$ is equal to the inner product of $X_{\rho\sigma(i)}$ and $X_{\rho\tau(j)}$. Therefore,

$$a_{ij} = \begin{cases} r & \text{if } \sigma(i) = \tau(j), \\ \lambda & \text{if } \sigma(i) \neq \tau(j) \end{cases}, \quad b_{ij} = \begin{cases} r & \text{if } \rho\sigma(i) = \rho\tau(j), \\ \lambda & \text{if } \rho\sigma(i) \neq \rho\tau(j). \end{cases}$$

Therefore, $XY^\top = (\rho X)(\rho Y)^\top$. Since

$$\sum_{\rho\in G} \rho Y = \sum_{\rho\in G} \rho\sigma X = \sum_{\rho\in G} \rho X = kJ,$$

the group G is a group of symmetries of $\mathcal{M}$. □

Remark 11.7.6. Any group of order v can be regarded as a sharply transitive group of permutations of a set of cardinality v and therefore may serve as a group of symmetries described in Proposition 11.7.5.

The following theorem introduces a method for constructing large quasi-derived designs from smaller quasi-derived designs.

Theorem 11.7.7. *Let Y be an incidence matrix of a (v, b, r, k, λ)-design* **D** *with $k \geq 2$ and let G be a sharply transitive group of permutations of rows of Y. For each positive integer m, let H_m be a $GH(G; v^{m-1})$. Let $Y_0 = Y$ and let matrices Y_m for $m \geq 1$ be recursively defined block matrices $Y_m = [H_m \otimes_G Y \; Y_{m-1} \otimes J_{v,1}]$. Then*:

(i) *Y_m is an incidence matrix of a $(v^{m+1}, b_m, r_m, v^m k, \lambda_m)$-design* $\mathbf{D}_m$ *with $r_m = \lambda(v^{m+1} - 1)/(k-1)$, $\lambda_m = \lambda(kv^m - 1)/(k-1)$, and $b_m = vr_m/k$;*

(ii) *if* **D** *is quasi-derived, then so is* $\mathbf{D}_m$;

(iii) *Y_m can be represented as a block matrix $Y_m = [Y_{ij}]$ so that each Y_{ij} is an $r \times (v-1)$ matrix and each row of each Y_{ij} is equal to a row of Y;*

(iv) *Y_m can be partitioned into $r \times (v-1)$ submatrices so that each of these submatrices can be obtained by permuting rows of Y and the sets of columns for any two of these submatrices are either the same or disjoint.*

Proof. By Proposition 11.7.7, the group G is a group of symmetries of set $\{\rho Y : \rho \in G\}$.

(i) We will prove this statement by induction on m. It is true for $m = 0$, so let $m \geq 1$ and let Y_{m-1} be an incidence matrix of a $(v^m, b_{m-1}, r_{m-1}, v^{m-1}k, \lambda_{m-1})$-design. For each $\sigma \in G$, the column sum of σY equals k, so the column sum of $H_m \otimes_G Y$ equals kv^m. The induction hypothesis implies that the column sum of $Y_{m-1} \otimes J_{v,1}$ is $v(kv^{m-1}) = kv^m$.

Let $H_m = [\eta_{ij}]$. For $i, h = 1, 2, \ldots, v^m$, let

$$P_{ih} = \sum_{j=1}^{v^m} (\eta_{ij} Y)(\eta_{hj} Y)^\top.$$

Then $P_{ii} = v^m YY^\top = v^m(r-\lambda)I + v^m \lambda J$ and, for $i \neq h$,

$$P_{ih} = \sum_{j=1}^{v^m} (\eta_{hj}^{-1} Y)Y^\top = v^{m-1}\left(\sum_{\sigma\in G} \sigma Y\right) Y^\top = v^{m-1}kJY^\top = v^{m-1}krJ.$$

Let $R_1, R_2, \ldots, R_{v^m}$ be consecutive rows of Y. Then $Y_{m-1} \otimes J_{v,1} = [S_1^\top \; S_2^\top \ldots S_{v^m}^\top]$ where $S_i = R_i \otimes J_{v,1}$. The induction hypothesis implies that, for $i, h = 1, 2, \ldots, v^m$,

$$S_i S_h^\top = \begin{cases} r_{m-1} J & \text{if } i = h, \\ \lambda_{m-1} J & \text{if } i \neq h. \end{cases}$$

Since $v^m r + r_{m-1} = r_m$ and $v^m \lambda + r_{m-1} = v^{m-1} kr + \lambda_{m-1} = \lambda_m$, we obtain that Y_m is an incidence matrix of a $(v^{m+1}, b_m, r_m, v^m k, \lambda_m)$-design.

(ii) If **D** is quasi-derived, then $\lambda = k - 1$. This implies that $\lambda_m = kv^m - 1$, so $\mathbf{D}_m$ is quasi-derived.

(iii) We will prove this statement by induction on m. It is trivial for $m = 0$, so let $m \geq 1$ and suppose that Y_{m-1} can be represented a block matrix of the required form. Then the statement is true for the matrix $Y_{m-1} \otimes J_{v,1}$. Since it is also true for the matrix $H_m \otimes_S Y = [\eta_{ij} Y]$, it is true for Y_m.

(iv) We will prove this statement by induction on m. It is trivial for $m = 0$, so let $m \geq 1$ and suppose that Y_{m-1} can be partitioned into submatrices of the required form. The submatrix $H_m \otimes_G Y$ is partitioned into submatrices $\eta_{ij} Y$ each of which is obtained by permuting rows of Y. We partition $Y_{m-1} \otimes J_{v,1}$ into v submatrices equal to Y_{m-1} and then apply the induction hypothesis to partition each of these submatrices into $r \times (v-1)$ submatrices of the required form. □

We will now describe a scheme for constructing an infinite family of locally decomposable symmetric designs. We begin with a symmetric (v, r, λ)-design **D** with r a prime power. Let B be a block of **D**. Then $\mathbf{D}^B$ is a residual design with replication number r. Let an incidence matrix X of $\mathbf{D}^B$ be contained in a set $\mathcal{M}$ of matrices having a cyclic group of symmetries S whose order divides $r - 1$. Then, for any positive integer m, there exists a balanced generalized weighing matrix W over S with parameters

$$\left(\frac{r^{m+1} - 1}{r - 1}, r^m, r^m - r^{m-1} \right). \tag{11.13}$$

By Theorem 11.7.3, $W \otimes_S X$ is an incidence matrix of a quasi-residual design with parameters

$$\left(\frac{(v - r)(r^{m+1} - 1)}{r - 1}, \frac{(v - 1)(r^{m+1} - 1)}{r - 1}, r^{m+1}, (r - \lambda) r^m, \lambda r^m \right). \tag{11.14}$$

Let Y be an incidence matrix of the derived design $\mathbf{D}_B$. Then Y is an $r \times (v - 1)$ matrix. Since r is a prime power, there exists, for every positive integer

m, a generalized Hadamard matrix GH(r, r^{m-1}) over a sharply transitive group of permutations of rows of Y. Then Theorem 11.7.7 produces a quasi-derived design with parameters

$$\left(r^{m+1}, \frac{(v-1)(r^{m+1}-1)}{r-1}, r^{m+1}-1, \lambda r^m, \lambda r^m - 1\right). \tag{11.15}$$

Note that (11.14) and (11.15) are the parameters of a residual and a derived design of a symmetric design with parameters

$$\left(1 + \frac{(v-1)(r^{m+1}-1)}{r-1}, r^{m+1}, \lambda r^m\right) \tag{11.16}$$

if such a symmetric design exists. Of course, the existence of designs (11.14) and (11.15) does not automatically imply the existence of a symmetric design (11.16). The next theorem gives a sufficient condition for combining designs (11.14) and (11.15) into a symmetric design.

Theorem 11.7.8. *Let* $\mathbf{D}$ *be a symmetric* (v, r, λ)*-design with* r *a prime power. Let* B *be a block of* $\mathbf{D}$ *and let* X *and* Y *be incidence matrices of designs* $\mathbf{D}^B$ *and* $\mathbf{D}_B$*, respectively. Let* $\mathcal{M}$ *be a set of* $(v-r) \times (v-1)$ *matrices that contains* X *and admits a cyclic group* S *of symmetries. Let* $|S|$ *divide* $r-1$ *and let* $(\sigma X)Y = \lambda J$ *for all* $\sigma \in S$*. Then, for any positive integer* m*, there exists a symmetric design with parameters* (11.16) *which is locally decomposable into symmetric* (v, r, λ)*-designs.*

Proof. Let m be a positive integer and let $W = [\omega_{ij}]$ be a $BGW(w, r^m, r^m - r^{m-1}; S)$ with $w = (r^{m+1}-1)/(r-1)$. By Theorem 11.7.3, $W \otimes_S X$ is an incidence matrix of a quasi-residual design with parameters (11.14). Let G be a sharply transitive elementary abelian group of permutations of the set of rows of Y. Let $H_m = [\eta_{ij}]$ be a $GH(G; r^{m-1})$. Then, in the notation of Theorem 11.7.7, Y_m is an incidence matrix of a design with parameters (11.15). Let matrix N be defined by

$$N = \begin{bmatrix} W \otimes_S X & \mathbf{0} \\ Y_m & \mathbf{j} \end{bmatrix}. \tag{11.17}$$

We claim that N is an incidence matrix of a symmetric design with parameters (11.16). Let $Y_m = [Y_{ij}]$ be a representation of Y_m of the form given by Theorem 11.7.7(iii). It suffices to show that, for $i = 1, 2, \ldots, w$ and $h = 1, 2, \ldots, r^m$,

$$\sum_{j=1}^{w} (\omega_{ij} X) Y_{hj}^{\top} = \lambda r^m J. \tag{11.18}$$

Since each row of each block Y_{ij} is a row of Y and since $(\sigma X)Y^\top = \lambda J$ for all $\sigma \in G$, we obtain that

$$(\omega_{ij}X)Y^\top = \begin{cases} \lambda J & \text{if } \omega_{ij} \neq 0, \\ O & \text{if } \omega_{ij} = 0. \end{cases}$$

Since the matrix W has exactly r^m nonzero entries, (11.18) follows.

Let **E** be the symmetric design with incidence matrix N. In order to show that **E** is locally decomposable, we partition $W \otimes_S X$ into blocks $\omega_{ij}X$ and let $Y_m = [Y_{ij}]$ be the partition given by Theorem 11.7.7(iv). For $j = 1, 2, \ldots, w$, there are r^m nonzero matrices $\omega_{ij}X$, each of which is an incidence matrix of a quasi-residual $(v - r, v - 1, r, r - \lambda, \lambda)$-design. Let these matrices be $X'_{1j}, X'_{2j}, \ldots, X'_{r^m j}$. Also, there are r^m matrices Y_{ij}, each of which is an incidence matrix of a quasi-derived $(r, v - 1, r - 1, \lambda, \lambda - 1)$-design. Since $(\sigma X)Y_{ij}^\top = \lambda J$ for all $\sigma \in S$ and all i and j, we obtain that every matrix

$$N_{ij} = \begin{bmatrix} X'_{ij} & \mathbf{0} \\ Y_{ij} & \mathbf{j} \end{bmatrix}$$

is an incidence matrix of a symmetric (v, r, λ)-design. These designs form a local decomposition of **E**. □

In the next section, we will give several realizations of Theorem 11.7.8.

11.8. Infinite families of locally decomposable symmetric designs

The following theorem presents two cases in which the conditions of Theorem 11.7.8 are satisfied.

Theorem 11.8.1. *Let r be a prime power and* **D** *a symmetric (v, r, λ)-design. If* **D** *is* (i) *a $PG_{d-1}(d, q)$ or* (ii) *a Hadamard 2-design, then, for any nonnegative integer m, there exists a locally decomposable symmetric design with parameters* (11.16).

Proof. Since r is a prime power, then, for any positive integer m, there exists a $BGW((r^{m+1} - 1)/(r - 1), r^m, r^m - r^{m-1})$ over any cyclic group S with $|S|$ dividing $r - 1$ and there exists a $GH(r, r^{m-1})$ over an elementary abelian group of order r.

Let B be a block of **D** and let X and Y be incidence matrices of $\mathbf{D}^B$ and $\mathbf{D}_B$, respectively, corresponding to the same ordering of the block set.

(i) Suppose $\mathbf{D}$ is a $PG_{d-1}(d, q)$ with $(q^d - 1)/(q - 1) = r$. Then $\mathbf{D}^B$ is $AG_{d-1}(d, q)$ and $\mathbf{D}_B$ is a q-fold multiple of $PG_{d-2}(d - 1, q)$. The columns of X are partitioned into r classes corresponding to parallel classes of $AG_{d-1}(d, q)$. The ith column of X and the jth column of X are in the same class if and only if the ith column of Y and the jth column of Y are equal to each other. Therefore, if ρ is a permutation of columns of X which permutes the columns of each parallel class without changing the classes, then $(\rho X)Y^\top = XY^\top = \lambda J$ with $\lambda = (q^{d-1} - 1)/(q - 1)$. Thus, if we assume that ρ cyclically permutes the columns of each class and denote by S the cyclic group of order q generated by ρ, then S is a group of symmetries of the set $\mathcal{M} = \{\sigma X : \sigma \in S\}$. Since q divides $r - 1$, Theorem 11.7.8 yields a locally decomposable symmetric design with parameters (11.16).

(ii) Suppose $\mathbf{D}$ is a Hadamard 2-design, i.e., a symmetric $(2r + 1, r, (r - 1)/2)$-design. Then the parameters of $\mathbf{D}^B$ are $(r + 1, 2r, r, (r + 1)/2, (r - 1)/2)$, the parameters of $\mathbf{D}_B$ are $(r, 2r, r - 1, (r - 1)/2, (r - 3)/2)$, and $XY^\top = \frac{r-1}{2}J$. Therefore, $(J - X)Y^\top = (r - 1)J - XY^\top = \frac{r-1}{2}J$. Let $\mathcal{M} = \{X, J - X\}$ and let S be the group of order 2 generated by the transposition τ acting on $\mathcal{M}$. Then S is a group of symmetries on $\mathcal{M}$, and again Theorem 11.7.8 yields a locally decomposable symmetric design with parameters (11.16). □

For case (i), we obtain the following symmetric designs.

Corollary 11.8.2. *Let q and $r = (q^d - 1)/(q - 1)$ be prime powers. Then, for any positive integer m, there exists a locally decomposable symmetric design with parameters*

$$\left(1 + \frac{qr(r^m - 1)}{r - 1}, r^m, \frac{r^{m-1}(r - 1)}{q}\right). \tag{11.19}$$

Remark 11.8.3. If $d = 2$ in the case (i) of Theorem 11.8.1, then $\mathbf{D}$ can be any projective plane of order q, not necessarily the desarguesian one.

For case (ii) , we obtain another family of symmetric designs.

Corollary 11.8.4. *Let q be an odd prime power. Then, for any positive integer m, there exists a locally decomposable symmetric design with parameters*

$$\left(1 + \frac{2q(q^m - 1)}{q - 1}, q^m, \frac{q^{m-1}(q - 1)}{2}\right). \tag{11.20}$$

If we apply case (i) of Theorem 11.8.1 to the Fano Plane, we obtain a symmetric $(25, 9, 3)$-design $\mathbf{D}$. The incidence matrices of its residual design and its derived design are the matrices X and Y introduced in the beginning of Section 11.7. The matrices $BGW(4, 3, 2)$ over $\mathbf{Z}_2 = \{1, -1\}$ and $GH(3, 1)$

over $\mathbf{Z}_3 = \{1, \rho, \rho^2\}$, which were used in the construction are, respectively,

$$\begin{bmatrix} 0 & 1 & 1 & 1 \\ 1 & 0 & 1 & - \\ 1 & - & 0 & 1 \\ 1 & 1 & - & 0 \end{bmatrix} \text{ and } \begin{bmatrix} 1 & 1 & 1 \\ 1 & \rho & \rho^2 \\ 1 & \rho^2 & \rho \end{bmatrix}.$$

Let $\mathbf{D}'$ be the complementary (25, 16, 10)-design. Matrices $J - Y$ and $J - X$ are incidence matrices of the residual design $\mathbf{D}'^A$ and the derived design $\mathbf{D}'_A$ corresponding to the same block A. Observe that the design with the incidence matrix $J - Y$ is 4-resolvable, i.e., its block set can be divided into classes $\mathcal{C}_1, \mathcal{C}_2, \mathcal{C}_3$, and $\mathcal{C}_4$ (each corresponding to consecutive columns of $J - Y$) so that each point is replicated four times in the blocks of each class. Furthermore, each class $\mathcal{C}_i$ can be divided into three subclasses $\mathcal{C}_{i1}, \mathcal{C}_{i2}, \mathcal{C}_{i3}$ of cardinality 2 so that each point occurs in all or none of the blocks of each subclass. Let us now divide and subdivide accordingly the block set of the design with the incidence matrix $J - X$. Then the number of occurrences of each point a of this design among the blocks of each subclass $\mathcal{C}_{ij}$ depends on i but does not depend on j. Therefore, if ρ is a permutation (of order 3) on the set of columns of $J - Y$, which permutes cyclically the subclasses $\mathcal{C}_{i1}, \mathcal{C}_{i2}$, and $\mathcal{C}_{i3}$, we obtain that $(\rho(J - Y))(J - X)^\top = (J - Y)(J - X)^\top = 10J$, a condition that is required by Theorem 11.7.8. Thus, the design $\mathbf{D}'$ may and, as the following theorem shows, does start an infinite family of locally decomposable symmetric designs.

Theorem 11.8.5. *Let $\mathbf{D} = (V, \mathcal{B})$ be a symmetric (v, r, λ)-design and let $A \in \mathcal{B}$. Suppose that the residual design $\mathbf{D}^A$ is α-resolvable with resolution classes $\mathcal{C}_1, \mathcal{C}_2, \ldots, \mathcal{C}_s$. Suppose further that each $\mathcal{C}_i$ admits a partition $\{\mathcal{C}_{i1}, \mathcal{C}_{i2}, \ldots, \mathcal{C}_{it}\}$ into subclasses of the same cardinality q so that, for each $a \in V$, for $i = 1, 2, \ldots, s$, and for $j = 1, 2, \ldots, t$,*

$$|\{B \in \mathcal{C}_{ij} : a \in B\}| = \begin{cases} 0 \text{ or } q & \text{if } a \notin A, \\ n_i(a) & \text{if } a \in A, \end{cases}$$

where $n_i(a)$ is an integer depending on i and a but not on j. If r is a prime power and t divides $r - 1$, then, for any positive integer m, there exists a locally decomposable symmetric design with parameters (11.16).

Proof. First note that $v - 1 = qst$. Let $\mathcal{M}$ be the set of all (0, 1)-matrices of size $(v - r) \times (v - 1)$. If $M \in \mathcal{M}$, we will assume that the columns of M are divided into s classes of consecutive columns, and each of these classes is divided into t subclasses, of q consecutive columns each. We will denote by $C_{ijh}(M)$ the hth column of the jth subclass of the ith class. Let $\rho \colon \mathcal{M} \to \mathcal{M}$

permute the set of columns of each matrix $M \in \mathcal{M}$ so that $C_{ijh}(\rho M) = C_{ij_1h}(M)$ with $j_1 \equiv j - 1 \pmod{t}$. In other words, ρ cyclically permutes subclasses within each class without changing the order of columns within the subclasses. Let S be the cyclic group generated by ρ. Then $|S| = t$. Since ρ is a permutation of columns, we have $(\sigma M)(\sigma N)^\top = MN^\top$ for all $M, N \in \mathcal{M}$ and all $\sigma \in S$.

We will now assume that the blocks of $\mathbf{D}$ are so ordered that the blocks of each resolution class $\mathcal{C}_i$ as well as the blocks of each subclass $\mathcal{C}_{ij}$ are consecutive blocks. Then the corresponding incidence matrix X of the residual design $\mathbf{D}^A$ is in $\mathcal{M}$. Let Y be the incidence matrix of $\mathbf{D}_A$ with respect to the same ordering, and let $\mathcal{M}_0 = \{\sigma X : \sigma \in S\}$. We claim that S is a group of symmetries on $\mathcal{M}_0$. Indeed, since the design $\mathbf{D}^A$ is α-resolvable and since each point of this design is replicated 0 or q times in each subclass $\mathcal{C}_{ij}$, each point is replicated q times in exactly α/q subclasses of each class $\mathcal{C}_i$. Therefore,

$$C_{ijh}\left(\sum_{\sigma \in S} \sigma X\right) = \sum_{\sigma \in S} C_{ijh}(\sigma X) = \sum_{j=1}^{t} C_{ijh}(X) = \frac{\alpha}{q} J.$$

We will represent any row $\mathbf{x}$ of length $v - 1$ as $\mathbf{x} = [\mathbf{x}_1\ \mathbf{x}_2 \ldots \mathbf{x}_s]$ where each $\mathbf{x}_i$ is a row of length $(v-1)/s$, and then represent each $\mathbf{x}_i$ as $\mathbf{x}_i = [\mathbf{x}_{i1}\ \mathbf{x}_{i2} \ldots \mathbf{x}_{it}]$, where each $\mathbf{x}_{ij}$ is a row of length q.

Let $\sigma \in S$ and let $\mathbf{x}$ and $\mathbf{y}$ be a row of X and a row of Y, respectively. Let $a \in A$ be the point of $\mathbf{D}$ corresponding to $\mathbf{y}$, and let $\mathbf{x}'$ be the row of σX corresponding to $\mathbf{x}$. If we show that $\mathbf{x}\mathbf{y}^\top = \mathbf{x}'\mathbf{y}^\top$, then $(\sigma X)Y^\top = XY^\top$ follows.

Each $\mathbf{x}_{ij}$ as well as each $\mathbf{x}'_{ij}$ is the all-one or the all-zero row of length q. Since the row sum of both $\mathbf{x}_i$ and $\mathbf{x}'_i$ is equal to α, the number of all-one rows $\mathbf{x}_{ij}$ for a fixed i as well as the number of the all-one rows $\mathbf{x}'_{ij}$ is equal to α/q. Since the row sum of each $\mathbf{y}_{ij}$ is equal to $n_i(a)$, we obtain that

$$\mathbf{x}_i\mathbf{y}_i^\top = \sum_{j=1}^{t} \mathbf{x}_{ij}\mathbf{y}_{ij}^\top = \frac{\alpha n_i(a)}{q}$$

and

$$\mathbf{x}'_i\mathbf{y}_i^\top = \sum_{j=1}^{t} \mathbf{x}'_{ij}\mathbf{y}_{ij}^\top = \frac{\alpha n_i(a)}{q}.$$

Since r is a prime power and $|S| = t$ divides $r - 1$, all the conditions of Theorem 11.7.8 are satisfied and we obtain a family of symmetric designs with parameters (11.16). □

Let $\mathbf{D}$ be the complement of the symmetric design obtained in Corollary 11.8.2. Then $\mathbf{D}$ is a symmetric (v, k, λ)-design with $v = 1 + qr(r^m - 1)/$

$(r-1)$, $k = q^d(r^m - 1)/(r-1)$, and $\lambda = q^d(q^{d-1}r^{m-1} - 1)/(r-1)$ where q and $r = (q^d - 1)/(q-1)$ are prime powers. If A is the block corresponding to the last column of the incidence matrix $J - S$, where S is the matrix (11.17), then the residual design $\mathbf{D}^A$ is q^d-resolvable. If each resolution class is divided into $t = r$ subclasses, each formed by q blocks corresponding to consecutive columns of the incidence matrix, then all the structural conditions of Theorem 11.8.5 are satisfied. Since $k \equiv q^d \pmod{t}$, t divides $k - 1$. Therefore, we obtain the following result.

Theorem 11.8.6. *Let d and e be positive integers. Let q be a prime power such that $p = (q^d - 1)/(q-1)$ and $r = q^d(p^{e+1} - 1)/(p-1)$ are prime powers. Then, for any nonnegative integer m, there exists a locally decomposable symmetric design with parameters* (11.16) *where $v = 1 + qp(p^{e+1} - 1)/(p-1)$ and $\lambda = q^d(q^{d-1}p^e - 1)/(p-1)$.*

The only realization of the conditions of Theorem 11.8.6 that we are aware of is $q = 2$, $p = 2^d - 1$ is a Mersenne prime, and $e = 1$, so $r = 2^{2d}$. Thus,

Corollary 11.8.7. *If $2^d - 1$ is a prime, then, for any positive integer m there exists a locally decomposable symmetric design with parameters*

$$\left(1 + \frac{2^{d+1}(2^{2dm} - 1)}{2^d + 1},\ 2^{2dm},\ 2^{2dm-d-1}(2^d + 1)\right). \tag{11.21}$$

Finally, we will describe one more approach to satisfying the conditions of Theorem 11.7.8 and thus obtaining an infinite family of locally decomposable symmetric designs. We begin with the following definition.

Definition 11.8.8. A 2-(v, k, λ) design is called cyclic if it has a *cyclic* automorphism group which is sharply transitive on the point set of the design.

By Corollary 9.1.21, the development of a cyclic difference set is a cyclic symmetric design. A multiple of a cyclic symmetric design is another example of a cyclic design. The following matrix is an incidence matrix of a cyclic 2-(13, 3, 1) design.

$$\begin{bmatrix}
1&0&0&0&0&0&0&0&0&0&0&1&1&\quad 1&0&0&0&0&0&1&0&0&0&0&1&0\\
1&1&0&0&0&0&0&0&0&0&0&0&1&\quad 0&1&0&0&0&0&0&1&0&0&0&0&1\\
0&1&1&0&0&0&0&0&0&0&0&0&0&\quad 1&0&1&0&0&0&0&0&1&0&0&0&0\\
1&0&1&1&0&0&0&0&0&0&0&0&0&\quad 0&1&0&1&0&0&0&0&0&1&0&0&0\\
0&1&0&1&1&0&0&0&0&0&0&0&0&\quad 0&0&1&0&1&0&0&0&0&0&1&0&0\\
0&0&1&0&1&1&0&0&0&0&0&0&0&\quad 0&0&0&1&0&1&0&0&0&0&0&1&0\\
0&0&0&1&0&1&1&0&0&0&0&0&0&\quad 0&0&0&0&1&0&1&0&0&0&0&0&1\\
0&0&0&0&1&0&1&1&0&0&0&0&0&\quad 1&0&0&0&0&1&0&1&0&0&0&0&0\\
0&0&0&0&0&1&0&1&1&0&0&0&0&\quad 0&1&0&0&0&0&1&0&1&0&0&0&0\\
0&0&0&0&0&0&1&0&1&1&0&0&0&\quad 0&0&1&0&0&0&0&1&0&1&0&0&0\\
0&0&0&0&0&0&0&1&0&1&1&0&0&\quad 0&0&0&1&0&0&0&0&1&0&1&0&0\\
0&0&0&0&0&0&0&0&1&0&1&1&0&\quad 0&0&0&0&1&0&0&0&0&1&0&1&0\\
0&0&0&0&0&0&0&0&0&1&0&1&1&\quad 0&0&0&0&0&1&0&0&0&0&1&0&1
\end{bmatrix}$$

Theorem 11.8.9. *Let* $\mathbf{D}$ *be a symmetric* (v, r, λ)*-design with* r *a prime power and* $v - r$ *dividing* $r - 1$. *If* $\mathbf{D}$ *has a cyclic residual design, then, for any positive integer* m, *there exists a locally decomposable symmetric design with parameters* (11.16).

Proof. Let $\mathbf{D}^A$ be a residual design of $\mathbf{D}$ having a cyclic automorphism group S which is sharply transitive on points. Let X and Y be the incidence matrices of the designs $\mathbf{D}^A$ and $\mathbf{D}_A$, respectively, corresponding to the same ordering of blocks. Let $\mathcal{M} = \{\sigma X : \sigma \in S\}$. Since the group S is sharply transitive on the rows of X, we have

$$\sum_{\sigma \in S} \sigma M = \sum_{\sigma \in S} \sigma X = kJ$$

for all $M \in \mathcal{M}$. For each $\sigma \in S$ and each $M \in \mathcal{M}$, σM can be obtained from M by a permutation of columns. Therefore, S is a group of symmetries of $\mathcal{M}$. Since each row of σX is a row of X, we have $(\sigma X)Y^\top = \lambda J$ for all $\sigma \in G$, and we apply Theorem 11.7.8. □

If $\mathbf{D}$ is the complement of $\mathrm{PG}_{d-1}(d, q)$ with $d \geq 2$ and q a prime power, then any residual design $\mathbf{D}^A$ is a q-fold multiple of the complement of the design $\mathrm{PG}_{d-2}(d - 1, q)$, which is cyclic. Therefore, $\mathbf{D}^A$ is also cyclic. The divisibility condition of Theorem 11.8.9 is also satisfied. The design that Theorem 11.8.9 yields in this case is the complement of $\mathrm{PG}_{(m+1)d-1}((m + 1)d, q)$. Thus, we have obtained the following result.

Corollary 11.8.10. *For any composite positive integer* d *and any prime power* q, *the complement of the design* $\mathrm{PG}_{d-1}(d, q)$ *is locally decomposable.*

We have now completed all constructions of symmetric designs that we intended to introduce in this book. The parameters of all currently known infinite families of symmetric designs are listed in Appendix. Among these families are the designs of points and hyperplanes of projective spaces, Hadamard 2-designs and Menon designs obtained from Hadamard matrices, and the designs obtained as developments of difference sets. All other parameters of known infinite families of symmetric designs arise as the parameters of either globally decomposable or locally decomposable symmetric designs.

11.9. An infinite family of designs with a nearly affine decomposition

In this section, we will apply BGW-matrices to obtain an infinite family of 2-designs admitting a nearly affine decomposition (cf. Definition 8.6.1). These designs are block-schematic and have three intersection numbers. The next theorem introduces a putative group of symmetries on a set of incidence matrices of 2-designs that increases the number of intersection numbers of the designs by at most one.

Theorem 11.9.1. *Let* **D** *be a* (v, b, r, k, λ)*-design with intersection numbers* $\alpha_1, \alpha_2, \ldots, \alpha_s$ *and let* N *be an incidence matrix of* **D**. *Let* G *be an abelian group of permutations of the set* $\mathcal{C}$ *of all columns of* N. *If* B *is a block of* **D** *corresponding to a column* C *of* N, *then, for each* $\sigma \in G$, *let* σB *be the block corresponding to the column* σC. *Suppose that the group* G, *as a group of bijections* $\mathcal{B} \to \mathcal{B}$, *satisfies the following conditions*:

(i) $|\sigma B_1 \cap \sigma B_2| = |B_1 \cap B_2|$ *for all* $B_1, B_2 \in \mathcal{B}$ *and all* $\sigma \in G$;
(ii) *there is an integer* t *such that, for any* $x \in X$ *and any* $B \in \mathcal{B}$, *the number of elements* $\sigma \in G$ *such that* $x \in \sigma(B)$ *is equal to* t.

For every $\sigma \in G$, *let* σN *be the incidence matrix of* **D** *obtained by replacing each column* C *of* N *with* σC. *Let* $\mathcal{N} = \{\sigma N : \sigma \in G\}$. *Then* G *is a group of symmetries of* $\mathcal{N}$. *Furthermore, if* W *is a* $BGW(w, l, \mu; G)$ *with* $rk\mu = v\lambda l$, *then* $\overline{W} \otimes_G N$ *is an incidence matrix of a* $(vw, bw, rl, kl, \lambda l)$*-design with intersection numbers* $\alpha_1 l, \alpha_2 l, \ldots, \alpha_s l$ *and* $\frac{v\lambda l}{b}$.

Proof. Since G is a group of permutations of columns of N, we have $(\sigma P)(\sigma Q)^\top = PQ^\top$ for all $P, Q \in \mathcal{N}$ and all $\sigma \in G$. In order to show that G is a group of symmetries of $\mathcal{N}$, we have to verify that $S = \sum_{\sigma \in G} \sigma N$ is a constant matrix. Let $x \in X$ and $B \in \mathcal{B}$. Then the entry of S, occupying the same position as the (x, B)-entry of N, is equal to the number of elements $\sigma \in G$ such that $x \in \sigma B$. By condition (ii), this number is t, and therefore $S = tJ_{v,b}$. Thus, G is a group of symmetries of $\mathcal{N}$. Note that the row sum of every matrix σN is r, and therefore, the row sum of S is $r|G|$. This implies that $r|G| = tb$, and therefore,

$$t = \frac{r|G|}{b} = \frac{k|G|}{v}. \tag{11.22}$$

Let $W = [\omega_{ij}]$ be a $BGW(w, l, \mu; G)$ with $rk\mu = v\lambda l$. By Theorem 11.7.3, $\overline{W} \otimes_G N$ is an incidence matrix of a $(vw, bw, rl, kl, \lambda l)$-design **E**. We will now determine the intersection numbers of **E**.

Let us assume that if ω_{ij} is a zero entry of W, then $\omega_{ij} B$ is the empty set for all $B \in \mathcal{B}$, so $\omega_{ij} B$ is now defined for all entries of W. Then the intersection numbers of **E** are precisely the sums

$$\beta_{jh}(A, B) = \sum_{i=1}^{w} |\overline{\omega_{ij}} A \cap \overline{\omega_{ih}} B|, \tag{11.23}$$

where A and B are blocks of $\mathbf{D}$ and $j, h \in \{1, 2, \dots, w\}$, with the only restriction that if $j = h$, then $A \neq B$.

Case 1: $j = h$ and $A \neq B$.

Since each column of $\overline{W}$ has l nonzero entries, we obtain that, for some $\sigma_1, \sigma_2, \dots, \sigma_l \in G$,

$$\beta_{jh}(A, B) = \sum_{i=1}^{l} |\sigma_i A \cap \sigma_i B| = l|A \cap B| \in \{\alpha_1 l, \alpha_2 l, \dots, \alpha_s l\}.$$

Case 2: $j \neq h$.

Then there exist $\sigma_1, \sigma_2, \dots, \sigma_\mu \in G$ and $\tau_1, \tau_2, \dots, \tau_\mu \in G$ such that

$$\beta_{jh}(A, B) = \sum_{i=1}^{\mu} |\sigma_i A \cap \tau_i B| = \sum_{i=1}^{\mu} |\tau_i^{-1}\sigma_i A \cap B| = \sum_{i=1}^{\mu} |\sigma_i \tau_i^{-1} A \cap B|.$$

Since the group G is abelian, $W^\top$ is a $BGW(w, l, \mu; G)$. Therefore,

$$\beta_{jh}(A, B) = \frac{\mu}{|G|} \sum_{\sigma \in G} |\sigma A \cap B|.$$

For each $x \in B$, there are exactly t elements $\sigma \in G$ such that $x \in \sigma A$. Therefore, we apply (11.22) and the condition $rk\mu = v\lambda l$ to obtain:

$$\beta_{jh}(A, B) = \frac{\mu k t}{|G|} = \frac{v\lambda l}{b}.$$

The proof is now complete. □

Remark 11.9.2. If W is a $BGW((q^{m+1} - 1)/(q - 1), q^m, q^{m-1}(q - 1))$, where q is a prime power, then the equality $kr\mu = v\lambda l$ is equivalent to $q = kr/(r - \lambda)$.

Remark 11.9.3. The design $\mathbf{E}$ constructed in Theorem 11.9.1 has the following property: its block set is partitioned into w subsets corresponding to the columns of $\overline{W}$ so that, for any distinct blocks P and Q of $\mathbf{E}$, $|P \cap Q| = v\lambda l/b$ if P and Q are not in the same subset; if P and Q are in the same subset, then $|P \cap Q| = \alpha_i l$ with $i \in \{1, 2, \dots, s\}$.

We will now obtain a family of 2-designs with nearly affine decomposition.

Theorem 11.9.4. *Let $\mathbf{D} = (X, \mathcal{B})$ be an affine α-resolvable (v, b, r, k, λ)-design. Suppose that $q = kr/(r - \lambda)$ is a prime power and $\alpha(r - \lambda)$ divides*

$k\lambda$. Then, for each positive integer m, there exists a

$$\left(\frac{(q^{m+1}-1)v}{q-1}, \frac{(q^{m+1}-1)b}{q-1}, q^m r, q^m k, q^m \lambda\right)\text{-design}$$

E *with at most three intersection numbers $\rho_1 = k^2 q^m / v$, $\rho_2 = (k + \lambda - r)q^m$, and $\rho_3 = v\lambda q^m / b$. Furthermore, the design* **E** *admits a nearly affine decomposition.*

Proof. Let $\mathfrak{R} = \{\mathcal{C}_1, \mathcal{C}_2, \dots, \mathcal{C}_{r/\alpha}\}$ be the affine resolution of **D**. Let $\mathcal{C}_i = \{B_{i1}, B_{i2}, \dots, B_{is}\}$ with $s = b\alpha/r = v\alpha/k$. Define a bijection $\sigma : \mathcal{B} \to \mathcal{B}$ as follows:

$$\sigma B_{ij} = \begin{cases} B_{i,j+1} & \text{if } j = 1, 2, , \dots, s-1, \\ B_{i1} & \text{if } j = s. \end{cases}$$

Let G be the cyclic group generated by σ. Then $|G| = s$ and G satisfies conditions (i) and (ii) of Theorem 11.9.1 with $t = \alpha$. Since

$$q - 1 = \frac{r(k-1) + \lambda}{r - \lambda} = \frac{v\lambda}{r - \lambda},$$

we obtain that

$$\frac{q-1}{|G|} = \frac{k\lambda}{\alpha(r - \lambda)},$$

which is an integer. Thus, $|G|$ divides $q - 1$, and then Theorem 10.2.5 implies that, for any positive integer m, there exists a $BGW((q^{m+1} - 1)/(q - 1), q^m, q^m - q^{m-1}; G)$. By Corollary 5.1.16, the intersection numbers of **D** are k^2/v and $k - r + \lambda$. We now complete the proof by applying Theorem 11.9.1. □

To apply this theorem, we let $\mathbf{D}_1$ be a Hadamard 2-$(2^n - 1, 2^{n-1}, 2^{n-2})$ design with $p = 2^n - 1$ a prime. Let $\mathbf{D}_2$ be the affine resolvable design $AG_{p-1}(2, p)$. The blocks of $\mathbf{D}_2$ are partitioned into 2^n classes of $2^n - 1$ blocks, each of cardinality $2^n - 1$. Theorem 5.3.10 then yields an affine 2^{n-1}-resolvable design **D** with parameters $((2^n - 1)^2, 2^n(2^n - 1), 2^{2n-1}, 2^{n-1}(2^n - 1), 2^{n-2}(2^n + 1))$. The parameters of **D** satisfy the conditions of Theorem 11.9.4 with $q = 2^n$ and $\alpha = 2^{n-1}$. Therefore, we obtain the following result.

Corollary 11.9.5. *Let $2^n - 1$ be a prime and let $q = 2^n$. Then for any positive integer m, there exists a $((q^{m+1} - 1)(2^n - 1), (q^{m+1} - 1) \cdot 2^n, q^m \cdot 2^{2n-1}, q^m \cdot 2^{n-1}(2^n - 1), q^m \cdot 2^{n-2}(2^n + 1))$-design admitting a nearly affine decomposition.*

Exercises

(1) Let X be an incidence matrix of a symmetric $(3, 2, 1)$-design, let $G = \langle \omega \rangle$ be a cyclic group of order 3, and let W be the matrix of Example 10.1.7. Construct the matrix $W \otimes_G X$. It is an incidence matrix of a symmetric $(15, 8, 4)$-design.

(2) Use Theorem 11.8.1 to construct an incidence matrix of a symmetric $(61, 25, 10)$-design.

(3) Prove that the graphs corresponding to the nearly affine decomposition obtained in Theorem 11.9.4 are complete multipartite graphs.

Notes

The notions of global and local decompositions of symmetric designs were introduced in Ionin and M. S. Shrikhande (2003) though, of course, many such designs were constructed long before that. The construction described in Theorem 11.1.1 is due to Bridges (1983). Designs with the same parameters were obtained earlier in van Trung (1982a). We follow Ionin and M. S. Shrikhande (2003) in the proofs of Theorems 11.2.6 and 11.2.8. Theorem 11.2.5 and Corollary 11.2.7 were obtained in Ionin (1998b). Symmetric designs from Theorems 11.3.3–11.3.8 were constructed in Ionin (1999b).

The notion of a productive regular Hadamard matrix was introduced in Ionin (2004) where Theorem 11.4.6 was proven. Theorem 11.4.3 to Kharaghani (2000) and Theorem 11.4.4 to Ionin and Kharaghani (2003b). Theorem 11.4.5 is proven in Behbahani and Kharaghani (2004).

Symmetric designs from Corollary 11.5.2 were constructed in S. S. Shrikhande and Singhi (1975). Ionin and M. S. Shrikhande (2003) showed that these designs were globally decomposable.

Constructions of strongly regular graphs and normally regular digraphs in Section 11.6. follow Ionin and Kharaghani (2003a, 2003b).

Theorem 11.7.7 is due to Rajkundlia (1983). Theorems 11.7.8, 11.8.1, and 11.8.5 are due to Ionin (2001). Symmetric designs with parameters (11.19) were constructed for $d = 2$ in Rajkundlia (1978, 1983) and Mitchell (1979), for $q = 2$ in Brouwer (1983), for $q = 8$ and $d = 3$ in Ionin (1999a), and in the general case in Ionin (2001). Symmetric designs with parameters (11.20) were constructed in Brouwer (1983). Symmetric designs of Theorem 11.8.6 were constructed in Fanning (1995) for $p = 3$ and in Ionin (1999a) for p a Mersenne prime. Theorem 11.8.9 is due to Ionin (2001). We are not aware of any realization of the condition of this theorem, except the one considered in Corollary 11.8.10.

The results of Section 11.9. are due to Ionin and M. S. Shrikhande (2000).

12

Subdesigns of symmetric designs

Substructures of an incidence structure may provide useful information about this incidence structure. Among interesting substructures of symmetric designs that are themselves symmetric designs are *tight subdesigns*, *Baer subdesigns*, *normal subdesigns*, and *Bruck subdesigns*. Partitioning the point set of a symmetric design into subsets, such as *M-arcs*, which are closely related to the structure of the design, could be used as a classification tool.

12.1. Tight subdesigns

We begin with the following specialization of Definition 2.1.4.

Definition 12.1.1. A symmetric design $\mathbf{D}_1 = (X_1, \mathcal{B}_1)$ is called a *symmetric subdesign* of a symmetric design $\mathbf{D} = (X, \mathcal{B})$ if (i) $X_1 \subseteq X$; (ii) $\mathcal{B}_1 \subseteq \mathcal{B}$; and (iii) for all $x \in X_1$ and $B \in \mathcal{B}_1$, the point x and the block B are incident in $\mathbf{D}_1$ if and only if they are incident in $\mathbf{D}$. If $\mathbf{D}_1$ is a symmetric (v_1, k_1, λ_1)-design, we will refer to it as a (v_1, k_1, λ_1)-subdesign of $\mathbf{D}$.

Remark 12.1.2. Throughout this chapter a subdesign of a symmetric design is always assumed to be a symmetric design.

Example 12.1.3. Let $\mathbf{D}$ be a symmetric design described in Example 1.3.3 and let $\mathbf{D}_1$ be the substructure formed by the four points of one row and the corresponding four blocks. Then $\mathbf{D}_1$ is a $(4, 3, 2)$-subdesign of $\mathbf{D}$.

Example 12.1.4. Let H be a regular Hadamard matrix of Bush type of order $4h^2$, $h > 0$, and let $N = \frac{1}{2}(H + J)$. Then N is an incidence matrix of a symmetric $(4h^2, 2h^2 + h, h^2 + h)$-design $\mathbf{D}$ and N has all-one submatrices of order $2h$. Therefore, $\mathbf{D}$ has $(2h, 2h, 2h)$-subdesigns.

Example 12.1.5. For $i = 1, 2$, let H_i be a regular Hadamard matrix of order $4h_i^2$ and let $N_i = \frac{1}{2}(J - H_i)$. Then N_i is an incidence matrix of a symmetric $(4h_i^2, 2h_i^2 - h_i, h_i^2 - h_i)$-design $\mathbf{D}_i$. Let $H = H_1 \otimes H_2$ and let $N = \frac{1}{2}(J - H)$. Then N is an incidence matrix of a symmetric $(16h_1^2h_2^2, 8h_1^2h_2^2 - 2h_1h_2, 4h_1^2h_2^2 - 2h_1h_2)$-design that has symmetric subdesigns isomorphic to $\mathbf{D}_1, \mathbf{D}_2$, and their complements.

Let a field K be a subfield of a field L and let K^n and L^n be the n-dimensional vector spaces over K and L, respectively. Then we can regard K^n as a subset of L^n. For $0 \le d \le n$, every d-dimensional subspace of K^n is contained in a unique d-dimensional subspace of L^n. These simple observations permit us to consider the projective geometry $PG(n, q)$ as embedded into the projective geometry $PG(n, q^m)$. Therefore, we have the following result.

Proposition 12.1.6. *Let q be a prime power and let m and n be positive integers. Then the design $PG_{n-1}(n, q^m)$ has a subdesign isomorphic to $PG_{n-1}(n, q)$.*

Remark 12.1.7. In particular, the projective plane $\mathbf{P}$ over $GF(q^2)$ has a *subplane* $\mathbf{P}_1$ that is a projective plane over $GF(q)$. In this case, the lines of the subplane are the lines of $\mathbf{P}$ that contain at least two points of $\mathbf{P}_1$. Let $\mathcal{L}_0$ be the set of all lines of $\mathbf{P}$ that are not lines of $\mathbf{P}_1$. Note that every point of $\mathbf{P}_1$ lies on $(q^2 + 1) - (q + 1) = q^2 - q$ lines of $\mathcal{L}_0$. Therefore, there are exactly $(q^2 + q + 1)(q^2 - q) = q^4 - q$ flags (x, L) where x is a point of $\mathbf{P}_1$ and $L \in \mathcal{L}_0$. Since $|\mathcal{L}_0| = (q^4 + q^2 + 1) - (q^2 + q + 1) = q^4 - q$, we obtain that each $L \in \mathcal{L}_0$ contains one point of $\mathbf{P}_1$. Therefore, the points of $\mathbf{P}_1$ and the lines of $\mathcal{L}_0$ form a $(q^2 + q + 1, q^4 - q, q^2 - q, 1, 0)$-design.

Definition 12.1.8. A symmetric subdesign $\mathbf{P}_1$ of a projective plane $\mathbf{P}$ is called a *Baer subplane of* $\mathbf{P}$ if $\mathbf{P}_1$ is a projective plane and every line of $\mathbf{P}$ that is not a line of $\mathbf{P}_1$ contains exactly one point of $\mathbf{P}_1$.

For the rest of this section, we will be interested in subdesigns of symmetric designs that have a similar property.

Definition 12.1.9. A (v_1, k_1, λ_1)-subdesign $\mathbf{D}_1 = (X_1, \mathcal{B}_1)$ of a symmetric (v, k, λ)-design $\mathbf{D} = (X, \mathcal{B})$ is called a *tight subdesign* if $v_1 < v$ and there is an integer k_2 such that $|B \cap X_1| = k_2$ for all $B \in \mathcal{B} \setminus \mathcal{B}_1$. If furthermore $\lambda_1 = \lambda$, then $\mathbf{D}_1$ is said to be a *Baer subdesign of* $\mathbf{D}$.

The following result is immediate.

Proposition 12.1.10. *If $\mathbf{D}_1 = (X_1, \mathcal{B}_1)$ is a tight (v_1, k_1, λ_1)-subdesign of a symmetric (v, k, λ)-design $\mathbf{D} = (X, \mathcal{B})$, then the complement of $\mathbf{D}_1$ is a* *tight subdesign of the complement of* $\mathbf{D}$ *and the substructure* $\mathbf{D}_2 = (X_1, \mathcal{B} \setminus \mathcal{B}_1)$ *of* $\mathbf{D}$ *is a* $(v_1, v - v_1, k - k_1, k_2, \lambda - \lambda_1)$*-design with* $k_2 = v_1(k - k_1)/(v - v_1)$.

The next theorem characterizes the parameters of Baer subdesigns.

Theorem 12.1.11. *Let* $\mathbf{D}_1$ *be a* (v_1, k_1, λ)*-subdesign of a symmetric* (v, k, λ)*-design* $\mathbf{D}$ *with* $v_1 < v$. *Then* $\mathbf{D}_1$ *is a Baer subdesign of* $\mathbf{D}$ *if and only if* $k - \lambda = (k_1 - 1)^2$. *Furthermore, if* $\mathbf{D}_1$ *is not a Baer subdesign of* $\mathbf{D}$, *then* $k - \lambda \geq k_1(k_1 - 1)$.

Proof. Let $\mathbf{D} = (X, \mathcal{B})$ and $\mathbf{D}_1 = (X_1, \mathcal{B}_1)$. We have $(v - 1)\lambda = k(k - 1)$ and $(v_1 - 1)\lambda = k_1(k_1 - 1)$, so

$$(v - v_1)\lambda = (k - k_1)(k + k_1 - 1). \tag{12.1}$$

Since $v > v_1$, we have $k > k_1$. If x and y are distinct points of $\mathbf{D}_1$, then no block of $\mathcal{B} \setminus \mathcal{B}_1$ contains $\{x, y\}$, i.e., every block of $\mathcal{B} \setminus \mathcal{B}_1$ contains at most one point of $\mathbf{D}_1$. Therefore, the subdesign $\mathbf{D}_1$ is a Baer subdesign of $\mathbf{D}$ if and only if every block of $\mathcal{B} \setminus \mathcal{B}_1$ contains exactly one point of $\mathbf{D}_1$. This is equivalent to $v - v_1 = v_1(k - k_1)$. We now apply (12.1) to obtain:

$$\begin{aligned} v - v_1 &= v_1(k - k_1) \Leftrightarrow (v - v_1)\lambda = v_1\lambda(k - k_1) \\ &\Leftrightarrow (k - k_1)(k + k_1 - 1) = v_1\lambda(k - k_1) \Leftrightarrow k + k_1 - 1 = v_1\lambda \\ &\Leftrightarrow (k - \lambda) + (k_1 - 1) = k_1(k_1 - 1) \Leftrightarrow k - \lambda = (k_1 - 1)^2. \end{aligned}$$

Suppose now that $\mathbf{D}_1$ is not a Baer subdesign of $\mathbf{D}$. Then there is a block $B_0 \in \mathcal{B} \setminus \mathcal{B}_1$ such that $B_0 \cap X_1 = \varnothing$. Then $|B_0 \cap (X \setminus X_1)| = k$. If B_1 and B_2 are distinct blocks of $\mathcal{B}_1$, then $|B_1 \cap B_2 \cap X_1| = \lambda$ and therefore, the λ-subsets $B_1 \cap B_0$ and $B_2 \cap B_0$ of $X \setminus X_1$ are disjoint. This implies that $k = |B_0 \cap (X \setminus X_1)| \geq v_1\lambda$, and then $k - \lambda \geq k_1(k_1 - 1)$. □

Corollary 12.1.12. *Let a projective plane* $\mathbf{P}_1$ *of order* m *be a subplane of a projective plane* $\mathbf{P}$ *of order* n. *If* $\mathbf{P}_1$ *is a Baer subplane of* $\mathbf{P}$, *then* $n = m^2$; *otherwise,* $n \geq m^2 + m$.

We will now characterize the parameters of tight subdesigns.

Theorem 12.1.13. *Let* $\mathbf{D}_1$ *be a* (v_1, k_1, λ_1)*-subdesign of a symmetric* (v, k, λ)*-design* $\mathbf{D}$ *with* $v_1 < v$ *and let* $k_2 = v_1(k - k_1)/(v - v_1)$. *Then*

$$k - \lambda \geq k_1^2 - v_1\lambda + k_2(k - k_1). \tag{12.2}$$

Moreover, the subdesign $\mathbf{D}_1$ *is tight if and only if*

$$k - \lambda = k_1^2 - v_1\lambda + k_2(k - k_1). \tag{12.3}$$

Proof. Let $\mathbf{D} = (X, \mathcal{B})$ and $\mathbf{D}_1 = (X_1, \mathcal{B}_1)$. Then $\mathbf{D}$ has an incidence matrix N of the form $N = \left[\begin{smallmatrix} N_1 & N_2 \\ N_3 & N_4 \end{smallmatrix}\right]$, where N_1 is an incidence matrix of $\mathbf{D}_1$.

For each $B \in \mathcal{B} \setminus \mathcal{B}_1$, let $\alpha_B = |B \cap X_1|$. Then counting in two ways flags (x, B) of $\mathbf{D}$ with $x \in X_1$ and $B \in \mathcal{B} \setminus \mathcal{B}_1$ yields

$$\sum_{B \in \mathcal{B} \setminus \mathcal{B}_1} \alpha_B = v_1(k - k_1) = (v - v_1)k_2. \tag{12.4}$$

Next count triples (x, y, B) with $B \in \mathcal{B} \setminus \mathcal{B}_1$ and $x, y \in B \cap X_1$ to obtain

$$\sum_{B \in \mathcal{B} \setminus \mathcal{B}_1} \alpha_B(\alpha_B - 1) = v_1(v_1 - 1)(\lambda - \lambda_1). \tag{12.5}$$

Hence we have

$$\begin{aligned} 0 \le \sum_{B \in \mathcal{B} \setminus \mathcal{B}_1} (k_2 - \alpha_B)^2 &= \sum_{B \in \mathcal{B} \setminus \mathcal{B}_1} \alpha_B^2 - k_2^2(v - v_1) \\ &= v_1(v_1 - 1)(\lambda - \lambda_1) + v_1(k - k_1) - k_2 v_1(k - k_1) \end{aligned}$$

and therefore we obtain (using $(v_1 - 1)\lambda_1 = k_1(k_1 - 1)$) the inequality (12.2).

The equality holds if and only if $\alpha_B = k_2$ for all $B \in \mathcal{B} \setminus \mathcal{B}_1$, i.e., if and only if $\mathbf{D}_1$ is tight. □

Corollary 12.1.14. *If a symmetric (v, k, λ)-design has a tight (v_1, k_1, λ_1)-subdesign, then $k - \lambda = (k_1 - k_2)^2$ with $k_2 = v_1(k - k_1)/(v - v_1)$.*

Proof. Let $\mathbf{D}_1$ be a tight (v_1, k_1, λ_1)-subdesign of a symmetric (v, k, λ)-design $\mathbf{D}$. Then $\mathbf{D}$ has an incidence matrix N of the form $N = \left[\begin{smallmatrix} N_1 & N_2 \\ N_3 & N_4 \end{smallmatrix}\right]$, where N_1 is an incidence matrix of $\mathbf{D}_1$. Choose a point $x_0 \in X \setminus X_1$ and denote by β the number of blocks of $\mathbf{D}_1$ containing x_0. Then counting the flags (x, B) such that $x \in X_1$, $B \in \mathcal{B}$, and $x_0 \in B$ yields

$$v_1\lambda = \beta k_1 + (k - \beta)k_2, \tag{12.6}$$

which implies that $(k_1 - k_2)\beta = v_1\lambda - kk_2$. If $k_1 = k_2$, then one derives that $k = \lambda$, which in turn implies $v = k$, a contradiction. Therefore, $k_1 \neq k_2$, so β does not depend on the choice of the point x_0. This means that N_3 has constant row sum β, and therefore N_4 has constant row sum $k - \beta$ and constant column sum $k - k_2$. Since N_4 is a square matrix, we obtain that $\beta = k_2$. Now (12.3) and (12.6) imply that $k - \lambda = (k_1 - k_2)^2$. □

We will state without proof the converse of the above result.

Theorem 12.1.15. *If a symmetric (v, k, λ)-design $\mathbf{D}$ has a symmetric (v_1, k_1, λ_1)-subdesign $\mathbf{D}_1$ and $k - \lambda = (k_1 - k_2)^2$ with $k_2 = v_1(k - k_1)/(v - v_1)$, then the subdesign $\mathbf{D}_1$ is tight.*

Remark 12.1.16. If the parameters (v, k, λ) and (v_1, k_1, λ_1) of two symmetric designs satisfy conditions of Theorem 12.1.13, this does not imply that there exists a symmetric (v, k, λ)-design with a tight (v_1, k_1, λ_1)-subdesign. For instance, let $(v, k, \lambda) = (36, 15, 6)$ and $(v_1, k_1, \lambda_1) = (15, 8, 4)$. In this case, $k_2 = 5$. If there is a symmetric (36, 15, 6)-design with a tight (15, 8, 4)-subdesign, then there exists a (15, 21, 7, 5, 2)-design. This design is quasi-residual and, by the Hall–Connor Theorem, it is embeddable in a symmetric (22, 7, 2)-design. However, the latter design does not exist (see Remark 2.4.11).

We will now derive necessary conditions on the parameters of a symmetric (v, k, λ)-design that admits a given (v_1, k_1, λ_1)-design as a tight subdesign.

Proposition 12.1.17. *If a nontrivial symmetric (v_1, k_1, λ_1)-design with $\lambda_1 \neq 0$ is a tight subdesign of a symmetric (v, k, λ)-design, then there exist positive integers d, t, u, and u_1 such that $du_1 = v_1$, $d \geq 2$, $t \leq d - 1$,*

$$u = \frac{(v_1 - 1)((k_1 - tu_1)^2 - (k_1 - \lambda_1))}{tu_1(d - t)}, \tag{12.7}$$

$v = d(u_1 + u)$, $k = k_1 + tu$, and $\lambda = k - (k_1 - tu_1)^2$.

Proof. Let a symmetric (v, k, λ)-design **D** have a nontrivial tight (v_1, k_1, λ_1)-subdesign $\mathbf{D}_1$ with $\lambda_1 \neq 0$. By Corollary 12.1.14, there exists a positive integer $k_2 = v_1(k - k_1)/(v - v_1)$ such that $k - \lambda = (k_1 - k_2)^2$.

Let d be the greatest common divisor of v and v_1 and let $v_1 = du_1$ and $v = d(u_1 + u)$. Then u divides $k - k_1$, so let $k = k_1 + tu$. Then $k_2 = tu_1$ and therefore $\lambda = k - (k_1 - tu_1)^2$. Since $k_2 \leq v_1$, we have $t \leq d$. Suppose $t = d$. Then $k_2 = v_1$ and therefore $v - v_1 = k - k_1$. Then $v - k = v_1 - k_1$ and therefore the complement of **D** has an incidence matrix N of the form $\left[\begin{smallmatrix} N_1 & O \\ O & N_2 \end{smallmatrix}\right]$ where N_1 is an incidence matrix of the complement of $\mathbf{D}_1$. This implies $\lambda = 0$ and then $\lambda_1 = 0$, a contradiction. Therefore, $t \leq d - 1$.

We have to show that u satisfies (12.7). The equation $(v - 1)\lambda = k(k - 1)$ implies that

$$(d(u_1 + u) - 1)((k_1 + tu) - (k_1 - k_2)^2) = (k_1 + tu)(k_1 + tu - 1).$$

This is a quadratic equation in u with the leading coefficient $t(d - t)$ and the free term that can be rewritten as $(v_1 - 1)((k_1 - \lambda_1) - (k_1 - k_2)^2)$. One can verify that $u = -u_1$ is a solution to this equation. Since it is negative, the only possible value of u is the other solution to this equation, and it is given by (12.7). □

Corollary 12.1.18. *A given nontrivial symmetric (v_1, k_1, λ_1)-design with $\lambda_1 \neq 0$ can be a tight subdesign of at most finitely many symmetric designs.*

Corollary 12.1.19. *If a symmetric* (v, k, λ)*-design has a tight* $(7, 3, 1)$*-subdesign, then* $(v, k, \lambda) = (21, 5, 1)$ *or* $(v, k, \lambda) = (56, 45, 36)$.

We will show in the next section that both possibilities suggested by Corollary 12.1.19 can be realized.

12.2. Examples of tight subdesigns

In this section we will give several examples of tight subdesigns of symmetric designs. The examples that initiated the notion of tight subdesign are the Baer subplanes of projective planes that are described in Remark 12.1.7. Proposition 12.1.10 implies that the complement of a Baer subplane of a projective plane **P** is a tight subdesign of the complement of **P**.

In fact, if **P** is the desarguesian projective plane of order q^2, then it admits a family of Baer subplanes (of order q) whose point sets partition the point sets of **P** and block sets partition the block set of **P**.

Definition 12.2.1. A *tight* (v_1, k_1, λ_1)*-partition* of a symmetric design $\mathbf{D} = (X, \mathcal{B})$ is a set of tight (v_1, k_1, λ_1)-subdesigns $\mathbf{D}_i = (X_i, \mathcal{B}_i)$, $1 \le i \le s$, of **D** such that $\{X_1, X_2, \dots, X_s\}$ is a partition of X and $\{\mathcal{B}_1, \mathcal{B}_2, \dots, \mathcal{B}_s\}$ is a partition of $\mathcal{B}$. If the designs $\mathbf{D}_i$ are Baer subdesigns of **D**, then the partition is called a *Baer partition*.

Theorem 12.2.2. *Let* q *be a prime power. The design* $PG_1(2, q^2)$ *admits a Baer partition into* $q^2 - q + 1$ *subdesigns isomorphic to* $PG_1(2, q)$.

Proof. Let M be the field of order q^6 and let F, K, L be the subfields of M of order q, q^3, and q^2, respectively. Let α be a primitive element of M. Then $\beta = \alpha^{q^3+1}$ is a primitive element of K and $\gamma = \alpha^{q^4+q^2+1}$ is a primitive element of L. Note that $K \cap L = F$.

Let $X = \{F, \beta F, \dots, \beta^{q^2+q} F\}$. Fix a two-dimensional subspace H of K over F and let $\mathcal{B} = \{H, \beta H, \dots, \beta^{q^2+q} H\}$. Then, by Proposition 3.6.8, the incidence structure $\mathbf{D} = (X, \mathcal{B})$ (with inclusion as the incidence relation) is $PG_1(2, F)$.

For $i = 0, 1, \dots, q^2 - q$, let $X_i = \{\alpha^i L, \alpha^i \beta L, \dots, \alpha^i \beta^{q^2+q} L\}$ and $\mathcal{B}_i = \{\alpha^i HL, \alpha^i \beta HL, \dots, \alpha^i \beta^{q^2+q} HL\}$. If $\alpha^i \beta^k L \subset \alpha^i \beta^l HL$, then $\beta^k L \subset \beta^l HL$, and therefore, $\beta^k = \beta^l xy$ with $x \in H$ and $y \in L$. Since $H \subset K$ and $K \cap L = F$, we obtain that $y \in F$, and then $\beta^k \in \beta^l H$. Therefore, for $i = 0, 1, \dots, q^2 - q$, the incidence structure $\mathbf{D}_i = (X_i, \mathcal{B}_i)$ is isomorphic to $PG_1(2, F)$.

We claim that, for $0 \le i < j \le q^2 - q$, $X_i \cap X_j = \varnothing$ and $\mathcal{B}_i \cap \mathcal{B}_j = \varnothing$.

To prove this claim, let $\alpha^i\beta^k L = \alpha^j\beta^l L$, for some $i, j \in \{0, 1, \ldots, q^2 - q\}$ and $k, l \in \{0, 1, \ldots, q^2 + q\}$. Then $\alpha^{i-j}\beta^{k-l} \in L$, $\alpha^{i-j+(q^3+1)(k-l)} \in L$, $i - j + (q^3 + 1)(k - l) \equiv 0 \pmod{q^4 + q^2 + 1}$. Since $q^4 + q^2 + 1 = (q^2 + q + 1)(q^2 - q + 1)$ and $q^3 + 1 = (q + 1)(q^2 - q + 1)$, we obtain that $i - j \equiv 0 \pmod{q^2 - q + 1}$, which implies $i = j$ and $(q + 1)(k - l) \equiv 0 \pmod{q^2 + q + 1}$. Since $q + 1$ and $q^2 + q + 1$ are relatively prime, we have $k - l \equiv 0 \pmod{q^2 + q + 1}$ and then $k = l$. Since HL is a two-dimensional subspace of M over L, Proposition 3.6.8 implies that if $\alpha^i\beta^k HL = \alpha^j\beta^l H$, then $\alpha^{i-j}\beta^{k-l} \in L$ and then again $i = j$ and $k = l$.

The elements of the set $Y = X_0 \cup X_1 \cup \ldots \cup X_{q^2-q}$ can be regarded as points of $PG_1(2, L)$ and the elements of the set $\mathcal{C} = \mathcal{B}_0 \cup \mathcal{B}_1 \cup \ldots \cup \mathcal{B}_{q^2-q}$ as blocks of $PG_1(2, L)$. Since $|Y| = |\mathcal{C}| = (q^2 - q + 1)(q^2 + q + 1) = q^4 + q^2 + 1$, we obtain that $(Y, \mathcal{C})$ is the $PG_1(2, L)$ admitting a Baer partition into subplanes isomorphic to $PG_1(2, F)$. □

Our next two examples come from regular Hadamard matrices.

Example 12.2.3. If H is a regular Hadamard matrix of Bush type with row sum $2h$, then the symmetric $(4h^2, 2h^2 + h, h^2 + h)$-design with incidence matrix $N = \frac{1}{2}(J + H)$ admits a tight partition into trivial $(2h, 2h, 2h)$-subdesigns.

Example 12.2.4. Let $h \neq 0$ be an integer and let H be a regular Hadamard matrix with row sum $2h$. Let K be a regular Hadamard matrix with row sum -2. Then $K \otimes H$ is a regular Hadamard matrix with row sum $-4h$. Let $N = \frac{1}{2}(J + (K \otimes H))$ Then the symmetric $(16h^2, 8h^2 - 2h, 4h^2 - 2h)$-design with incidence matrix N admits a tight partition into four isomorphic $(4h^2, 2h^2 + h, h^2 + h)$-subdesigns.

The following proposition gives another infinite family of symmetric designs with a trivial tight subdesign.

Proposition 12.2.5. *Let q be a prime power and d a positive integer. The design $PG_{2d}(2d + 1, q)$ has a tight partition into $(\lambda, \lambda, \lambda)$-subdesigns with $\lambda = (q^{d+1} - 1)/(q - 1)$.*

Proof. By Theorem 3.6.13, the projective space $PG(2d + 1, q)$ can be partitioned into d-dimensional subspaces. If C is any of these subspaces, then Proposition 3.6.2 implies that the substructure of $PG_{2d}(2d + 1, q)$ formed by the points of C and the hyperplanes containing C is a symmetric $(\lambda, \lambda, \lambda)$-subdesign with $\lambda = (q^{d+1} - 1)/(q - 1)$. Since every hyperplane that does not contain C intersects C in a $(d - 1)$-dimensional subspace, we obtain that the blocks of $PG_{2d}(2d+1,q)$ that are not blocks of the subdesign meet C in the same number of points. Therefore, the subdesign is tight. □

Remark 12.2.6. Proposition 12.2.5 yields a Baer partition if and only if $d = 1$.

Stanton–Sprott difference sets provide us with one more family of symmetric designs with a trivial tight subdesign.

Proposition 12.2.7. *Let q and $q+2$ be odd prime powers. Then there exists a symmetric $(q^2+2q, (q^2+2q-1)/2, (q^2+2q-3)/4)$-design with a tight (q,q,q)-subdesign.*

Proof. Let $\mathbf{D} = dev(D)$ where D is the Stanton–Sprott difference set described in Theorem 9.2.3. Let D_1 be the subset of D consisting of all pairs $(x,0)$ with $x \in GF(q)$ and let $\mathbf{D}_1 = dev(D_1)$. Then $\mathbf{D}_1$ is a symmetric (q,q,q)-subdesign of $\mathbf{D}$ and it is tight by Theorem 12.1.13. □

Remark 12.2.8. Proposition 12.2.7 yields a Baer subdesign if and only if $q = 3$.

Global decomposition of symmetric designs may produce designs with a tight partition.

Theorem 12.2.9. *Suppose a symmetric design $\mathbf{D}$ admits a uniform global decomposition, i.e., $\mathbf{D}$ admits an incidence matrix $N = [N_{ij}]$ such that each N_{ij} is an incidence matrix of a symmetric $(v_{ij}, k_{ij}, \lambda_{ij})$-design. Suppose further that there exist parameters (v_1, k_1, λ_1) and (v_2, k_2, λ_2) such that, for all i and j, $(v_{ij}, k_{ij}, \lambda_{ij})$ is equal to either (v_1, k_1, λ_1) or (v_2, k_2, λ_2) and every row and every column of blocks of the block matrix $N = [N_{ij}]$ contains exactly one incidence matrix of a symmetric (v_1, k_1, λ_1)-design. Then the design $\mathbf{D}$ admits a tight (v_1, k_1, λ_1)-partition.*

Proof. Without loss of generality, assume that N_{11} is an incidence matrix of a symmetric (v_1, k_1, λ_1)-design $\mathbf{D}_1 = (X_1, \mathcal{B}_1)$. Then every block of $\mathbf{D}$ that is not in $\mathcal{B}_1$ meets X_1 in k_2 points. Therefore, $\mathbf{D}_1$ is a tight (v_1, k_1, λ_1)-subdesign of $\mathbf{D}$. □

Corollary 12.2.10. *Let a symmetric design $\mathbf{D}$ admit an incidence matrix $N = W \otimes_S X$ where X is an incidence matrix of a symmetric (v_2, k_2, λ_2)-design, S is a group of symmetries of a set of matrices containing X, and W is a generalized conference matrix over S. Then the complement of $\mathbf{D}$ admits a tight (v_2, v_2, v_2)-partition.*

Corollary 12.2.10 in conjunction with constructions of globally decomposable symmetric design in Sections 11.3 and 11.4 yields several infinite families of symmetric designs with a tight partition into trivial subdesigns.

We will now turn our attention to subdesigns with incidence matrix $J - I$. The first such example is given by the construction of a symmetric (66, 26, 10)-design in Section 11.1.

Proposition 12.2.11. *There exists a symmetric* (66, 40, 24)*-design with a tight partition into* (11, 10, 9)*-subdesigns.*

Proof. Let **D** be the complement of the symmetric (66, 26, 10)-design constructed in Theorem 11.1.1. This construction shows that **D** satisfies the conditions of Theorem 12.2.9 with $(v_1, k_1, \lambda_1) = (11, 10, 9)$ and $(v_2, k_2, \lambda_2) = (11, 6, 3)$. □

Our next example comes from the symmetric (56, 45, 36)-design constructed in Theorem 6.6.1.

Proposition 12.2.12. *There exists a symmetric* (56, 45, 36)*-design with a tight* (16, 15, 14)*-subdesign.*

Proof. Let a be a point of the 3-(22, 6, 1) design $\mathbf{W}_{22}$ and let $\mathcal{A}$ be the set of all blocks of $\mathbf{W}_{22}$ that do not contain a. Then $\mathbf{D} = (\mathcal{A}, \mathcal{A})$ with blocks $B_1, B_2 \in \mathcal{A}$ incident if and only if $|B_1 \cap B_2| = 2$ is a symmetric (56, 45, 36)-design (Theorem 6.6.1). Fix a block A of $\mathbf{W}_{22}$ that contains the point a and let $\mathcal{A}_1$ be the set of 16 blocks of $\mathbf{W}_{22}$ that are disjoint from A (Theorem 6.5.13). Then $\mathcal{A}_1 \subset \mathcal{A}$. We claim that the substructure $\mathbf{D}_1 = (\mathcal{A}_1, \mathcal{A}_1)$ of **D** is a symmetric (16, 15, 14)-design. It suffices to show that $|B_1 \cap B_2| = 2$ for any distinct blocks $B_1, B_2 \in \mathcal{A}_1$.

Fix a block $B \in \mathcal{A}_1$ and, for $i, j \in \{0, 2\}$, denote by n_{ij} the number of blocks C of $\mathbf{W}_{22}$ such that $|C \cap A| = i$ and $|C \cap B| = j$. Theorem 6.5.13 implies that $n_{00} + n_{02} = n_{00} + n_{20} = 15$ and $n_{20} + n_{22} = n_{02} + n_{22} = 60$. Counting triples (x, y, C) where $x \in A$, $y \in B$, and C is a block of $\mathbf{W}_{22}$ that contains $\{x, y\}$ yields, by Proposition 6.5.12, the equation $4n_{22} = 36 \cdot 5$. Therefore, $n_{22} = 45$, $n_{02} = n_{20} = 15$, and $n_{00} = 0$. Therefore, no two blocks of $\mathcal{A}_1$ are disjoint. This completes the proof of the claim.

Theorem 12.1.13 now implies that the subdesign $\mathbf{D}_1$ is tight. □

Suppose there exists an affine-resolvable (v, b, r, k, λ)-design, and let $N = [N_{ij}]$ be the incidence matrix of a symmetric $(v(r + 1), kr, \lambda r)$-design constructed in Theorem 5.3.15. Since the diagonal entries of matrices $M_1, M_2, \ldots, M_r$ in this proof are equal to 1 and since $M_{r+1} = O$, we obtain that the submatrix of N formed by the $(1, 1)$-entries of all blocks N_{ij} is an incidence matrix of a symmetric $(r+1, r, r-1)$-design. Theorem 12.1.13 implies that this design is a tight subdesign of the symmetric $(v(r+1), kr, \lambda r)$-design. Thus, we have the following result.

Theorem 12.2.13. *If there exists an affine resolvable (v, b, r, k, λ)-design, then there exists a symmetric $(v(r+1), kr, \lambda r)$-design with a tight $(r+1, r, r-1)$-subdesign.*

Using the known affine resolvable designs, that is, the designs $AG_{d-1}(d, q)$ and Hadamard 3-designs, we obtain the following two corollaries of this theorem.

Corollary 12.2.14. *For any prime power q and positive integer d, there exists a symmetric (v, k, λ)-design with a tight $(r+1, r, r-1)$-design, where $r = (q^{d+1}-1)/(q-1)$, $v = (r+1)q^{d+1}$, $k = rq^d$, and $\lambda = (r-1)q^{d-1}$.*

Remark 12.2.15. Corollary 12.2.14 yields a Baer subdesign if and only if $d = 1$.

Corollary 12.2.16. *If $4n$ is the order of a Hadamard matrix, then there exists a symmetric $(16n^2, 8n^2-2n, 4n^2-2n)$-design with a tight $(4n, 4n-1, 4n-2)$-subdesign.*

Another approach to finding subdesigns with incidence matrix $J - I$ is through (v, k, λ)-graphs. The following proposition is immediate.

Proposition 12.2.17. *Let N be an adjacency matrix of a (v, k, λ)-graph Γ. If Γ contains a clique of cardinality v_1, then the submatrix of N corresponding to the vertices of this clique is an incidence matrix of a tight (v_1, v_1-1, v_1-2)-subdesign of the symmetric (v, k, λ)-design with incidence matrix N.*

Theorems 8.2.25 and 12.1.13 now imply the next result.

Theorem 12.2.18. *Let k be a positive integer. If there exists a 2-$(2k^2-k, k, 1)$ design, then there exists a symmetric $(4k^2-1, 2k^2, k^2)$-design with a tight $(2k+1, 2k, 2k-1)$-subdesign.*

Such symmetric designs exist for $k \leq 8$ (cf. Remark 8.2.30).

Let $\mathbf{N} = (X, \mathcal{L})$ be an $(2n, n)$-net, let Γ be the associated strongly regular $(4n^2, 2n^2-n, n^2-n)$-graph, and let $\mathbf{D} = (X, \mathcal{B})$ be the corresponding symmetric $(4n^2, 2n^2-n, n^2-n)$-design (see Proposition 7.4.16). Let $\mathcal{C}$ be a parallel class of $\mathbf{N}$. Then $\mathcal{C}$ can be regarded as a partition of the vertex set X of Γ into $2n$ cliques of cardinality $2n$. The points and the blocks of $\mathbf{D}$ corresponding to each of these cliques form a symmetric $(2n, 2n-1, 2n-2)$-subdesign of $\mathbf{D}$ which is tight by Theorem 12.1.13. Thus, we have the following result.

Theorem 12.2.19. *If there exists an $(2n, n)$-net, then there exists a symmetric $(4n^2, 2n^2 - n, n^2 - n)$-design admitting a tight $(2n, 2n-1, 2n-2)$-partition.*

Remark 12.2.20. The existence of a $(2n, n)$-net is immediate if n is an even prime power. Such a net is also known to exist for $n = 6$.

In most examples of symmetric designs with tight subdesigns in this section, the subdesign has I, J, O, or $J - I$ as incidence matrix. This makes the next two examples especially interesting. We will begin with the following lemma.

Lemma 12.2.21. *Let $\mathcal{B}$ be the block set of the 3-$(22, 6, 1)$ design $\mathbf{W}_{22}$. Let blocks $A_0, A_1 \in \mathcal{B}$ be such that $|A_0 \cap A_1| = 2$ and let $a \in A_0 \setminus A_1$. Let*

$$\mathcal{C}_1 = \{C \in \mathcal{B}\colon |(A_0 \setminus A_1) \cap C| = |(A_1 \setminus A_0) \cap C| = 2\},$$
$$\mathcal{C}_2 = \{C \in \mathcal{C}_1 : a \notin C\},$$
$$\mathcal{C}_3 = \{C \in \mathcal{B}\colon |A_0 \cap C| = |A_1 \cap C| = 2 \text{ and } |A_0 \cap A_1 \cap C| = 1\},$$
$$\mathcal{C}_4 = \{C \in \mathcal{B}\colon |A_0 \cap C| = 2 \text{ and } A_1 \cap C = \varnothing\},$$
$$\mathcal{C}_5 = \{C \in \mathcal{B}\colon |A_1 \cap C| = 2 \text{ and } A_0 \cap C = \varnothing\},$$
$$\mathcal{C}_6 = \{C \in \mathcal{B}\colon (A_0 \cup A_1) \cap C = \varnothing\}.$$

Then $|\mathcal{C}_1| = 12$, $|\mathcal{C}_2| = 6$, $|\mathcal{C}_3| = 32$, $|\mathcal{C}_4| = |\mathcal{C}_5| = 12$, and $|\mathcal{C}_6| = 4$.

Proof. Counting in two ways quadruples (x, y, z, C) with $x, y \in A_0 \setminus A_1$, $z \in A_1 \setminus A_0$, and C a block from $\mathcal{C}_1$ that contains $\{x, y, z\}$ yields $4 \cdot 3 \cdot 4 \cdot 1 = |\mathcal{C}_1| \cdot 2 \cdot 2 \cdot 1$, so $|\mathcal{C}_1| = 12$. If we also require $x \neq a$ and $y \neq a$, the same counting yields $|\mathcal{C}_2| = 6$. For $x \in A_0 \setminus A_1$, $y \in A_1 \setminus A_0$, and $z \in A_0 \cap A_1$, there is a unique block $C \in \mathcal{B}$ that contains $\{x, y, z\}$. This yields $|\mathcal{C}_3| = 32$. The design $\mathbf{W}_{22}$ has 60 blocks that are not disjoint from A_0 (Theorem 6.5.13). This set of 60 blocks is the disjoint union $\mathcal{C}_1 \cup \mathcal{C}_3 \cup \mathcal{C}_4 \cup \mathcal{C}_0$ where $\mathcal{C}_0$ is the set of the four blocks other than A_0 that contain $A_0 \cap A_1$. Therefore, $|\mathcal{C}_4| = 12$. Similarly, $|\mathcal{C}_5| = 12$. The design $\mathbf{W}_{22}$ has 16 blocks disjoint from A_0 (Theorem 6.5.13). Since the set of these 16 blocks is the disjoint union of $\mathcal{C}_5$ and $\mathcal{C}_6$, we obtain that $|\mathcal{C}_6| = 4$. □

Theorem 12.2.22. *The symmetric $(56, 45, 36)$-design of Theorem 6.6.1 has a tight $(7, 3, 1)$-subdesign.*

Proof. Let a be a point of the 3-$(22, 6, 1)$ design $\mathbf{W}_{22}$ and let $\mathcal{A}$ be the set of all blocks of $\mathbf{W}_{22}$ that do not contain a. Let $\mathbf{D} = (\mathcal{A}, \mathcal{A})$ be the symmetric $(56, 45, 36)$-design of Theorem 6.6.1.

Fix a block A_0 of $\mathbf{W}_{22}$ that contains a and let b and c be distinct points of $A_0 \setminus \{a\}$. Let A_1, A_2, A_3, A_4 be the four blocks of $\mathbf{W}_{22}$ other than A_0 that contain $\{b, c\}$. Let X be the set of seven points of $\mathbf{D}$ that consists of A_1 and the six elements of the set $\mathcal{C}_2$ of Lemma 12.2.21. Let $\mathcal{C}$ be the set of seven blocks of $\mathbf{D}$ that consists of A_2, A_3, A_4, and the four elements of the set $\mathcal{C}_6$ of Lemma 12.2.21. We claim that the substructure $(X, \mathcal{C})$ of $\mathbf{D}$ is a symmetric $(7, 3, 1)$-design.

Observe that the 2-subset $\{b, c\}$ and the 4-subsets $Z_i = A_i \setminus \{b, c\}$, $i = 0, 1, 2, 3, 4$, partition the point set of the design $\mathbf{W}_{22}$. Let $A_0 = \{a, b, c, x_1, x_2, x_3\}$ and $A_1 = \{b, c, y_1, y_2, y_3, y_4\}$. There is a unique $B_1 \in X$ such that $x_1, x_2, y_1 \in B_1$. Let $A_1 \cap B_1 = \{y_1, y_2\}$. There is a unique $B_2 \in X$ such that $x_1, x_2, y_3 \in B_2$. Then $A_1 \cap B_2 = \{y_3, y_4\}$. Therefore, if $B \in X \setminus \{A_1, B_1, B_2\}$, then B contains exactly one of the points x_1, x_2, exactly one of the points y_1, y_2, and exactly one of the points y_3, y_4. This implies that the intersection of any two distinct blocks of the set $X \setminus \{A_1\}$ is contained in $Z_0 \cup Z_1$. Therefore, the intersections of these six blocks with $Z_2 \cup Z_3 \cup Z_4$ partition the set $Z_2 \cup Z_3 \cup Z_4$ into six 2-subsets. Since each of the six elements of $X \setminus \{A_1\}$ meets each of the blocks B_2, B_3, B_4 in zero or two points, each of the six 2-subsets that partition $Z_2 \cup Z_3 \cup Z_4$ is contained in Z_2 or Z_3 or Z_4. Let these 2-subsets be Z_{ij} with $i = 2, 3, 4$ and $j = 1, 2$ and let $Z_i = Z_{i1} \cup Z_{i2}$.

Thus, each $B \in X \setminus \{A_1\}$ contains one of the sets Z_{ij} and each Z_{ij} is contained in one block $B \in X \setminus \{A_1\}$. Let $C \in \mathcal{C}$, $B \in X \setminus \{A_1\}$, and $Z_{ij} \subset B$. If $C \cap B = \varnothing$, then $C \cap Z_{ij} = \varnothing$; if $|C \cap B| = 2$, then $C \cap B = Z_{ij}$. Therefore, each of the four blocks $C \in \mathcal{C}_6$ is the union of three 2-subsets Z_{ij}. Since any two of these four blocks meet in at most two points, each 2-subset Z_{ij} is contained in two blocks $C \in \mathcal{C}_6$.

We are now ready to prove that $(X, \mathcal{C})$ is a symmetric $(7, 3, 1)$-design.

Let $B \in X \setminus \{A_1\}$ and let $Z_{ij} \subset B$. Then B is incident with exactly three blocks $C \in \mathcal{C}$, namely, A_i and the two blocks of $\mathcal{C}_6$ that contain Z_{ij}. Since A_1 is incident with A_2, A_3, and A_4 and with no block $C \in \mathcal{C}_6$, the incidence structure $(X, \mathcal{C})$ has constant replication number 3. Since each $C \in \mathcal{C}_6$ is the union of three 2-subsets Z_{ij}, C is incident with three points $B \in X \setminus \{A_1\}$ and C is not incident with A_1. For $i = 1, 2, 3$, the block A_i is incident with A_1 and the two points $B \in X \setminus \{A_1\}$ that contain Z_{i1} and Z_{i2}, respectively. Thus, the incidence structure $(X, \mathcal{C})$ has constant block size 3. Let B_1 and B_2 be distinct elements of $X \setminus \{A_1\}$ and let $Z_{ij} \subset B_1$ and $Z_{kl} \subset B_2$. Since any block $C \in \mathcal{C}$ that is incident with both B_1 and B_2 must contain the 4-subset $Z_{ij} \cup Z_{kl}$, there is at most one such block. The only block of $\mathcal{C}$ incident with both A_1 and B_i is A_i. Thus, any two distinct points of X are incident with at most one block of $\mathcal{C}$. Proposition 2.3.14 now implies that $(X, \mathcal{C})$ is a $(7, 3, 1)$-subdesign of $\mathbf{D}$ and then Theorem 12.1.13 implies that it is a tight subdesign. □

Theorem 12.2.22 and Proposition 12.1.10 imply

Corollary 12.2.23. *The symmetric* $(56, 11, 2)$*-design of Theorem 6.6.1 has a Baer* $(7, 4, 2)$*-subdesign.*

Another example comes from the symmetric $(176, 50, 14)$-design constructed in Section 6.6.

Proposition 12.2.24. *The symmetric* $(176, 50, 14)$*-design of Theorem 6.6.2 has a tight symmetric* $(16, 10, 6)$*-subdesign.*

Proof. Let a and b be distinct points of the Witt design $\mathbf{W_{24}}$. Let $\mathcal{A}$ be the set of all blocks of $\mathbf{W_{24}}$ that contain a and do not contain b. Let $\mathcal{B}$ be the set of all blocks of $\mathbf{W_{24}}$ that contain b and do not contain a. By Theorem 6.6.2, the incidence structure $\mathbf{H} = (\mathcal{A}, \mathcal{B}, I)$ with $(A, B) \in I$ if and only if $|A \cap B| \neq 2$ is a symmetric $(176, 50, 14)$-design. It suffices to show that the complementary symmetric $(176, 126, 90)$-design $\mathbf{H}'$ admits a tight $(16, 6, 2)$-subdesign. Note that, for $A \in \mathcal{A}$ and $B \in \mathcal{B}$, (A, B) is a flag of $\mathbf{H}'$ if and only if $|A \cap B| = 2$.

Fix distinct points c and d of $\mathbf{W_{24}}$ such that $\{c, d\} \cap \{a, b\} = \varnothing$ and let $\mathcal{A}_0 = \{A \in \mathcal{A} \colon c, d \in A\}$ and $\mathcal{B}_0 = \{B \in \mathcal{B} \colon c, d \in B\}$. We claim that the substructure $\mathbf{D} = (\mathcal{A}_0, \mathcal{B}_0)$ of $\mathbf{H}'$ is a symmetric $(16, 6, 2)$-design. Theorem 12.1.13 would then imply that $\mathbf{D}$ is a tight subdesign of $\mathbf{H}'$.

For $i = 3, 4, 5$, let λ_i denote the number of blocks of $\mathbf{W_{24}}$ containing any given set of i points. By Proposition 6.5.12, $\lambda_5 = 1$, $\lambda_4 = 5$, and $\lambda_3 = 21$. We have $|\mathcal{A}_0| = |\mathcal{B}_0| = \lambda_3 - \lambda_4 = 16$.

Let $A \in \mathcal{A}_0$. For $i = 2$ and 4, let m_i denote the number of blocks $B \in \mathcal{B}_0$ such that $|A \cap B| = i$. Then $m_2 + m_4 = |\mathcal{B}_0| = 16$. Counting in two ways flags (x, B) of $\mathbf{W_{24}}$ with $x \in A \setminus \{a, c, d\}$ yields another equation: $2m_4 = 5(\lambda_4 - \lambda_5) = 20$, so $m_4 = 10$ and $m_2 = 6$. Thus, every element of $\mathcal{B}_0$ is incident in $\mathbf{H}'$ with exactly six elements of $\mathcal{A}_0$.

Let $A_1, A_2 \in \mathcal{A}_0$, $A_1 \neq A_2$. By Proposition 2.3.14, it suffices to show that there are at most two blocks $B \in \mathcal{B}_0$ such that $|A_1 \cap B| = |A_2 \cap B| = 2$. Since $a, c, d \in A_1 \cap A_2$, we have $|A_1 \cap A_2| = 4$, so let $S = A_1 \cap A_2 = \{a, c, d, e\}$, $T_1 = A_1 \setminus S$, and $T_2 = A_2 \setminus S$. Let C be the unique block of $\mathbf{W_{24}}$ containing $\{a, b, c, d, e\}$ and let $U = C \setminus S$. The design $\mathbf{W_{24}}$ has five blocks containing $\{a, c, d, e\}$. Three of them are A_1, A_2, and C. Let A_3 and A_4 be the remaining two, and let $T_3 = A_3 \setminus S$ and $T_4 = A_4 \setminus S$. Then $S \cup T_1 \cup T_2 \cup T_3 \cup T_4 \cup U$ is the entire point set of $\mathbf{W_{24}}$.

Let $B \in \mathcal{B}_0$ be such that $|A_1 \cap B| = |A_2 \cap B| = 2$. Then $A_1 \cap B = A_2 \cap B = \{c, d\}$. Therefore, $B \cap S = \{c, d\}$ and $B \cap T_1 = B \cap T_2 = \varnothing$. Since $b, c, d \in B \cap C$, we have $|B \cap C| = 4$, so $|B \cap U| = 2$. Thus, B contains two points of S and two points of U, and the remaining four points of B are elements of $T = T_3 \cup T_4$. Suppose there are three distinct blocks $B_1, B_2, B_3 \in \mathcal{B}_0$ such that $|A_i \cap B_j| = 2$ for $i = 1, 2$ and $j = 1, 2, 3$. Since $|B_j \cap T| = 4$ for $j = 1, 2, 3$ and $|T| = 8$, we obtain that there exists a 2-subset $\{t_1, t_2\}$ of T which is contained in at least two of the blocks B_1, B_2, B_3. Since $b, c, d \in B_1 \cap B_2 \cap B_3$, we have found two blocks of $\mathbf{W}_{24}$ that share at least five points, b, c, d, t_1, and t_2, a contradiction. Therefore, there are at most two blocks of $\mathcal{B}_0$ that are incident in $\mathbf{H}'$ with both A_1 and A_2. The proof is now complete. □

In fact, the following stronger result is true. Its proof is outlined in Exercise 2.

Theorem 12.2.25. *The symmetric* $(176, 50, 14)$*-design of Theorem 6.6.2 admits a tight* $(16, 10, 6)$*-partition.*

We will now return to Theorem 12.1.11 and consider a (v_1, k_1, λ)-subdesign $\mathbf{D}_1$ of a symmetric (v, k, λ)-design $\mathbf{D}$. If $\mathbf{D}_1$ is not a Baer subdesign of $\mathbf{D}$, then $k - \lambda \geq k_1(k_1 - 1)$.

Definition 12.2.26. A (v_1, k_1, λ)-subdesign of a symmetric (v, k, λ)-design is called a *Bruck subdesign* if $k - \lambda = k_1(k_1 - 1)$.

No example of a Bruck subdesign is known for projective planes. To the best of our knowledge, the next theorem gives the only known family of symmetric designs with Bruck subdesigns.

Theorem 12.2.27. *Let q and $q - 1$ be prime powers. Then there exists a symmetric $(q^3 - q + 1, q^2, q)$-design $\mathbf{D}$ with a Bruck (q, q, q)-subdesign.*

Proof. The definition of Bruck subdesigns implies that any (q, q, q)-subdesign of a symmetric $(q^3 - q + 1, q^2, q)$-design is a Bruck subdesign.

Consider the construction of symmetric designs given in Theorem 11.8.1 and let $\mathbf{D}$ be a projective plane of order $q - 1$. For $m = 1$, this construction utilizes matrices $BGW(q + 1, q, q - 1)$ and $GH(q, 1)$, and we assume that these matrices are normalized. Let B be a block of $\mathbf{D}$ and let X and Y be incidence matrices of $\mathbf{D}^B$ and $\mathbf{D}_B$, respectively, corresponding to the same order of blocks. The resulting design is a symmetric $(q^3 - q + 1, q^2, q)$-design $\mathbf{E}$ with an incidence matrix N that can be represented as a block matrix $N = [N_{ij}]$ with $i = 1, 2, \ldots, 2q + 1$ and $j = 1, 2, \ldots, q + 2$ satisfying the following conditions: (i) $N_{1j} = X$ for $j = 2, 3, \ldots, q + 1$, (ii) $N_{q+2,j} = Y$ for $j = 1, 2, \ldots, q$, and (iii) the first row of $N_{q+2,q+1}$ is the same as the first row of Y.

Let x be the point of $\mathbf{D}$ corresponding to the first row of Y and let A be a block of $\mathbf{D}$, other than B, that contains x. Let $R_1, R_2, \ldots, R_{q-1}$ be the rows of X corresponding to the points of $A \setminus B$. For $i = 1, 2, \ldots, q - 1$, let R_i' be the row of N containing row R_i of N_{12}. Let R_q' be the row of N containing the first row of $N_{q+2,1}$. For $j = 2, 3, \ldots, q+1$, let C_j be the column of N_{1j} corresponding to the block A of $\mathbf{D}^B$ and let C'_j be the column of N containing C_j. Then the submatrix of N formed by rows $R'_1, R'_2, \ldots, R'_q$ and columns $C'_2, C'_3, \ldots, C'_{q+1}$ is the all-one matrix of order q. The corresponding substructure of $\mathbf{E}$ is a Bruck (q, q, q)-subdesign. □

12.3. Normal subdesigns

We defined normal subdesigns in Section 5.4. We will now frame this definition in the context of subdesigns of symmetric designs.

Definition 12.3.1. A nontrivial symmetric subdesign $\mathbf{D_1} = (X_1, \mathcal{B}_1)$ of a symmetric design $\mathbf{D} = (X, \mathcal{B})$ is called a *normal subdesign* of $\mathbf{D}$ if X_1 is a proper subset of X and, for every block $B \in \mathcal{B}$, $B \cap X_1$ is either a block of $\mathbf{D}_1$ or the empty set.

We begin with two classical examples of normal subdesigns.

Example 12.3.2. Let q be a prime power and let $\mathbf{D}$ be the complement of $PG_{n-1}(n, q)$ with $n \geq 3$. Let B be a block of $\mathbf{D}$. Propositions 2.4.16 and 3.6.10 imply that the residual design $\mathbf{D}^B$ is a q-fold multiple of a design $\mathbf{D}_1$ isomorphic to the complement of $PG_{n-2}(n-1, q)$. Thus, $\mathbf{D}_1$ is a normal subdesign of $\mathbf{D}$.

Example 12.3.3. Let H be a normalized Hadamard matrix of order $4n$ and let $K = \left[\begin{smallmatrix} H & H \\ H & -H \end{smallmatrix}\right]$. Let C be the core of K and let $N = \frac{1}{2}(J - C)$. Then N is an incidence matrix of a symmetric $(8n-1, 4n, 2n)$-design $\mathbf{D}$ and the submatrix of N formed by the first $4n-1$ rows and columns of N is an incidence matrix of a normal $(4n-1, 2n, n)$-subdesign of $\mathbf{D}$.

The next theorem imposes restrictions on the parameters of normal subdesigns.

Theorem 12.3.4. *If a symmetric (v, k, λ)-design has a normal (v_1, k_1, λ_1)-subdesign, then k_1 divides λ and there is a positive integer q such that $k = k_1 q$ and $\lambda = \lambda_1 q$.*

Proof. Let $\mathbf{D}_1 = (X_1, \mathcal{B}_1)$ be a normal (v_1, k_1, λ_1)-subdesign of a symmetric (v, k, λ)-design $\mathbf{D} = (X, \mathcal{B})$. Let $\mathcal{B}_0$ be the multiset of all nonempty intersections $B \cap X_1$ with $B \in \mathcal{B}$. Then $\mathcal{D}_0 = (X_1, \mathcal{B}_0)$ is a quasi-symmetric 2-(v_1, k_1, λ) design with replication number k and block intersection numbers k_1 and λ_1. By Proposition 8.2.5, $\mathbf{D}$ is isomorphic to the q-fold multiple of $\mathbf{D}_1$ for some q, and then $k = k_1 q$ and $\lambda = \lambda_1 q$.

Theorem 5.4.4 now implies that there exists a positive integer α such that $\mathbf{D}(X \setminus X_1)$ is a PBD that admits an affine resolution with one resolution class $\mathcal{C}_0$ of replication number α, consisting of all blocks of $\mathbf{D}$ that are disjoint from X_1, and the other resolution classes $\mathcal{C}_1, \mathcal{C}_2, \ldots, \mathcal{C}_q$ of cardinality q and replication number $q - \alpha$. Blocks B_1 and B_2 of $\mathbf{D}$ are in the same resolution class of this PBD if and only if $B_1 \cap X_1 = B_2 \cap X_1$. Let $x \in X_1$ and $y \in X \setminus X_1$. For $i = 1, 2, \ldots, q$, if there is a block of $\mathcal{C}_i$ that contains x, then x is contained in every block of $\mathcal{C}_i$ and therefore x and y occur together in exactly $q - \alpha$ blocks of $\mathcal{C}_i$. Since x occurs in k_1 blocks of $\mathbf{D}_1$, there are exactly k_1 classes $\mathcal{C}_i$ whose blocks contain x. Therefore, $k_1(q - \alpha) = \lambda$. This implies that k_1 divides λ. □

Examples 12.3.2 and 12.3.3 give infinite sequences of symmetric designs, in which every design is a normal subdesign of the next design of the sequence. This motivates the following definition.

Definition 12.3.5. An infinite sequence $\{\mathbf{D}_m\}_{m=1}^{\infty}$ of symmetric designs is called a *normal series of symmetric designs* if $\mathbf{D}_{m-1}$ is a normal subdesign of $\mathbf{D}_m$ for each $m \geq 2$.

Examples 12.3.2 and 12.3.3 give normal series of symmetric designs with the following parameters:

(i) $\mathbf{D}_m$ is a symmetric $((q^{m+1} - 1)/(q - 1), q^m, q^m - q^{m-1})$-design where q is a prime power;

(ii) $\mathbf{D}_m$ is a Hadamard 2-$(2^{m+1}n - 1, 2^m n, 2^{m-1}n)$ design where $4n$ is the order of a Hadamard matrix.

Globally decomposable symmetric designs form several normal series.

Proposition 12.3.6. *Let N_1 be an incidence matrix of a symmetric design $\mathbf{D}_1$ and let $\{W_m\}_{m=1}^{\infty}$ be a sequence of normalized BGW-matrices over a group S. Suppose there exists an integer $q \geq 2$ and a sequence $\{C_m\}_{m=1}^{\infty}$ of $(0, S)$-matrices such that, for $m \geq 2$,*

$$W_m = \begin{bmatrix} \mathbf{0}^\top & \mathbf{j}^\top \\ W_{m-1} \otimes J_{q,1} & C_{m-1} \end{bmatrix}.$$

Suppose further that, for all $m \geq 1$, the matrix $W_m \otimes_S N_1$ is defined and is an incidence matrix of a symmetric design $\mathbf{D}_{m+1}$. Then $\{\mathbf{D}_m\}_{m=1}^{\infty}$ is a normal series of symmetric designs.

Proof. For $m \geq 1$, let W_m be a $BGW(w_m, l_m, \mu_m; S)$ and let $N_{m+1} = W_m \otimes_S N_1$. Let v be the order of N_1. For $m \geq 2$, the submatrix of W_m formed by all the rows and the first w_{m-1} columns can be split into a zero submatrix and q submatrices equal to W_{m-1}. Therefore, the submatrix of N_{m+1} formed by all the rows and the first vw_{m-1} columns can be split into a zero submatrix and q submatrices equal to N_m. Since W_1 is normalized, the submatrix of N_2 formed by all the rows and the first v columns can be split into a zero submatrix and l_1 submatrices equal to N_1. Therefore, for all $m \geq 1$, $\mathbf{D}_m$ is a normal subdesign of $\mathbf{D}_{m+1}$. □

If q is a prime power, then Corollary 10.4.28 yields a family of matrices

$$BGW\left(\frac{q^{m+1}-1}{q-1}, q^m, q^m - q^{m-1}; S\right),$$

where S is a cyclic group whose order divides $q-1$. All these matrices satisfy the condition of Proposition 12.3.6. Therefore, the families of symmetric designs obtained in Theorems 11.3.3–11.3.8 and in Theorem 11.4.2 can be obtained as normal series of symmetric designs.

Several more normal series are formed by locally decomposable symmetric designs.

Let $N_1 = \left[\begin{smallmatrix} X & \mathbf{0} \\ Y & \mathbf{j} \end{smallmatrix}\right]$ be an incidence matrix of a symmetric (v, r, λ)-design with r a prime power and let N_{m+1} be the matrix (11.17) described in Theorem 11.7.8. We will assume that the matrix W_m from the proof of this theorem is given by Corollary 10.4.28 (with $q = r$). We will also assume that, for $m \geq 2$, the matrix H_m from the same proof is equal to $H \otimes H_{m-1}$ where H is a normalized $GH(v, 1)$, and that $H_1 = H$. Then, as in the proof of Proposition 12.3.6, we obtain that the submatrix of N_{m+1} formed by all the rows, the first $(v-1)(r^{m+1}-1)/(r-1)$ columns and the last column can be split into a zero submatrix and several submatrices equal to N_m. Therefore, if $\mathbf{D}_m$ is a symmetric design with incidence matrix N_m, then $\{\mathbf{D}_m\}_{m=1}^{\infty}$ is a normal series of symmetric designs. In particular, the constructions of Section 11.8 yield three normal series with parameters (11.19), (11.20), and (11.21).

Remark 12.3.7. Let $\mathbf{D}_1 = (X_1, \mathcal{B}_1)$ and $\mathbf{D}_2 = (X_2, \mathcal{B}_2)$ be symmetric designs with parameters (11.20) with $m = 1$ and $m = 2$, respectively. Then $\mathbf{D}_1$ is a Hadamard 2-$(2q+1, q, (q-1)/2)$ design and $\mathbf{D}_2$ is a symmetric $(2q^2+2q+1, q^2, q(q-1)/2)$-design. The construction of $\mathbf{D}_2$ can be carried on so that $X_1 \subset X_2$ and $\mathbf{D}_2(X_1)$ is a q-fold multiple of $\mathbf{D}_1$. Theorem 5.4.4 implies that $\mathbf{D}_2(X_2 \setminus X_1)$ is a PBD that admits an affine resolution with one resolution class of cardinality $q+1$ and replication number $(q+1)/2$ and $2q+1$ resolution classes of cardinality q and replication number $(q-1)/2$. The complement of this PBD is an affine $(\frac{q+1}{2})$-resolvable PBD embedded in a symmetric $(2q^2+2q+1, (q+1)^2, q(q+1)/2)$-design. This example realizes Case (ii) of Theorem 5.4.11 with $n = 0$ and $2\alpha - 1 = q$, a prime power.

12.4. Symmetric designs with M-arcs

The complement of a block of a symmetric $(4n-1, 2n, n)$-design meets every block of this design in either 0 or n points.

Definition 12.4.1. A proper subset P of the point set of a symmetric (v, k, λ)-design is called an *M-arc* if $|P \cap B| \in \{0, \lambda\}$ for every block B of the design.

Existence of an M-arc imposes restrictions on the parameters of a symmetric design.

Proposition 12.4.2. *Let $\mathbf{D} = (X, \mathcal{B})$ be a symmetric (v, k, λ)-design with $\lambda \geq 1$ having an M-arc P of cardinality m. Let $\mathcal{A}$ be the set of all blocks of $\mathbf{D}$ that meet P in λ points and $\mathcal{A}'$ the set of all blocks of $\mathbf{D}$ disjoint from P. Let $a = |\mathcal{A}|$. Then:*

(i) *the substructure $(P, \mathcal{A})$ of $\mathbf{D}$ is an $(m, a, k, \lambda, \lambda)$-design;*
(ii) *there exists a positive integer s such that $k = \lambda s$, $v = \lambda s^2 - s + 1$, and $m = \lambda s - s + 1$;*
(iii) *the substructure $(X \setminus P, \mathcal{A}')$ of $\mathbf{D}$ is the dual of an $((s-1)^2, (s-1)k, k, s-1, \lambda)$-design.*

Proof. The statement (i) is straightforward, and it implies that $mk = a\lambda$ and $(m-1)\lambda = k(\lambda - 1)$. The latter equation implies that λ divides k, so let $k = \lambda s$. Then $m = \lambda s - s + 1$ and $a = ms$. Since $(v-1)\lambda = k(k-1)$, we obtain that $v = \lambda s^2 - s + 1$.

To prove (iii), fix a point $x \in X \setminus P$ and denote by α the number of blocks of $\mathcal{A}'$ that contain x. Counting in two ways flags (y, B) of $\mathbf{D}$ with $y \in P$ and $x \in B$ yields $m\lambda = (k - \alpha)\lambda$, so $\alpha = k - m = s - 1$. Since $|\mathcal{A}'| = v - a = v - ms = (s-1)^2$, (iii) follows. □

The complement of a block of a symmetric $(4n-1, 2n, n)$-design is an M-arc. Our next example involves a family of locally decomposable symmetric designs.

Theorem 12.4.3. *Let q and $q+1$ be prime powers and let m be a positive integer. Then there exists a symmetric $((q+1)^{m+1} - q, (q+1)^m, (q+1)^{m-1})$-design with an M-arc.*

Proof. Let X be an incidence matrix of an affine plane $\mathbf{A}$ of order q and let the blocks of each parallel class of $\mathbf{A}$ correspond to consecutive columns of X.

Let

$$Y = \begin{bmatrix} 11\dots1 & 00\dots0 & \dots & 00\dots0 \\ 00\dots0 & 11\dots1 & \dots & 00\dots0 \\ \dots & \dots & \dots & \dots \\ 00\dots0 & 00\dots0 & \dots & 11\dots1 \end{bmatrix}$$

be an incidence matrix of a $(q+1, q^2+q, q, 1, 0)$-design. Let S be the group of symmetries considered in the proof of Theorem 11.8.1(i) with $d = 2$. Let m be a positive integer and let $W = [\omega_{ij}]$ be a $BGW(w, r^m, r^m - r^{m-1}; S)$ with $r = q+1$. The proofs of Theorem 11.7.7 and Theorem 11.8.1(i) show that there exists matrix Y_m such that

$$N = \begin{bmatrix} W \otimes_S X & \mathbf{0} \\ Y_m & \mathbf{j} \end{bmatrix}$$

is an incidence matrix of a symmetric $(r^{m+2} - r + 1, r^{m+1}, r^m)$-design $\mathbf{E}$.

Let C be the block of $\mathbf{E}$ corresponding to the last column of N. Then $W \otimes_S X$ is an incidence matrix of the residual design $\mathbf{E}^C$. We represent this matrix as a block matrix $W \otimes_S X = [M_1\ M_2 \dots M_w]$ with each block formed by q^2+q consecutive columns of $W \otimes_S X$. We then represent each M_j as a block matrix $M_j = [M_{j1}\ M_{j2} \dots M_{j,q+1}]$ with each block formed by q consecutive columns of M_j. For $1 \le j \le w$, $1 \le k \le q+1$, and $1 \le l \le q$, let A_{jkl} be the block of $\mathbf{E}^C$ corresponding to the lth column of M_{jk}. Let B_{jkl} be the block of $\mathbf{E}$ containing A_{jkl} and let $C_{jkl} = B_{jkl} \cap C$. For $i = 1, 2, \dots, w$, let Q_i be the set of all points of $\mathbf{E}^C$ corresponding to rows of matrices $\omega_{ij} X$. The action of S described in the proof of Theorem 11.8.1(i) implies that, if $\omega_{ij} \ne 0$, then $\omega_{ij} X$ is an incidence matrix of an affine plane isomorphic to $\mathbf{A}$. The parallel classes of this affine plane are sets $\{A_{jkl} \cap Q_i : 1 \le l \le q\}$ where $k = 1, 2, \dots, q+1$.

Let A_{jkl} and $A_{j'k'l'}$ be distinct blocks of $\mathbf{E}^C$ and let $\alpha = |A_{jkl} \cap A_{j'k'l'}|$. For $i = 1, 2, \dots, w$, let $L_i = A_{jkl} \cap Q_i$ and $L'_i = A_{j'k'l'} \cap Q_i$.

If $j' = j$, $k' = k$, and $l' \ne l$, then L_i and L'_i are either empty sets or distinct parallel lines. Therefore, $\alpha = 0$.

If $j' = j$ and $k' \ne k$, then L_i and L'_i are either empty sets or intersecting lines depending on whether $\omega_{ij} = 0$ or $\omega_{ij} \ne 0$. Therefore, $\alpha = r^m$.

Let $j' \ne j$ and let R be the set of all $i \in \{1, 2, \dots, w\}$ with $\omega_{ij} \ne 0$ and $\omega_{ij'} \ne 0$. Then $|R| = r^m - r^{m-1} = r^{m-1}q$. If $k' \ne k$, then, for each $i \in R$, L_i and L'_i correspond to intersecting lines of $\mathbf{A}$. Therefore, in this case, $\alpha = |R| = r^{m-1}q$. If $k' = k$, then, for $i \in R$, lines L_i and L'_i correspond to parallel lines of $\mathbf{A}$. Let σ be the unique element of S that transforms the lth line of the kth parallel class of $\mathbf{A}$ into the $(l')^{th}$ line of the same parallel class. Then $L_i = L'_i \ne \varnothing$ if and only if $\omega_{ij'}^{-1}\omega_{ij} = \sigma$. Since $W^\top$ is a BGW-matrix with the same parameters as

W, there are exactly $(r^m - r^{m-1})/q = r^{m-1}$ indices $i \in R$ with $L_i = L_i' \neq \varnothing$, and again $\alpha = r^{m-1}q$.

Fix a block B_{jkl} of $\mathbf{E}$, let $Q = \bigcup_{\omega_{ij}=0} Q_i$, and let $P = Q \cup C_{jkl}$. We claim that P is an M-arc of $\mathbf{E}$.

If $j' = j$, $k' = k$, and $l' = l$, then $|B_{j'k'l'} \cap P| = |C_{jkl}| = r^m$.

If $j' = j$, $k' = k$, and $l' \neq l$, then $|B_{j'k'l'} \cap P| = |C_{j'k'l'} \cap C_{jkl}| = |B_{j'k'l'} \cap B_{jkl}| - |A_{j'k'l'} \cap A_{jkl}| = r^m$.

If $j' = j$ and $k' \neq k$, then $|B_{j'k'l'} \cap P| = r^m - |A_{j'k'l'} \cap A_{jkl}| = 0$.

Let $j' \neq j$ and let $R' = \{i \in \{1, 2, \ldots, w\} : \omega_{ij} = 0, \omega_{ij'} \neq 0\}$. Then $|R'| = r^{m-1}$. Therefore, $|B_{j'k'l'} \cap Q| = r^{m-1}q$, and then $|B_{j'k'l'} \cap P| = r^{m-1}q + r^m - |A_{j'k'l'} \cap A_{jkl}| = r^m$.

The proof is now complete. □

We will now consider symmetric designs with the point set partitioned into M-arcs.

Proposition 12.4.4. *If the point set of a symmetric (v, k, λ)-design can be partitioned into M-arcs, then $v = \lambda^3 + 2\lambda^2$ and $k = \lambda^2 + \lambda$.*

Proof. Suppose the point set of a symmetric (v, k, λ)-design can be partitioned into M-arcs. Let m and s be the same as in Proposition 12.4.2. Then $m = \lambda s - s + 1$ divides $v = \lambda s^2 - s + 1$. Therefore, m divides $v - m = \lambda s(s-1)$. Since m and s are relatively prime, we obtain that m divides $\lambda(s-1)$. Then $m \leq \lambda(s-1)$. This implies $s \geq \lambda + 1$, and then $m = (\lambda - 1)s + 1 \geq \lambda^2$.

On the other hand, m divides $v - sm = (s-1)^2 = (\lambda s - m)^2$. Since m and s are relatively prime, m divides λ^2. Therefore, $m = \lambda^2$ and then $s = \lambda + 1$. This implies the required values of k and v. □

For q a prime power, symmetric designs with these parameters were constructed in Theorem 3.8.3.

Theorem 12.4.5. *For any prime power q, there exists a symmetric $(q^3 + 2q^2, q^2 + q, q)$-design whose point set can be partitioned into M-arcs.*

Proof. Let q be a prime power and let $\mathbf{D} = (X, \mathcal{B})$ be the symmetric design of Theorem 3.8.3 with $d = 1$. Let Latin square L, sets F_{ij}, parameter s, and block matrix $M = [M(F_{ij})]$ be the same as in the proof of Theorem 3.8.3. For $i = 1, 2, \ldots, q+2$, let X_i be the set of all points of $\mathbf{D}$ corresponding to the rows of $M(F_{i1})$. Then $\{X_1, X_2, \ldots, X_{q+2}\}$ is a partition of X. For $j = 1, 2, \ldots, q+2$, if B is a block of $\mathbf{D}$ corresponding to a column of $M(F_{1j})$, then

$$|B \cap X_i| = \begin{cases} 0 & \text{if } L(i, j) = s, \\ q & \text{if } L(i, j) \neq s. \end{cases}$$

Therefore, each X_i is an M-arc of $\mathbf{D}$. □

Exercises

(1) Symmetric subdesigns $\mathbf{D}_1 = (X_1, \mathcal{B}_1)$ and $\mathbf{D}_2 = (X_2, \mathcal{B}_2)$ of a symmetric design $\mathbf{D}$ are said to be *independent* if $X_1 \cap X_2 = \varnothing$ and $\mathcal{B}_1 \cap \mathcal{B}_2 = \varnothing$. Prove that the symmetric (56, 45, 36)-design of Theorem 6.6.1 has four pairwise independent (7, 3, 1)-subdesigns.

(2) Let $\mathbf{W}_{24} = (Q, \mathcal{W})$ be the Witt 5-design and let $a, b \in Q$, $a \neq b$. Let $\mathcal{A} = \{A \in \mathcal{W} : a \in A, b \notin A\}$ and $\mathcal{B} = \{B \in \mathcal{W} : b \in B, a \notin B\}$. Let $\mathbf{H}' = (\mathcal{A}, \mathcal{B}, I)$ be the incidence structure with $(A, B) \in I$ if and only if $|A \cap B| = 2$. Then $\mathbf{H}'$ is a symmetric (176, 126, 90)-design (Theorem 6.6.2).

(a) Prove that the set $Q \setminus \{a, b\}$ can be partitioned into 2-subsets P_i, $1 \leq i \leq 11$, so that $\mathbf{W}_{24}$ has the following 13 blocks:

$U_i = \{a, b\} \cup P_1 \cup P_{2i} \cup P_{2i+1}$, for $i = 1, 2, 3, 4, 5$,
$\{a, b\} \cup P_2 \cup P_i \cup P_{i+2}$, for $i = 4, 5, 8, 9$,
$\{a, b\} \cup P_3 \cup P_i \cup P_{i+1}$, for $i = 5, 9$,
$\{a, b\} \cup P_3 \cup P_i \cup P_{i+3}$, for $i = 4, 8$.

(b) For $1 \leq i \leq 7$, let $\mathcal{A}_i = \{A \in \mathcal{A} : P_i \subset A\}$ and $\mathcal{B}_i = \{B \in \mathcal{A} : P_i \subset B\}$. Then the substructure $\mathbf{D}_i = (\mathcal{A}_i, \mathcal{B}_i)$ of $\mathbf{H}'$ is a tight (16, 6, 2)-subdesign (Proposition 12.2.24). Prove that the subdesigns $\mathbf{D}_i$, $1 \leq i \leq 7$, are pairwise independent.

(c) For any block C of $\mathbf{W}_{24}$ containing $\{a, b\}$ and for any 2-subset P of $C \setminus \{a, b\}$, let $\mathcal{A}(C, P) = \{A \in \mathcal{A} : |A \cap C| = 4 \text{ and } A \cap P = \varnothing\}$ and $\mathcal{B}(C, P) = \{B \in \mathcal{B} : |B \cap C| = 4 \text{ and } B \cap P = \varnothing\}$. Prove that the substructure $\mathbf{E}(C, P) = (\mathcal{A}(C, P), \mathcal{B}(C, P))$ of $\mathbf{H}'$ is a tight (16, 6, 2)-subdesign.

(d) Let $\mathbf{D}_8 = \mathbf{E}(U_4, P_1)$ and $\mathbf{D}_9 = \mathbf{E}(U_5, P_1)$. Prove that the subdesigns $\mathbf{D}_i$, $1 \leq i \leq 9$, are pairwise independent.

(c) Prove that the complement of $\mathbf{H}'$ admits a tight (16, 10, 6)-partition.

Notes

The notion of Baer subplane was first studied in Baer (1946). The classical result on the possible orders of subplanes of finite projective planes (Corollary 12.1.12) is due to Bruck (1955). Kantor (1969a, Lemma 9.5) generalized it to symmetric designs (Theorem 12.1.11) though the term *Baer subdesigns* first appears in Bose and S. S. Shrikhande (1976). Theorem 12.2.2 is due to Bruck (1960). For further results on Baer partitions of projective geometries, see Baker, Dover, Ebert and Wantz (2000).

The notion of tight design was introduced in Haemers and M. S. Shrikhande (1979). We use an equivalent definition proposed in Jungnickel (1982a). Theorem 12.1.13 is due to Jungnickel (1982a). Theorem 12.1.15 is in Haemers and M. S. Shrikhande (1979). The proof relies on the technique of interlacing of eigenvalues developed earlier in Haemers (1978).

Proposition 12.2.5 and 12.2.12 and Theorem 12.2.13 can be found in Jungnickel (1982a). Proposition 12.2.18 and Theorem 12.2.19 are due to Baartmans and M. S. Shrikhande (1981). Our proof of Theorem 12.2.22 is different from the one in Haemers and M.S. Shrikhande (1979). Proposition 12.2.24 and Theorem 12.2.25 are proven by Klee and Yates (2004).

The generalization of Baer partitions to symmetric designs is due to Cron and Mavron (1983). This paper also mentions that the symmetric (56, 11, 2)-design of Hall, Lane and Wales (which admits a Baer (7, 4, 2)-subdesign by Corollary 12.2.23) does not admit a Baer partition into such subdesigns.

The notion of Bruck subdesign was introduced in Baker (1982) where the properties and constructions of such subdesigns were considered. In particular, Theorem 12.2.27 is proved there, though our construction is different.

Normal subdesigns were introduced as *strong subdesigns* in Ionin (1999a) where Theorem 12.3.4 is proved. Normal subdesigns corresponding to difference sets are considered in Jungnickel and Tonchev (1999a).

For the latest survey on symmetric subdesigns of symmetric designs, see M. S. Shrikhande (2002).

The notion of M-arc was introduced (as λ-*arc*) by Mavron (1988) as a special case of an *arc of a symmetric design* introduced by Sane, S.S. Shrikhande and Singhi (1985). All theorems of Section 12.4. are due to Mavron (1988) though our proofs are sometimes different.

13
Non-embeddable quasi-residual designs

Quasi-residual designs can be useful in constructing symmetric designs. However, there are quasi-residual designs that cannot be embedded in a symmetric design. This may happen because the corresponding symmetric design does not exist. Another reason for nonembeddability could be the existence of several blocks whose intersection sizes do not allow them to be extended to blocks of a symmetric design. Also, a quasi-residual design may have a substructure preventing it from embeddability in a symmetric design.

13.1. Quasi-residuals of non-existing symmetric designs

Recall that a (v, b, r, k, λ)-design is called *quasi-residual* if $r = k + \lambda$. This condition is satisfied by any residual design of a symmetric $(v + r, r, \lambda)$-design. If a quasi-residual design is isomorphic to a residual of a symmetric design **D**, it is said to be *embeddable in* **D**. In this section we will construct several families of quasi-residual (v, b, r, k, λ)-designs for which a symmetric $(v + r, r, \lambda)$-design does not exist.

To obtain the first family of such designs, let **D** be the $(v - 1)$ fold multiple of the complete symmetric $(v, v - 1, v - 2)$-design. Then **D** is a $(v, v(v - 1), (v - 1)^2, v - 1, (v - 1)(v - 2))$-design, so **D** is quasi-residual. If **D** is embeddable in a symmetric design, then the complement of this symmetric design is a projective plane of order $v - 1$. Now the Bruck–Ryser Theorem gives an infinite family of non-embeddable quasi-residual designs.

Proposition 13.1.1. *If there is no projective plane of order* $v - 1$, *then there exists a non-embeddable quasi-residual* $(v, v(v - 1), (v - 1)^2, v - 1, (v - 1)(v - 2))$*-design.*

Next consider the complete design on v points with block size 2. Its complement is a quasi-residual $(v, v(v-1)/2, (v-1)(v-2)/2, v-2, (v-2)(v-3)/2)$-design. If it is embeddable in a symmetric design, the complement of this symmetric design is a biplane of order $v-2$. Proposition 2.4.10 and the Bruck–Ryser–Chowla Theorem provide another infinite family of nonembeddable quasi-residual designs.

Proposition 13.1.2. *If there is no biplane of order $v-2$, then there exists a non-embeddable quasi-residual $(v, v(v-1)/2, (v-1)(v-2)/2, v-2, (v-2)(v-3)/2)$-design.*

For any $k \leq 5$ and for any r such that $r(r+1) \equiv 0 \pmod{k}$, there exists an $(r+1, r(r+1)/k, r, k, k-1)$-design (see Remark 2.3.11). Its complement is a quasi-residual design. If this quasi-residual design is embeddable, then there exists a symmetric design with parameters

$$\left(1 + \frac{r(r+1)}{k}, \frac{r(r+1-k)}{k}, \frac{(r+1-k)(r-k)}{k}\right). \tag{13.1}$$

However, for infinitely many values of r such a design does not exist due to the Bruck–Ryser–Chowla Theorem or Proposition 2.4.10. This gives another infinite family of non-embeddable quasi-residual designs.

Proposition 13.1.3. *Let positive integers k and r be such that $k \leq 5$ and $r(r+1) \equiv 0 \pmod{k}$. If there is no symmetric design with parameters (13.1), then there exists a non-embeddable quasi-residual design with parameters*

$$\left(r+1, \frac{r(r+1)}{k}, \frac{r(r+1-k)}{k}, r+1-k, \frac{(r+1-k)(r-k)}{k}\right).$$

The following theorem uses resolvable unitals to obtain non-embeddable quasi-residual designs.

Theorem 13.1.4. *If $q \neq 3$ is an odd prime power such that $q-1$ is a prime power, then there exists a non-embeddable quasi-residual design with parameters*

$$(q^3+1, q^2(q^2-q+1), q^2(q-1)^2, (q+1)(q-1)^2, (q-1)^2(q^2-q-1)). \tag{13.2}$$

Proof. Let q and $q-1$ be prime powers. By Theorem 5.3.9, there exists a resolvable design $\mathbf{D}_1$ with parameters $(q^3+1, q^2(q^2-q+1), q^2, q+1, 1)$. Let $\mathbf{D}_2$ be the complement of a projective plane of order $q-1$, so $\mathbf{D}_2$ is a symmetric $(q^2-q+1, (q-1)^2, (q-1)(q-2))$-design. Since q^2-q+1 is the cardinality of a parallel class of $\mathbf{D}_1$, we can apply Theorem 5.3.10 to obtain a design with parameters (13.2), which is quasi-residual. If this design is embeddable, then there exists a symmetric design **S** with parameters

$$(q^2(q^2-q+1)+1, q^2(q-1)^2, (q-1)^2(q^2-q-1)).$$

However, if q is odd, then the number of points of **S** is even, while the order of **S** is equal to $(q-1)^2(q+1)$, and it is not a square unless $q=3$. Therefore, if q is odd and $q \neq 3$, such a design does not exist by Proposition 2.4.10. □

13.2. Linear non-embeddability conditions

In Section 2.4 we gave an example of a non-embeddable quasi-residual 2-(16, 6, 3) design **D** corresponding to a symmetric (25, 9, 3)-design (Example 2.4.18). In this example, **D** has a pair of blocks meeting in four points, and therefore, this design cannot be embedded in a symmetric (v, k, λ)-design with $\lambda = 3$. In general, if **D** is a quasi-residual 2-(v, k, λ) design with distinct blocks B_1 and B_2 such that $|B_1 \cap B_2| - \lambda > 0$, then **D** cannot be embedded in a symmetric design. In this section we will develop several non-embeddability conditions that are expressed as linear inequalities involving block intersection numbers and the parameters k and λ of a quasi-residual 2-(v, k, λ) design.

Definition 13.2.1. Let $\mathcal{P}_m$ be the set of all linear polynomials

$$f = a + a_0x_0 + \sum_{i=1}^{m}\sum_{j=1}^{m} a_{ij}x_{ij}$$

in $1+m^2$ variables x_0, x_{ij} with integer coefficients a_0, a_{ij} and the free term a equal to 0 or 1. Let F be a subset of $\mathcal{P}_m$. We will say that a 2-(v, k, λ) design **D** satisfies the inequality $F > 0$ if D has m distinct blocks $B_1, B_2, \ldots, B_m$ such that, if $x_0 = \lambda$ and $x_{ij} = |B_i \cap B_j|$ for $i, j = 1, 2, \ldots, m$, then the value of each polynomial $f \in F$ is positive.

We will call the set F an *m block non-embeddability condition* if every quasi-residual design which satisfies the inequality $F > 0$ is not embeddable in a symmetric design.

Remark 13.2.2. If $a = 0$, then the polynomial f of this definition is homogeneous; if $a = 1$, then $f - 1$ is homogeneous. Since all values of f are integers, the requirement that $a = 0$ or $a = 1$ is equivalent to considering inequalities $f > 0$ or $f \geq 0$, respectively, with a homogeneous polynomial f. Also note that the condition $F > 0$ involves the parameter k of the design **D** as the value of x_{ii}.

Example 13.2.3 (Two block non-embeddability condition). Let $F = \{-x_0 + x_{12}\}$. Suppose a quasi-residual 2-(v, k, λ) design **D** satisfies the inequality

$F > 0$, i.e., the design has distinct blocks B_1 and B_2 such that $|B_1 \cap B_2| - \lambda > 0$. This is precisely the condition that warrants the non-embeddability of the design of Example 2.4.18.

The next two propositions give m block non-embeddability conditions for $m = 3$ and $m = 5$, respectively.

Proposition 13.2.4. *The singleton* $F = \{-x_0 - x_{11} + x_{12} + x_{13} + x_{23}\}$ *is a three block non-embeddability condition.*

Proof. Suppose that a quasi-residual (v, b, r, k, λ)-design has distinct blocks B_1, B_2, and B_3 satisfying $F > 0$. Then

$$|B_1 \cap B_2| + |B_1 \cap B_3| + |B_2 \cap B_3| > r. \tag{13.3}$$

Suppose that this design is embeddable in a symmetric $(v + r, r, \lambda)$-design **D**. For $i = 1, 2, 3$, let A_i be the block of **D** that contains B_i and let $C_i = A_i \setminus B_i$. Then, for $i \neq j$, $|C_i \cap C_j| = \lambda - |B_i \cap B_j|$. Therefore,

$$\begin{aligned} r &\geq |C_1 \cup C_2 \cup C_3| \geq |C_1| + |C_2| + |C_3| \\ &\quad - (|C_1 \cap C_2| + |C_1 \cap C_3| + |C_2 \cap C_3|) \\ &= 3(r - k) - 3\lambda + (|B_1 \cap B_2| + |B_1 \cap B_3| + |B_2 \cap B_3|) \\ &= |B_1 \cap B_2| + |B_1 \cap B_3| + |B_2 \cap B_3| > r, \end{aligned}$$

a contradiction. □

We will now give three examples of quasi-residual designs satisfying the condition of Proposition 13.2.4.

Example 13.2.5. Let $X = \{1, 2, 3, \dots, 12\}$ and let $\mathcal{B} = \{B_1, B_2, \dots, B_{22}\}$ be the set of the following 22 subsets of X:

$$\begin{array}{ll}
B_1 = \{3, 5, 6, 8, 10, 11\} & B_2 = \{1, 2, 6, 8, 10, 11\} \\
B_3 = \{2, 3, 6, 8, 11, 12\} & B_4 = \{1, 2, 3, 7, 8, 9\} \\
B_5 = \{4, 5, 6, 7, 8, 12\} & B_6 = \{1, 2, 4, 7, 8, 10\} \\
B_7 = \{1, 2, 5, 7, 11, 12\} & B_8 = \{3, 4, 6, 7, 8, 9\} \\
B_9 = \{3, 4, 5, 7, 10, 11\} & B_{10} = \{1, 3, 4, 9, 10, 11\} \\
B_{11} = \{2, 5, 6, 7, 9, 10\} & B_{12} = \{1, 3, 5, 8, 9, 12\} \\
B_{13} = \{2, 4, 6, 9, 10, 12\} & B_{14} = \{1, 3, 6, 7, 10, 12\} \\
B_{15} = \{2, 4, 5, 8, 9, 11\} & B_{16} = \{1, 4, 5, 8, 10, 12\} \\
B_{17} = \{1, 4, 6, 9, 11, 12\} & B_{18} = \{2, 3, 5, 9, 10, 12\} \\
B_{19} = \{1, 5, 6, 7, 9, 11\} & B_{20} = \{2, 3, 4, 7, 11, 12\} \\
B_{21} = \{1, 2, 3, 4, 5, 6\} & B_{22} = \{7, 8, 9, 10, 11, 12\}
\end{array}$$

It can be checked that $\mathbf{D} = (X, \mathcal{B})$ is a 2-(12, 6, 5) design. Blocks B_1, B_2, and B_3 satisfy the condition of Proposition 13.2.4. Therefore, the design **D** is not embeddable.

Example 13.2.6. Let $X = \{1, 2, 3, \ldots, 16\}$ and let $\mathcal{B} = \{B_1, B_2, \ldots, B_{30}\}$ be the set of the following 30 subsets of X:

$B_1 = \{2, 3, 5, 7, 11, 12, 14, 15\}$ $B_2 = \{1, 2, 3, 4, 11, 12, 14, 16\}$
$B_3 = \{1, 2, 3, 4, 11, 13, 14, 15\}$ $B_4 = \{5, 6, 7, 8, 11, 12, 14, 16\}$
$B_5 = \{5, 6, 7, 8, 11, 13, 14, 15\}$ $B_6 = \{1, 2, 3, 4, 5, 6, 7, 8\}$
$B_7 = \{9, 10, 11, 12, 13, 14, 15, 16\}$ $B_8 = \{1, 2, 5, 6, 9, 10, 11, 12\}$
$B_9 = \{3, 4, 7, 8, 9, 10, 11, 12\}$ $B_{10} = \{1, 2, 5, 7, 9, 10, 11, 13\}$
$B_{11} = \{3, 4, 6, 8, 9, 10, 11, 14\}$ $B_{12} = \{1, 2, 6, 8, 9, 10, 14, 15\}$
$B_{13} = \{3, 4, 5, 7, 9, 10, 14, 15\}$ $B_{14} = \{1, 2, 7, 8, 9, 12, 14, 16\}$
$B_{15} = \{3, 4, 5, 6, 9, 12, 13, 16\}$ $B_{16} = \{1, 3, 5, 6, 9, 14, 15, 16\}$
$B_{17} = \{2, 4, 7, 8, 9, 13, 15, 16\}$ $B_{18} = \{1, 3, 5, 8, 9, 11, 13, 16\}$
$B_{19} = \{2, 4, 6, 7, 9, 11, 15, 16\}$ $B_{20} = \{1, 3, 6, 7, 9, 12, 13, 15\}$
$B_{21} = \{2, 4, 5, 8, 9, 12, 13, 14\}$ $B_{22} = \{1, 3, 7, 8, 10, 12, 15, 16\}$
$B_{23} = \{2, 4, 5, 6, 10, 12, 15, 16\}$ $B_{24} = \{1, 4, 5, 7, 10, 13, 14, 16\}$
$B_{25} = \{2, 3, 6, 8, 10, 13, 14, 16\}$ $B_{26} = \{1, 4, 5, 8, 10, 11, 15, 16\}$
$B_{27} = \{2, 3, 6, 7, 10, 11, 13, 16\}$ $B_{28} = \{1, 4, 6, 7, 10, 12, 13, 14\}$
$B_{29} = \{2, 3, 5, 8, 10, 12, 13, 15\}$ $B_{30} = \{1, 4, 6, 8, 11, 12, 13, 15\}$

It can be checked that $\mathbf{D} = (X, \mathcal{B})$ is a 2-(16, 8, 7) design. Blocks B_1, B_2, and B_3 satisfy the condition of Proposition 13.2.4. Therefore, the design $\mathbf{D}$ is not embeddable.

Example 13.2.7. Let $X = \{1, 2, 3, \ldots, 20\}$ and let $\mathcal{B} = \{B_1, B_2, \ldots, B_{38}\}$ be the set of the following 38 subsets of X:

$B_1 = \{2, 5, 6, 7, 9, 13, 14, 15, 17, 20\}$ $B_2 = \{1, 2, 5, 7, 9, 13, 14, 16, 17, 20\}$
$B_3 = \{1, 5, 6, 7, 10, 12, 14, 17, 18, 20\}$ $B_4 = \{1, 2, 3, 4, 5, 11, 12, 13, 14, 15\}$
$B_5 = \{6, 7, 8, 9, 10, 11, 12, 13, 14, 15\}$ $B_6 = \{1, 7, 8, 9, 10, 11, 12, 13, 16, 20\}$
$B_7 = \{2, 3, 4, 5, 6, 11, 12, 13, 16, 20\}$ $B_8 = \{1, 2, 3, 7, 8, 11, 12, 13, 17, 18\}$
$B_9 = \{4, 5, 6, 9, 10, 11, 12, 13, 17, 18\}$ $B_{10} = \{1, 4, 5, 9, 1, 11, 12, 14, 15, 19\}$
$B_{11} = \{2, 3, 6, 7, 8, 11, 12, 14, 19, 20\}$ $B_{12} = \{1, 2, 3, 9, 10, 11, 13, 16, 17, 19\}$
$B_{13} = \{4, 5, 6, 7, 8, 11, 13, 16, 17, 19\}$ $B_{14} = \{1, 4, 5, 7, 8, 11, 14, 17, 18, 19\}$
$B_{15} = \{2, 3, 6, 9, 10, 11, 14, 17, 18, 19\}$ $B_{16} = \{1, 2, 4, 6, 7, 11, 14, 15, 16, 18\}$
$B_{17} = \{3, 5, 8, 9, 10, 11, 14, 16, 18, 20\}$ $B_{18} = \{1, 3, 5, 6, 8, 11, 15, 16, 18, 20\}$
$B_{19} = \{2, 4, 7, 9, 10, 11, 15, 16, 18, 20\}$ $B_{20} = \{1, 3, 5, 6, 9, 11, 15, 17, 19, 20\}$
$B_{21} = \{2, 4, 7, 8, 10, 11, 15, 17, 19, 20\}$ $B_{22} = \{1, 2, 4, 6, 10, 12, 13, 18, 19, 20\}$
$B_{23} = \{3, 5, 7, 8, 9, 12, 13.15, 18, 19\}$ $B_{24} = \{2, 3, 4, 8, 9, 12, 14, 17, 18, 20\}$
$B_{25} = \{1, 4, 6, 8, 9, 12, 14, 16, 19, 20\}$ $B_{26} = \{2, 3, 5, 7, 10, 12, 14, 15, 16, 19\}$
$B_{27} = \{1, 3, 6, 7, 10, 12, 15, 16, 17, 18\}$ $B_{28} = \{2, 4, 5, 8, 9, 12, 15, 16, 17, 18\}$
$B_{29} = \{1, 2, 6, 8, 9, 12, 15, 16, 17, 19\}$ $B_{30} = \{3, 4, 5, 7, 10, 12, 16, 17, 19, 20\}$
$B_{31} = \{1, 3, 4, 8, 10, 13, 14, 15, 17, 20\}$ $B_{32} = \{3, 4, 6, 8, 10, 13, 14, 15, 16, 17\}$
$B_{33} = \{1, 2, 5, 8, 10, 13, 15, 18, 19, 20\}$ $B_{34} = \{3, 4, 6, 7, 9, 13, 15, 18, 19, 20\}$
$B_{35} = \{1, 3, 4, 7, 9, 13, 14, 16, 18, 19\}$ $B_{36} = \{2, 5, 6, 8, 10, 13, 14, 16, 18, 19\}$
$B_{37} = \{1, 2, 3, 4, 5, 6, 7, 8, 9, 10\}$ $B_{38} = \{11, 12, 13, 14, 15, 16, 17, 18, 19, 20\}$

It can be checked that $\mathbf{D} = (X, \mathcal{B})$ is a 2-(20, 10, 9) design. Blocks B_1, B_2, and B_3 satisfy the condition of Proposition 13.2.4. Therefore, the design $\mathbf{D}$ is not embeddable.

Proposition 13.2.8. *The set of linear polynomials* $F = \{1 + x_{12} - x_{13} - x_{23}, 1 + x_{12} - x_{14} - x_{24}, 1 + x_{34} - x_{35} - x_{45}, 1 - x_{12} + x_{15}, x_{25}\}$ *is a five-block non-embeddability condition.*

Proof. Let $\mathbf{D}$ be a quasi-residual 2-(v, k, λ) design that satisfies the inequality $F > 0$. This means that $\mathbf{D}$ has five distinct blocks B_1, B_2, B_3, B_4, and B_5, such that $|B_i \cap B_j| \geq |B_i \cap B_h| + |B_j \cap B_h|$ for $(i, j, h) = (1, 2, 3), (1, 2, 4)$, and $(3, 4, 5)$, $|B_1 \cap B_2| \leq |B_1 \cap B_5|$, and $B_2 \cap B_5 \neq \varnothing$.

Suppose that $\mathbf{D}$ is embeddable in a symmetric design $\mathbf{E}$. For $i = 1, 2, 3, 4, 5$, let B_i' be the block of $\mathbf{E}$ which contains B_i, and let $C_i = B_i' \setminus B_i$. Then $|C_i| = \lambda$ and, for $1 \leq i < j \leq 5$, $|C_i \cap C_j| = \lambda - |B_i \cap B_j|$.

Claim. Let (i, j, h) be any of the triples $(1, 2, 3)$, $(1, 2, 4)$, and $(3, 4, 5)$. Then

(i) $|B_i \cap B_j| = |B_i \cap B_h| + |B_j \cap B_h|$;
(ii) $C_i \cap C_j \subseteq C_h \subseteq C_i \cup C_j$;
(iii) $|(C_i \setminus C_j) \cap C_h| = |B_j \cap B_h|$;
(iv) $|(C_j \setminus C_i) \cap C_h| = |B_i \cap B_h|$.

To prove the claim, let $x_1 = |C_i \cap C_j \cap C_h|$, $x_2 = |(C_i \setminus C_j) \cap C_h|$, $x_3 = |(C_j \setminus C_i) \cap C_h|$, and $x_4 = |C_h \setminus (C_i \cup C_j)|$. Then

$$x_1 + x_2 = |C_i \cap C_h| = \lambda - |B_i \cap B_h|,$$
$$x_1 + x_3 = |C_j \cap C_h| = \lambda - |B_j \cap B_h|,$$
$$x_1 + x_2 + x_3 + x_4 = |C_h| = \lambda.$$

These equations imply

$$x_1 - x_4 = \lambda - |B_i \cap B_h| - |B_j \cap B_h| \geq \lambda - |B_i \cap B_j| = |C_i \cap C_j|.$$

Since, on the other hand, $x_1 \leq |C_i \cap C_j|$, we obtain that $x_4 = 0$ and $x_1 = |C_i \cap C_j|$, and then (i) and (ii) follow. Furthermore, $x_2 = \lambda - |B_i \cap B_h| - x_1 = |B_j \cap B_h|$ and $x_3 = \lambda - |B_j \cap B_h| - x_1 = |B_i \cap B_h|$, so the claim is true.

Let $S_0 = C_1 \cap C_2$ and $S = C_1 \cup C_2$. By the above claim, $S_0 \subseteq C_3 \subseteq S$ and $S_0 \subseteq C_4 \subseteq S$. We partition the set $S \setminus S_0$ into subsets $S_1 = (C_1 \setminus C_2) \cap C_3$, $S_2 = (C_1 \setminus C_2) \setminus C_3$, $S_3 = (C_2 \setminus C_1) \cap C_3$, and $S_4 = (C_2 \setminus C_1) \setminus C_3$. For $i = 1, 2, 3, 4$, let $y_i = |S_i \cap C_4|$. Since $C_3 \cap C_4 = S_0 \cup (S_1 \cap C_4) \cup (S_3 \cap C_4)$, we have $\lambda - |B_3 \cap B_4| = \lambda - |B_1 \cap B_2| + y_1 + y_3$, so

$$y_1 + y_3 = |B_1 \cap B_2| - |B_3 \cap B_4|. \tag{13.4}$$

Let $T_0 = C_3 \cap C_4$ and $T = C_3 \cup C_4$. Then $S_0 \subseteq T_0$, $|C_1 \cap T_0| = |S_0| + y_1$, and $|C_2 \cap T_0| = |S_0| + y_3$. By the above claim, $T_0 \subseteq C_5 \subseteq T$. We partition the set $T \setminus T_0$ into subsets $T_1 = (C_1 \cap C_3) \setminus (C_4 \cup C_2)$, $T_2 = (C_1 \cap C_4) \setminus (C_3 \cup C_2)$, $T_3 = (C_2 \cap C_3) \setminus (C_4 \cup C_1)$, and $T_4 = (C_2 \cap C_4) \setminus (C_3 \cup C_1)$. For $i = 1, 2, 3, 4$, let $z_i = |T_i \cap C_5|$. Since $C_1 \cap C_5 = (C_1 \cap T_0) \cup (T_1 \cap C_5) \cup (T_2 \cap C_5)$, we have $\lambda - |B_1 \cap B_5| = \lambda - |B_1 \cap B_2| + y_1 + z_1 + z_2$, so

$$y_1 + z_1 + z_2 = |B_1 \cap B_2| - |B_1 \cap B_5| \leq 0,$$

which implies $y_1 = z_1 = z_2 = 0$. Therefore, $C_3 \cap C_5 = T_0 \cup (T_3 \cap C_5)$ and $C_4 \cap C_5 = T_0 \cup (T_4 \cap C_5)$. Then we have $z_3 = |C_3 \cap C_5| - |C_3 \cap C_4| = |B_3 \cap B_4| - |B_3 \cap B_5| = |B_4 \cap B_5|$ and, similarly, $z_4 = |B_3 \cap B_5|$. Therefore,

$$z_3 + z_4 = |B_4 \cap B_5| + |B_3 \cap B_5| = |B_3 \cap B_4|. \tag{13.5}$$

Since $C_2 \cap C_5 = (C_2 \cap T_0) \cup (T_3 \cap C_5) \cup (T_4 \cap C_5)$, we obtain that

$$\lambda - |B_2 \cap B_5| = \lambda - |B_1 \cap B_2| + y_3 + z_3 + z_4.$$

Since $y_1 = 0$, (13.4) now implies that $z_3 + z_4 = |B_3 \cap B_4| - |B_2 \cap B_5|$, and then (13.5) implies that $B_2 \cap B_5 = \varnothing$, a contradiction. □

Example 13.2.9. In this example we describe a quasi-residual 2-(27, 9, 4) design $\mathbf{D}$ satisfying the conditions of Proposition 13.2.8. The point set V of $\mathbf{D}$ is the set of points of the three-dimensional vector space over $GF(3) = \{0, 1, 2\}$, i.e., the set of triples (x, y, z) with $x, y, z \in GF(3)$. A plane in V is given by an equation $ax + by + cz = d$ with coefficients a, b, c, and d equal to 0, 1, 2 and not equal to 0 simultaneously. To have a unique equation for each plane, we assume the following: $a = 0$ or 1; if $a = 0$, then $b = 0$ or 1; if $a = b = 0$, then $c = 1$. We will denote by $P(a, b, c, d)$ the plane with this equation and by $A(a, b, c, d)$ the subset of $P(a, b, c, d)$ consisting of all points with the third coordinate not equal to 2. If we select the set of planes $P(a, b, c, d)$ as the block set, we obtain the design $AG_2(3, 3)$, which is, of course, embeddable in a symmetric (40, 13, 4)-design. To obtain the desired non-embeddable design, we will keep the part $A(a, b, c, d)$ of each plane $P(a, b, c, d)$, except the planes $z = 0$, $z = 1$, and $z = 2$, and change the remaining three points. We put $B(0, 0, 1, d) = P(0, 0, 1, 2)$ for $d = 0, 1, 2$. If $a \neq 0$ or $b \neq 0$, we define $B(a, b, c, d) = A(a, b, c, d) \cup C(a, b, c, d)$ where sets $C(a, b, c, d)$ are the subsets of cardinality 3 of the plane $z = 2$ that are described below. The third coordinate of every point of sets $C(a, b, c, d)$ is equal to 2, the first coordinate

x and the second coordinate y are given as xy in the following table:

abcd	$C(a, b, c, d)$	*abcd*	$C(a, b, c, d)$	*abcd*	$C(a, b, c, d)$
0100	10, 11, 12	0101	10, 20, 12	0102	00, 10, 12
0110	20, 21, 22	0111	00, 21, 22	0112	11, 21, 22
0120	00, 01, 02	0121	01, 11, 02	0122	20, 01, 02
1000	02, 12, 22	1001	02, 12, 22	1002	02, 12, 22
1010	10, 21, 02	1011	10, 21, 02	1012	10, 21, 02
1020	10, 01, 22	1021	10, 01, 22	1022	10, 01, 22
1100	01, 11, 21	1101	20, 01, 21	1102	00, 01, 21
1110	20, 01, 12	1111	00, 01, 12	1112	01, 11, 12
1120	00, 21, 12	1121	11, 21, 12	1122	20, 21, 12
1200	00, 10, 20	1201	00, 10, 11	1202	10, 20, 11
1210	00, 11, 22	1211	20, 11, 22	1212	00, 20, 22
1220	20, 11, 02	1221	00, 20, 02	1222	00, 11, 02

It can be checked that the design **D** with point set V and blocks $B(a, b, c, d)$ is indeed a 2-(27, 9, 4) design. Let $B_1 = B(1, 1, 0, 0)$, $B_2 = B(0, 1, 1, 2)$, $B_3 = B(1, 2, 0, 0)$, $B_4 = B(0, 1, 0, 0)$, and $B_5 = B(0, 1, 2, 1)$. The (i, j)-entry of the following 5×5 matrix is equal to $|B_i \cap B_j|$:

$$\begin{bmatrix} 9 & 4 & 2 & 3 & 4 \\ 4 & 9 & 2 & 1 & 1 \\ 2 & 2 & 9 & 3 & 2 \\ 3 & 1 & 3 & 9 & 1 \\ 4 & 1 & 2 & 1 & 9 \end{bmatrix}$$

One can verify that blocks B_i $(i = 1, 2, 3, 4, 5)$ satisfy Proposition 13.2.8, and therefore **D** is non-embeddable.

13.3. BGW-matrices and non-embeddability

In this section we apply balanced generalized weighing matrices to obtain several infinite families of non-embeddable quasi-residual designs.

We begin with the following construction.

Proposition 13.3.1. *Let* **D** *be a* 2-$(2\lambda + 2, \lambda + 1, \lambda)$ *design satisfying the inequality* $F > 0$ *with* $F = \{-x_0 - x_{11} + x_{12} + x_{13} + x_{23}\}$. *Let* X *be an incidence matrix of* **D**. *Then the block matrix*

$$N = \begin{bmatrix} X & X & \mathbf{j} & \mathbf{0} \\ X & J - X & \mathbf{0} & \mathbf{j} \end{bmatrix}$$

is an incidence matrix of a 2-$(4\lambda+4, 2\lambda+2, 2\lambda+1)$ *design satisfying the inequality* $F > 0$.

Proof. It is straightforward to show that N is an incidence matrix of a 2-$(4\lambda+4, 2\lambda+2, 2\lambda+1)$ design **E**. Without loss of generality, we assume that the blocks B_1, B_2, and B_3 of **D** corresponding to the first three columns of X satisfy the inequality (13.3) with $r = 2\lambda+1$. Let C_1, C_2, and C_3, be the blocks of **E** corresponding to the first three columns of N. Then

$$\begin{aligned}&|C_1 \cap C_2| + |C_1 \cap C_3| + |C_2 \cap C_3|\\&\quad = 2(|B_1 \cap B_2| + |B_1 \cap B_3| + |B_2 \cap B_3|) \geq 2(2\lambda+2) > 4\lambda+3,\end{aligned}$$

so **E** satisfies the inequality $F > 0$. □

Proposition 13.3.1 and Examples 13.2.5, 13.2.6, and 13.2.7 now yield the following infinite families of non-embeddable quasi-residual designs.

Theorem 13.3.2. *For any integer $n \geq 2$, there exist non-embeddable quasi-residual designs with parameters* $(3 \cdot 2^n, 3 \cdot 2^{n-1}, 3 \cdot 2^{n-1} - 1)$, $(2^{n+2}, 2^{n+1}, 2^{n+1} - 1)$, *and* $(5 \cdot 2^n, 5 \cdot 2^{n-1}, 5 \cdot 2^{n-1} - 1)$.

Remark 13.3.3. Let X be an incidence matrix of a 2-$(2\lambda+2, \lambda+1, \lambda)$ design. Then $J - X$ is an incidence matrix of a 2-design with the same parameters. Let $S = \{1, \sigma\}$ be a group of order 2 acting on the set $\mathcal{M} = \{X, J - X\}$ (with $\sigma X = J - X$). Then S is a group of symmetries of $\mathcal{M}$ and $H = \left[\begin{smallmatrix} 1 & 1 \\ 1 & \sigma \end{smallmatrix}\right]$ is a $GH(S; 1)$. The matrix N constructed in Proposition 13.3.1 can be represented as $N = \left[H \otimes_S X \quad \left[\begin{smallmatrix} 1 & 0 \\ 0 & 1 \end{smallmatrix}\right] \otimes J_{2\lambda+2,1}\right]$. The following two theorems use BGW-matrices to produce more infinite families of non-embeddable quasi-residual designs.

Theorem 13.3.4. *Let r be an odd prime power and let* **D** *be an* $(r+1, 2r, r, (r+1)/2, (r-1)/2)$*-design satisfying an m block non-embeddability condition. Then, for any positive integer n, there exists a quasi-residual design with parameters*

$$\left(\frac{(r+1)(r^n-1)}{r-1}, \frac{2r(r^n-1)}{r-1}, r^n, \frac{(r+1)r^{n-1}}{2}, \frac{(r-1)r^{n-1}}{2}\right) \tag{13.6}$$

satisfying the same non-embeddability condition.

Proof. Let X be an incidence matrix of **D** and let $\mathcal{M} = \{X, J - X\}$. Let $S = \{1, \sigma\}$ be a group of order 2 acting on $\mathcal{M}$ (with $\sigma X = J - X$). Then S is a group of symmetries of $\mathcal{M}$. Since r is an odd prime power, Theorem 10.2.5

implies that, for any positive integer n,there exists a

$$BGW\left(\frac{r^{n+1}-1}{r-1}, r^n, r^n - r^{n-1}; S\right).$$

Let W be such a matrix, and we assume that W is normalized. By Theorem 11.7.3, $N = W \otimes_S X$ is an incidence matrix of a design $\mathbf{E}$ with parameters (13.6). Let $F > 0$ be an m block non-embeddability condition satisfied by $\mathbf{D}$ and let $B_1, B_2, \ldots, B_m$ be the corresponding m blocks of $\mathbf{D}$. Without loss of generality, we assume that these blocks correspond to the first m columns of X. Let $B'_1, B'_2, \ldots, B'_m$ be the blocks of $\mathbf{E}$ corresponding to the first m columns of N. Since W is normalized, we obtain that, for all $i, j \in \{1, 2, \ldots, m\}$, $|B'_i \cap B'_j| = r^n|B_i \cap B_j|$. Therefore, $\mathbf{E}$ satisfies the condition $F > 0$. □

All designs obtained in Theorem 13.3.2 have parameters of the form $(r+1, 2r, r, (r+1)/2, (r-1)/2)$ and satisfy a three block non-embeddability condition. Therefore, as soon as the replication number of any of these designs is an odd prime power, it produces an infinite family of non-embeddable quasi-residual designs with parameters (13.6). Thus, we obtain the following result.

Corollary 13.3.5. *Let m and d be positive integers and let $r = 2^d - 1$ or $r = 3 \cdot 2^d - 1$ or $r = 5 \cdot 2^d - 1$. If r is a prime power and $r \geq 11$, then there exists a non-embeddable quasi-residual design with parameters*

$$\left(\frac{(r+1)(r^m-1)}{r-1}, \frac{2r(r^m-1)}{r-1}, r^m, \frac{(r+1)r^{m-1}}{2}, \frac{(r-1)r^{m-1}}{2}\right).$$

Remark 13.3.6. Symmetric designs corresponding to all quasi-residual designs of Corollary 13.3.5 exist (see Corollary 11.8.4).

We will now use BGW-matrices to obtain an infinite family of non-embeddable quasi-residual designs that has the design of Example 13.2.9 as a starter. We begin with a certain regularity condition that is satisfied by the design of Example 13.2.9.

Definition 13.3.7. Let X be a (0, 1)-matrix with n columns and a constant row sum r. We will call the ratio $\frac{r}{n}$ the *density* of X and denote it by $d(X)$.

Definition 13.3.8. Let q be a positive integer. A matrix of type $(q, 1)$ is a (0, 1)-matrix of size $1 \times q^2$ which can be represented as $[A\ A\ \ldots\ A]$ where A is a $1 \times q$ matrix. A matrix of type $(q, 2)$ is a (0, 1)-matrix which can be represented as $[A_1\ A_2\ \ldots\ A_q]$ where $A_1, A_2, \ldots, A_q$ are $1 \times n$ matrices with the same n and $A_1 + A_2 + \ldots + A_q = \alpha J_{1,n}$ for some integer α.

Remark 13.3.9. Any $1 \times q$ matrix is of type $(q, 2)$ (with $n = 1$).

Definition 13.3.10. Let $n_1, n_2, \ldots, n_s$, and q be positive integers. A matrix X is said to be of type $(n_1, n_2, \ldots, n_s; q)$ (or simply of type q) if X can be represented as a block-matrix $X = [X_{ij}]$, $1 \le j \le s$, where each X_{ij} is a $1 \times n_j$ matrix of type $(q, 1)$ or $(q, 2)$ and all X_{ij} have the same density.

Remark 13.3.11. Observe the following immediate consequences of this definition.

(i) If d is the density of all the matrices X_{ij}, then d is the density of X.
(ii) If, for some j, at least one of the matrices X_{ij} is of type $(q, 1)$, then $n_j = q^2$.
(iii) The all-one and all-zero matrices with $n \equiv 0 \pmod{q}$ columns are of type q.
(iv) Any design $AG_{n-1}(n, q)$ has an incidence matrix of type q.

Remark 13.3.12. Let **D** be the 2-(27, 9, 4) design constructed in Example 13.2.9. We will order the block set of **D** lexicographically, i.e., $B(a, b, c, d)$ precedes $B(a', b', c', d')$ if and only if $a < a'$ or $a = a'$ and $b < b'$ or $a = a'$, $b = b'$, and $c < c'$ or $a = a'$, $b = b'$, $c = c'$, and $d < d'$. Let X be the incidence matrix of D with respect to this ordering of blocks. Then X is a matrix of type $(3, 9, 9, 9, 9; 3)$.

The following proposition is straightforward.

Proposition 13.3.13. *Let $M = [M_{ij}]$ be a block-matrix. If, for all i, every block M_{ij} is a matrix of type $(n_{1j}, n_{2j}, \ldots, n_{s_j j}; q)$ and all blocks M_{ij} are of the same density, then M is a matrix of type q.*

We will now describe an action of a group of order q on matrices of type q.

Let $G = \{g_1, g_2, \ldots, g_q\}$ be a group of order q. If $A = [a_1\ a_2\ \ldots\ a_q]$ is a $1 \times q$ matrix, then, for $m = 1, 2, \ldots, q$, define $g_m A = [b_1\ b_2\ \ldots\ b_q]$ with $b_i = a_j$ if and only if $g_m g_i = g_j$. If $X = [A\ A\ \ldots\ A]$ is a matrix of type $(q, 1)$, define $g_m X = [g_m A\ g_m A\ \ldots\ g_m A]$. Clearly, $g_m X$ is a matrix of type $(q, 1)$ and

$$\sum_{m=1}^{q} g_m X - qd(X)J.$$

If $X = [A_1\ A_2\ \ldots\ A_q]$ is a matrix of type $(q, 2)$, define $g_m X = [B_1\ B_2\ \ldots\ B_q]$ with $B_i = A_j$ if and only if $g_m g_i = g_j$. Then $g_m X$ is a matrix of type $(q, 2)$ and

$$\sum_{m=1}^{q} g_m X = qd(X)J.$$

Let $X = [X_{ij}]$ be a matrix of type $(n_1, n_2, \ldots, n_s; q)$. Define, for $m = 1, 2, \ldots, q$, $g_m X = [g_m X_{ij}]$. Then X is a matrix of type $(n_1, n_2, \ldots, n_s; q)$,

and

$$\sum_{m=1}^{q} g_m X = qd(X)J. \tag{13.7}$$

Proposition 13.3.14. *Let G be a group of order q and let X and Y be matrices of type $(n_1, n_2, \ldots, n_s; q)$. Then, for any $g \in G$, $(gX)(gY)^\top = XY^\top$.*

Proof. Let $X = [X_{ij}]$ and $Y = [Y_{ij}]$, $j = 1, 2, \ldots, s$ where X_{ij} and Y_{ij} are $1 \times n_j$ matrices of type $(q, 1)$ or $(q, 2)$. It suffices to verify that $(gX_{ij})(gY_{ij})^\top = X_{ij}Y_{ij}^\top$ for $j = 1, 2, \ldots, s$. It is true if X_{ij} and Y_{ij} are of the same type, because in this case g acts as the same permutation of their entries. Suppose $X_{ij} = [A\ A\ \ldots\ A]$ is of type $(q, 1)$ and $Y_{ij} = [A_1\ A_2\ \ldots\ A_q]$ is of type $(q, 2)$. Then $A_1, A_2, \ldots, A_q$ are $1 \times q$ matrices and $A_1 + A_2 + \cdots + A_q = qd(Y)J$. Therefore,

$$X_{ij}Y_{ij}^\top = A\sum_{m=1}^{q} A_m^\top = [q^2 d(X)d(Y)]$$

and

$$(gX_{ij})(gY_{ij})^\top = (gA)\sum_{m=1}^{q} A_m^\top = [q^2 d(gX)d(Y)] = [q^2 d(X)d(Y)].$$

□

Through the remainder of this section, for $n \equiv 0 \pmod{q}$, $\mathcal{M}(m, n, q)$ denotes the set of all (0, 1)-matrices of size $m \times n$ and of type q. Let G be a group of order q. Then each element g of G acts as a bijection $g\colon \mathcal{M}(m, n, q) \to \mathcal{M}(m, n, q)$. Equation (13.7) and Proposition 13.3.14 imply that G is a group of symmetries of $\mathcal{M}(m, n, q)$.

Theorem 13.3.15. *Let v, b, and q be positive integers, $v \equiv 0 \pmod{q}$, and let a group G of order q act on the set $\mathcal{M}(v, b, q)$. Let H be a $GH(G; 1)$. Suppose that $X \in \mathcal{M}(v, b, q)$ is an incidence matrix of a quasi-residual 2-(qk, k, λ) design **D** and let $Y = [H \otimes X \quad I_q \otimes J_{v,1}]$. Then Y is an incidence matrix of a quasi-residual 2-$(q^2k, qk, q\lambda + 1)$ design **E** and $Y \in \mathcal{M}(qv, qb + q, q)$. Furthermore, if **D** satisfies a non-embeddability condition $F > 0$ and the matrix H is normalized, then **E** satisfies the condition $F > 0$ and thus is non-embeddable.*

Proof. Let $H = [\eta_{ij}]$ $(i, j = 1, 2, \ldots, q)$ and $Y = [Y_{ij}]$ $(i = 1, 2, \ldots, q;$ $j = 1, 2, \ldots, q + 1)$ so that each Y_{ij} with $j \neq q + 1$ is a $v \times b$ matrix and each

$Y_{i,q+1}$ is a $q \times q$ matrix. For $i, h = 1, 2, \ldots, q$, let

$$P_{ih} = \sum_{j=1}^{q+1} Y_{ij} Y_{hj}^T.$$

Then, by Proposition 13.3.14,

$$P_{ii} = \sum_{j=1}^{q} (\eta_{ij} X)(\eta_{ij} X)^T + J = \sum_{j=1}^{q} XX^T + J = qXX^T + J = qkI + (q\lambda + 1)J.$$

If $i \neq h$, we apply Proposition 13.3.14 again to obtain:

$$\begin{aligned} P_{ih} &= \sum_{j=1}^{q} (\eta_{ij} X)(\eta_{hj} X)^T = \sum_{j=1}^{q} (\eta_{hj}^{-1} \eta_{ij} X) X^T = \left(\sum_{g \in G} gX \right) X^T \\ &= \frac{q(k+\lambda)}{b} JX^T = \frac{qk}{v} JX^T = (k+\lambda)J. \end{aligned}$$

Since **D** is quasi-residual, we have $(qk-1)\lambda = (k+\lambda)(k-1)$, which implies $k + \lambda = q\lambda + 1$. By Proposition 2.3.13, Y is an incidence matrix of a 2-$(q^2k, qk, q\lambda + 1)$ design **E** with replication number $r = qk + q\lambda + 1$.

Since each matrix $\eta_{ij} X$ is obtained by permuting rows of X, all these matrices are of the same type and density as X. Since X has $b = qr$ columns, $d(X) = 1/q$. Since also $d(Y_{i,q+1}) = 1/q$ for all i, Proposition 13.3.13 implies that Y is of type q, so $Y \in \mathcal{M}(qv, qb + q, q)$.

Suppose now that **D** satisfies an m block non-embeddability condition $F > 0$, and let $B_1, B_2, \ldots, B_m$ be the m blocks of **D** such that, for any $f \in F$, $f > 0$ whenever $x_0 = \lambda$ and $x_{ij} = |B_i \cap B_j|$. If the free term of $f \in F$ is equal to 1, we replace the inequality $f > 0$ by an equivalent (for integer values of variables) inequality $f^* \geq 0$ where $f^* = f - 1$. Without loss of generality, we assume that the blocks $B_1, B_2, \ldots, B_m$ correspond to the first m columns of matrix X.

Suppose further that the matrix H is normalized. Let $C_1, C_2, \ldots, C_m$ be the blocks of **E** corresponding to the first m columns of Y. Then, for $i, j = 1, 2, \ldots, m$, $|C_i \cap C_j| = q|B_i \cap B_j|$. Therefore, if inequality $f > 0$ for a polynomial f with the free term 0 or $f^* \geq 0$ for a polynomial f with the free term 1 is satisfied by the blocks $B_1, B_2, \ldots, B_m$, it is satisfied by the blocks $C_1, C_2, \ldots, C_m$, i.e., the design **E** satisfies the inequality $F > 0$. □

Since the design **E**, constructed in Theorem 13.3.15, satisfies the conditions of this theorem (with v and b replaced by qv and $qb + b$, respectively) we obtain by induction the following result.

Corollary 13.3.16. *Let q be a prime power. Suppose there exists a quasi-residual 2-(qk, k, λ) design, which satisfies a non-embeddability condition*

$F > 0$ and has an incidence matrix of type q. Then, for any positive integer m, there exists a quasi-residual 2-$(q^m k, q^{m-1}k, q^{m-1}\lambda + (q^{m-1} - 1)/(q-1))$ design, which satisfies condition $F > 0$ and has the incidence matrix of type q.

Applying this result to the 2-(27, 9, 4) design of Example 13.2.9 yields

Corollary 13.3.17. *For any positive integer $m \geq 3$, there exists a non-embeddable quasi-residual 2-$(3^m, 3^{m-1}, (3^{m-1} - 1)/2)$ design.*

Remark 13.3.18. The corresponding symmetric designs are designs having the parameters of $PG_{m-1}(m, 3)$.

We will now apply BGW-matrices to the designs obtained in Corollary 13.3.16 to obtain more infinite families of non-embeddable quasi-residual designs.

Theorem 13.3.19. *Let q be a prime power. Suppose there exists a quasi-residual 2-(qk, k, λ) design, which satisfies a non-embeddability condition $F > 0$ and has an incidence matrix of type q. If, for a positive integer m, $r_m = q^m\lambda + (q^m - 1)/(q-1)$ is a prime power, then, for any positive integer n, there exists a quasi-residual 2-$(q^m k(r_m^n - 1)/(r-1), q^{m-1}kr_m^{n-1}, (r_m - q^{m-1}k)r_m^{n-1})$ design, which satisfies the condition $F > 0$.*

Proof. Note that r_m is the replication number of the 2-design obtained in Corollary 13.3.16. Let $\mathbf{D}_m$ be this design and let X be its incidence matrix of type q. Then $X \in \mathcal{M}(v, b, q)$ where v and b are the number of points and the number of blocks of $\mathbf{D}_m$, respectively. Let S be a cyclic group of order q acting on the set $\mathcal{M}(v, b, q)$. Since r_m is a prime power and $r_m \equiv 1 \pmod q$, there exists a

$$BGW\left(\frac{r_m^{n+1} - 1}{r_m - 1}, r_m^n, r_m^n - r_m^{n-1}; S\right).$$

Let W be a normalized BGW matrix with these parameters. Then the set of matrices $\mathcal{M}(v, b, q)$ and matrices W and X satisfy the conditions of Theorem 11.7.3, which implies that $W \otimes_S X$ is an incidence matrix of the required quasi-residual design. □

Applying this theorem to the designs of Corollary 13.3.17 yields the next result.

Corollary 13.3.20. *If $m \geq 3$ and $(3^m - 1)/2$ is a prime power, then, for any positive integer n, there exists a non-embeddable quasi-residual*

$$2\text{-}\left(\frac{3^{m-1}((3^m - 1)^n - 2^n)}{2^{n-1}(3^{m-1} - 1)}, \frac{3^{m-1}(3^m - 1)^{n-1}}{2^{n-1}}, \frac{(3^{m-1} - 1)(3^m - 1)^{n-1}}{2^n}\right)$$

design.

Remark 13.3.21. Symmetric designs corresponding to all quasi-residual designs of Corollary 13.3.5 exist (see Corollary 11.8.2).

13.4. Non-embeddable quasi-derived designs

A residual design of a symmetric design is isomorphic to a derived design of the complementary symmetric design (Proposition 2.4.16). A derived design of a nontrivial symmetric (v, k, λ)-design has parameters $(k, v - 1, k - 1, \lambda, \lambda - 1)$. This motivates the following definition.

Definition 13.4.1. A (v, b, r, k, λ)-design is called *quasi-derived* if $k = \lambda + 1$. A quasi-derived design is *embeddable* if it is isomorphic to a derived design of a nontrivial symmetric design. Otherwise, a quasi-derived design is called *non-embeddable*.

The following proposition is immediate.

Proposition 13.4.2. *A (v, b, r, k, λ)-design* **D** *is quasi-derived if and only if its complement is quasi-residual; a quasi-derived design is embeddable if and only the complementary quasi-residual design is embeddable.*

In this section we will obtain two families of non-embeddable quasi-derived designs.

Definition 13.4.3. A substructure $\mathbf{D}_1$ of a (v, b, r, k, λ)-design **D** is called a *subdesign of* **D** if $\mathbf{D}_1$ is a $(v_1, b_1, r_1, k_1, \lambda_1)$-design.

Remark 13.4.4. In Chapter 12, we considered symmetric subdesigns of symmetric designs. In this section, neither the design, nor the subdesign are assumed to be symmetric.

The non-embeddability of quasi-derived designs in subsequent constructions will be based on the following result.

Theorem 13.4.5. *Let* **D** *be a symmetric (v, k, λ)-design with $\lambda \geq 1$ and let* $\mathbf{D}_1$ *be a derived design of* **D**. *If* $\mathbf{D}_1$ *contains a 2-$(k_0, \lambda, \lambda - 1)$ subdesign, then*

$$k_0(k_0 - 1)^2(k_0 - \lambda) + 2\lambda(v - k) \geq 2k_0(k_0 - 1)(k - \lambda).$$

Proof. Let $\mathbf{D} = (X, \mathcal{B})$ and $\mathbf{D}_1 = (X_1, \mathcal{B}_1)$. Let $\mathbf{D}_1$ contain a 2-$(k_0, \lambda, \lambda - 1)$ subdesign $\mathbf{D}_2 = (X_2, \mathcal{B}_2)$. Then $\mathbf{D}_1$ is a $(k, v - 1, k - 1, \lambda, \lambda - 1)$-design and $\mathbf{D}_2$ is a $(k_0, b_0, k_0 - 1, \lambda, \lambda - 1)$-design with $b_0 = k_0(k_0 - 1)/\lambda$.

Observe that if $x \in X_1 \setminus X_2$ and $B \in \mathcal{B}_2$, then x and B are not incident. For $i = 0, 1, \ldots, v$, let n_i be the number of points $x \in X \setminus X_1$ that are contained in exactly i blocks of $\mathcal{B}_2$. Then

$$\sum_{i=0}^{v} n_i = |X \setminus X_1| = v - k. \tag{13.8}$$

Counting in two ways flags (x, B) with $x \in X \setminus X_1$ and $B \in \mathcal{B}_2$ yields

$$\sum_{i=0}^{v} i n_i = b_0(k - \lambda). \tag{13.9}$$

Counting in two ways triples (x, B, C) with $B, C \in \mathcal{B}_2$, $B \neq C$, and $x \in B \cap C$, yields

$$\sum_{i=0}^{v} i(i-1)n_i + k_0(k_0 - 1)(k_0 - 2) = b_0(b_0 - 1)\lambda. \tag{13.10}$$

From equations (13.8)–(13.10),

$$\begin{aligned}\sum_{i=0}^{v}(i-1)(i-2)n_i &= \sum_{i=0}^{v} i(i-1)n_i - 2\sum_{i=0}^{v} i n_i + 2\sum_{i=0}^{v} n_i \\ &= b_0(b_0 - 1)\lambda - k_0(k_0 - 1)(k_0 - 2) - 2b_0(k - \lambda) + 2(v - k) \\ &= b_0(k_0 - 1)(k_0 - \lambda) - 2b_0(k - \lambda) + 2(v - k).\end{aligned}$$

Since $(i-1)(i-2) \geq 0$ for any integer i, the required inequality follows. □

Taking $k_0 = \lambda + 1$ yields the following corollary.

Corollary 13.4.6. *Let* **D** *be a symmetric* (v, k, λ)*-design with* $\lambda \geq 1$ *and let* $\mathbf{D}_1$ *be a derived design of* **D**. *If* $\mathbf{D}_1$ *contains a* 2-$(\lambda + 1, \lambda, \lambda - 1)$ *subdesign, then*

$$2(v - k) \geq (\lambda + 1)(2k - 3\lambda).$$

In order to apply Corollary 13.4.6, we need the following lemma.

Lemma 13.4.7. *For* $n \geq 3$, *any* $(n-1)$*-resolvable* 2-$(n^2 - n - 1, n - 1, n - 1)$ *design is embeddable in a* 2-$(n^2, n, n - 1)$ *design having a* 2-$(n + 1, n, n - 1)$ *subdesign.*

Proof. Let **D** be an $(n-1)$-resolvable 2-$(n^2 - n - 1, n - 1, n - 1)$ design. Then it has $n + 1$ resolution classes. Let $\mathfrak{R} = \{\mathcal{C}_1, \mathcal{C}_2, \ldots, \mathcal{C}_{n+1}\}$ be the resolution of **D**. Let $Y = \{y_1, y_2, \ldots, y_{n+1}\}$ be a set disjoint from the point set X of **D**. Form an incidence structure $\mathbf{E} = (X \cup Y, \mathcal{A} \cup \mathcal{B})$ where

$$\mathcal{A} = \bigcup_{i=1}^{n+1} \{A \cup \{y_i\} : A \in \mathcal{C}_i\}$$

and $\mathcal{B}$ is the set of all n-subsets of Y. Then $\mathbf{E}$ is a 2-$(n^2, n, n-1)$ design and $\mathbf{D}_1 = (Y, \mathcal{B})$ is a 2-$(n+1, n, n-1)$ subdesign of $\mathbf{E}$. □

Theorem 13.4.8. *If $n \geq 3$ is an integer and $q = n^2 - n - 1$ is a prime power, then there exists a non-embeddable quasi-derived 2-$(n^2, n, n-1)$ design.*

Proof. Let $n \geq 3$ and let $q = n^2 - n - 1$ be a prime power. Since $q - 1 = (n+1)(n-2)$, the multiplicative group of the field $GF(q)$ has a subgroup H of order $n-2$. By Theorem 5.3.12, there exists an $(n-1)$-resolvable 2-$(n^2 - n - 1, n-1, n-1)$ design. By Lemma 13.4.7, this design can be embedded in a quasi-derived 2-$(n^2, n, n-1)$ design having a 2-$(n+1, n, n-1)$ subdesign. Corollary 13.4.6 implies that this quasi-residual design is non-embeddable. □

Our next construction of non-embeddable quasi-derived designs uses the complements of affine planes.

Theorem 13.4.9. *Let $n \geq 3$ and $q = n^2(n^2 - n - 2) + 1$ be prime powers. Then there exists a non-embeddable quasi-derived 2-$(n^2(n^2 - n - 1) + 1, n^2 - n, n^2 - n - 1)$ design.*

Proof. Let $\mathbf{D}_0 = (X_0, \mathcal{B}_0)$ be the complement of an affine plane of order n and let $X_0 = \{x_1, x_2, \ldots, x_{n^2}\}$. Let H be the subgroup of order $n^2 - n - 1$ of the multiplicative group of the field $GF(q)$. By Theorem 5.3.12, there exists an $(n^2 - n - 1)$-resolvable 2-$(q, n^2 - n - 1, n^2 - n - 1)$ design $\mathbf{D}_2 = (X_2, \mathcal{B}_2)$. It has exactly n^2 resolution classes $\mathcal{C}_i$, $i = 1, 2, \ldots, n^2$. We assume that the sets X_0 and X_2 are disjoint and let $X_1 = X_0 \cup X_2$. Let

$$\mathcal{B}_1 = \bigcup_{i=1}^{n^2} \{B \cup \{x_i\} : B \in \mathcal{C}_i\}.$$

It is straightforward to verify that the incidence structure $\mathbf{D}_1 = (X_1, \mathcal{B}_1)$ is a 2-$(n^2(n^2 - n - 1) + 1, n^2 - n, n^2 - n - 1)$ design. The parameters of the quasi-derived design $\mathbf{D}_1$ and its subdesign $\mathbf{D}_2$ do not satisfy the condition of Theorem 13.4.5. Therefore, $\mathbf{D}_1$ is non-embeddable. □

Remark 13.4.10. The two smallest values of n satisfying the condition of Theorem 13.4.9 are $n = 3$ ($q = 37$) and $n = 8$ ($q = 3457$).

Exercises

(1) Determine the smallest possible number of points of a non-embeddable quasi-residual design.

(2) Determine the smallest possible number of points of a non-embeddable quasi-derived design.

Notes

The first example of a non-embeddable quasi-residual design (Example 2.4.18) was obtained in Bhattacharya (1944b). For a long time this was the only known parameter set of a non-embeddable quasi-residual (v, b, r, k, λ)-design with $k \leq v/2$, until the paper by van Lint (1978), in which two new designs were constructed. The case $k > v/2$ was considered in Parker (1967) and van Lint and Tonchev (1984).

Our presentation of Section 13.1. follows van Trung (1990). The notion of a linear non-embeddability condition was introduced in Ionin and Mackenzie-Fleming (2002). Sections 13.2. and 13.3. are based on this paper. All results of Section 13.4. are due to van Lint and Tonchev (1993).

More parameters of non-embeddable quasi-residual and quasi-derived designs can be found in Kageyama and Miao (1996), Mackenzie-Fleming (2000a, 2000b), Mackenzie-Fleming and Smith (1998), Tonchev (1986c, 1992), van Lint and Tonchev (1984, 1993), van Lint, Tonchev and Landgev (1990), and van Trung (1986, 1990).

14

Ryser designs

If X is a set of cardinality v and $\mathcal{B}$ is a set of nonempty subsets of X, any two of which meet in the same number of points, then the set $\mathcal{B}$ consists of at most v subsets of X. If $|\mathcal{B}| = v$, then either $\mathcal{B}$ is the block set of a symmetric design on the point set X or $(X, \mathcal{B})$ is an incidence structure known as a Ryser design. If A is a block of a symmetric (v, k, λ)-design with $k \neq 2\lambda$, then replacing every other block of this design by its symmetric difference with A produces a Ryser design. In this chapter, we introduce evidence in support of the conjecture that all Ryser designs arise from symmetric designs in this manner.

14.1. Basic properties of Ryser designs

We recall the definition of Ryser designs given in Section 1.3.

Definition 14.1.1. Let v and λ be positive integers. A *Ryser design of index* λ *on* v *points* is an incidence structure $\mathbf{D} = (X, \mathcal{B})$ where X is a set of cardinality v and $\mathcal{B}$ is a set of v subsets of X (blocks) such that

(i) $|A \cap B| = \lambda$ for all distinct $A, B \in \mathcal{B}$;
(ii) $|B| > \lambda$ for all $B \in \mathcal{B}$; and
(iii) there are $A, B \in \mathcal{B}$ such that $|A| \neq |B|$.

We will also recall the Ryser–Woodall Theorem.

Theorem 14.1.2 (The Ryser–Woodall Theorem). *For any Ryser design* $\mathbf{D}$ *on* v *points there are distinct integers* r *and* r^* *greater than* 1 *such that* $r + r^* = v + 1$ *and every point of* $\mathbf{D}$ *has replication number* r *or* r^*.

The following definition introduces the set of incidence structures and specific notations that will be used throughout this chapter.

Definition 14.1.3. Let X be a set of cardinality v and let r be an integer such that $2 \le r \le v-1$ and $r \ne (v+1)/2$. Let $\mathfrak{D}_r(X)$ denote the set of all incidence structures $\mathbf{D} = (X, \mathcal{B})$ where $\mathcal{B}$ is a set of subsets of X and $\mathbf{D}$ is a Ryser design with replication numbers r and $r^* = v + 1 - r$ or a symmetric design with block size r or r^*. For all $\mathbf{D} \in \mathfrak{D}_r(X)$, we will denote by ρ and ρ^* the fractions $\rho = (r-1)/(r^*-1)$ and $\rho^* = (r^*-1)/(r-1)$ and by g the greatest common divisor of $r-1$ and r^*-1. We will also use the following notations: $\lambda(\mathbf{D})$ is the cardinality of the intersection of two distinct blocks of $\mathbf{D}$; $E(\mathbf{D})$ is the set of all points of $\mathbf{D}$ with replication number r and $E^*(\mathbf{D})$ is the set of all points of $\mathbf{D}$ with replication number r^*; $e(\mathbf{D}) = |E(\mathbf{D})|$ and $e^*(\mathbf{D}) = |E^*(\mathbf{D})|$; for any block B of $\mathbf{D}$, we denote by $k_B(\mathbf{D})$ and $k_B^*(\mathbf{D})$ the cardinalities of the sets $B \cap E(\mathbf{D})$ and $B \cap E^*(\mathbf{D})$, respectively.

Remark 14.1.4. If there is no confusion, we will use the above notations without reference to the design $\mathbf{D}$.

The following is immediate.

Proposition 14.1.5. *An incidence structure $\mathbf{D} \in \mathfrak{D}_r(X)$ is a symmetric design if and only if $E(\mathbf{D})$ equals X or $\varnothing$.*

We will now introduce the operation of *block complementation* on $\mathfrak{D}_r(X)$.

Definition 14.1.6. For any $\mathbf{D} = (X, \mathcal{B}) \in \mathfrak{D}_r(X)$ and for any block A of $\mathbf{D}$, we denote by $\mathbf{D} * A$ the incidence structure $\mathbf{D} * A = (X, \mathcal{C})$ where

$$\mathcal{C} = \{A\} \cup \{B \triangle A : B \in \mathcal{B} \setminus \{A\}\}.$$

The incidence structure $\mathbf{D} * A$ is called the *block complementation of* $\mathbf{D}$ *with respect to* A.

The proof of the following proposition is straightforward.

Proposition 14.1.7. *Let $\mathbf{D} \in \mathfrak{D}_r(X)$ and let A be a block of $\mathbf{D}$. Then $\mathbf{D} * A \in \mathfrak{D}_r(X)$ and the following conditions hold:*

(i) $(\mathbf{D} * A) * A = \mathbf{D}$;
(ii) *if B is a block of $\mathbf{D} * A$ and $B \ne A$, then $A \triangle B$ is a block of $\mathbf{D}$ and* $(\mathbf{D} * A) * B = \mathbf{D} * (A \triangle B)$;
(iii) $\lambda(\mathbf{D} * A) = |A| - \lambda(\mathbf{D})$;
(iv) $E(\mathbf{D} * A) = E(\mathbf{D}) \triangle A$;
(v) $e(\mathbf{D} * A) = e(\mathbf{D}) - k_A(\mathbf{D}) + k_A^*(\mathbf{D})$;
(vi) *$\mathbf{D} * A$ is a symmetric design if and only if $A = E(\mathbf{D})$ or $A = E^*(\mathbf{D})$.*

We will now derive several useful relations between parameters of a Ryser design.

Let $\mathbf{D} = (X, \mathcal{B})$ be a Ryser design of index λ on v points with replication numbers r and r^* and let B be a block of $\mathbf{D}$. Then $e + e^* = v$ and $k_B + k_B^* = |B|$. Counting in two ways triples (x, A, B) with $A, B \in \mathcal{B}$, $A \neq B$, and $x \in A \cap B$ yields

$$er(r-1) + e^* r^*(r^* - 1) = \lambda v(v-1), \tag{14.1}$$

which can be transformed into the following three equations:

$$e = \frac{\lambda(\rho+1)^2 - (v+\rho)}{\rho^2 - 1}, \tag{14.2}$$

$$(r - r^*)e = (v-1)\lambda - r^*(r^* - 1), \tag{14.3}$$

$$(e - r^*)(r - e^*) = \lambda(v-1) - ee^*. \tag{14.4}$$

Since $e - r^*$ and $r - e^*$ are two consecutive integers, the last equation implies

$$ee^* \leq \lambda(v-1). \tag{14.5}$$

Fixing $B \in \mathcal{B}$ and counting in two ways flags (x, A) with $x \in B$ and $A \neq B$ yields

$$k_B(r-1) + k_B^*(r^* - 1) = (v-1)\lambda, \tag{14.6}$$

which can be rewritten in the following three forms:

$$(k_B - \lambda)\rho = \lambda - k_B^*, \tag{14.7}$$

$$(r-1)(|B| - 2k_B) = (v-1)(|B| - k_B - \lambda), \tag{14.8}$$

$$(r^* - 1)(|B| - 2\lambda) = (r - r^*)(\lambda - k_B). \tag{14.9}$$

The last equation immediately implies the following result.

Proposition 14.1.8. *Let* $\mathbf{D} = (X, \mathcal{B})$ *be a Ryser design of index* λ *and let* $r > r^*$. *Then:*

(i) *for any* $A, B \in \mathcal{B}$, $|A| = |B|$ *if and only if* $k_A = k_B$;
(ii) *for* $B \in \mathcal{B}$, *if* $|B| > 2\lambda$, *then* $k_B < \lambda < k_B^*$;
(iii) *for* $B \in \mathcal{B}$, *if* $|B| < 2\lambda$, *then* $k_B^* < \lambda < k_B$;
(iv) *for* $B \in \mathcal{B}$, *if* $|B| = 2\lambda$, *then* $k_B = k_B^* = \lambda$.

Equation (14.9) implies

$$\frac{r^* - 1}{g}(|B| - 2\lambda) = \frac{r - r^*}{g}(\lambda - k_B).$$

Since $(r^* - 1)/g$ and $(r - r^*)/g$ are relatively prime, we obtain that, for every block B, there is an integer l_B such that $\lambda - k_B = l_B(r^* - 1)/g$. The integers

l_B then satisfy the following equations:

$$k_B = \lambda - \frac{r^* - 1}{g} l_B, \tag{14.10}$$

$$k_B^* = \lambda + \frac{r - 1}{g} l_B, \tag{14.11}$$

$$|B| = 2\lambda + \frac{r - r^*}{g} l_B. \tag{14.12}$$

We will now derive another useful result from these equations.

Proposition 14.1.9. *For any block B of a Ryser design,*

$$|B| = r^* + \frac{(r - r^*)(e - k_B)}{r^* - 1}. \tag{14.13}$$

Furthermore, if $k_B < e$ and $r > r^*$, then $|B| \geq r^* + (r - r^*)/g$.

Proof. Let $(r-1)/g = c$ and $(r^* - 1)/g = c^*$. From (14.3), we derive that

$$\lambda - e = \frac{(r^* - 1)(r^* - 2\lambda)}{r - r^*} = \frac{c^*(r^* - 2\lambda)}{c - c^*},$$

so $(\lambda - e)(c - c^*) = c^*(r^* - 2\lambda)$. Since $c - c^*$ and c^* are relatively prime, we obtain that $c - c^*$ divides $r^* - 2\lambda$. Then (14.10) implies

$$l_B = \frac{(\lambda - k_B)g}{r^* - 1} = \frac{(\lambda - e)g}{r^* - 1} + \frac{(e - k_B)g}{r^* - 1} = \frac{r^* - 2\lambda}{c - c^*} + \frac{(e - k_B)g}{r^* - 1}.$$

Now (14.13) follows from (14.12).

Since $c - c^*$ divides $r^* - 2\lambda$, we obtain that $(e - k_B)g/(r^* - 1)$ is an integer. Therefore, if $k_B < e$, then (14.13) implies that $|B| \geq r^* + (r - r^*)/g$. □

We will now obtain the following inequalities.

Proposition 14.1.10. *Let $\mathbf{D} = (X, \mathcal{B})$ be a Ryser design of index λ and let $r > r^*$. Then* (i) *$e < 4\lambda$ and* (ii) *for any block B of $\mathbf{D}$, $k_B < 2\lambda$.*

Proof. (i) Since $r > r^*$, we have $\rho > 1$. Therefore, by (14.2), it suffices to show that $f(\rho) > 0$ where

$$f(t) = 3\lambda t^2 - (2\lambda - 1)t + (v - 5\lambda).$$

Note that $f(1) = v - 4\lambda + 1$ and $(2\lambda - 1)/(6\lambda) < 1$. Therefore, if $f(1) > 0$, then $f(\rho) > 0$, and so $e < 4\lambda$. If $f(1) \leq 0$, then $v < 4\lambda$, and therefore again $e < 4\lambda$.

(ii) If, for some block B, $k_B \geq 2\lambda$, then (14.7) implies that $\lambda\rho \leq \lambda$ which contradicts $\rho > 1$. Therefore, $k_B < 2\lambda$ for every block B. □

Remark 14.1.11. If r and r^* are the replication numbers of a Ryser design $\mathbf{D} = (X, \mathcal{B})$, then the sets $\mathfrak{D}_r(X)$ and $\mathfrak{D}_{r^*}(X)$ are the same. Therefore, one can switch r with r^*, e with e^*, k_B with k_B^* in identities (14.2), (14.3), (14.7)–(14.13) to obtain "dual" identities.

Let $\mathbf{D} = (X, \mathcal{B})$ be a Ryser design of index λ on v points with replication numbers r and r^*. We will regard X as a set of v indeterminates and denote by $Pol(\mathbf{D})$ the set of all linear polynomials $\sum_{x \in X} a_x x + b$ in these indeterminates with rational coefficients. We will regard $Pol(\mathbf{D})$ as a vector space over $\mathbb{Q}$. Clearly, $\dim Pol(\mathbf{D}) = v + 1$. For $f = \sum_{x \in X} a_x x + b$ and any subset Y of X, we let

$$f(Y) = \sum_{x \in Y} a_x + b.$$

Then $(f + g)(Y) = f(Y) + g(Y)$ and $(af)(Y) = af(Y)$ for $a \in \mathbb{Q}$. With each block $B \in \mathcal{B}$, we associate the polynomial $f_B = \sum_{x \in B} x - \lambda$. Then, for any blocks $A, B \in \mathcal{B}$,

$$f_B(A) = \begin{cases} 0 & \text{if } A \neq B, \\ |A| - \lambda & \text{if } A = B. \end{cases}$$

It was shown in the proof of the Ryser–Woodall Theorem that the set $\{f_B : B \in \mathcal{B}\} \cup \{1\}$ of linear polynomials is a basis of $Pol(\mathbf{D})$. We will restate equations (1.6), (1.9), (1.10), and (1.11) as parts of the following theorem.

Theorem 14.1.12. *Let $\mathbf{D} = (X, \mathcal{B})$ be a Ryser design of index λ on v points with replication numbers r and r^*. Let $x, y \in X$, $x \neq y$, and let $r(x) = r$. Then the polynomial $(v - 1)x \in Pol(\mathbf{D})$ has the following expansion:*

$$(v - 1)x = (r^* - 1) \sum_{B \ni x} \frac{f_B}{|B| - \lambda} - (r - 1) \sum_{B \not\ni x} \frac{f_B}{|B| - \lambda} + (r - 1). \quad (14.14)$$

Furthermore,

$$\sum_{B \in \mathcal{B}} \frac{1}{|B| - \lambda} = \frac{(v - 1)^2}{(r - 1)(r^* - 1)} - \frac{1}{\lambda}, \quad (14.15)$$

$$\sum_{\substack{B \in \mathcal{B} \\ B \ni x}} \frac{1}{|B| - \lambda} = \frac{v - 1}{r^* - 1} = \rho + 1, \quad (14.16)$$

$$\sum_{\substack{B \in \mathcal{B} \\ B \not\ni x}} \frac{1}{|B| - \lambda} = \frac{v - 1}{r - 1} - \frac{1}{\lambda} = \frac{\rho + 1}{\rho} - \frac{1}{\lambda}, \quad (14.17)$$

$$\sum_{\substack{B \in \mathcal{B} \\ B \ni x, y}} \frac{1}{|B| - \lambda} = \begin{cases} \rho & \text{if } r(y) = r, \\ 1 & \text{if } r(y) = r^*, \end{cases} \quad (14.18)$$

$$\sum_{\substack{B \in \mathcal{B} \\ B \cap \{x, y\} = \varnothing}} \frac{1}{|B| - \lambda} = \begin{cases} \frac{1}{\rho} - \frac{1}{\lambda} & \text{if } r(y) = r, \\ 1 - \frac{1}{\lambda} & \text{if } r(y) = r^*. \end{cases} \quad (14.19)$$

Proof. Equation (14.14) follows from (1.6) if we take into account $\beta_i = (r-1)/(v-1)$. Equations (14.15), (14.16), and (14.17) are equivalent to (1.11), (1.9), and (1.10), respectively.

To obtain (14.18), we apply both sides of (14.14) to the singleton $\{y\}$. Let S be the left hand side of (14.18). If $y \in E$, we obtain

$$0 = (r^* - 1)(1-\lambda)S - (r^* - 1)\lambda\left(\frac{v-1}{r^*-1} - S\right) - (r-1)(1-\lambda)$$
$$\times\left(\frac{v-1}{r^*-1} - S\right) + (r-1)\lambda\left(\frac{v-1}{r-1} - \frac{1}{\lambda} - \frac{v-1}{r^*-1} + S\right) + r - 1.$$

If $y \in E^*$, we obtain

$$0 = (r^* - 1)(1-\lambda)S - (r^* - 1)\lambda\left(\frac{v-1}{r^*-1} - S\right)$$
$$-(r-1)(1-\lambda)\left(\frac{v-1}{r-1} - S\right) + (r-1)\lambda\left(-\frac{1}{\lambda} + S\right) + r - 1.$$

Solving these equations for S yields (14.18).

Now,

$$\sum_{\substack{B\in\mathcal{B}\\ B\cap\{x,y\}=\varnothing}} \frac{1}{|B|-\lambda} = \sum_{B\in\mathcal{B}} \frac{1}{|B|-\lambda} - \sum_{\substack{B\in\mathcal{B}\\ B\ni x}} \frac{1}{|B|-\lambda} - \sum_{\substack{B\in\mathcal{B}\\ B\ni y}} \frac{1}{|B|-\lambda} + \sum_{\substack{B\in\mathcal{B}\\ B\ni x,y}} \frac{1}{|B|-\lambda},$$

and (14.19) follows. □

We can now classify certain Ryser designs.

Theorem 14.1.13 (The de Bruijn–Erdös Theorem). *Let $v \geq 4$ be an integer and let $\mathcal{B}$ be a set of v subsets of a v-set X such that any two distinct members of $\mathcal{B}$ have exactly one element in common. Then one of the following situations occurs:*

(i) *$\mathcal{B}$ consists of a singleton and all 2-subsets of X containing it;*
(ii) *$\mathcal{B}$ consists of a $(v-1)$-subset and all 2-subsets of X that this $(v-1)$-subset does not contain, i.e., $\mathcal{B}$ is a pencil;*
(iii) *B is the set of all lines of a projective plane with point set X.*

Proof. If all elements of $\mathcal{B}$ have the same cardinality k, then $\mathcal{B}$ is the block set of a symmetric $(v, k, 1)$-design, i.e., the set of lines of a projective plane.

If $\mathcal{B}$ contains a singleton $\{x\}$, then every element of $\mathcal{B}$ contains x and therefore the set $\mathcal{P} = \{B \setminus \{x\} \colon B \in \mathcal{B}, B \neq \{x\}\}$ partitions $X \setminus \{x\}$. Since $|X \setminus \{x\}| = |\mathcal{P}| = v - 1$, the set $\mathcal{P}$ consists of $v-1$ singletons and therefore, $\mathcal{B}$ consists of $\{x\}$ and all 2-subsets of X containing x.

Suppose $|B| \geq 2$ for all $B \in \mathcal{B}$ and not all elements of $\mathcal{B}$ are of the same cardinality. Then $\mathbf{D} = (X, \mathcal{B})$ is a Ryser design of index 1. Let $x, y \in X$ and let $r(x) \neq r(y)$. Since any two distinct blocks of $\mathbf{D}$ meet in one point, there is at most one block containing $\{x, y\}$. By (14.18),

$$\sum_{\substack{B \in \mathcal{B} \\ B \ni x, y}} \frac{1}{|B| - 1} = 1.$$

Therefore, there is a unique block $B \in \mathcal{B}$ containing $\{x, y\}$ and $|B| = 2$. Thus, any pair of points of $\mathbf{D}$ of different replication numbers forms a block and there are ee^* such blocks. Let $e^* \geq e$. Then $e^* \geq v/2$. If $e \geq 2$, then, by (14.18), there is a block containing two points with replication number r. Therefore, in the case, $\mathbf{D}$ has at least $1 + ee^* > v$ blocks, a contradiction. Therefore, $e = 1$. Let x be the only point of $\mathbf{D}$ with replication number r. Then $\mathbf{D}$ has $v - 1$ blocks $\{x, y\}$ with $y \in X \setminus \{x\}$. The only other subset of X of cardinality greater than 1 that meets each of these blocks in one point is $X \setminus \{x\}$, so $\mathcal{B}$ consists of $X \setminus \{x\}$ and all 2-subsets containing x. □

Corollary 14.1.14. *The only Ryser designs of index* 1 *are pencils.*

As the proof of the de Bruijn–Erdös Theorem shows, we have $e = 1$ or $e^* = 1$ in any Ryser design of index 1. We will now prove the converse theorem.

Theorem 14.1.15. *Let* $\mathbf{D}$ *be a Ryser design of index* λ. *If all but one point of* $\mathbf{D}$ *have the same replication number, then* $\lambda = 1$, *and therefore* $\mathbf{D}$ *is a pencil.*

Proof. Let $\mathbf{D} = (X, \mathcal{B})$, $v = |X|$, and let r and r^* be the replication numbers of $\mathbf{D}$. Let x be the only point of $\mathbf{D}$ with replication number r. Let $\mathcal{B}_1$ be the set of all blocks of $\mathbf{D}$ that contain x and let $\mathcal{B}_0$ be the set of the remaining blocks of $\mathbf{D}$. Then $|\mathcal{B}_1| = r$ and $|\mathcal{B}_0| = v - r = r^* - 1$. Note that $k_B = i$ for $B \in \mathcal{B}_i$. Let $\rho = (r - 1)/(r^* - 1)$. Eq. (14.13) implies that, for $B \in \mathcal{B}$,

$$|B| = \begin{cases} \lambda(\rho + 1) & \text{if } B \subset \mathcal{B}_0, \\ r^* = \lambda\rho + \lambda - \rho + 1 & \text{if } B \in \mathcal{B}_1. \end{cases} \tag{14.20}$$

Let $B \in \mathcal{B}_0$. Then $k_B = 0$ and Proposition 14.1.8 implies that $|B| > 2\lambda$. Now (14.20) implies that $\rho > 1$.

Let y be a point of $\mathbf{D}$ other than x. Since all blocks containing $\{x, y\}$ are in $\mathcal{B}_1$, (14.18) implies that there are exactly $\lambda\rho - \rho + 1$ blocks containing $\{x, y\}$. Therefore, there are exactly $r^* - (\lambda\rho - \rho + 1) = \lambda$ blocks containing y but not x.

Fix a block $A \in \mathcal{B}_0$ and a point $y \in A$.

Suppose first that $A = E^*$. Then Proposition 14.1.7(vi) implies that $\mathbf{D} * A$ is a symmetric design and therefore, $|B \triangle A| = |A|$ for $B \in \mathcal{B}_1$, and then (14.20) implies that $(\lambda - 1)(\rho - 1) = 0$. Since $\rho \neq 1$, we have $\lambda = 1$.

Suppose now that $A \neq E^*$ and let $z \in E^* \setminus A$. For $i = 0, 1$, let λ_i denote the number of blocks $B \in \mathcal{B}_i$ containing y and z. Then the equation "dual" to (14.18) implies that

$$\frac{\lambda_0}{\lambda\rho} + \frac{\lambda_1}{\lambda\rho - \rho + 1} = \frac{1}{\rho}. \tag{14.21}$$

We rewrite (14.21) as

$$\lambda\rho(\lambda_0 + \lambda_1) = \lambda^2\rho + (\rho - 1)(\lambda_0 - \lambda).$$

Since $\rho > 1$ and $\lambda_0 < \lambda$, this equation implies that $\lambda_0 + \lambda_1 < \lambda$ and therefore

$$\lambda_0 + \lambda_1 + 1 \leq \lambda.$$

We now rewrite (14.21) as

$$\rho(\lambda^2 - \lambda(\lambda_0 + \lambda_1 + 1) + \lambda_0) = \lambda_0 - \lambda.$$

Since $\lambda_0 - \lambda < 0$, we obtain that

$$\lambda^2 + \lambda_0 < \lambda(\lambda_0 + \lambda_1 + 1) \leq \lambda^2,$$

a contradiction. Therefore, the case $A \neq E^*$ is not possible and the proof is now complete. □

Corollary 14.1.16. *If* $\mathbf{D}$ *is a Ryser design with* $g = 1$, *then* $\mathbf{D}$ *is a pencil.*

Proof. If $g = \gcd(r - 1, r^* - 1) = 1$, then $r - 1$ and $v - 1 = (r - 1) + (r^* - 1)$ are relatively prime. Then (14.8) implies that $v - 1$ divides $|B| - 2k_B$ for any block B. If $|B| \neq 2k_B$, then we obtain that $|B| \geq (v - 1) + 2k_B$, which implies $k_B = 0$ and $|B| = v - 1$. Then $e^* = v - 1$, so $e = 1$, and $\mathbf{D}$ is a pencil. If $|B| = 2k_B$ for every block B, then Proposition 14.1.8 implies that all blocks of $\mathbf{D}$ have cardinality 2λ, a contradiction. □

We will now obtain bounds for the parameter $\rho = (r - 1)/(r^* - 1)$ of a Ryser design of index λ.

Theorem 14.1.17. *For any Ryser design of index* $\lambda \geq 2$ *having replication numbers* $r > r^*$,

$$\frac{\lambda}{\lambda - 1} \leq \rho \leq \lambda$$

and $\rho \notin (\lambda - 1, \lambda)$.

Proof. Let $\mathbf{D} = (X, \mathcal{B})$ be a Ryser design of index $\lambda \geq 2$ having replication numbers $r > r^*$. Let $g = \gcd(r-1, r^*-1)$, and let $(r-1)/g = c$ and $(r^*-1)/g = c^*$. The inequality $\rho \leq \lambda$ follows immediately from (14.19). To prove that $\rho \geq \lambda/(\lambda-1)$, note that (14.6) implies that, for any block B,

$$c(k_B - \lambda) = c^*(\lambda - k_B^*).$$

Since c and c^* are relatively prime, we obtain that c^* divides $k_B - \lambda$. Proposition 14.1.10 implies that $-\lambda \leq k_B - \lambda < \lambda$. Let B be a block of cardinality other than 2λ. Then, by Proposition 14.1.8, $k_B - \lambda \neq 0$ and therefore $c^* \leq \lambda$. If $c^* \leq \lambda - 1$, then

$$\rho = \frac{c}{c^*} \geq \frac{c + \lambda - 1 - c^*}{\lambda - 1} \geq \frac{\lambda}{\lambda - 1}.$$

Suppose $c^* = \lambda$. Then λ divides $k_B - \lambda$ for any block B. Let $\mathcal{B}_1 = \{B \in \mathcal{B} \colon |B| = 2\lambda\}$ and $\mathcal{B}_2 = \mathcal{B} \setminus \mathcal{B}_1$. Then $k_B = 0$ for all $B \in \mathcal{B}_2$. Let $x \in E$ and $y \in E^*$. Then every block B containing $\{x, y\}$ is in $\mathcal{B}_1$ and therefore its cardinality is 2λ. Now (14.18) implies that there are exactly λ blocks containing $\{x, y\}$. Fix $y \in E^*$ and count in two ways flags (x, B) with $x \in E$ and $B \ni y$. Since every block containing $\{x, y\}$ meets E in λ points, we obtain $e\lambda = r^*\lambda$, so $e = r^*$. Let $x_1, x_2 \in E$, $x_1 \neq x_2$. Since all blocks containing $\{x_1, x_2\}$ are in $\mathcal{B}_1$, (14.18) implies that the number of such blocks is $\mu = \lambda\rho$. Therefore, the substructure $\mathbf{D}_1 = (E, \mathcal{B}_1)$ of $\mathbf{D}$ is a $(e, |\mathcal{B}_1|, r, \lambda, \mu)$-design. Then $(e-1)\mu = r(\lambda - 1)$, so $(r^* - 1)\mu = r(\lambda - 1)$, which implies $r = \lambda$. Then $r^* = \lambda g + 1 > r$, a contradiction. Thus, $c^* \leq \lambda - 1$, and therefore, we have proved that $\rho \geq \lambda/(\lambda - 1)$.

Suppose $\lambda - 1 < \rho < \lambda$. Since $\lambda \geq 2$, Theorem 14.1.15 implies that $e \geq 2$. Let $x_1, x_2 \in E$, $x_1 \neq x_2$. Then (14.19) implies that

$$\sum_{\substack{B \in \mathcal{B} \\ B \cap \{x_1, x_2\} = \varnothing}} \frac{1}{|B| - \lambda} = \frac{1}{\rho} - \frac{1}{\lambda} = \frac{\lambda - \rho}{\rho\lambda} < \frac{1}{\rho\lambda}.$$

Since this sum is not equal to 0, there is a block B such that $|B| - \lambda > \rho\lambda$. If B is such a block, then, by (14.7), $|B| - \lambda = k_B + k_B^* - \lambda = k_B + (\lambda - k_B)\rho \leq \lambda\rho$, a contradiction. Therefore, $\rho \notin (\lambda - 1, \lambda)$. □

In the course of the above proof, we showed that $c^* \leq \lambda - 1$ and therefore obtained the following result.

Proposition 14.1.18. *Let r^* be the smaller replication number of a Ryser design of index $\lambda \geq 2$. Then $r^* - 1 \leq (\lambda - 1)g$.*

We can now prove the following important result.

Theorem 14.1.19. *The number of points of a Ryser design of index* $\lambda > 1$ *does not exceed* $\lambda^3 + 2$ *and therefore, for any fixed* $\lambda > 1$*, there are only finitely many Ryser designs of index* λ.

Proof. Let $\mathbf{D}$ be a Ryser design of index $\lambda \geq 2$ on v points. By Theorem 14.1.15, $e \geq 2$. Therefore, (14.2) implies $\lambda(\rho + 1)^2 - (v + \rho) \geq 2\rho^2 - 2$, which is equivalent to $v \leq (\lambda - 2)\rho^2 + (2\lambda - 1)\rho + \lambda + 2$. By Theorem 14.1.17, $\rho \leq \lambda$ and therefore $v \leq \lambda^3 + 2$. □

The next theorem gives another useful inequality involving parameters of Ryser designs.

Theorem 14.1.20. *If* r *and* r^* *are the replication numbers of a Ryser design of index* $\lambda \geq 2$*, then* $r - r^* \leq (\lambda - 1)g$ *where* $g = \gcd(r - 1, r^* - 1)$.

Proof. Let $\mathbf{D} = (X, \mathcal{B})$ be a Ryser design of index $\lambda \geq 2$. Let $(r - 1)/g = c$ and $(r^* - 1)/g = c^*$. Theorem 14.1.15 implies that $e \geq 2$. Let $x, y \in E, x \neq y$. Then (14.16) and (14.18) imply that

$$\sum_{\substack{B \in \mathcal{B} \\ B \ni x, B \not\ni y}} \frac{1}{|B| - \lambda} = 1. \tag{14.22}$$

Let $B \in \mathcal{B}$ be a block that contains x and does not contain y. Then $k_B < e$ and Proposition 14.1.9 implies that $|B| \geq r^* + c - c^*$. Therefore, (14.22) implies that there are at least $r^* + c - c^* - \lambda$ blocks $B \in \mathcal{B}$ that contain x and do not contain y. On the other hand, there are exactly $v - r = r^* - 1$ blocks that do not contain y. Therefore, $r^* + c - c^* - \lambda \leq v - r$, so $c - c^* \leq \lambda - 1$, i.e., $r - r^* \leq (\lambda - 1)g$. □

14.2. Type-1 Ryser designs

Proposition 14.1.7 implies that if $\mathbf{S}$ is a symmetric (v, r, μ)-design with $v \neq 2r - 1$ and A is a block of $\mathbf{S}$, then the block complementation of $\mathbf{S}$ with respect to A is a Ryser design of index $\lambda = r - \mu$. Ryser designs that can be obtained in this way are called *type-1 Ryser designs.*

Definition 14.2.1. A *type-1 Ryser design* is a Ryser design that can be obtained as $\mathbf{S} * A$ where $\mathbf{S}$ is a symmetric design and A is a block of $\mathbf{S}$.

Remark 14.2.2. All known Ryser designs are type-1. In this and following sections we will obtain results towards the following *Ryser–Woodall conjecture*: all Ryser designs are type-1.

We will often use the following simple result.

Proposition 14.2.3. *Let* $\mathbf{D}$ *be a Ryser design and let* A *be a block of* $\mathbf{D}$. *If* $\mathbf{D} * A$ *is a symmetric design or a type-1 Ryser design, then* $\mathbf{D}$ *is a type-1 Ryser design.*

Proof. If $\mathbf{D} * A$ is a symmetric design, then, by Proposition 14.1.7(i), $\mathbf{D} = (\mathbf{D} * A) * A$ is type-1. If $\mathbf{D} * A$ is a type-1 Ryser design, i.e., $\mathbf{D} * A = \mathbf{S} * B$ where $\mathbf{S}$ is a symmetric design and B is a block of $\mathbf{S}$, then A is a block of $\mathbf{S} * B$. If $A = B$, then $\mathbf{D} = \mathbf{S}$, which is not the case. Therefore, $B \neq A$, and then $A \triangle B$ is a block of $\mathbf{S}$ and, by Proposition 14.1.7(ii), $\mathbf{D} = \mathbf{S} * (A \triangle B)$, so $\mathbf{D}$ is type-1. □

The following properties of type-1 Ryser designs are straightforward.

Proposition 14.2.4. *Let* $\mathbf{S} = (X, \mathcal{A})$ *be a nontrivial symmetric* (v, k, μ)*-design with* $v \neq 2k - 1$ *and let* A *be a block of* $\mathbf{S}$. *Then* $\mathbf{D} = \mathbf{S} * A$ *is a type-1 Ryser design of index* $\lambda = k - \mu$ *on* v *points with replication numbers* $r = k$ *and* $r^* = v - k + 1$. *The design* $\mathbf{D}$ *satisfies the following conditions*:

(i) $E(\mathbf{D}) = X \setminus A$ *and* $E^*(\mathbf{D}) = A$, *so* $E^*(\mathbf{D})$ *is a block of* $\mathbf{D}$;
(ii) $r(r - 1)/(v - 1)$ *is an integer*;
(iii) $e(\mathbf{D}) = r^* - 1$ *and* $e^*(\mathbf{D}) = r$;
(iv) $e(\mathbf{D})e^*(\mathbf{D}) = (v - 1)\lambda$;
(v) *for distinct* $x, y \in X$, *the number of blocks of* $\mathbf{D}$ *containing* $\{x, y\}$ *depends only on whether* $r(x) = r(y) = r$ *or* $r(x) = r(y) = r^*$ *or* $r(x) \neq r(y)$.

In this section, we will show that if a Ryser design $\mathbf{D}$ satisfies any one of the conditions (i)–(v), then $\mathbf{D}$ is type-1. We begin with condition (i).

Proposition 14.2.5. *If* E *or* E^* *is a block of a Ryser design* $\mathbf{D}$, *then* $\mathbf{D}$ *is type-1.*

Proof. Suppose E is a block of a Ryser design $\mathbf{D}$. Then, by Proposition 14.1.7(iv), $E(\mathbf{D} * E) = \varnothing$, i.e., all points of $\mathbf{S} = \mathbf{D} * E$ have the same replication number. Then $\mathbf{S}$ is a symmetric design. By Proposition 14.1.7, $\mathbf{D} = \mathbf{S} * E$, so $\mathbf{D}$ is type-1. □

We will now turn our attention to condition (ii) of Proposition 14.2.4. The proof that a Ryser design satisfying this condition has to be type-1 is significantly more involved. We begin with the following lemma. Recall that, for any distinct points x and y of an incidence structure, $\lambda(x, y)$ denotes the number of blocks containing $\{x, y\}$.

Lemma 14.2.6. *Let* $\mathbf{D} = (X, \mathcal{B})$ *be a Ryser design of index* λ *with replication numbers* r *and* r^*. *Let* $x \in E$ *and let*

$$U(x) = \sum_{\substack{B \in \mathcal{B} \\ B \ni x}} (|B| - \lambda) - \frac{r^2(r^* - 1)}{v - 1}.$$

Then $U(x) \geq 0$. *Moreover,* $U(x) = 0$ *if and only if all blocks containing* x *are of the same cardinality. Furthermore,*

$$\sum_{y \in E^*} \lambda(x, y) = \frac{r - 1}{r - r^*} U(x) + \frac{e^* r(r^* - 1)}{v - 1}, \tag{14.23}$$

$$\sum_{y \in E \setminus \{x\}} \lambda(x, y) = \frac{r(r - 1)(e - 1)}{v - 1} - \frac{r^* - 1}{r - r^*} U(x), \tag{14.24}$$

$$\sum_{y \in X \setminus \{x\}} (\lambda(x, y))^2 = U(x) + r^2 \left(\lambda - \frac{r - 1}{v - 1} \right). \tag{14.25}$$

Proof. The inequality between the arithmetic and harmonic means implies

$$\left(U(x) + \frac{r^2(r^* - 1)}{v - 1} \right) \sum_{\substack{B \in \mathcal{B} \\ B \ni x}} \frac{1}{|B| - \lambda} \geq r^2 \tag{14.26}$$

with equality if and only if $|B| - \lambda$ is the same for all blocks B containing x. Equation (14.16) implies that (14.26) is equivalent to $U(x) \geq 0$.

To prove (14.23), observe that

$$\sum_{y \in E^*} \lambda(x, y) = \sum_{\substack{B \in \mathcal{B} \\ B \ni x}} k_B^*.$$

From the equation dual to (14.9), we obtain

$$\lambda - k_B^* = \frac{(r - 1)(|B| - 2\lambda)}{r^* - r},$$

$$k_B^* = \frac{(r - 1)(|B| - \lambda)}{r - r^*} - \frac{\lambda(r^* - 1)}{r - r^*}.$$

Therefore,

$$\sum_{y \in E^*} \lambda(x, y) = \frac{r - 1}{r - r^*} U(x) + \frac{r(r^* - 1)((r(r - 1) - \lambda(v - 1))}{(r - r^*)(v - 1)},$$

and then (14.3) implies (14.23).

To prove (14.24), observe that

$$\sum_{y \in E \setminus \{x\}} \lambda(x, y) = \sum_{\substack{B \in \mathcal{B} \\ B \ni x}} (k_B - 1).$$

From (14.9),

$$k_B = \frac{\lambda(r-1)}{r-r^*} - \frac{r^*-1}{r-r^*}(|B|-\lambda).$$

Therefore,

$$\sum_{y\in E\setminus\{x\}} \lambda(x,y) = \frac{\lambda r(r-1)}{r-r^*} - r - \frac{r^*-1}{r-r^*}\left(U(x) + \frac{r^2(r^*-1)}{v-1}\right),$$

and after routine manipulations we obtain (14.24).

To prove (14.25), observe that

$$\sum_{y\in X\setminus\{x\}} (\lambda(x,y))^2 = \sum_{\substack{B\in\mathcal{B}\\ B\ni x}} \sum_{y\in B\setminus\{x\}} \lambda(x,y).$$

If B is a block containing x, then

$$\sum_{y\in B\setminus\{x\}} \lambda(x,y) = \sum_{\substack{A\in\mathcal{B}\\ A\ni x}} (|A\cap B|-1) = |B| - \lambda + r(\lambda-1).$$

Therefore,

$$\sum_{y\in X\setminus\{x\}} (\lambda(x,y))^2 = \sum_{\substack{B\in\mathcal{B}\\ B\ni x}} (|B|-\lambda+r(\lambda-1)) = U(x) + r^2\left(\lambda - \frac{r-1}{v-1}\right).$$

□

Theorem 14.2.7. *Let* $\mathbf{D} = (X, \mathcal{B})$ *be a Ryser design of index* λ *on* v *points with replication numbers* r *and* r^*. *If* $r(r-1)/(v-1)$ *or* $r^*(r^*-1)/(v-1)$ *is an integer, then* $\mathbf{D}$ *is type-1.*

Proof. Without loss of generality, assume that $n = r(r-1)/(v-1)$ is an integer. For $x \in E$, define $U(x)$ as in Lemma 14.2.6 and also define

$$S(x) = \sum_{y\in E\setminus\{x\}} (\lambda(x,y)-n)^2 + \sum_{y\in E^*} (\lambda(x,y)+n-r)(\lambda(x,y)+n-r-1)$$

and

$$S^*(x) = \sum_{y\in E^*} (\lambda(x,y)+n-r)^2 + \sum_{y\in E\setminus\{x\}} (\lambda(x,y)-n)(\lambda(x,y)-n-1).$$

We claim that

$$S(x) = -\frac{r^*-1}{r-r^*}U(x) \tag{14.27}$$

and

$$S^*(x) = \frac{r-1}{r-r^*}U(x). \tag{14.28}$$

We have

$$S^*(x) = \sum_{y \in X\setminus\{x\}} (\lambda(x,y))^2 + 2(n-r)\sum_{y\in E^*}\lambda(x,y)$$
$$-(2n+1)\sum_{y\in E\setminus\{x\}}\lambda(x,y) + e^*(n-r)^2 + (e-1)n(n+1).$$

We now apply (14.23)–(14.25) and use (14.3) and the equation dual to (14.3) to express $S(x)$ in terms of $U(x)$, r, and r^* and derive (14.27). Equation (14.28) is obtained in a similar manner.

Since the product of two consecutive integers is nonnegative, we obtain that $S(x) \geq 0$ and $S^*(x) \geq 0$. On the other hand, by Lemma 14.2.6, $U(x) \geq 0$. If $r > r^*$, then (14.27) implies $S(x) \leq 0$; if $r < r^*$, then (14.28) implies $S^*(x) \leq 0$. Therefore, either $S(x) = 0$ or $S^*(x) = 0$, and in either case $U(x) = 0$. Lemma 14.2.6 now implies that, for any $x \in E$, all blocks containing x are of the same cardinality m. Since $U(x) = 0$, we have

$$\sum_{\substack{B\in\mathcal{B}\\ B\ni x}}(|B|-\lambda) = \frac{r^2(r^*-1)}{v-1},$$

so $m = \lambda + r(r^*-1)/(v-1) = \lambda + r - n$.

Thus, for any Ryser design of index λ on v points with a replication number r such that $n = r(r-1)/(v-1)$ is an integer, all blocks that contain at least one point with replication number r have cardinality $\lambda + r - n$. Let A be a block of $\mathbf{D}$ of cardinality $\lambda + r - n$. If $\mathbf{D} * A$ is a symmetric design, then $\mathbf{D}$ is type-1. Suppose $\mathbf{D} * A$ is a Ryser design. By Proposition 14.1.7(iii), $\lambda(\mathbf{D} * A) = r - n$. Let B be another block of $\mathbf{D}$ of cardinality $\lambda + r - n$ (the design $\mathbf{D}$ has at least r such blocks). Then $|B \triangle A| = |B| + |A| - 2\lambda = 2(r-n)$. Thus, the Ryser design $\mathbf{D} * A$ of index $r - n$ has a block $B\triangle A$ of cardinality $2(r-n)$. Proposition 14.1.8(iv) implies that this block is not disjoint from the set $E(\mathbf{D} * A)$. Let $\mathcal{C}$ be the set of all blocks of $\mathbf{D} * A$ that are not disjoint from $E(\mathbf{D} * A)$. Since r is a replication number of $\mathbf{D} * A$ and $n = r(r-1)/(v-1)$ is an integer, we conclude that every block of $\mathcal{C}$ has cardinality $2(r-n)$. By Proposition 14.1.8(iv), $|C \cap (E^*(\mathbf{D} * A))| = r - n$ for all $C \in \mathcal{C}$. Let F be a block of $\mathbf{D} * A$ that is disjoint from $E(\mathbf{D} * A)$, i.e., $F \subseteq E^*(\mathbf{D} * A)$. Since $|F \cap C| = r - n$ for all $C \in \mathcal{C}$, we obtain that

$$F \supseteq \bigcup_{C\in\mathcal{C}}(C \cap (E^*(\mathbf{D} * A))).$$

By (14.18), for any $x \in E(\mathbf{D} * A)$ and any $y \in E^*(\mathbf{D} * A)$, there is a block $C \in \mathcal{C}$ that contains $\{x, y\}$. Therefore,

$$\bigcup_{C\in\mathcal{C}}(C \cap (E^*(\mathbf{D} * A))) = E^*(\mathbf{D} * A),$$

and then $F = E^*(\mathbf{D} * A)$. Thus, $E^*(\mathbf{D} * A)$ is a block of $\mathbf{D} * A$, and therefore, by Proposition 14.2.5, $\mathbf{D} * A$ is type-1. Proposition 14.2.3 now implies that $\mathbf{D}$ is type-1. □

Using Theorem 14.2.7, it is easy to show that a Ryser design satisfying condition (iii) or (iv) of Proposition 14.2.4 has to be type-1.

Theorem 14.2.8. *Let* $\mathbf{D}$ *be a Ryser design with replication numbers* r *and* r^*. *If* $e(\mathbf{D}) = r^*$ *or* $e^*(\mathbf{D}) = r$, *then* $\mathbf{D}$ *is type-1.*

Proof. Let $\mathbf{D}$ have v points and λ be the index of $\mathbf{D}$. Without loss of generality, we assume that $e = e(\mathbf{D}) = r^*$. Then (14.1) implies

$$r^*(v + 1 - r^*)(v - r^*) + (v - r^*)r^*(r^* - 1) = \lambda v(v - 1).$$

Therefore,

$$r^*(2 - r^*)(1 - r^*) + (1 - r^*)r^*(r^* - 1) \equiv 0 \pmod{v - 1},$$
$$-r^*(r^* - 1) \equiv 0 \pmod{v - 1}.$$

Therefore, $r^*(r^* - 1)/(v - 1)$ is an integer, and $\mathbf{D}$ is type-1 by Theorem 14.2.7. □

Theorem 14.2.8 and (14.4) immediately imply the following result.

Corollary 14.2.9. *Let* $\mathbf{D}$ *be a Ryser design with replication numbers* r *and* r^* *and let* $e(\mathbf{D})e^*(\mathbf{D}) = (v - 1)\lambda$. *Then* $\mathbf{D}$ *is type-1.*

We will now show that any Ryser design satisfying condition (v) of Proposition 14.2.4 is type-1.

Theorem 14.2.10. *Let* $\mathbf{D} = (X, \mathcal{B})$ *be a Ryser design with replication numbers* r *and* r^*. *Suppose there exist integers* μ, μ^*, *and* $\overline{\mu}$ *such that, for any distinct points* $x, y \in X$,

$$\lambda(x, y) = \begin{cases} \mu & \text{if } r(x) = r(y) = r, \\ \mu^* & \text{if } r(x) = r(y) = r^*, \\ \overline{\mu} & \text{if } r(x) \neq r(y). \end{cases}$$

Then $\mathbf{D}$ *is type-1.*

Proof. Let $|X| = v$ and let λ be the index of $\mathbf{D}$. Our first goal is to show that $\mathbf{D}$ has exactly two distinct block sizes. In the following we extend the notation $\lambda(x, y)$ to the case $x = y$ assuming that $\lambda(x, x) = r(x)$ for all $x \in X$.

For any block A of $\mathbf{D}$, let

$$L(A) = \sum_{x \in E} \sum_{y \in A} \lambda(x, y)$$

and

$$L^*(A) = \sum_{x \in E^*} \sum_{y \in A} \lambda(x, y).$$

Observe that, for $x \in E$,

$$\sum_{y \in A \cap E} \lambda(x, y) = \begin{cases} (k_A - 1)\mu + r & \text{if } x \in A, \\ k_A \mu & \text{if } x \notin A. \end{cases}$$

and

$$\sum_{y \in A \cap E^*} \lambda(x, y) = k_A^* \overline{\mu}.$$

Therefore,

$$L(A) = \sum_{x \in E} \sum_{y \in A \cap E} \lambda(x, y) + \sum_{x \in E} \sum_{y \in A \cap E^*} \lambda(x, y)$$
$$= k_A((k_A - 1)\mu + r) + (e - k_A)k\mu + ek_A^* \overline{\mu} = ((e-1)\mu + r)k_A + ek_A^* \overline{\mu}.$$

Using the "dual" expression for $L^*(A)$, we obtain that

$$L(A) + L^*(A) = sk_A + s^* k_A^*,$$

where $s = (e-1)\mu + e^* \overline{\mu} + r$ and $s^* = (e^* - 1)\mu^* + e\overline{\mu} + r^*$.

On the other hand,

$$L(A) + L^*(A) = \sum_{x \in X} \sum_{y \in A} \lambda(x, y),$$

and we will evaluate this sum by counting in two ways triples (x, y, B) with $B \in \mathcal{B}$, $x \in B$, and $y \in A \cap B$:

$$\sum_{x \in X} \sum_{y \in A} \lambda(x, y) = \sum_{B \in \mathcal{B}} |B \cap A| \cdot |B| = |A|^2 + \lambda \sum_{B \in \mathcal{B} \setminus \{A\}} |B|.$$

Since $\sum_{B \in \mathcal{B}} |B| = er + e^* r^*$, we obtain that

$$|A|^2 + \lambda(er + e^* r^* - |A|) = sk_A + s^* k_A^*. \tag{14.29}$$

From (14.9) and the "dual" equation, we express k_A and k_A^* in terms of $|A|$ and transform (14.29) into a quadratic equation for $|A|$. Thus, $|A|$ may have at most two distinct values and, since **D** is a Ryser design, it has exactly two distinct values, which we denote by n_1 and n_2. For $i = 1, 2$, let $\mathcal{B}_i = \{B \in \mathcal{B} \colon |B| = n_i\}$, $k_i = |B \cap E|$ for $B \in \mathcal{B}_i$, and $k_i^* = |B \cap E^*|$ for $B \in \mathcal{B}_i$. Then $k_i + k_i^* = n_i$.

Fix $i \in \{1, 2\}$, a point $x \in E$, and a block $A \in \mathcal{B}_i$ and count in two ways pairs (y, B) with $B \in \mathcal{B}$, $x \in B$, and $y \in A \cap B$. If $x \notin A$, we obtain

$$k_i \mu + k_i^* \overline{\mu} = r\lambda. \tag{14.30}$$

If $x \in A$, we obtain

$$r + (k_i - 1)\mu + k_i^* \overline{\mu} = n_i + (r - 1)\lambda. \tag{14.31}$$

If $k_i \neq 0$ and $k_i \neq e$, we subtract (14.31) from (14.30) to obtain that $\mu - r = \lambda - n_i$.

Thus, for $i = 1$ and for $i = 2$, one of the following three conditions is satisfied: (i) $k_i = 0$; (ii) $k_i = e$; (iii) $\mu - r = \lambda - n_i$. Also, for each i, one of the three "dual" conditions is satisfied: (i) $k_i^* = 0$; (ii) $k_i^* = e^*$; (iii) $\mu^* - r^* = \lambda - n_i$.

Assume first that $k_1 = 0$. Then $k_1^* \neq 0$. If $k_1^* = e^*$, then E^* is a block of **D**, and then **D** is type-1 by Proposition 14.2.5. Suppose $\mu^* - r^* = \lambda - n_1$. Since $n_1 \neq n_2$, we have $\mu^* - r^* \neq \lambda - n_2$, and therefore $k_2^* \in \{0, e^*\}$. If $k_2^* = 0$, then any block $A \in \mathcal{B}_1$ and any block $B \in \mathcal{B}_2$ are disjoint, which is not the case. Therefore, $k_2^* = e^*$, i.e., $E^* \subseteq B$ for all $B \in \mathcal{B}_2$. Since all blocks of $\mathcal{B}_1$ are disjoint from E, we obtain that $k_1^* = \lambda$. But then, by Proposition 14.1.8(iv), $k_1 = \lambda$, which is not the case. Thus, if $k_1 = 0$, then **D** is type-1.

Assume now that $k_1 = e$. If $k_1^* = 0$, then E is a block of **D**, and then **D** is type-1 by Proposition 14.2.5. If $k_1^* = e^*$, then $n_1 = e + e^* = v$, which is not possible. Suppose $\mu^* - r^* = \lambda - n_1$. Then $\mu^* - r^* \neq \lambda - n_2$, and therefore $k_2^* \in \{0, e^*\}$. If $k_2^* = 0$, then, for $A \in \mathcal{B}_1$ and $B \in \mathcal{B}_2$, $\lambda = |A \cap B| = |E \cap B| = k_2$. Then Proposition 14.1.8(iv) implies that $k_2^* = \lambda$, which is not the case. Therefore, $k_2^* = e^*$. Since $k_1 = e$, Proposition 14.1.9 implies that $n_1 = r^*$. Since $k_2^* = e^*$, the "dual" proposition implies that $n_2 = r$. Therefore, for $A \in \mathcal{B}_1$ and $B \in \mathcal{B}_2$, $|A \cup B| = n_1 + n_2 - \lambda = r + r^* - \lambda = v + 1 - \lambda$. On the other hand, since $E \subseteq A$ and $E^* \subseteq B$, we have $|A \cup B| = v$. Therefore, $\lambda = 1$, and then **D** is type-1. Thus, if $k_1 = e$, then **D** is type-1.

In the last two paragraphs, we showed that if $k_1 \in \{0, e\}$, then **D** is type-1. Similar reasoning would show that if $k_2 \in \{0, e\}$, then **D** is type-1. If $k_1, k_2 \notin \{0, e\}$, then $\mu - r = \lambda - n_i$ for $i = 1$ and $i = 2$, so $n_1 = n_2$. Since this is not the case, the proof is now complete. □

We will state without proof the following strengthening of Theorem 14.2.10.

Theorem 14.2.11. *Let* $\mathbf{D} = (X, \mathcal{B})$ *be a Ryser design. Suppose there exist* $x \in X$ *and integers* μ *and* $\overline{\mu}$ *such that, for any point* y *of* **D**, *other than* x,

$$\lambda(x, y) = \begin{cases} \mu & \text{if } r(y) = r(x), \\ \overline{\mu} & \text{if } r(y) \neq r(x). \end{cases}$$

Then **D** *is type-1.*

We will state without proof another characterization of type-1 Ryser designs.

Theorem 14.2.12. *A Ryser design* $\mathbf{D}$ *of index* λ *is type-1 if and only if one of the following conditions is satisfied:*

(i) *there exists* $x \in E(\mathbf{D})$ *such that*

$$\sum_{y \in E^*(\mathbf{D})} \lambda(x, y) = \lambda e^*(\mathbf{D});$$

(ii) *there exists* $y \in E^*(\mathbf{D})$ *such that*

$$\sum_{x \in E(\mathbf{D})} \lambda(x, y) = \lambda e(\mathbf{D}).$$

14.3. Ryser designs of prime index

In this section we will show that all Ryser designs of prime index are type-1. We begin with two technical lemmas.

Lemma 14.3.1. *Let* $\mathbf{D} = (X, \mathcal{B})$ *be a Ryser design of index* λ *with replication numbers* r *and* r^*. *Then, for any* $x \in E$,

$$\sum_{y \in E \setminus \{x\}} \left(\lambda(x, y) - \frac{r(r-1)}{v-1}\right)\left(\lambda(x, y) - \frac{r(r-1)}{v-1} + \frac{r - r^*}{v-1}\right) \tag{14.32}$$
$$+ \sum_{y \in E^*} \left(\lambda(x, y) - \frac{r(r^*-1)}{v-1}\right)\left(\lambda(x, y) - \frac{r(r^*-1)}{v-1} - \frac{r - r^*}{v-1}\right) = 0.$$

Proof. The left-hand side of (14.32) can be represented as

$$\sum_{y \in X \setminus \{x\}} (\lambda(x, y))^2 + \alpha \sum_{y \in E \setminus \{x\}} \lambda(x, y) + \beta \sum_{y \in E^*} \lambda(x, y) + \gamma,$$

where α, β, and γ are certain rational function of r and r^*. Equation (14.3) and the "dual" equation allow us to represent e and e^* as rational functions of r, r^*, and λ. We then use (14.23), (14.24), and (14.25) to express the left hand side of (14.32) as a rational function of r, r^*, λ, and $U(x)$ and then show (manually or with help of a computer algebra system such as MAPLE) that this rational function identically equals zero. □

Lemma 14.3.2. *Let* $\mathbf{D} = (X, \mathcal{B})$ *be a Ryser design of index* λ *with replication numbers* r *and* r^*. *Let* $g = \gcd(r-1, r^*-1)$, $c = (r-1)/g$, *and* $c^* = (r^* - 1)/g$. *Let* $h = \gcd(c - c^*, \lambda)$. *If* $h = 1$ *or* 2, *then, for any distinct* $x, y \in X$, *there exists an integer* $t(x, y)$ *such that*

$$\lambda(x, y) = \begin{cases} \dfrac{r(r-1)}{v-1} + \dfrac{r-r^*}{v-1}t(x, y) & \text{if } r(x) = r(y) = r, \\ \dfrac{r^*(r^*-1)}{v-1} + \dfrac{r-r^*}{v-1}t(x, y) & \text{if } r(x) = r(y) = r^*, \\ \dfrac{r(r^*-1)}{v-1} + \dfrac{r-r^*}{v-1}t(x, y) & \text{if } r(x) \neq r(y). \end{cases}$$

Proof. Note that (14.3) implies that

$$(c - c^*)e = (c + c^*)\lambda - r^*c^*.$$

Since c^* and $c - c^*$ are relatively prime and since $r \equiv r^* \pmod{c - c^*}$, we obtain

$$r \equiv r^* \equiv 2\lambda \pmod{c - c^*}. \tag{14.33}$$

Let $d = (c - c^*)/h$.

Case 1: $x, y \in E, x \neq y$.

Then (14.18) and (14.12) imply

$$\frac{hc}{c^*} = \sum_{\substack{B \in \mathcal{B} \\ B \ni x, y}} \frac{1}{\frac{\lambda}{h} + dl_B}.$$

Multiplying both sides of this equation by the product of all $1 + \lambda(x, y)$ denominators and reducing the resulting equation modulo d yields

$$hc(\lambda/h)^{\lambda(x,y)} \equiv c^*(\lambda/h)^{\lambda(x,y)-1}\lambda(x, y) \pmod{d}.$$

Since $c \equiv c^* \pmod{d}$ and λ/h and d are relatively prime as well as c and d, we obtain

$$\lambda(x, y) \equiv \lambda \pmod{d}. \tag{14.34}$$

Let

$$t(x, y) = \left(\lambda(x, y) - \frac{r(r-1)}{v-1}\right) \cdot \frac{v-1}{r-r^*} = \frac{(c + c^*)\lambda(x, y) - rc}{c - c^*}.$$

Suppose first that $h = 1$, i.e., $d = c - c^*$. Since $(c + c^*)\lambda(x, y) \equiv 2c\lambda(x, y) \pmod{d}$, (14.33) and (14.34) imply that $t(x, y)$ is an integer.

Suppose now that $h = 2$. Then (14.33) implies that r is even and

$$\frac{r}{2} \equiv \lambda \pmod{d}. \tag{14.35}$$

Since

$$t(x, y) = \frac{c(\lambda(x, y) - \frac{r}{2})}{d} - \lambda(x, y),$$

(14.34) and (14.35) imply that $t(x, y)$ is an integer.

Case 2: $x \in E, y \in E^*$.

Then (14.18) and (14.12) imply

$$h = \sum_{\substack{B \in \mathcal{B} \\ B \ni x, y}} \frac{1}{\frac{\lambda}{h} + dl_B}$$

and clearing the denominators again yields (14.34). Let

$$t(x, y) = \left(\lambda(x, y) - \frac{r(r^* - 1)}{v - 1}\right) \cdot \frac{v - 1}{r - r^*}$$

$$= \frac{(c + c^*)\lambda(x, y) - rc^*}{c - c^*} = \frac{c^*(\lambda(x, y) - \frac{r}{2})}{(c - c^*)/2} + \lambda(x, y).$$

If $h = 1$, then (14.33) and (14.34) imply that $t(x, y)$ is an integer. If $h = 2$, then (14.34) and (14.35) imply that $t(x, y)$ is an integer.

The case $x, y \in E^*$ is "dual" to Case 1. □

We are now ready to prove the main result of this section.

Theorem 14.3.3. *Let* $\mathbf{D} = (X, \mathcal{B})$ *be a Ryser design of index* λ *with replication numbers* r *and* r^*. *Let* $g = \gcd(r - 1, r^* - 1)$, $c = (r - 1)/g$, $c^* = (r^* - 1)/g$, *and* $h = \gcd(c - c^*, \lambda)$. *If* $h = 1$ *or* 2, *then* $\mathbf{D}$ *is type-1.*

Proof. Suppose $h = 1$ or 2. Let $x \in E$. Then Lemma 14.3.2 allows us to rewrite (14.32) as follows:

$$\frac{r - r^*}{v - 1}\left(\sum_{y \in E \setminus \{x\}} t(x, y)(t(x, y) + 1) + \sum_{y \in E^*} t(x, y)(t(x, y) - 1)\right) = 0.$$

Since all $t(x, y)$ are integers and the product of two consecutive integers is nonnegative, we obtain that $t(x, y) \in \{0, -1\}$ for all $y \in E \setminus \{x\}$ and $t(x, y) \in \{0, 1\}$ for all $y \in E^*$. Using in the same manner the equation "dual" to (14.32) would imply a similar conclusion for $x \in E^*$.

Observe that the values of each expression for $\lambda(x, y)$, given by Lemma 14.3.2, corresponding to two consecutive values of $t(x, y)$, differ by $(r - r^*)/(v - 1)$. Since this is not an integer, at most one of these values can be actually equal to $\lambda(x, y)$. Therefore, as soon as it is known whether $x, y \in E$

or $x, y \in E^*$ or $x \in E$, $y \in E^*$, the value of $\lambda(x, y)$ is uniquely determined. Theorem 14.2.9 then implies that **D** is a type-1 Ryser design. □

Theorems 14.3.3 and 14.1.20 immediately imply the following important result.

Corollary 14.3.4. *All Ryser designs of prime index are type-1.*

Let **D** be a Ryser design of index $\lambda = 2p$ where p is a prime. Then the greatest common divisor h of $c - c^*$ and $\lambda = 1, 2$, or p. If $h = 1$ or 2, then Theorem 14.3.3 implies that **D** is type-1. The case $h = p$ is significantly more involved. We state the following result without proof.

Theorem 14.3.5. *Let* $\mathbf{D} = (X, \mathcal{B})$ *be a Ryser design of index* λ *with replication numbers* r *and* r^*. *Let* $g = \gcd(r - 1, r^* - 1)$, $c = (r - 1)/g$, *and* $c^* = (r^* - 1)/g$. *If* $\gcd(c - c^*, \lambda) = \lambda/2$, *then* **D** *is type-1.*

Corollary 14.3.6. *If* p *is a prime, then all Ryser designs of index* $2p$ *are type-1.*

Proof. Let **D** be a Ryser design of index $2p$ and let $h = \gcd(c - c^*, \lambda)$. Then $h \in \{1, 2, p, 2p\}$. If $d = 1$ or 2, then **D** is type-1 by Theorem 14.3.3. If $d = p$, then **D** is type-1 by Theorem 14.3.5. Finally, Theorem 14.1.20 rules out the case $h = 2p$. □

14.4. Ryser designs of small index

In this and subsequent sections, we will often use the *p-order function* ν_p defined, for every prime p, on the set $\mathbb{Q}^*$ of all nonzero rational numbers as follows.

Definition 14.4.1. Let $r = \frac{a}{b}$ be a nonzero rational number with relatively prime integers a and b. Let m be the highest power of p dividing a and n the highest power of p dividing b. Then $\nu_p(r) = m - n$.

For instance, $\nu_3(-9/5) = 2$, $\nu_2(3/8) = -3$, and $\nu_5(11/3) = 0$.

Proof of the following proposition is left as an exercise.

Proposition 14.4.2. *For any* $x, y \in \mathbb{Q}^*$ *and any prime* p,

(i) $\nu_p(xy) = \nu_p(x) + \nu_p(y)$;
(ii) $\nu_p(x/y) = \nu_p(x) - \nu_p(y)$;
(iii) *if* $\nu_p(x) > \nu_p(y)$, *then* $\nu_p(x \pm y) = \nu_p(y)$;
(iv) *if* $\nu_p(x) = \nu_p(y)$, *then* $\nu_p(x \pm y) \geq \nu_p(y)$.

We will now prove that all Ryser designs of index $\lambda < 10$, are type-1. Due to Corollaries 14.1.16 and 14.3.4, it suffices to consider $\lambda = 4$, 6, 8, and 9. Theorem 14.1.15 implies that whenever $\lambda > 1$, we have $e \geq 2$ and $e^* \geq 2$. In the subsequent proofs, we will also use the following two simple results.

Lemma 14.4.3. *Let* $\mathbf{D}$ *be a Ryser design of index* λ *with a block* A *such that* $|A| < 2\lambda$. *If all Ryser designs of index less than* λ *are type-1, then* $\mathbf{D}$ *is type-1.*

Proof. By Proposition 14.1.7, $\mathbf{D} * A$ is either a symmetric design or a Ryser design of index $\lambda(\mathbf{D} * A) = |A| - \lambda < \lambda$. Therefore, if all Ryser designs of index less than λ are type-1, then $\mathbf{D}$ is type-1. □

Proposition 14.4.4. *Let* $\mathbf{D} = (X, \mathcal{B})$ *be a Ryser design of index* λ *on* v *points with replication numbers* r *and* r^*. *If* $|B| \geq 2\lambda$ *for all* $B \in \mathcal{B}$, *then* $r(r^* - 1) \geq (v - 1)\lambda$ *and* $r^*(r - 1) \geq (v - 1)\lambda$.

Proof. Apply (14.16) and the "dual" equation. □

Theorem 14.4.5. *Any Ryser design of index* 4 *is type-1.*

Proof. Let $\mathbf{D} = (X, \mathcal{B})$ be a Ryser design of index 4 with replication numbers $r > r^*$. Let $g = \gcd(r - 1, r^* - 1)$, $c = (r - 1)/g$, and $c^* = (r^* - 1)/g$. Proposition 14.1.18 and Theorem 14.1.20 imply that $\gcd(c - c^*, 4)$ equals 1 or 2, and then Theorem 14.3.3 implies that $\mathbf{D}$ is type-1. □

Theorem 14.4.6. *All Ryser designs of index 6 are type-1.*

Proof. Let $\mathbf{D} = (X, \mathcal{B})$ be a Ryser design of index 6 on v points with replication numbers $r > r^*$. Let $g = \gcd(r - 1, r^* - 1)$. Theorems 14.1.20 and 14.3.3 allow us to split the proof into the following two cases: (i) $r^* - 1 = g$, $r - 1 = 4g$; (ii) $r^* - 1 = 2g$, $r - 1 = 5g$. Lemma 14.4.3 allows us to assume that $|B| \geq 12$ for all $B \in \mathcal{B}$, and then Proposition 14.1.7 implies that $k_B \leq 6 \leq k_B^*$. For $0 \leq i \leq 6$, let $\mathcal{B}_i = \{B \in \mathcal{B} \colon k_B = i\}$. Note that if $i > e$, then $\mathcal{B}_i = \varnothing$. For distinct $x, y \in X$, let $\mu_i(x, y)$ be the number of blocks $B \in \mathcal{B}_i$ disjoint from $\{x, y\}$ and let $r_i(x)$ be the number of blocks $B \in \mathcal{B}_i$ that contain x.

Case 1: $r^* - 1 = g$, $r - 1 = 4g$.

Then $v - 1 = 5g$ and (14.3) implies that $3e = 29 - g$. Therefore, $g \equiv 2 \pmod 3$ and $g \leq 23$. Proposition 14.4.4 implies that $g \geq 8$. If $g \equiv \pm 1 \pmod 5$, then $\mathbf{D}$ is type-1 by Theorem 14.2.7. Therefore, we assume that $g \in \{8, 17, 20, 23\}$. From (14.6), $4k_B + k_B^* = 30$, so $|B| = 30 - 3k_B$ for all $B \in \mathcal{B}$. Let $x, y \in E$, $x \neq y$.

Suppose first that $g = 8$. Then $r = 33$, $r^* = 9$, $e = 7$, and $\rho = (r-1)/(r^*-1) = 4$. Therefore, $k_B \le 7$ for all $B \in \mathcal{B}$. We have $\mu_i(x, y) = 0$ for $i \ge 6$, and (14.19) implies that

$$\sum_{i=0}^{5} \frac{\mu_i(x, y)}{24 - 3i} = \frac{1}{12},$$

$$\sum_{i=0}^{5} \frac{\mu_i(x, y)}{8 - i} = \frac{1}{4}.$$

Since every fraction on the left hand side does not exceed $1/4$, we obtain that $\mu_5(x, y) = 0$. Applying functions ν_3, ν_5, and ν_7 to both sides of this equation, we obtain that $\mu_2(x, y) \equiv 0 \pmod 3$, $\mu_3(x, y) \equiv 0 \pmod 5$, and $\mu_1(x, y) \equiv 0 \pmod 7$. Again, since every fraction on the left hand side does not exceed $1/4$, we obtain that $\mu_i(x, y) = 0$ for $i = 1, 2, 3$. Therefore, we have

$$\frac{\mu_0(x, y)}{8} + \frac{\mu_4(x, y)}{4} = \frac{1}{4},$$

and so either $\mu_0(x, y) = 2$ and $\mu_4(x, y) = 0$ or $\mu_0(x, y) = 0$ and $\mu_4(x, y) = 1$. Since for $0 \le i \le 5$, every block $B \in \mathcal{B}_i$ is disjoint from at least one 2-subset $\{x, y\}$ of E, we obtain that $\mathcal{B}_i = \varnothing$ for $i = 1, 2, 3, 5$.

If $\mu_0(x, y) = 2$ and $\mu_4(x, y) = 0$, then $\mathcal{B}_4 = \varnothing$, and then (14.16) implies that, for all $x \in E$,

$$\frac{r_6(x)}{2} + r_7(x) = 3(\rho + 1) = 15.$$

Since also $r_6(x) + r_7(x) = r = 33$, we obtain that $r_7(x) = -3$, a contradiction. Therefore, $\mu_0(x, y) = 0$ (which implies $\mathcal{B}_0 = \varnothing$) and $\mu_4(x, y) = 1$. Since this is true for any 2-subset $\{x, y\}$ of E, we obtain that the complement of the substructure $(E, \mathcal{B}_4)$ of $\mathbf{D}$ is a 2-(7, 3, 1) design, i.e., the Fano Plane. Therefore, $r_4(x) = 4$ for all $x \in E$. Now (14.16) implies that

$$1 + \frac{r_6(x)}{2} + r_7(x) = 15.$$

Since also $r_6(x) + r_7(x) = r - 4 = 29$, we obtain that $r_7(x) = -1$, a contradiction.

Now let $g \in \{17, 20, 23\}$. Then $e \le 4$. Therefore, $\mu_i(x, y) = 0$ for $3 \le i \le 6$ and then (14.19) implies that

$$\frac{\mu_0(x, y)}{8} + \frac{\mu_1(x, y)}{7} + \frac{\mu_2(x, y)}{6} = \frac{1}{4}.$$

Applying functions ν_3 and ν_7 to both sides of this equation and taking into account that no term on the left hand side exceeds 1/4, we obtain that $\mu_1(x, y) = 0$ and $\mu_2(x, y) = 0$, and then $\mu_0(x, y) = 2$, i.e., $|\mathcal{B}_0| = 2$. If $B \in \mathcal{B}_i$ with $i \leq e - 2$, then B is disjoint from some 2-subset of E. Therefore, $\mathcal{B}_i = \varnothing$ if $i \in \{1, 2\}$ and $i \leq e - 2$.

Let $\mathcal{B}_0 = \{A_1, A_2\}$ and let $z \in A_1 \setminus A_2$. Then (14.16) implies that

$$\frac{r_{e-1}(z)}{9-e} + \frac{r_e(z)}{8-e} = \frac{119}{8}.$$

For $e = 2$, 3, and 4, $\nu_2(r_{e-1}(z)/(9-e)) \geq -1$ and $\nu_2(r_e(z)/(8-e)) \geq -2$, while $\nu_2(119/8) = -3$, a contradiction.

Thus, Case 1 is ruled out.

Case 2: $r^* - 1 = 2g, r - 1 = 5g$.

Then $v - 1 = 7g$ and (14.3) implies that $3e = 40 - 4g$. Therefore, $g \equiv 1 \pmod 3$ and $g \leq 8$. Proposition 14.4.4 implies that $g \geq 4$. If $g = 4$, then **D** is type-1 by Theorem 14.2.7. Therefore, we assume that $g = 7$. Then $e = 4$. From (14.6), $5k_B + 2k_B^* = 42$, so $k_B \in \{0, 2, 4\}$ and $2|B| = 42 - 3k_B$ for all $B \in \mathcal{B}$.

Let $x, y \in E, x \neq y$. Then (14.19) implies that

$$\frac{\mu_0(x, y)}{5} + \frac{\mu_2(x, y)}{4} = \frac{7}{10}.$$

The only solution to this equation is $\mu_0(x, y) = 1$ and $\mu_2(x, y) = 2$. Therefore, $|\mathcal{B}_0| = 1$. Let $\mathcal{B}_0 = \{A\}$ and let $z \in E^* \setminus A$. Then (14.16) implies that

$$\frac{r_2(z)}{4} + \frac{r_4(z)}{3} = \frac{21}{5}.$$

Then the values of ν_5 on the left hand side and the right hand side of this equation are different, a contradiction.

Thus, all Ryser designs of index 6 are type-1. □

Theorem 14.4.7. *All Ryser designs of index 8 are type-1.*

Proof. Let $\mathbf{D} = (X, \mathcal{B})$ be a Ryser design of index 8 on v points with replication numbers $r > r^*$. Let $g = \gcd(r - 1, r^* - 1)$. Proposition 14.1.20 and Theorem 14.3.3 allow us to split the proof into the following two cases: (i) $r^* - 1 = g, r - 1 = 5g$; (ii) $r^* - 1 = 3g, r - 1 = 7g$. Lemma 14.4.3 allows us to assume that $|B| \geq 16$ for all $B \in \mathcal{B}$, and then Proposition 14.1.7 implies that $k_B \leq 8 \leq k_B^*$. For $0 \leq i \leq 8$, let $\mathcal{B}_i = \{B \in \mathcal{B} \colon k_B = i\}$. If $i > e$, then $\mathcal{B}_i = \varnothing$. For distinct $x, y \in X$, let $\mu_i(x, y)$ be the number of blocks $B \in \mathcal{B}_i$ disjoint from $\{x, y\}$ and let $r_i(x)$ be the number of blocks $B \in \mathcal{B}_i$ that contain x.

Case 1: $r^* - 1 = g, r - 1 = 5g$.

Then $v - 1 = 6g$ and (14.3) implies that $4e = 47 - g$. Therefore, $g \equiv 3 \pmod 4$ and $g \leq 39$. Proposition 14.4.4 implies that $g \geq 10$. If $g \equiv \pm 1 \pmod 6$, then **D** is type-1 by Theorem 14.2.7. Therefore, we assume that $g \in \{15, 27, 39\}$. Then $e \leq 8$. From (14.6), $5k_B + k_B^* = 48$, so $|B| = 48 - 4k_B$ for all $B \in \mathcal{B}$.

Let $x, y \in E, x \neq y$. Then (14.19) implies that

$$\sum_{i=0}^{6} \frac{\mu_i(x, y)}{10 - i} = \frac{3}{10},$$

which we rewrite as

$$\sum_{i=1}^{4} \frac{\mu_i(x, y)}{10 - i} + \frac{\mu_6(x, y)}{4} = \frac{3 - \mu_0(x, y) - 2\mu_5(x, y)}{10}.$$

Applying ν_5 to both sides of this equation yields

$$\mu_0(x, y) + 2\mu_5(x, y) = 3 \tag{14.36}$$

and then

$$\mu_i(x, y) = 0 \text{ for } 1 \leq i \leq 4 \text{ and } i = 6. \tag{14.37}$$

Suppose $g = 15$. Then $e = 8$ and therefore (14.37) implies that $\mathcal{B}_i = \varnothing$ for $1 \leq i \leq 4$ and for $i = 6$. Since there is a 2-subset $\{x, y\}$ of E such that $\mu_5(x, y) \geq 1$ and since $\mu_0(x, y) = |\mathcal{B}_0|$ is the same for all distinct $x, y \in E$, we obtain from (14.36) that $\mu_0(x, y) = \mu_5(x, y) = 1$ for all distinct $x, y \in E$. Counting triples (x, y, B) where $x, y \in E, x \neq y, B \in \mathcal{B}_5$, and $B \cap \{x, y\} = \varnothing$ yields $8 \cdot 7 = |\mathcal{B}_5| \cdot 3 \cdot 2$, a contradiction.

Suppose $g = 27$. Then $e = 5$ and therefore $\mu_5(x, y) = 0$ for any 2-subset $\{x, y\}$ of E. Now (14.36) implies that $|\mathcal{B}_0| = 3$ and (14.37) implies that $\mathcal{B} = \mathcal{B}_0 \cup \mathcal{B}_4 \cup \mathcal{B}_5$. Let $\mathcal{B}_0 = \{A_1, A_2, A_3\}$. Since $|A_1| \cap (A_2 \cup A_3)| \leq 16$, we have $A_1 \setminus (A_2 \cup A_3) \neq \varnothing$. Let $z \in A_1 \setminus (A_2 \cup A_3)$. Then (14.16) implies that

$$\frac{1}{10} + \frac{r_4(z)}{6} + \frac{r_5(z)}{5} = \frac{24}{5}.$$

Since $r_4(z) + r_5(z) = r^* - 1 = 27$, we obtain that $r_5(z) < 0$, a contradiction.

Suppose $g = 39$. Then $e = 2$. Therefore, (14.36) implies that $|\mathcal{B}_0| = 3$. Let again $\mathcal{B}_0 = \{A_1, A_2, A_3\}$ and let $z \in A_1 \setminus (A_2 \cup A_3)$. Then (14.16) implies that

$$\frac{r_1(z)}{9} + \frac{r_2(z)}{8} = \frac{47}{10}.$$

Applying ν_5 to both sides of this equation yields a contradiction.

Case 2: $r^* - 1 = 3g, r - 1 = 7g$.

Then $v - 1 = 10g$ and (14.3) implies that $4e = 77 - 9g$. Therefore, $g \equiv 1 \pmod 4$ and $g \le 7$. Proposition 14.4.4 implies that $g \ge 4$, so $g = 5$ and $e = 8$. From (14.6), $7k_B + 3k_B^* = 80$ for all $B \in \mathcal{B}$, so $3|B| = 80 - 4k_B$, and then $k_B \equiv 2 \pmod 3$. Therefore, $k_B \in \{2, 5, 8\}$.

For $z \in E^*$, (14.16) implies that

$$\frac{r_2(z)}{16} + \frac{r_5(z)}{12} + \frac{r_8(z)}{8} = \frac{40}{7},$$

and applying ν_7 to both sides of this equation yields a contradiction. □

Theorem 14.4.8. *All Ryser designs of index 9 are type-1.*

Proof. Let $\mathbf{D} = (X, \mathcal{B})$ be a Ryser design of index 9 on v points with replication numbers $r > r^*$. Let $g = \gcd(r - 1, r^* - 1)$. Proposition 14.1.20 and Theorem 14.3.3 allow us to split the proof into the following four cases: (i) $r^* - 1 = g, r - 1 = 4g$; (ii) $r^* - 1 = g, r - 1 = 7g$; (iii) $r^* - 1 = 2g, r - 1 = 5g$; (iv) $r^* - 1 = 4g, r - 1 = 7g$. Lemma 14.4.3 allows us to assume that $|B| \ge 18$ for all $B \in \mathcal{B}$, and then Proposition 14.1.7 implies that $k_B \le 9 \le k_B^*$. For $0 \le i \le 9$, let $\mathcal{B}_i = \{B \in \mathcal{B} \colon k_B = i\}$. If $i > e$, then $\mathcal{B}_i = \varnothing$. For distinct $x, y \in X$, let $\mu_i(x, y)$ be the number of blocks $B \in \mathcal{B}_i$ disjoint from $\{x, y\}$, let $r_i(x)$ be the number of blocks $B \in \mathcal{B}_i$ that contain x, and $\alpha_i(x, y)$ be the number of blocks $B \in \mathcal{B}_i$ that contain y but do not contain x.

Case 1: $r^* - 1 = g, r - 1 = 4g$.

Then $v - 1 = 5g$ and (14.3) implies that $3e = 44 - g$. Therefore, $g \equiv 2 \pmod 3$ and $g \le 38$. Proposition 14.4.4 implies that $g \ge 11$. If $g \equiv \pm 1 \pmod 5$, then $\mathbf{D}$ is type-1 by Theorem 14.2.7. Therefore, we assume that $g \in \{17, 20, 23, 32, 35, 38\}$. Then $e \le 9$. From (14.6), $4k_B + k_B^* = 45$, so $|B| = 45 - 3k_B$ for all $B \in \mathcal{B}$.

Let $x, y \in E, x \ne y$. Then (14.19) implies that

$$\sum_{i=0}^{e-2} \frac{\mu_i(x, y)}{12 - i} = \frac{5}{12}. \tag{14.38}$$

Applying ν_{11}, ν_7, and ν_5 to both sides of (14.38) yields the following conditions: $\mu_1(x, y) \equiv 0 \pmod{11}$, $\mu_5(x, y) \equiv 0 \pmod 7$, and $\mu_2(x, y) + 2\mu_7(x, y) \equiv 0 \pmod 5$. Since every term on the left hand side of (14.38) does not exceed $5/12$, we obtain that

$$\mu_1(x, y) = \mu_2(x, y) = \mu_5(x, y) = \mu_7(x, y) = 0, \tag{14.39}$$

so

$$\frac{\mu_3(x,y)}{9}+\frac{\mu_4(x,y)}{8}=\frac{5-\mu_0(x,y)-2\mu_6(x,y)}{12}. \tag{14.40}$$

Applying ν_3 to both sides of (14.40) yields $\mu_3(x,y)\equiv 0 \pmod 3$. Since $\mu_3(x,y)<6$, we obtain

$$\mu_3(x,y)\in\{0,3\}. \tag{14.41}$$

Suppose first that $g\in\{32,35,38\}$. Then $e\le 4$ and therefore $\mu_i(x,y)=0$ for $i\ge 3$. Then (14.40) implies that $\mu_0(x,y)=5$, so $|\mathcal{B}_0|=5$ and $\mathcal{B}=\mathcal{B}_0\cup\mathcal{B}_{e-1}\cup\mathcal{B}_e$. For $x\in E$, $z\in E^*$, and $i\in\{0,e-1\}$, (14.16) and (14.18) imply that

$$\frac{\alpha_0(x,z)}{12}+\frac{\alpha_{e-1}(x,z)}{13-e}=\frac{3}{4}.$$

Note that $\alpha_0(x,z)=r_0(z)$, so $0\le r_0(z)\le 5$ and

$$\alpha_{e-1}(x,z)=\frac{(9-r_0(z))(13-e)}{12}.$$

If $e=2$, then $\alpha_{e-1}(x,z)$ is not an integer. If $e=3$, then $\alpha_{e-1}(x,z)$ is an integer if and only if $r_0(z)=3$. This means that any point $z\in E^*$ is contained in three blocks of $\mathcal{B}_0$. This cannot be the case, because any three blocks of $\mathcal{B}_0$ meet in at most nine points, while $e^*>9\cdot\binom{5}{3}=90$. If $e=4$, then $\alpha_{e-1}(x,z)$ is an integer if and only if $r_0(z)\in\{1,5\}$. This means that every point $z\in E^*$ is contained in all or one block of $\mathcal{B}_0$. Since the blocks of $\mathcal{B}_0$ have cardinality 45, this implies that $e^*=9+4\cdot 36=153$. However, if $e=4$, then $g=32$, $v=161$, and $e^*=157$, a contradiction.

Suppose next that $g\in\{17,20,23\}$ and $\mathcal{B}_3=\varnothing$. Then $7\le e\le 9$ and (14.40) implies

$$\mu_4(x,y)=\mu_0(x,y)+2\mu_6(x,y)=2. \tag{14.42}$$

Therefore, the incidence structure complementary to the substructure $(E,\mathcal{B}_4)$ of **D** is a 2-$(e,e-4,2)$ design. The basic parameter relations imply that there is no 2-(9, 5, 2) design and there is no 2-(8, 4, 2) design. Thus $e=7$ and $|\mathcal{B}_4|=14$. For $x\in E$ and $z\in E^*$, (14.16) and (14.18) imply that

$$\frac{r_0(z)}{12}+\frac{\alpha_4(x,z)}{8}+\frac{\alpha_6(x,z)}{6}=\frac{3}{4}. \tag{14.43}$$

Since $e=7$, we have $\mu_6(x,y)=0$, and then (14.42) implies that $\mu_0(x,y)=2$. Let $\mathcal{B}_0=\{A_1,A_2\}$. For $z\in A_1\setminus A_2$, $r_0(z)=1$ and then (14.43) implies that $\alpha_4(x,z)=4$ and $\alpha_6(x,z)=1$. For $z\in A_1\cap A_2$, $r_0(z)=1$ and (14.43) implies that $\alpha_4(x,z)=\alpha_6(x,z)=2$. Counting in two ways triples (x,z,B) where $x\in$

$E, z \in A_1, B \in \mathcal{B}_4, x \notin B$, and $z \in B$ yields $7 \cdot 36 \cdot 4 + 7 \cdot 9 \cdot 2 = 14 \cdot 9 \cdot 3$, a contradiction.

Suppose finally that $g \in \{17, 20, 23\}$ and $\mathcal{B}_3 \neq \varnothing$. Then (14.41) implies that $\mu_3(x, y) = 3$ for all distinct $x, y \in E$. Therefore, the incidence structure complementary to the substructure $(E, \mathcal{B}_3)$ of $\mathbf{D}$ is a 2-$(e, e-3, 3)$ design. However, for $7 \leq e \leq 9$, there is no 2-design with these parameters.

Case 2: $r^* - 1 = g, r - 1 = 7g$.

Then $v - 1 = 8g$ and (14.3) implies that $6e = 71 - g$. Therefore, $g \equiv -1 \pmod 6$ and $g \leq 59$. Proposition 14.4.4 implies that $g \geq 11$. If $g \equiv \pm 1 \pmod 8$, then $\mathbf{D}$ is type-1 by Theorem 14.2.7. Therefore, we assume that $g \in \{11, 29, 35, 53, 59\}$. Then $e \in \{10, 7, 6, 3, 2\}$. From (14.6), $7k_B + k_B^* = 72$, so $|B| = 72 - 6k_B$ for all $B \in \mathcal{B}$.

Let $x, y \in E, x \neq y$. Then (14.19) implies that

$$\sum_{i=0}^{e-2} \frac{\mu_i(x, y)}{21 - 2i} = \frac{2}{21}.$$

Therefore, $\mu_i(x, y) = 0$ for $i \geq 1$ and $\mu_0(x, y) = 2$. This implies that $\mathcal{B} = \mathcal{B}_0 \cup \mathcal{B}_{e-1} \cup \mathcal{B}_e$ and $\mathcal{B}_0 = 2$. Let $\mathcal{B}_0 = \{A_1, A_2\}$ and let $z \in A_1 \setminus A_2$. Then (14.16) and (14.18) imply that

$$\frac{\alpha_{e-1}(x, z)}{23 - 2e} = \frac{2}{21}.$$

For $e \in \{2, 3, 6, 7, 10\}$, $\nu_7(23 - 2e) = 0$, while $\nu_7(2/21) = -1$, a contradiction.

Case 3: $r^* - 1 = 2g, r - 1 = 5g$.

Then $v - 1 = 7g$ and (14.3) implies that $3e = 61 - 4g$. Therefore, $g \equiv 1 \pmod 3$ and $g \leq 13$. Proposition 14.4.4 implies that $g \geq 7$. If $g = 10$, then $\mathbf{D}$ is type-1 by Theorem 14.2.7. Therefore, we assume that $g \in \{7, 13\}$. From (14.6), $5k_B + 2k_B^* = 63$, so k_B is odd and $|B| = 63 - 3k_B$ for all $B \in \mathcal{B}$.

Let $x, y \in E, x \neq y$. Then (14.19) implies that

$$\sum_{i=0}^{e-2} \frac{\mu_i(x, y)}{18 - i} = \frac{13}{15}. \tag{14.44}$$

If $g = 13$, then $e = 3$ and (14.44) implies that $\mu_1(x, y) = 221/15$, a contradiction. Let $g = 7$, so $e = 11$. For $z \in E^*$, (14.16) implies that

$$\frac{r_1(z)}{17} + \frac{r_3(z)}{15} + \frac{r_5(z)}{13} + \frac{r_7(z)}{11} + \frac{r_9(z)}{9} = \frac{21}{5}. \tag{14.45}$$

On the other hand, $r_1(z) + r_3(z) + r_5(z) + r_7(z) + r_9(z) = r^* = 15$ and therefore

$$\frac{r_1(z)}{9} + \frac{r_3(z)}{9} + \frac{r_5(z)}{9} + \frac{r_7(z)}{9} + \frac{r_9(z)}{9} = \frac{5}{3}. \tag{14.46}$$

Clearly, the left hand side of (14.45) does not exceed the left hand side of (14.46), a contradiction.

Case 4: $r^* - 1 = 4g, r - 1 = 7g$.

Then $v - 1 = 11g$ and (14.3) implies that $3e = 95 - 16g$. Therefore, $g \equiv 2 \pmod 3$ and $g \le 4$, i.e., $g = 2$. On the other hand, Proposition 14.4.4 implies that $g \ge 4$, a contradiction.

Case 5: $r^* - 1 = 5g, r - 1 = 8g$.

Then $v - 1 = 13g$ and (14.3) implies that $3e = 112 - 25g$. Therefore, $g \equiv 2 \pmod 3$ and $g \le 4$, i.e., $g = 2$. Since Proposition 14.4.4 implies that $g \ge 3$, the proof is now complete. □

14.5. Ryser designs of small gcd

We showed in Section 14.3. that there are infinitely many values of λ such that every Ryser design of index λ is type-1. In this section, we will show that there are infinitely many values of v such that every Ryser design on v points is type-1.

Let $\mathbf{D} = (X, \mathcal{B})$ be a Ryser design of index λ on v points with replication numbers r and r^*. As before, we will denote by g the greatest common divisor of $r - 1$ and $r^* - 1$. Since $v - 1 = (r - 1) + (r^* - 1)$, g is the greatest common divisor of $v - 1$ and $r - 1$, and we let $q = (v - 1)/g$ throughout this section. Equations (14.8), (14.9), and (14.10) imply that

$$|B| - 2k_B = ql_B \tag{14.47}$$

for all $B \in \mathcal{B}$. Since $||B| - 2k_B| = |k_B^* - k_B| \le v - 1$, we obtain that $|l_B| \le g$ for all $B \in \mathcal{B}$. For $-g \le i \le g$, let

$$\mathcal{A}_i = \{B \in \mathcal{B} : l_B = i\}, \quad a_i = |\mathcal{A}_i|.$$

It follows from (14.47) that all blocks of each set $\mathcal{A}_i$ are of the same cardinality.

From (14.12),

$$|B| - \lambda = \frac{g\lambda + (r - r^*)l_B}{g}.$$

and we rewrite (14.15) as

$$\sum_{i=-g}^{g} \frac{ga_i}{g\lambda + (r - r^*)i} = \frac{g^2q^2}{(r-1)(r^*-1)} - \frac{1}{\lambda}. \tag{14.48}$$

Counting in two ways all flags (x, B) and the flags with $x \in E$ yields $\sum_{B\in\mathcal{B}} |B| = er + e^*r^*$ and $\sum_{B\in\mathcal{B}} k_B = er$. Let $s = \sum_{B\in\mathcal{B}} l_B$. Then, (14.47) implies that

$$s = \sum_{B\in\mathcal{B}} l_B = \frac{1}{q}(e^*r^* - er) = \frac{1}{q}((gq + 1 - e)(gq + 2 - r) - er),$$

giving

$$sq = gq(gq - e - r + 3) - (2e + r - 2). \tag{14.49}$$

Therefore, $2e + r - 2 \equiv 0 \pmod{q}$, and we introduce the integral parameter $m = m(\mathbf{D})$ by

$$2e + r - 2 = mq \tag{14.50}$$

and the "dual" parameter m^* by

$$2e^* + r^* - 2 = m^*q. \tag{14.51}$$

Then

$$m + m^* = 3g. \tag{14.52}$$

Since $s = \sum_{B\in\mathcal{B}} l_B = \sum_{i=-g}^{g} ia_i$, (14.49), (14.50) and (14.52) imply that

$$\sum_{i=-g}^{g} ia_i = g^2q - g(e + r) + m^*. \tag{14.53}$$

In this section, we will regard λ, q, g, and m as the main parameters of a Ryser design. In the following two lemmas, we express r, r^*, and e in terms of these parameters and give bounds for the integers l_B.

Lemma 14.5.1. *Let $\mathbf{D} = (X, \mathcal{B})$ be a Ryser design of index λ on v points with replication numbers r and r^*. Let $g = \gcd(r-1, r^*-1)$, $q = (v-1)/g$, and $e = e(\mathbf{D})$. Let m and m^* be defined by (14.50) and (14.51). Then $m = m^*$ if and only if $v = 4\lambda - 1$. Furthermore, if $v \neq 4\lambda - 1$, then*

$$r = \frac{(2g - m)(gq + 2) - 2g\lambda}{3g - 2m}, \tag{14.54}$$

$$r^* = \frac{2g\lambda + (g - m)(gq + 2)}{3g - 2m}, \tag{14.55}$$

and

$$e = \frac{g\lambda - (g-m)^2 q + (g-m)}{3g - 2m}. \tag{14.56}$$

Proof. From (14.50) and (14.51), $(m^* - m)q = 2(e^* - e) - (r - r^*)$. From (14.3) and the "dual" equation, $(r - r^*)(e^* - e) = r(r-1) + r^*(r^* - 1) - 2\lambda(v-1)$. Therefore,

$$(r - r^*)(m^* - m)q = 2(r - r^*)(e^* - e) - (r - r^*)^2 = (v-1)(v+1-4\lambda),$$

and then

$$(r - r^*)(m^* - m) = g(v + 1 - 4\lambda). \tag{14.57}$$

Therefore, $m = m^*$ if and only if $v = 4\lambda - 1$.

Suppose $v \neq 4\lambda - 1$. Since $r^* = gq + 2 - r$ and $m^* - m = 3g - 2m \neq 0$, we use (14.6) to express r in terms of λ, q, g, and m and obtain (14.54). Since $r^* = gq + 2 - r$, we derive (14.55), and we use (14.50) to obtain (14.56). □

Remark 14.5.2. If **D** is a type-1 Ryser design of index λ on v points, then Proposition 2.4.12 implies that $v \geq 4\lambda - 1$. It is not known whether this inequality holds for all Ryser designs. If r is the larger replication number of **D**, then (14.57) implies that $v \geq 4\lambda - 1$ if and only if $r - r^* \leq 2(e^* - e)$.

Lemma 14.5.3. *Let* $\mathbf{D} = (X, \mathcal{B})$ *be a Ryser design of index* λ *on* v *points with replication numbers* r *and* r^*. *Let* $g = \gcd(r-1, r^*-1)$, $q = (v-1)/g$, *and* $e = e(\mathbf{D})$. *Let* m *and* m^* *be defined by* (14.50) *and* (14.51). *If* $\lambda \neq 1$, *then, for all* $B \in \mathcal{B}$,

$$|l_B| \leq g - 1. \tag{14.58}$$

If $v \neq 4\lambda - 1$, *then, for all* $B \in \mathcal{B}$,

$$g - m \leq l_B < \frac{(3g - 2m)\lambda}{2\lambda - 1 + (g-m)q}. \tag{14.59}$$

Furthermore, $l_B = g - m$ *if and only if* $B \supseteq E$ *and the upper bound in* (14.59) *is attained if and only if* $B \subseteq E^*$.

Proof. If $\lambda \neq 1$, then, by Theorem 14.1.15, $e \geq 2$ and $e^* \geq 2$. Therefore, for any $B \in \mathcal{B}$, $||B| - 2k_B| < v - 1$, and then (14.47) implies (14.58).

Suppose $v \neq 4\lambda - 1$. We have $0 \leq k_B \leq e$ for all $B \in \mathcal{B}$, and then (14.10) implies that

$$\frac{\lambda - e}{(r^* - 1)/g} \leq l_B \leq \frac{\lambda}{(r^* - 1)/g}.$$

From (14.55),

$$\frac{r^* - 1}{g} = \frac{2\lambda - 1 + (g - m)q}{3g - 2m}.$$

From (14.56),

$$\lambda - e = \frac{(g - m)(2\lambda - 1 + (g - m)q)}{3g - 2m},$$

and (14.59) follows. □

Remark 14.5.4. Let $\mathbf{D} = (X, \mathcal{B})$ be a symmetric (v, r, λ)-design with $1 < r < v$ and $r \neq (v + 1)/2$. Let $e = v$, $e^* = 0$, and $r^* = v + 1 - r$. Let $g = \gcd(r - 1, r^* - 1)$ and $q = (v - 1)/g$. Then, as in the case of Ryser designs, $m = (2e + r - 2)/q$ is a positive integer. For a symmetric (v, r, λ)-design, $m = 2g + gr/(v - 1)$.

Proposition 14.1.7(v) implies that if a Ryser (or symmetric) design $\mathbf{D}$ is replaced by $\mathbf{D} * A$ where A is a block of $\mathbf{D}$, then the parameter m changes as follows:

$$m(\mathbf{D} * A) = m(\mathbf{D}) + 2l_A. \tag{14.60}$$

The smaller the value of m, the better is the lower bound on l_B given by (14.59). This motivates the following notion.

Definition 14.5.5. A design $\mathbf{D} \in \mathfrak{D}_r(X)$ is said to be *extremal* if $|m(\mathbf{D}) - m^*(\mathbf{D})| \geq |m(\mathbf{D} * A) - m^*(\mathbf{D} * A)|$ for every block A of $\mathbf{D}$.

Proposition 14.5.6. *Every symmetric design* $\mathbf{D} \in \mathfrak{D}_r(X)$ *is extremal.*

Proof. Let $\mathbf{D}$ be a symmetric (v, r, λ)-design and A a block of $\mathbf{D}$. Let $r^* = v + 1 - r$ and $g = \gcd(r - 1, r^* - 1)$. Then, by Remark 14.5.4, $m(\mathbf{D}) = 2g + gr/(v - 1)$ and then $m^*(\mathbf{D}) = 3g - m(\mathbf{D}) = g - gr/(v - 1)$, so $|m(\mathbf{D}) - m^*(\mathbf{D})| = g + 2gr/(v - 1)$. Since $k_A^* = 0$, (14.11) implies that $l_A = -g\lambda/(r - 1)$. Therefore, $m(\mathbf{D} * A) = m(\mathbf{D}) + 2l_A = 2g + gr/(v - 1) - 2g\lambda/(r - 1)$ and $m^*(\mathbf{D} * A) = g - gr/(v - 1) + 2g\lambda/(r - 1)$, so $|m(\mathbf{D} * A) - m^*(\mathbf{D} * A)| = |g + 2gr/(v - 1) - 4g\lambda/(r - 1)|$. Since $r/(v - 1) = \lambda/(r - 1)$, this implies $|m(\mathbf{D}) - m^*(\mathbf{D})| > |m(\mathbf{D} * A) - m^*(\mathbf{D} * A)|$, and therefore the design $\mathbf{D}$ is extremal. □

Remark 14.5.7. Let $\mathbf{D} = (X, \mathcal{B})$ be a Ryser design. Choose a block A of $\mathbf{D}$ so that $|m(\mathbf{D} * A) - m^*(\mathbf{D} * A)| \geq |m(\mathbf{D} * B) - m^*(\mathbf{D} * B)|$ for all $B \in \mathcal{B}$. Proposition 14.1.8(ii) implies that $\mathbf{D} * A$ is extremal. Therefore, if we want to show that a Ryser design $\mathbf{D}$ is type-1, we can assume without loss of generality that $\mathbf{D}$ is extremal and then either show that $\mathbf{D}$ is type-1 or obtain a contradiction.

For an extremal Ryser design, the parameters l_B satisfy the following inequality.

Lemma 14.5.8. *Let* **D** *be an extremal Ryser design with* $m(\mathbf{D}) \le m^*(\mathbf{D})$. *Then* $0 \le l_B \le (m^*(\mathbf{D}) - m(\mathbf{D}))/2$, *for every block* B *of* **D**.

Proof. For a block B of $\mathbf{D}$, we have $m(\mathbf{D} * B) = m(\mathbf{D}) + 2l_B$ and $m^*(\mathbf{D} * B) = m^*(\mathbf{D}) + 2l_B^* = m^*(\mathbf{D}) - 2l_B$. Therefore, $|m(\mathbf{D} * B) - m^*(\mathbf{D} * B)| = |m(\mathbf{D}) - m^*(\mathbf{D}) + 4l_B|$. Since $\mathbf{D}$ is extremal,

$$m(\mathbf{D}) - m^*(\mathbf{D}) \le m(\mathbf{D}) - m^*(\mathbf{D}) + 4l_B \le m^*(\mathbf{D}) - m(\mathbf{D}),$$

and therefore, $0 \le l_B \le (m^*(\mathbf{D}) - m(\mathbf{D}))/2$. □

We will now investigate Ryser designs with small values of $g = \gcd(r-1, r^*-1)$. Note that Theorem 14.1.16 implies that all Ryser designs with $g = 1$ are type-1. In the subsequent proofs we will often use the fact that all Ryser designs of index 9 or less are type-1.

Theorem 14.5.9. *Let* **D** *be a Ryser design with replication numbers* r *and* r^*. *If* $\gcd(r-1, r^*-1) = 2$, *then* **D** *is type-1.*

Proof. Let $\mathbf{D} = (X, \mathcal{B})$. Without loss of generality, we assume that $\mathbf{D}$ is extremal and that $m \le m^*$. Since r is odd, (14.50) implies that m is odd. By (14.52), $m + m^* = 6$, so $m = 1$ or 3.

If $m = 1$, then (14.58) and (14.59) imply that $l_B = 1$ for all $B \in \mathcal{B}$. If $m = 3$, then Lemma 14.5.8 implies that $l_B = 0$ for all $B \in \mathcal{B}$. In either case, all blocks of $\mathbf{D}$ have the same cardinality, a contradiction. □

Theorem 14.5.10. *Let* **D** *be a Ryser design with replication numbers* r *and* r^*. *If* $\gcd(r-1, r^*-1) = 3$, *then* **D** *is type-1.*

Proof. Let $\mathbf{D} = (X, \mathcal{B})$, let $v = |X|$, and let λ be the index of $\mathbf{D}$. Without loss of generality, we assume that $\mathbf{D}$ is extremal and that $m \le m^*$. Then (14.52) implies that $1 \le m \le 4$. If $m = 1$, then (14.58) and (14.59) imply that $l_B = 2$ for all $B \in \mathcal{B}$. If $m = 4$, then Lemma 14.5.8 implies that $l_B = 0$ for all $B \in \mathcal{B}$. In either case, all blocks of $\mathbf{D}$ have the same cardinality, a contradiction.

Suppose $m = 2$. Then (14.59) and Lemma 14.5.8 imply that $1 \le l_B \le 2$ for all $B \in \mathcal{B}$. Since not all blocks of $\mathbf{D}$ are of the same cardinality, there is a block $B \in \mathcal{A}_1$. By Lemma 14.5.3, $B \supseteq E$ and (14.59) implies that $5\lambda \ge 2(2\lambda - 1 + q)$, so $\lambda \ge 2q - 2$. Now (14.54) implies that $r = (12q - 6\lambda + 8)/5$ and therefore $r \le 4$. Since $r \equiv 1 \pmod 3$ and $r > 1$, we obtain that $r = 4$. Therefore, $\lambda = 2q - 2$, and then (14.56) implies that $e = q - 1$. Since $B \in \mathcal{A}_1$, we have $k_B = e$, and then (14.47) implies that $|B| = 3q - 2$. If $B_1, B_2 \in \mathcal{A}_1$ and

$B_1 \neq B_2$, then $|B_1 \cup B_2| = 2(3q-2) - \lambda = 4q - 2$. If $q \leq 3$, then $\lambda \leq 4$, so **D** is type-1. If $q \geq 4$, then $4q - 2 > v$, and therefore, $a_1 \leq 1$. Since $\mathcal{A}_1 \neq \varnothing$, we obtain that $a_1 = 1$ and then (14.53) implies that $a_2 = 3q - 3/2$, a contradiction.

Suppose $m = 3$. Then (14.54), (14.55), and (14.56) imply that $r = -2\lambda + 3q + 2$, $r^* = 2\lambda$, and $e = \lambda$. Lemma 14.5.8 implies that $0 \leq l_B \leq 1$ for all $B \in \mathcal{B}$, and then (14.53) implies that $a_1 = 3\lambda$. Therefore, $a_0 = v - a_1 = 3q - 3\lambda + 1$. Since all blocks of $\mathcal{A}_0$ contain E and meet E in λ points, Proposition 14.1.8 implies that all these blocks meet E^* in pairwise disjoint subsets of cardinality λ. Therefore, $a_0\lambda \leq e^*$, i.e., $(3q - 3\lambda + 1)\lambda \leq 3q - \lambda + 1$, so

$$3q(\lambda - 1) \leq (\lambda - 1)(3\lambda + 1) + 2. \tag{14.61}$$

Since all Ryser designs of index $\lambda \leq 3$ are type-1, we assume that $\lambda > 3$, and then (14.61) implies that $3q \leq 3\lambda + 1$, so $q \leq \lambda$. Since $a_0 \geq 1$, we have $q \geq \lambda$, and therefore, $q = \lambda$. Since $r^* \equiv 1 \pmod 3$, we obtain that $\lambda \equiv 2 \pmod 3$, and then $r^*(r^* - 1)/(v - 1) = 2(2\lambda - 1)/3$ is an integer. Thus, **D** is type-1 by Theorem 14.2.7.

The proof is now complete. □

Theorem 14.5.11. *Let* **D** *be a Ryser design with replication numbers r and r^*. If* $\gcd(r-1, r^*-1) = 4$, *then* **D** *is type-1.*

Proof. Let $\mathbf{D} = (X, \mathcal{B})$, let $v = |X|$, and let λ be the index of **D**. Without loss of generality, we assume that **D** is extremal and that $m \leq m^*$. Then (14.52) implies that $m \leq 6$. Since $r \equiv 1 \pmod 4$, r is odd and (14.50) implies that m is odd. Therefore, $m \in \{1, 3, 5\}$.

If $m = 1$, then (14.58) and (14.59) imply that $l_B = 3$, so all blocks of **D** have the same cardinality, a contradiction.

Suppose $m = 3$. Then (14.54) implies that $r = (10q - 4\lambda + 5)/3$. Since $r \equiv 1 \pmod 4$ and $r > 1$, we have $r \geq 5$, which implies that $q \geq (2\lambda + 5)/5$. Therefore,

$$\frac{6\lambda}{2\lambda - 1 + q} \leq \frac{5}{2},$$

and then (14.59) implies that $1 \leq l_B \leq 2$ for all $B \in \mathcal{B}$. Therefore, $a_1 + a_2 = 4q + 1$ and (14.53) implies that $a_1 + 2a_2 = (10q + 8\lambda + 5)/3$. These equations imply that $a_1 = (14q - 8\lambda + 1)/3$ and $a_2 = (8\lambda - 2q + 2)/3$. Now (14.48) can be transformed into the following equation:

$$(2\lambda - 2q - 1)^2(5\lambda + 34 - (7\lambda - 10)(7q - 4\lambda - 2)) = 0.$$

Since $2\lambda - 2q - 1 \neq 0$, we obtain that $(7\lambda - 10)(7q - 4\lambda - 2) = 5\lambda + 34$. If $7\lambda - 10 = 5\lambda + 34$, then $\lambda = 22$ and $q = 13$. This gives $r = 47/3$, a contradiction. Therefore, $2(7\lambda - 10) \leq 5\lambda + 34$, which gives $\lambda \leq 6$, so **D** is type-1.

Suppose now that $m = 5$. Then (14.54), (14.55), and (14.56) imply that $r = 6q - 4\lambda + 3$, $r^* = 4\lambda - 2q - 1$, and $e = (4\lambda - q - 1)/2$. Lemma 14.5.8 implies that $0 \leq l_B \leq 1$ for all $B \in \mathcal{B}$. Now (14.53) implies that $a_1 = 8\lambda - 6q + 3$, and then $a_0 = v - a_1 = 10q - 8\lambda + 4$. Now (14.48) can be transformed into $(2\lambda - 2q - 1)^2 f(q) = 0$ where $f(q) = 15q^2 - 2(16\lambda - 10)q + (16\lambda^2 - 16\lambda + 5)$. The discriminant of the quadratic polynomial $f(q)$ has to be a square. This implies $(4\lambda - 10)^2 - x^2 = 75$ with a positive integer x. The solutions of the equation are $\lambda = x = 5$, $\lambda = 6$ and $x = 11$, $\lambda = 12$ and $x = 37$. The last solution does not yield an integer q, so $\lambda \leq 6$, and therefore **D** is type-1. The proof is now complete. □

Theorem 14.5.12. *Let* **D** *be a Ryser design with replication numbers r and r^*. If* $\gcd(r - 1, r^* - 1) = 5$, *then* **D** *is type-1.*

Proof. Let $\mathbf{D} = (X, \mathcal{B})$, let $v = |X|$, and let λ be the index of **D**. Without loss of generality, we assume that **D** is extremal and that $m \leq m^*$. Then (14.52) implies that $1 \leq m \leq 7$. If $m = 1$, then (14.58) and (14.59) imply that $l_B = 4$ for all $B \in \mathcal{B}$. If $m = 7$, then Lemma 14.5.8 implies that $l_B = 0$ for all $B \in \mathcal{B}$. In either case, all blocks of **D** have the same cardinality, a contradiction. Therefore, we assume that $2 \leq m \leq 6$.

Case 1: $m = 2$.

Since not all blocks of **D** are of the same cardinality, (14.59) implies that $11\lambda/(2\lambda - 1 + 3q) \geq 4$, so $3\lambda \geq 12q - 4$, and then

$$\lambda \geq 4q - 1. \tag{14.62}$$

From (14.54), $r = (40q - 10\lambda + 16)/11$. Since $r \equiv 1 \pmod 5$ and $r > 1$, we have $r \geq 6$, which implies $\lambda \leq 4q - 5$, in contradiction with (14.62).

Case 2: $m = 3$.

From (14.54), $r = (35q - 10\lambda + 14)/9$. Since $r \geq 6$, we obtain $q \geq (2\lambda + 8)/7$. Therefore,

$$\frac{9\lambda}{2\lambda - 1 + 2q} \leq \frac{63\lambda}{18\lambda + 11} < 4,$$

and then (14.59) implies that $2 \leq l_B \leq 3$ for all $B \in \mathcal{B}$. Therefore, $a_2 + a_3 = 5q + 1$ and, by (14.53), $2a_2 + 3a_3 = (70q + 25\lambda + 28)/9$. These equations give $a_2 = (65q - 25\lambda - 1)/9$ and $a_3 = (25\lambda - 20q + 10)/9$. Now (14.48) implies after routine manipulations that

$$(5q - 4\lambda + 2)^2(4q - 5\lambda - 2)((13\lambda - 21)q - 5\lambda^2 + 4\lambda - 3) = 0.$$

Since not all blocks of **D** are of the same cardinality, we have $a_3 \geq 1$. Therefore, $4q - 5\lambda - 2 = -a_3/5 \neq 0$. If $5q - 4\lambda + 2 = 0$, then $r = 2\lambda$ and $v = 4\lambda - 1$, and therefore $r = r^*$, a contradiction. Hence, $(13\lambda - 21)q = 5\lambda^2 - 4\lambda + 3$, which can be rewritten as $(13q - 5\lambda - 4)(13\lambda - 21) = \lambda + 123$. Thus, $13\lambda - 21$ divides $\lambda + 123$. If $13\lambda - 21 = \lambda + 123$, then $\lambda = 12, q = 5$, and $r = 23/3$, a contradiction. Thus, $2(13\lambda - 21) \leq \lambda + 123$, which gives $\lambda \leq 6$, and therefore **D** is type-1.

Case 3: $m = 4$.

From (14.54) and (14.56), $r = (30q - 10\lambda + 12)/7$ and $e = (5\lambda - q + 1)/7$. Since $r \geq 6$, we obtain that $3q \geq \lambda + 3$, and then

$$\frac{7\lambda}{2\lambda - 1 + q} \leq 3.$$

Therefore, (14.59) implies that $1 \leq l_B \leq 3$ for all $B \in \mathcal{B}$.

Assume first that $\mathcal{A}_3 \neq \varnothing$. Then $7\lambda/(2\lambda - 1 + q) = 3$. Then $\lambda = 3q - 3$, $r = 6$, and $e = 2q - 2$. By Lemma 14.5.3, any block $B \in \mathcal{A}_1$ contains E and then, by (14.13), $|B| = r^* = 5q - 4$. If $B_1, B_2 \in \mathcal{A}_1$ and $B_1 \neq B_2$, then $|B_1 \cup B_2| = 2(5q - 4) - \lambda = 7q - 5$. If $q \leq 2$, then $\lambda \leq 4$, so **D** is type-1. If $q \geq 3$, then $|B_1 \cup B_2| > v$, so $a_1 \leq 1$. If $B \in \mathcal{A}_2$, then $k_B = \lambda - 2(q - 1) = q - 1$, so, by (14.47), $|B| = 4q - 2$. If $B \in \mathcal{A}_3$, then $k_B = 0$ and $|B| = 3q$. Therefore, (14.48) implies that

$$\frac{a_1}{2q - 1} + \frac{a_2}{q + 1} + \frac{a_3}{3} = q + 1 + \frac{2}{3(q - 1)}. \tag{14.63}$$

We have $a_1 + a_2 + a_3 = 5q + 1$ and, from (14.53), $a_1 + 2a_2 + 3a_3 = 15q - 9$. Since $a_1 \in \{0, 1\}$, we obtain that (a_1, a_2, a_3) equals $(0, 12, 5q - 11)$ or $(1, 10, 5q - 10)$. In either case, the left hand side of (14.63) is greater than $10/(q + 1) + (5q - 11)/3$, while the right hand side is less than $q + 2$. Then we have

$$q + 2 > \frac{10}{q + 1} + \frac{5q - 11}{3},$$

which implies that $q \leq 6$. One can check that, for $q = 5$ or 6, neither triple (a_1, a_2, a_3) satisfies (14.63). Therefore, $q \leq 4$, and then $\lambda \leq 9$, so **D** is type-1.

Assume now that $\mathcal{A}_3 = \varnothing$. Then $a_1 + a_2 = 5q + 1$ and $a_1 + 2a_2 = 15q - 9$, so $a_1 = (40q - 25\lambda + 2)/7$ and $a_2 = (25\lambda - 5q + 5)/7$. Then

(14.48) implies that

$$(5q - 4\lambda + 2)^2(5\lambda - q + 1)(4q(3 - 2\lambda) + 5\lambda^2 - 3\lambda + 2) = 0.$$

Since not all blocks of **D** are of the same cardinality, we have $a_2 \geq 1$. Therefore, $5\lambda - q + 1 \neq 0$. If $5q - 4\lambda + 2 = 0$, then $r = 2\lambda$ and $v = 4\lambda - 1$, and therefore $r = r^*$, a contradiction. Therefore, $4q(3 - 2\lambda) + 5\lambda^2 - 3\lambda + 2 = 0$, which we rewrite as $(8q - 5\lambda - 4)(2\lambda - 3) = \lambda + 16$. Thus, $2\lambda - 3$ divides $\lambda + 16$. If $2\lambda - 3 = \lambda + 16$, then $\lambda = 19$ and $q = 25/2$, a contradiction. Hence, $2(2\lambda - 3) \leq \lambda + 16$. This gives $\lambda \leq 7$, so **D** is type-1.

Case 4: $m = 5$.

Then (14.59) and Lemma 14.5.8 imply that $0 \leq l_B \leq 2$ for all $B \in \mathcal{B}$. Note that if $5q - 4\lambda + 2 = 0$, then $r = 2\lambda$ and $v = 4\lambda - 1$, and therefore $r = r^*$, a contradiction. Therefore, we assume that $5q - 4\lambda + 2 \neq 0$.

Subcase $a_0 = 0$.

In this case $a_1 + a_2 = 5q + 1$ and $a_1 + 2a_2 = 25q + 10 - 5(\lambda + 5q - 2\lambda + 2) = 5\lambda$. Thus, $a_1 = 10q - 5\lambda + 2$ and $a_2 = 5\lambda - 5q - 1$, and (14.48) implies that

$$50\lambda q^2 + 5q(-15\lambda^2 + 11\lambda + 2) + 25\lambda^3 - 22\lambda^2 + 7\lambda + 2 = 0$$

The discriminant $D = 625\lambda^4 - 3850\lambda^3 + 125\lambda^2 + 700\lambda + 100$ of this quadratic (in q) equation must be a square. Observe that for $\lambda \geq 378$, $(25\lambda^2 - 77\lambda - 117)^2 < D < (25\lambda^2 - 77\lambda - 116)^2$, so D is not a square. One can check that, for $2 \leq \lambda \leq 377$, D is not a square either.

Subcase $a_0 = 1$.

In this case $a_1 + a_2 = 5q$ and $a_1 + 2a_2 = 5\lambda$. Thus, $a_0 = 1, a_1 = 5(2q - \lambda)$ and $a_2 = 5(\lambda - q)$ and (14.48) implies that

$$50\lambda q^2 - 5q(15\lambda^2 - 7\lambda - 4) + 25\lambda^3 - 14\lambda^2 - \lambda + 4 = 0$$

Therefore, $D = 625\lambda^4 - 2450\lambda^3 - 1575\lambda^2 + 600\lambda + 400$ is a square. Observe that for $\lambda \geq 303$, $(25\lambda^2 - 49\lambda - 80)^2 < D < (25\lambda^2 - 49\lambda - 79)^2$. One can check that, for $2 \leq \lambda \leq 302$, D is a square if and only if $\lambda \in \{5, 8\}$, and therefore, **D** is type-1.

Subcase $a_2 = 0$.

In this case $a_0 + a_1 = 5q + 1$ and $a_1 = 5\lambda$. Thus, $a_0 = 5q - 5\lambda + 1$ and (14.48) implies that $5q(\lambda - 1) - 5\lambda^2 + 2\lambda - 1 = 0$, which can be rewritten as

$(5q - 5\lambda - 3)(\lambda - 1) = 4$. Therefore, $\lambda - 1$ divides 4. This implies that $\lambda \le 5$, and therefore **D** is type-1.

Subcase $a_2 = 1$

In this case $a_0 + a_1 = 5q$ and $a_1 = 5\lambda - 2$. Thus, $a_0 = 5q - 5\lambda + 2$ and (14.48) implies that

$$50q^2(\lambda - 1) - 5q(13\lambda^2 - 15\lambda + 8) + 15\lambda^3 - 34\lambda^2 + 19\lambda - 6 = 0.$$

Therefore, $D = 49\lambda^4 + 2\lambda^3 + 9\lambda^2 - 40\lambda + 16$ is a square. Observe that, for $\lambda \ge 41$, $(49\lambda^2 + \lambda + 4)^2 < D < (49\lambda^2 + \lambda + 5)^2$. One can check that, for $2 \le \lambda \le 40$, D is not a square either.

Subcase $a_0 \ge 2$ and $a_2 \ge 2$

Then there exist distinct blocks B_1 and B_2 such that $l_{B_1} = l_{B_2} = 0$ and distinct blocks C_1 and C_2 such that $l_{C_1} = l_{C_2} = 2$.

For blocks B_1 and B_2, we have $k_{B_1} = k_{B_2} = e = \lambda$. This implies $|B_1| = |B_2| = 2\lambda$ and $B_1 \cap B_2 \cap E^* = \varnothing$. Let $E^* \cap B_1 = E_1^*$ and $E^* \cap B_2 = E_2^*$.

For blocks C_1 and C_2, we have $k_{C_1} = k_{C_2} = (\lambda + 2)/5$. Since $|C_1 \cap B_1| = \lambda$, we have $|C_1 \cap E_1^*| = \lambda - (\lambda + 2)/5 = (4\lambda - 2)/5$. Similarly, $|C_2 \cap B_1| = \lambda$ and $|C_2 \cap E_1^*| = (4\lambda - 2)/5$. Thus, $|C_1 \cap C_2 \cap E_1^*| \ge |C_1 \cap E_1^*| + |C_2 \cap E_2^*| - |E_1^*| = (3\lambda - 4)/5$. By similar reasons, $|C_1 \cap C_2 \cap E_2^*| \ge (3\lambda - 4)/5$.

Now, $|C_1 \cap C_2| = \lambda$. On the other hand, $|C_1 \cap C_2| \ge |C_1 \cap C_2 \cap E_1^*| + |C_1 \cap C_2 \cap E_2^*| \ge (6\lambda - 8)/5$. Therefore, $(3\lambda - 4)/5 \le (6\lambda - 8)/5$. This implies $\lambda \le 8$, and therefore **D** is type-1.

Case 5: $m = 6$.

Then Lemma 14.5.8 implies that $0 \le l_B \le 1$ for all $B \in \mathcal{B}$. Therefore, from (14.53), $a_1 = (25\lambda - 20q - 8)/3$, and then $a_0 = v - a_1 = (35q - 25\lambda + 11)/3$. Now (14.48) implies that

$$(5q - 4\lambda + 2)^2(28q^2 - 5q(11\lambda - 7) + 25\lambda^2 - 22\lambda + 7) = 0.$$

If $5q - 4\lambda + 2 = 0$, then (14.53) implies that $r = 2\lambda$. Since $v = 5q + 1 = 4\lambda - 1$, we obtain that $r^* = r$, a contradiction. Therefore, $28q^2 - 5q(11\lambda - 7) + 25\lambda^2 - 22\lambda + 7 = 0$, and therefore $D = 25\lambda^2 - 154\lambda + 49$ is a square. However, for $\lambda \ge 35$, $(5\lambda - 16)^2 < D < (5\lambda - 15)^2$. One can check that D is not a square for $10 \le \lambda \le 34$, except $\lambda = 15$. In this case, $q = 11$, $r = 26$, and $r^* = 31$. Therefore, $(r - r^*)/g$ and λ are relatively prime, so **D** is type-1 by Theorem 14.3.3. If $\lambda \le 9$, then **D** is type-1 by the results of Section 14.4. This completes the proof of this theorem. □

In a similar manner but with more involved analysis, one can prove the next theorem.

Theorem 14.5.13. *Let* **D** *be a Ryser design with replication numbers r and r^* and let $g = \gcd(r-1, r^*-1)$. If $g = 6$ or $g = 8$, then* **D** *is type-1.*

We will now obtain several infinite families of values of v such that every Ryser design on v points is type-1. We begin with the following result.

Theorem 14.5.14. *Let* **D** *be a Ryser design with replication numbers r and r^*. If $\gcd(r - r^*, r + r^*) = 1$ or 2, then* **D** *is type-1.*

Proof. Let $d = \gcd(r - r^*, r + r^*)$ and $g = \gcd(r-1, r^*-1)$. Let v and λ be the number of points and the index of **D**, respectively. If $d \le 2$, then, since $r + r^* = v + 1$, we have $\gcd((r - r^*)/g, v+1) \le 2$. Then (14.57) implies that $\gcd((r - r^*)/g, 4\lambda) \le 2$ and therefore $\gcd((r - r^*)/g, \lambda) \le 2$. Now Theorem 14.3.3 implies that **D** is type-1. □

We will now consider Ryser designs on $v = np + 1$ points where p is a prime.

Theorem 14.5.15. *Let p be a prime and let $v = np + 1$ where n is a positive integer. In each of the following cases, every Ryser design on v points is type-1:*

(i) $n \in \{1, 2, 3, 4, 6\}$;
(ii) $n = 5$ *and* $p \not\equiv 2, 8 \pmod{15}$;
(iii) $n = 8$ *and* $p \not\equiv 5, 11 \pmod{24}$.

Proof. Let **D** be a Ryser design of index λ on v points with replication numbers $r > r^*$. Let $d = \gcd(r - r^*, r + r^*)$ and $g = \gcd(r-1, r^*-1)$. Then $g < (v-1)/2$. Since all Ryser designs with $g \le 6$ and $g = 8$ are type-1, we will assume that $g > 6$ and $g \ne 8$. Then g does not divide the given values of n. Since g divides $v - 1$, we obtain that $g = tp$ where t divides n. If we show that $d = 1$ or 2, then **D** is type-1 by Theorem 14.5.14.

(i) if $n = 1$ or 2, then $p \ge (v-1)/2$, so the statement of the theorem is true. Suppose $n = 3$. Then $g = p$, so $r^* = p + 1$ and $r = 2p + 1$. Therefore, $d = 1$. Suppose $n = 4$. Then $g = p$, $r^* = p + 1$, and $r = 3p + 1$. Therefore, $d = 2$.

Suppose $n = 6$. Then $g = p$ or $2p$. If $g = p$, then $r^* = p + 1$, and $r = 5p + 1$. If $p \equiv 1 \pmod 6$, then $r(r-1)/(v-1)$ is an integer. If $p \equiv 5 \pmod 6$, then $r^*(r^*-1)/(v-1)$ is an integer. In either case, **D** is type-1 by Theorem 14.2.7. If $g = 2p$, then $r^* = 2p + 1$ and $r = 4p + 1$, so $d = 2$.

(ii) Suppose $n = 5$. Then $g = p$ and $r^* = p + 1$ or $2p + 1$. If $r^* = 2p + 1$, then $r = 3p + 1$ and $d = 1$. Let $r^* = p + 1$. Then $r = 4p + 1$. If $p \not\equiv 2 \pmod 3$, then $r + r^* \not\equiv 0 \pmod 3$. Therefore $d = 1$. If $p \equiv 2 \pmod 3$, then, since $p \not\equiv 2, 8 \pmod{15}$, we obtain that $p \equiv \pm 1 \pmod 5$, and therefore **D** is type-1 by Theorem 14.2.7.

(iii) Suppose $n = 8$. Then $g = p$ or $2p$. If $g = 2p$, then $r^* = 2p + 1$, $r = 6p + 1$, and $d = 2$. If $g = p$, then $r^* = p + 1$ or $3p + 1$. If $r^* = 3p + 1$, then $r = 5p + 1$ and $d = 2$. If $r^* = p + 1$, then $r = 7p + 1$. If $p \not\equiv 2 \pmod 3$, then $r + r^* \not\equiv 0 \pmod 3$. Therefore, $d = 2$. If $p \equiv 2 \pmod 3$, then, since $p \not\equiv 5, 11 \pmod{24}$, we obtain that $p \equiv \pm 1 \pmod 8$. Hence, **D** is type-1 by Theorem 14.2.7. □

Exercises

(1) Prove Proposition 14.4.2.

(2) Let **D** be a Ryser design with replication numbers r and r^* such that $0 < r - r^* \le 20$. Prove that **D** is type-1.

(3) Let p be an odd prime and let $\mathcal{F}$ be a set of $2p + 1$ subsets of a set X of cardinality $2p + 1$. Suppose there exists $\lambda \neq 0$ such that $|A \cap B| = \lambda$ for any distinct $A, B \in \mathcal{F}$. Prove that one of the following is true:
   (i) $|A| = 2p$ for all $A \in \mathcal{F}$;
   (ii) there exists $x \in X$ such that $\mathcal{F}$ consists of $\{x\}$ and all 2-subsets of X containing x;
   (iii) there exists $x \in X$ such that $\mathcal{F}$ consists of $X \setminus \{x\}$ and all 2-subsets of X containing x;
   (iv) there exists a Hadamard 2-design $\mathbf{D} = (X, \mathcal{B})$ with a block $A \in \mathcal{B}$ such that $\mathcal{F} = \mathcal{B}$ or $\mathcal{F} = \{X \setminus B : B \in \mathcal{B}\}$ or $\mathcal{F} = \{A \triangle B : B \in \mathcal{B}, B \neq A\} \cup \{A\}$.

Notes

Ryser designs were first introduced in Ryser (1968) as λ-designs. The term *Ryser design* is taken from Stanton (1997). We prefer to use this term to avoid confusion with the common usage of such terms as 2-design, t-design, etc. in design theory.

The Ryser–Woodall Theorem was independently proven by Ryser (1968) and Woodall (1970). They both conjectured that all Ryser designs can be obtained from symmetric designs by block complementation. To the best of our knowledge, the term type-1 for such Ryser designs was introduced in Bridges (1970).

The de Bruijn–Erdös Theorem was published in de Bruijn and Erdös (1948). There are many different proofs of this classical result. See, for instance, Batten (1986), Beth, Jungnickel and Lenz (1999), or Ionin and M. S. Shrikhande (1996a). Two short proofs can be found in Chapter 9 of Aigner and Ziegler (2004).

The Ryser–Woodall conjecture was proven for $\lambda = 2$ in Ryser (1968), for $\lambda = 3$ in Bridges and Kramer (1970), for $\lambda = 4$ in Bridges (1970), and for $5 \le \lambda \le 9$ in Kramer (1969, 1974). Weisz (1995) extended these results to $\lambda \le 34$.

Singhi and S. S. Shrikhande (1976) proved the conjecture for Ryser designs of prime index. A more general Theorem 14.3.3 was proven in S. S. Shrikhande and Singhi (1984) and extended to Theorem 14.3.5 in Seress (2001). In the latter paper, the Ryser–Woodall conjecture is proven for Ryser designs of index $2p$ where p is a prime.

Theorem 14.1.19 was obtained in Woodall (1970). Our proof follows Kramer (1974). It is interesting to note that the corresponding result for symmetric designs (the M. Hall Jr. conjecture) is still open. Theorem 14.1.15 was established in Bridges (1970) and Woodall (1970). In Bridges (1970), it was also shown that there is no Ryser design with $e = 2$. Wu (1989) proved a similar result for $e = 3$. In that paper, the upper bound $v \leq \lambda^3 - 2\lambda^2 + 4$ for Ryser designs of index $\lambda \geq 3$ on v points was obtained.

Theorems 14.2.7 and 14.2.8 were announced in Woodall (1971). We follow the proof given in Seress (1989). Theorems 14.2.11 and 14.2.12 are due to Seress (1989). This paper contains other characterizations of type-1 Ryser designs. Theorem 14.2.10 was proven in Kramer (1969) and Woodall (1970).

A technique for proving the Ryser–Woodall conjecture for Ryser designs with small values of $\gcd(r - 1, r^* - 1)$ was developed in Ionin and M. S. Shrikhande (1996a, 1996b) where Theorems 14.1.16, 14.5.9, 14.5.10, and 14.5.11 were proved. Theorem 14.5.12 is due to Hein and Ionin (2002), Theorem 14.5.13 is due to Fiala (2002, 2003). The result of Exercise 3 is due to Ionin and M. S. Shrikhande (1994a).

Appendix

In this appendix, we list the parameters (v, k, λ) of all known symmetric designs up to complementation. We will combine most of these parameters into 23 families and then conclude the list by giving the parameters of the remaining known symmetric designs. If a difference set or a strongly regular graph or a normally regular digraph with the parameters of a symmetric design is known to exist, we provide this information. We also mention some non-existence results for (v, k, λ)-graphs.

Natural Series. Let q be a prime power and d a positive integer. Then there exists a symmetric design with

$$v = \frac{q^{d+1} - 1}{q - 1}, \quad k = \frac{q^d - 1}{q - 1}, \quad \lambda = \frac{q^{d-1} - 1}{q - 1}. \tag{15.1}$$

These parameters are realized by designs $PG_{d-1}(d, q)$. See Corollary 3.6.7.

There exists a difference set with parameters (15.1) (Theorem 9.2.1). There is no (v, k, λ)-graph with these parameters (Corollary 7.4.13). If d is odd, then there exists a (v, k, λ)-graph with parameters of the complement of $PG_{d-1}(d, q)$ (Corollary 10.4.17). If d is even and q is not a power of 4, then such a graph does not exist (Corollary 7.4.14). There exists an $NRD(q^2 + q + 1, q + 1, 1)$ whenever there exists a projective plane of order q (Theorem 7.5.9).

Hadamard Series. For a positive integer n, a symmetric design with parameters

$$v = 4n - 1, \quad k = 2n - 1, \quad \lambda = n - 1 \tag{15.2}$$

exists if and only if there exists a Hadamard matrix of order $4n$. (See Proposition 4.1.7). The Hadamard conjecture states that a Hadamard matrix of order $4n$ exists for every positive integer n. The smallest open case is $n = 167$. (See

Proposition 4.2.4, Theorem 4.2.5, and Corollary 4.3.3 for different constructions of Hadamard matrices and Hadamard 2-designs). By Theorem 4.3.24, Hadamard matrices can be obtained from symmetric and skew-symmetric conference matrices. For different constructions of symmetric and skew-symmetric conference matrices (see Proposition 4.3.5 and Theorems 4.3.13, 4.3.14, and 4.6.4).

Difference sets with parameters (15.2) are known to exist if $4n-1$ is a prime power (Example 9.1.24) or $n=(q+1)^2/4$ where q and $q+2$ are prime powers (Theorem 9.2.3). There is no (v,k,λ)-graph with parameters (15.2) (Corollary 7.4.13). A (v,k,λ)-graph with the complementary parameters is known to exist in the following cases: (i) there is a Hadamard matrix of order n (Theorem 7.4.6); (ii) $n=2^{s+t+u+1}\cdot 3^s\cdot 5^t\cdot 7^u m$ and there is a Hadamard matrix of order $4m$ (Theorem 8.2.29); (iii) $2n-1$ and $2n+1$ are prime powers (Theorem 4.5.1, Remark 4.5.2, and Theorem 7.4.23).

Series 1. For a positive integer n, a symmetric design with parameters

$$v=4n^2,\quad k=2n^2-n,\quad \lambda=n^2-n \tag{15.3}$$

exists if and only if there exists a regular Hadamard matrix of order $4n^2$ (Theorem 4.4.5). For different constructions of regular Hadamard matrices, see Proposition 4.4.9 and Theorems 4.4.11, 4.4.16, 4.4.18, 4.5.1, and 4.6.11. Theorem 8.2.29 gives regular Hadamard matrices obtained from strongly regular graphs. For regular Hadamard matrices related to Hadamard difference sets, see Theorem 9.5.10 and Beth, Jungnickel and Lenz (1999), Theorem VI.12.15.

A (v,k,λ)-graph with parameters (15.3) is known to exist in the following cases: (i) there is a Hadamard matrix of order $2n$ (Theorem 7.4.6); (ii) $n=2^{s+t+u+1}\cdot 3^s\cdot 5^t\cdot 7^u m$ and there is a Hadamard matrix of order $4m$ (Theorem 8.2.29), and (iii) n is a square (Corollary 7.4.8). Note that a (v,k,λ)-graph with parameters (15.3) exists if and only if there exists a (v,k,λ)-graph with parameters $v=4n^2$, $k=2n^2+n$, $\lambda=n^2+n$.

An $NRD(v,k,\lambda,\lambda)$ with parameters (15.3) is known to exist when there exists a Hadamard matrix of order $2n$ (Theorem 7.5.10).

Series 2. Let h be an integer (not necessarily positive) such that $q=(2h-1)^2$ is a prime power. If there exists a productive regular Hadamard matrix of order $4h^2$, then, for any positive integer m, there exists a symmetric design with parameters

$$v=\frac{4h^2(q^{m+1}-1)}{q-1},\quad k=(2h^2-h)q^m,\quad \lambda=(h^2-h)q^m. \tag{15.4}$$

(See Theorem 11.4.2.) For the notion of productive regular Hadamard matrices, see Definition 11.4.1. For different constructions of productive regular Hadamard matrices, see Theorems 11.4.3, 11.4.4, 11.4.5, and 11.4.6.

Series 3. Let q be a prime power and d a positive integer. Then there exists a symmetric design with parameters

$$v = \frac{q^{d+1}(q^{d+1}+q-2)}{q-1}, \quad k = \frac{q^d(q^{d+1}-1)}{q-1}, \quad \lambda = \frac{q^d(q^d-1)}{q-1}. \tag{15.5}$$

(See Theorem 3.8.3).

For parameters (15.5), there exists a difference set (Theorem 9.5.2), a (v, k, λ)-graph (Theorem 7.4.18), and an $NRD(v, k, \lambda, \lambda)$ (Theorem 7.5.11). A (v, k, λ)-graph with the complementary parameters exists if and only if qd is even (Theorem 7.4.26).

Series 4. Let d be a positive integer. Then there exists a symmetric design with parameters

$$v = \frac{3^{d+1}(3^{d+1}-1)}{2}, \quad k = \frac{3^d(3^{d+1}+1)}{2}, \quad \lambda = \frac{3^d(3^d+1)}{2}. \tag{15.6}$$

(See Theorem 3.8.5).

For parameters (15.6), there exists a difference set (Theorem 9.5.5) and a (v, k, λ)-graph (Theorem 7.4.25). A (v, k, λ)-graph with the complementary parameters exists if and only if d is odd (Theorem 7.4.27).

Series 5. Let q be a prime power such that q^2+q+1 is also a prime power. Then there exists a symmetric design with parameters

$$v = (q+1)^3 + q + 2, \quad k = (q+1)^2 + 1, \quad \lambda = q + 1. \tag{15.7}$$

(See Theorem 11.5.1). There is no (v, k, λ)-graph with parameters (15.7) (Exercise 19 of Chapter 7).

Series 6. Let q be an odd prime power and d a positive integer. Then there exists a symmetric design with parameters

$$v = 1 + \frac{2q(q^d-1)}{q-1}, \quad k = q^d, \quad \lambda = \frac{q^{d-1}(q-1)}{2}. \tag{15.8}$$

(See Corollary 11.8.4.) There is no (v, k, λ)-graph with parameters (15.8) (Exercise 20 of Chapter 7).

Series 7. Let a prime power q and a positive integer d be such that $r = (q^d - 1)/(q - 1)$ is also a prime power. Let m be a positive integer. Then there exists a symmetric design with parameters

$$v = 1 + \frac{qr(r^m - 1)}{r - 1}, \quad k = r^m, \quad \lambda = \frac{r^{m-1}(r - 1)}{q}. \tag{15.9}$$

(See Corollary 11.8.2.) There is no (v, k, λ)-graph with parameters (15.9) (Exercise 21 of Chapter 7).

Series 8. Let a prime power q and a positive integer d be such that $r = (q^{d+1} - 1)/(q - 1)$ is also a prime power. Let m be a positive integer. Then there exists a symmetric design with parameters

$$v = \frac{q^{d+1}(r^{2m} - 1)}{r - 1}, \quad k = r^{2m-1}q^d, \quad \lambda = (r - 1)r^{2m-2}q^{d-1}. \tag{15.10}$$

(See Theorem 11.3.3.) If q is odd and m is even, then there exists a (v, k, λ)-graph with parameters (15.10) (Theorem 11.6.6). If q is even and m is even, then there exists an $NRD(v, k, \lambda, \lambda)$ with parameters (15.10) (Theorem 11.6.13).

Series 9. Let a prime power q and a positive integer d be such that $r = q^{d+1} + q - 1$ is also a prime power. Let m be a positive integer. Then there exists a symmetric design with parameters

$$v = \frac{q^d(r^{2m} - 1)}{(q - 1)(q^d + 1)}, \quad k = q^d r^{2m-1}, \quad \lambda = q^d(q^d + 1)(q - 1)r^{2m-2}. \tag{15.11}$$

(See Theorem 11.3.4.) If q is odd and m is even, then there exists a (v, k, λ)-graph with parameters (15.11) (Theorem 11.6.7).

Series 10. Let a positive integer d be such that $q = (3^{d+1} + 1)/2$ is a prime power. Let m be a positive integer. Then there exists a symmetric design with parameters

$$v = \frac{2 \cdot 3^d(q^{2m} - 1)}{3^d + 1}, \quad k = 3^d q^{2m-1}, \quad \lambda = \frac{3^d(3^d + 1)q^{2m-2}}{2}. \tag{15.12}$$

(See Theorem 11.3.5.) If m is even, then there exists a (v, k, λ)-graph with parameters (15.12) (Theorem 11.6.8).

Series 11. Let a positive integer d be such that $q = 3^{d+1} - 2$ is a prime power. Let m be a positive integer. Then there exists a symmetric design with parameters

$$v = \frac{3^d(q^{2m}-1)}{2(3^d-1)}, \quad k = 3^d q^{2m-1}, \quad \lambda = 2 \cdot 3^d(3^d-1)q^{2m-2}. \tag{15.13}$$

(See Theorem 11.3.6.) If m is even, then there exists a (v, k, λ)-graph with parameters (15.13) (Theorem 11.6.9).

Series 12. Let a prime power q and positive integers d and e be such that $p = (q^d - 1)/(q - 1)$ and $r = q^d(p^{e+1} - 1)/(p - 1)$ are also prime powers. Let m be a positive integer. Then there exists a symmetric design with parameters

$$v = 1 + qp(p^{e+1} - 1)/(p - 1), \quad k = r^{m+1}, \quad \lambda = q^d(q^{d-1}p^e - 1)/(p - 1).$$

(See Theorem 11.8.6.) The only known realization of these conditions is $q = 2$, $p = 2^d - 1$ is a Mersenne prime, and $e = 1$, so $r = 2^{2d}$. This gives the parameters

$$v = 1 + \frac{2^{d+1}(2^{2dm}-1)}{2^d+1}, \quad k = 2^{2dm}, \quad \lambda = 2^{2dm-d-1}(2^d+1).$$

Series 13. Let d be a nonnegative integer. Then there exists a symmetric design with parameters

$$v = \frac{2^{2d+4}(2^{2d+2}-1)}{3}, \quad k = \frac{2^{2d+1}(2^{2d+3}+1)}{3}, \quad \lambda = \frac{2^{2d+1}(2^{2d+1}+1)}{3}.$$

These designs correspond to the Davis–Jedwab difference sets. (See Theorem 9.5.7 and Beth, Jungnickel and Lenz (1999), Theorem VI.9.4.)

Series 14. Let d be a positive integer and let q be the square of an odd prime or a power of 2 or 3. Then there exists a symmetric design with parameters

$$v = \frac{4q^{2d}(q^{2d}-1)}{q^2-1}, \quad k = q^{2d-1}\left(\frac{2(q^{2d}-1)}{q+1} + 1\right),$$
$$\lambda = \frac{q^{2d-1}(q-1)(q^{2d-1}+1)}{q+1}.$$

These designs correspond to the Chen difference sets. (See Beth, Jungnickel and Lenz (1999), Theorem VI.9.5.)

Series 15. Let d be a positive integer such that $q = (2^{2d+3} + 1)/3$ is a prime power. Let m be a positive integer. Then there exists a symmetric design with parameters

$$v = \frac{2^{2d+3}(q^{2m}-1)}{q+1}, \quad k = 2^{2d+1}q^{2m-1}, \quad \lambda = 2^{2d-1}(q+1)q^{2m-2}.$$

(See Theorem 11.3.7.)

Series 16. Let d be a positive integer such that $q = 2^{2d+3} - 3$ is a prime power. Let m be a positive integer. Then there exists a symmetric design with parameters

$$v = \frac{2^{2d+3}(q^{2m}-1)}{3(q-1)}, \quad k = 2^{2d+1}q^{2m-1}, \quad \lambda = 3 \cdot 2^{2d-1}(q-1)q^{2m-2}.$$

(See Theorem 11.3.8.)

Series 17. Let t be an odd positive integer such that $q = 4t^2 + 1$ is a prime power. Then there exists a difference set with parameters

$$v = q, \quad k = t^2, \quad \lambda = \frac{t^2-1}{4}.$$

(See Beth, Jungnickel and Lenz (1999), Theorem VI.8.11 or Hall (1986), Section 11.6.)

Series 18. Let t be an odd positive integer such that $q = 4t^2 + 9$ is a prime power. Then there exists a difference set with parameters

$$v = q, \quad k = t^2 + 3, \quad \lambda = \frac{t^2+3}{4}.$$

(See Beth, Jungnickel and Lenz (1999), Theorem VI.8.11 or Hall (1986), Section 11.6.)

Series 19. Let t and u be odd positive integers such that $q = 8t^2 + 1 = 64u^2 + 9$ is a prime. Then there exists a difference set with parameters

$$v = q, \quad k = t^2, \quad \lambda = u^2.$$

(See Beth, Jungnickel and Lenz (1999), Theorem VI.8.11 or Hall (1986), Section 11.6).

Series 20. Let t be an odd positive integer and u an even integer such that $q = 8t^2 + 49 = 64u^2 + 441$ is a prime. Then there exists a difference set with parameters

$$v = q, \quad k = t^2 + 6, \quad \lambda = u^2 + 7.$$

(See Beth, Jungnickel and Lenz (1999), Theorem VI.8.11 or Hall (1986), Section 11.6.)

Series 21. Let p be a prime such that $q = 3p + 2$ is a prime and $(pq - 1)/4$ is an odd square. Then there exists a difference set with parameters

$$v = pq, \quad k = \frac{pq - 1}{4}, \quad \lambda = \frac{pq - 5}{16}.$$

(See Whiteman (1962)). The only known realizations of these parameters are $p = 17, q = 53$ and $p = 46817, q = 140453$.

Sporadic parameters. Symmetric design with the following parameters are known to exist. They do not belong to any of the above 23 series. If a design with given parameters is constructed in this book, we refer to the appropriate theorem. In this case, a reference to the original paper can be found in the notes to the corresponding chapter. If no construction of a design is given in this book, we refer to other sources.

v	k	λ	Reference
41	16	6	Theorem 2.7.1; Bridges, Hall and Hayden (1981), van Trung (1982a)
49	16	5	Brouwer and Wilbrink (1984)
56	11	2	Theorem 6.6.1; Hall, Lane and Wales (1970)
66	26	10	Theorem 11.1.1; van Trung (1982a)
70	24	8	Janko and van Trung (1984)
71	15	3	Theorem 8.7.8
71	21	6	Janko and van Trung (1985a)
78	22	6	Janko and van Trung (1985b); Tonchev (1987)
79	13	2	Theorem 2.8.1; Aschbacher (1971)
105	40	15	Janko (1999)
176	50	14	Theorem 6.6.2
189	48	12	Janko (1997)

For more detailed tables of small symmetric designs, see van Trung (1996) and Beth, Jungnickel and Lenz (1999). For tables of difference sets, see Jungnickel and Pott (1996) and Beth, Jungnickel and Lenz (1999). For nonexistence results on difference sets, see Pott (1995) and Beth, Jungnickel and Lenz (1999).

References

Abel, R. J. R., Brouwer, A. E., Colbourn, C. J. and Dinitz, J. H. (1996) Mutually orthogonal Latin squares (MOLS), in: *The CRC Handbook of Combinatorial Designs*, C. J. Colbourn and J. H. Dinitz (eds.). Boca Raton: CRC Press, pp. 111–42.

Abel, R. J. R. and Furino, S. C. (1996) Resolvable and near resolvable designs, in: *The CRC Handbook of Combinatorial Designs*, C. J. Colbourn and J. H. Dinitz (eds.). Boca Raton: CRC Press, pp. 87–94.

Agaian, S. S. (1985) *Hadamard Matrices and their Applications*. Berlin: Springer.

Aigner, M. and Ziegler, G. M. (2004) *Proofs from The Book*, 3rd edn. New York: Springer.

Alon, N., Babai, L. and Suzuki, H. (1991) Multilinear polynomials and Frankl–Ray Chaudhuri–Wilson type intersection theorems, *J. Combin. Theory Ser. A*, **58**, 165–80.

Anderson, I. (1987) *Combinatorics of Finite Sets*. Oxford: Oxford University Press.

André, J. (1954) Über nicht-Desarguessche Ebenen mit transitiver Translationsgruppe, *Math. Z.*, **60**, 156–86.

Arasu, K. T., Dillon, J. F., Leung, K. H. and Ma, S. L. (2001) Cyclic relative difference sets with classical parameters, *J. Combin. Theory Ser. A*, **94**, 118–26.

Artin, E. (1957) *Geometric Algebra*. New York: Wiley.

Aschbacher, M. (1971) On the collineation groups of symmetric designs, *J. Combin. Theory Ser. A*, **11**, 272–81.

Assmus, E. F. and Key, J. D. (1992) *Designs and their Codes*. Cambridge: Cambridge University Press.

Assmus, E. F. and Mattson, H. F. Jr. (1967) On tactical configurations and error-correcting codes, *J. Combin. Theory*, **2**, 243–57.

Baartmans, A. H. and Shrikhande, M. S. (1981) Tight subdesigns of symmetric designs, *Ars Combin.*, **12**, 303–10.

— (1985) A characterization of the extensions of symmetric designs, *Discrete Math.*, **57**, 301–5.

Baartmans, A. H., Shrikhande, M. S. and Tonchev, V. D. (1994) Residual designs of affine designs, preprint.

Babai, L. (1987) On the nonuniform Fisher inequality, *Discrete Math.*, **66**, 303–7.

Baer, R. (1946) Projectivities with fixed points on every line of the plane, *Bull. Amer. Math. Soc.* **52**, 273–86.

— (1952) *Linear Algebra and Projective Geometry*. New York: Academic Press.

Bailey, R. A. (2003) *Association Schemes*. Cambridge: Cambridge University Press.

Baker, R. D. (1977) Elliptic semi-planes I. Existence and classification, *Congr. Numer.*, **19**, 61–73.

— (1982) Symmetric designs with Bruck subdesigns, *Combinatorica*, **2**, 103–9.

Baker, R. D., Dover, J. M., Ebert, G. L. and Wantz, K. L. (2000) Baer subgeometry partitions, *J. Geom.*, **67**(1–2), 23–34.

Bannai, E. (1977) On tight designs, *Quart. J. Math.* (Oxford) **28**, 433–48.

Bannai, E. and Ito, T. (1984) *Algebraic Combinatorics: Association Schemes*. New York: Benjamin.

Bannai, E. and Sawano, M. (2002) Symmetric designs attached to four-weight spin models, *Des. Codes Cryptog.*, **25**, 73–90.

Batten, L. M. (1986) *Combinatorics of Finite Geometries*. Cambridge: Cambridge University Press.

Batten, L. M. and Beutelspacher, A. (1993) *The Theory of Finite Linear Spaces: Combinatorics of Points and Lines*. Cambridge: Cambridge University Press.

Baumert, L. D. (1971) *Cyclic Difference Sets, Lecture Notes in Mathematics*, **182**. Berlin: Springer.

Behbahani, M. and Kharaghani, H. (2004) On a new class of productive regular Hadamard matrices, preprint.

Beker, H. and Haemers, W. (1980) 2-designs with an intersection number $k - n$, *J. Combin. Theory Ser. A*, **28**, 64–81.

Bekker, B., Ionin, Y. J. and Shrikhande, M. S. (1998) Embeddability and construction of affine α-resolvable pairwise balanced designs, *J. Combin. Des.*, **6**, 111–29.

Belevitch, V. (1950) Theory of $2n$-terminal networks with applications to conference telephony, *Electrical Communication*, **27**, 231–44.

Berman, G. (1977) Weighing matrices and group divisible designs determined by $EG(t, p^n)$, $t > 2$, *Util. Math.*, **12**, 183–92.

— (1978) Families of generalized weighing matrices, *Canad. J. Math.*, **30**, 1016–28.

Best, M. R. (1977) The excess of a Hadamard matrix, *Indag. Math.*, **39**, 357–61.

Beth, T. and Jungnickel, D. (1981) Mathieu groups, Witt designs, and Golay codes, in: *Geometries and Groups*, M. Aigner and D. Jungnickel (eds.). Berlin: Springer, pp. 157–59.

Beth, T., Jungnickel, D. and Lenz, H. (1999) *Design Theory*, Vols. I and II. Cambridge: Cambridge University Press.

Beutelspacher, A. (1996) Classical geometries, in: *The CRC Handbook of Combinatorial Designs*, C. J. Colbourn and J. H. Dinitz (eds.). Boca Raton: CRC Press, pp. 694–708.

Beutelspacher, A. and Rosenbaum, U. (1998) *Projective Geometry*. Cambridge: Cambridge University Press.

Bhagwandas and Shrikhande, S. S. (1968) Seidel-equivalence of strongly regular graphs, *Sankhyā, Ser. A*, **30**, 359–68.

Bhaskar Rao, M. (1966) Group divisible family of PBIB designs, *J. Indian Statist. Assoc.*, **4**, 14–28.

— (1970) Balanced orthogonal designs and their applications in the construction of some BIB and group divisible designs, *Sankhyā Ser. A*, **32**, 439–48.

Bhattacharya, K. N. (1944a) A new symmetrical balanced incomplete block design, *Bull. Calcutta Math. Soc.*, **36**, 91–6.

— (1944b) A new balanced incomplete block design, *Science and Culture*, **9**, 508.

— (1946) A new solution in symmetrical balanced incomplete block designs ($v = b = 31, r = k = 10, \lambda = 3$), *Sankhyā Ser*, **7**, 423–24.

Biggs, N. L. (1993) *Algebraic Graph Theory* (2nd edn.). Cambridge: Cambridge University Press.

Biliotti, M., Jha, V. and Johnson, N. L. (2001) *Foundations of Translation Planes*. Marcel Dekker, Inc., New York.

Blokhuis, A. (1993) Polynomials in finite geometries and combinatorics, in: *Surveys in Combinatorics*, K. Walter (ed.). Cambridge: Cambridge University Press, pp. 35–52.

Bollobás, B. (1986) *Combinatorics*. Cambridge: Cambridge University Press.

— (1998) *Modern Graph Theory*. New York: Springer.

Bose, R. C. (1938) On the application of the properties of Galois fields to the problem of construction of hyper-Graeco-Latin squares, *Sankhyā*, **3**, 323–38.

— (1939) On the construction of balanced incomplete block designs, *Ann. Eugenics*, **9**, 353–99.

— (1942) A note on the resolvability of incomplete block designs, *Sankhyā Ser.*, **6**, 105–10.

— (1955) Versuche in unvollstandiger Blöcken, *Gastvarlesung Universitat Frankfurt/ M., Naturwissnschaftliche Fakultät.*

— (1958–9) On the application of finite projective geometry for deriving a certain series of balanced Kirkman arrangements, in: *Calcutta Math. Soc. Golden Jubilee Commemoration*, Vol. II, pp. 341–56.

— (1963) Strongly regular graphs, partial geometries, and partially balanced designs, *Pacific J. Math.*, **13**, 389–419.

Bose, R. C. and Mesner, D. M. (1959) On linear associative algebras corresponding to association schemes of partially balanced designs, *Ann. Math. Statist.*, **30**, 21–38.

Bose, R. C. and Nair, K. R. (1939) Partially balanced incomplete block designs, *Sankhyā Ser.*, **4**, 337–72.

Bose, R. C. and Shimamoto, T. (1952) Classification and analysis of partially balanced incomplete block designs with two associate classes, *J. Amer. Statist. Assoc.*, **47**, 151–84.

Bose, R. C. and Shrikhande, S. S. (1959a) A note on a result in the theory of code construction, *Information and Control*, **2**, 183–94.

— (1959b) On the composition of balanced incomplete block designs, *Canad. J. Math.*, **12**, 177–88.

— (1959c) On the falsity of Euler's conjecture about the non-existence of two orthogonal Latin squares of order $4t + 2$, *Proc. Natl. Acad. Sci. USA*, **45**, 734–37.

— (1960) On the construction of sets of mutually orthogonal Latin squares and the falsity of a conjecture of Euler, *Trans. Amer. Math. Soc.*, **95**, 191–209.

— (1970) Graphs in which each pair of vertices is adjacent to d other vertices, *Studia Sci. Math. Hungar.* **5**, 181–95.

— (1971) Some further constructions for $G_2(d)$ graphs, *Studia Sci. Math. Hunga.*, **6**, 127–32.

— (1976) Baer subdesigns of symmetric balanced incomplete block designs, in: *Essays in Probability and Statistics*, S. Ikeda, *et al.* (eds.) Tokyo: Shinko Tsusho, pp. 1–16.

Bose, R. C., Shrikhande, S. S. and Parker, E. T. (1959) Further results on the construction of sets of mutually orthogonal Latin squares and the falsity of Euler's conjecture, *Canad. J. Math.*, **12**, 189–203.

Bose, R. C., Shrikhande, S. S. and Singhi, N. M. (1976) Edge regular multigraphs and partial geometric designs with an application to the embedding of quasi-residual designs, *Atti Dei Convegni Lincei*, **17**, 49–81.

Bracken, C. (2004) A counterexample to a question of Cameron, preprint.

Bracken, C. and McGuire, G. (2002) Characterization of SDP designs that yield certain spin models, preprint.

— (2003) On quasi-3 designs and spin models, preprint.

Brauer, R. (1941) On the connections between the ordinary and modular characters of groups of finite order, *Ann. Math.*, **42**, 926–35.

Bremner, A. (1979) A diophantine equation arising from tight 4-designs, *Osaka J. Math.*, **16**, 353–6.

Bridges, W. G. (1970) Some results on λ-designs, *J. Combin. Theory Ser. A*, **8**, 350–60.

— (1983) A (66, 26, 10) design, *J. Combin. Theory Ser. A*, **35**, 360.

Bridges, W. G., Hall, M. Jr. and Hayden, J. L. (1981) Codes and designs, *J. Combin. Theory Ser. A*, **31**, 155–74.

Bridges, W. G. and Kramer, E. S. (1970) The determination of all λ-designs with $\lambda = 3$, *J. Combin. Theory Ser. A*, **8**, 343–9.

Broughton, W. and McGuire, G. (1999) Some observations on quasi-3 designs and Hadamard matrices, *Des. Codes Cryptog.*, **18**, 55–61.

— (2003) On the non-existence of quasi-3 designs, *Discrete Math.*, **262**, 79–87.

Brouwer, A. E. (1982) Polarities of G. Higman's symmetric design and a strongly regular graph on 176 vertices, *Aequationes Math.*, **25**, 77–82.

— (1983) An infinite series of symmetric designs, *Report ZW 202/83, Math.* Centrum, Amsterdam.

— (1995) Block Designs, in: *Handbook of Combinatorics*, R. Graham, M. Grötschel, and L. Lovász (eds.). Amsterdam: Elsevier Science BV, pp. 693–745.

— (1996) Strongly regular graphs, in: *The CRC Handbook of Combinatorial Designs*, C. J. Colbourn and J. H. Dinitz (eds.). Boca Raton: CRC Press, pp. 667–85.

Brouwer, A. E., Cohen, A. M. and Neumaier, A. (1989) *Distance-Regular Graphs*. New York: Springer-Verlag.

Brouwer, A. E. and Haemers, W. H. (1995) Association Schemes, in: *Handbook of Combinatorics*, R. Graham, M. Grötschel and L. Lovász (eds.). Amsterdam: Elsevier Science BV, pp. 747–71.

Brouwer, A. E. and van Lint, J. H. (1994) Strongly regular graphs and partial geometries, in: *Enumeration and Design*, D. M. Jackson and S. A. Vanstone (eds.). New York: Academic Press, pp. 85–122.

Brouwer, A. E. and Wilbrink, H. A. (1984) A symmetric design with parameters 2 – (49, 16, 5), *J. Combin. Theory Ser. A*, **37**, 193–4.

Brown, E. and Reid, K. B. (1972) Doubly regular tournaments are equivalent to skew-Hadamard matrices, *J. Combin. Theory Ser. A*, **12**, 332–8.

Bruck, R. H. (1951) Finite nets I. Numerical invariants, *Canad. J. Math.*, **3**, 94–107.
— (1955) Difference sets in a finite group, *Trans. Amer. Math. Soc.*, **78**, 464–81.
— (1960) Quadratic extensions of cyclic planes, *Proc. Sympos. Appl. Math.*, **10**, 15–44.
— (1963) Finite nets II, *Pacific J. Math.*, **13**, 421–57.
Bruck, R. H. and Ryser, H. J. (1949) The non-existence of certain finite projective planes, *Canad. J. Math.*, **1**, 88–93.
Burnside, W. (1911) Theory of groups of finite order (2nd edn). Cambridge (Dover, 1955).
Bush, K. A. (1971) Unbalanced Hadamard matrices and finite projective planes of even order, *J. Combin. Theory Ser. A*, **11**, 38–44.
Bussemaker, F. C. and Seidel, J. J. (1970) Symmetric Hadamard matrices of order 36, *Annals of New York Academy of Sciences*, **175**, 66–79.
Butson, A. T. (1962) Generalized Hadamard matrices, *Proc. Amer. Math. Soc.*, **13**, 894–8.
— (1963) Relations among generalized Hadamard matrices, relative difference sets and maximal length recurring sequences, *Canad. J. Math.*, **15**, 42–8.
Calderbank, A. R. and Morton, P. (1990) Quasi-symmetric 3-designs and elliptic curves, *SIAM J. Discrete Math.*, **3**, 178–96.
Cameron, P. J. (1973a) Extending symmetric designs, *J. Combin. Theory A*, **14**, 215–20.
— (1973b) Near-regularity conditions for designs, *Geom. Dedicata*, **2**, 213–23.
— (1973c) Biplanes, *Math. Z.*, **131**, 85–101.
— (1975) Two remarks on Steiner systems, *Geom. Dedicata*, **4**, 403–18.
— (1976) *Parallelisms of Complete Designs*. Cambridge: Cambridge University Press.
Cameron, P. J., Delsarte, P. and Goethals, J. M. (1979) Hemisystems, orthogonal configurations and dissipative conference matrices, *Philips J. Res.*, **34**, 147–62.
Cameron, P. J. and Seidel, J. J. (1973) Quadratic forms over GF(2), *Proc. Kon. Nederl. Akad. Wetensch.(A)*, **76**, 1 – 8 (= *Indag. Math.*, **35**, 1–8).
Cameron, P. J. and van Lint, J. H. (1991) *Designs, Graphs, Codes, and Their Links*. Cambridge: Cambridge University Press.
Chang, L. C. (1960a) The uniqueness and non-uniqueness of the triangular association scheme, *Sci. Record Peking Math. (new series)*, **3**, 604–13.
— (1960b) Association schemes of partially balanced block designs with parameters $v = 28, n_1 = 12, n_2 = 15$ and $p_{11}^2 = 4$, *Sci. Record Peking Math. (new series)*, **4**, 12–18.
Chen, Y. Q. (1998) A construction of difference sets, *Des. Codes Cryptog.*, **13**, 247–50.
Chouinard II, L. G., Jajcay, R. and Magliveras, S. S. Finite groups and designs, in: *The CRC Handbook of Combinatorial Designs*, C. J. Colbourn and J. H. Dinitz (eds.). Boca Raton: CRC Press, pp. 587–615.
Chowla, S. and Ryser, H. J. (1950) Combinatorial problems, *Canad. J. Math.*, **2**, 93–9.
Christian, J. D. and Shader, B. L. (2004) Nonexistence results for Hadamard-like matrices, *The Electronic Journal of Combinatorics*, **11**, Section N1.
Colbourn, C. J. and Dinitz, J. H. (eds.) (1996) *The CRC Handbook of Combinatorial Designs*. Boca Raton: CRC Press.
— (2001) Mutually orthogonal latin squares: a brief survey of constructions, *J. Statist. Plann. Inf.*, **95**, 9–48.
Connor, W. S. (1952) On the structure of balanced incomplete block designs, *Ann. Math. Stat.*, **23**, 57–71.

— (1958) The uniqueness of the triangular association scheme, *Ann. Math. Stat.*, **29**, 262–6.

Conwell, G. M. (1910) The 3-space $PG(3, 2)$ and its group, *Annals of Math.*, **11**, 60–76.

Coxeter, H. S. M. (1963) *Regular Polytopes* (2nd edn.). New York: Macmillan.

Craigen, R. (1996a) Hadamard matrices and designs, in: *The CRC Handbook of Combinatorial Designs*, C. J. Colbourn and J. H. Dinitz (eds.). Boca Raton: CRC Press, pp. 370–7.

— (1996b) Weighing matrices and conference matrices, in: *The CRC Handbook of Combinatorial Designs*, C. J. Colbourn and J. H. Dinitz (eds.). Boca Raton: CRC Press, pp. 496–504.

Craigen, R. and Wallis, W. D. (1993) Hadamard matrices: 1893–1993. *Congr. Numer.*, **97**, 99–129.

Cron, N. J. and Mavron, V. C. (1983) On intersections of symmetric subdesigns of symmetric designs, *Archiv der Mathematik*, **40**, 475–80.

Davis, J. A. and Jedwab, J. (1996) A survey of Hadamard difference sets, in: *Groups, Difference Sets, and the Monster*, K. T. Arasu, J. F. Dillon, K. Harada, S. Sehgal and R. Solomon, Walter de Gruyter (eds.). Berlin, pp. 145–56.

— (1997) A unifying construction for difference sets, *J. Combin. Theory Ser. A*, **80**, 13–78.

— (1998) Some recent developments in difference sets, in: *Combinatorial Designs and their Applications*, K. Quinn, B. Webb, F. Holyroyd, and C. Rowley (eds.). Pittman Research Notes in Mathematics, Addison Wesley Longman.

Dawson, J. (1985) A construction for the generalized Hadamard matrices $GH(4q; EA(q))$, *J. Statist. Plann. Inf.*, **11**, 103–10.

de Bruijn, N. G. and Erdös, P. (1948) On a combinatorial problem, *Indag. Math.(N.S)*, **10**, 421–3.

de Launey, W. (1986) A survey of generalized Hadamard matrices, difference matrices $D(k, \lambda; G)$ with large k, *Util. Math.*, **30**, 5–29.

— (1989) Square GBRDs over non-abelian groups, *Ars Combin.*, **27**, 42–9.

— (1992a) Generalized Hadamard matrices which are developed modulo a group, *Discrete Math.*, **104**, 49–65.

— (1996) Bhaskar Rao designs, in: *The CRC Handbook of Combinatorial Designs*, C. J. Colbourn and J. H. Dinitz (eds.). Boca Raton: CRC Press, pp. 241–6.

de Launey, W. and Dawson, J. E. (1992) A note on the construction of GH($4tq$; EA(q)) for $t = 1, 2$, *Australas. J. Combin.*, **6**, 177–86.

— (1994) An asymptotic result on the existence of generalized Hadamard matrices, *J. Combin. Theory Ser. A*, **65**, 158–63.

de Launey, W. and Sarvate, D. G. (1983) Non-existence of certain GBRDs, *Ars Combin.*, **18**, 5–20.

de Resmini, M. J. (1996) Projective planes, nondesarguesian, in: *The CRC Handbook of Combinatorial Designs*, C. J. Colbourn and J. H. Dinitz (eds.). Boca Raton: CRC Press, pp. 708–18.

Delsarte, P. (1968) Orthogonal matrices over a group and related tactical configurations, *M.B.L.E. Report R90*.

— (1973a) An algebraic approach to the association schemes of coding theory, *Philips Research Report Suppl.*, **10**.

— (1973b), Four fundamental parameters of a code and their combinatorial significance, *Information and Control*, **23**(5), 407–38.

Delsarte, P., Goethals, J.-M. and Seidel, J. J. (1971) Orthogonal matrices with zero diagonal II, *Canad. J. Math.*, **23**, 816–32.

— (1977) Spherical codes and designs, *Geom. Dedicata*, **6**, 363–88.

Dembowski, P. (1958) Verallgemeinerungen von Transitivitätsklassen endlicher projektiver Ebenen, *Math. Z.*, **69**, 59–89.

— (1968) *Finite Geometries*. Berlin-Heidelberg: Springer.

Dembowski, P. and Piper, F. C. (1967) Quairegular collineation groups of finite projective planes, *Math. Z.*, **99**, 53–75.

Dembowski, P. and Wagner, A. (1960) Some characterizations of finite projective spaces, *Arch. Math.*, **11**, 465–9.

Denés, J. and Keedwell, A. D. (1974) *Latin Squares and Their Applications*. New York: Academic Press.

Denniston, R. H. F. (1982) Enumeration of symmetric designs (25, 9, 3), *Ann. Disc. Math.*, **15**, 111–27.

Deza, M. (1973) Une propriété extrémale des plans projectifs finis dans une classe de codes équidistants, *Discr. Math.*, **6**, 343–52.

Dickson, L. E. (1905a) Definitions of a group and a field by independent postulates, *Trans. Amer. Math. Soc.*, **6**, 198–204.

— (1905b) On finite algebras, *Nachr. Akad. Wiss. Göttingen Math.-Phys.K.1*, 358–93.

Dillon, J. F. and Schatz, J. R. (1987) Block-designs with the symmetric difference property, *Proc. NSA Mathematical Sciences Meeting Ft. George Meade, Maryland*, 159–64.

Dougherty, S. T. (1994) A coding theoretic solution to the 36 officers problem, *Des. Codes Cryptog.*, **4**, 123–8.

Drake, D. A. (1979) Partial λ-geometries and generalized Hadamard matrices over groups, *Canad. J. Math.*, **31**, 617–27.

Drake, D. A. and Jungnickel, D. (1978) Klingenberg structures and partial designs II. Regularity and uniformity, *Pacific J. Math.*, **77**, 389–415.

Elliott, J. E. H. and Butson, A. T. (1966) Relative difference sets, *Illinois J. Math.*, **10**, 517–31.

Enomoto, H., Ito, N. and Noda, R. (1979) Tight 4-designs, *Osaka J. Math.*, **16**, 39–43.

Euler, L. (1782) Recherches sur une nouvelie espéce de quarres magiques, *Verhandlingen Zeeuwach Genootschap Wetenschappen Vlissengen*, **9**, 85–239.

Fanning, J. D. (1995) A family of symmetric designs, *Discrete Math.*, **146**, 307–12.

Fano, G. (1892) Sui postulati fondamentali della geometria proiettiva, *Giornale di Matematiche*, **30**, 106–32.

Fiala, N. C. (2002), λ-designs on $6p + 1$ points, in: *Codes and Designs – Proceedings of Conference Honoring Prof. D. K. Ray-Chaudhuri on the Occasion of his 65th birthday*, K. T. Arasu and A. Seress (eds.). Ohio State University Mathematical Research Institute Publications, **10**. Berlin: Walter de Gruyter, pp. 109–24.

— (2003), λ-designs on $8p + 1$ points, *Ars Combin.*, **68**, 17–32.

Fiedler, M., Klin, M. and Muzychuk, M. (2002) Small vertex-transitive directed strongly regular graphs, *Discrete Math.*, **255**, 87–115.

Fisher, R. A. (1940) An examination of the different possible solutions of a problem in incomplete block designs, *Ann. Eugenics*, **10**, 363–88.

— (1949) *The Design of Experiments* (5th edn.). Edinburgh.

Fisher, R. A. and Yates, F. (1934) The 6×6 Latin squares, *Proc. Cambridge Philos. Soc.*, **30**, 492–507.

— (1963) *Statistical Tables for Biological, Agricultural, and Medical Research* (6th edn.). New York: Hafner.

Frankl, P. and Wilson, R. M. (1981) Intersection theorems with geometric consequences, *Combinatorica*, **1**, 357–68.

Fried, E. and Sós, V. (1975) Weakly associative lattices and projective planes, *Algebra Universalis*, **5**, 114–19.

Frolov, M. (1890) Recherches sur le permutations carrées, *J. Math. Spéc*, **4**, 8–11.

Furino, S., Miao, Y. and Yin, J. (1996) *Frames and Resolvable Designs*. Boca Raton: CRC Press.

Ganley, M. J. (1976) Direct product difference sets, *J. Combin. Theory Ser. A*, **27**, 93–8.

Ge, G., Greig, M. and Seberry, J. (2003) Generalized Bhaskar Rao designs with block size 4 signed over elementary abelian groups, *J. Combin. Math. Combin. Comput.*, **46**, 3–45.

Ghinelli, D. and Jungnickel, D. (2003) Finite projective planes with a large abelian group, in: *Survey in Combinatorics*, C. D. Wensley (ed.). Cambridge: Cambridge University Press, pp. 175–237.

Gibbons, P. B. and Mathon, R. A. (1987a) Group signings of symmetric balanced incomplete block designs, *Ars Combin.*, **A23**, 123–34.

— (1987b) Construction methods for Bhaskar Rao and related designs, *J. Austral. Math. Soc. (A)*, **42**, 5–30.

Glynn, D. (1978) *Finite Projective Planes and Related Combinatorial Systems*, Ph.D. thesis, Adelaide, Australia: University of Adelaide.

Godsil, C. (1993) *Algebraic Combinatorics*. New York: Chapman and Hall.

— (1996) Association Schemes, in: *The CRC Handbook of Combinatorial Designs*, C. J. Colbourn and J. H. Dinitz (eds.). Boca Raton: CRC Press, pp. 229–33.

Godsil, C. and Royle, G. (2001) *Algebraic Graph Theory*. New York: Springer.

Goethals, J. M. and Seidel, J. J. (1970) Strongly regular graphs derived from combinatorial designs, *Canad. J. Math.*, **22**, 597–614.

Golay, M. J. E. (1949) Notes on digital coding, *Proc. IEEE*, **37**, 657.

Goldbach, R. W. and Claasen, H. L. (1998) 3-class association schemes and Hadamard matrices of a certain block form, *European J. Combin.*, **19**, 943–51.

Gordon, B., Mills, W. H. and Welch, L. R. (1962) Some new difference sets, *Canad. J. Math.*, **14**, 614–25.

Gropp, H. (1992) The birth of a mathematical theory in British India, *Coll. Math. Soc. Janos Bolyai*, **60**, 315–27.

Guo, H. and Huang, T. (2001) Some classes of four-weight spin models, *J. Statist. Plann. Inf.*, **94**, 231–47.

Hadamard, J. (1893) Résolution d́une question relative aux déterminants, *Bull. Sci. Math.*, **17**(2), 240–6.

Haemers, W. H. (1978) A generalization of the Higman–Sims technique, *Indag. Math. (NS)*, **40**, 445–7.

— (1995) Interlacing eigenvalues and graphs, *Linear Algebra Appl.*, **227–8**, 593–616.

Haemers, W. H. and Shrikhande, M. S. (1979) Some remarks on subdesigns of symmetric designs, *J. Statist. Plann. Inf.*, **3**, 361–6.

Hall, M., Jr. (1943) Projective planes, *Trans. Amer. Math. Soc.*, **54**, 229–77.

— Jr. (1947), Cyclic projective planes, *Duke J. Math.*, **14**, 1079–90.

— Jr. (1986), *Combinatorial Theory* (2nd edn.). New York: Wiley.

Hall, Jr. and Connor, W. S. (1954) An embedding theorem for balanced incomplete block designs *Canad. J. Math.*, **6**, 35–41.

Hall, M., Jr., Lane, R. and Wales, D. (1970) Designs derived from permutation groups, *J. Combin. Theory*, **8**, 12–22.

Hall, M., Jr. and Ryser, H. J. (1951) Cyclic incidence matrices, *Canad. J. Math.*, **III**(4), 495–502.

Hamada, N. and Ohmori, H. (1975) On the BIB design having minimum p-rank, *J. Combin. Theory Ser. A*, **18**, 131–40.

Hanani, H. (1975) Balanced incomplete block designs and related designs, *Discrete Math.*, **11**, 255–369.

Hedayat, A. S. and Kageyama, S. (1980) The family of t-designs – Part I, *J. Statist. Plann. Inf.*, **4**, 173–212.

Hedayat, A. S. and Wallis, W. D. (1978) Hadamard matrices and their applications, *Ann. Statist.*, **6**, 1184–238.

Hein, D. W. and Ionin, Y. J. (2002) On the λ-design conjecture for $v = 5p + 1$ points, in: *Codes and Designs – Proceedings of Conference Honoring Prof. D. K. Ray-Chaudhuri on the Occasion of his 65th birthday*, K. T. Arasu and A. Seress (eds.). Ohio State Univ. Math. Res. Inst. Publ. **10**. Berlin: Walter de Gruyter, pp. 145–56.

Higman, G. (1969) On the simple group of D. G. Higman and C. C. Sims, *Illinois J. Math.*, **13**, 74–80.

Hill, R. (1986) *A First Course in Coding Theory*. Oxford: Oxford University Press.

Hirschfeld, J. W. P. (1985) *Finite Projective Spaces of Three Dimensions*. Oxford: Oxford University Press.

— (1998) *Projective Geometries Over Finite Fields* (2nd edn.). Oxford: Oxford University Press.

Hirschfeld, J. W. P. and Thas, J. A. (1991) *General Galois Geometries*. Oxford: Oxford University Press.

Hoffman, A. J. (1960) On the uniqueness of the triangular association scheme, *Ann. Math. Stat.*, **31**, 492–7.

— (1961) On the exceptional case in a characterization of the arcs of a complete graph, *IBM J. Res. Develop.*, **4**, 487–96.

— (1963) On the polynomial of a graph, *Amer. Math. Monthly*, **70**, 30–6.

— (1970) On eigenvalues and colourings of graphs, in: *Graph Theory and Its Applications*, B. Harris (ed.). New York: Academic Press, pp. 79–91.

Holton, D. A. and Sheehan, J. (1993) *The Petersen Graph*. Cambridge: Cambridge University Press.

Hubaut, X. (1974) Strongly regular graphs, *Discrete Math.*, **13**, 357–81.

Hughes, D. R. (1957a) Collineations and generalized incidence matrices, *Trans. Amer. Math. Soc.*, **86**, 284–96.

— (1957b) A class of non-Desarguesian projective planes, *Canad. J. Math.*, **9**, 378–88.

— (1965) On t-designs and groups, *Amer. J. Math.*, **87**(4), 761–78.

Hughes, D. R. and Piper, F. C. (1976) On resolutions and Bose's theorem, *Geom. Dedicata*, **5**, 129–31.

— (1982) *Projective Planes* (2nd edn.). Berlin: Springer.

— (1985) *Design Theory*. Cambridge: Cambridge University Press.

Humphrey, J. F. (1996) *A Course in Group Theory*, Oxford: Oxford University Press.

Ionin, Y. J. (1998) A technique for constructing symmetric designs, *Des. Codes Cryptog.*, **14**, 147–58.

— (1999a) Symmetric subdesigns of symmetric designs, *J. Combin. Math. Combin. Comput.*, **29**, 65–78.

— (1999b) Building symmetric designs with building sets, *Des. Codes Cryptog.*, **17**, 159–75.

— (2000) A technique for constructing divisible difference sets, *J. Geom.*, **67**, No. 1/2, 164–172.

— (2001) Applying balanced generalized weighing matrices to construct block designs, *The Electronic Journal of Combinatorics*, **8**, #R12.

— (2004) Regular Hadamard matrices generating infinite families of symmetric designs, *Des. Codes Cryptog.*, **32**, 227–33.

— (2005) Generalized conference matrices of index 1, preprint.

Ionin, Y. J. and Kharaghani, H. (2003a) New families of strongly regular graphs, *J. Comb. Des.*, **11**, 208–17.

— (2003b) Doubly regular digraphs and symmetric designs, *J. Combin. Theory Ser. A*, **101**, 35–48.

— (2005) A recursive construction for new symmetric designs, *Des. Codes Cryptog.*, **35**, 303–10.

Ionin, Y. J. and Mackenzie-Fleming, K. (2002) A technique for constructing non-embeddable quasi-residual designs, *J. Combin. Des.*, **10**, 160–72.

Ionin, Y. J. and Shrikhande, M. S. (1993) $(2s-1)$-designs with s intersection numbers, *Geom. Dedicata*, **48**, 247–65.

— (1994a) Maximal intersecting families with constant intersection size (preprint).

— (1994b) On a conjecture of Jungnickel and Tonchev for quasi-symmetric designs, *J. Comb. Des.*, **2**, 49–59.

— (1995a) 5-designs with three intersection numbers, *J. Combin. Theory Ser. A*, **69**, 36–50.

— (1995b) Equidistant families of sets, *Linear Algebra Appl.*, **226–228**, 223–35.

— (1996a) On the λ-design conjecture, *J. Combin. Theory Ser. A*, **74**, 100–14.

— (1996b) λ-designs on $4p+1$ points, *J. Combin. Math. Combin. Comput.*, **22**, 135–42.

— (1998) Resolvable pairwise balanced designs, *J. Statist. Plann. Inf.*, **72**, 393–405.

— (2000) Strongly regular graphs and designs with three intersection numbers, *Des. Codes Cryptog.*, **21**, 113–25.

— (2002) Fisher's inequality for designs on regular graphs, *J. Statist. Plann. Inf.*, **100**, 185–90.

— (2003) Decomposable symmetric designs, *Discrete Math.* (To appear.)

Ito, N. (1975) On tight 4-designs, *Osaka J. Math.*, **12**, 493–522.

Ito, N. (1978) (Corrections and Supplements, *Osaka J. Math.*, **15**, 693–7.

Janko, Z. (1997) The existence of symmetric designs with parameters (189, 48, 12), *J. Combin. Theory Ser. A*, **80**(2), 334–8.

— (1999) The existence of symmetric designs with parameters (105, 40, 15), *J. Combin. Des.*, **7**, 17–19.

— (2001) The existence of a Bush-type Hadamard matrix of order 36 and two new infinite classes of symmetric designs, *J. Combin. Theory Ser. A*, **95**(2), 360–4.

Janko, Z. and Kharaghani, H. (2002) A block negacyclic Bush-type Hadamard matrix and two strongly regular graphs, *J. Combin. Theory Ser. A*, **98**(1), 118–26.

Janko, Z., Kharaghani, H. and Tonchev, V. D. (2001a) The existence of a Bush-type Hadamard matrix of order 324 and two new infinite classes of symmetric designs, *J. Combin. Des.*, **9**, 72–8.

— (2001b) A Bush-type Hadamard matrix of order 100 and two new infinite classes of symmetric designs, *Des. Codes Cryptog.*, **24**, 225–32.

Janko, Z. and Trung, van (1984) The existence of a symmetric block design for (70, 24, 8), *Mitt. Math. Sem. Giessen*, **165**, 17–18.

— (1985a) Construction of symmetric block designs for (71, 21, 6), *Discrete Math.*, **55**, 327–8.

— (1985b) Construction of a new symmetric block design for (78, 22, 6) with the help of tactical decompositions, *J. Combin. Theory Ser. A*, **40**, 451–5. **55**, 327–8.

— (1986) A new biplane of order 9 with a small automorphism group, *J. Combin. Theory Ser. A*, **42**, 305–9.

Johnsen, E. G. (1966) Integral solutions to the incidence equations for finite projective plane cases of order $n = 2 \pmod 4$), *Pacific J. Math.*, **17**, 97–120.

Jones, B. W. (1950) *The Arithmetic Theory of Quadratic Forms*. New York: Wiley.

Jónsson, W. (1973) The (56, 11, 2)-design of Hall, Lane and Wales, *J. Combin. Theory Ser. A*, **14**, 113–18.

Jørgensen, L. (1994) On Normally Regular Digraphs, Preprint R-94-2023. Aalborg University: Institute for Electronic Systems.

Jungnickel, D. (1979) On difference matrices, resolvable transversal designs, and generalized Hadamard matrices, *Math. Z.*, **167**, 49–60.

— (1982a) On subdesigns of symmetric designs, *Math. Z.*, **181**, 383–93.

— (1982b) On automorphism groups of divisible designs, *Canad. J. Math.*, **34**, 257–97.

— (1990a) Latin squares, their geometries and their groups. A survey, in: *Coding Theory and Design Theory, Part II, IMA Volumes in Mathematics and Its Applications*, Vol. 21, D. K. Ray-Chaudhuri (ed.). New York: Springer-Verlag, pp. 166–225.

— (1992) Difference sets, in: *Contemporary Design Theory: A Collection of Surveys*, J. H. Dinitz and D. R. Stinson (eds.). New York: Wiley, 241–324.

— (1993) *Finite Fields*. B. I. Wissenschaftsverlag, Mannheim.

— (2005) Balanced generalized weighing matrices and related structures, *Quaderni Elettronici del Seminario di Geometria Combinatoria*, **16E**.

Jungnickel, D. and Kharaghani, H. (2004) Balanced generalized weighing matrices and their applications, *Le Matematiche*, submitted.

Jungnickel, D. and Pott, A. (1996) Difference sets: abelian, in: *The CRC Handbook of Combinatorial Designs*, C. J. Colbourn and J. H. Dinitz (eds.). Boca Raton: CRC Press, pp. 297–307.

Jungnickel, D. and de Resmini, M. J. (2002) Another case of the prime power conjecture for finite projective planes, *Adv. Geom.*, **2**, pp. 215–18.

Jungnickel, D. and Tonchev, V. D. (1991a) Intersection numbers of quasi-multiples of symmetric designs, in: *Advances in Finite Geometries and Designs, Proceedings of the Third Isle of Thorns Conference 1990*, J. W. P. Hirschfeld, D. R. Hughes, and J. A. Thas (eds.). Oxford: Oxford University Press, pp. 227–36.

— (1991b) Exponential number of quasi-symmetric SDP-designs and codes meeting the Grey–Rankin bound, *Des. Codes Cryptog.*, **1**, 247–53.

— (1992) On symmetric and quasi-symmetric designs with the symmetric difference property and their codes, *J. Combin. Theory Ser. A*, **59**, 42–50.

— (1999a) Decompositions of difference sets, *J. Algebra*, **217**, 21–39.

— (1999b) Perfect codes and balanced generalized weighing matrices, *Finite Fields Appl.*, **5**, 294–300.

— (2002) Perfect codes and balanced generalized weighing matrices II, *Finite Fields Appl.*, **8**, 155–65.

Kageyama, S. and Hedayat, A. S. (1983) The family of t-designs – Part II, *J. Statist. Plann. Inf.*, **7**, 257–87.

Kageyama, S. and Miao, Y. (1996) Some non-embeddable quasi-derived designs, *J. Combin. Theory Ser. A*, **76**, 295–304.

Kantor, W. M. (1969a) 2-transitive symmetric designs, *Trans. Amer. Math. Soc.*, **146**, 1–28.

— (1969b) Characterization of finite projective and affine spaces, *Canad. J. Math.*, **21**, 64–75.

— (1975) Symplectic groups, symmetric designs and line ovals, *J. Algebra*, **33**, 43–58.

— (2001) Note on GMW-designs, *European J. Combin.*, **22**, 63–69.

Key, J. D. and Tonchev, V. D. (1997) Computational results for the known biplanes of order 9, in: *Geometry, Combinatorial Designs and Related Structures*, J. W. P. Hirschfeld, S. S. Magliveras and M. J. de Resmini (eds.). Cambridge: Cambridge University Press, pp. 113–22.

Kharaghani, H. (1985) New classes of weighing matrices, *Ars Combin.*, **19**, 69–72.

— (2000) On the twin designs with the Ionin-type parameters, *Electron. J. Combin.*, **7**(7), R1.

— (2003) On a class of symmetric balanced generalized weighing matrices, *Des. Codes Cryptog.*, **30**, 139–49.

Kharaghani, H. and Tayfeh-Rezaie, B. A. (2005) A Hadamard matrix of order 428, *J. Comb. Des.*, **13**, 435–40.

Kirkman, T. P. (1847) On a problem in combinatorics, *Cambridge and Dublin Math. J.*, **2**, 191–204.

Klee, S. and Yates, L. (2004) Tight subdesigns of the Higman–Sims design, *Rose-Hulman Undergraduate Math. J.*, **5**(2).

Klin, M., Munemasa, A., Muzychuk, M. and Zieschang, P.-H. (2004) Directed strongly regular graphs obtained from coherent algebras, *Linear Algebra Appl.*, **377**, 83–109.

Koornwinder, T. H. (1976) A note on the absolute bound for systems of lines, *Proc. Kon. Nederl. Akad. Wetensch. Ser. A =Indag. Math.*, **38**, 152–3.

Kramer, E. S. (1969) On λ-designs, Ph.D. dissertation. Michigan: University of Michigan.

— (1974) On λ-designs, *J. Combin. Theory Ser. A*, **16**, 57–75.

Kreher, D. L. (1996) t-Designs, $t \geq 3$, in: *The CRC Handbook of Combinatorial Designs*, C. J. Colbourn and J. H. Dinitz (eds.). Boca Raton: CRC Press, pp. 47–66.

Lam, C. W. H. (1991) The search for a finite projective plane of order 10, *Amer. Math. Monthly*, **98**, 305–18.

Lam, C. W. H., Thiel, L. and Swiercz, S. (1989) The non-existence of finite projective planes of order 10, *Canad. J. Math.*, **XLI**(6), 1117–23.

Lam, C. W. H., Thiel, L., Swiercz, S. and McKay, J. (1983) The non-existence of ovals in the projective plane of order 10, *Discrete Math.*, **45**, 319–22.

Lander, E. S. (1983) *Symmetric Designs: An Algebraic Approach*. Cambridge: Cambridge University Press.

Laywine, C. F. and Mullen, G. L. (1998) *Discrete Mathematics Using Latin Squares*. John Wiley & Sons, New York.

Lenz, H. (1965) *Vorlesungen über projektive Geometrie*, Akad. Leipzig: Verlagsgesellschaft.

Lenz, H. and Jungnickel, D. (1979) On a class of symmetric designs, *Arch. Math.*, **33**, 590–2.

Leung, K. H., Ma, S. L. and Schmidt, B. (2002) Constructions of relative difference sets with classical parameters and circulant weighing matrices, *J. Combin. Theory Ser. A*, **99**, 111–27.

Levi, F. W. (1942) *Finite Geometrical Systems*. Calcutta: University of Calcutta.

Lidl, R. and Niederreiter, H. (1997) *Finite Fields*. Cambridge: Cambridge University Press.

Lüneburg, H. (1969) *Transitive Erweiterungen endlicher Permutationsgruppen*, *Lecture Notes in Math.*, **84**. Berlin: Springer-Verlag.

— (1980) *Translation Planes*. Berlin: Springer-Verlag.

MacInnes, C. R. (1907) Finite planes with less than eight points on a line, *Amer. Math. Monthly*, **14**, 171–4.

Mackenzie-Fleming, K. (2000a) An infinite family of non-embeddable quasi-residual Hadamard designs, *J. Geom.*, **67**, 173–9.

— (2000b) An infinite family of non-embeddable Hadamard designs, *Electron. J. Combin.*, **6**, R24.

Mackenzie-Fleming, K. and Smith, K. W. (1998) An infinite family of non-embeddable quasi-residual designs, *J. Statist. Plann. Inf.*, **73**, 77–83.

MacNeish, H. F. (1922) Euler squares, *Ann. Math.*, **23**, 221–7.

MacWilliams, F. J. and Sloane, N. J. A. (1977) *The Theory of Error-correcting Codes*. Amsterdam: North-Holland.

Majumdar, K. N. (1953) On some theorems in combinatorics related to incomplete block designs, *Ann. Math. Statist.*, **24**, 377–89.

— (1962) On the parameters and intersection of blocks of balanced incomplete block designs, *Ann. Math. Statist.*, **33**, 1200–5.

Mann, H. B. (1965a) *Addition Theorems: The Addition Theorems of Group Theory and Number Theory*. New York: Wiley.

— (1965b) Difference sets in elementary abelian groups, *Illinois J. Math.*, **9**, 212–19.

Marrero, O. (1972) A property of (v, k, λ)-designs, *Israel J. Math.*, **12**, 277–8.

Masuyama, M. (1957) On difference sets for constructing orthogonal arrays of index 2 and of strength 2, *Rep. Stat. Appl. Res. JUSE*, **5**, 27–34.

Mathon, R. (1975) 3-class association schemes, in: *Proceedings Conference on Algebraic Aspects of Combinatorics*, D. Corneil and E. Mendelsohn (eds.). Utilitas Mathematica Pub., Winnipeg, pp. 123–55.

— (1978) Symmetric conference matrices of order $pq^2 + 1$, *Canad. J. Math.*, **30**, 321–31.

— (1987) On a new divisible semiplane. Announcement at the Eleventh British Combinatorial Conference.

Mathon, R. and Rosa, A. (1996) 2-(v, k, λ) designs of small order, in: *The CRC Handbook of Combinatorial Designs*, C. J. Colbourn and J. H. Dinitz (eds.). Boca Raton: CRC Press, pp. 1–41.

Mavron, V. C. (1972) Parallelisms in designs, *J. Lond. Math. Soc.*, **4**, 682–4.

— (1973) A class of symmetric designs, *Geom. Dedicata*, **1**, 468–78.

— (1988) Symmetric designs and λ-arcs, *European J. Combin.*, **8**, 507–16.

Mavron, V. C., McDonough, T. P. and Pallikaros, C. A. (2001) A difference matrix construction and a class of balanced generalized weighing matrices, *Arch. Math.*, **76**, 259–64.

McDonough, T. P., Mavron, V. C. and Pallikaros, C. A. (2000) Generalized Hadamard matrices and translations, Special issue in honor of Professor Ralph Stanton, *J. Statist. Plann. Inference*, **86**(2), 527–33.

McFarland, R. L. (1973) A family of difference sets in non-cyclic groups, *J. Combin. Theory Ser. A*, **15**, 1–10.

— (1974) On (v, k, λ)-configurations with $v = 4p^e$, *Glasgow J. Math.*, **15**, 180–3.

McGuire, G. and Ward, H. N. (1998) Characterization of certain minimal rank designs, *J. Combin. Theory Ser. A*, **83**, 42–56.

Menon, P. K. (1960) Difference sets in abelian groups, *Proc. Amer. Math. Soc.*, **11**, 368–76.

— (1962a) Certain Hadamard designs, *Proc. Amer. Math. Soc.*, **13**, 524–39.

— (1962b) On difference sets whose parameters satisfy a certain relation, *Proc. Amer. Math. Soc.*, **13**, 739–45.

Mesner, D. M. (1956) *An Investigation of Certain Combinatorial Properties of Partially Balanced Incomplete Block Designs and Association Schemes, with a Detailed Study of Designs of the Latin Square and Related Types*. Michigan: Michigan State Univ., doctoral thesis.

— (1967) A new family of partially balanced incomplete block designs with some Latin square design properties, *Ann. Math. Statist.*, **38**, 571–81.

Metsch, K. (1991) Improvement of Bruck's completion theorem, *Des. Codes Cryptog.*, **1**, 99–116.

Mitchell, C. J. (1979) An infinite family of symmetric designs, *Discrete Math.*, **26**, 247–50.

Motose, K. (2001) On finite Dickson near fields, *Bull. Fac. Sci. Technol. Hirosaki Univ.*, **2**, 69–78.

Mukhopadhay, A. C. (1978) Some infinite classes of Hadamard matrices, *J. Combin. Theory Ser. A*, **25**, 128–41.

Mullin, R. C. and Gronau, H.-D. O. F. (1996) PBDs and GDDs: the basics, in: *The CRC Handbook of Combinatorial Designs*, C. J. Colbourn and J. H. Dinitz (eds.). Boca Raton: CRC Press, pp. 185–93.

Mullin, R. C. and Stanton, R. G. (1975a) Group matrices and balanced weighing designs, *Util. Math.*, **8**, 277–301.

— (1975b) Balanced weighing matrices and group divisible designs, *Util. Math.*, **8**, 303–10.

Muzychuk, M. and Klin, M. (1998) On graphs with three eigenvalues, *Discrete Math.*, **189**, 191–207.

Muzychuk, M. and Xiang, Q. (2005) Symmetric Bush-type Hadamard matrices of order $4m^4$ exist for all odd m, preprint.

Paley, R. E. A. C. (1933) On orthogonal matrices, *J. Math. Phys.*, **12**, 311–20.

Parker, E. T. (1957) On collineations of symmetric designs, *Proc. Amer. Math. Soc.*, **8**, 350–1.

— (1959) Construction of some sets of pairwise orthogonal Latin squares, *Proc. Amer. Math. Soc.*, **10**, 946–9.

— (1960) Orthogonal latin squares, *Proc. Nat. Acad. Sci. USA*, **21**, 859–62.

— (1967) A result in balanced incomplete block designs, *J. Combin. Theory Ser.*, **3**, 283–5.

Pawale, R. M. and Sane, S. S. (1991) A short proof of a conjecture on quasi-symmetric 3-designs, *Discrete Math.*, **96**, 71–4.

Petersen, J. (1898) Sur le théorème de Tait, *L'Intermédiare des Mathématiciens*, **5**, 225–7.

Peterson, C. (1977) On tight 6-designs, *Osaka J. Math.*, **14**, 417–35.

Petrenjuk, A. J. (1968) On Fisher's Inequality for tactical configurations (Russian). *Mat. Zametki*, **4**, 417–25.

Pilz, G. (1983) *Near-Rings*. Amsterdam: North-Holland.

Plackett, R. L. and Burman, J. P. (1946) Designs of optimal multifactorial experiments, *Biometrika*, **33**, 305–25.

Pless, V. (1989) *Introduction to the Theory of Error-correcting Codes*. New York: John Wiley & Sons.

Pless, V. S. and Huffman, W. C. (eds.) (1998) *Handbook of Coding Theory*, vols. 1 and 2. Amsterdam: North-Holland.

Plotkin, M. (1960) Binary codes with specified minimum distances, *IEEE Trans. Inform. Theory*, **6**, 445–50.

Plücker, J. (1839) *Theorie der algebraischen Curven, gegrundet auf eine neue Behandlungweise der analytischen Geometrie*. Bonn: Marcus.

Pott, A. (1994) On projective planes admitting elations and homologies, *Geom. Dedicata*, **52**, 181–93.

— (1995) Finite geometry and character theory, *Lecture Notes in Mathematics*, **1601**, Springer-Verlag.

Raghavarao, D. (1971) *Constructions and Combinatorial Problems in Design of Experiments*. New York: John Wiley (reprinted Dover Paperback Series, 1989).

Rajkundlia, D. (1978) Some techniques for constructing infinite families of BIBDs, Ph.D. dissertation. Kingston, Canada: Queen's University.

— (1983) Some techniques for constructing infinite families of BIBDs, *Discrete Math.*, **44**, 61–96.

Ray-Chaudhuri, D. K. and Wilson, R. M. (1973) Solution of Kirkman's schoolgirl problem, *Symp. Pure Math.*, **19**, 187–203.

— (1975) On t-designs, *Osaka J. Math.*, **12**, 737–44.

Room, T. G. and Kirkpatrick, P. B. (1971) *Miniquaternion Geometry*. Cambridge: Cambridge University Press.

Ryser, H. J. (1963) *Combinatorial Mathematics*. New York: John Wiley.

— (1968) An extension of a theorem of de Bruijn and Erdös, *J. Algebra*, **10**, 246–61.

Sane, S. S. (2000) Quasi–multiple quasi–symmetric designs corresponding to biplanes, *Congr. Numer.*, **146**, 213–16.

— (2001) A proof of the Jungnickel–Tonchev conjecture on quasi–multiple quasi–symmetric designs, *Des. Codes Cryptog.*, **23**, 291–6.

Sane, S. S. and Shrikhande, M. S. (1987) Quasi-symmetric 2, 3, 4-designs, *Combinatorica*, **7**(3), 291–301.

Sane, S. S., Shrikhande, S. S. and Singhi, N. M. (1985) Maximal arcs in designs, *Graphs and Combinatorics*, **1**, 97–106.

Schmidt, B. (2002) Cyclotomic integers and finite geometry, *Lecture Notes in Mathematics*, **1797**. Berlin: Springer-Verlag.

Schützenberger, M. P. (1949) A non-existence theorem for an infinite family of symmetrical block designs, *Ann. Eugenics*, **14**, 286–7.

Seberry, J. (1979a) Some remarks on generalized Hadamard matrices, in: *Combinatorial Mathematics*, D. A. Holtan and J. Seberry (eds.). Berlin: Springer, pp. 275–81.

— (1979b) Some remarks on generalized Hadamard matrices and theorems of Rajkundlia on SBIBD's, in: *Combinatorial Mathematics VI*, A. F. Horadam and W. D. Wallis (eds.). Berlin: Springer, pp. 154–64.

Seberry, J. and Whiteman, A. L. (1988) New Hadamard matrices and conference matrices obtained via Mathon's construction, *Graphs and Combinatorics*, **4**, 355–77.

Seberry, J. and Yamada, M. (1992) Hadamard matrices, sequences, and block designs, in: *Contemporary Design Theory, A Collection of Surveys*, J. H. Dinitz and D. R. Stinson (eds.). New York: Wiley, pp. 431–560.

Seberry, J. and Zhang, X. (1993) Regular sets of matrices and applications, *Graphs and Combinatorics*, **9**, 185–95.

Segre, B. (1961) *Lectures on Modern Geometry*. Rome: Cremonese.

Seidel, J. J. (1967) Strongly regular graphs of L_2 type and triangular type, *Indag. Math.*, **29**, 188–96.

— (1968) Strongly regular graphs with $(-1, 1, 0)$ adjacency matrix having eigenvalue 3, *Linear Algebra Appl.*, **1**, 281–98.

— (1976) A survey two-graphs, *Atti Dei Convegni Lincei*, **17**, 481–511.

Seidel, J. J. and Taylor, D. E. (1981) Two-graphs, a second survey, in: *Algebraic Methods in Graph Theory*, L. Lovaász and V. T. Sós (eds.). *Coll. Math. Soc. J. Bolyai*, **25**, 689–711.

Semakov, N. V. and Zinoviev, V. A. (1968) Equidistant q-ary codes with maximal distance and resolvable balanced incomplete block designs, *Probl. Inf. Transm.*, **4**, 1–7.

Seress, A. (1989) Some characterizations of type-1 λ-designs, *J. Combin. Theory Ser. A*, **52**, 288–300.

— (2001) All lambda-designs with $\lambda = 2p$ are type-1, *Des. Codes Cryptog.*, **22**, 5–17.

Shrikhande, M. S. (1996) Quasi-symmetric designs, in: *The CRC Handbook of Combinatorial Designs*, C. J. Colbourn and J. H. Dinitz (eds.). Boca Raton: CRC Press, pp. 430–4.

— (2002) Subdesigns of symmetric designs, in: *Codes and Designs – Proceedings of Conference Honoring Prof. D. K. Ray-Chaudhuri on the Occasion of his 65th birthday*, K. T. Arasu and A. Seress (eds.). Ohio State Univ. Math. Res. Inst. Publ. **10**. Berlin: Walter de Gruyter, pp. 237–57.

Shrikhande, M. S. and Sane, S. S. (1991) *Quasi-symmetric Designs*. Cambridge: Cambridge University Press.

Shrikhande, S. S. (1950) The impossibility of certain symmetric balanced incomplete block designs, *Ann. Math. Stat.*, **21**, 106–111.

— (1951) On the non-existence of affine resolvable balanced incomplete block designs, *Sankhyā*, **11**, 185–6.
— (1952) On the dual of some balanced incomplete block designs, *Biometrics*, **8**, 66–72.
— (1959a) On a characterization of the triangular association scheme, *Ann. Math. Stat.*, **30**, 39–47.
— (1959b) On the uniqueness of L_2 association scheme, *Ann. Math. Stat.*, **30**, 781–98.
— (1960) Relations between certain incomplete block designs, in: *Contributions to Probability and Statistics: Essays in Honor of Harold Hotelling*, I. Olkin *et al.* (eds.). Stanford: Stanford University Press, pp. 388–95.
— (1964) Generalized Hadamard matrices and orthogonal arrays of strength two, *Canad. J. Math.*, **16**, 736–40.
— (1976) Affine resolvable balanced incomplete block designs: a survey, *Aequationes. Math.*, **14**, 251–69.
Shrikhande, S. S. and Bhagwandas (1965) Duals of incomplete block designs, *J. Indian Statist. Assoc.*, **3**, 30–7.
Shrikhande, S. S. and Raghavarao, D. (1963) A method of construction of incomplete block designs, *Sankhyā, Ser. A*, **25**, 399–402.
— (1964) A note on the non-existence of symmetric balanced incomplete block designs, *Sankhyā, Ser. A*, **26**, 91–2.
Shrikhande, S. S. and Singhi, N. M. (1975) Construction of geometroids, *Util. Math.*, **8**, 187–92.
— (1984) Some combinatorial problems, in: *Combinatorics and its Applications*, K. S. Vijayan and N. M. Singhi (eds.). Calcutta: Indian Statistical Institute, pp. 340–49.
Singer, J. (1938) A theorem in finite projective geometry and some applications to number theory, *Trans. Amer. Math. Soc.*, **43**, 377–85.
Singhi, N. M. and Shrikhande, S. S. (1973) Embeddings of quasi-residual designs with $\lambda = 3$, *Util. Math.*, **4**, 35–53.
— (1974) Embeddings of quasi-residual designs, *Geom. Dedicata*, **2**, 509–17.
— (1976) On the λ-design conjecture, *Util. Math.*, **9**, 301–18.
Smith, K. W. (1995) Non-abelian Hadamard difference sets, *J. Combin. Theory Ser. A*, **70**, 144–56.
— (1996) Difference sets: nonabelian, in: *The CRC Handbook of Combinatorial Designs*, C. J. Colbourn and J. H. Dinitz (eds.). Boca Raton: CRC Press, pp. 308–12.
Smith, M. (1976a) On the isomorphism of two simple groups of order 44,352,000, *J. Algebra*, **41**, 172–4.
— (1976b) A combinatorial configuration associated with the Higman–Sims simple group, *J. Algebra*, **41**, 175–95.
Snevily, H. S. (1994) On generalizations of the de Bruijn and Erdös Theorem , *J. Combin. Theory Ser. A*, **68**, 232–8.
Spence, E. (1977) A family of difference sets in non-cyclic groups, *J. Combin. Theory Ser. A*, **22**, 103–6.
— (1996) Two-graphs, in: *The CRC Handbook of Combinatorial Designs*, C. J. Colbourn and J. H. Dinitz (eds.). Boca Raton: CRC Press, pp. 686–94.
Stanton, R. G. (1997) Ryser Designs, *Ars Combinatoria*, **46**, 133–44.
Stanton, R. G. and Kalbfleisch, J. G. (1968) Quasi-symmetric balanced incomplete block designs, *J. Combin. Theory Ser. A*, **4**, 391–6.

Stanton, R. G. and Sprott, D. A. (1958) A family of difference sets, *Canad. J. Math.*, **10**, 73–7.

Steiner, J. (1853) Combinatoriesche Aufgabe, *J. für die reine und angewandte Mathematik*, **45**, 181–2.

Stinson, D. R. (1982) Applications and generalizations of the variance method in combinatorial designs, *Util. Math.*, **22**, 323–33.

— (1984) A short proof of the non-existence of a pair of orthogonal Latin squares of order 6, *J. Combin. Theory Ser. A*, **36**, 373–6.

Stinson, D. R. and van Rees, G. H. J. (1984) The equivalence of certain equidistant codes and symmetric designs, *Combinatorica*, **4**, 357–62.

Storer, T. (1967) *Cyclotomy and Difference Sets*. Chicago: Markham.

Street, A. P. (1979) Generalized Hadamard matrices, orthogonal arrays and F-squares, *Ars Combin.*, **8**, 131–41.

Street, A. P. and Street, D. J. (1987) *Combinatorics of Experimental Design*. Oxford: Oxford University Press.

Street, A. P. and Wallis, W. D. (1977) *Combinatorial Theory: An Introduction*. Winnipeg: CRBC Press.

Szekeres, E. and Szekeres, G. (1965) On a problem of Schütte and Erdös, *Math. Gas.*, **49**, 290–93.

Tarry, G. (1900) Le problème des 36 officiers, *C. R. Assoc. Franc. Avanc. Sci. Nat.*, **1**, 122–3.

Taylor, D. E. (1971) Some Topics in the Theory of Finite Groups, Ph.D. thesis. Oxford: Oxford University.

— (1977) Regular 2-graphs, *Proc. London Math. Soc.*, **35**(3), 257–74.

Teirlinck, L. (1987) Non-trivial t-designs without repeated blocks exist for all t, *Discrete Math.*, **65**, 301–11.

Tits, J. (1964) Sur les systèmes de Steiner associés aux trois "grands" groupes de Mathieu. *Rend. Math. e Appl. (5)*, **23**, 166–84.

Todd, J. A. (1933) A combinatorial problem, *J. Math. Phys.*, **12**, 321–33.

Tonchev, V. D. (1987a) Embedding of the Witt system $S(3, 6, 22)$ in a symmetric $2-(78, 22, 6)$-design, *Geom. Dedicata.*, **22**, 49–75.

— (1987b), Symmetric $2-(31, 10, 3)$-designs with automorphism of order seven, *Ann. Disc. Math.*, **34**, 461–4.

— (1988) *Combinatorial Configurations*. Pittman Monographs and Surveys in Pure and Applied Mathematics **40**. New York: Wiley.

— (1992) Some small non-embeddable designs, *Discrete Math.*, **106/107**, 489–92.

— (1996) Codes, in: *The CRC Handbook of Combinatorial Designs*, C. J. Colbourn and J. H. Dinitz (eds.). Boca Raton: CRC Press, pp. 517–43.

— (1998) Codes and designs, in: *Handbook of Coding Theory*, V. S. Pless and W. C. Huffman (eds.), vol. 2. North-Holland, pp. 1229–67.

Tsuzuku, T. (1982) *Finite Groups and Finite Geometries*. Cambridge: Cambridge University Press.

Turyn, R. J. (1965) Character sums and difference sets, *Pacific J. Math.*, **15**, 319–46.

— (1971) On C-matrices of arbitrary powers, *Canad. J. Math.*, **XXIII**, 531–5.

ud Din, S. and Mavron, V. C. (1992) Generalized Hadamard matrices and nets, *Sankhyā*, **54**, 503–9.

van Dam, E. R. (1998) Nonregular graphs with three eigenvalues, *J. Combin. Theory Ser. B*, **73**, 101–18.
van Lint, J. H. (1973) A theorem on equidistant codes, *Discrete Math.*, **6**, 353–8.
— (1978) Non-embeddable quasi-residual designs, *Indag. Math.*, **40**, 269–75.
— (1982) *Introduction to Coding Theory*, New York: Springer-Verlag.
— (1984a) On ovals in $PG(2, 4)$ and the McLaughlin graph, in: *Papers dedicated to J. J. Seidel*, P. J. Doelder, J. de Graaf and J. H. van Lint (eds.). Eindhoven: Eindhoven University of Technology Press, pp. 234–54.
— (1984b) On equidistant binary codes of length $n = 4k + 1$ with distance $d = 2k$, *Combinatorica*, **4**(4), 321–3.
van Lint, J. H. and Seidel, J. J. (1966) Equilateral point sets in elliptic geometry, *Indag. Math.*, **28**, 335–48.
van Lint, J. H. and Tonchev, V. D. (1984) Non-embeddable quasi-residual designs with large k, *J. Combin. Theory Ser. A*, **37**, 359–62.
— (1993) A class of non-embeddable designs, *J. Combin. Theory Ser. A*, **62**, 252–60.
van Lint, J. H., Tonchev, V. D. and Landgev, I. N. (1990) A new design, in: *Coding Theory and Design Theory, Part II*, D. K. Ray-Chaudhuri (ed.). The IMA Volumes in Mathematics and its Applications, **21**. New York: Springer-Verlag, pp. 251–56.
van Lint, J. H. and Wilson, R. M. (1993) *A Course in Combinatorics*. Cambridge: Cambridge University Press.
van Trung, T. (1982a) The existence of symmetric block designs with parameters (41, 16, 6) and (66, 26, 10), *J. Combin. Theory Ser. A*, **33**, 201–204.
— (1986) Non-embeddable quasi-residual designs with $k < \frac{v}{2}$, *J. Combin. Theory Ser. A*, **40**, 133–7.
— (1990) Nonembeddable quasi-residual designs, *Contemp. Math.*, **111**, 237–78.
— (1996) Symmetric designs, in: *The CRC Handbook of Combinatorial Designs*, C. J. Colbourn and J. H. Dinitz (eds.). Boca Raton: CRC Press, pp. 75–87.
Vanstone, S. A. (1979) Resolvable (r, λ)-designs and the Fisher inequality, *J. Aust. Math. Soc. Ser. A*, **28**, 471–8.
Veblen, O. and Bussey, W. H. (1906) Finite projective geometries, *Trans. Amer. Math. Soc.*, **7**, 241–59.
Veblen, O. and Young, J. W. (1916), *Projective Geometry* (2 vols.). Boston: Ginn & Co.
Wähling, H. (1987) *Theorie der Fastkörper*. Thales Monographs **1**. Essen: Thales-Verlag.
Wallis, W. D. (1971) Construction of strongly regular graphs using affine designs, *Bull. Aust. Math. Soc.*, **4**, 41–9.
— (1972) On a problem K. A. Bush concerning Hadamard matrices, *Bull. Aust. Math. Soc.*, **6**, 321–6.
Wallis, W. D., Street, A. P. and Wallis, J. S. (1972) *Combinatorics: Room Squares, Sum-Free Sets, Hadamard Matrices*. Berlin: Springer.
Weisz, I. (1995) Lambda-designs with small lambda are type-1. Ph.D. dissertation. Ohio: Ohio State University.
Whiteman, A. L. (1962) A family of difference sets, *Illinois J. Math.*, **6**, 107–21.
Wilson, R. M. (1973) The necessary conditions for t-designs are sufficient for something, *Util. Math.*, **4**, 207–15.

— (1984) On the theory of T-designs, in: *Enumeration and Design*, D. M. Jackson and S. A. Vanstone (eds.). New York: Academic Press, pp. 19–49.

Witt, E. (1938a) Die 5-fach transitiven Gruppen von Mathieu, *Abh. Math. Sem. Univ. Hamburg*, **12**, 256–64.

— (1938b) Über Steinersche Systeme, *Abh. Math. Sem. Univ. Hamburg*, **12**, 265–75.

Woodall, D. R. (1970) Square λ-linked designs, *Proc. London Math. Soc.*, **20**, 669–87.

— (1971) Square λ-linked designs: a survey, in: *Combinatorial Mathematics and Its Applications*. New York/London: Academic Press, pp. 349–55.

Wu, Q. (1989) Results on λ-designs, *Ars Combin.*, **28**, 231–41.

Yates, F. (1935) Complex experiments, *J. R. Stat. Soc.*, **B2**, 181–223.

— (1936) Incomplete randomized blocks, *Ann. Eugenics*, **7**, 121–40.

Zassenhaus, H. (1935/36) Über endliche Fastkörper, *Abh. Math. Sem. Univ. Hamburg*, **11**, 187–220.

Index